Sustainability Sciences in Asia and Africa

Sustainable Agriculture and Food Security

Sustainability as a scientific endeavor requires broad understanding of the interconnections in our global environment. It is therefore positioned at the crossroads of research in natural sciences, technology and human sciences. This book series supports the global efforts by providing timely coverage of the progress, opportunities, and challenges towards achieving sustainable development goals (SDGs) in Asia and Africa. It fosters the research in transdisciplinary academic fields spanning across agriculture and global food security, health and diseases, access to clean water and affordable energy, climate resilience and effects of global climate change on biodiversity, water and land. This series is planned to narrate the stories of scientific development and progression towards SDGs from the regions of Asia and Africa. The overall aim of the series is to provide a comprehensive publication platform and act as a knowledge engine in the growth of sustainability sciences with a special focus on developing nations. The series constitute of four sub-series each focusing on a specific theme closely linked to a set of SDGs. Each sub-series is leaders and experts in agricultural sciences, environmental sciences, biomedicine and medical sciences respectively. Books in this series are useful for environmentalists, biotechnologists, plant scientists, medical practitioners, conservation biologists, policy makers, NGOs and scientific bodies working in the areas of sustainability sciences in the developing regions.

Series Editor

Rajeev K. Varshney, Perth, WA, Australia

This book series support the global efforts towards sustainability by providing timely coverage of the progress, opportunities, and challenges of sustainable food production and consumption in Asia and Africa. The series narrates the success stories and research endeavors from the regions of Africa and Asia on issues relating to SDG 2: Zero hunger. It fosters the research in transdisciplinary academic fields spanning across sustainable agriculture systems and practices, post- harvest and food supply chains. It will also focus on breeding programs for resilient crops, efficiency in crop cycle, various factors of food security, as well as improving nutrition and curbing hunger and malnutrition. The focus of the series is to provide a comprehensive publication platform and act as a knowledge engine in the growth of sustainability sciences with a special focus on developing nations. The series will publish mainly edited volumes but some authored volumes. These volumes will have chapters from eminent personalities in their area of research from different parts of the world.

Anandita Singh • Sandip Das
Rajeev K. Varshney
Editors

Sustainable Crop Production

New Research Paradigms in Plant Sciences

Springer

Editors
Anandita Singh
Department of Biotechnology
TERI School of Advanced Studies
New Delhi, Delhi, India

Sandip Das
Department of Botany
University of Delhi
New Delhi, Delhi, India

Rajeev K. Varshney
Murdoch University
Food Futures Institute
Perth, WA, Australia

ISSN 2730-6771 ISSN 2730-678X (electronic)
Sustainability Sciences in Asia and Africa
ISSN 2730-6798 ISSN 2730-6801 (electronic)
Sustainable Agriculture and Food Security

ISBN 978-981-95-7843-6 ISBN 978-981-95-7844-3 (eBook)
https://doi.org/10.1007/978-981-95-7844-3

This work was jointly supported by University of Essex and UK Research and Innovation.

This Springer imprint is published by the registered company Springer Nature Singapore Pte Ltd.
The registered company address is: 152 Beach Road, #21-01/04 Gateway East, Singapore 189721, Singapore

Foreword

Sustainable crop production is one of the cornerstones of the UN Charter on Sustainable Development Goals (SDG2). SDG2 envisions guaranteed access to better nutrition and food security by 2030 for all people worldwide. Hunger and chronic malnutrition remain serious threats to human prosperity and social justice, despite the tireless efforts of agriculturists, applied and basic research scientists, and policymakers. In addition, ongoing and accelerated climate change will amplify the pressing requirement for reinventing sustainable crop production and empowering farming communities.

The book titled *Sustainable Crop Production: New Research Paradigms in Plant Sciences* is a timely and vital contribution to international agricultural research community. The 15 chapters cover subjects that span both foundational and leading edge ideas: reconfiguring developmental pathways for trait manipulation, building resilience to abiotic and biotic stresses, bioengineering carbon capture mechanisms, application of Q-dots to improve photosynthetic efficiency, deciphering how the rhizobiome can contribute to abiotic stress tolerances, and developing novel strategies for heterosis breeding based on molecular insights. One chapter envisions agriculture systems in space, a future scenario that should not be ignored.

This rich collection of information and insights will not only teach but also motivate the next generation of scientists to reimagine the future of food production. As this volume makes eminently clear, combining plant science research with technological innovations will be the strategy that carries agriculture forward to new levels of success worldwide. The editors have curated a comprehensive collection of knowledge that will help researchers and policymakers bridge the gap between scientific discoveries and translation to sustainable crop production. I would like to congratulate all Editors—Anandita Singh, Sandip Das, and Rajeev K. Varshney, for bringing out such a nice volume.

Our planet faces unprecedented and interwoven crises, especially rapid climate change, habitat loss, skyrocketing demand for better qualities of life, global ecological fragility, and massive population growth in the poorest parts of the world. The book *Sustainable Crop Production: New Research Paradigms in Plant Sciences* elegantly supports a call to action for global cooperation toward increased production, equity, and resilience in worldwide agriculture.

Jeff Bennetzen
US National Academy of Science Member
Norman and Doris Giles Professor Georgia
Research Alliance Eminent Scholar
University of Georgia
Athens, GA, USA

Series Preface

The UN's second Sustainable Development Goal (SDG) sets out to end hunger, achieve food security and improved nutrition, and promote sustainable agriculture by 2030. Yet despite the tireless efforts of the farmers, researchers, and policymakers the world over, over 700 million people still suffer from undernourishment. To address this food security gap, global food production systems will need to expand their capacities and produce almost twice their existing levels to feed the world's ever-growing population.

It is a significant challenge that is further exacerbated by the increasing prevalence of extreme weather events, unpredictability in rainfall patterns, and pest-pathogen dynamics associated with global climate change, which continues to have a profoundly negative impact on agricultural productivity and farm incomes worldwide.

To address these challenges, it is clear that we need to open up pathways toward sustainable crop production, particularly in resource-constrained agricultural settings, as well as more inclusive farming practices that encourage the participation of women in agriculture.

Efforts to implement these sustainable innovations have resulted in improvements in food access, farmer autonomy, and the stabilization of agriculture's global carbon footprint, yet the challenge to reduce hunger remains pronounced in the Global South, particularly Asia and Africa, which house the largest proportion of people suffering from malnutrition and other nutrition-related issues. Furthermore, the agri-food systems in Asia and Africa are severely constrained by the subsistence nature of farming, declining land and other agricultural resources, increasing environmental pollution, soil and biodiversity degradation, and climate change.

This book series, Sustainable Agriculture and Food Security, aims to support global efforts toward sustainability by providing timely coverage of the progress, opportunities, and challenges of sustainable food production and consumption in Asia and Africa. The series narrates the success stories and research endeavors from the regions of Africa and Asia on issues relating to SDG 2: Zero Hunger. It fosters research in transdisciplinary academic fields spanning sustainable agriculture systems and practices, pre-seed and post-harvest innovations, and food supply chains.

The focus of the series is to provide a comprehensive publication platform and to act as a repository of knowledge in sustainability sciences with a particular focus on developing nations.

Agriculture's contribution to global greenhouse gas emissions is significant, and as the world's population continues to rise, so too will agriculture's contribution to climate change. This ultimately results in a negative feedback loop, whereby farmers need more land and more chemical inputs to offset lower yields and productivity. Therefore, to successfully meet growing demand without exacerbating the negative impacts of climate change, it is essential that we embrace sustainable crop production approaches. However, it is also vital that efforts to deliver sustainable global food systems do not come at the cost of crop productivity and yield gains, that they do not come at the detriment of farmers, and that they factor in the everyday realities of participants from across the agriculture value chain.

With that in mind, this book, *Sustainable Crop Production: New Research Paradigms in Plant Sciences*, edited by my co-editors, Anandita Singh, Sandip Das, and myself, Rajeev K. Varshney, serves as a timely and important resource for new research approaches and farming practices that enable more sustainable and environmentally conscious agricultural systems.

The book presents 15 chapters of diverse, multidisciplinary research that has been loosely collected into five broad themes, covering new approaches in established research areas, such as photosynthesis research, to more futuristic research fields, such as astrobiology. Contributing to these 15 chapters are subject matter experts and practitioners from across the globe, each containing up-to-date and important insights into this broad, yet critically important topic. I would like to thank all contributing authors for their valuable insights, and for their continued efforts to deliver a more equitable, sustainable, and nutritious food future.

I am sure this book is a great resource for students, researchers, farmers, educators and extension agents, policymakers, and other stakeholders engaged in transforming agriculture around the world.

I wish to extend my sincere thanks and gratitude to the Springer staff, particularly Sanchi Bhimrajika, Associate Editor (Books, Medicine and Life Sciences), and Nick Melchior, Editorial Director (Books, Asia Medicine and Life Sciences), for their constant support in the accomplishment of this compendium. The cooperation received from my supervisors and research leaders, such as Prof. Andrew Deeks (Vice Chancellor), Prof Peter Eastwood (Deputy Vice Chancellor- Research) and Prof. Peter Davies (Pro Vice Chancellor, Food Futures Institute) from Murdoch University (Australia), is also gratefully acknowledged. I would like to thank my family members—Monika Varshney, Prakhar Varshney, and Preksha Varshney, for their love and support in discharging my duties as Series Editor.

Perth, WA, Australia Rajeev K. Varshney

Preface

Science and innovation continue to steer global efforts toward eradication of hunger and poverty, and, in achieving the shared goal of food sovereignty. The Green Revolution of mid-twentieth century is a historical landmark illustrating the power of technology in enhancing food production. Despite these developments, world continues to grapple with hunger, chronic malnutrition, and economic inequalities.

The largescale climate disruption, ecological vulnerabilities, and depletion of natural resources pose constant challenge to crop production. The global goal of United Nations aims to ensure access to safe, nutritious, and sufficient food. Under current trends, this is unlikely to be met by 2030. Harnessing research leads from modern plant sciences is therefore not only aligned with SDG2, but is also essential for bolstering yields of crop varieties that may no longer remain resilient in agroecologies under altered climatic conditions. Agricultural targets must be achieved while balancing the conflicting demands of agriculture and biodiversity conservation. Sustainable crop production must thus include adoption of holistic multidisciplinary approaches that are grounded on fundamental principles of plant sciences. To realize this vision, academics and researchers need to be exposed to topical themes in plant research that have the potential to transform agriculture. This book was conceived to inspire curiosity and stimulate creative scientific pursuits, and, for enable researchers to address critical bottlenecks in sustainable crop production.

The compilation brings together diverse research themes, offering a holistic view of current developments and emerging opportunities. By offering a repository of ideas, the editors aim to motivate future researchers to tackle the vexing research challenges that hinder effective translation of modern plant science into crop improvement. The book is organized into 15 chapters spanning five broad research domains.

The first theme explores how molecular insights from plant development can be harnessed to drive trait engineering. Understanding molecular mechanisms governing flowering time, formation of reproductive organs, fruit, seed, and overall plant architecture, is crucial for enhancing crop performance. Chapters 1, 2, 3, and 4 illustrate how lessons can be drawn from natural morphological innovations shaped by development and evolutionary processes to achieve trait manipulation. For example, modifying flowering time helps crops evade environmental stresses and enables inclusion of multiple cropping cycles within a single growing season whereas engineering optimal root architecture enhances resource efficiency in degrades soils.

The second theme, covered in Chaps. 5, 6, and 7, addresses enhancement of photosynthetic efficiency for increased biomass production and yield. The proposed strategies include Rubisco modification for efficient carbon fixation, reduced photorespiration, engineering carbon concentrating mechanisms, integration of photorespiratory bypasses, bioengineering of new-to-nature synthetic pathways and metabolic retrosynthesis, enhancing light use efficiency, and, synthesizing chloroplast mimics. Modification of chloroplast structure, distribution and membrane composition to boost photon capture, electron transport, and ATP synthesis are novel approaches. Deploying nanoparticles such as Quantum dots to improve light use efficiency demonstrates the interdisciplinary approach required for increasing crop production.

The third theme, covered in Chaps. 8, 9, 10, 11, 12, and 13, examines plant response to combined environmental stresses including interactions with pathogens. While molecular mechanisms underlying individual stresses are well-understood, it is important to uncover how multiple stresses operate simultaneously and impact crop performance. The range of topics include hormonal cross-talk, priming plant defense using immunity inducers, novel pattern recognition systems, rhizobiome engineering, root exudates influencing rhizobiomes, and rhizobial symbiosis for engineering biological N_2 fixation.

The fourth theme, Chap. 14, addresses heterosis breeding and hybrid seed production while emphasizing the importance of deciphering developmental basis of CMS, environmental sensitivity of male sterility and fertility restorer. Authors reflect upon insights from autoimmune-induced hybrid necrosis to support more robust hybrid development programs.

The final emerging theme is astrobiology. Chapter 15 explores an exciting but nascent research domain of growing crops beyond Earth. Futuristic as it may seem, human space habitation and extraterrestrial agriculture may soon become a reality. The authors discuss contemporary research aimed at testing feasibility of practicing agriculture in micro-gravity environments. The topics discuss how microgravity affects developmental regulation, hormone signaling, and overall crop productivity to enable future agriculture in space.

The compilation represents only a fraction of knowledge needed for achieving the goal of sustainable agriculture and food security. Several equally important topics could not be included current collection. These include apomixis, autoimmunity and hybrid necrosis, metabolic engineering, and synthetic biology. Likewise, the use of structural biology for designing novel trait alleles is still in its early stages.

A total of 60 researchers have authored 15 chapters and have summarized key actionable points. The editors express their deep sense of appreciation to all contributing authors who readily agreed to this collaborative academic endeavor. Their invaluable scholarship have shaped the contours of this book. A special word of appreciation to 13 domain experts whose careful evaluations and critical insights guided the review process. Their patience and constructive feedback have refined the overall quality of the book. It has been a sheer delight and honor to collaborate with such talented and dedicated academicians who authored and reviewed these contributions.

AS expresses heartfelt gratitude to TERI (The Energy and Resources Institute) and TERI SAS (TERI School of Advanced Studies), the two institutions that served as the cradle of sustainability science. Over last three decades, numerous mentors from TERI and TERI SAS shaped her perspectives on impacts of climate change, sustainable development of which sustainable agriculture is one component. Special gratitude to Late Dr. RK Pachauri, former Director General of TERI and former Chancellor of TERI SAS under whose able leadership, TERI and TERI SAS became agents of a massive global movement advocating pathways toward a more sustainable world. Sincere appreciation to co-editors. A word of gratitude to all teachers, collaborators, and students for constant support.

SD would like to express his gratitude toward University of Delhi, Delhi, for providing all necessary academic, infrastructure, and administrative support for research and teaching activities. SD is extremely grateful to all teachers, and past and present research scholars of his laboratory for constant motivation, support, and academic achievements.

RKV is thankful to his supervisors and research leaders, such as Prof. Andrew Deeks (Vice Chancellor), Prof Peter Eastwood (Deputy Vice Chancellor-Research) and Prof. Peter Davies (Pro Vice Chancellor, Food Futures Institute) as well as his colleagues from Murdoch University (Australia) for their support and cooperation.

We, the, editors hope that the themes compiled in this book inspire the present and future generations of students, scholars, scientists, and policymakers. It is our collective goal that the book brings all stakeholders together to achieve the challenges of sustainable agriculture for food and nutrition security.

New Delhi, Delhi, India — Anandita Singh
New Delhi, Delhi, India — Sandip Das
Perth, VA, Australia — Rajeev K. Varshney

List of Reviewers

Surekha Katiyar Agarwal
Department of Plant Molecular Biology,
University of Delhi, South Campus
Benito Juarez Marg, New Delhi, India

Joshua Kaste
Department of Civil and Environmental Engineering
University of Illinois Urbana-Champaign
Champaign, USA

Ashish Kumar
ICAR-National Institute for Plant Biotechnology,
Indian Agricultural Research Institute
New Delhi, India

Megan L. Matthews
Department of Civil and Environmental Engineering
University of Illinois Urbana-Champaign
Champaign, USA

H.P. Meena
ICAR—Indian Institute of Oilseeds Research
Hyderabad, India

Ashis Kumar Nandi
School of Life Sciences
Jawaharlal Nehru University
New Delhi, India

Sudhakar Reddy Palakolanu
International Crops Research Institute for the Semi-Arid Tropics (ICRISAT)
Hyderabad, India

Debabrata Panda
Department of Biodiversity & Conservation of Natural Resources
Central University of Odisha
Landiguda, Koraput, Odisha, India

Nirala Ramchiary
School of Life Sciences
Jawaharlal Nehru University
New Delhi, India

Julian Schroeder
Department of Cell and Developmental Biology
University of California- San Diego,
La Jolla, San Diego, USA

Bhaskar Sharma
Chemical and Environmental Engineering, Botany and Plant Sciences
Marlan and Rosemary Bourns College of Engineering
University of California, Riverside 350B Bourns Hall of Engineering
Riverside, California, USA

Shilpi Sharma
Department of Biochemical Engineering and Biotechnology,
Indian Institute of Technology-Delhi
New Delhi, India

Sudhir K Sopory
International Centre for Genetic Engineering and Biotechnology (ICGEB)
Aruna Asaf Ali Marg
New Delhi, India

Contents

Series Editor and Editorial Board Members

About the Series Editor

Rajeev K. Varshney, FRS, is a renowned agricultural scientist with over two decades of experience in genomics and molecular breeding. Currently, he holds multiple prestigious positions at Murdoch University in Australia. He has made a significant impact on global food security by creating genomic resources for major "orphan" tropical crops, using these resources to identify genetic loci and candidate genes for drought and pest tolerance in staple crops in sub-Saharan Africa and India, and spearheading international programs that deliver superior crop varieties to some of the world's most impoverished farmers. He is an accomplished author, with more than 600 publications to his name, and has been recognized as a Clarivate Highly Cited Researcher for ten consecutive years (2014–2023).

Food Futures Institute, Centre for Crop and Food Innovation, State Agricultural Biotechnology Centre, Agriculture and Food Security, Murdoch University, Perth, WA, Australia.

Editorial Board Members

About the Editors and Contributors

About the Editors

Anandita Singh is currently a Professor at the Department of Biotechnology, TERI School of Advanced Studies, New Delhi, India. She teaches courses in molecular genetics and mentors doctoral and postdoctoral researchers. Her special research interest lies in decoding the regulatory logic underlying developmental traits and translating fundamental insights into actionable strategies for oilseed mustard improvement. With over two decades of research experience, including a postdoctoral award from the Max Planck Institute of Developmental Biology, Tuebingen, Germany, she actively contributes to cross-disciplinary collaborations in functional genomics. Her work on early flowering transgenic Indian mustard was recognized among the top 15 scientific achievements in India.

Sandip Das a Professor at the Department of Botany, University of Delhi, is a plant molecular biologist with over 30 years of teaching and research experience. He earned his academic degrees from Jamia Hamdard and TERI, New Delhi (PhD), and from the University of Delhi (MPhil, MSc), specializing in crop genetics and plant breeding. He received postdoctoral training at the Max Planck Institute for Developmental Biology, Tuebingen, Germany, and Cornell University, Ithaca, USA. His research focuses on functional and regulatory diversification, regulatory RNAs, and organization and evolution of the polyploid genome, particularly in Brassicaceae. His group integrates comparative genomics, evolutionary developmental biology, and functional genomics to understand plant development and adaptation for trait modification and crop improvement. He has published 55+ research articles in leading international journals and has supervised a number of PhD, MPhil, and postdoctoral researchers.

Rajeev K. Varshney, FRS is a renowned agricultural scientist with over two decades of experience in genomics and molecular breeding. Currently, he holds multiple prestigious positions at Murdoch University in Australia, which are

Professor, Food Futures Institute; Director, Centre for Crop & Food Innovation; Director, WA State Agricultural Biotechnology Centre; and International Chair in Agriculture & Food Security. He has made a significant impact on global food security by creating genomic resources for major "orphan" tropical crops, using these resources to identify genetic loci and candidate genes for drought and pest tolerance in staple crops in Sub-Saharan Africa and India, and spearheading international programs that deliver superior crop varieties to some of the world's most impoverished farmers. He is an accomplished author, with more than 600 publications to his name, and has been recognized as a Clarivate Highly Cited Researcher for ten consecutive years (2014–2024).

Contributors

Mohana Sai Akula Amity Institute of Biotechnology, Amity University of Maharashtra, Mumbai, Maharashtra, India

Aryaman National Institute of Plant Genome Research, New Delhi, India

Prakhar Awasthi National Institute of Plant Genome Research, New Delhi, India

Sureshkumar Balasubramanian School of Biological Sciences, Monash University Clayton, Clayton, VIC, Australia

Debjyoti Bandhu Banerjee Molecular Plant & Microbiology Laboratory (MPML), Department of Botany (Post Graduate), Ramakrishna Mission Vivekananda Centenary College, Kolkata, India

Anirban Bhar Molecular Plant & Microbiology Laboratory (MPML), Department of Botany (Post Graduate), Ramakrishna Mission Vivekananda Centenary College, Kolkata, India

Abhishek Bohra ICAR-Indian Institute of Pulses Research (IIPR), Kanpur, Uttar Pradesh, India

Halidev Krishna Botta National Institute of Plant Genome Research, New Delhi, India

Jordyn Coutts School of Biological Sciences, Monash University Clayton Campus, Clayton, VIC, Australia

Alok Das ICAR-Indian Institute of Pulses Research (IIPR), Kanpur, Uttar Pradesh, India

Sandip Das Department of Botany, University of Delhi, Delhi, India

Proharsha Dey Molecular Plant & Microbiology Laboratory (MPML), Department of Botany (Post Graduate), Ramakrishna Mission Vivekananda Centenary College, Kolkata, India

Febina Fernandez Plant Biotechnology and Disease Biology Division, BRIC-Rajiv Gandhi Centre for Biotechnology (BRIC-RGCB), Thiruvananthapuram, Kerala, India

Ramya Ganesan School of Biological Sciences, Monash University Clayton Campus, Clayton, VIC, Australia

Sumanti Gupta Department of Botany, Rabindra Mahavidyalaya, Hooghly, West Bengal, India

Zeik Hargreaves School of Biological Sciences, Monash University Clayton, Clayton, VIC, Australia

Özge İbiş Department of Chemistry, İstanbul Technical University, İstanbul, Türkiye

Ayushi Jaiswal Department of Botany, University of Delhi, Delhi, India

Mouesanao K. Kandjoze School of Life Sciences, University of Essex, Colchester, UK

Abhishek Kanojia Department of Botany, University of Delhi, Delhi, India

Kundan Kumar Department of Biological Sciences, Birla Institute of Technology & Science-Pilani, K. K. Birla Goa Campus, Sancoale, Goa, India

Mukesh Kumar National Institute of Plant Genome Research, New Delhi, India

Anish Kundu Plant Biotechnology and Disease Biology Division, BRIC-Rajiv Gandhi Centre for Biotechnology (BRIC-RGCB), Thiruvananthapuram, Kerala, India

Ashverya Laxmi National Institute of Plant Genome Research, New Delhi, India

Sangram K. Lenka Gujarat Biotechnology University, Gandhinagar, Gujarat, India

Manjima Mohanan Plant Biotechnology and Disease Biology Division, BRIC-Rajiv Gandhi Centre for Biotechnology (BRIC-RGCB), Thiruvananthapuram, Kerala, India

Yashwanti Mudgil Department of Botany, University of Delhi, Delhi, India

Sourav Mukherjee School of Biological Sciences, Monash University Clayton, Clayton, VIC, Australia

Sofiya Muniyasamy School of Biological Sciences, Monash University Clayton, Clayton, VIC, Australia

Sheeba Naaz National Institute of Plant Genome Research, New Delhi, India

Yogesh Negi Department of Biological Sciences, Birla Institute of Technology & Science-Pilani, K. K. Birla Goa Campus, Sancoale, Goa, India

Anjali Pande National Institute of Plant Genome Research, New Delhi, India

Shalini Pareek ICAR-Indian Institute of Pulses Research (IIPR), Kanpur, Uttar Pradesh, India

Ekta Pokhriyal Department of Botany, University of Delhi, Delhi, India

Chandan Kumar Pradhan School of Biological Sciences, Monash University Clayton Campus, Clayton, VIC, Australia

Indian Institute of Technology-Bombay, Mumbai, Maharashtra, India

Susithra Priyadarshan Mugunthan Amity Institute of Biotechnology, Amity University of Maharashtra, Mumbai, Maharashtra, India

Tilahun Rabuma Department of Biotechnology, College of Natural and Computational Science, Wolkite University, Wolkite, Ethiopia

Plant RNAi Biology Group, International Centre for Genetic Engineering and Biotechnology, New Delhi, India

Shiva Sai Ravva Department of Biological Sciences, Birla Institute of Technology & Science-Pilani, K. K. Birla Goa Campus, Sancoale, Goa, India

Sanjay Singh Rawat National Institute of Plant Genome Research, New Delhi, India

Harshita Bharti Saksena National Institute of Plant Genome Research, New Delhi, India

Neeti Sanan-Mishra Plant RNAi Biology Group, International Centre for Genetic Engineering and Biotechnology, New Delhi, India

Shital Sandhya National Institute of Plant Genome Research, New Delhi, India

Aishwarye Sharma National Institute of Plant Genome Research, New Delhi, India

Richa Shukla Department of Botany, University of Delhi, Delhi, India

Gurbir Kaur Sidhu TERI School of Advanced Studies, New Delhi, India

Pallavi Singh School of Life Sciences, University of Essex, Colchester, UK

Shivam Singh Department of Biological Sciences, Birla Institute of Technology & Science-Pilani, K. K. Birla Goa Campus, Sancoale, Goa, India

Penna Suprasanna Amity Center for Nuclear Biotechnology, Amity Institute of Biotechnology, Amity University of Maharashtra, Mumbai, Maharashtra, India

Sridevi Sureshkumar School of Biological Sciences, Monash University Clayton Campus, Clayton, VIC, Australia

Kajal Taneja Department of Botany, University of Delhi, Delhi, India

Caner Ünlü Department of Chemistry, İstanbul Technical University, İstanbul, Türkiye

Poonam Yadav Department of Botany, University of Delhi, Delhi, India

Ritu Yadav Department of Botany, University of Delhi, Delhi, India

Chapter 1
Flowering Time: A Target Trait to Enhance Agricultural Productivity

Sourav Mukherjee, Ramya Ganesan, Zeik Hargreaves, and Sureshkumar Balasubramanian

Abstract Flowering time is an important phenotype linked with plant adaptation. In the context of food crops, flowering time has a significant impact on agricultural productivity. Climate change has negative impacts on agricultural productivity that poses a threat to the global food security, particularly with the growing population. Thus, development of climate-friendly crops is essential for sustainable agriculture, and flowering time is an important trait in that context. Flowering time regulation at the molecular level is well studied across different plants, including crops. Major insights on the molecular pathways regulating flowering time have come from studies on the model plant *Arabidopsis thaliana*. Study of natural variation provides insights of evolution of plants, crop domestication and plant breeding. These insights, combined with recent advances in genetic tools, offer new opportunities for improving crop yield and agricultural productivity. Incorporating genomics together with predictive modelling and artificial intelligence, flowering time could emerge as a key trait for the development of climate-smart crops.

Keywords Flowering · Climate change · Genomics · CRISPR · Photoperiod

Editor's Notes

Floral transition is a key life-history trait that fosters adaptation and survivability in plants under fluctuating environments. Plants perceive, relay, and integrate multiple environmental cues and endogenous signals by way of complex regulatory pathways to take the crucial decision on flowering. This maximizes reproductive success, growth, and survival of the progeny. In crops, unfavorable local environments manifest as severe yield penalties. Thus understanding control of flowering is essential for manipulating flowering time. Particularly, the generation of early flowering

Authors Sourav Mukherjee and Ramya Ganesan have equally contributed to this chapter.

S. Mukherjee · R. Ganesan · Z. Hargreaves · S. Balasubramanian (✉)
School of Biological Sciences, Monash University, Clayton, VIC, Australia
e-mail: mb.suresh@monash.edu

A. Singh et al. (eds.), *Sustainable Crop Production*, Sustainability Sciences in Asia and Africa, https://doi.org/10.1007/978-981-95-7844-3_1

crop varieties has been a long-standing breeder's goal for mitigating crop losses. Shorter life cycles associated with earliness in flowering allow crops to evade environmental stresses.

This chapter by Mukherjee et al. provides a comprehensive account on how multiple input pathways, epistatic interactions, chromatin remodeling factors, and RNA processing, intricately control flowering time in context to plant adaptations. The authors build focus on thermo-sensory pathways for regulation of flowering time given its significance under global warming scenario. A major area relates to exploiting natural variation in flowering time by genomic selection for introgression of useful alleles from crop wild relatives. The authors encourage scholars and breeders to leverage modern technologies to identify and characterize functional orthologs, investigate underlying GRNs, dissect genotype-phenotype complexities, and study the trait trade-offs in the context of natural variation. All these are required to translate knowledge from model plants to crops.

Actionable Points

1. Flowering time is highly sensitive to environmental cues, with minor changes in photoperiod or temperature significantly impacting yield and quality. While early flowering can be advantageous in certain conditions, it may reduce vegetative biomass and ultimately lower yields in leafy vegetables as well as in some crops such as maize and sorghum. In rice, alterations in photoperiod affect grain quality, while early flowering in maize can affect biomass and yield.
2. Comparative genomics has shown that *FT* genes function similarly across many crop species. For example, the discovery of *FT* homologues in tomato and soybean has enabled options for precisely regulating their flowering. Improved sequencing technologies enable researchers to study wild relatives of domesticated crops more effectively, helping to uncover natural genetic variations that contribute to valuable traits like drought resistance.
3. Integrating marker-assisted selection (MAS) and genomic selection (GS) with multi-environment high-throughput phenotyping enables precise identification of climate-resilient alleles while maintaining genetic diversity essential for biotic and abiotic stress tolerance. Precise introgression of beneficial alleles while reducing linkage drag can be achieved with MAS and GS.
4. CRISPR/Cas9 revolutionises plant breeding by enabling precise edits to flowering genes without introducing foreign DNA, increasing regulatory acceptance; however, off-target effects remain a key limitation, potentially causing unintended mutations. CRISPR-mediated knockouts of floral repressors like *TFL1* in tomato and *Ghd7* in rice have led to earlier flowering and enhanced yield.
5. Combining early flowering alleles with heat-responsive genes in wheat has led to varieties that perform well in high-temperature conditions. AI-driven integration of genomic, phenotypic and environmental data can be useful in enabling user-friendly platforms to identify desirable flowering traits linked with stress-resilient alleles, facilitating the development of climate-smart, high-yielding crops tailored to specific environments.

1.1 Introduction

The advent of agriculture is one of the key milestones of human progress. Domestication of crops over the years has resulted in crop varieties that are suitable and profitable from a human context. As such, natural life history traits of plants have been influenced, modified and selected upon over time to enhance agricultural productivity. For example, seed shattering, a key requirement for seed dispersal that enables plants to propagate and thrive in nature, needs to be substantially reduced to mitigate losses in human-aided agriculture. Therefore, such traits have been selected during domestication of crops for agriculture. For example, domestication of rice involved selection for varieties with less seed-shattering properties (Konishi et al. 2006).

Agricultural productivity is also heavily dependent on environmental and genetic factors (Ma et al. 2022). Growth environment, in particular predation and climate (temperature, humidity, rainfall, etc.), has a significant influence on strategies that plants utilise for their success (Abbass et al. 2022; War et al. 2012). Plant adaptations are among strategies that involve gradual or radical modifications in key life history traits. Such strategies are used by plants to adapt and manage hostile environments. Remodelling of life history traits allows them to develop into morphotypes that favour fitness in an altered environment, which is crucial for plant success especially owing to their general lack of motility (Howe et al. 2005). Several of these life history traits, including flowering time (Carlquist 2012), are relevant to crop productivity and sustainable agriculture (Cruz et al. 2021; Eckardt 2010).

The International Panel on Climate Change portends that agriculture is among the key sectors that will be adversely impacted by global warming (Masson-Delmotte et al. 2018). Climate change-associated alterations will likely outpace the natural diversification of plants (Franks and Weis 2008; Habib-ur-Rahman et al. 2022; Martén-Rodríguez et al. 2025). Heat stress, especially at the flowering and grain filling stage, can have massive impacts on seed set. In addition, even small changes in temperature at growth temperatures are predicted to have significant impacts of crop productivity. With every 1 °C rise in temperature, significant drop in agricultural productivity is predicted in all major food crops (Challinor et al. 2014; Rezaei et al. 2023; Zhao et al. 2017). A net result of all of these factors poses a major threat to food security and sustainable agriculture, especially in the context of ever-growing global population.

Plant growth is primarily regulated at the shoot and root meristems. The shoot meristem starts producing leaves, and at some point in its life cycle, it makes a switch and forms flowers instead of leaves. The timing of this transition, referred to as “flowering time”, is one of the critical decisions that a plant makes in its life cycle. Flowering essentially means that the plants will make the flowers, set seeds and complete its life. Thus, a decision to flower is a key life history trait for a plant, and unsurprisingly, flowering time is highly regulated. Flowering time is a quintessential adaptive trait for plant adaptation and has major influences on agricultural productivity. Flowering at a suitable time in coordination with local environmental conditions is critical for plant's reproductive success and seed set (Blümel et al. 2015). In out crossing plants, asynchronous flowering, for example, can reduce the success of seed set (Liu et al. 2021). The seasonal fluctuation in photoperiod and temperature during winter and summer significantly influences the transition of plants from a vegetative stage to reproductive stage by adapting to rapid cycling or delaying flowering until the completion of winter to synchronise with favourable spring conditions (Prevéy 2020). A variety of cues, including day length, temperature and abiotic stress, are all integrated to fine-tune appropriate timing of flowering.

From an agronomic perspective, flowering time is of particular importance for several reasons. The success of plant breeding involves crossing, and therefore, availability of varieties and cultivars with stable flowering time is highly desirable (Alter et al. 2016). In crops, the edible parts of the crops vary across species. Flowering marks the culmination of the vegetative phase, when plants cease leaf production and begin floral development to facilitate seed set. This has varied implications based on which parts of the crops are edible. For crops where the seeds are the edible parts (rice, maize, wheat, canola, mustard) often having earlier flowering would indicate multiple generations to be planted within a year. Conversely, in leafy vegetables, premature flowering leads to reduced vegetative yield, as the plant diverts energy away from leaf development. Furthermore, stress factors such as herbivory can induce flowering prematurely, compromising both the quality and quantity of the edible parts, whether seeds or leaves. Thus, the timing of flowering directly influences the quality, yield and sustainability of crop production systems (Chen et al. 2019; Mutasa-Göttgens et al. 2010; Sujariya et al. 2023). Germplasm collections of plants fortuitously harbour natural variation in flowering time, which present a valuable genetic reservoir for dissecting the underlying genetic mechanisms of floral transition and addressing how environmental factors influence the time of flowering (Fournier-Level et al. 2022).

In summary, flowering time is a critical life history trait that can be manipulated to bolster crop yields. Thus, it becomes essential to understand the genes, pathways and mechanisms underlying the regulation of flowering time. In this context, the following sections will examine the major genetic pathways and key genes involved in flowering time regulation, analyse insights gained from the evolution of flowering through natural variation and conclude with current challenges, limitations, the use of CRISPR-based genome editing to advance sustainable agriculture and food security.

1.2 Flowering Time Pathways

Molecular genetic analysis with the model organism *Arabidopsis thaliana* resulted in uncovering multiple genes that upon mutation lead to a flowering time defects. A combinatorial analysis of these genes has allowed us to infer relationships between these genes and group them into different pathways. Based on these grouping, one could say that there are four main pathways that regulate flowering time. While other pathways also influence flowering time, focusing on these four provides a useful framework for understanding the process and help us decipher an instructive framework of major floral genes. These pathways are photoperiod pathways, vernalisation pathway, autonomous pathway and gibberellic acid (GA) pathway. These pathways integrate environmental cues (day length and temperature) with developmental signals from within the organism. Given the importance of temperature in the context of climate change, this chapter will also briefly discuss thermal regulation of flowering. There are multiple genes that have been identified to play a role in these pathways, which eventually converge on a set of genes, which are referred to as floral pathway integrators to promote floral transition. The discovery of these genes and pathways in *Arabidopsis* then further facilitates the discovery of flowering pathway genes in diverse range of plants. At a molecular level, various mechanisms, such as chromatin remodelling, RNA processing and hormonal signalling are involved in these pathways to optimise flowering in response to environmental and internal cues. This chapter will discuss few essential genes associated with flowering and mechanisms associated with each of these pathways that regulate flowering time.

1.2.1 Photoperiod Pathway

The quantum of light a plant experiences (day length/photoperiod) in a day has a significant effect on its flowering behaviour. This behaviour is due to a complex process that integrates circadian rhythms with defined light environments and includes various signalling to coordinate flowering for reproductive success. Based on the photoperiod responsiveness, plants are categorised into long-day plants, short-day plants or day-neutral plants. Arabidopsis is a facultative long day plant, in which long days promote flowering, but flowering can also occur in short days. Genetic screens resulted in the isolation of mutants in which late flowering was observed only in long days but not in short days (Koornneef et al. 1991, 1998), which indicated that these genes are required for promoting flowering specifically in long days. Subsequent map-based cloning and further molecular analysis revealed that the underlying genes operate at different steps to confer this long-day-specific phenotypic response. First, as one would expect, several genes associated with the perception of light form the first step of recognising day length. Second, genes that are associated with the circadian clock, which is again dependent on the light, form another group of proteins in this cascade. Third, there are proteins whose activity itself is influenced by light that plays a role in day-length-dependent flowering response (Fig. 1.1).

Red light receptors phytochromes (PHY) and blue light receptor cryptochromes (CRY) provide cues on light perception and subsequent signalling, and they form the first part of the photoperiodic pathway. There are at least five different phytochromes (PHYA, PHYB, PHYC, PHYD and PHYE) and two different cryptochromes (CRY1 and CRY2) known from *Arabidopsis* (Casal 2000). Mutations in all of them show flowering phenotypes, but they are not specific only to long days, since the light perception is also required to measure other photoperiods (e.g. short days). In the second set, there are genes such as *TIMING OF CAB EXPRESSION 1 (TOC1)*, *LONG HYPOCOTYL (LHY1)*, and *CIRCADIAN CLOCK ASSOCIATED 1 (CCA1)* that play a role in maintaining the circadian clock that also feed onto this photoperiodic pathway. The clock can be entrained both by light and temperature, and these have an impact on the perception of the plant as to how they have experienced day length, which is reflected in their physiological responses (Hayama and Coupland 2003).

The key protein of the photoperiodic pathway is CONSTANS (CO). Mutations in CO result in plants that show late flowering only in long-day conditions (Koornneef et al. 1991, 1998). CO is connected to the circadian clock via two other proteins. The FLAVIN-BINDING KELCH REPEAT F-BOX 1 (FKF1) protein is stabilised by the GIGANTEA (GI) protein under long days. GI also stabilises CO. FKF1 degrades CYCLING DOF FACTORS (CDFs) to repress CONSTANS (CO) transcription (Imaizumi et al. 2005; Sawa et al. 2007). Under long days, CO accumulation correlates with light exposure, which is stabilised by PHYTOCHROME A (PHYA) and CRYPTOCHROME 1/2 (CRY1/2) (Valverde et al. 2004). CO integrates the photoperiodic pathway by activating the floral integrator FLOWERING LOCUS T (FT), which is also the mobile signal, florigen, that is produced in the leaf vasculature and moves to shoot apical meristem (SAM) to induce flowering.

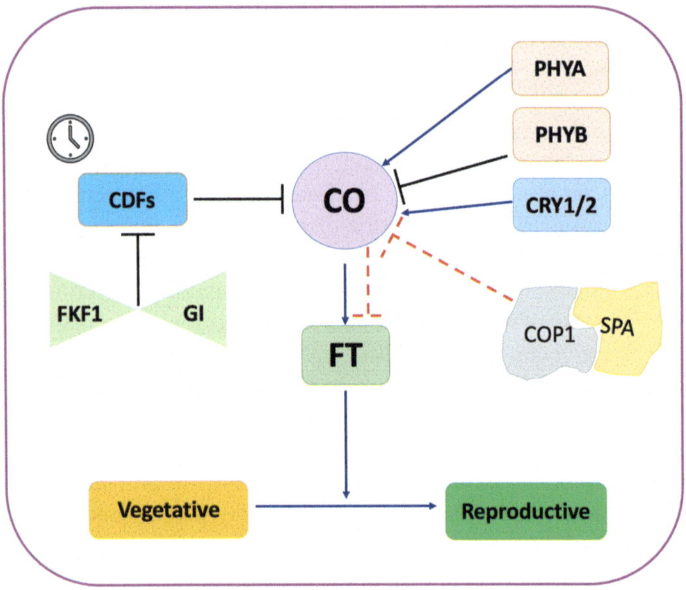

Fig. 1.1 *CONSTANS* is the key player in the photoperiod pathway. The signals of photoperiod are integrated by photoreceptors (PHY/CRY). Day length also influences CO stability through the COP1/SPA complex. CO also is affected by the circadian clock. CO affects *FT* levels, which in turn mediate floral transition

Flowering is promoted at the SAM by FT's interaction with FLOWERING LOCUS D (FD), which activates floral meristem identity genes such as APETALA1 (AP1) and SUPPRESSOR OF OVEREXPRESSION OF CO 1 (SOC1) (Abe et al. 2005; Wigge et al. 2005). In short days, the CONSTITUTIVE PHOTOMORPHOGENIC 1 (COP1)–SUPPRESSOR OF PHYA (SPA) complex degrades CO mRNA, which is transcribed outside of the light phase and inhibits FT activation and flowering (Jang et al. 2008; Laubinger et al. 2006).

1.2.2 Vernalisation Pathway

Flowering at an inappropriate time can be detrimental to the plants. For example, flowering during peak snow-covered winters is not ideal for seed set. Therefore, plants have evolved mechanisms that prevent them to flower during cold winters. Some plants can only flower after they experience winter conditions or vernalisation. Vernalisation is an essential epigenetic process that allows plants to sense and "remember" extended cold exposure, guaranteeing that flowering takes place under optimal conditions (Chouard 1960). By preventing early flowering in the fall, this adaptive mechanism increases the possibility of successful reproduction the following spring. *Arabidopsis* strains differ in their vernalisation requirement for flowering with rapid cyclers not requiring vernalisation and summer annuals presenting a need for vernalisation before flowering (Johanson et al. 2000; Lempe et al. 2005). Natural variation at two loci, FRIGIDA (FRI) and FLOWERING LOCUS C (FLC), explains most of the variation in the vernalisation requirement in the model plant *Arabidopsis thaliana* (Koornneef et al. 1994; Lempe et al. 2005; Napp-Zinn 1987; Shindo et al. 2005; Werner et al. 2005). Flowering in fall is delayed by FRI's upregulation of FLC, a MADS-box transcription factor that inhibits floral integrators like FLOWERING LOCUS T (FT) and SUPPRESSOR OF OVEREXPRESSION OF CONSTANS 1 (SOC1) (Michaels and Amasino 1999; Rouse et al. 2002).

FLC proved to be an excellent system, and studies on *FLC* have led to the discovery of some of the basic principles of epigenetic gene regulation in plants (Fig. 1.2). FLC has served as a key model for understanding epigenetically mediated gene regulation in plants. It appears that as a function of cold temperatures, through histone modifications, chromatin remodelling eventually silence FLC in winter. This requires the polycomb repressive complex 2 (PRC2)-like complex which consists of SWINGER (SWN), VERNALIZATION 2 (VRN2), FERTILIZATION-INDEPENDENT ENDOSPERM (FIE) and MULTICOPY SUPPRESSOR OF IRA 1 (MSI1) and causes the trimethylation of histone H3 lysine 27 (H3K27me3) at the *FLC* locus, thereby locking *FLC* in a repressed state (Finnegan and Dennis 2007; Yuan et al. 2016). VERNALIZATION-INSENSITIVE 3 (VIN3), a plant homeodomain (PHD)-finger protein that initiates the vernalisation response by histone deacetylation, further aids in this silencing (Qüesta et al. 2016; Sung and Amasino 2004). The recruitment of VRN1 and VRN2, which sustains FLC repression after vernalisation, is preceded by cold-dependent VIN3 induction (Heo and Sung 2011;

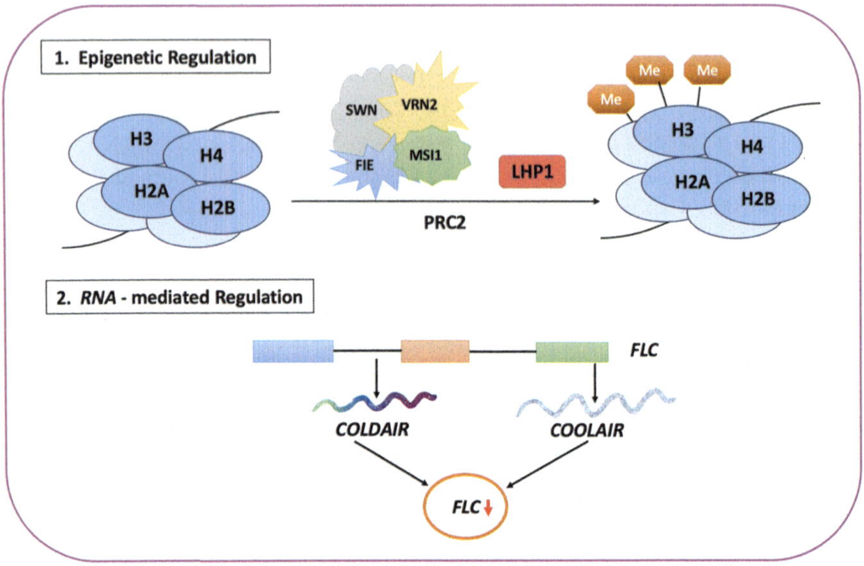

Fig. 1.2 Regulation of *FLC* is the key gene of the vernalisation pathway. Regulation of *FLC* occurs through epigenetic regulation involving the PRC2 complex and the LHP1 protein. This epigenetic regulation is also through the involvement of anti-sense RNAs *COLDAIR* and *COOLAIR*, which in turn lead to the epigenetic silencing of *FLC upon vernalisation*

Li et al. 2018, 2020). LIKE-HETEROCHROMATIN PROTEIN 1 (LHP1) stabilises FLC silencing by recognising histone marks like H3K27me3 which ensures that vernalisation memory is retained until embryogenesis resets it in the next generation (Feng and Lu 2017).

There is also a role for noncoding RNA in this epigenetic silencing. FLC is modulated by noncoding RNAs, specifically COOLAIR and COLD-ASSISTED INTRONIC NONCODING RNA (COLDAIR), which interact with PRC2-like complexes to mediate repression of transcription (Heo and Sung 2011). An anti-sense transcript called COOLAIR, which arises from the 3' region of FLC, helps remodel chromatin, and interacts with RNA 3'-processing factors like CstF64 and CstF77 to change the expression of FLC (Li et al. 2018), while COLDAIR, which is transcribed from an intronic region of FLC facilitates PRC2 recruitment, ensures H3K27me3 deposition for stable repression (Qüesta et al. 2016). During vegetative growth, PRC2-like complex consisting of CURLY LEAF (CLF), EMBRYONIC FLOWER 2 (EMF2), and FIE represses FT indicating that a larger regulatory network combines temperature cues with flowering control. This suggests that the role of chromatin modifiers in vernalisation goes beyond FLC (Li et al. 2018). Also, FLC physically interacts with the floral repressor protein SHORT VEGETATIVE PHASE (SVP) to strengthen its suppression of SOC1 and FT (Michaels et al. 2003). LHP1 and chromatin remodelling factors are recruited during SVP-mediated repression, which strengthens the transcriptional blockade in *Arabidopsis* accessions that need vernalisation (Yuan et al. 2016). The complex mechanisms of temperature-dependent floral transition are highlighted by the interaction of FLC, SVP and epigenetic regulators. Vernalisation is mostly dependent on chromatin remodelling, while gibberellin pathways and photoperiodic signalling also interact with vernalisation-induced flowering to provide the optimal seasonal synchronisation (Tilmes et al. 2019).

1.2.3 Autonomous Pathway

Flowering time mutants in which flowering was delayed irrespective of their day length were referred to be acting through the autonomous pathway. This pathway basically ensures that the floral repressor *FLC* is kept in check (Fig. 1.3). Given the impact of vernalisation on *FLC*, the autonomous pathway mutants also showed a strong response to vernalisation (Koornneef et al. 1991; Martinez-Zapater and Somerville 1990). The floral transition in autonomous pathway is facilitated by the downregulation of FLC expression via mechanisms involving RNA processing and chromatin modifications. Major components of this pathway include *FCA*, *FPA*, *FY*, *FLOWERING LOCUS K (FLK)*, *LUMINIDEPENDENS (LD)*, *FLOWERING LOCUS D (FLD)*, *FVE* and *RELATIVE OF EARLY FLOWERING 6*, which function to repress *FLC* and thereby permit the activation of flowering-promoting genes (Simpson 2004).

The primary mechanism of the autonomous pathway includes RNA-mediated regulation of *FLC* expression. *FCA* and *FPA* have been demonstrated to encode RNA-binding proteins that control the alternative polyadenylation of an antisense

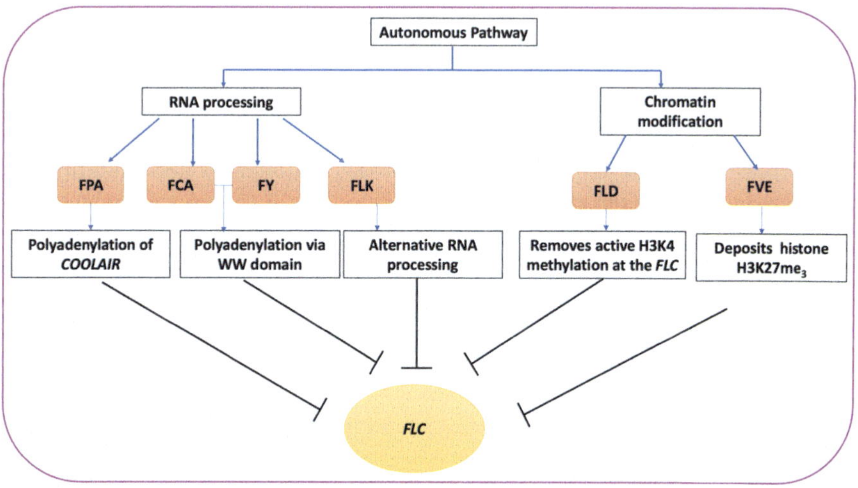

Fig. 1.3 Autonomous pathway genes regulate *FLC* expression levels through RNA processing and chromatin modification

non-coding RNA-*COOLAIR*, at the *FLC* locus. Both FCA and FPA facilitate proximal polyadenylation of COOLAIR, leading to a repressive chromatin state at *FLC* and reduced *FLC* expression (Liu et al. 2010). Further, FCA interacts with FY, a polyadenylation factor via its WW domain; this complex modulates *FLC* mRNA levels through RNA 3′ end processing (Henderson et al. 2005). *FLK*, another RNA-binding protein, has been suggested in alternative RNA processing events at *FLC*, although the precise molecular mechanism by which *FLK* acts is not fully clear (Lim et al. 2004). These RNA-based regulatory mechanisms demonstrate that post-transcriptional control plays a pivotal role in autonomous pathway function.

Other important regulatory aspects of the autonomous pathway are the chromatin modifications. *FLD*, a histone demethylase, removes transcriptionally active H3K4 methylation marks at the *FLC* locus, leading to its repression. While *FVE*, a component of the PRC2 complex, promotes the deposition of repressive histone marks such as H3K27me3 on *FLC*, securing transcriptional silencing of *FLC* (Ausín et al. 2004). Further genetic studies confirm that all autonomous pathway components converge on the same regulator *FLC*, as mutations in these genes result in increased *FLC* expression and significant delays in flowering. Further studies confirm that the late-flowering phenotype of autonomous-pathway mutants is fully suppressed in an *flc*-null background (Michaels and Amasino 2001; Sheldon et al. 1999). Though all genes of the autonomous pathway led to *FLC* repression, their molecular function is diverse, including RNA processing, histone modification and chromatin remodelling. The interplay of these mechanisms maintains strong repression of FLC and appropriate timing of floral induction. It is worth noting that some of the autonomous pathway genes, such as *FCA* and *FPA*, are also known to play a role in general developmental processes like seed germination and transposon silencing (Bäurle et al. 2007), suggesting functions extending beyond flowering time regulation.

1.2.4 Gibberellic Acid: GA Pathway

Gibberellins (GAs) are important phytohormones that control flowering time by integrating environmental and endogenous signals (Fig. 1.4). GA signalling is particularly important in promoting flowering under noninductive conditions, such as short days in *Arabidopsis thaliana* (Andres and Coupland 2012). A key set of molecules involved in this pathway are the DELLA proteins that repress GA responses. DELLA proteins are degraded in the GA signalling pathway. The GA receptor GID1 (GIBBERELLIN INSENSITIVE DWARF1) undergoes a conformational shift in the presence of bioactive GAs, which facilitates its interaction with DELLA proteins to degrade them (Sun 2011). GA serves as the predominant flowering inducer under short-day conditions, and it directly activates SOC1 expression in the shoot apical meristem (SAM) that triggers flowering (Mutasa-Gottgens and Hedden 2009). DELLA proteins act as negative regulators of SOC1, sequestering the SOC1-activating transcription factors that already bind with SOC1. GA-mediated DELLA

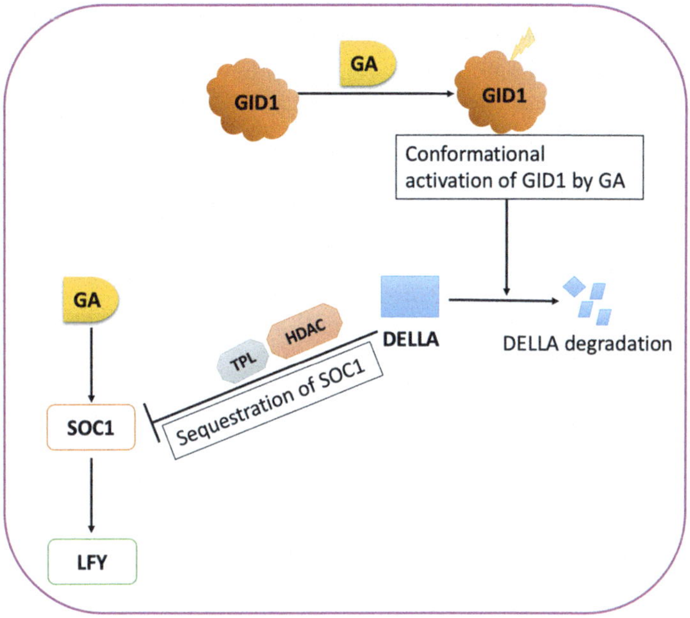

Fig. 1.4 GA regulates flowering by activating SOC1 and by removing the repression by degradation involving DELLA proteins

degradation releases this repression, enabling SOC1 to induce the expression of floral meristem identity genes such as LEAFY (LFY) (Davière and Achard 2013). GAs also cross talk with the vernalisation pathway to regulate flowering time.

1.2.5 Thermal Regulation of Flowering

Temperature controls flowering through multiple, distinct pathways involving a wide range of genes dispersed across various signalling cascades, including circadian clock components (ELF3), chromatin-level regulators (e.g. H2A.Z, PWR) and hormonal signalling pathways (BZR1). In general, high ambient temperatures promote flowering, while low temperatures delay it (Lee et al. 2023). PIF4, transcriptional factor, acts as the central integrator in regulating flowering by responding to thermal cues, whose expression and activity are elevated at higher temperatures. PIF4 directly influences the expression of FT (Kumar et al. 2012). Upstream regulators of PIF4 such as BZR1 and TCP5 are observed to increase the expression of PIF4 at elevated temperatures (Han et al. 2019; Ibañez et al. 2018). BZR1 mediates brassinosteroid (BR) signalling pathway. BR signalling is thermosensitive as BZR1 nuclear localisation is promoted at higher temperatures leading to increased PIF4 expression (Ibañez et al. 2018).

The circadian clock also integrates temperature signals to modulate flowering. ELF3, a component of the Evening Complex (EC), plays a critical role in temperature-dependent transcriptional repression of PIF4. At lower temperatures, ELF3 is diffusely localised and actively represses PIF4, both transcriptionally and through direct protein interaction (Nusinow et al. 2011; Yadav et al. 2025). However, at elevated temperatures, ELF3 forms nuclear speckles with ELF4 and LUX, resulting in its inactivation and loss of repression on PIF4 (Jung et al. 2020). Moreover, high temperature promotes ubiquitin-mediated proteasomal degradation of ELF3, further diminishing its repressive activity (Ezer et al. 2017). While the thermal sensitivity of ELF3 was initially attributed to its prion-like domain containing polyglutamine (PolyQ) repeats, subsequent studies have shown that temperature responsiveness is retained even in the absence of this domain (Zhu et al. 2024), indicating other contributing elements.

At the chromatin level, the histone variant H2A.Z plays a critical role. H2A.Z binds DNA more tightly than canonical H2, limiting transcription factor access and leading to reduced expression of target genes (Lei and Berger 2020). Importantly, its incorporation into nucleosomes is temperature-sensitive: High temperatures evict H2A.Z, particularly near the FT promoter, correlating with increased FT expression (Kumar and Wigge 2010). Similarly, histone deacetylation contributes to chromatin-level control of flowering. POWERDRESS (PWR), a SANT-domain protein, functions in association with HISTONE DEACETYLASE9 (HDA9) to deacetylate histone H3K9 at the +1 nucleosome of PIF4. This epigenetic modification is also temperature-sensitive, linking PWR activity to high-temperature-induced flowering (Tasset et al. 2018). Temperature-specific regulation of flowering also involves RNA processing, particularly alternative splicing as seen in FLM (negative regulator of FT) and posttranscriptional control of key regulators.

Together, these findings underscore a multilayered regulatory framework in which transcriptional, posttranscriptional and chromatin-based mechanisms work in combination to interpret ambient temperature changes. All of these regulatory layers ultimately converge on FT, establishing it as the central output node for temperature-mediated flowering control in *Arabidopsis*.

1.2.6 Influence of Other Signals for Flowering Regulation

In addition to these major pathways, there are other signals that have a significant impact of flowering time. This is not surprising, given that flowering is the most important decision in the life of a plant and any physiological changes will naturally have an influence on flowering time. Thus, mutations that lead to changes in flowering time are not necessarily "flowering time mutants" but rather it is an output that is expected to be affected by a variety of means. For example, light quality, ambient growth temperature, shade, nutritional status, various types of biotic and abiotic stress and the developmental age of the plants, etc. do influence the timing of flowering. Genes that play a role in mediating the impacts of these signals in flowering too typically converge on the same set of genes. In interest of brevity, we shall not discuss additional pathways, and readers are advised to refer to some excellent reviews that have been published (Freytes et al. 2021; Maple et al. 2024; Niu et al. 2024; Park et al. 2016; Zhao et al. 2025). Light quality also affects FT expression; PHYB encourages CO degradation, whereas CRY1/2 and PHYA stabilise CO during long days (Valverde et al. 2004). (FT-FD complexes activate floral meristem identity genes after FT transport to the SAM.) (Wigge et al. 2005; Yu et al. 2006).

1.2.7 Regulatory Interactions Among FT, SOC1 and FLC in Floral Induction

FLC, a repressor of flowering, which delays flowering in non-vernalised plants, can repress *SOC1* and *FT* expression. There is almost a perfect negative correlation between *FLC* expression and *SOC1* expression in *Arabidopsis* strains (Anandita Singh, personal communication). *FLC* is also a strong repressor of *FT*. Plants show high levels of *FT* expression in long days, and in short days, its expression is reduced.

FLC, FT and *SOC1* become an integrative point for all of these pathways, and a threshold level of FT essentially becomes the inflection point for a decision to flower (Fig. 1.5). The photoperiodic component CO induces *FT* expression, but this induction cannot occur when there are high levels of *FLC* expression. So, upon vernalisation, *FLC* levels decline, and the *FT* levels increase with the arrival of spring to ensure flowering. Similarly, the repression of *SOC1* can be counteracted by the application of GA which acts by reducing DELLA-mediated stabilisation of *FLC*-associated chromatin modifiers (Li et al. 2015, 2016). Additionally, gene regulation during floral transition is regulated by the combined action of GA signalling with chromatin remodelling factors. Inhibition of *SOC1* and *LFY* expression by repressor complexes like *TOPLESS (TPL)* and histone deacetylase (HDACS) can be mediated by DELLA proteins. These repressor complexes are disrupted by GA-dependent DELLA degradation, which opens up the chromatin, conducive to floral activation (Davière and Achard 2013). Thus, GA signalling regulates flowering by fine-tuning the transcriptional network through DELLA degradation, interactions with key flowering regulators and chromatin modifications. This pathway is therefore pivotal in ensuring that plants flower under suboptimal photoperiodic conditions by integrating multiple endogenous and environmental cues.

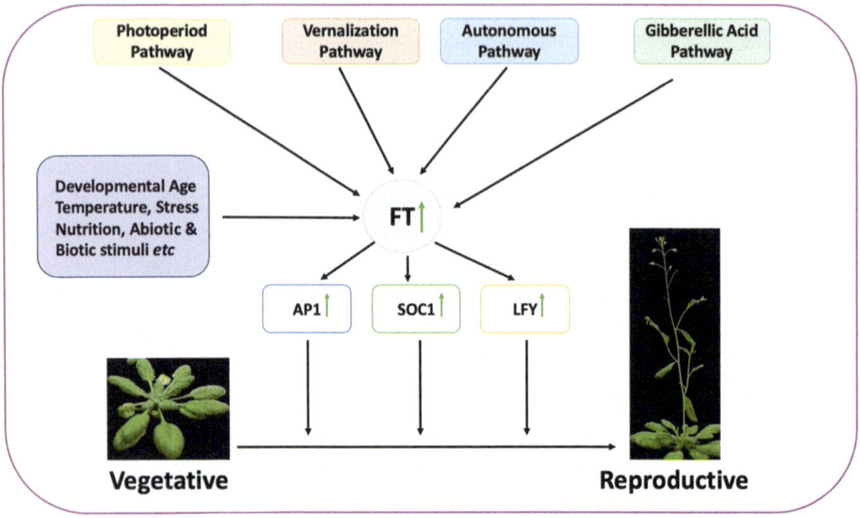

Fig. 1.5 A summary of flowering time pathways. Almost all pathways appear to integrate at the level of *FT*, and once the level of *FT* crosses a threshold, the plants make the transition from the vegetative phase to the flowering phase

1.3 Natural Variation and Evolution of Flowering

Flowering time is a highly variable trait that has evolved in response to diverse environmental cues such as photoperiod, temperature and water availability. There is a multilayered control system, and it appears that majority of these pathways or the principles of these pathways are evolutionarily conserved, while adaptive on a species level, and provide a framework to develop breeding strategies to control flowering time in crops leading to increased yield and adaptation to climate change. Natural variation in flowering time is observed across and within species, reflecting adaptations to local climates and ecological niches. *A. thaliana* exhibits extensive natural variation in flowering time across its geographic range. Early-flowering ecotypes, such as L*er* and Col-0, are adapted to warmer climates, while late-flowering ecotypes, such as Co-4, Ge-2 and Zu-0, are suited to cooler regions (Sanda et al. 1997). Genome-wide association studies (GWAS) and quantitative trait locus (QTL) mapping have further revealed the polygenic nature of flowering time variation, with numerous small-effect loci contributing to the trait (Atwell et al. 2010). *A. thaliana* has more than 1100 accessions identified and maintained so far, representing a rich resource for understanding the genetic basis of flowering and its adaptation to different environments.

Studies on natural variation in *Arabidopsis* also reveal that there are specific nodes in the flowering time pathways that appear to be selected for introducing natural variation in flowering. For example, common allelic variation at *FLC*, *FRI* and *PHYC* seem to contribute to natural variation in flowering time (Balasubramanian et al. 2006; Caicedo et al. 2004; Johanson et al. 2000; Lempe et al. 2005; Shindo et al. 2006). However, there are rare variations seen in several flowering time genes (Maple et al. 2024).

1.3.1 Evolution of Flowering

The way flowering has evolved across plants is remarkable. Analysis of flowering time in diverse species presents as excellent window to look into these pathways. Comparative studies in crop species and *A. thaliana* revealed how photoperiodic flowering have been optimised to ensure the life history of plants. For example, a pseudo-response regulator (PRR) that promotes early flowering under long days is encoded by photoperiod H1 (Ppd-H1) in barley. Middle Eastern strains of barley are adaptive to this trait to avoid drought. Mutant Ppd-H1 alleles found in Northern European barley varieties postpone flowering in order to prolong vegetative growth and boost yield (Turner et al. 2005). The domestication of rice, maize and other crops also relies on similar photoperiodic adaptations (Hayama et al. 2003; Izawa et al. 2002).

Comparative studies in rice, which is a short-day plant in which short days promote flowering, show how the CO-pathway has been optimised to achieve this response. Rice HEADING DATE 1 (Hd1), a homolog of CO, is induced in short days compared to long days in *Arabidopsis*. This in turn activates HEADING DATE 3a (Hd3a), the FT homolog to induce flowering in response to short days (Turck et al. 2008; Yano et al. 2000). The CO-FT module in Poplar regulates dormancy depending on the duration of the daylight, indicating that this pathway is both conserved and flexible (Böhlenius et al. 2006).

Similarly, the analysis of the vernalisation response in wheat revealed that they harbour another MADS-box gene (VRN2) that appears to regulate flowering time. FLC-clade is restricted to Brassicaceae. However, it appears that the same principles, such as a long first intron that has signatures of insertions, epigenetic silencing in response to cold, etc, make it clear that the principles of gene regulation for the vernalisation response may be similar despite having a completely different sets of genes for this pathway (Distelfeld et al. 2009; Dubcovsky et al. 2006; Trevaskis et al. 2006; Yan et al. 2004).

Flowering pathways are also optimised to for other responses. For example, FT homologs control other seasonal reactions beyond flowering. The FT homolog SP6A mediates tuberisation in *Solanum tuberosum* under short days, whereas SP3D encourages flowering (Navarro et al. 2011; Rodriguez-Falcon et al. 2006).

1.3.2 Evolutionary Shaping of Flowering Time Through Crop Domestication and Selection

Historically, farmers selectively cultivated plants that were well-suited to their local environmental conditions and agricultural practices. This process has significantly influenced flowering time and associated traits during the domestication of crops. Domestication often involved the selection of early flowering varieties to reduce the growing season and enable multiple harvests per year. For example, in cereal crops like wheat and barley, early flowering was preferred to escape drought and heat stress in arid regions. Similarly, in case of rice, the selection of photoperiod-insensitive varieties allowed for cultivation across globe. Studies reveal that these changes were driven by mutations in key regulatory genes of flowering time, such as *Ghd7* in rice (Lu et al. 2012) and *VRN1* in wheat (Dubcovsky et al. 1998), which were selected by farmers initially. Additionally, traits which influence flowering time, such as day-length sensitivity (photoperiodism), vernalisation requirement and inflorescence architecture, were also targeted during domestication (Nakamichi 2015). The domestication of flowering traits not only enhanced yield and adaptability but also set the foundation for modern breeding strategies.

1.3.3 Natural Variation in Plant Breeding

Natural variation in flowering time has long been fundamental to plant breeding, facilitating the cultivars development suited to diverse environmental conditions. Traditional breeding methods, such as cross-hybridisation and selection, have successfully harnessed natural alleles to improve flowering-related traits (Neri et al. 2018). A notable example is the incorporation of photoperiod-insensitive alleles from tropical maize into temperate varieties, broadening the crop's cultivation range (Coles et al. 2010). Advances in genomic technologies have enhanced the use of natural variation, with marker-assisted selection (MAS) and genomic selection (GS) enabling breeders to identify and incorporate beneficial alleles from wild relatives or landraces into elite cultivars. For instance, targeting the Hd1 gene, which regulates flowering time in rice, has led to early-maturing varieties adapted to short growing seasons (Vicentini et al. 2023).

Beyond flowering time, natural variation has also been instrumental in enhancing traits like inflorescence architecture, flower number and flowering synchrony. This approach has proven particularly valuable in adapting crops to climate change, helping mitigate challenges such as unpredictable growing seasons and the rising frequency of extreme weather events.

1.3.4 Genetic and Environmental Perturbations for Precision Breeding

The future of flowering time breeding hinges on the integration of genetic advancements with environmental adjustments to achieve precise and rapid improvements. Transgenic approaches, such as the overexpression or knockdown of flowering genes, offer effective strategies for fine-tuning flowering time. For example, overexpressing *FT* homologs in crops like tomato and potato has been shown to accelerate flowering and reduce the time to harvest (Plantenga et al. 2019; Xu et al. 2012).

Integrating genetic modifications with advanced cultivation techniques, such as speed breeding (Kabade et al. 2024), offers a powerful strategy for accelerating breeding cycles. Speed breeding leverages extended photoperiods and controlled temperatures to induce rapid flowering and seed set (Kabade et al. 2024). This has been shown to work in diverse crops (Choi et al. 2023; Gangashetty et al. 2024; Rodrmguez et al. 2023). When combined with CRISPR-Cas9 genome editing, this approach facilitates the precise introduction and evaluation of targeted mutations in flowering genes. For instance, CRISPR-mediated knockout of flowering repressors like *TFL1* in apples and pears has resulted in earlier flowering and a more compact growth habit (Ramirez-Torres et al. 2021).

Additionally, leveraging natural alleles through genomic approaches, alongside environmental adjustments, can optimise flowering time. Identifying alleles that enhance flowering plasticity could enable plants to adapt their flowering in response to temperature fluctuations, which can strengthen resilience to climate variability (Zhu et al. 2023). This integrative strategy, combining genetic insights with environmental control, presents a promising avenue for developing crops with precisely tailored flowering behaviours to meet the challenges of a changing climate.

1.4 Challenges, Limitations and Future Directions

Significant progress has been made in understanding and manipulating flowering time in plants, yet several challenges and limitations persist. Addressing these issues is crucial for advancing crop improvement and ensuring global food security in the face of climate change.

1.4.1 Translating Knowledge to Diverse Crops and Species

One of the major challenges in plant breeding is applying insights gained from model species, such as *Arabidopsis thaliana*, to a wide range of crop species. While *Arabidopsis* has been helpful in exploring the genetic regulation of flowering, many crop species exhibit distinct flowering mechanisms and reproductive strategies. For instance, perennial crops like apple and grapevine have complex flowering behaviours that differ substantially from annual crops such as wheat and rice (Rehman et al. 2023). However, studies on orthologous genes reveal that there is some level of conservation in function, which can be exploited for further understanding of flowering time in the species of interest (Rehman et al. 2023).

Comparative genomics and functional studies in non-model species are necessary for bridging this gap. Recent advancement in sequencing technologies has facilitated genome assembly for various crops, assisting in the discovery of orthologous flowering-related genes. For example, the discovery of *FT* homologs in tomato and soybean has allowed specific regulation of flowering time (Jiang et al. 2022; Liu et al. 2018). Wild relatives of domesticated crops also harbour essential genetic diversity for traits like drought tolerance and disease resistance (Kapazoglou et al. 2023). Integrating knowledge from model systems with crop-specific studies can lead to tailored breeding strategies that optimise flowering and other agronomic traits across diverse species.

1.4.2 Challenges and Lessons Learned

Altering flowering time in crops presents multiple challenges. This includes complexity of genetic regulatory networks, genotype-by-environment (G x E) interactions and trade-offs with other important agronomic traits. For example, while early flowering can be beneficial in some situation, it may reduce vegetative biomass and eventually negatively impact yield in crops like maize and sorghum (Farré and Faci 2006; Murphy et al. 2014). Likewise, selection for photoperiod insensitivity in rice has sometime led to reductions in grain quality (Sujariya et al. 2023). To address these challenges, plant breeders have combined traditional breeding approaches with molecular and computational tools. Marker-assisted selection (MAS) and genomic selection (GS) have facilitated the precise introgression of beneficial alleles while minimising linkage drag (Jacobsen and van der Vossen 2009). Multi-environment trials and high-throughput phenotyping have further refined our understanding of G x E interactions, supporting the development of robust cultivars. A major lesson from past breeding efforts is the importance of maintaining genetic diversity (Salgotra and Chauhan 2023). Over-reliance on a narrow genetic base has left many crops vulnerable to biotic and abiotic stresses (Kissoudis et al. 2014). Maintaining diverse germplasm is key for upgrading crop resilience and adaptability.

1.4.3 Gene Duplication and Polyploidy: Challenges and Opportunities

Gene duplication and polyploidy create both obstacles and opportunities in the investigation and regulation of flowering time. Many staple crops, including wheat, cotton and canola, are polyploids (Weeks 2017). This genomic complexity can make it harder to target and alter specific genes, as duplicated copies may exhibit functional redundancy or divergence. Flowering time analysis in canola as well as other mustard varieties reveals that it is feasible to assess the functional implications of specific orthologous genes. However, challenges remain in mechanistic elucidation of these genes (Raman et al. 2013, 2015, 2019). For instance, hexaploid wheat carries three copies of the *VRN1* gene, each contributing to vernalisation requirements (Konopatskaia et al. 2016). Disrupting a single copy may have little effect due to redundancy. However, polyploidy also promotes phenotypic diversity and evolutionary adaptation (Balao et al. 2011). In Brassica species, polyploidy has been correlated to increased variability in flowering time, enhancing adaptability across various environments. To overcome these complexities, scientists are utilising advanced genomic tools such as CRISPR-Cas9 to precisely target multiple gene copies (Lopez-Obando et al. 2016). Comparative analyses between diploid and polyploid species also uncover meaningful insights into the evolutionary history of gene duplication, influencing future breeding strategies.

1.4.4 CRISPR: Opportunities and Challenges

CRISPR-Cas9 genome editing has transformed plant breeding by establishing precise alteration to flowering genes and other traits. Unlike traditional transgenic approaches, CRISPR introduces targeted mutations without incorporating foreign DNA, making it more socially acceptable and less hindered by regulatory restrictions (Doudna and Charpentier 2014; Hamdan and Tan 2024). For example, CRISPR-mediated knockouts of flowering repressors such as *TFL1* in tomato (Sheng et al. 2023) and *Ghd7* in rice (Wang et al. 2020) have resulted in earlier flowering and improved yield.

Despite its advantages, CRISPR presents several challenges. Off-target effects can introduce unintended mutations with unpredictable consequences. Additionally, genome-editing efficiency varies across species and tissue types, necessitating crop-specific optimisation (Javaid and Choi 2021). Nevertheless, CRISPR offers huge potential for accelerating breeding cycles and enhancing climate resilience. By integrating CRISPR with speed breeding, scientists can quickly introduce and assess targeted mutations, strikingly reducing the time needed to develop improved crop cultivars.

1.4.5 Future Climate-Proofing Solutions

Climate change introduces an alarming challenge to global agriculture and food security (Mirzabaev et al. 2023). Climate change can be characterised by increasing earth temperatures, shifting growing seasons and increased frequency of extreme weather events (Asefa Bogale 2023). To upgrade climate resilience, breeders must develop crop varieties with greater phenotypic plasticity, allowing plants to adjust their flowering time in response to various environmental cues. One promising approach involves developing "climate-smart" crops that integrate early flowering with traits such as drought tolerance and high temperature resistance. For example, in wheat, the incorporation of early-flowering alleles with heat-responsive genes has produced varieties that perform well under high-temperature conditions (Wang et al. 2023). Similarly, in rice, submergence-tolerant varieties with adaptive flowering times have enhanced resilience to flooding (Singh et al. 2017).

Predictive modelling and artificial intelligence also have higher potential for climate-proofing crops. By integrating genomic, proteomic, phenotypic and environmental data, machine learning algorithms can be useful to discover beneficial genotype for specific conditions.

Acknowledgements We thank Sridevi Sureshkumar from Monash University for comments on this manuscript. SB lab group has been supported primarily by funds from the Australian Research Council and the National Health and Medical Research Council over the years. SM and RG are recipients of the Monash Graduate Scholarship. We attempted to summarise findings and presented in a simplified manner, and we apologise to all our colleagues whom we could not cite due to space limitations and attempted simplification of this large area of research.

References

Abbass K, Qasim MZ, Song H, Murshed M, Mahmood H, Younis I (2022) A review of the global climate change impacts, adaptation, and sustainable mitigation measures. Environ Sci Pollut Res 29(28):42539–42559

Abe M, Kobayashi Y, Yamamoto S, Daimon Y, Yamaguchi A, Ikeda Y, Ichinoki H, Notaguchi M, Goto K, Araki T (2005) FD, a bZIP protein mediating signals from the floral pathway integrator FT at the shoot apex. Science 309(5737):1052–1056. http://www.ncbi.nlm.nih.gov/entrez/query.fcgi?cmd=Retrieve&db=PubMed&dopt=Citation&list_uids=16099979

Alter P, Bircheneder S, Zhou L-Z, Schlüter U, Gahrtz M, Sonnewald U, Dresselhaus T (2016) Flowering time-regulated genes in maize include the transcription factor *ZmMADS1*. Plant Physiol 172(1):389–404

Andres F, Coupland G (2012) The genetic basis of flowering responses to seasonal cues [Research Support, Non-U.S. Gov't Review]. Nat Rev Genet 13(9):627–639. https://doi.org/10.1038/nrg3291

Asefa Bogale G (2023) Analysis the characterization of climate change and its impacts on smallholder farmers in Eastern Ethiopia. Heliyon 9(10):e20293. https://doi.org/10.1016/j.heliyon.2023.e20293

Atwell S, Huang YS, Vilhjalmsson BJ, Willems G, Horton M, Li Y, Meng D, Platt A, Tarone AM, Hu TT, Jiang R, Muliyati NW, Zhang X, Amer MA, Baxter I, Brachi B, Chory J, Dean C,

Debieu M, de Meaux J, Ecker JR, Faure N, Kniskern JM, Jones JD, Michael T, Nemri A, Roux F, Salt DE, Tang C, Todesco M, Traw MB, Weigel D, Marjoram P, Borevitz JO, Bergelson J, Nordborg M (2010) Genome-wide association study of 107 phenotypes in *Arabidopsis thaliana* inbred lines [Research Support, N.I.H., Extramural Research Support, U.S. Gov't, Non--P.H.S.]. Nature 465(7298):627–631. https://doi.org/10.1038/nature08800

Ausín I, Alonso-Blanco C, Jarillo JA, Ruiz-García L, Martínez-Zapater JM (2004) Regulation of flowering time by FVE, a retinoblastoma-associated protein. Nat Genet 36(2):162–166

Balao F, Herrera J, Talavera S (2011) Phenotypic consequences of polyploidy and genome size at the microevolutionary scale: a multivariate morphological approach. New Phytol 192(1):256–265. https://doi.org/10.1111/j.1469-8137.2011.03787.x

Balasubramanian S, Sureshkumar S, Agrawal M, Michael TP, Wessinger C, Maloof JN, Clark R, Warthmann N, Chory J, Weigel D (2006) The *PHYTOCHROME C* photoreceptor gene mediates natural variation in flowering and growth responses of *Arabidopsis thaliana* [Research Support, N.I.H., Extramural Research Support, Non-U.S. Gov't]. Nat Genet 38(6):711–715. https://doi.org/10.1038/ng1818

Bäurle I, Smith L, Baulcombe DC, Dean C (2007) Widespread role for the flowering-time regulators FCA and FPA in RNA-mediated chromatin silencing. Science 318(5847):109–112

Blümel M, Dally N, Jung C (2015) Flowering time regulation in crops—what did we learn from Arabidopsis? Curr Opin Biotechnol 32:121–129

Böhlenius H, Huang T, Charbonnel-Campaa L, Brunner AM, Jansson S, Strauss SH, Nilsson O (2006) *CO/FT* regulatory module controls timing of flowering and seasonal growth cessation in trees. Science 312(5776):1040–1043

Caicedo AL, Stinchcombe JR, Olsen KM, Schmitt J, Purugganan MD (2004) Epistatic interaction between *Arabidopsis FRI* and *FLC* flowering time genes generates a latitudinal cline in a life history trait. Proc Natl Acad Sci USA 101(44):15670–15675. http://www.ncbi.nlm.nih.gov/entrez/query.fcgi?cmd=Retrieve&db=PubMed&dopt=Citation&list_uids=15505218

Carlquist S (2012) How wood evolves: a new synthesis. Botany 90(10):901–940

Casal JJ (2000) Phytochromes, cryptochromes, phototropin: photoreceptor interactions in plants [Research Support, Non-U.S. Gov't Review]. Photochem Photobiol 71(1):1–11. http://www.ncbi.nlm.nih.gov/pubmed/10649883

Challinor AJ, Watson J, Lobell DB, Howden SM, Smith DR, Chhetri N (2014) A meta-analysis of crop yield under climate change and adaptation. Nat Clim Chang 4:287–291

Chen C, Huang W, Hou K, Wu W (2019) Bolting, an important process in plant development, two types in plants. J Plant Biol 62:161–169

Choi H, Back S, Kim GW, Lee K, Venkatesh J, Lee HB, Kwon JK, Kang BC (2023) Development of a speed breeding protocol with flowering gene investigation in pepper (*Capsicum annuum*). Front Plant Sci 14:1151765. https://doi.org/10.3389/fpls.2023.1151765

Chouard P (1960) Vernalization and its relations to dormancy. Annu Rev Plant Physiol 11(1):191–238

Coles ND, McMullen MD, Balint-Kurti PJ, Pratt RC, Holland JB (2010) Genetic control of photoperiod sensitivity in maize revealed by joint multiple population analysis. Genetics 184(3):799–812. https://doi.org/10.1534/genetics.109.110304

Cruz M, Arbelaez JD, Loaiza K, Cuasquer J, Rosas J, Graterol E (2021) Genetic and phenotypic characterization of rice grain quality traits to define research strategies for improving rice milling, appearance, and cooking qualities in Latin America and the Caribbean. Plant Genome 14(3):e20134

Davière J-M, Achard P (2013) Gibberellin signaling in plants. Development 140(6):1147–1151

Distelfeld A, Tranquilli G, Li C, Yan L, Dubcovsky J (2009) Genetic and molecular characterization of the *VRN2* loci in tetraploid wheat. Plant Physiol 149(1):245–257. https://doi.org/10.1104/pp.108.129353

Doudna JA, Charpentier E (2014) Genome editing. The new frontier of genome engineering with CRISPR-Cas9. Science 346(6213):1258096. https://doi.org/10.1126/science.1258096

Dubcovsky J, Lijavetzky D, Appendino L, Tranquilli G (1998) Comparative RFLP mapping of *Triticum monococcum* genes controlling vernalization requirement. Theor Appl Genet 97(5):968–975. https://doi.org/10.1007/s001220050978

Dubcovsky J, Loukoianov A, Fu D, Valarik M, Sanchez A, Yan L (2006) Effect of photoperiod on the regulation of wheat vernalization genes *VRN1* and *VRN2*. Plant Mol Biol 60(4):469–480. http://www.ncbi.nlm.nih.gov/entrez/query.fcgi?cmd=Retrieve&db=PubMed&dopt=Citation&list_uids=16525885

Eckardt NA (2010) Evolution of the domesticated bread wheat. Plant Cell 22(4):993

Ezer D, Jung J-H, Lan H, Biswas S, Gregoire L, Box MS, Charoensawan V, Cortijo S, Lai X, Stöckle D (2017) The evening complex coordinates environmental and endogenous signals in *Arabidopsis*. Nat Plants 3(7):1–12

Farré I, Faci JM (2006) Comparative response of maize (*Zea mays* L.) and sorghum (*Sorghum bicolor* L. Moench) to deficit irrigation in a Mediterranean environment. Agric Water Manag 83(1):135–143. https://doi.org/10.1016/j.agwat.2005.11.001

Feng J, Lu J (2017) *LHP1* could act as an activator and a repressor of transcription in plants. Front Plant Sci 8:2041

Finnegan EJ, Dennis ES (2007) Vernalization-induced trimethylation of histone H3 lysine 27 at *FLC* is not maintained in mitotically quiescent cells. Curr Biol 17(22):1978–1983

Fournier-Level A, Taylor MA, Paril JF, Martínez-Berdeja A, Stitzer MC, Cooper MD, Roe JL, Wilczek AM, Schmitt J (2022) Adaptive significance of flowering time variation across natural seasonal environments in *Arabidopsis thaliana*. New Phytol 234(2):719–734

Franks S, Weis A (2008) A change in climate causes rapid evolution of multiple life-history traits and their interactions in an annual plant. J Evol Biol 21(5):1321–1334

Freytes SN, Canelo M, Cerdan PD (2021) Regulation of flowering time: when and where? Curr Opin Plant Biol 63:102049. https://doi.org/10.1016/j.pbi.2021.102049

Gangashetty PI, Belliappa SH, Bomma N, Kanuganahalli V, Sajja SB, Choudhary S, Gaviyappanavar R, Bomireddy D, Anil Kumar V, Pranati J, Sharma M, Pandey MK (2024) Optimizing speed breeding and seed/pod chip based genotyping techniques in pigeonpea: a way forward for high throughput line development. Plant Methods 20(1):27. https://doi.org/10.1186/s13007-024-01155-w

Habib-ur-Rahman M, Ahmad A, Raza A, Hasnain MU, Alharby HF, Alzahrani YM, Bamagoos AA, Hakeem KR, Ahmad S, Nasim W (2022) Impact of climate change on agricultural production: issues, challenges, and opportunities in Asia. Front Plant Sci 13:925548

Hamdan MF, Tan BC (2024) Genetic modification techniques in plant breeding: a comparative review of CRISPR/Cas and GM technologies. Hortic Plant J 11(5):1807–1829. https://doi.org/10.1016/j.hpj.2024.02.012

Han X, Yu H, Yuan R, Yang Y, An F, Qin G (2019) *Arabidopsis* transcription factor TCP5 controls plant thermomorphogenesis by positively regulating PIF4 activity. Iscience 15:611–622

Hayama R, Coupland G (2003) Shedding light on the circadian clock and the photoperiodic control of flowering. Curr Opin Plant Biol 6(1):13–19. http://www.ncbi.nlm.nih.gov/entrez/query.fcgi?cmd=Retrieve&db=PubMed&dopt=Citation&list_uids=12495746

Hayama R, Yokoi S, Tamaki S, Yano M, Shimamoto K (2003) Adaptation of photoperiodic control pathways produces short-day flowering in rice. Nature 422(6933):719–722. http://www.ncbi.nlm.nih.gov/entrez/query.fcgi?cmd=Retrieve&db=PubMed&dopt=Citation&list_uids=12700762

Henderson IR, Liu F, Drea S, Simpson GG, Dean C (2005) An allelic series reveals essential roles for FY in plant development in addition to flowering-time control. Development 132(16):3597–3607. http://www.ncbi.nlm.nih.gov/entrez/query.fcgi?cmd=Retrieve&db=PubMed&dopt=Citation&list_uids=16033802

Heo JB, Sung S (2011) Vernalization-mediated epigenetic silencing by a long intronic noncoding RNA. Science 331(6013):76–79

Howe GT, Brunner AM, De la Barrera E, Andrade JL (2005) An evolving approach to understanding plant adaptation. New Phytol 167(1):1–5

Ibañez C, Delker C, Martinez C, Bürstenbinder K, Janitza P, Lippmann R, Ludwig W, Sun H, James GV, Klecker M (2018) Brassinosteroids dominate hormonal regulation of plant thermomorphogenesis via BZR1. Curr Biol 28(2):303–310. e303

Imaizumi T, Schultz TF, Harmon FG, Ho LA, Kay SA (2005) FKF1 F-box protein mediates cyclic degradation of a repressor of *CONSTANS* in *Arabidopsis*. Science 309(5732):293–297

Izawa T, Oikawa T, Sugiyama N, Tanisaka T, Yano M, Shimamoto K (2002) Phytochrome mediates the external light signal to repress *FT* orthologs in photoperiodic flowering of rice. Genes Dev 16(15):2006–2020. http://www.ncbi.nlm.nih.gov/entrez/query.fcgi?cmd=Retrieve&db=PubMed&dopt=Citation&list_uids=12154129

Jacobsen E, van der Vossen EAG (2009) Plant disease resistance: breeding and transgenic approaches. In: Schaechter M (ed) Encyclopedia of microbiology, 3rd edn. Academic, pp 597–612. https://doi.org/10.1016/B978-012373944-5.00342-4

Jang S, Marchal V, Panigrahi KC, Wenkel S, Soppe W, Deng XW, Valverde F, Coupland G (2008) *Arabidopsis* COP1 shapes the temporal pattern of CO accumulation conferring a photoperiodic flowering response. EMBO J 27(8):1277–1288

Javaid N, Choi S (2021) CRISPR/Cas system and factors affecting its precision and efficiency. Front Cell Dev Biol 9:761709. https://doi.org/10.3389/fcell.2021.761709

Jiang X, Lubini G, Hernandes-Lopes J, Rijnsburger K, Veltkamp V, de Maagd RA, Angenent GC, Bemer M (2022) *FRUITFULL*-like genes regulate flowering time and inflorescence architecture in tomato. Plant Cell 34(3):1002–1019. https://doi.org/10.1093/plcell/koab298

Johanson U, West J, Lister C, Michaels S, Amasino R, Dean C (2000) Molecular analysis of *FRIGIDA*, a major determinant of natural variation in *Arabidopsis* flowering time. Science 290(5490):344–347. http://www.ncbi.nlm.nih.gov/entrez/query.fcgi?cmd=Retrieve&db=PubMed&dopt=Citation&list_uids=11030654

Jung J-H, Barbosa AD, Hutin S, Kumita JR, Gao M, Derwort D, Silva CS, Lai X, Pierre E, Geng F (2020) A prion-like domain in ELF3 functions as a thermosensor in *Arabidopsis*. Nature 585(7824):256–260

Kabade PG, Dixit S, Singh UM, Alam S, Bhosale S, Kumar S, Singh SK, Badri J, Varma NRG, Chetia S, Singh R, Pradhan SK, Banerjee S, Deshmukh R, Singh SP, Kalia S, Sharma TR, Singh S, Bhardwaj H, Kohli A, Kumar A, Sinha P, Singh VK (2024) SpeedFlower: a comprehensive speed breeding protocol for indica and japonica rice. Plant Biotechnol J 22(5):1051–1066. https://doi.org/10.1111/pbi.14245

Kapazoglou A, Gerakari M, Lazaridi E, Kleftogianni K, Sarri E, Tani E, Bebeli PJ (2023) Crop wild relatives: a valuable source of tolerance to various abiotic stresses. Plants (Basel) 12(2):328. https://doi.org/10.3390/plants12020328

Kissoudis C, van de Wiel C, Visser RG, van der Linden G (2014) Enhancing crop resilience to combined abiotic and biotic stress through the dissection of physiological and molecular crosstalk. Front Plant Sci 5:207. https://doi.org/10.3389/fpls.2014.00207

Konishi S, Izawa T, Lin SY, Ebana K, Fukuta Y, Sasaki T, Yano M (2006) An SNP caused loss of seed shattering during rice domestication. Science 312(5778):1392–1396. http://www.ncbi.nlm.nih.gov/entrez/query.fcgi?cmd=Retrieve&db=PubMed&dopt=Citation&list_uids=16614172

Konopatskaia I, Vavilova V, Kondratenko EY, Blinov A, Goncharov NP (2016) *VRN1* genes variability in tetraploid wheat species with a spring growth habit. BMC Plant Biol 16(Suppl 3):244. https://doi.org/10.1186/s12870-016-0924-z

Koornneef M, Hanhart CJ, van der Veen JH (1991) A genetic and physiological analysis of late flowering mutants in *Arabidopsis thaliana*. Mol Gen Genet 229(1):57–66. http://www.ncbi.nlm.nih.gov/entrez/query.fcgi?cmd=Retrieve&db=PubMed&dopt=Citation&list_uids=1896021

Koornneef M, Blankestijn-de Vries H, Hanhart C, Soppe W, Peeters T (1994) The phenotype of some late-flowering mutants is enhanced by a locus on chromosome 5 that is not effective in the Landsberg erecta wild-type. Plant J 6(6):911–919

Koornneef M, Alonso-Blanco C, Blankestijn-de Vries H, Hanhart CJ, Peeters AJ (1998) Genetic interactions among late-flowering mutants of *Arabidopsis*. Genetics 148(2):885–892. http://www.ncbi.nlm.nih.gov/entrez/query.fcgi?cmd=Retrieve&db=PubMed&dopt=Citation&list_uids=9504934

Kumar SV, Wigge PA (2010) H2A. Z-containing nucleosomes mediate the thermosensory response in *Arabidopsis*. Cell 140(1):136–147

Kumar SV, Lucyshyn D, Jaeger KE, Alós E, Alvey E, Harberd NP, Wigge PA (2012) Transcription factor PIF4 controls the thermosensory activation of flowering. Nature 484(7393):242–245

Laubinger S, Marchal V, Gentilhomme J, Wenkel S, Adrian J, Jang S, Kulajta C, Braun H, Coupland G, Hoecker U (2006) *Arabidopsis* SPA proteins regulate photoperiodic flowering and interact with the floral inducer CONSTANS to regulate its stability. Development 133(16):3213–3222

Lee Z, Kim S, Choi SJ, Joung E, Kwon M, Park HJ, Shim JS (2023) Regulation of flowering time by environmental factors in plants. Plants 12(21):3680

Lei B, Berger F (2020) H2A variants in *Arabidopsis*: versatile regulators of genome activity. Plant Commun 1(1):100015

Lempe J, Balasubramanian S, Sureshkumar S, Singh A, Schmid M, Weigel D (2005) Diversity of flowering responses in wild *Arabidopsis thaliana* strains. PLoS Genet 1(1):109–118. https://doi.org/10.1371/journal.pgen.0010006

Li K, Gao Z, He H, Terzaghi W, Fan L-M, Deng XW, Chen H (2015) Arabidopsis DET1 represses photomorphogenesis in part by negatively regulating DELLA protein abundance in darkness. Mol Plant 8(4):622–630

Li K, Yu R, Fan L-M, Wei N, Chen H, Deng XW (2016) DELLA-mediated PIF degradation contributes to coordination of light and gibberellin signalling in Arabidopsis. Nat Commun 7(1):11868

Li H, Fan Y, Yu J, Chai L, Zhang J, Jiang J, Cui C, Zheng B, Jiang L, Lu K (2018) Genome-wide identification of flowering-time genes in *Brassica* species and reveals a correlation between selective pressure and expression patterns of vernalization-pathway genes in *Brassica napus*. Int J Mol Sci 19(11):3632

Li Y, Xiong H, Guo H, Zhou C, Xie Y, Zhao L, Gu J, Zhao S, Ding Y, Liu L (2020) Identification of the vernalization gene *VRN-B1* responsible for heading date variation by QTL mapping using a RIL population in wheat. BMC Plant Biol 20:1–15

Lim MH, Kim J, Kim YS, Chung KS, Seo YH, Lee I, Kim J, Hong CB, Kim HJ, Park CM (2004) A new *Arabidopsis* gene, FLK, encodes an RNA binding protein with K homology motifs and regulates flowering time via *FLOWERING LOCUS C*. Plant Cell 16(3):731–740. http://www.ncbi.nlm.nih.gov/entrez/query.fcgi?cmd=Retrieve&db=PubMed&dopt=Citation&list_uids=14973162

Liu F, Marquardt S, Lister C, Swiezewski S, Dean C (2010) Targeted 3′ processing of antisense transcripts triggers *Arabidopsis FLC* chromatin silencing. Science 327(5961):94–97

Liu W, Jiang B, Ma L, Zhang S, Zhai H, Xu X, Hou W, Xia Z, Wu C, Sun S, Wu T, Chen L, Han T (2018) Functional diversification of *Flowering Locus T* homologs in soybean: *GmFT1a* and *GmFT2a/5a* have opposite roles in controlling flowering and maturation. New Phytol 217(3):1335–1345. https://doi.org/10.1111/nph.14884

Liu F, Gao C, Chen M, Tang G, Sun Y, Li K (2021) The impacts of flowering phenology on the reproductive success of the narrow endemic *Nouelia insignis* Franch. (Asteraceae). Ecol Evol 11(14):9396–9409

Lopez-Obando M, Hoffmann B, Gery C, Guyon-Debast A, Teoule E, Rameau C, Bonhomme S, Nogue F (2016) Simple and efficient targeting of multiple genes through CRISPR-Cas9 in *Physcomitrella patens*. G3 (Bethesda) 6(11):3647–3653. https://doi.org/10.1534/g3.116.033266

Lu L, Yan W, Xue W, Shao D, Xing Y (2012) Evolution and association analysis of *Ghd7* in rice. PLoS One 7(5):e34021. https://doi.org/10.1371/journal.pone.0034021

Ma L, Zhang Y, Chen S, Yu L, Zhu Y (2022) Environmental effects and their causes of agricultural production: evidence from the farming regions of China. Ecol Indic 144:109549

Maple R, Zhu P, Hepworth J, Wang JW, Dean C (2024) Flowering time: from physiology, through genetics to mechanism. Plant Physiol 195(1):190–212. https://doi.org/10.1093/plphys/kiae109

Martén-Rodríguez S, Cristobal-Pérez EJ, de Santiago-Hernández MH, Huerta-Ramos G, Clemente-Martínez L, Krupnick G, Taylor O, Lopezaraiza-Mikel M, Balvino-Olvera FJ, Sentíes-Aguilar

EM (2025) Untangling the complexity of climate change effects on plant reproductive traits and pollinators: a systematic global synthesis. Glob Chang Biol 31(2):e70081

Martinez-Zapater JM, Somerville CR (1990) Effect of light quality and vernalization on late-flowering mutants of *Arabidopsis thaliana*. Plant Physiol 92(3):770–776. http://www.ncbi.nlm.nih.gov/entrez/query.fcgi?cmd=Retrieve&db=PubMed&dopt=Citation&list_uids=16667348

Masson-Delmotte V, Zhai P, Poertner HO, Roberts D, Skea J, Shukla PR, Pirani A, Moufouma-Okia W, Pean C, Pidcock R, Connors S, Mathews JBR, Chen Y, Zhou X, Gomis MI, Lonnoy E, Maycock T, Tignor M, Waterfield T (2018) IPCC, 2018: summary for policymakers. Cambridge University Press. https://doi.org/10.1017/9781009157940.001

Michaels SD, Amasino RM (1999) *FLOWERING LOCUS C* encodes a novel MADS domain protein that acts as a repressor of flowering. Plant Cell 11(5):949–956. http://www.ncbi.nlm.nih.gov/entrez/query.fcgi?cmd=Retrieve&db=PubMed&dopt=Citation&list_uids=10330478

Michaels SD, Amasino RM (2001) Loss of *FLOWERING LOCUS C* activity eliminates the late-flowering phenotype of *FRIGIDA* and autonomous pathway mutations but not responsiveness to vernalization. Plant Cell 13(4):935–941. http://www.ncbi.nlm.nih.gov/entrez/query.fcgi?cmd=Retrieve&db=PubMed&dopt=Citation&list_uids=11283346

Michaels SD, He Y, Scortecci KC, Amasino RM (2003) Attenuation of *FLOWERING LOCUS C* activity as a mechanism for the evolution of summer-annual flowering behavior in *Arabidopsis*. Proc Natl Acad Sci USA 100(17):10102–10107. http://www.ncbi.nlm.nih.gov/entrez/query.fcgi?cmd=Retrieve&db=PubMed&dopt=Citation&list_uids=12904584

Mirzabaev A, Bezner Kerr R, Hasegawa T, Pradhan P, Wreford A, Tirado C, von der Pahlen M, Gurney-Smith H (2023) Severe climate change risks to food security and nutrition. Clim Risk Manag 39:100473. https://doi.org/10.1016/j.crm.2022.100473

Murphy RL, Morishige DT, Brady JA, Rooney WL, Yang S, Klein PE, Mullet JE (2014) *Ghd7 (Ma)* represses sorghum flowering in long days: *Ghd7* alleles enhance biomass accumulation and grain production. Plant Genome 7(2):plantgenome2013.2011.0040. https://doi.org/10.3835/plantgenome2013.11.0040

Mutasa-Gottgens E, Hedden P (2009) Gibberellin as a factor in floral regulatory networks. J Exp Bot 60(7):1979–1989. https://doi.org/10.1093/jxb/erp040

Mutasa-Göttgens ES, Qi A, Zhang W, Schulze-Buxloh G, Jennings A, Hohmann U, Müller AE, Hedden P (2010) Bolting and flowering control in sugar beet: relationships and effects of gibberellin, the bolting gene B and vernalization. AoB Plants 2010:plq012

Nakamichi N (2015) Adaptation to the local environment by modifications of the photoperiod response in crops. Plant Cell Physiol 56(4):594–604. https://doi.org/10.1093/pcp/pcu181

Napp-Zinn K (1987) Vernalization-environmental and genetic regulation. In: Atherton JG (ed) Manipulation of flowering. Butterworths, London, pp 123–132

Navarro C, Abelenda JA, Cruz-Oró E, Cuéllar CA, Tamaki S, Silva J, Shimamoto K, Prat S (2011) Control of flowering and storage organ formation in potato by *FLOWERING LOCUS T*. Nature 478(7367):119–122

Neri J, Wendt T, Palma-Silva C (2018) Natural hybridization and genetic and morphological variation between two epiphytic bromeliads. AoB Plants 10(1):plx061. https://doi.org/10.1093/aobpla/plx061

Niu F, Rehmani MS, Yan J (2024) Multilayered regulation and implication of flowering time in plants. Plant Physiol Biochem 213:108842. https://doi.org/10.1016/j.plaphy.2024.108842

Nusinow DA, Helfer A, Hamilton EE, King JJ, Imaizumi T, Schultz TF, Farré EM, Kay SA (2011) The ELF4–ELF3–LUX complex links the circadian clock to diurnal control of hypocotyl growth. Nature 475(7356):398–402

Park HJ, Kim WY, Pardo JM, Yun DJ (2016) Molecular interactions between flowering time and abiotic stress pathways. Int Rev Cell Mol Biol 327:371–412. https://doi.org/10.1016/bs.ircmb.2016.07.001

Plantenga FDM, Bergonzi S, Bachem CWB, Visser RGF, Heuvelink E, Marcelis LFM (2019) High light accelerates potato flowering independently of the *FT*-like flowering signal *StSP3D*. Environ Exp Bot 160:35–44. https://doi.org/10.1016/j.envexpbot.2019.01.004

Prevéy JS (2020) Climate change: flowering time may be shifting in surprising ways. Curr Biol 30(3):R112–R114

Qüesta JI, Song J, Geraldo N, An H, Dean C (2016) *Arabidopsis* transcriptional repressor *VAL1* triggers Polycomb silencing at *FLC* during vernalization. Science 353(6298):485–488

Raman H, Raman R, Eckermann P, Coombes N, Manoli S, Zou X, Edwards D, Meng J, Prangnell R, Stiller J, Batley J, Luckett D, Wratten N, Dennis E (2013) Genetic and physical mapping of flowering time loci in canola (*Brassica napus* L.). Theor Appl Genet 126(1):119–132. https://doi.org/10.1007/s00122-012-1966-8

Raman H, Raman R, Coombes N, Song J, Prangnell R, Bandaranayake C, Tahira R, Sundaramoorthi V, Killian A, Meng J, Dennis ES, Balasubramanian S (2015) Genome-wide association analyses reveal complex genetic architecture underlying natural variation for flowering time in canola. Plant Cell Environ 39(6):1228–1239. https://doi.org/10.1111/pce.12644

Raman H, Raman R, Qiu Y, Yadav AS, Sureshkumar S, Borg L, Rohan M, Wheeler D, Owen O, Menz I, Balasubramanian S (2019) *GWAS* hints at pleiotropic roles for *FLOWERING LOCUS T* in flowering time and yield-related traits in canola. *BMC* Genomics 20(1):636. https://doi.org/10.1186/s12864-019-5964-y

Ramirez-Torres F, Ghogare R, Stowe E, Cerda-Bennasser P, Lobato-Gomez M, Williamson-Benavides BA, Giron-Calva PS, Hewitt S, Christou P, Dhingra A (2021) Genome editing in fruit, ornamental, and industrial crops. Transgenic Res 30(4):499–528. https://doi.org/10.1007/s11248-021-00240-3

Rehman S, Bahadur S, Xia W (2023) An overview of floral regulatory genes in annual and perennial plants. Gene 885:147699. https://doi.org/10.1016/j.gene.2023.147699

Rezaei EE, Webber H, Asseng S, Boote KJ, Durand JL, Ewert F, Martre P, MacCarthy DS (2023) Climate change impact on crop yields. Nat Rev Earth Environ 4:831–846

Rodriguez-Falcon M, Bou J, Prat S (2006) Seasonal control of tuberization in potato: conserved elements with the flowering response. Annu Rev Plant Biol 57:151–180. http://www.ncbi.nlm.nih.gov/entrez/query.fcgi?cmd=Retrieve&db=PubMed&dopt=Citation&list_uids=16669759

Rodrmguez EPB, Morante N, Salazar S, Hyde PT, Setter TL, Kulakow P, Aparicio JS, Zhang X (2023) Flower-inducing technology facilitates speed breeding in cassava. Front Plant Sci 14:1172056. https://doi.org/10.3389/fpls.2023.1172056

Rouse DT, Sheldon CC, Bagnall DJ, Peacock WJ, Dennis ES (2002) *FLC*, a repressor of flowering, is regulated by genes in different inductive pathways. Plant J 29(2):183–191. http://www.ncbi.nlm.nih.gov/entrez/query.fcgi?cmd=Retrieve&db=PubMed&dopt=Citation&list_uids=11851919

Salgotra RK, Chauhan BS (2023) Genetic diversity, conservation, and utilization of plant genetic resources. Genes (Basel) 14(1):174. https://doi.org/10.3390/genes14010174

Sanda S, John M, Amasino R (1997) Analysis of flowering time in ecotypes of *Arabidopsis thaliana*. J Hered 88(1):69–72. http://www.ncbi.nlm.nih.gov/entrez/query.fcgi?cmd=Retrieve&db=PubMed&dopt=Citation&list_uids=9048445

Sawa M, Nusinow DA, Kay SA, Imaizumi T (2007) FKF1 and GIGANTEA complex formation is required for day-length measurement in *Arabidopsis*. Science 318(5848):261–265. http://www.ncbi.nlm.nih.gov/entrez/query.fcgi?cmd=Retrieve&db=PubMed&dopt=Citation&list_uids=17872410

Sheldon CC, Burn JE, Perez PP, Metzger J, Edwards JA, Peacock WJ, Dennis ES (1999) The *FLF* MADS box gene: a repressor of flowering in *Arabidopsis* regulated by vernalization and methylation. Plant Cell 11(3):445–458. http://www.ncbi.nlm.nih.gov/entrez/query.fcgi?cmd=Retrieve&db=PubMed&dopt=Citation&list_uids=10072403

Sheng X, Mahendra RA, Wang CT, Brunner AM (2023) CRISPR/Cas9 mutants delineate roles of *Populus FT* and *TFL1/CEN/BFT* family members in growth, dormancy release and flowering. Tree Physiol 43(6):1042–1054. https://doi.org/10.1093/treephys/tpad027

Shindo C, Aranzana MJ, Lister C, Baxter C, Nicholls C, Nordborg M, Dean C (2005) Role of *FRIGIDA* and *FLOWERING LOCUS C* in determining variation in flowering time of

Arabidopsis. Plant Physiol 138(2):1163–1173. http://www.ncbi.nlm.nih.gov/entrez/query.fcgi?cmd=Retrieve&db=PubMed&dopt=Citation&list_uids=15908596

Shindo C, Lister C, Crevillen P, Nordborg M, Dean C (2006) Variation in the epigenetic silencing of *FLC* contributes to natural variation in *Arabidopsis* vernalization response. Genes Dev 20(22):3079–3083. http://www.ncbi.nlm.nih.gov/entrez/query.fcgi?cmd=Retrieve&db=PubMed&dopt=Citation&list_uids=17114581

Simpson GG (2004) The autonomous pathway: epigenetic and post-transcriptional gene regulation in the control of *Arabidopsis* flowering time. Curr Opin Plant Biol 7(5):570–574. http://www.ncbi.nlm.nih.gov/entrez/query.fcgi?cmd=Retrieve&db=PubMed&dopt=Citation&list_uids=15337100

Singh A, Septiningsih EM, Balyan HS, Singh NK, Rai V (2017) Genetics, physiological mechanisms and breeding of flood-tolerant rice (*Oryza sativa* L.). Plant Cell Physiol 58(2):185–197. https://doi.org/10.1093/pcp/pcw206

Sujariya S, Jongdee B, Fukai S (2023) Estimation of flowering time and its effect on grain yield of photoperiod sensitive varieties in rainfed lowland rice in Northeast Thailand. Field Crop Res 302:109075

Sun T-P (2011) The molecular mechanism and evolution of the GA–GID1–DELLA signaling module in plants. Curr Biol 21(9):R338–R345

Sung S, Amasino RM (2004) Vernalization in *Arabidopsis thaliana* is mediated by the PHD finger protein VIN3. Nature 427(6970):159–164. http://www.ncbi.nlm.nih.gov/entrez/query.fcgi?cmd=Retrieve&db=PubMed&dopt=Citation&list_uids=14712276

Tasset C, Singh Yadav A, Sureshkumar S, Singh R, van der Woude L, Nekrasov M, Tremethick D, van Zanten M, Balasubramanian S (2018) POWERDRESS-mediated histone deacetylation is essential for thermomorphogenesis in *Arabidopsis thaliana*. PLoS Genet 14(3):e1007280

Tilmes V, Mateos JL, Madrid E, Vincent C, Severing E, Carrera E, López-Díaz I, Coupland G (2019) Gibberellins act downstream of *Arabis* PERPETUAL FLOWERING1 to accelerate floral induction during vernalization. Plant Physiol 180(3):1549–1563

Trevaskis B, Hemming MN, Peacock WJ, Dennis ES (2006) *HvVRN2* responds to daylength, whereas *HvVRN1* is regulated by vernalization and developmental status. Plant Physiol 140(4):1397–1405. http://www.ncbi.nlm.nih.gov/entrez/query.fcgi?cmd=Retrieve&db=PubMed&dopt=Citation&list_uids=16500994

Turck F, Fornara F, Coupland G (2008) Regulation and identity of florigen: *FLOWERING LOCUS T* moves center stage. Annu Rev Plant Biol 59:573–594. https://doi.org/10.1146/annurev.arplant.59.032607.092755

Turner A, Beales J, Faure S, Dunford RP, Laurie DA (2005) The pseudo-response regulator *Ppd-H1* provides adaptation to photoperiod in barley. Science 310(5750):1031–1034. http://www.ncbi.nlm.nih.gov/entrez/query.fcgi?cmd=Retrieve&db=PubMed&dopt=Citation&list_uids=16284181

Valverde F, Mouradov A, Soppe W, Ravenscroft D, Samach A, Coupland G (2004) Photoreceptor regulation of CONSTANS protein in photoperiodic flowering. Science 303(5660):1003–1006. http://www.ncbi.nlm.nih.gov/entrez/query.fcgi?cmd=Retrieve&db=PubMed&dopt=Citation&list_uids=14963328

Vicentini G, Biancucci M, Mineri L, Chirivi D, Giaume F, Miao Y, Kyozuka J, Brambilla V, Betti C, Fornara F (2023) Environmental control of rice flowering time. Plant Commun 4(5):100610. https://doi.org/10.1016/j.xplc.2023.100610

Wang G, Wang C, Lu G, Wang W, Mao G, Habben JE, Song C, Wang J, Chen J, Gao Y, Liu J, Greene TW (2020) Knockouts of a late flowering gene via CRISPR-Cas9 confer early maturity in rice at multiple field locations. Plant Mol Biol 104(1–2):137–150. https://doi.org/10.1007/s11103-020-01031-w

Wang X, Zhang J, Mao W, Guan P, Wang Y, Chen Y, Liu W, Guo W, Yao Y, Hu Z, Xin M, Ni Z, Sun Q, Peng H (2023) Association mapping identifies loci and candidate genes for grain-related traits in spring wheat in response to heat stress. Plant Sci 331:111676. https://doi.org/10.1016/j.plantsci.2023.111676

War AR, Paulraj MG, Ahmad T, Buhroo AA, Hussain B, Ignacimuthu S, Sharma HC (2012) Mechanisms of plant defense against insect herbivores. Plant Signal Behav 7(10):1306–1320

Weeks DP (2017) Gene editing in polyploid crops: wheat, camelina, canola, potato, cotton, peanut, sugar cane, and citrus. Prog Mol Biol Transl Sci 149:65–80. https://doi.org/10.1016/bs.pmbts.2017.05.002

Werner JD, Borevitz JO, Uhlenhaut NH, Ecker JR, Chory J, Weigel D (2005) FRIGIDA-independent variation in flowering time of natural *Arabidopsis thaliana* accessions. Genetics 170(3):1197–1207. http://www.ncbi.nlm.nih.gov/entrez/query.fcgi?cmd=Retrieve&db=PubMed&dopt=Citation&list_uids=15911588

Wigge PA, Kim MC, Jaeger KE, Busch W, Schmid M, Lohmann JU, Weigel D (2005) Integration of spatial and temporal information during floral induction in *Arabidopsis* [Research Support, Non-U.S. Gov't]. Science 309(5737):1056–1059. https://doi.org/10.1126/science.1114358

Xu F, Rong X, Huang X, Cheng S (2012) Recent advances of *Flowering Locus T* gene in higher plants. Int J Mol Sci 13(3):3773–3781. https://doi.org/10.3390/ijms13033773

Yadav AS, Sureshkumar S, Sinha AK, Balasubramanian S (2025) Dispersed components drive temperature sensing and response in plants. Science 388(6752):1161–1166

Yan L, Loukoianov A, Blechl A, Tranquilli G, Ramakrishna W, SanMiguel P, Bennetzen JL, Echenique V, Dubcovsky J (2004) The wheat *VRN2* gene is a flowering repressor down-regulated by vernalization. Science 303(5664):1640–1644. http://www.ncbi.nlm.nih.gov/entrez/query.fcgi?cmd=Retrieve&db=PubMed&dopt=Citation&list_uids=15016992

Yano M, Katayose Y, Ashikari M, Yamanouchi U, Monna L, Fuse T, Baba T, Yamamoto K, Umehara Y, Nagamura Y (2000) *Hd1*, a major photoperiod sensitivity quantitative trait locus in rice, is closely related to the *Arabidopsis* flowering time gene *CONSTANS*. Plant Cell 12(12):2473–2483

Yu X, Klejnot J, Lin C (2006) Florigen: one found, more to follow? J Integr Plant Biol 48(6):617–621

Yuan W, Luo X, Li Z, Yang W, Wang Y, Liu R, Du J, He Y (2016) A cis cold memory element and a trans epigenome reader mediate Polycomb silencing of *FLC* by vernalization in Arabidopsis. Nat Genet 48(12):1527–1534

Zhao C, Liu B, Piao S, Wang X, Lobell DB, Huang Y, Huang M, Yao Y, Bassu S, Ciais P, Durand JL, Elliott J, Ewert F, Janssens IA, Li T, Lin E, Liu Q, Martre P, Muller C, Peng S, Penuelas J, Ruane AC, Wallach D, Wang T, Wu D, Liu Z, Zhu Y, Zhu Z, Asseng S (2017) Temperature increase reduces global yields of major crops in four independent estimates. Proc Natl Acad Sci USA 114(35):9326–9331. https://doi.org/10.1073/pnas.1701762114

Zhao B, Zhai D, Wang JW (2025) Flowering time regulation through the lens of evolution. Curr Opin Plant Biol 85:102734. https://doi.org/10.1016/j.pbi.2025.102734

Zhu Z, Esche F, Babben S, Trenner J, Serfling A, Pillen K, Maurer A, Quint M (2023) An exotic allele of barley *EARLY FLOWERING 3* contributes to developmental plasticity at elevated temperatures. J Exp Bot 74(9):2912–2931. https://doi.org/10.1093/jxb/erac470

Zhu Z, Trenner J, Delker C, Quint M (2024) Tracing the evolutionary history of the temperature-sensing prion-like domain in *EARLY FLOWERING 3* highlights the uniqueness of *At ELF3*. Mol Biol Evol 41(10):msae205

Chapter 2
Remodelling Plant Development for Trait Engineering: Lessons from Morphological Innovations—Post-translational Modifications in Regulating Flowering Time

Jordyn Coutts, Chandan Kumar Pradhan, Ramya Ganesan, Sofiya Muniyasamy, and Sridevi Sureshkumar

Abstract The continuously growing global population and intensifying climatic change pose increasing challenges to plants, especially crop species, in adapting to and surviving within their environments. Crop productivity and resilience are determined by key agronomics traits such as the flowering time, fruit or seed production, and responses to abiotic and biotic stresses. Among these, flowering time is particularly significant, as it integrates a wide range of environmental cues, starting from seasonal fluctuations to metal-induced stresses. Post-translational modifications (PTMs) of proteins governing these traits are tightly controlled and play crucial roles in modulating protein function. PTMs are thus emerging as crucial modulators in plant adaptive responses. In this chapter, we explore the significance of these traits in the production of sustainable crops, capable of withstanding environmental challenges. We highlight the importance of PTMs in manipulating or enhancing agriculturally important traits. Additionally, we discuss the current advances in modern techniques used to detect and characterise PTMs associated with particular traits as well as efforts to manipulate PTMs in various crops to enhance the yield and quality. We also address potential trade-off and unintended consequences of targeting PTMs. Finally, we emphasise about the relevance of proteomics as a powerful tool for advancing crop improvement strategies under environmental stress.

Jordyn Coutts and Chandan Kumar Pradhan: Shared first authors

---

J. Coutts · R. Ganesan · S. Muniyasamy · S. Sureshkumar (✉)
School of Biological Sciences, Monash University, Clayton, VIC, Australia
e-mail: sridevi.sureshkumar@monash.edu

C. K. Pradhan
School of Biological Sciences, Monash University, Clayton, VIC, Australia

Indian Institute of Technology Bombay, Mumbai, Maharashtra, India

A. Singh et al. (eds.), *Sustainable Crop Production*, Sustainability Sciences in Asia and Africa, https://doi.org/10.1007/978-981-95-7844-3_2

Keywords Post-translational modifications · Flowering · Crop improvement · Proteomics

Abbreviations

AAP1	Amino acid permease 1
ABA	Abscisic acid
ABFs	ABA-RESPONSIVE ELEMENT BINDING PROTEINS
ABI5	Abscisic acid-insensitive 5
AL3	ALPHIN-LIKE 3
Asp	Aspartic acid
ATX1	ARABIDOPSIS HOMOLOG OF TRITHORAX1
BisChIP-seq	Bisulfide-treated chromatin immunoprecipitation
CBL	Calcineurin B-like proteins
CCA1	CIRCADIAN CLOCK-ASSOCIATED 1
ChIP	Chromatin Immunoprecipitation
CID	Collision-induced dissociation
CIPK	CBL-interacting protein kinases
CO	CONSTANS
COP1	Constitutive photomorphogenesis 1
CRY	Cryptochromes
CUT & Tag	Cleavage under targets and tagmentation
DELLA	Asp-Glu-Leu-Leu-Ala
E1, E2, and E3	Ubiquitin-protein ligases
ESD4	EARLY IN SHORT DAY 4
ESI	Electron-spray ionisation
FBH4	Flowering BHLH4
FLC	FLOWERING LOCUS C
FLD	FLOWERING LOCUS D
FRI	FRIGIDA
FT	FLOWERING LOCUS T
FUG-1	FOURTH ULP GENE CLASS 1
GA	Gibberellin
GI	GIGANTEA
H3K27me3	Trimethylation of histone H3 lysine 27
H3K4me3	Trimethylation of histone H3 lysine 4
H3K9ac	Histone 3 lysine 9 acetylation
HATs	Histone acetyltransferases
HD1	Heading date 1
HD3A	HEADING DATE 3A
HD6	Head date 6
HDACs	Histone deacetylases

HMTs	Histone methyltransferases
HOS1	HIGH EXPRESSION OF OSMOTICALLY RESPONSIVE GENES 1
HPY2	HIGH PLOIDY 2
HSP	Heat shock proteins
LHP-1	LIKE HETEROCHROMATIN 1
MALDI	Matrix-assisted laser desorption/ionisation
MAPKs	Mitogen-activated protein kinases
MAS	Marker-assisted selection
MLKs	MuTP9-like kinases
MS	Mass spectrometry
OTS1	OVERLAY TOLERANT TO SALT-1
PHYA	PHYTOCHROME A
PHYB	PHYTOCHROME B
PHYC	PHYTOCHROME C
PHYD	PHYTOCHROME D
PIFs	PHYTOCHROME INTERACTING FACTORs
PPD1	PHOTOPERIOD 1
PRC2	POLYCOMB REPESSIVE COMPLEX 2
PRRs	Pathogen recognition receptors
PTMs	Post-translational modifications
PUB13	PLANT U-BOX 13
RBOH1	Respiratory burst oxidase homolog 1
RLK	Receptor-like kinases
RLPs	Receptor-like proteins
ROS	Reactive oxygen species
ROS1	Repressor of silencing 1
SEC	SECRET AGENT
Ser	Serine
SIZ1	Salt-induced zinc finger protein 1
SnRK1	SNF1-RELATED PROTEIN KINASE 1
SOC1	SUPPRESSOR OF OVEREXPRESSION OF CONSTANS 1
SOS	Salt overlay sensitivity
SPA1	Suppressor of PHYA-105
SSR	SUMO stress response
SUMO	Small ubiquitin-like modifier
Thr	Threonine
Tyr	Tyrosine
UBPs	Ubiquitin-specific protease
UPS	Ubiquitin-proteasome system
VIN3	VERNALIZATION INSENSITIVE 3
VRN1	VERNALIZATION 1
VRN2	VERNALIZATION 2

Editor's Notes

In this chapter, intended for research scholars, authors present the exciting and untapped potential of post-translational modifications (PTMs) in remodeling crop traits. As crucial integrators of cell signaling, PTMs intricately control fundamental cellular processes by regulating protein function, stability, localization, and interactions. Since the dynamic nature of PTMs enables plants to swiftly respond to changing and challenging environments, PTMs are emerging as key determinants for manipulating adaptive traits. As opposed to genome modification-based methods, PTMs promise trait modification in a more nuanced manner without offsetting the fine balance of other agronomic traits. The authors quote *"PTMs based strategies are rapidly evolving and transformative, they hold immense promise for research applications and the production of sustainable crops."* The authors propose students and scholars to examine the potential of following strategies for designing agronomically important traits.

Actionable Points

1. Kinase pathways like SnRK2, prone to phosphorylation, can be targeted for ameliorating environmental stresses like drought. Kinases are primarily associated with inducing ABA signaling.
2. PTMs in plant pathogen receptors, such as PRRs and PUB22, can be targeted to produce disease-resistant crop varieties.
3. Plant growth could be enhanced by knocking out negative regulators of growth and development. Some of the best examples are the DELLA protein, which targets GA signaling, and histone acetyltransferases (e.g., GCN5), which influence flower and root development.
4. Nutrient utilization in plants can also be controlled by regulating PTMs of nutrient receptors and channels to increase the quality and quantity of crops.
5. Identification of PTM-related genes related to flowering regulators, such as Heading Date, for higher yield.

2.1 Introduction

Climate change poses a major threat to global food security, disrupting agricultural productivity through rising temperatures, unpredictable rainfall, and more frequent extreme weather events. These changes reduce crop yields, distort growing seasons, and exacerbate pest and disease outbreaks (Bibi and Rahman 2023). Small farmers are particularly vulnerable, as they rely on stable climates (Habtemariam et al. 2017). As the world's population grows and food demand rises (Molotoks et al. 2021), climate-induced stress can have a devastating effect on food security (Aslebagh et al. 2019; Balasubramanian et al. 2006; Bandeira et al. 2007). Crop growth and development is directly affected by the increased weather variability caused by rising global temperatures (Rummukainen 2012), in particular, the timing of important reproductive transitions such as flowering (Thompson 1975; Almaraz et al. 2008).

Flowering time, an essential determinant of crop yield and resilience, influences grain filling, fruit development, and the timing of reproductive phases with suitable environmental conditions (Wu et al. 2023a). Flowering time must be carefully regulated by seasonal cues such as temperature, day length, and water availability to ensure plants transition to reproduction under favourable conditions. If plants flower too early or too late, they may expose growing buds and fruits to harsh weather, such as heat, drought, or frost; miss the active season of pollinators, not have enough resources for fruit production; and compromise the overall survivability of the plant (Talukder et al. 2014). The precise regulation of flowering time is essential for maximising yield and improving crop resilience to environmental stress (Kopeć 2024). With extreme and unpredictable weather due to climate change, the precise regulation of flowering time becomes even more important.

Increased weather variability threatens food security (Berhane 2018), as variation can disrupt flowering time regulation amongst crops (Craufurd and Wheeler 2009). Even temperature variation of a few degrees can trigger early or delayed flowering (Balasubramanian et al. 2006). Heatwaves during critical flowering periods can lead to floral sterility (Parker et al. 2020), while unexpected cold snaps can damage buds and stunt development (Inouye 2000; Thakur et al. 2010). These disruptions reduce grain filling, fruit set, and overall yield stability (Craufurd and Wheeler 2009). Changes to flowering times can also disrupt farming management practices, making it more difficult to decide harvesting time and when to apply pest control measures, further reducing efficiency and profits. Developing crops that can tolerate accelerated climate change—through targeted biotechnological interventions—will become pivotal in achieving robust agricultural productivity and global food security (Blümel et al. 2015).

Flowering time regulation is influenced by transcriptional regulation and post-translational modifications (PTMs) (Niu et al. 2024). PTMs allow plants to alter flowering time in response to environmental stress by modifying protein stability, activity, and interactions within flowering time pathways (Qazi et al. 2019; Martí-Guillén et al. 2022; Sharma et al. 2023), offering a promising avenue of research for overcoming climate-change induced challenges within agriculture. This chapter explores the many types of PTMs, their roles in flowering time, and their potential for modification in developing climate-resilient crops. Harnessing PTMs can allow researchers and agriculturalists to engineer crops with improved adaptability, enhancing sustainability, and food security in the face of increasingly unpredictable growing conditions (Han et al. 2022).

2.2 PTMs in Flowering Time

PTMs are chemical modifications made to already synthesised proteins. These changes can affect protein folding, stability, activity, localisation, and protein-protein interactions. PTMs act as regulatory mechanisms that allow cells to rapidly adapt to environmental stress. In parallel, other cellular responses, such as the induction of heat shock proteins and activation of stress-responsive transcription factors, also contribute to maintaining proteome stability. PTMs play a crucial role in fine-tuning flowering time by activating key floral regulators. Flowering time is regulated by PTMs through their effect on signalling, epigenetic modifications, and activity changes (Si et al. 2021). A set of enzymes, referred to as writers and erasers, dynamically regulate PTMs by adding and removing different chemical groups, respectively, to target proteins. Additionally, proteins that can recognise and bind to specific PTMs are known as readers (Chang et al. 2021). Together, these enzymes help the cell to interpret and react to environmental cues, such as light and temperature. Interpreting these cues is crucial, as they determine when a plant should flower to ensure optimal growth and reproduction. This precise timing is essential for agricultural activities and crop production, as it directly affects yield and quality.

PTMs affect several flowering regulatory proteins, such as FLOWERING LOCUS C (FLC), FLOWERING LOCUS T (FT), FRIGIDA (FRI), and CONSTANS (CO), altering their activity in flowering time regulation (Kwak et al. 2017). "When to flower" is decided through a complex and intertwining network of genetic and epigenetic regulators which are influenced by both environmental and endogenous signals (Lee et al. 2023). The major PTMs associated with flowering time regulation include phosphorylation, ubiquitination, SUMOylation, histone acetylation, histone and DNA methylation, and glycosylation (Niu et al. 2024).

2.2.1 Phosphorylation

Phosphorylation involves the addition of a phosphate group (PO_4^{3-}) to a protein. The typical targets of phosphorylation are serine (Ser), threonine (Thr), tyrosine (Tyr), and aspartic acid (Asp). The phosphate group is added by kinase enzymes and removed by phosphatases, a process carefully regulated for healthy plant cell growth and development (Beltrao et al. 2013). Approximately 30–50% of proteins are phosphorylated at any point of time during the life cycle of plants, such as rice, wheat, soybean, and maize (Zhang et al. 2023).

Photoperiod, the duration of light and darkness in a 24-h period, is a major environmental cue determining plant reproductive success. Photoreceptor proteins undergo phosphorylation to regulate the transition from vegetative to reproductive phases. Phosphorylation modulates phytochrome activity, with studies showing that phosphorylation of Ser-86 in PHYTOCHROME B (PHYB) directly affects flowering time (Medzihradszky et al. 2013; Nito et al. 2013; Hajdu et al. 2015). Further analysis revealed that the early flowering phenotype resulting from this modification is not due to changes in the circadian clock protein GIGANTEA (GI) but rather reflects PHYB's associated role in flowering regulation. Tyrosine phosphorylation also plays a major role in PHYB light and temperature signalling. Specifically, phosphorylation of Tyr-104 disrupts PHYB's interaction with PHYTOCHROME INTERACTING FACTOR 3 (PIF3), preventing nuclear body formation (Nito et al.

2013). This disruption alters plant architecture, causing pale leaves and early flowering, highlighting the importance of PTMs in light signalling and flowering regulation, and raising questions about similar mechanisms in other phytochromes (PHYA, PHYC, PHYD). The selection of phosphorylation sites is crucial for integrating environmental signals, with hierarchical PTM order ensuring precise control over PHYB's function in flowering time regulation.

2.2.2 Ubiquitination and SUMOylation

Ubiquitination and SUMOylation are both PTMs which involve small protein molecules being covalently attached to target proteins. In ubiquitination, also known as ubiquitylation, small ubiquitin proteins are attached to the lysine residues of substrate proteins in a three-enzyme cascade catalysed by ubiquitin-protein ligases E1, E2, and E3 (Pickart 2001). The number of ubiquitin proteins added, and the way they are added to a protein impacts the effect the modification has on the substrate protein. For example, mono-ubiquitination and multi-ubiquitination, the addition of a single ubiquitin proteins (one-per-lysine), tend to affect protein function and localisation (Haglund and Dikic 2005). However, since the ubiquitin protein itself contains lysine residues, ubiquitin proteins can be ubiquitinated, resulting in long chains of ubiquitin, known as polyubiquitination (Komander 2009). This often targets the modified protein for degradation by the ubiquitin proteasome system (UPS) (Fig. 2.1) (Hochstrasser 1996). This PTM is implicated in a variety of biochemical processes in plants, particularly in plant immunity and stress response (Trujillo and Shirasu 2010; Marino et al. 2012; Zhou and Zeng 2017; Gao et al. 2022).

SUMOylation is a rapidly emerging PTM in both animals and plants. Small ubiquitin-like modifier (SUMO), a ubiquitin-like peptide, is covalently attached to lysine residues of target proteins, either as mono- or poly-molecules, and plays a significant role in regulating flowering time in *Arabidopsis thaliana* (Kurepa et al. 2003). Approximately 5–10% of proteins in *A. thaliana* undergo SUMO modifications. The SUMOylation process involves a cascade of enzymatic steps mediated by E1, E2, and E3 ligases, which attach SUMO to target substrates (Gareau and Lima 2010). This modification is reversible, with SUMO proteases removing SUMO moieties to ensure dynamic regulation and cellular homeostasis. Mutations in SUMO proteins or SUMO proteases have been shown to alter flowering in *A. thaliana* (Kong et al. 2017). For instance, mutations in the SUMO protease EARLY IN SHORT DAY 4 (ESD4) promote early flowering under both inductive (long day) and non-inductive (short day) photoperiodic conditions (Murtas et al. 2003), highlighting SUMOylation's role in floral regulation. The reduced FLC expression in *esd4* mutants leads to early flowering and explains the role of ESD4 in integrating SUMO-mediated protein regulation into the genetic network responsible for regulating flowering time (Reeves et al. 2002; Murtas et al. 2003).

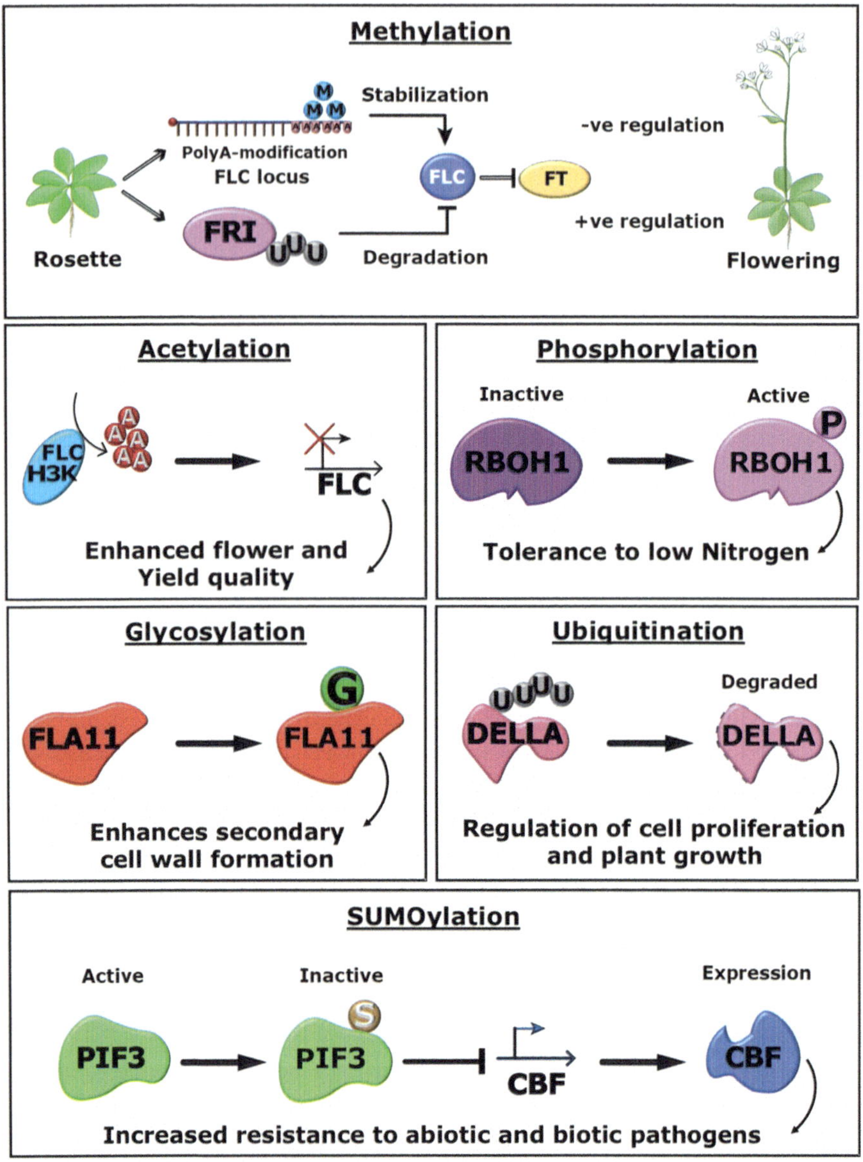

Fig. 2.1 Role of PTMs in crop improvement. PTMs play crucial roles in various signalling pathways that influence numerous crop traits. By targeting specific pathways, PTMs can be leveraged to modify crops, enhancing desirable traits such as resistance to both biotic and abiotic environmental stresses. Enhancing stress tolerance through PTMs not only improves crop strength but also contributes to overall crop quality

2.2.3 Acetylation and Methylation

Acetylation and methylation alter chromatin structure to regulate flowering time (Shi et al. 2022). Histone acetyltransferases (HATs) add an acetyl group (CH_3CO^-) to lysine residues in histone monomers, typically leading to a loosening of chromatin structure and promoting gene expression. This is seen in the regulation of flowering time genes such as *FT* (Kumar et al. 2021) and *SUPPRESSOR OF OVEREXPRESSION OF CONSTANS 1 (SOC1)* (Liang et al. 2025). Histone deacetylases (HDACs) remove these acetyl groups to encourage a compact chromatin formation and gene silencing (Ma et al. 2013). During vernalisation, histones bound to the FLC gene undergo deacetylation, thereby silencing FLC expression and allowing flowering to occur after cold exposure has ended (Xu and Chong 2018). This suggests that the regulation of flowering through acetylation can be influenced by changes in light and temperature conditions (Boss et al. 2004; Proietti et al. 2022).

Methylation adds methyl groups (CH_3) to histones or DNA, catalysed by histone methyltransferases (HMTs) or DNA methyltransferases, respectively (Vanyushin 2006). The effect methylation has on chromatin function depends on the specific histone residue modified. For instance, tri-methylation of histone H3 lysine 4 (H3K4me3) (Howe et al. 2017) promotes the expression of flowering-related genes such as FT, SOC1, and FLC (Jeong et al. 2015; Ibrar et al. 2023; Haider and Farrona 2024), whereas histone H3 lysine 27 tri-methylation (H3K27me3) associates with their silencing (Hou et al. 2014; Costa and Dean 2019). DNA methylation at promoter regions can further reinforce gene silencing, as seen in FLC, where cytosine methylation restricts transcription. Additionally, the m^6A-methylation of FLC mRNA is known to stabilise FLC expression (Fig. 2.1) (Xu et al. 2021). For example, in *Chrysanthemum lavandulifolium*, a species of flowering plant, the DNA demethylase gene *ROS1* (*REPRESSOR OF SILENCING 1*) plays a vital role in regulating flowering time. Studies show that *ROS1*-mediated demethylation promotes flowering by decreasing methylation at the *CONSTANTS* gene promoter, thereby increasing its expression (Zhang et al. 2018a; Shi et al. 2024). DNA methylation and demethylation play key roles in regulating flowering time in plants (Zhang et al. 2018a). DNA methylation dynamics are influenced by temperature fluctuations and drought stress, which can shift flowering time (Sun et al. 2022).

2.2.4 Glycosylation

N-glycosylation involves the attachment of glycans, sugar molecules, to asparagine residues within a target protein. N-glycosylation is involved in protein folding and quality control within the endoplasmic reticulum, where it prevents the premature export of misfolded proteins, helping to maintain overall protein stability (Nagashima et al. 2018). This is the most common PTM found in plant secretory pathways, including secondary cell wall formation (e.g. glycosylation of FLA11) (Fig. 2.1)

(Hebert et al. 2014; Ma et al. 2022). Glycosylation has recently been implicated in plant stress response (Takano et al. 2015). In *A. thaliana*, an estimated 1–2% of proteins undergo glycosylation. However, despite its significance, glycosylation remains understudied. Given its limited exploration, glycosylation warrants further research in plant biology.

2.3 PTMs in Environmental Stress Response

Various changes to the regulation of flowering occur during stress to integrate stress signals into developmental timing (Hashiguchi and Komatsu 2016; Liu et al. 2019; Zhang et al. 2022). For example, the stability and activity of photoperiod transcription factor CO are influenced by PTMs under oxidative stress (Valverde 2011). Cryptochromes (CRY1 and CRY2) and phytochromes (PHYA and PHYB) inhibit the activity of COP1-SPA1 (CONSTITUTIVE PHOTOMORPHOGENESIS 1 and SUPPRESSOR OF PHYA-105) in long-day conditions, preventing the ubiquitination of CO, allowing CO to accumulate and activate FT, ultimately promoting flowering (Liu et al. 2008). However, during high-intensity red light stress, E3 ubiquitin ligase HOS1 (HIGH EXPRESSION OF OSMOTICALLY RESPONSIVE GENES 1) targets CO for degradation and delays flowering (Lazaro et al. 2015). PTMs can also help plants balance stress tolerance with reproductive timing. During non-stressful conditions, gibberellin (GA) promotes flowering, and DELLA (Asp-Glu-Leu-Leu-Ala) proteins, GA repressors, are phosphorylated and ubiquitinated, leading to their degradation (Fig. 2.1) (Hauvermale et al. 2012). Under drought and heat stress, DELLA proteins are stabilised and repress GA (Blanco-Touri et al. 2020), delaying flowering until favourable reproductive conditions are met. Flowering time is often co-regulated by several different pathways, involving multiple PTMs acting in coordination.

2.3.1 PTM Crosstalk and the Co-regulation of Traits

Multiple PTMs can function within the same pathways at once, increasing the complexity of trait regulation. For example, as previously mentioned, CO is subject to ubiquitination in the photoperiod pathway, but CO is also phosphorylated to modulate its activity (Sarid-Krebs et al. 2015). Crosstalk is also seen in the GA pathway, where DELLA proteins are mostly regulated by several PTMs. DELLA proteins are not only regulated by ubiquitination-related degradation (Blanco-Touri et al. 2020); they can be phosphorylated to modulate their interaction with other proteins, such as bZIP23, bHLH089, bHLH094, and OSH1. In addition, DELLA proteins are SUMOylated during their role in the regulation GA biosynthesis under salt stress (Fernandes et al. 2024). Some DELLA proteins are also O-GlcNAcylated and O-fucosylated which promotes their activity with PIF3 and PIF4 (Zentella et al. 2017).

VERNALIZATION INSENSITIVE 3 (VIN3) plays a role in flowering time regulation by mediating vernalisation and interacting with the autonomous pathway. Under prolonged cold exposure, VIN3 forms a complex with VERNALIZATION 1 (VRN1) and VERNALIZATION 2 (VRN2), acting as part of the POLYCOMB REPRESSIVE COMPLEX 2 (PRC2) (Gendall et al. 2001; Levy et al. 2002; De Lucia et al. 2008) to add repressive histone marks (H3K27me3) to FLC, leading to flowering after the cold period has ended (Bastow et al. 2004; Sung and Amasino 2004).

The many pathways that influence flowering time work together through different PTMs. PTMs are integral not only to immediate environmental response but also play a key role in epigenetic memory, allowing plants, and future generations to "remember" past stressors. The PTMs made during stress response can have impacts that carry across generations. Epigenetic regulation allows plants to quickly regulate flowering response when stress is recurring (Oberkofler et al. 2021). The novel SUMO protease FUG-1 (FOURTH ULP GENE CLASS 1) recovered from thermal screens has the potential to reshape chromatin by interacting with histone reader AL3 (ALPHIN-LIKE protein 3) and chromatin remodeller LHP-1 (LIKE HERTEROCHROMATIN 1) (Sureshkumar et al. 2024).

2.3.2 PTMs Role in Epigenetic Memory and Trans-generational Stress Response

PTMs establish and maintain epigenetic memory across development and into subsequent generations, allowing plants to adapt their flowering time based on historic environmental conditions. Epigenetic memory refers to DNA methylation patterns, histone modifications, and chromatin states which are maintained over a long period of time, and even across generations, with the purpose of making future stress responses quicker and more efficient (D'Urso and Brickner 2014). This memory is actioned through histone modifications, such as methylation, acetylation, ubiquitination, and phosphorylation (Turgut-Kara et al. 2020). Histone-modifying enzymes, such as histone methyltransferases/demethylases, acetyltransferases/deacetylases, help maintain histone marks through successive cell divisions (Zhao et al. 2019), including meiosis (Nguyen et al. 2022).

One of the best studied examples of a flowering time gene influenced by trans-generational stress responses is FLC (Berry and Dean 2015; Iwasaki 2015; Costa and Dean 2019). Drought and heat stress during growth can influence FLC expression in later progeny. These stress conditions lead to a decrease in active H3K4me3 and acetylation of histone 3 lysine 9 (H3K9ac) marks and an increase in repressive H3K27me3 marks. This leads to stable repression of FLC in subsequent generations, promoting early flowering as an adaptive strategy to avoid future stress (Fig. 2.1).

2.3.3 Examples in Arabidopsis and Model Crops

A. thaliana has been instrumental in uncovering the role of PTMs in flowering regulation. These modifications influence key regulatory proteins, allowing plants to integrate environmental signals such as light and temperature into their developmental programs. FLC, a transcriptional repressor, delays flowering by inhibiting floral integrator genes. Phosphorylation of FLC influences its stability and activity, while FT undergoes phosphorylation to regulate its stability and interaction with proteins, ultimately influencing floral induction (Niu et al. 2024). SUMOylation also plays a role in stabilising FLC, with SUMO E3 ligases salt-induced zinc finger protein 1 (SIZ1) and HIGH PLOIDY 2 (HPY2) ensuring its functional stability. O-GlcNAcylation by SECRET AGENT (SEC) on ARABIDOPSIS HOMOLOG OF TRITHORAX1 (ATX1) modifies chromatin and influences FLC expression (Xing et al. 2018).

Light perception and circadian regulation are also modulated by PTMs. The COP1-SPA1 complex ubiquitinates CO, marking it for degradation, but this is counteracted by PHYB phosphorylation, which stabilises CO and promotes FT expression (Sarid-Krebs et al. 2015). The circadian clock regulator CIRCADIAN CLOCK-ASSOCIATED 1 (CCA1) undergoes phosphorylation, linking photoperiod sensing with flowering (Portolés and Zhang 2014). PTMs similarly regulate flowering across species. In rice (*Oryza sativa*), Heading Date 1 (HD1) and GI undergo phosphorylation, affecting photoperiodic flowering (Dai and Xue 2010). In maize (*Zea mays*), mitogen-activated protein kinases (MAPKs) and SNF1-RELATED PROTEIN KINASE 1 (SnRK1) integrate stress signals with floral transition (Forestan et al. 2020), and in barrel medick (*Medicago truncatula*), phosphorylation regulates both nitrogen fixation and flowering time, linking nutrient availability to developmental transitions (Ma and Chen 2021).

2.3.4 Links Between PTMs and Stress Signalling Pathways

PTMs induced by environmental stress responses regulate flowering by modifying proteins involved in developmental transitions. SnRK1, a metabolic sensor, phosphorylates Flowering BHLH 4 (FBH4) under low nitrogen conditions, accelerating flowering (Sanagi et al. 2021; Niu et al. 2024). Additionally, the abscisic acid (ABA) signalling pathway mediates flowering responses to drought via ABA-RESPONSIVE ELEMENT BINDING PROTEINS (ABFs) phosphorylation and ABSCISIC ACID-INSENSITIVE 5 (ABI5) ubiquitination, enabling drought escape flowering (Chirivì and Betti 2023).

SUMOylation also regulates flowering under stress. SIZ1 SUMOylates FLOWERING LOCUS D (FLD) to increase FLC expression, delaying flowering. HPY2 stabilises FLC, prolonging vegetative growth (Elrouby 2017). MAPKs phosphorylate WRKY and MYB transcription factors to modulate flowering under osmotic stress (Li et al. 2024). Additionally, reactive oxygen species (ROS) induce PTMs such as oxidation and S-nitrosylation, modifying GI and CO, ultimately influencing FT expression (Chirivì and Betti 2023).

The ubiquitin-proteasome system regulates stress-induced flowering by targeting floral repressors for degradation. COP1, HOS1, and PLANT U-BOX 13 (PUB13) ubiquitinate CO and GI, refining flowering responses (Shu and Yang 2017). The ubiquitin-specific proteases or UBP12 and UBP13 de-ubiquitinases counteract degradation, stabilising CO to promote early flowering (Cui et al. 2013). In cereals, PTMs modulate PHOTOPERIOD 1 (PPD1), VRN1, and HEADING DATE 3A (HD3A—rice ortholog of FT), synchronising flowering with seasonal variations (Chirivì and Betti 2023).

2.3.5 Techniques to Study PTMs in Crops

The classical approaches for detecting PTMs are Edman degradation, Western blotting, and mass spectrometry (MS) (Edman et al. 1950). Edman degradation-based protein sequencing uses peptide labelling, N-terminal residue cleavage, and detection for protein sequence less than ~50 amino acids in length. The modified peptide residues inhibit further sequencing (Restrepo-Pérez et al. 2018). Western blotting techniques (or immunoblotting) use a gel electrophoresis approach to separate protein samples either in their native or degraded form, followed by detection of PTMs using targeted antibodies. Western blotting can be further aided with advanced detection techniques like immunoprecipitation/co-immunoprecipitation assays to detect the subsequent interacting proteins (Xie et al. 2017). In MS, the PTMs are detected by enzymatic degradation of proteins and detection of peptide fragments. Here, a detector captures the ion signals according to their mass-to-charge (m/z) ratio in the form of a spectrum (Aslebagh et al. 2019). But MS analysis is usually assisted with different types of ionisation techniques, such as matrix-assisted laser desorption/ionisation (MALDI) or electron-spray ionisation (ESI) with HPLC-MS/MS, LC-MS, or FT-ICR-MS for higher accuracy (Bandeira et al. 2007; Picotti and Aebersold 2012). Additionally, specific dissociation methods like collision-induced dissociation or CID is used in the detection of glycopeptides (Bobalova et al. 2023). To overcome detection of less abundant proteins in crops, immuno-enrichment and affinity chromatography can be used for the enrichment of PTMs such as phosphorylation (Li et al. 2015), acetylation (Nallamilli et al. 2014), ubiquitination (Xie et al. 2015), and glycosylation (Catalá et al. 2011).

Cutting-edge molecular techniques like chromatin immunoprecipitation (ChIP), including ChIP-qPCR and ChIP-seq, are widely used for investigating PTMs on histones, allowing researchers to map modification sites and analyse their influence on gene expression (Sharma et al. 2023). Beyond chromatin, mass spectrometry, particularly top-down and bottom-up proteomics, is essential for detecting diverse PTMs, such as phosphorylation, acetylation, methylation, and ubiquitination, by analysing peptide fragments or intact proteins (Taylor and Young 2021). Cleavage under targets and tagmentation (CUT & Tag) offers a high-resolution alternative to ChIP, enabling efficient profiling of PTM-associated proteins with minimal input material (Kaya-Okur et al. 2019). Bisulfite-treated chromatin immunoprecipitation (BisChIP-seq) integrates DNA methylation analysis with PTM profiling, revealing epigenetic crosstalk (Tollefsbol 2020). CRISPR/dCas9-based epigenetic editing allows targeted manipulation of PTMs, helping elucidate functional roles in gene regulation, signal transduction, and disease mechanisms (Zhang et al. 2024). Integrating these approaches—ChIP-seq for mapping, MS for precise identification, and CRISPR for functional validation—provides a comprehensive framework for studying PTMs across various cellular processes, including stress responses, differentiation, and disease progression.

2.4 Using PTMs in Crop Improvement

Climatic changes over past years have shown to be a serious threat to agricultural practices, as plants struggle to cope with the changing environment. Stressful environments contribute to changes in metabolic processes, which in turn are associated with food production and yield (Teixeira et al. 2013). PTMs help plants survive these stresses by modulating protein stability, function, localisation, and interaction with other cellular components (Millar et al. 2019). PTMs influence crop response to biotic and abiotic stresses like cold, heat, drought, high salinity, pathogenic attacks, and nutrient deficiency. For example, in a high salt environment, the salt overlay sensitivity (SOS) pathway becomes activated due to phosphorylation, nitrosylation, and ubiquitination, activating Na^+/H^+ antiporters membrane bound transport proteins, such as calcineurin B-like protein (CBL) SOS3 and CBL-interacting protein kinases (CIPKs) SOS2 (Mahajan et al. 2008), to remove excessive Na^+ ions generated. PTMs are also tightly regulated in biotic stresses, where phosphorylation and ubiquitination are mostly known for their role in activating/deactivating certain defence-related signalling cascades. Plant immune response genes undergo alterations after infection signals are perceived (Schulze et al. 2010). For example, phosphorylation of RBOH1 (respiratory burst oxidase homolog 1) activates the protein to increase low-nitrogen tolerance. SUMOylation similarly activates PIF3, increasing resistance to pathogens (Jiang et al. 2017; Xie et al. 2019; Zheng et al. 2025) (Fig. 2.1). By effectively targeting PTMs that are involved in stress tolerance in crops, researchers have been successful in engineering crops that tolerate extreme environments (Pandey and Gayen 2024). Hence, a proper understanding of PTMs is essential in regulating plant responses toward these stress conditions, thereby leading a way to reduce the usage of chemical pesticides for stress tolerance and promoting sustainable crop production by utilising PTMs to improve crop immunity.

2.4.1 Proteomics: Identifying PTM Signatures Associated with Desirable Traits

The study of whole proteomes within organisms has modernised the pipeline identifying targeted PTMs under a particular stress condition, which is involved in shaping crop traits. For instance, phosphorylation of serine and threonine residues plays a crucial role in biological functions like cell cycle, growth and development, gene expression, apoptosis, and cellular communication (Ytterberg and Jensen 2010). Protein activity in turn can be activated or deactivated by switching the phases between phosphorylation and dephosphorylation (Whitmarsh and Davis 2000). Similarly, SUMOylation of lysine residues in targeted proteins increases under high temperature conditions as a stress response, a process also known as the SUMO stress response or SSR (Kurepa et al. 2003). Under high temperatures, plants also tend to increase hypocotyl length, extend petioles, show leaf hyponasty, and flower early as morphological changes (Quint et al. 2016). Secretory pathway proteins are modified by N-glycosylation to facilitate the effective folding of proteins, with the assistance of heat shock proteins (HSPs) in osmotic and/or drought stresses (Williams 2006). In tomato (*Solanum lycopersicum*) crops, heat shock proteins are overexpressed to maintain proper protein folding in order to increase its tolerance to salt stress (Fu et al. 2016). Likewise, rice crops become more susceptible to low heat and osmotic stress, in case of a defective β1,2-xylosyltransferase (incorporates xylose into N-glycans), which obstructs aerenchyma formation in roots (Takano et al. 2015). Therefore, identification of PTMs that are linked to certain traits provides a powerful tool to enhance crop yield and breeding.

2.4.2 Breeding and Biotech Approaches to Modify PTM in Desirable Ways

Tools of modern biotechnology involve transgenic plants and genetically modified crops to gain explicit control over PTMs. Genetic modification in plants can be achieved at various levels including overexpression, knockout, or single nucleotide base editing. PTMs such as SUMOylation induce specific traits in crops such as early flowering, biotic and abiotic stress tolerance, and crop yields (Rosa and Abreu 2019). Overexpression of the associated genes that regulate favourable PTMs has been shown to be beneficial in sustainable crop production. For example, overexpressing the *SIZ1* gene in rice and tomato leads to a higher tolerance to abiotic stress (Li et al. 2013; Zhang et al. 2018b). Similarly, genetically modified sugarcane (*Saccharum officinarum*) with overexpression of the SUMO protease *OVERLY TOLERANT TO SALT-1* (*OTS1*) gene of *A. thaliana* showed improved resistance to water deficit and drought conditions (Conti et al. 2008; Masoabi et al. 2023). Apart from overexpressing beneficial PTM containing genes, knockouts or targeted modifications to negatively regulate PTMs can also be helpful. The CRISP/Cas9-based targeting approach is now being widely used to elucidate the roles of functional PTMs in crops. This technique is also used in improving the physical characteristics of crops to produce higher-quality crops (Liu et al. 2021). However, synthetic biology approaches offer an innovative way to modify PTM sites within proteins to provide better stability and function, increasing durability as a trait. This involves the design of synthetically programmed biological systems that regulate PTMs only at the post-translational level without altering their genetic code. Researchers have developed synthetic molecules such as microproteins that can alter PTM patterns in response to stress stimuli (Abdeeva et al. 2024).

2.4.3 Marker-Assisted Selection: Breeding Crops with Natural PTM Variants Associated with Stress Tolerance

PTMs play a crucial role in modulating protein function, making them valuable targets for marker-assisted selection, aimed to design crops that are resilient to change in temperatures, pest resistance, and adaptable to other environmental stresses. Harnessing natural variants in PTM-related genes through marker-assisted selection (MAS) offers a strategic avenue for breeding stress-tolerant crops (Tariq et al. 2022). MAS is a breeding technique that utilises molecular markers linked to desirable traits, enabling the selection of plants carrying specific genetic variants without relying solely on phenotypic assessments (Hasan et al. 2021). In the context of PTMs, natural allelic variations affecting protein modifications can be identified and tracked using MAS to develop stress-resilient crop varieties. For instance, phosphorylation plays a critical role in stress signal transduction. Variation in genes encoding protein kinases or phosphatases can alter phosphorylation patterns, impacting stress responses (Damaris and Yang 2021). By identifying molecular markers associated with favourable alleles of these genes, breeders can select plants with enhanced stress tolerance. Similarly, ubiquitination-related genes, which govern protein degradation pathways, may harbour natural variants that confer improved stress resilience (Wu et al. 2023b). MAS facilitates the incorporation of these beneficial alleles into elite cultivars.

2.4.4 Precise Reprogramming of PTMs to Create Smart Crops

Advancements in genome editing and synthetic biology have revolutionised our ability to modify PTMs, offering precise tools to enhance stress tolerance in crops. The CRISPR/Cas9 system enables targeted editing of specific genes. By designing guide RNAs that direct Cas9 to precise genomic locations, researchers can introduce mutations or insertions to modify PTM-related genes or perform epigenomic editing (Alves et al. 2024; Batra et al. 2024). For example, editing genes encoding

E3 ubiquitin ligases can alter protein ubiquitination patterns, affecting protein stability and stress responses, or alternatively, residues within proteins can be edited, opting for a more protein specific effect. This targeted approach allows for the manipulation of protein function. Apart from CRISPR, synthetic biology techniques can be utilised, which involve the design and construction of novel genetic circuits and pathways to reprogram cellular functions.

The first ever genome editing application based on CRISPR/Cas9 for the purpose of crop improvement was utilised over rice and wheat (*Triticum aestivum*). After which has been an exponentially growing area, now incorporating most widely grown crops such as potato, tomato, maize, soyabean, sugarcane, and cotton (Li et al. 2021). CRISPR-based editing techniques have shown to be a meticulous tool for plant breeding to produce tolerant lines against biotic and abiotic stresses. One such example is plant receptors for the recognition of biotic pathogens, such as receptor-like kinases or RLK, which work as pathogen recognition receptors (PRRs). The intracellular kinase domain of RLKs turns on the defence signalling cascade by phosphorylation after receiving molecular cues from a pathogen (Burkart and Stahl 2017). In this scenario, enhanced expression of RLKs and knockdowns of receptor-like proteins or RLPs, would help in boosting plant pathogen response. Similarly, N-glycosylation of the GPP protein from a nematode inside a rice host promotes parasitism and helps in supressing host immune response (Chen et al. 2017), whereas overexpression of a RING-type E3 ligase *CaAIP1* from capsicum helps in gaining resistance to salt tolerance. This gene was recognised to be a stress responsive ubiquitin E3 ligase and has functions in osmotic stress tolerance (Park et al. 2016). Another study in tomatoes shows that silencing of the *SlWRKY 81* gene reduced transpiration loss through stomata and increased tolerance to drought conditions through various signalling pathways (Birhanu 2020). In a similar way, the glycosylation of soybean (*Glycine max*) proteins was found to be more frequent in flooding conditions, which is further linked to the degradation of those proteins (Mustafa and Komatsu 2014).

Flowering time is well studied in *A. thaliana*, and *FLC* is known to be the master regulator of flowering (Srikanth and Schmid 2011). Similarly, the *GHD7* gene in rice and maize and the *VRN2* gene in wheat and barley (*Hordeum vulgare*) are homologs of *FLC*, both working to repress FT-like genes. During exposure to cold environments, phosphorylation of VRN2 increases significantly, which leads to the accumulation of phosphorylated protein in the shoots and tips of young leaves (Xiao et al. 2014; Woods et al. 2016). Flowering time in plants is also determined by several other factors, such as temperature and light intensity. One group of such proteins found in *A. thaliana* is the PIFs, which are negative regulators of FT. But under high-intensity light, phytochromes tend to phosphorylate PIFs which target them for degradation (de Lucas and Prat 2014). Alternatively, SUMOylation of PHYB increases stability and induces the degradation of PIFs, resulting in the acceleration of flowering (Bernula et al. 2021). In rice, *HEAD DATE 6* (*HD6*) was found to encode a CK2α subunit that represses flowering time under long-day conditions by phosphorylation. However, replacing one serine residue of CK2 with glutamate rescues this phenomenon without changing the plant's biological clock (Ogiso et al.

2010). With knowledge about these PTMs holding a primary role in regulating flowering time in crops, CRISPR/Cas9 can be used to target modifications to achieve the desired flowering traits under circumstances.

2.4.5 Considering Which Features to Enhance

Modifying the desired PTM sites in the plant genome, improving crop yields, can be achieved through modern agricultural techniques. Different proteins in plants are recruited for certain functions, and their modification could result in distinct plant growth, development, and responses. For crop sustainability and improvement, the traits selected for modification should be considered carefully. In crops like rice, wheat, barley, and maize, where a majority portion of the yield is destroyed due to pathogens, a desirable pathogen-resistant feature should be incorporated followed by the improvement of seed quality and size. To meet the global demand of cereal crops, seed characteristics are also an essential factor to improve the grain quality while improving the flowering time of the crops. Shorter flowing time not only improves the harvest time but also reduces crop life cycle, which decreases the load on agricultural landmass. Whereas in crops like potato, peanut, and radish, the focus should be on increasing the yield by enhancing the vegetative growth of the plant. Targeted features for improvement should be primarily decided based on the major challenges that crop experiences based on cultivar location and secondarily by the strategy to improve production.

2.5 Sustainability Considerations

In crop breeding practices, agricultural biologists work to improve crop resistance and yield. However, in certain scenarios, the regulation of a particular target trait could also be involved in controlling multiple pathways in the plant and might affect other regulatory mechanisms if altered. One such example is the modification of amino acid transporters in crops, which is associated with improving resistance to disease and the nutritional values of the crop. However, the desired trait might also lead to reduction in overall crop yield and improper plant growth and development (Dhatterwal et al. 2024). In rice, when *OsAAP1* and *OsAAP4* were overexpressed, due to increased transportation of neutral amino acids, the crop was able to use nitrogen more effectively without resulting in any other defects in crop development and growth (Ji et al. 2020; Fang et al. 2021). In contrast, overexpression of AMINO ACID PERMEASE 1 (*AAP1*) in *Pisum sativum* resulted in an increase of grain nitrogen by 10–25%, with a complementary reduction of seed weight and nutritional values (Weigelt et al. 2008) (Fig. 2.2). While this specific overexpression helped increasing the yield, in other ways also led to the downfall of other essential traits that were adjusted with the desired phenotype. A similar trade-off has also

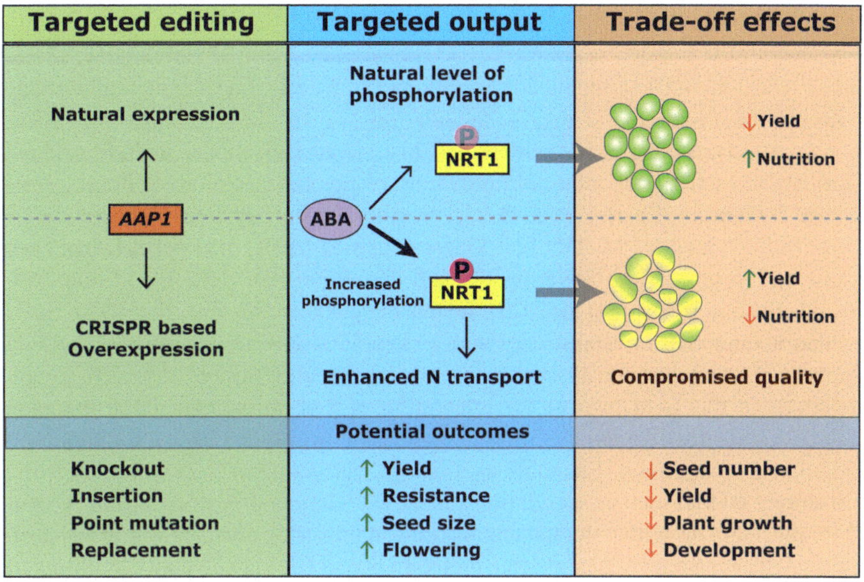

Fig. 2.2 Targeted PTM and associated off-target effects. This figure demonstrates the application of advanced gene editing tools, for generating targeted knockouts, mutations, and overexpressing crop lines. The targeted manipulation of genes known to enhance PTM levels in crops can lead to significant improvements in crop yield. However, these modifications may also have unintended consequences, potentially compromising other desirable traits

been reported with the overexpression of genes for improved pathogen resistance which in turn shows compromised plant growth and development (Yang et al. 2014), highlighting another scenario where making the crops stress resistance comes with the penalty of reduced crop yield and quality (Srivastava and Sahoo 2021).

While considering modifications in crops, the environmental, ecological, and consumer impacts should also be evaluated. Modifying a particular protein in economically important crops could also potentially hamper its interaction with pollinators and essential soil microbiota. A non-targeted trait being affected by modification can also result in long-term environmental impacts. For example, if the goal is to improve nutrient quality in produce, this also leads to quickening nutrient depletion of the growing soil. On the other hand, consumer satisfaction is arguably the most important trait to consider, as even the most nutritious, resilient, and sustainable crop is effectively worthless if it does not sell. As an extra layer of complexity, genetically modified crops are usually perceived with conflicts around safety, ethics, and potential associated hazards to human health which impacts the acceptance of the trait enhancement (Ishii and Araki 2016). Hence, biologists are suggested to produce modified cultivars that could satisfy consumer needs by checking on the possible off-target effects of the modification. Proper, detailed information related to modified crops should be communicated to consumers to have clear knowledge about food safety (Siddiqui et al. 2022).

2.6 Future Directions in PTM-Based Intervention

Future research on PTM-mediated regulation of flowering time should aim to refine molecular interventions for precise modulation of plant developmental processes. Proteomic advancements have identified histone modifications, including phosphorylation, methylation, and ubiquitination, as pivotal regulators of floral transition in *A. thaliana* (Sharma et al. 2023). Notably, phosphorylation of histone H2A at Ser95, mediated by MuTP9-like kinases such as MLK3 and MLK4, has been implicated in promoting flowering through chromatin remodelling and transcriptional activation (Sharma et al. 2023). Increasing evidence suggests that PTM-based regulation integrates metabolic cues with endogenous flowering pathways, enabling plants to respond dynamically to environmental stimuli (Giese et al. 2023). Future directions in this field involve the integration of high-throughput PTM mapping with CRISPR-based genome editing to engineer flowering pathways for enhanced agricultural productivity (Pramanik et al. 2021). Furthermore, elucidating the cross-talk among PTMs, such as acetylation and SUMOylation, may uncover novel regulatory mechanisms with potential applications in flowering control (Marquardt et al. 2020). Advancements in quantitative proteomics will facilitate dynamic profiling of PTMs across developmental stages and environmental conditions, providing deeper mechanistic insights into floral regulation. Extending this research to diverse plant species holds promise for the development of climate-resilient crops with optimised flowering times, addressing challenges posed by global climate variability. Some of the post-translational modifications on target genes in the various essential crops are currently under consideration in modern agricultural practices (Table 2.1).

2.7 Conclusion

Post-translational modifications often influence target protein functions to maintain cellular homeostasis. This chapter covers the most well-known PTMs and their roles in controlling flowering in model plants and crops. Thanks to cutting-edge technologies, new insights have emerged into understanding the dynamics of PTMs and revealing their critical role in linking environmental signals to plant development. As sessile organisms, plants depend heavily on environmental conditions for their growth and development. Adverse climatic conditions directly affect plant growth, correlating with food supply and prompting significant efforts to create climate-proof sustainable crops. Despite challenges like transient interactions and biologically low abundance, PTMs act as precise signal integrators, offering potential biomarkers for selective breeding. Furthermore, by modulating PTMs, crops could be engineered for accelerated growth and reduced resource demands (e.g. water-intensive rice varieties) while optimising yield without compromising productivity. In agronomy, there is significant potential to utilise PTMs as biomarkers, which would be less error-prone, more precise, and provide better protein-level

Table 2.1 Post-translational modifications reported and successfully implemented in plant species

Plant species	PTM type	Targeted protein	Effect of PTM	References
Solanum lycopersicum	Phosphorylation	RBOH1	Tolerance to low nitrogen	Zheng et al. (2025)
Saccharum officinarum	SUMOylation	OTS1	Increased drought tolerance	Masoabi et al. (2023)
Oryza sativa	Deacetylation	HDAC	Tolerance to salt	Cheng et al. (2018)
Triticum aestivum	Glycosylation	RLK	Cell wall modulation in drought condition	Chang et al. (2021)
Oryza sativa	Acetylation	STRK1	Tolerance to salt and heat	Tian et al. (2024)
Arabidopsis thaliana	Glycosylation	FLA11	Enhanced cell wall formation	Ma et al. (2022)
Zea mays	Acetylation	GCN5	Tolerance to salinity	Li et al. (2014)
Oryza sativa	SUMOylation	DELLA	Improves yield under stress	Fernandes et al. (2024)
Capsicum annuum	Methylation	SWC4	Tolerance to heat	Huang et al. (2023)
Gossypium herbaceum	Ubiquitination	MYB	Resistance to pathogen	Guang et al. (2024)
Oryza sativa	SUMOylation	SIZ1	Tolerance to heat	Mishra et al. (2018)
Oryza sativa	Glycosylation	ABI	Promotes growth during flooding	He et al. (2023)
Hordeum vulgare	Acetylation	MYST	Tolerance to drought	Papaefthimiou et al. (2010)
Gossypium herbaceum	Ubiquitination	RDUF4D	Resistance to fungus	Zhao et al. (2021)
Poncirus trifoliata	Phosphorylation	APX	Reduced ROS accumulation	Meng et al. (2020)
Setaria italica	Acetylation	HAT	Tolerance to abiotic stress and viral infection	Xing et al. (2022)

control without compromising the original functions. Unlike traditional GMO-dependent technologies, PTMs-based strategies allow target protein modifications and reduce unintended effects. These strategies are rapidly evolving and transformative; they hold immense promise for research applications and the production of sustainable crops. In conclusion, PTMs bridge the gap between molecular biology and agricultural practices, offering promising sustainable solutions for climate-resilient crops, high crop yield, and better crop selection. PTMs offer a powerful tool to engineer sustainable crops to offset adverse environmental conditions.

Acknowledgements We thank Arnab Singha Mahapatra, a visitor to Monash University, for the comments on the book chapter. We sincerely apologise to our colleagues whose contributions could not be cited due to the limited focus of this article and space limitations. The funding support: JC is supported by the Research Training Program (RTP) Scholarship, Monash University.

CKP is supported by the IITB-Monash Research Academy Fellowship. RG is supported by the Monash Graduate Scholarship (MGS), Monash University. SS is supported by the Australian Research Council Future Fellowship FT190100403.

Author Contributions

Conceptualisation: SS
Supervision: SS
Writing—original draft: JC, CKP, RG, SM, SS
Writing—review and editing: JC, CKP, SS

Competing Interests The authors declare that they have no competing interests.

References

Abdeeva IA, Panina YS, Maloshenok LG (2024) Synthetic biology approaches to posttranslational regulation in plants. Biochem Mosc 89(Suppl 1):S278–S289

Almaraz JJ, Mabood F, Zhou X, Gregorich EG, Smith DL (2008) Climate change, weather variability and corn yield at a higher latitude locale: Southwestern Quebec. Clim Chang 88(2):187–197

Alves J, Schwinn M, Machleidt T, Goueli SA, Cali JJ, Zegzouti H (2024) Monitoring phosphorylation and acetylation of CRISPR-mediated HiBiT-tagged endogenous proteins. Sci Rep 14(1):2138

Aslebagh R, Wormwood KL, Channaveerappa D, Wetie AGN, Woods AG, Darie CC (2019) Identification of posttranslational modifications (PTMs) of proteins by mass spectrometry. In: Advancements of mass spectrometry in biomedical research. Springer, pp 199–224

Balasubramanian S, Sureshkumar S, Lempe J, Weigel D (2006) Potent induction of *Arabidopsis thaliana* flowering by elevated growth temperature. PLoS Genet 2(7):e106

Bandeira N, Clauser KR, Pevzner PA (2007) Shotgun protein sequencing: assembly of peptide tandem mass spectra from mixtures of modified proteins. Mol Cell Proteomics 6(7):1123–1134

Bastow R, Mylne JS, Lister C, Lippman Z, Martienssen RA, Dean C (2004) Vernalization requires epigenetic silencing of *FLC* by histone methylation. Nature 427(6970):164–167

Batra SS, Cabrera A, Spence JP, Goell J, Anand SS, Hilton IB, Song YS (2024) Predicting the effect of CRISPR/Cas9-based epigenome editing. bioRxiv:2023.2010.2003.560674

Beltrao P, Bork P, Krogan NJ, van Noort V (2013) Evolution and functional cross-talk of protein post-translational modifications. Mol Syst Biol 9(1):714

Berhane A (2018) Climate change and variability impacts on agricultural productivity and food security. Clim Weather Forecast 6(240):2

Bernula P, Pettkó-Szandtner A, Hajdu A, Kozma-Bognár L, Josse EM, Ádám É, Nagy F, Viczián A (2021) SUMOylation of PHYTOCHROME INTERACTING FACTOR 3 promotes photomorphogenesis in *Arabidopsis thaliana*. New Phytol 229(4):2050–2061

Berry S, Dean C (2015) Environmental perception and epigenetic memory: mechanistic insight through *FLC*. Plant J 83(1):133–148

Bibi F, Rahman A (2023) An overview of climate change impacts on agriculture and their mitigation strategies. Agriculture 13(8):1508

Birhanu MW (2020) Transgenic approaches of improving tomato (*Solanum lycopersicum*) to salt stress tolerance. J Biol Agric Healthc 10:1–10

Blanco-Touri N, Serrano-Mislata A, Alabad D (2020) Regulation of DELLA proteins by post-translational modifications. Plant Cell Physiol 61(11):1891–1901

Blümel M, Dally N, Jung C (2015) Flowering time regulation in crops—what did we learn from *Arabidopsis*? Curr Opin Biotechnol 32:121–129

Bobalova J, Strouhalova D, Bobal P (2023) Common post-translational modifications (PTMs) of proteins: analysis by up-to-date analytical techniques with an emphasis on barley. J Agric Food Chem 71(41):14825–14837

Boss PK, Bastow RM, Mylne JS, Dean C (2004) Multiple pathways in the decision to flower: enabling, promoting, and resetting. Plant Cell 16:S18–S31

Burkart RC, Stahl Y (2017) Dynamic complexity: plant receptor complexes at the plasma membrane. Curr Opin Plant Biol 40:15–21

Catalá C, Howe KJ, Hucko S, Rose JK, Thannhauser TW (2011) Towards characterization of the glycoproteome of tomato (*Solanum lycopersicum*) fruit using Concanavalin A lectin affinity chromatography and LC-MALDI-MS/MS analysis. Proteomics 11(8):1530–1544

Chang Y, Zhu D, Duan W, Deng X, Zhang J, Ye X, Yan Y (2021) Plasma membrane N-glycoproteome analysis of wheat seedling leaves under drought stress. Int J Biol Macromol 193:1541–1550

Chen J, Lin B, Huang Q, Hu L, Zhuo K, Liao J (2017) A novel *Meloidogyne graminicola* effector, MgGPP, is secreted into host cells and undergoes glycosylation in concert with proteolysis to suppress plant defenses and promote parasitism. PLoS Pathog 13(4):e1006301

Cheng X, Zhang S, Tao W, Zhang X, Liu J, Sun J, Zhang H, Pu L, Huang R, Chen T (2018) INDETERMINATE SPIKELET1 recruits histone deacetylase and a transcriptional repression complex to regulate rice salt tolerance. Plant Physiol 178(2):824–837

Chirivì D, Betti C (2023) Molecular links between flowering and abiotic stress response: a focus on Poaceae. Plants 12(2):331

Conti L, Price G, O'Donnell E, Schwessinger B, Dominy P, Sadanandom A (2008) Small ubiquitin-like modifier proteases OVERLY TOLERANT TO SALT1 and-2 regulate salt stress responses in *Arabidopsis*. Plant Cell 20(10):2894–2908

Costa S, Dean C (2019) Storing memories: the distinct phases of Polycomb-mediated silencing of *Arabidopsis FLC*. Biochem Soc Trans 47(4):1187–1196

Craufurd PQ, Wheeler TR (2009) Climate change and the flowering time of annual crops. J Exp Bot 60(9):2529–2539

Cui X, Lu F, Li Y, Xue Y, Kang Y, Zhang S, Qiu Q, Cui X, Zheng S, Liu B (2013) Ubiquitin-specific proteases UBP12 and UBP13 act in circadian clock and photoperiodic flowering regulation in *Arabidopsis*. Plant Physiol 162(2):897–906

D'Urso A, Brickner JH (2014) Mechanisms of epigenetic memory. Trends Genet 30(6):230–236

Dai C, Xue HW (2010) Rice *early flowering 1*, a CKI, phosphorylates DELLA protein SLR1 to negatively regulate gibberellin signalling. EMBO J 29(11):1916–1927

Damaris RN, Yang P (2021) Protein phosphorylation response to abiotic stress in plants. In: Plant phosphoproteomics: methods and protocols. Springer US, New York, pp 17–43

de Lucas M, Prat S (2014) PIF s get BR right: PHYTOCHROME INTERACTING FACTOR s as integrators of light and hormonal signals. New Phytol 202(4):1126–1141

De Lucia F, Crevillen P, Jones AME, Greb T, Dean C (2008) A PHD-polycomb repressive complex 2 triggers the epigenetic silencing of *FLC* during vernalization. Proc Natl Acad Sci 105(44):16831–16836

Dhatterwal P, Prasad M, Mehrotra S, Mehrotra R (2024) Yield loss and growth-defense trade-offs: impact of engineering amino acid transporters. Nucleus 67:1–5

Edman P, Högfeldt E, Sillén LG, Kinell P-O (1950) Method for determination of the amino acid sequence in peptides. Acta Chem Scand 4(7):283–293

Elrouby N (2017) Regulation of plant cellular and organismal development by SUMO. In: SUMO regulation of cellular processes. Springer, pp 227–247

Fang Z, Wu B, Ji Y (2021) The amino acid transporter OsAAP4 contributes to rice tillering and grain yield by regulating neutral amino acid allocation through two splicing variants. Rice 14:1–17

Fernandes T, Gonçalves NM, Matiolli CC, Rodrigues MAA, Barros PM, Oliveira MM, Abreu IA (2024) SUMOylation of rice DELLA SLR1 modulates transcriptional responses and improves yield under salt stress. Planta 260(6):136

Forestan C, Farinati S, Zambelli F, Pavesi G, Rossi V, Varotto S (2020) Epigenetic signatures of stress adaptation and flowering regulation in response to extended drought and recovery in *Zea mays*. Plant Cell Environ 43(1):55–75

Fu C, Liu X, Yang W, Zhao C, Liu J (2016) Enhanced salt tolerance in tomato plants constitutively expressing heat-shock protein in the endoplasmic reticulum. Genet Mol Res 15(2):10.4238

Gao C, Tang D, Wang W (2022) The role of ubiquitination in plant immunity: fine-tuning immune signaling and beyond. Plant Cell Physiol 63(10):1405–1413

Gareau JR, Lima CD (2010) The SUMO pathway: emerging mechanisms that shape specificity, conjugation and recognition. Nat Rev Mol Cell Biol 11(12):861–871

Gendall AR, Levy YY, Wilson A, Dean C (2001) The *VERNALIZATION 2* gene mediates the epigenetic regulation of vernalization in *Arabidopsis*. Cell 107(4):525–535

Giese J, Eirich J, Walther D, Zhang Y, Lassowskat I, Fernie AR, Elsässer M, Maurino VG, Schwarzländer M, Finkemeier I (2023) The interplay of post-translational protein modifications in *Arabidopsis* leaves during photosynthesis induction. Plant J 116(4):1172–1193

Guang H, Xiaoyang G, Zhian W, Ye W, Peng W, Linfang S, Bingting W, Anhong Z, Fuguang L, Jiahe W (2024) The cotton *MYB33* gene is a hub gene regulating the trade-off between plant growth and defense in *Verticillium dahliae* infection. J Adv Res 61:1–17

Habtemariam LT, Abate Kassa G, Gandorfer M (2017) Impact of climate change on farms in smallholder farming systems: yield impacts, economic implications and distributional effects. Agric Syst 152:58–66

Haglund K, Dikic I (2005) Ubiquitylation and cell signaling. EMBO J 24(19):3353–3359

Haider S, Farrona S (2024) Decoding histone 3 lysine methylation: insights into seed germination and flowering. Curr Opin Plant Biol 81:102598

Hajdu A, Ádám É, Sheerin DJ, Dobos O, Bernula P, Hiltbrunner A, Kozma-Bognár L, Nagy F (2015) High-level expression and phosphorylation of phytochrome B modulates flowering time in Arabidopsis. Plant J 83(5):794–805

Han D, Yu Z, Lai J, Yang C (2022) Post-translational modification: a strategic response to high temperature in plants. aBIOTECH 3(1):49–64

Hasan N, Choudhary S, Naaz N, Sharma N, Laskar RA (2021) Recent advancements in molecular marker-assisted selection and applications in plant breeding programmes. J Genet Eng Biotechnol 19(1):128

Hashiguchi A, Komatsu S (2016) Impact of post-translational modifications of crop proteins under abiotic stress. Proteome 4(4):42

Hauvermale AL, Ariizumi T, Steber CM (2012) Gibberellin signaling: a theme and variations on DELLA repression. Plant Physiol 160(1):83–92

He Y, Sun S, Zhao J, Huang Z, Peng L, Huang C, Tang Z, Huang Q, Wang Z (2023) UDP-glucosyltransferase OsUGT75A promotes submergence tolerance during rice seed germination. Nat Commun 14(1):2296

Hebert DN, Lamriben L, Powers ET, Kelly JW (2014) The intrinsic and extrinsic effects of N-linked glycans on glycoproteostasis. Nat Chem Biol 10(11):902–910

Hochstrasser M (1996) Protein degradation or regulation: Ub the judge. Cell 84(6):813–815

Hou X, Zhou J, Liu C, Liu L, Shen L, Yu H (2014) Nuclear factor Y-mediated H3K27me3 demethylation of the *SOC1* locus orchestrates flowering responses of *Arabidopsis*. Nat Commun 5(1):4601

Howe FS, Fischl H, Murray SC, Mellor J (2017) Is H3K4me3 instructive for transcription activation? BioEssays 39(1):1–12

Huang Y, Cai W, Lu Q, Lv J, Wan M, Guan D, Yang S, He S (2023) PMT6 is required for SWC4 in positively modulating pepper thermotolerance. Int J Mol Sci 24(5):4849

Ibrar D, Ahmad R, Hasnain Z, Gul S, Rais A, Khan S (2023) Epigenetic-based control of flowering and seed development in plants: a review. Plant Breed 142(6):732–744

Inouye DW (2000) The ecological and evolutionary significance of frost in the context of climate change. Ecol Lett 3(5):457–463

Ishii T, Araki M (2016) Consumer acceptance of food crops developed by genome editing. Plant Cell Rep 35:1507–1518

Iwasaki M (2015) Chromatin resetting mechanisms preventing transgenerational inheritance of epigenetic states. Front Plant Sci 6:380

Jeong HJ, Yang J, Yi J, An G (2015) Controlling flowering time by histone methylation and acetylation in arabidopsis and rice. J Plant Biol 58(4):203–210

Ji Y, Huang W, Wu B, Fang Z, Wang X (2020) The amino acid transporter AAP1 mediates growth and grain yield by regulating neutral amino acid uptake and reallocation in *Oryza sativa*. J Exp Bot 71(16):4763–4777

Jiang B, Shi Y, Zhang X, Xin X, Qi L, Guo H, Li J, Yang S (2017) PIF3 is a negative regulator of the *CBF* pathway and freezing tolerance in *Arabidopsis*. Proc Natl Acad Sci 114(32):E6695–E6702

Kaya-Okur HS, Wu SJ, Codomo CA, Pledger ES, Bryson TD, Henikoff JG, Ahmad K, Henikoff S (2019) CUT&Tag for efficient epigenomic profiling of small samples and single cells. Nat Commun 10(1):1930

Komander D (2009) The emerging complexity of protein ubiquitination. Biochem Soc Trans 37(5):937–953

Kong X, Luo X, Qu GP, Liu P, Jin JB (2017) *Arabidopsis* SUMO protease ASP1 positively regulates flowering time partially through regulating *FLC* stability. J Integr Plant Biol 59(1):15–29

Kopeć P (2024) Climate change—the rise of climate-resilient crops. Plants 13(4):490

Kumar V, Thakur JK, Prasad M (2021) Histone acetylation dynamics regulating plant development and stress responses. Cell Mol Life Sci 78(10):4467–4486

Kurepa J, Walker JM, Smalle J, Gosink MM, Davis SJ, Durham TL, Sung D-Y, Vierstra RD (2003) The small ubiquitin-like modifier (SUMO) protein modification system in *Arabidopsis*: accumulation of SUMO1 and-2 conjugates is increased by stress. J Biol Chem 278(9):6862–6872

Kwak JS, Son GH, Song JT, Seo HS (2017) Post-translational modifications of FLOWERING LOCUS C modulate its activity. J Exp Bot 68(3):383–389

Lazaro A, Mouriz A, Piñeiro M, Jarillo JA (2015) Red light-mediated degradation of CONSTANS by the E3 ubiquitin ligase HOS1 regulates photoperiodic flowering in *Arabidopsis*. Plant Cell 27(9):2437–2454

Lee Z, Kim S, Choi SJ, Joung E, Kwon M, Park HJ, Shim JS (2023) Regulation of flowering time by environmental factors in plants. Plants 12(21):3680

Levy YY, Mesnage S, Mylne JS, Gendall AR, Dean C (2002) Multiple roles of *Arabidopsis VRN1* in vernalization and flowering time control. Science 297(5579):243–246

Li Z, Hu Q, Zhou M, Vandenbrink J, Li D, Menchyk N, Reighard S, Norris A, Liu H, Sun D (2013) Heterologous expression of *Os SIZ 1*, a rice SUMO E 3 ligase, enhances broad abiotic stress tolerance in transgenic creeping bentgrass. Plant Biotechnol J 11(4):432–445

Li H, Yan S, Zhao L, Tan J, Zhang Q, Gao F, Wang P, Hou H, Li L (2014) Histone acetylation associated up-regulation of the cell wall related genes is involved in salt stress induced maize root swelling. BMC Plant Biol 14:1–14

Li M, Yin X, Sakata K, Yang P, Komatsu S (2015) Proteomic analysis of phosphoproteins in the rice nucleus during the early stage of seed germination. J Proteome Res 14(7):2884–2896

Li C, Brant E, Budak H, Zhang B (2021) CRISPR/Cas: a Nobel Prize award-winning precise genome editing technology for gene therapy and crop improvement. J Zhejiang Univ Sci B 22(4):253–284

Li S, Khoso MA, Xu H, Zhang C, Liu Z, Wagan S, Dinislam K, Liu L (2024) WRKY transcription factors (TFs) as key regulators of plant resilience to environmental stresses: current perspective. Agronomy 14(10):2421

Liang Z, Huang Y, Hao Y, Song X, Zhu T, Liu C, Li C (2025) The HISTONE ACETYLTRANSFERASE 1 interacts with CONSTANS to promote flowering in Arabidopsis. J Genet Genomic/Yi chuan xue bao 52(12):1627–1637

Liu L-J, Zhang Y-C, Li Q-H, Sang Y, Mao J, Lian H-L, Wang L, Yang H-Q (2008) COP1-mediated ubiquitination of CONSTANS is implicated in cryptochrome regulation of flowering in *Arabidopsis*. Plant Cell 20(2):292–306

Liu Y, Lu S, Liu K, Wang S, Huang L, Guo L (2019) Proteomics: a powerful tool to study plant responses to biotic stress. Plant Methods 15(1):135
Liu Q, Yang F, Zhang J, Liu H, Rahman S, Islam S, Ma W, She M (2021) Application of CRISPR/Cas9 in crop quality improvement. Int J Mol Sci 22(8):4206
Ma Y, Chen R (2021) Nitrogen and phosphorus signaling and transport during legume–rhizobium symbiosis. Front Plant Sci 12:683601
Ma X, Lv S, Zhang C, Yang C (2013) Histone deacetylases and their functions in plants. Plant Cell Rep 32(4):465–478
Ma Y, MacMillan CP, de Vries L, Mansfield SD, Hao P, Ratcliffe J, Bacic A, Johnson KL (2022) FLA11 and FLA12 glycoproteins fine-tune stem secondary wall properties in response to mechanical stresses. New Phytol 233(4):1750–1767
Mahajan S, Pandey GK, Tuteja N (2008) Calcium-and salt-stress signaling in plants: shedding light on SOS pathway. Arch Biochem Biophys 471(2):146–158
Marino D, Peeters N, Rivas S (2012) Ubiquitination during plant immune signaling. Plant Physiol 160(1):15–27
Marquardt S, Leng X, Thomas Q, Rasmussen S (2020) AG (enomic) P (ositioning) S (ystem) for plant RNAPII transcription. Preprints. https://doi.org/10.20944/preprints202003.0290.v1
Martí-Guillén JM, Pardo-Hernández M, Martínez-Lorente SE, Almagro L, Rivero RM (2022) Redox post-translational modifications and their interplay in plant abiotic stress tolerance. Front Plant Sci 13:1027730
Masoabi M, Burger N, Botha AM, Le Roux M, Vlok M, Snyman S, Van der Vyver C (2023) Overexpression of the Small Ubiquitin-Like Modifier protease *OTS1* gene enhances drought tolerance in sugarcane (*Saccharum* spp. hybrid). Plant Biol 25(7):1121–1141
Medzihradszky M, Bindics J, Ádám É, Viczián A, Klement É, Lorrain S, Gyula P, Mérai Z, Fankhauser C, Medzihradszky KF (2013) Phosphorylation of phytochrome B inhibits light-induced signaling via accelerated dark reversion in Arabidopsis. Plant Cell 25(2):535–544
Meng L, Zhang Q, Yang J, Xie G, Liu J-H (2020) PtrCDPK10 of *Poncirus trifoliata* functions in dehydration and drought tolerance by reducing ROS accumulation via phosphorylating PtrAPX. Plant Sci 291:110320
Millar AH, Heazlewood JL, Giglione C, Holdsworth MJ, Bachmair A, Schulze WX (2019) The scope, functions, and dynamics of posttranslational protein modifications. Annu Rev Plant Biol 70(1):119–151
Mishra N, Srivastava AP, Esmaeili N, Hu W, Shen G (2018) Overexpression of the rice gene *OsSIZ1* in Arabidopsis improves drought-, heat-, and salt-tolerance simultaneously. PLoS One 13(8):e0201716
Molotoks A, Smith P, Dawson TP (2021) Impacts of land use, population, and climate change on global food security. Food Energy Secur 10(1):e261
Murtas G, Reeves PH, Fu Y-F, Bancroft I, Dean C, Coupland G (2003) A nuclear protease required for flowering-time regulation in *Arabidopsis* reduces the abundance of SMALL UBIQUITIN-RELATED MODIFIER conjugates. Plant Cell 15(10):2308–2319
Mustafa G, Komatsu S (2014) Quantitative proteomics reveals the effect of protein glycosylation in soybean root under flooding stress. Front Plant Sci 5:627
Nagashima Y, von Schaewen A, Koiwa H (2018) Function of N-glycosylation in plants. Plant Sci 274:70–79
Nallamilli BRR, Edelmann MJ, Zhong X, Tan F, Mujahid H, Zhang J, Nanduri B, Peng Z (2014) Global analysis of lysine acetylation suggests the involvement of protein acetylation in diverse biological processes in rice (*Oryza sativa*). PLoS One 9(2):e89283
Nguyen NH, Vu NT, Cheong J-J (2022) Transcriptional stress memory and transgenerational inheritance of drought tolerance in plants. Int J Mol Sci 23(21):12918
Nito K, Wong CC, Yates JR, Chory J (2013) Tyrosine phosphorylation regulates the activity of phytochrome photoreceptors. Cell Rep 3(6):1970–1979
Niu F, Rehmani MS, Yan J (2024) Multilayered regulation and implication of flowering time in plants. Plant Physiol Biochem 213:108842

Oberkofler V, Pratx L, Bäurle I (2021) Epigenetic regulation of abiotic stress memory: maintaining the good things while they last. Curr Opin Plant Biol 61:102007

Ogiso E, Takahashi Y, Sasaki T, Yano M, Izawa T (2010) The role of casein kinase II in flowering time regulation has diversified during evolution. Plant Physiol 152(2):808–820

Pandey A, Gayen D (2024) Decoding post-translational modifications for understanding stress tolerance in plant. Crop Design 3(4):100077

Papaefthimiou D, Likotrafiti E, Kapazoglou A, Bladenopoulos K, Tsaftaris A (2010) Epigenetic chromatin modifiers in barley: III. Isolation and characterization of the barley GNAT-MYST family of histone acetyltransferases and responses to exogenous ABA. Plant Physiol Biochem 48(2–3):98–107

Park C, Lim CW, Lee SC (2016) The pepper RING-Type E3 ligase, CaAIP1, functions as a positive regulator of drought and high salinity stress responses. Plant Cell Physiol 57(10):2202–2212

Parker LE, McElrone AJ, Ostoja SM, Forrestel EJ (2020) Extreme heat effects on perennial crops and strategies for sustaining future production. Plant Sci 295:110397

Pickart CM (2001) Mechanisms underlying ubiquitination. Annu Rev Biochem 70(1):503–533

Picotti P, Aebersold R (2012) Selected reaction monitoring–based proteomics: workflows, potential, pitfalls and future directions. Nat Methods 9(6):555–566

Portolés S, Zhang D-P (2014) ABA signaling and circadian clock. In: Abscisic acid: metabolism, transport and signaling. Springer, Dordrecht, pp 385–407

Pramanik D, Shelake RM, Kim MJ, Kim J-Y (2021) CRISPR-mediated engineering across the central dogma in plant biology for basic research and crop improvement. Mol Plant 14(1):127–150

Proietti S, Scariot V, De Pascale S, Paradiso R (2022) Flowering mechanisms and environmental stimuli for flower transition: bases for production scheduling in greenhouse floriculture. Plants 11(3):432

Qazi HA, Jan N, Ramazan S, John R (2019) Protein modification in plants in response to abiotic stress. Academic, an imprint of Elsevier, London

Quint M, Delker C, Franklin KA, Wigge PA, Halliday KJ, Van Zanten M (2016) Molecular and genetic control of plant thermomorphogenesis. Nat Plants 2(1):1–9

Reeves PH, Murtas G, Dash S, Coupland G (2002) early in short days 4, a mutation in *Arabidopsis* that causes early flowering and reduces the mRNA abundance of the floral repressor *FLC*. Development 129(23):5349–5361

Restrepo-Pérez L, Joo C, Dekker C (2018) Paving the way to single-molecule protein sequencing. Nat Nanotechnol 13(9):786–796

Rosa MT, Abreu IA (2019) Exploring the regulatory levels of SUMOylation to increase crop productivity. Curr Opin Plant Biol 49:43–51

Rummukainen M (2012) Changes in climate and weather extremes in the 21st century. Wiley Interdiscip Rev 3(2):115–129

Sanagi M, Aoyama S, Kubo A, Lu Y, Sato Y, Ito S, Abe M, Mitsuda N, Ohme-Takagi M, Kiba T (2021) Low nitrogen conditions accelerate flowering by modulating the phosphorylation state of FLOWERING BHLH 4 in *Arabidopsis*. Proc Natl Acad Sci 118(19):e2022942118

Sarid-Krebs L, Panigrahi KC, Fornara F, Takahashi Y, Hayama R, Jang S, Tilmes V, Valverde F, Coupland G (2015) Phosphorylation of CONSTANS and its COP_1-dependent degradation during photoperiodic flowering of *Arabidopsis*. Plant J 84(3):451–463

Schulze B, Mentzel T, Jehle AK, Mueller K, Beeler S, Boller T, Felix G, Chinchilla D (2010) Rapid heteromerization and phosphorylation of ligand-activated plant transmembrane receptors and their associated kinase BAK1. J Biol Chem 285(13):9444–9451

Sharma M, Sidhu AK, Samota MK, Gupta M, Koli P, Choudhary M (2023) Post-translational modifications in histones and their role in abiotic stress tolerance in plants. Proteome 11(4):38

Shi M, Wang C, Wang P, Zhang M, Liao W (2022) Methylation in DNA, histone, and RNA during flowering under stress condition: a review. Plant Sci 324:111431

Shi Z, Zhao W, Li C, Tan W, Zhu Y, Han Y, Ai P, Li Z, Wang Z (2024) Overexpression of the *Chrysanthemum lavandulifolium* ROS1 gene promotes flowering in *Arabidopsis thaliana* by reducing the methylation level of CONSTANS. Plant Sci 342:112019

Shu K, Yang W (2017) E3 ubiquitin ligases: ubiquitous actors in plant development and abiotic stress responses. Plant Cell Physiol 58(9):1461–1476

Si S, Zhang M, Hu Y, Wu C, Yang Y, Luo S, Xiao X (2021) *BrcuHAC1* is a histone acetyltransferase that affects bolting development in Chinese flowering cabbage. J Genet 100:1–11

Siddiqui SA, Asif Z, Murid M, Fernando I, Adli DN, Blinov AV, Golik AB, Nugraha WS, Ibrahim SA, Jafari SM (2022) Consumer social and psychological factors influencing the use of genetically modified foods—a review. Sustainability 14(23):15884

Srikanth A, Schmid M (2011) Regulation of flowering time: all roads lead to Rome. Cell Mol Life Sci 68:2013–2037

Srivastava R, Sahoo L (2021) Balancing yield trade-off in legumes during multiple stress tolerance via strategic crosstalk by native NAC transcription factors. J Plant Biochem Biotechnol 30(4):708–729

Sun M, Yang Z, Liu L, Duan L (2022) DNA methylation in plant responses and adaption to abiotic stresses. Int J Mol Sci 23(13):6910

Sung S, Amasino RM (2004) Vernalization in *Arabidopsis thaliana* is mediated by the PHD finger protein VIN3. Nature 427(6970):159–164

Sureshkumar S, Bandaranayake C, Lv J, Dent CI, Bhagat PK, Mukherjee S, Sarwade R, Atri C, York HM, Tamizhselvan P (2024) SUMO protease FUG1, histone reader AL3 and chromodomain protein LHP1 are integral to repeat expansion-induced gene silencing in *Arabidopsis thaliana*. Nat Plants 10(5):749–759

Takano S, Matsuda S, Funabiki A, Furukawa J-i, Yamauchi T, Tokuji Y, Nakazono M, Shinohara Y, Takamure I, Kato K (2015) The rice *RCN11* gene encodes β1, 2-xylosyltransferase and is required for plant responses to abiotic stresses and phytohormones. Plant Sci 236:75–88

Talukder A, McDonald GK, Gill GS (2014) Effect of short-term heat stress prior to flowering and early grain set on the grain yield of wheat. Field Crop Res 160:54–63

Tariq A, Javaid A, Farhat F, Aslam MM, Subhani MM, Raja S, Naheed R, Tabbasum HN, Zulfiqar S (2022) Proteomic and genomic approaches for antioxidant enzyme-mediated defense analyses in higher plants. In: Antioxidant defense in plants: molecular basis of regulation. Springer, pp 57–70

Taylor BC, Young NL (2021) Combinations of histone post-translational modifications. Biochem J 478(3):511–532

Teixeira EI, Fischer G, Van Velthuizen H, Walter C, Ewert F (2013) Global hot-spots of heat stress on agricultural crops due to climate change. Agric For Meteorol 170:206–215

Thakur P, Kumar S, Malik JA, Berger JD, Nayyar H (2010) Cold stress effects on reproductive development in grain crops: an overview. Environ Exp Bot 67(3):429–443

Thompson LM (1975) Weather variability, climatic change, and grain production. Science 188(4188):535–541

Tian Y, Zeng H, Wu J-C, Dai G-X, Zheng H-P, Liu C, Wang Y, Zhou Z-K, Tang D-Y, Deng G-F (2024) The zinc finger protein DHHC09 S-acylates the kinase STRK1 to regulate H2O2 homeostasis and promote salt tolerance in rice. Plant Cell 36(4):919–940

Tollefsbol TO (2020) Epigenetics methods. Academic

Trujillo M, Shirasu K (2010) Ubiquitination in plant immunity. Curr Opin Plant Biol 13(4):402–408

Turgut-Kara N, Arikan B, Celik H (2020) Epigenetic memory and priming in plants. Genetica 148(2):47–54

Valverde F (2011) CONSTANS and the evolutionary origin of photoperiodic timing of flowering. J Exp Bot 62(8):2453–2463

Vanyushin B (2006) DNA methylation in plants. In: DNA methylation: basic mechanisms. Springer, Berlin/Heidelberg, pp 67–122

Weigelt K, Küster H, Radchuk R, Müller M, Weichert H, Fait A, Fernie AR, Saalbach I, Weber H (2008) Increasing amino acid supply in pea embryos reveals specific interactions of N and C metabolism, and highlights the importance of mitochondrial metabolism. Plant J 55(6):909–926

Whitmarsh A, Davis R (2000) Regulation of transcription factor function by phosphorylation. Cell Mol Life Sci 57:1172–1183

Williams DB (2006) Beyond lectins: the calnexin/calreticulin chaperone system of the endoplasmic reticulum. J Cell Sci 119(4):615–623

Woods DP, McKeown MA, Dong Y, Preston JC, Amasino RM (2016) Evolution of *VRN2/Ghd7*-like genes in vernalization-mediated repression of grass flowering. Plant Physiol 170(4):2124–2135

Wu L, Marcelis LFM, Kong F, Zhu Y (2023a) Editorial: flowering time control in agricultural and horticultural crops. Front Plant Sci 14:1116197

Wu M, Musazade E, Yang X, Yin L, Zhao Z, Zhang Y, Lu J, Guo L (2023b) ATL protein family: novel regulators in plant response to environmental stresses. J Agric Food Chem 71(51):20419–20440

Xiao J, Xu S, Li C, Xu Y, Xing L, Niu Y, Huan Q, Tang Y, Zhao C, Wagner D (2014) *O*-GlcNAc-mediated interaction between VER2 and TaGRP2 elicits *TaVRN1* mRNA accumulation during vernalization in winter wheat. Nat Commun 5(1):4572

Xie X, Kang H, Liu W, Wang G-L (2015) Comprehensive profiling of the rice ubiquitome reveals the significance of lysine ubiquitination in young leaves. J Proteome Res 14(5):2017–2025

Xie Y, Kang R, Tang D (2017) Assessment of posttranslational modifications of ATG proteins. Methods Enzymol (Elsevier) 587:171–188

Xie Z, Nolan TM, Jiang H, Yin Y (2019) AP2/ERF transcription factor regulatory networks in hormone and abiotic stress responses in *Arabidopsis*. Front Plant Sci 10:228

Xing L, Liu Y, Xu S, Xiao J, Wang B, Deng H, Lu Z, Xu Y, Chong K (2018) *Arabidopsis O*-Glc NA c transferase SEC activates histone methyltransferase ATX 1 to regulate flowering. EMBO J 37(19):e98115

Xing G, Jin M, Qu R, Zhang J, Han Y, Han Y, Wang X, Li X, Ma F, Zhao X (2022) Genome-wide investigation of histone acetyltransferase gene family and its responses to biotic and abiotic stress in foxtail millet (*Setaria italica* [L.] P. Beauv). BMC Plant Biol 22(1):292

Xu S, Chong K (2018) Remembering winter through vernalisation. Nat Plants 4(12):997–1009

Xu C, Wu Z, Duan H-C, Fang X, Jia G, Dean C (2021) R-loop resolution promotes co-transcriptional chromatin silencing. Nat Commun 12(1):1790

Yang H, Postel S, Kemmerling B, Ludewig U (2014) Altered growth and improved resistance of *Arabidopsis* against *Pseudomonas syringae* by overexpression of the basic amino acid transporter AtCAT1. Plant Cell Environ 37(6):1404–1414

Ytterberg AJ, Jensen ON (2010) Modification-specific proteomics in plant biology. J Proteome 73(11):2249–2266

Zentella R, Sui N, Barnhill B, Hsieh W-P, Hu J, Shabanowitz J, Boyce M, Olszewski NE, Zhou P, Hunt DF, Sun T-P (2017) The *Arabidopsis* O-fucosyltransferase SPINDLY activates nuclear growth repressor DELLA. Nat Chem Biol 13(5):479–485

Zhang H, Lang Z, Zhu J-K (2018a) Dynamics and function of DNA methylation in plants. Nat Rev Mol Cell Biol 19(8):489–506

Zhang S, Wang S, Lv J, Liu Z, Wang Y, Ma N, Meng Q (2018b) SUMO E3 ligase SlSIZ1 facilitates heat tolerance in tomato. Plant Cell Physiol 59(1):58–71

Zhang H, Zhu J, Gong Z, Zhu J-K (2022) Abiotic stress responses in plants. Nat Rev Genet 23(2):104–119

Zhang WJ, Zhou Y, Zhang Y, Su YH, Xu T (2023) Protein phosphorylation: a molecular switch in plant signaling. Cell Rep 42(7):112729

Zhang X, Bhattacharya A, Pu C, Dai Y, Liu J, Rao L, Tian C (2024) A programmable CRISPR/dCas9-based epigenetic editing system enabling loci-targeted histone citrullination and precise transcription regulation. J Genet Genomics 51(12):1485–1493

Zhao T, Zhan Z, Jiang D (2019) Histone modifications and their regulatory roles in plant development and environmental memory. J Genet Genomics/Yi chuan xue bao 46(10):467–476

Zhao Y-P, Shen J-L, Li W-J, Wu N, Chen C, Hou Y-X (2021) Evolutionary and characteristic analysis of RING-DUF1117 E3 ubiquitin ligase genes in *Gossypium* discerning the role of *GhRDUF4D* in *Verticillium dahliae* resistance. Biomolecules 11(8):1145

Zheng X, Yang H, Zou J, Jin W, Qi Z, Yang P, Yu J, Zhou J (2025) SnRK1α1-mediated RBOH1 phosphorylation regulates reactive oxygen species to enhance tolerance to low nitrogen in tomato. Plant Cell 37(1):koae321

Zhou B, Zeng L (2017) Conventional and unconventional ubiquitination in plant immunity. Mol Plant Pathol 18(9):1313–1330

Chapter 3
Hormonal and Molecular Regulation of Root Traits in *Arabidopsis* in Response to Abiotic Stress: A Way Forward to Climate-Smart Crop Improvement Program

Abhishek Kanojia, Poonam Yadav, Ayushi Jaiswal, Kajal Taneja, Ritu Yadav, and Yashwanti Mudgil

Abstract In the context of climate change and anthropogenic activities, escalating abiotic stresses, such as extreme temperatures, drought, salinity, exposure to genotoxic substances, and heavy metals, pose significant challenges to plant health, impacting growth and productivity. The survival of plants under these fluctuating conditions relies upon their ability to perceive and respond to various abiotic stresses, which collectively constitute significant constraints on plant growth and agricultural productivity worldwide. Focusing on roots in the context of abiotic stress is crucial, given their pivotal role in water and nutrient uptake. Roots exhibit remarkable plasticity, allowing them to adapt their development in response to environmental cues. Understanding the molecular mechanisms underlying root responses to stress is essential for improving crop resilience. *Arabidopsis*, a widely used model organism, serves as an ideal system to investigate the molecular pathways that regulate root responses to abiotic stresses. This chapter examines how different types of abiotic stresses influence root development by activating signaling pathways in *Arabidopsis*. These pathways control the expression of stress-responsive genes and shape root architecture, enabling plants to cope with environmental challenges. Comprehending these molecular mechanisms provides potential strategies for enhancing crop resilience in the face of climate change and a foundation for engineering stress-tolerant crops.

Keywords Anthropogenic activities · Abiotic stresses · *Arabidopsis* · Stress-tolerant crops

A. Kanojia · P. Yadav · A. Jaiswal · K. Taneja · R. Yadav · Y. Mudgil (✉)
Stress Signaling and Molecular Physiology Laboratory, Department of Botany, University of Delhi, Delhi, India

A. Singh et al. (eds.), *Sustainable Crop Production*, Sustainability Sciences in Asia and Africa, https://doi.org/10.1007/978-981-95-7844-3_3

Editor's Notes

Roots, the below ground structures of plants, play a crucial role in exploration of water and mineral nutrients. Roots display an extraordinary ability to dynamically change form and shape in response to availability of resources in soils, a term defined as "root plasticity." Depending upon the soil features, root system architecture (RSA) may vary with respect to length of primary and lateral roots, branching patterns, and, root growth angles. In the last decade, understanding the RSA has become a research priority from a crop adaptation perspective. Agronomists are aiming to decipher the fine control of root phenotypes to develop crop varieties adapted to marginal, water and nutrient-depleted soil types with more efficient systems for resource foraging.

In this chapter, Kanojia et al. reinforce the critical role of root systems in ameliorating abiotic stresses and enhancing nutrient uptake. The authors provide a comprehensive overview on root development, architecture, and plasticity under various abiotic stresses while highlighting extensive cross-talks among molecular regulators, and hormones, triggered by external and internal environmental factors. However, the large body of available information is only tip of the iceberg. The authors argue that understanding root development and deciphering molecular network in model plants such as *Arabidopsis thaliana* needs to be taken up by researchers on an urgent basis. The deployment of new, noninvasive advanced imaging technologies is proving to be a boon to researchers for understanding RSA under complex environments. Integrating such technologies will hasten translation from model plants to crops for realizing the goal of sustainable crop production.

Actionable Points

1. Understanding how root systems develop in the model plant *Arabidopsis* under various environmental conditions is crucial, as this knowledge could directly inform and guide research in optimizing root architecture in diverse crop species.
2. For molecular breeding of drought-resistant crops with improved root systems, regulatory genes such as *NAC41*, *RoLe1*, *AGAP*, *RRS1* (*MYB* family), and *IAA3* are promising targets and potential of these must be examined.
3. Breeders can put efforts into manipulating alleles of key regulatory genes (e.g., those controlling auxin homeostasis/transport, MAP kinases, specific transcription factors, etc.) that mediate beneficial root architectural changes under stress.
4. A reciprocal cross-species approach must be adopted to study root response with respect to multiple genes and environmental conditions.
5. Crop management must focus on studying root growth of both natural and genetically modified plants, and these must be analyzed directly in the natural field conditions to capture gene and environmental effects.

3.1 Introduction

As sessile organisms, plants necessitate adaptive strategies to navigate the ever--evolving environmental landscape, and the plasticity of their root systems is a critical component of this adaptation. Roots play a pivotal role in mediating responses to abiotic stressors, owing to their essential functions in water and nutrient acquisition and their integral interactions with the soil microbiome.

With respect to roots, the external environment modulates root growth and development, directly affecting plant health. Externally, the root can be divided into different parts like primary root, lateral root, root hair, root tip, root cap, and internally into vascular cylinder, cortex, epidermis, and endodermis. One of the prominent characteristic features of root architecture adaptation is the modulation of lateral root development (Benková and Bielach 2010). Lateral roots (LRs) originate from the pericycle of primary roots and help in the expansion of the root surface area for water and nutrient acquisition. Like under drought conditions, *Arabidopsis* enhances its primary root growth and reduces lateral root branching (Osakabe et al. 2014). Meanwhile, under limited soil, a localized nutrient-based experiment shows root growth toward the nutrient (López-Bucio et al. 2003).

The perception of stress and the subsequent signal transduction involve a complex network of interconnected signaling molecules, including proteins, kinases, phosphatases, transcription factors, and hormones. These components ultimately activate downstream signaling pathways, leading to alterations in gene expression and directing morphophysiological changes in plant parts (Shinozaki and Yamaguchi-Shinozaki 2007). Notably, one of the primary mitogen-activated protein kinase (MAPK) cascades serves as a crucial signal transduction pathway, facilitating the transmission of stress-induced signals and thereby regulating root developmental processes. Similarly, transcription factor families, particularly WRKY and MYB, play a significant regulatory role, influencing the expression of genes involved in root growth and various stress tolerance mechanisms.

Several hormones also mediate stress-related developmental adjustments (Sabagh et al. 2021, 2022). For example, abscisic acid (ABA) coordinates root responses to water deficit (Rosales et al. 2019), and a rapid increase in ABA levels in response to various stresses regulates plant growth and stress responses by affecting downstream signaling pathways (Tuteja 2007; Cutler et al. 2010). ABA also acts as an essential component that promotes stress tolerance by influencing root cell division, elongation, and branching, which regulates water uptake. Under stress conditions, especially during salinity and drought, ABA plays a crucial role in controlling root responses to stress (Osakabe et al. 2014). Auxin is a key regulator of root development functions in a concentration-dependent manner in cells and plant parts, modulating root tip growth, cell elongation, lateral root initiation, and gravitropism, and it also plays a role under stress (Richter et al. 2010). Abiotic stress results in high levels of endogenous ethylene (Lequeux et al. 2010; Asgher et al. 2014; Schellingen et al. 2014; Arteca and Arteca 2007). Ethylene is known to influence root growth mediated by cross talk with auxin, where ethylene can alter auxin biosynthesis, transport, and signaling within the root tip, leading to altered cell proliferation and elongation affecting the root phenotypes (Vaseva et al. 2018). Cytokinin (CK) regulates lateral root initiation, nodulation, vascular differentiation, and meristematic activity and is involved in various abiotic and biotic stress responses (Frugier et al. 2008; Argueso et al. 2010; Perilli et al. 2010). In *Arabidopsis*, cytokinins exist in two major forms: trans-zeatin (TZ) and isopentenyladenine (IP) (Li et al. 2021a, b). Cytokinin promotes shoot growth and negatively regulates root development. CK can accumulate in areas with lower water potential, influencing root curvature (Karlova et al. 2021). Exogenous application of CK inhibited the elongation of primary roots (Beemster and Baskin 1998), whereas reduced endogenous cytokinin resulted in increased elongation of primary roots (Werner et al. 2003; Miyawaki et al. 2006). JA signaling is also crucial for stress adaptation and frequently interacts with other hormonal pathways, including ethylene and auxin, to fine-tune root development and root hair elongation (Xu et al. 2020). It is noted that BRs generally promote root growth at low concentrations by influencing both cell division in the root apical meristem and cell elongation in the elongation zone. However, at higher concentrations, they inhibit root growth.

Root system architecture (RSA) is the distribution of roots in the soil and is primarily determined by the length, thickness, number of branches, and growth angle (Khan et al. 2016). RSA is important to plants for the uptake of water and nutrients from the soil. The structural traits of RSA include length, density, volume, depth, growth angle, diameter, root surface area, and root-shoot ratio (Germon et al. 2020; Siddiqui et al. 2021). Exploring the molecular mechanisms and identifying key regulatory factors will enhance crop engineering capabilities for better root systems. This will ensure agricultural productivity under climate change and anthropogenically influenced environmental conditions.

This review synthesizes current knowledge regarding the hormonal and molecular mechanisms governing root trait modulation in *Arabidopsis* under abiotic stress, highlighting the integrated roles of diverse phytohormones and associated signaling pathways in regulating stress-responsive gene expression and root architecture.

3.2 Root Architecture Under Salt Stress

Plant roots serve as the primary organ that experiences the detrimental effects of soil salinity. Elevated salt concentrations in the soil environment adversely impact crop cultivation, reducing yield and nutritional value. These unfavorable conditions trigger a signaling cascade that disrupts normal physiological and metabolic processes. Under salt stress, plants exhibit decreased water absorption capacity, leading to sodium (Na^+) osmotic imbalance. Increased Na^+ levels in plants result in potassium (K^+) imbalance, which impairs critical metabolic and physiological functions. These malfunctions diminish key processes, such as photosynthetic capacity, enzymatic activities, and the ability to synthesize essential compounds. Furthermore, excessive salt triggers the production of reactive oxygen species (ROS), which act as messengers in the signaling cascade. ROS further induce lipid membrane peroxidation and electrolyte leakage, disrupting cellular homeostasis. To mitigate these effects, plants employ various counteractive processes, including the production of protective osmolytes, metabolites, and proteins. Additionally, a protective enzyme system, comprising oxidoreductases, catalases, superoxide dismutase (SOD), peroxidases, and glutathione S-transferases (GSTs), is activated to counterbalance the effects of ROS. Other mechanisms involved in protecting against the harmful effects of sodium include a wide array of membrane proteins, such as antiporters, cation channels, anion transporters, ABC-type transporters, sodium and potassium transporters, plasma membrane and vacuolar ATPases, and aquaporins, which are crucial for maintaining sodium ion homeostasis in plants.

From an experimental perspective, studies involving *Arabidopsis* help to provide insights into root development. Where we can easily visualize the phenotypical changes under normal or stressed conditions, genetic modification approaches, which include mutating or overexpressing the genes, help elucidate the gene's effect on plant growth and development. Root hair density and root hair length are found to decrease under salt stress conditions. Exposure to high salinity levels results in alterations to root anatomy, specifically leading to a diminished cell count in defined outer periclinal cortical cell walls and a concurrent increase in the size of individual root cells (Wang et al. 2008). These changes indicate the plant's adaptive strategies under stress, reflecting modifications at the cellular and molecular levels to maintain functionality (Abou-Shlell et al. 2020). Salt stress induces a range of morphological, physiological, biochemical, and molecular changes in plants, impacting growth and cell metabolism (Soltabayeva et al. 2021).

Auxin-related genes play a pivotal role in modulating the root phenotype in response to stress conditions. Research has established that IAA-CONJUGATE--RESISTANT 4 (IAR4) is critical for the growth of primary roots under saline stress conditions. The *iar4* mutants exhibited heightened sensitivity to salt stress, characterized by inhibited root growth and diminished survival rates. Furthermore, *iar4* plants demonstrated an accumulation of Na^+, resulting in an elevated Na^+/K^+ ratio and revealing increased levels of ROS due to compromised scavenging mechanisms (Fu et al. 2019). Similarly, the expression of *PIN-2* was disrupted and diminished under salt stress, subsequently impacting root growth (Smolko et al. 2021). Moreover, salt stress induces a specific elevation in indole-3-acetic acid (IAA) levels, which is primarily achieved through enhanced IAA biosynthesis via the activation of associated biosynthetic genes. The gene *NITRILASE 2* (*NIT2*), which is responsible for the conversion of indole-3-acetonitrile (IAN) to IAA, exhibited increased activity under conditions of salt stress (Cackett et al. 2022). Additionally, salt stress has been observed to inhibit root meristem size, likely due to the suppression of PIN gene expression, which further contributes to the reduction in auxin levels (Liu et al. 2015). It is very interesting to note that the auxin biosynthesis genes are upregulated while the *PIN* genes are downregulated. This change in distribution and local auxin availability highlights a multifaceted response toward salt stress in plants with other mediator effectors. Another auxin signaling-related mutant, *axr1-24*, demonstrated improved resilience to salt stress across various concentrations (Tiryaki 2007). The salt-induced promotion of LR initiation was absent in *aux1-7*, a mutant with disrupted auxin transport. This observation implies that salt stress stimulates auxin accumulation in developing primordia, thus promoting their exit from preemergence to emergence (Zolla et al. 2010). Mutants with deficiencies in either auxin transport (*pin3,4,7, aux1-7*) or auxin signaling (*tir1-1, arf10-16*) exhibited significantly stunted root growth when subjected to salt stress (Yang et al. 2023). The overexpression of *AUXIN SIGNALING F-BOX 3* (*AFB3*) receptor enhances salt tolerance in roots and germination processes. The *SALT-INDUCIBLE ZINC FINGER PROTEIN 1 (SZF1)* has been identified as crucial for the auxin-dependent response to salt stress by regulating the No Apical Meristem (NAM)/*Arabidopsis* transcription activation factor (ATAF)/cup-shaped cotyledon (CUC) NAC4, indicating that auxin plays a modulatory role in root architecture under saline stress conditions (Garrido-Vargas et al. 2020). *CDK8* positively regulates salt tolerance; overexpression lines show longer roots, whereas its mutation results in shorter roots in salt stress (Guo et al. 2025).

Furthermore, mutants in ethylene signaling components, such as *ein2-1*, *ein4-1*, and *etr1-1*, exhibited reduced root lengths, underscoring their significance in signaling pathways related to salt stress (Cao et al. 2007). The *EIN3*ox seedlings consistently displayed an augmented capacity to withstand salt stress, characterized by superior survival rates, enhanced fresh weight, and increased root length compared to wild-type seedlings (Peng et al. 2014).

Other genes, including *Ferrochelatase-1* (*FC1*), which is induced by salt, demonstrated that mutants are sensitive to salinity, evidenced by shorter root lengths, while overexpression lines exhibited resistance to salt stress. Further analyses indicated modulation in the expression of genes associated with salt-induced oxidative stress responses, monovalent cation-proton (Na^+/H^+) exchange, and Na^+ detoxification processes (Zhao et al. 2017). Additionally, mutants in the *Responsive to Desiccation* (*RD29*) gene family, which encodes hydrophilic proteins significant for plant tolerance toward drought, cold, and high salinity, specifically *rd29a* and *rd29b*, demonstrated increased root length when exposed to salt stress (Msanne et al. 2011). The mutants of *SaGa-associated Factor* (*sgf29a-1*), which facilitates the binding of TFs to DNA and acts in histone acetylation and deubiquitylation, a critical step for activating or repressing gene expression, also exhibited resistance to salt stress in terms of root length. In contrast, the *ALTERATION/DEFICIENCY IN ACTIVATION 2* (*ada2b-1*) mutant displayed hypersensitivity, indicated by reduced root lengths. Analysis of the double mutant (*sgf29a-1*; *ada2b-1*) revealed a phenotype analogous to that of *ada2b-1* plants. This finding suggests that the function of SGF29a may have been obscured by ADA2b under these conditions, indicating that SGF29a may play a supplementary role in the regulatory mechanisms associated with ADA2b (Kaldis et al. 2011). Also, class I trehalose-6-phosphate synthase (TPS) mutant, *attps9*, showed decreased root length under salt stress conditions, indicating its positive role in controlling root length under salt stress (Vishal et al. 2024). It was also found that SPIRRIG plays a crucial role in maintaining BRICK1 stability and facilitating the developmental plasticity of root hairs in response to salt stress (Zhang et al. 2024). Likewise, the small GTPase gene *AtRABA1b/BEX5* (a RAS gene from RAT BRAIN A1b homolog) mutation in plants increased sensitivity to saline conditions in terms of root length (Rajappa et al. 2024).

SALT OVERLY SENSITIVE (SOS) mutants also exhibit hypersensitivity to salt stress, resulting in shorter root lengths and fewer root hairs. Further experimental investigations revealed that one of the primary causes of this phenotype is ionic imbalance and potentially an adaptive mechanism that reduces ion uptake (Wang et al. 2008). Studies on the *U-box E3 Ub ligase* mutant (*Atpub30-1,2*) demonstrate an increased root length compared to the wild type under salt stress, suggesting that *At*PUB30 may function as a negative regulator in high salinity conditions during the seedling phase (Hwang et al. 2015). Similarly, experiments conducted with varying concentrations of salt stress highlight the sensitivity of *Pollen-Pistil Incompatibility 2* (*pop-2*) mutants, particularly concerning root length. Additional research suggests that the root phenotype could be linked to an accumulation of amino acids and a reduction in carbohydrate levels (Renault et al. 2010). Overexpression of *NAM*, *ATAF1-2*, and *CUC2* (*NAC2*) also promotes lateral root development under normal and salt stress conditions (He et al. 2005). Seifert and co-workers (2014) also noted the phenotypic effects of salt stress across different genotypes. They discovered that *FASCICLIN-LIKE ARABINOGALACTAN PROTEIN 4* (*fla4*) mutants exhibited abnormal growth when exposed to NaCl. Double mutants *abi4 fla4* and *cpl3 fla4* also displayed unusual growth patterns in the presence of NaCl. The phenotype is also altered by ABA modulation. Another study on the *cyp79b2 cyp79b3* double mutant revealed shorter and fewer lateral roots than wild-type plants (Julkowska et al. 2017). During osmotic and salt stress, the *wrky46* mutants exhibited a significant reduction in lateral root growth. In contrast, overexpressing *WRKY46* in plants encouraged the development of more lateral roots (Ding et al. 2015). Whereas, *wrky45* showed enhanced tolerance to salt stress regarding root length compared to wild-type plants (Zhou et al. 2025). Meanwhile, mutations in *WRKY1* increase the plants' sensitivity to salt stress (Wu et al. 2022). Additionally, the overexpression of *miR393 transport inhibitor response protein 1* (*mTIR1*) in plants enhances their tolerance to salt stress regarding root elongation. The mTIR1 overexpression lines accumulate higher levels of proline and anthocyanins, contributing to their resilience against stress conditions (Iglesias et al. 2014). A linker histone variant (HIS1-3) mutation promotes root length under salt stress, while overexpression increases sensitivity (Wu et al. 2022).

Mutation in MAD box-related genes like *XAANTAL2* (*XAL2*), double mutation in *XAL2 AGL24* (*AGAMOUS-LIKE 24*), while triple mutation in the genes *XAL2 AGL24 SOC1* (*SUPPRESSOR OF OVEREXPRESSION OF CONSTANS 1*) leads to a decrease in root length, indicating a role in root growth and development (Castañon-Sua'rez et al. 2024). Likewise, mutation in outward-rectifying potassium channels-GUARD CELL OUTWARD RECTIFIER K^+ (GORK) and STELAR K^+ OUTWARD RECTIFIER (SKOR) leads to decrease in root length (Hiya et al. 2024). The *1,3,4-trisphosphate 5/6-kinase 4* (*ITPK4*) mutation leads to a greater density of root hairs and an enhanced accumulation of reactive oxygen species under saline conditions (Ren et al. 2024a, b). Moreover, the constitutive emission of isoprene from roots helped plant roots to grow under salt stress. It was also found that it helps in regulating or priming hormone production and signaling and influences the expression of genes essential for salt stress defense (Bellucci et al. 2024). The study also highlights that the gravitropic response of Col-0 roots became compromised at 75 mM NaCl concentration, with the most significant impairment noted at 150 mM NaCl (Zheng et al. 2024). Moreover, the *sr3g-5* (*salt root: shoot ratio regulator gene*) mutant displayed a notably reduced growth rate of lateral roots under 75 mM NaCl (Ishka et al. 2025).

In conclusion, soil salinity severely influences plant root development, activating a cascade of detrimental physiological and metabolic changes. Plants exhibit remarkable resilience and retaliate by diverse cellular and molecular strategies to cope with these challenging conditions. These internal modifications are also observable through morphological and anatomical changes, often mediated by key hormones like auxin (Fig. 3.1).

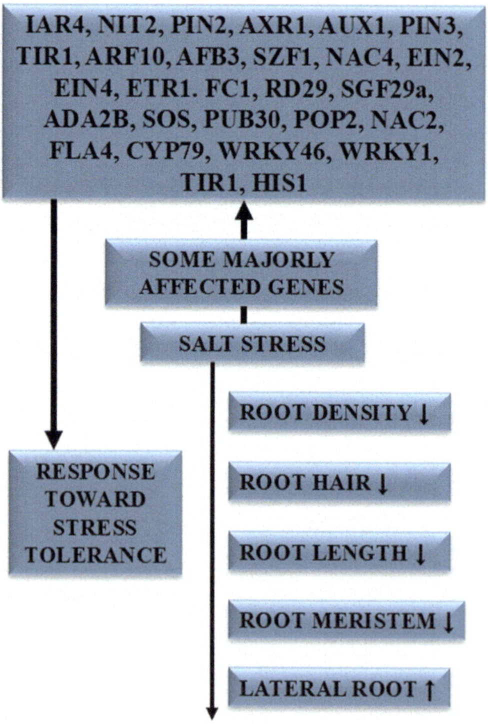

Fig. 3.1 Genes regulated under salt stress and induced root architecture remodeling in *Arabidopsis*. Salt stress triggers changes in the expression of multiple genes involved in hormone signaling, stress response, and root development. These altered gene activities lead to significant modifications in root density, root hair development, root length, and root meristem size, collectively influencing the plant's ability to tolerate saline conditions

3.3 Root Architecture Under Thermal Stress

Optimal temperature is essential for a plant's growth, reproduction, and overall functioning at its best. Both excessively high (heat stress) and excessively low (cold stress) temperatures lead to thermomorphogenic changes. One notable thermomorphogenic alteration is the modified architecture of the root system. When exposed to elevated temperatures (30 °C), the primary root growth of *Arabidopsis* accelerates, exhibiting increased cell elongation rates in the root elongation zone (Yang et al. 2017). Additionally, higher temperatures in *Arabidopsis* enhance the gravitropic response of lateral roots, prompting them to grow deeper and take on a more vertical alignment (Rellan Alvarez et al. 2015). Shoot-to-root communication plays a dynamic role in regulating root growth under heat stress, as shoots are the first organ to sense the heat stress. At early seedling stages, roots detect the heat stress signals and don't rely on shoot signals. This has been shown by studies where roots detached from shoots show the root growth phenotype under warm conditions, and mutants of known shoot thermosensors such as phyB, PIF4, or ELF3 do not display altered root growth under warm conditions (Ai et al. 2023; Bellstaedt et al. 2019; Borniego et al. 2022). Nonetheless, the soil buffering capacity in full-grown plants makes these temperature changes less pronounced to roots. Hence, root thermomorphogenesis relies on the shoot-sensed temperature signals. Recent studies propose that mobile signals such as oscillatory auxin (IAA) effluxes elicited by light and controlled temperature sensed by the shoot affect lateral root density (Ren et al. 2024a, b preprint). Apart from this, the response of root traits to thermal stress is governed by a complex interaction of hormones, including auxin, cytokinin, ABA, and Brassinosteroids. At elevated temperatures, the concentration of HSP90 proteins rises, which stabilizes the transport inhibitor response1/auxin response F-box (TIR1/AFB) auxin co-receptors by safeguarding them from proteasomal degradation. This stabilization leads to increased auxin levels, ultimately promoting lateral root development in *Arabidopsis* under heat stress (Wang et al. 2016). Cheng et al. (2011) and Fei et al. (2017) studied the T-DNA insertion mutants of *GRXS17 (MONOTHIOL GLUTAREDOXIN)* and *CKRC1 (CYTOKININ INDUCED ROOT CURLING 1)*, respectively, and found that under heat stress, these mutants exhibit elevated ROS levels, reduced root length, and impaired gravitropism due to disrupted polar auxin transport. This suggests that both genes regulate auxin signaling and transport, thus ensuring proper cell division in root cells and gravitropism. Moreover, the auxin maxima in the root tip at high temperatures can be sustained through a particular *SORTING NEXIN1* (*SNX-1*)-dependent endosomal trafficking pathway that directs PIN2 to the plasma membrane from late endosomes (Hanzawa et al. 2013). Elevated temperatures stimulate primary root growth by downregulating the BRASSINOSTEROID INSENSITIVE1 (BRI1) receptor levels and its downstream signaling pathways (Martins et al. 2017). Similarly, root growth is promoted at high temperatures through the BR-auxin crosstalk pathway, where BR signaling inhibits the accumulation of the PIN-LIKES 6 (PILS6) protein in the endoplasmic reticulum. This increases auxin's nuclear abundance, promoting root growth through auxin signaling (Sun et al. 2020). Under high temperatures, SPA1 (suppressor of Phy A-105) kinase phosphorylates HY5 (ELONGATED HYPOCOTYL 5) at the serine 36 residue, stabilizing it to facilitate BR-mediated root elongation (Lee et al. 2021). ABA-treated *Arabidopsis* roots display elevated expression of *HSP90* family member *HSP81*, underscoring ABA's role in thermo-tolerance (Yabe et al. 1994). The overactivation of certain genes, namely, *Nodulin--related protein 1* (*NRP1*) (Fu et al. 2010) and *stress-associated protein 10* (*SAP10*) (Dixit and Dhankher 2011), influenced by ABA signaling, enhances the capacity for heat stress tolerance in *Arabidopsis* roots. Additionally, heat stress can trigger the activation of ONSEN, a Ty1/copia type long terminal repeat (LTR) retrotransposon, which can impact root phenotypes during stress. The transposition of ONSEN into ABA-responsive genes such as *ABSCISIC ACID INSENSITIVE 4* and *ABSCISIC ACID INSENSITIVE 5* (*ABI4* and *ABI5*) resulted in an elongated root phenotype during heat stress (Ito et al. 2016). The study by Leal et al. (2022) shows that combined heat and osmotic stress brings about zone-specific architectural remodeling in *Arabidopsis* roots, largely mediated by suberization. The suberin deposition on old primary roots significantly increases to develop a protective barrier, while young primary roots are largely unaffected. This diverse response implies a strategic adjustment to reduce the stress on the root function. The study by Dard et al. (2023) has established that glutathione is vital in regulating thermomorphogenesis and acute heat stress tolerance responses in *Arabidopsis*. By quickly restoring redox balance, glutathione effectively reduces oxidative damage in the cytoplasm and nucleus, which supports the necessary gene expression changes for overall plant survival under thermal stress. Heat stress disrupts phosphate (Pi) homeostasis, leading to Pi starvation-like symptoms that impair root architecture, including reduced primary root growth and lateral root emergence. Transcriptome analysis further discovered that many of the genes activated by heat stress were similar to those regulated by P-starvation leading to a decrease in *PHR1* (*phosphate starvation response 1*), *PHL1* (*PHR1-like 1*), and *PHO2* (*phosphate overaccumulator 2*) and increase in Pi transporters active levels reflecting a compensatory Pi uptake response. In mutants, it was confirmed further that the *phr1 phl1* mutant, which is already known to be deficient in Pi starvation signaling, displayed low recovery under heat. On the other hand, the *pho2* mutant, which accumulates more Pi in shoots, has shown improved root elongation and biomass recovery under heat when roots were exposed to a cooler temperature gradient using the TGRooZ system (González-García et al. 2023).

Plant roots produce small peptides (SP), which are less than 120 amino acids, and act like signaling molecules; they bind to the cell surface receptors and activate downstream signaling. They participate in root development, stress resilience, and influencing RSA to improve nutrient acquisition. CLE is a well-noted peptide playing a role in stem cell maintenance in RAM and LR development; other reported SPs are CEP, RALF, PSY1, and CIF. Study of root-derived peptides is a growing new field and holds a lot of potential, promising applications in plant biology (Keerthana et al. 2025).

Cold stress (4 °C) inhibits root growth by reducing the meristem size and cell number. But cell elongation remains unaffected (Zhu et al. 2015). Also, the cell cycle progression is affected by the downregulation of *CYCB1;1*, *CDKA1*, *CYCB1;2*, and *CYCA1;1* (Zhao et al. 2014). Shibasaki and co-workers (2009) reported the inhibition of auxin transport via PIN2 and PIN3 transporters and IAA2-GUS (an auxin-responsive marker) in low temperature stress. Reduced trafficking of PIN2 transporter inhibits the basipetal transport of auxin, causing reduced root length and inhibited gravitropic response at low temperature. This has been further supported by reduced translation of GNOM (a guanine nucleotide exchange factor that acts on ADP ribosylation factor (ARF)-type G proteins) under low temperature (Ashraf and Rahman 2019). GNOM is found to be involved in the intracellular trafficking of PIN2. Zhu et al. (2015) examined *Arabidopsis response regulator* (*ARR*) mutants, specifically *arr1-3* and *arr12-1*, which are involved in cytokinin signaling. They discovered that these mutants displayed higher levels of *PIN1/3* expression and were less affected by cold temperatures in terms of root length and meristem cell count. This suggests the involvement of cytokinin signaling in the RAS development modulation during cold stress. This has been further supported by a study that reports the involvement of *CRF2* and *CRF3* (*CYTOKININ RESPONSE FACTORS*) in promoting lateral root initiation in a cytokinin-dependent manner during cold stress (Jeon et al. 2016). The morphology of *Arabidopsis* roots is affected at low temperature conditions when auxin and cytokinin balance is disturbed. The regulation involves the two-component signaling pathway of cytokinin, which consists of histidine kinases (AHKs), histidine phosphotransfer proteins (AHPs), and response regulators (ARRs). Cytokinin response factors (CRFs) working within this path specifically bind to the PCRE (PIN cytokinin-responsive element) located on the promoter regions of PIN auxin efflux carrier gene family members, including *PIN1* and *PIN7*. By regulating cold stress-induced polar auxin accumulation pattern via modulating transcription of *PIN* genes, cytokinin signaling precisely determines root development at low temperature (Tiwari et al. 2023). Both high and low temperatures significantly alter root phenotype in plants like *Arabidopsis* through complex hormonal cross talk and gene regulation, highlighting the critical role of temperature in root development (Fig. 3.2).

3.4 Effect of Heavy Metal Stress on Root Architecture

Roots are the first organs that come in contact with heavy metal stress under soil contamination conditions. Heavy metals are potentially toxic metallic elements essential to life only when present in trace amounts, while their excess causes cellular damage (Jalmi et al. 2018). Some of the common heavy metals that affect both animal and plant kingdom include lead (Pb), cadmium (Cd), iron (Fe), chromium (Cr), arsenite (AsIII), arsenate (AsV), mercury (Hg), copper (Cu), aluminum (Al), etc. (Tchounwou et al. 2012). The mechanism underlying damage caused by heavy

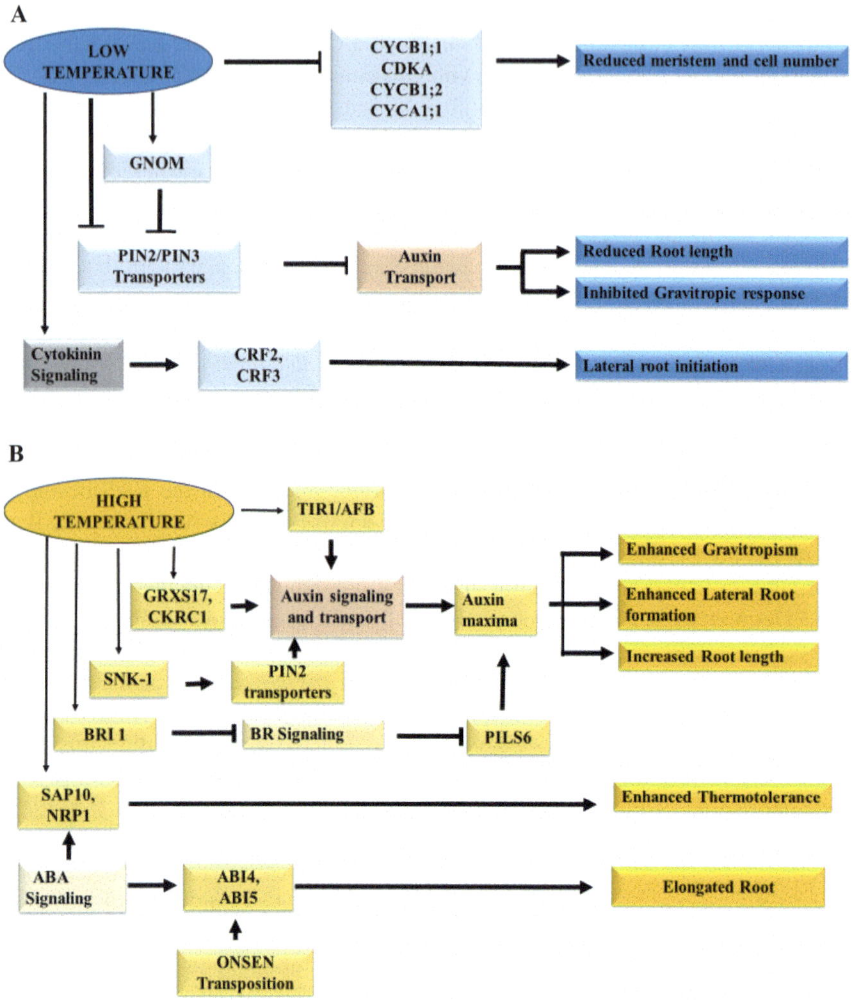

Fig. 3.2 Temperature-dependent signaling pathways regulating root development in *Arabidopsis*. Plants integrate environmental temperature cues through hormonal and transcriptional networks to adjust root development and stress resilience. (**a**) Low temperature downregulates the expression of key cell cycle regulators (CYCB1;1, CDKA, CYCB1;2, and CYCA1;1), leading to reduced meristem size and cell number. It also disrupts GNOM-dependent trafficking, impairing the activity of PIN2/PIN3 transporters, and thereby decreasing auxin transport. This leads to shortened root length and inhibited gravitropic response. Simultaneously, CRF2 and CRF3 are found to be involved in enhancing lateral root initiation (**b**) during high temperature, auxin signaling and transport are improved via the TIR1/AFB auxin receptors and redox regulators GRXS17 and CKRC1, increasing auxin maxima through PIN2 transporters. SNK-1 supports auxin transport, while Brassinosteroid (BR) signaling represses the accumulation of PILS6, further promoting auxin activity. These interactions collectively lead to increased root length, enhanced lateral root formation, and improved gravitropism. SAP10 and NRP1 genes involved in ABA signaling, also activated by heat, lead to enhanced thermotolerance independently of auxin. Through ABI4 and ABI5, ABA signaling contributes to root elongation and is further influenced by ONSEN transposition, a heat-responsive transposable element

metal stress includes ROS production, which further induces oxidative stress (Stohs and Bagchi 1995).

In *Arabidopsis*, metal stress reduces primary root elongation and increases lateral root density (Staňová et al. 2012). To lessen and/or avoid heavy metal toxicity, roots employ a variety of strategies, which includes avoiding heavy metal uptake, activation of antioxidative mechanisms, regulating heavy metal levels through transport proteins, sequestration of heavy metals into vacuole, callose synthesis, and deposition especially in root hair and root epidermis that acts as a barrier and restricts further uptake of heavy metals (Wang and Yang 2021). These strategies enhance the phenotypic plasticity of roots, such as reduced root length, altered lateral root density, reducing the area of contact and thus, blocking the entry of heavy metals into the roots or sequestering them into vacuoles, therefore making them inactive (Shahid et al. 2017).

Root architecture plays a vital role in plant growth in metal-contaminated areas, as altering the architecture of the root might serve as a strategy to adapt to heavy metal stress. Auxin transport directly stimulates root growth in response to various environmental stresses, including heavy metals, as it plays an essential role in the primary root development and lateral root initiation (De Smet et al. 2015). In *Arabidopsis*, copper treatment results in increased DR5 expression, inhibiting elongation of primary roots by decreasing expression of *PIN1*, resulting in reduced division of meristematic cells (Lequeux et al. 2010; Yuan et al. 2013; Yuan and Huang 2016). Aluminum shows auxin-mediated inhibition of root elongation by altering *AUX1* and *PIN2* (Sun et al. 2010). Cadmium alters the expression of various genes involved in biosynthesis and catabolism of auxin, *AUX1*, and *PINs* and results in increased density of lateral roots; however, *pin2-1* mutants failed to show increased cadmium-induced LR density (Hu et al. 2013). Cd stress results in upregulation of auxin biosynthetic gene *NIT*, which increases IAA concentration in roots of *Arabidopsis*, thus promoting growth of lateral roots and protecting the roots from cadmium stress (Vitti et al. 2013). Cadmium-induced nitric oxide accumulation inhibited primary root meristem growth by stabilizing AUX/IAA proteins, which repressed auxin transport and signaling (Yuan and Huang 2016). Aluminum inhibits PIN2 vesicle transport from the plasma membrane to the endosomes, disturbing the synthesis of IAA in the apical buds, thus creating an imbalance of transportation and distribution of IAA in roots and subsequently inhibiting root growth (Shen et al. 2008). Higher concentration of copper resulted in auxin-mediated inhibition of primary root length by decreasing the expression of *PIN1* protein in *Arabidopsis* (Yuan et al. 2013). Copper stress reduced root length by altering auxin levels, interfering with NO functioning in *Arabidopsis* (Zhang et al. 2018). Arsenic stress resulted in oxidative stress, leading to downregulation of auxin carrier AUX1, hindering root growth in *Arabidopsis* seedlings (Krishnamurthy and Rathinasabapathi 2013). Aluminum stress results in upregulation of PIN (PIN1, PIN3, PIN4, PIN7) and AUX/LAX (AUX1, LAX1, LAX2) in the transition zone (TZ) of the root, disrupting local auxin distribution and accumulation and inhibiting root growth (Li et al. 2021a, b). Exposure to 5 μM lanthanum (La) and ytterbium (Yb) was enough to inhibit primary and lateral root growth and alter root architecture in *Arabidopsis*

(Grosjean et al. 2024). Exposure to La leads to downregulation of genes involved in loosening of the cell wall and root hair development, while the genes involved in lateral root formation were upregulated (Grosjean et al. 2024). Cytokinin downregulates *PINs* expression and inhibits the initiation of lateral roots (Li et al. 2006; Laplaze et al. 2007). In *Arabidopsis*, copper stress results in increased accumulation of cytokinin and inhibition of the growth of the primary root (Lequeux et al. 2010). Cadmium induced upregulation of *CKX5* (*CYTOKININ OXIDASE 5*), a cytokinin turnover gene in the roots of *Arabidopsis*, resulting in enlarged root meristem (Aloni et al. 2006). Cadmium stress resulted in enhanced cytokinin signaling in root tips and meristems of *Arabidopsis* to upregulate *SHORT HYPOCOTYLS 2* (SHY2) gene via *ARABIDOPSIS* RESPONSE REGULATOR 1 (ARR 1), ARR12, and *ARABIDOPSIS* HISTIDINE KINASES 3 (AHK3) enzymes that reduced the toxic effects of cadmium on root elongation (Bruno et al. 2017). SHY2 downregulates *PIN 1*, *3*, and *7* and hinders auxin transport to the root tips and meristem, emphasizing the role of auxin-cytokinin cross talk under heavy metal stress (Bruno et al. 2017). GAs play important role to protect the plant from cadmium stress by reducing the changes induced by cadmium. GA was found to suppress cadmium-induced upregulation of *IRT1*, a transporter involved in cadmium uptake. GA also reduced cadmium-induced NO accumulation (Zhu et al. 2012). Exogenous application of ABA significantly reduced levels of nickel, cadmium, and zinc in the roots and shoots of *Arabidopsis* as compared to ABA-free control plants by inhibiting expression of *IRT1*, *NRAMP1*, *NRAMP4*, and *HMA3* genes, involved in heavy metal uptake (Zhu et al. 2024). Exogenous application of ABA in *Arabidopsis* also mitigates damage induced by cadmium and promotes root growth via (ABA)-importing transporter 1 (AIT1)-mediated reduced cadmium uptake (Pan et al. 2020). Exogenous application of ABA resulted in downregulation of genes involved in uptake and translocation of cadmium and upregulation of genes involved in cadmium chelation and efflux, thus inhibiting its accumulation (Meng et al. 2022).

Cadmium stress in *Arabidopsis* resulted in upregulation of ethylene-responsive genes *ACS2* and *ACS6*, resulting in enhanced ethylene production (Schellingen et al. 2014; Arteca and Arteca 2007). *Arabidopsis ein2–1* mutants showed a significant reduction in expression of genes involved in oxidative stress, emphasizing the role of ethylene in response to cadmium stress (Schellingen et al. 2015). Ethylene was shown to reverse zinc-induced photosynthesis inhibition by altering the photosystem II activity (Khan and Khan 2014). Aluminum also induces enhanced ethylene accumulation, resulting in inhibition of root elongation (Yoon et al. 2010; Sun et al. 2010). ANAC017 and MYB103 regulate cell wall modification in response to aluminum stress in *Arabidopsis*. ANAC017, along with XTH31, functioning downstream, alters the composition of the cell wall and influences Al tolerance. MYB103 regulates xyloglucan O-acetylation via its target TBL27, modulating cell wall composition and managing Al toxicity (Thiébaut et al. 2025). In *Arabidopsis*, uranium stress leads to its accumulation in root cell membranes and cell walls, notably changing protein expression, impacting the vesicular trafficking of proteins involved in metal transport, indicating adjustments at the root level in response to stress (Przybyla-Toscano et al. 2025).

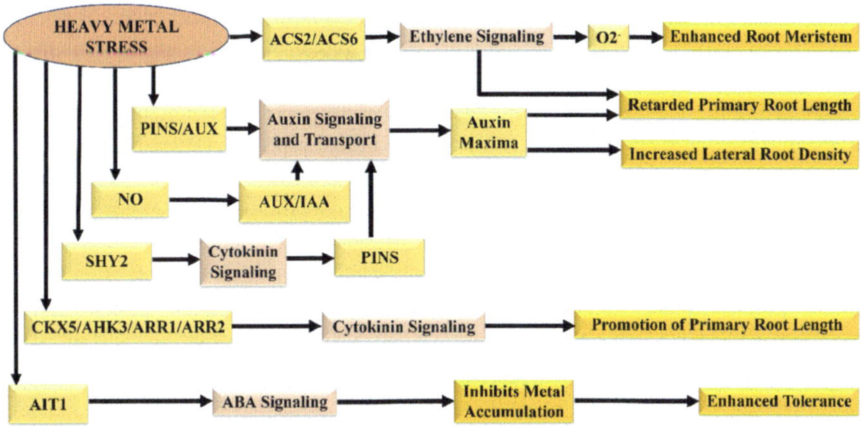

Fig. 3.3 Phytohormone-mediated remodeling of *Arabidopsis* root architecture under heavy metal stress. Heavy metal stress reduces the elongation of primary root length and increases lateral root density by activating auxin, cytokinin, ethylene, and ABA signaling pathways, and their cross talk with other signaling molecules plays an important role in reshaping root architecture in response to heavy metal stress. Heavy metal stress alters the expression of AUX/PINs and NO accumulation, which stabilizes AUX/IAA proteins, thus repressing auxin transport and signaling, disturbing auxin maxima, and affecting root phenotypes. Heavy metals also enhance cytokinin signaling that upregulates CKX5/AHK3/ARR1/ARR2 genes that reduce the toxic effects of metal and SHY2 gene that downregulates PIN expression, affecting auxin transport and signaling. ABA application results in activation of AIT1 that promotes root growth by reducing cadmium uptake, thus limiting its accumulation. Cadmium stress results in upregulation of ACS2 and ACS6 resulting in enhanced ethylene production and retarded root elongation. Ethylene interacts with metal-induced superoxide radical ($O2^-$), which results in enhanced root meristem

Heavy metal stress reduces the elongation of primary root length and increases lateral root density by decreasing the expression of PIN proteins, as altering the architecture of the root might serve as a strategy to adapt to heavy metal stress (De Smet et al. 2015). Auxin, cytokinin, and ethylene are the plant hormones that influence root architecture and patterning (Vanstraelen and Benková 2012) and also regulate lateral roots development (De Smet et al. 2015) (Fig. 3.3). Phytohormones and their cross talk with other signaling molecules play important role in reshaping root architecture in response to heavy metal stress (De Smet et al. 2015).

3.5 Root Architecture Under Drought Conditions

Drought stress is a major abiotic stress that limits the growth of the plant due to impairment of physiological, biochemical, and metabolic processes in plants (Blum 2009; Hu and Xiong 2014; Lorts and Lasky 2020). Roots are the first organs to sense drought stress and respond at the cellular as well as whole root structure (Dinneny 2019). When roots sense drought stress, the meristem and vasculature, along with stem cells of the root, coordinate with each other (Gupta et al. 2020). As a result, the entire root system architecture undergoes structural changes in the morphology of the root in order to improve the ability to absorb water (Dinneny 2019). Under drought stress, deep rooting facilitates the uptake of water, nutrients, and nitrogen from deeper layers of soil, thus maintaining crop yield (Wasson et al. 2012; Lynch 2013; Uga et al. 2013). Various key regulatory networks and molecular pathways are known to modulate root architecture in response to drought stress in *Arabidopsis*. Auxin is reported to regulate various developmental processes, deciding root structure, such as elongation of the root, lateral root development, and gravitropism (Pierik and Testerink 2014). Auxin is reported to control the initiation of lateral roots on sides (which are in contact with water), thus determining the direction of lateral root formation in soil (Orosa-Puente et al. 2018). Further, variation in water availability in the soil is found to stimulate the root growth toward water. Root tips sense soil moisture gradients and bend in the direction of high-water potential, resulting in stimulated root growth toward water, which is termed hydrotropism (Iwata et al. 2013; Lucob-Augustin et al. 2021). It is an auxin-dependent response that enables water acquisition under drought stress (Takahashi et al. 2002; Dietrich 2018). Experiments suggest antagonistic functioning of gravitropism and hydrotropism (Li et al. 2020a, b). However, auxin plays a major role in supporting hydrotropism, which is supported by a recent study revealing the foraging of roots for water (Zhang et al. 2025). According to this report, under drought conditions, MIZU-KUSSEI 1 (MIZ1), a key player in hydrotropism, is involved in mediating stress-induced lateral diffusion of PINs across the PM. MIZ1 interferes with the maintenance of PIN polarity in root cells, resulting in defective gravity-induced auxin redistribution. Thus, MIZ1 mediates the inhibition of gravitropism and promotion of hydrotropism (Zhang et al. 2025). In *Arabidopsis*, *EXOCYST70A3*, a gene that causes variation in root structure by controlling the auxin pathway, has been identified using natural variation and genome-wide association mapping. *EXO70A3* regulates PIN4 auxin efflux carrier protein levels in columella tissue and modulates the auxin level in root tips. It also influences the distribution of the PIN4 proteins. Genetic variation and perturbation of EXOCYST70A3 alter the root growth orientation, resulting in different root structure depth profiles. Moreover, the allelic distribution of EXOCYST70A3 is also found to be related to patterns of rainfall and drought resistance (Ogura et al. 2019). Several AUX/IAA-ARF modules coordinate different stages of LR development (De Smet et al. 2010). Recent studies have shown that rhythmic changes in ARF abundance in the root, which plays a pivotal role in LR formation, are regulated by autophagy. Ebstrup and co-workers (2024) have reported the role of the neighbor of BRCA1 gene 1 (NBR1)-mediated autophagy in auxin-mediated ARF7 degradation.

In rice, a GWAS study has identified *root length 1* (*RoLe1*), playing an important role in controlling root development and drought resistance. Further, this study showed interaction between RoLe1 and ARF-GTPase-activating protein (*Os*AGAP), which regulates auxin-dependent root development. This interaction interferes with *Os*AGAP's function of modulating polar auxin transport. TF NAC41 binds to the *RoLe1* promoter and upregulates the expression of *RoLe1*. Under moderate drought stress, *RoLe1* is found to increase the seed setting rate, resulting in increased crop yield. Further in this study, using genomic evolutionary analysis, they identified a new favorable allelic variant *proRoLe1*$^{-526T}$ retained in upland rice (during domestication) (Han et al. 2024). Together, this study indicates that the NAC1-RoLe1--AGAP module promotes root development and drought resistance in rice. Another gene, *Robust root system 1* (*RRS1*), encoding R2R3-type MYB family TF acts as a negative regulator of root growth and development, knocking down this gene, resulted in enhanced root growth and increased resistance to drought. This negative regulation of rice root development by *RRS1* occurs through activation of *IAA3*, which is involved in auxin signaling (Gao et al. 2023).

Hormones like cytokinin and ABA generally suppress LR formation, while brassinosteroids promote it, often by interacting with auxin signaling (Fukaki and Tasaka 2009). ABA and drought stress exert similar inhibitory effects on the primary root elongation, LR formation, and branching (De Smet et al. 2006). Under drought stress, ABA plays the most critical role in the hormone signaling network (Soma et al. 2021). ABA affects roots by modulating other hormone signaling pathways, like auxin and cytokinin signaling, under drought stress conditions. Ethylene also acts as an inhibitor of root growth, and during drought, ABA inhibits the synthesis of ethylene, thus promoting root growth (Rowe et al. 2016; Sharp and Lenoble 2002). ABA is sensed by a family of cytosolic receptors known as PYR (PYRABACTIN RESISTANCE 1), PYL (PYRABACTIN RESISTANCE 1-LIKE), and RCAR (REGULATORY COMPONENT OF ABA RECEPTOR 1). In *Arabidopsis*, the mutants of *PYL8* and *PYL9* receptors exhibited prolonged ABA--induced quiescence of lateral root growth and reduced sensitivity to ABA during primary and lateral root formation. This reduced sensitivity was reversed by the exogenous application of auxin (IAA), highlighting a functional interaction between ABA and auxin signaling. Thus, PYL8 and PYL9 have a role in mediating the cross talk between ABA and auxin during lateral root formation (Xing et al. 2016). In *Arabidopsis*, ABA inhibits the elongation of roots by regulating the expression of *PIN1* and *PIN2*. The altered expression of *PIN1* modulates auxin transport in the root and size of the root meristem and structure of the root (Sun et al. 2018).

In *Arabidopsis*, when ABA and lateral root development 2 (LRD2) interact with auxin, it results in repressed LR formation under drought (Deak and Malamy 2005). During drought, MYB96 acts as a regulator of lateral root meristem initiation through ABA-auxin signaling. Under drought stress, ABA negatively regulates the expression of respiratory burst oxidase homolog (RBOH) genes and reduces the accumulation of ROS; as a result, primary root length increases (Sharp et al. 1994, 2004). In *Arabidopsis*, WRKY family transcription factors are also involved in regulating root structure. In an expression study, a decrease in *WRKY75* expression resulted in increased length and number of LR as well as number of root hairs (Devaiah et al. 2007). Further, under drought stress, *WRKY46* controls lateral root formation via regulation of ABA signaling and auxin homeostasis in *Arabidopsis* (Ding et al. 2015).

Cytokinin modulates root morphology and metabolism during drought stress in different plant species. In *Arabidopsis* and tobacco, overexpression of cytokinin oxidase (CKX) induced the degradation of cytokinin specifically in the root, resulting in longer primary roots, increased branching, and a larger root system (Werner et al. 2010).

Ethylene is also involved in modulating various aspects of root development, including primary root elongation, lateral root initiation, and radial expansion (Ullah et al. 2018; Feng et al. 2017). These modifications in roots, driven by ethylene, are often adaptive, enabling plants to better cope with drought. Transcription factors *EIN3* (ethylene insensitive) and *EIL1* (ethylene insensitive like 1) act as central regulators in the ethylene signaling pathway, orchestrating the activation of ethylene-responsive genes (Feng et al. 2017; Dolgikh et al. 2019). Under drought stress, the *edt1* (enhanced drought tolerance 1) mutant of *Arabidopsis* showed increased tolerance and improved root structure. This extensive root system results from increased expression of transcription factor *HDG11* (homeodomain glabrous 11) (Yu et al. 2008). Also, the transcription factor *ERF109* (ethylene response factor 109) plays a key role in mediating the improved root structure.

JA signaling regulates roots via cross talk with ABA, where ABA receptor PYL forms a complex with the Jasmonate ZIM domain (JAZ), resulting in transcriptional activation of MYC2. However, JA and ABA act antagonistically in the regulation of the root system. Jasmonoyl isoleucine (JA-Ile) binds to COI1 and mediates jasmonate-dependent signaling (Fonseca et al. 2009). Under water stress, early JA and jasmonoyl isoleucine (JA-Ile) increase were found necessary for the rise of ABA level. Also, water stress-induced JA-Ile accumulation through COI1 is required for the complete activation of ABA biosynthesis in roots of *Arabidopsis*. Overall, these results suggest an interaction between jasmonates, and ABA occurs at the biosynthetic level (de Ollas et al. 2015).

Brassinosteroids (BRs) are an important part of processes like cell division, elongation, photomorphogenesis, senescence, and response to various changes in environment and biotic stresses (Clouse and Sasse 1998; Mussig et al. 2003; Clouse 2011; Yang et al. 2011). Thus, BRs also participate in root development and drought response (Gonzalez-Garcia et al. 2011; Chaiwanon and Wang 2015; Nolan et al. 2017). BR regulates the quiescent center's ability to self-renewal and expression of cell cycle-related genes (Vilarrasa-Blasi et al. 2014; Planas-Riverola et al. 2019). BR signaling coordinates root growth and drought tolerance by suppressing the expression of *PLETHORA1* and *PLETHORA2* (PLT1 and PLT2). Plants overexpressing *PLT1* and *PLT2* showed enhanced drought tolerance than wild type. Brassinazole-resistant 1 (BZR1) is found to directly bind the promoter of *PLT1* and repress its expression level. The treatment with BR reduced the expression of *PLT1* and *PLT2* in roots. However, during primary root growth, the *PLT2*-overexpressing plants were less sensitive to BR in comparison to wild type (Zhao et al. 2023).

ROS acts as a signaling molecule influencing root growth under different stress conditions. NADPH oxidase and oxalate oxidases produce ROS signaling molecules and induce cell growth in the root by loosening the cell wall. Transcription factors like zinc finger protein *Zat12* and *WRKY25* act as regulators of ROS signaling in roots, and their expression is induced under oxidative stress, such as drought stress (Davletova et al. 2005). RBOH activity generates ROS gradients in the root, which induces directional growth in root hairs (Takeda et al. 2008; Nestler et al. 2014; Kim et al. 2019). In order to regulate plant response to environmental stresses, different TFs cross-talk with each other (Singh et al. 2013; Sharma et al. 2019; Goel et al. 2019). A class of transcription factors named PLTHORA (PLTs), containing AP2/ERF (Apetala2/ethylene response factor) domain, are involved in root development (Aida et al. 2004; Galinha et al. 2007; Mahonen et al. 2014). During drought stress, expression of *PLT1* and *PLT2* and their *AP2/ERF* domain is upregulated in root tissue. Another member of the *AP2/ERF* transcription factor family, the *HARDY* gene from *Arabidopsis*, was overexpressed in rice. During drought stress, the transgenic plants displayed induced water use efficiency, primarily attributed to an increase in root biomass (Karaba et al. 2007). *RDH6* (*root hair defective 6*) is a member of the bHLH transcription factor family. The activity of *RDH6* depends on *RSL4* (*root hair defective 6-like 4*), a direct target of *RHD6*, and promotes root hair elongation by altering cell wall composition (Feng et al. 2017). Overexpression of *RDH6* leads to a greater number and longer root hairs, thereby enhancing the plant's ability to uptake water (Shibata and Sugimoto 2019). A recent study has examined gene expression and chromatin accessibility in *Arabidopsis* root tip nuclei and reconstructed a cell-type-specific transcriptional regulatory network (TRN) (Liu et al. 2024). These TRNs regulate the development of the root tip under the condition of osmotic stress. The study found that different cell types showed varied responses toward osmotic stress. These responses were regulated by specific TRNs or by cis-regulatory elements linked with stress-related genes (gl-cCREs) specifically. Additionally, in the trichoblast regulatory network, WRKY11 transcription factor was identified, and its expression level was also higher in osmotic stress--treated samples. Similar to *WRKY11*, *WRKY17* also showed induced expression level in osmotic stress-treated root tips (Liu et al. 2024). Under drought stress, *Arabidopsis* employs a complex network involving hormonal cross talk, transcriptional regulation, and ROS homeostasis to adapt its root architecture (Fig. 3.4). Unraveling these mechanisms provides valuable insights for engineering drought-resilient crops, ensuring sustainable agricultural productivity in water-limited environments.

3.6 Root Architecture Under Genotoxic Stress

Genome integrity is consistently under threat from genotoxic agents such as harmful radiation and chemical mutagens, which severely undermine genome stability by causing cross-links in DNA strands, RNA pyrimidine dimers, chromosomal DNA damage, and the initiation of oxidative stress. While all genotoxic agents modify the structure and chemistry of DNA, only a select few can induce mutations. To put it differently, "Every mutagen is genotoxic, but not every genotoxic agent qualifies as a mutagen." Types of radiation, like ultraviolet (UV), ionizing radiation (IR), and gamma radiation, provoke intricate alterations in root morphology and physiology, primarily through complex molecular pathways. These radiations

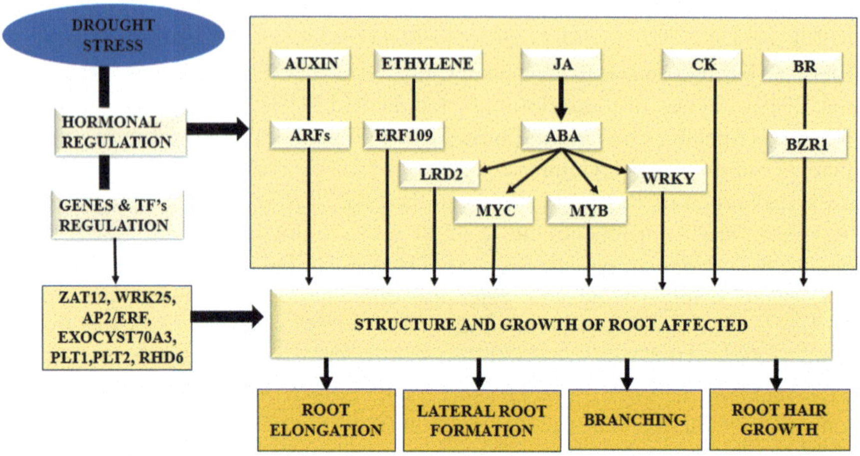

Fig. 3.4 Regulatory mechanism of *Arabidopsis* root architecture under drought stress. Drought stress triggers hormones like auxin, ethylene, jasmonic acid, and ABA, and their cross-talk modulates root morphological structure and growth under drought stress. Regulatory genes and transcription factors activated under drought stress to modify the root traits of root system architecture are-ZAT12, WRKY25, AP2/ERF, EXOCYST70A3, PLT1 and PLT2, RDH6, ARF, ERF109, LRD2, MYC, and MYB

predominantly influence auxin dynamics, lead to the depolymerization of actin filaments, cause the accumulation of flavonoids and ROS, and trigger DNA damage, ultimately disrupting cell division and resulting in cell cycle arrest. Wan et al. (2018) discovered that exposure to UV-B in *Arabidopsis* prompts root bending driven by a flavonoid-regulated auxin pathway. This process involves the suppression of auxin signaling due to enhanced stabilization of Aux/IAA repressors in the roots, while ultraviolet resistance 8 (UVR8)-mediated flavonoid accumulation leads to positive phototropic root bending induced by UV-B radiation. The AUX1 and PIN2 transporters exhibited elevated expression and auxin accumulation on the non-illuminated side, contrasting with the UV-B illuminated side, indicating their UV stress-regulated expression and localization (Wan et al. 2018). UV-B exposure also disturbs the actin cytoskeleton, potentially through the accumulation of hydrogen peroxide or via UVR8-mediated signaling, resulting in thickened actin bundles in the roots that interfere with root development (Du et al. 2020). Exposure to UV-B radiation suppresses primary root elongation in the meristematic zone mediated by growth regulating factor 3 (GRF3), while the elongation zone remains unaffected. This is driven by the generation of ROS and nitric oxide along with DNA damage, which collectively lead to reduced cell proliferation and meristem cell death in *Arabidopsis* roots (Sheridan et al. 2022). CeO_2 nanoparticles (CeO_2-NP) reduce UV-B induced damage in *Arabidopsis* roots by limiting oxidative stress. CeO_2-NP reduces ROS accumulation by mimicking the activity of antioxidant enzymes like CAT and SOD in root tips, as well as boosting plant's own antioxidant system, thereby reducing DNA damage and preventing cell death in plants. Under UV-B stress, CeO_2-NP also enhances PIN protein expression, polar auxin transport, auxin production, and accumulation and stabilizes cell cycle regulation. As a result, cell division activity in RAM is preserved leading to improved root growth. Overall, under UV-B exposure, CeO_2-NP significantly reduces cellular damage and preserves root structure and function by reducing oxidative damage and maintaining hormonal and cellular stability (Sun et al. 2025).

It has also been observed that UV-B downregulates genes associated with cell wall and lipid biosynthesis and metabolism, thereby influencing root plasticity and growth (Zhang et al. 2020). Research has shown that different cell types exhibit varying responses to double-strand breaks (DSB). The epidermis and cortex demonstrate increased cell volume along with inhibition of G2/M cell cycle progression and endoreduplication (Adachi et al. 2011), whereas stem cells at the root tip undergo programmed cell death (PCD) upon zeocin treatment, a DSB-inducing agent (Hirakawa and Matsunaga 2019). Stem cell initials and actively dividing meristematic cells in roots are highly susceptible to DNA-damaging agents such as bleomycin, UV-B, and X-rays, while QC cells exhibit strong resistance to these factors (Fulcher and Sablowski 2009; Cruz-Ramírez et al. 2013). Timilsina et al. (2019) observed that the frequency of QC cell division increases as the root ages, and the susceptibility of QC cells to genotoxic stresses also escalates in older roots.

Suppressor of gamma radiation 1 (SOG1), a TF that functions as the master regulator of DNA damage repair (DDR) in plants, is a functional counterpart of mammalian p53. SOG1 operates as a vital regulator of the DNA damage response by activating genes that are involved in DDR and cell cycle control, as well as regulating root meristem recovery after stress withdrawal. DNA damage caused by genotoxic stresses activates the plant's DNA damage response, which initiates various downstream pathways. In response to DSBs and single-strand breaks (SSBs), SOG1 gets activated through phosphorylation by ATM and ATR, respectively. Activated, SOG1 regulates the transcription of genes involved in various cellular responses, including cell cycle arrest, DDR, initiating endoreduplication, apoptosis, PCD, and hormonal regulation. In the root tip, SOG1 initiates cell cycle arrest and PCD during DNA damage. Once the stress withdraws, SOG1 helps in regenerating root meristem activity and restoring root growth (Li and Qian 2023). Zeocin treatment in *Arabidopsis* causes DNA damage such as DSBs; in response, SOG1 inhibits lateral root formation (Davis et al. 2016). This inhibition is associated with cytokinin signaling, as mutants deficient in cytokinin biosynthesis or signaling exhibit less LR inhibition under DNA-damaging conditions. Under zeocin stress, cytokinin biosynthesis genes are induced, and cytokinin levels are enhanced in LR primordia. The intricate process reveals that when faced with genotoxic stress, SOG1 activates and upregulates the expression of genes responsible for cytokinin biosynthesis. Enhanced cytokinin levels also activate SHY/IAA3, an inhibitor of AUX/IAA, resulting in downregulation of auxin signaling. Another crucial role of SOG1 is to inhibit PIN proteins. This implies that increased cytokinin levels signal the plant to focus on DNA repair rather than primary and lateral root (LR) development (Davis et al. 2016; Li and Qian 2023). RAD54 is an ATPase and chromatin remodeling factor crucial for homologous recombination. RAD54 foci form and accumulate in the nuclei of *Arabidopsis* root cells under gamma radiation stress, predominantly observed in cells during the G1 and G2 phases. Plant-specific internuclear membrane proteins, crowded nuclei 1 (CRWN1) and CRWN2, play a role in anchoring RAD54 foci to the nuclear periphery. This anchoring aids in nuclear envelope-mediated homologous recombination repair and helps to maintain genome stability in response to DNA double-strand breaks in plants (Hirakawa and Matsunaga 2019). Ribosomal protein mutant, *ars27A*, exhibits significant growth suppression and developmental issues when exposed to 100 ppm methyl methane sulfonate (MMS), a concentration that can be easily tolerated by wild type. Instead of forming auxiliary roots, *ars27A* develops tumor-like structures at the site of auxiliary root primordia in response to MMS, indicating a unique role for ARS27A in handling genotoxic stress (Revenkova et al. 1999). RAD51 is associated with the S and G2 phases of the cell cycle, essential for homologous recombination and DDR in *Arabidopsis* roots (Goldfarb and Lichten 2010). However, during genotoxic stress, elevated RAD51 expression leads to arrest at the G1/S and G2/M checkpoints, with a significant accumulation of cells at the G1/S transition. Kutashev et al. (2024) demonstrated that root stele cells exhibit different RAD51 transcriptional and cell cycle dynamics in response to increased DNA damage compared to other root cell types. Specifically, stele cells show stronger RAD51 activation and a shorter cell cycle duration (around 15 h) than cortex (approximately 23 h) and epidermis cells (approximately 24 h) (Kutashev et al. 2024). Furthermore, MEDIATOR 18 (MED18), a transcription factor involved in floral development and defense responses along with the ERF115, a cell regeneration stimulating factor and AUX/IAA4-mediated auxin signaling pathway, plays a vital role in root regeneration and stem cell replenishment under various genotoxic stresses (cisplatin, bleomycin, zeocin, etc.) (Ruiz-Aguilar et al. 2020). *MED18* mutation, which enhances root susceptibility to DNA-damaging agents combined with chromate, causes root meristem splitting, resulting in twin root formation even though pro-vasculature cells in *med18* root meristem undergo PCD in response to genotoxic stress (Ruiz-Aguilar et al. 2020). Tyrosyl-DNA phosphodiesterase (TDP1) enzymes, TDPα1 and TDPβ1, also function as DNA repair enzymes under genotoxic stress. *TDP1α* mutants show high sensitivity toward camptothecin (CPT, inhibitor of topoisomerase 1), while TDP1β mutants are sensitive to cisplatin and zeocin, along with impaired root growth. Pagano et al. (2023) show that TDP1β functions in root development mediated by gibberellin and brassinosteroid pathways. Additionally, *TDP1β* expression is enhanced in replication-related gene mutants *e2fb*, *pol2a*, and *atr*, supporting its role in replication stress and root growth (Pagano et al. 2024). In response to genotoxic stress, *Argonaute 2* (*AGO*) is another key component of the DNA damage response, particularly under gamma irradiation and zeocin exposure. In *Arabidopsis*, *AGO2* expression is significantly upregulated under these conditions (Kim et al. 2025). *AGO2* contributes to DSBs repair by recruiting small RNAs (diRNAs), modifying chromatin structure, and promoting repair protein complex assembly (Oliver et al. 2014). Plants with a mutation in the N-terminal domain of AGO2 show increased sensitivity to DNA damage and growth defects, including increased root cell death and growth inhibition. Overexpressing *AGO2* can alleviate these phenotypes, suggesting its role in limiting overall nuclear DNA damage under genotoxic stress (Kim et al. 2025).

These findings reveal plants' complex, cell-specific strategies to preserve root growth and genome stability under genotoxic stress. The orchestration of genes contributing to DNA repair and hormonal signaling, primarily driven by auxin and cytokinin, and structural remodeling form a dynamic defense system and visible structural alteration in *Arabidopsis* roots (Fig. 3.5). Such multilayered responses not only reflect the remarkable adaptability of plants but also offer valuable insights into enhanced resilience mechanisms. Applying this insight paves the way for developing crops that are better equipped to withstand genotoxic environments.

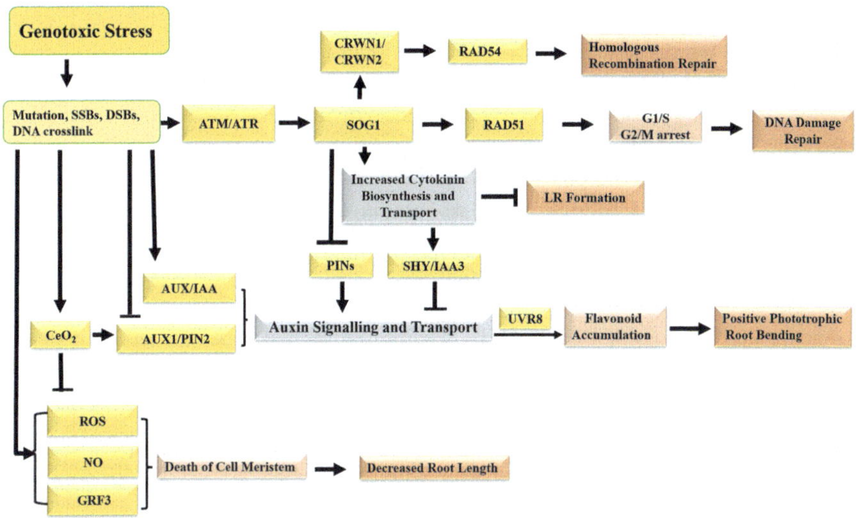

Fig. 3.5 Transcriptional and morphological responses of *Arabidopsis* roots under genotoxic stress. Genotoxic stress causes DNA damage and oxidative stress, which disrupt hormone signaling, alter root architecture, and trigger cell death. This DNA damage is detected by the ATM/ATR transcription factors, which activate downstream RAD-dependent repair pathways mediated via suppressor of gamma response 1 (SOG1). SOG1 is an essential signaling component that initiates DNA damage responses, activates cell cycle checkpoints, inhibits PIN proteins, and enhances cytokinin biosynthesis and transport in lateral roots, inhibiting their formation and suppressing auxin signaling and transport. Additionally, genotoxic damage disrupts auxin signaling and transport by inhibiting AUX/PIN2 and PINS, activating AUX/IAA, leading to positive phototropic root bending. Oxidative stress and NO accumulation further reduce primary root length by causing cell death in the meristematic zone. CeO_2 application limits ROS accumulation and activates AUX1/PIN2 proteins that preserve root architecture by reducing oxidative activity and maintaining hormonal stability

3.7 Conclusion

As the primary interface between a plant and its soil, the root system exhibits considerable plasticity, undergoing cellular and structural modifications to enhance survival and growth in challenging conditions. These adaptive responses are orchestrated by signaling molecules like auxin, cytokinin, ABA, JA, and ethylene. The interplay between these hormones and regulatory genes (such as those encoding MAP kinases and transcription factors) governs the morphophysiological adjustments in root development under stress. For example, salinity can alter auxin homeostasis and transport, affecting root elongation and lateral root development. At the same time, heavy metal stress induces changes in root architecture, such as reduced elongation of the primary root and increased lateral root density to minimize uptake and toxicity. Heavy metal stress disrupts root development and auxin signaling by altering the expression of AUX/PIN/IAA and enhancing the production of cytokinin and ethylene, reducing heavy metal-induced phenotypic defects. At the same time, ABA limits the uptake of heavy metals. Thermal stress prompts thermomorphogenesis, influenced by hormone signaling and protein stabilization, and drought leads to architectural changes that improve water foraging.

Further research into the specific roles of genes and hormone interactions under combined stresses is needed to translate these discoveries into agricultural solutions. Moreover, new emerging techniques that allow deep phenotyping of different crops are helping to visualize the root phenotypes in real time. These noninvasive techniques involve automated systems, robotics, and advanced imaging techniques using MRI, X-ray, micro-CT scans, and automated rhizotrons to measure root parameters like length, surface area, diameter, and branching patterns and produce high-throughput assessment of RSA in complex soil environments. By analyzing 3D root growth and dynamics under controlled or field-simulated stress conditions, it can help to identify stress-specific superior root traits – "ideotypes" that effectively respond to different stress dynamics, which will definitely aid in breeding resilient crops for climate-smart precision agriculture.

Acknowledgments Research in Y.M Lab is currently funded by a grant from the ANRF-SERB (project CRG/2023/007304). A.K., P.Y., and R.Y. thank the University Grants Commission (UGC), while A.J. and K.T. acknowledge financial support from CSIR-UGC INDIA for their JRF/SRF fellowships.

References

Abou-Shlell MK, El-Emary FA, Khalifa AA (2020) Effect of nanoparticle on growth, biochemical and anatomical characteristics of moringa plant (*Moringa oleifera* L.) under salinity stress condition. Arch Agric Sci J 3(3):186–213

Adachi S, Minamisawa K, Okushima Y, Inagaki S, Yoshiyama K, Kondou Y, Kaminuma E, Kawashima M, Toyoda T, Matsui M, Kurihara D (2011) Programmed induction of endoreduplication by DNA double-strand breaks in *Arabidopsis*. Proc Natl Acad Sci 108(24):10004–10009

Ai H, Bellstaedt J, Bartusch KS, Eschen-Lippold L, Babben S, Balcke GU, Tissier A, Hause B, Andersen TG, Delker C, Quint M (2023) Auxin-dependent regulation of cell division rates governs root thermomorphogenesis. EMBO J 42(11):e111926

Aida M, Beis D, Heidstra R, Willemsen V, Blilou I, Galinha C, Nussaume L, Noh YS, Amasino R, Scheres B (2004) The *PLETHORA* genes mediate patterning of the *Arabidopsis* root stem cell niche. Cell 119(1):109–120. https://doi.org/10.1016/j.cell.2004.09.018

Aloni R, Aloni E, Langhans M, Ullrich CI (2006) Role of cytokinin and auxin in shaping root architecture: regulating vascular differentiation, lateral root initiation, root apical dominance and root gravitropism. Ann Bot 97(1):883–893

Argueso CT, Raines T, Kieber JJ (2010) Cytokinin signaling and transcriptional networks. Curr Opin Plant Biol 13(5):533–539

Arteca RN, Arteca JM (2007) Heavy-metal-induced ethylene production in *Arabidopsis thaliana*. J Plant Physiol 164(11):1480–1488

Asgher M, Khan NA, Khan MIR, Fatma M, Masood A (2014) Ethylene production is associated with alleviation of cadmium-induced oxidative stress by sulfur in mustard types differing in ethylene sensitivity. Ecotoxicol Environ Saf 106(2014):54–61

Ashraf MA, Rahman A (2019) Cold stress response in *Arabidopsis thaliana* is mediated by GNOM ARF-GEF. Plant J 97(3):500–516

Beemster GT, Baskin TI (1998) Analysis of cell division and elongation underlying the developmental acceleration of root growth in *Arabidopsis thaliana*. Plant Physiol 116(4):1515–1526

Bellstaedt J, Trenner J, Lippmann R, Poeschl Y, Zhang X, Friml J, Quint M, Delker C (2019) A mobile auxin signal connects temperature sensing in cotyledons with growth responses in hypocotyls. Plant Physiol 180(2):757–766

Bellucci M, Mostofa MG, Weraduwage SM, Xu Y, Abdelrahman M, De Gara L, Loreto F, Sharkey TD (2024) The effect of constitutive root isoprene emission on root phenotype and physiology under control and salt stress conditions. Plant Direct 8(7):e617

Benková E, Bielach A (2010) Lateral root organogenesis-from cell to organ. Curr Opin Plant Biol 13(6):677–683

Blum A (2009) Effective use of water (EUW) and not water-use efficiency (WUE) is the target of crop yield improvement under drought stress. Field Crop Res 112(2–3):119–123. https://doi.org/10.1016/j.fcr.2009.03.009

Borniego MB, Costigliolo Rojas MC, Casal JJ (2022) Shoot thermosensors do not fulfil the same function in the root. New Phytol 236:9–14

Bruno L, Pacenza M, Forgione I, Lamerton LR, Greco M, Chiappetta A, Bitonti MB (2017) In *Arabidopsis thaliana* cadmium impact on the growth of primary root by altering *SCR* expression and auxin-cytokinin cross-talk. Front Plant Sci 8:1323

Cackett L, Cannistraci CV, Meier S, Ferrandi P, Pěnčík A, Gehring C, Novák O, Ingle RA, Donaldson L (2022) Salt-specific gene expression reveals elevated auxin levels in *Arabidopsis thaliana* plants grown under saline conditions. Front Plant Sci 13:804716

Cao WH, Liu J, He XJ, Mu RL, Zhou HL, Chen SY, Zhang JS (2007) Modulation of ethylene responses affects plant salt-stress responses. Plant Physiol 143(2):707–719

Castañón-Suárez CA, Arrizubieta M, Castelán-Muñoz N, Sánchez-Rodríguez DB, Caballero-Cordero C, Zluhan-Martínez E, Patiño-Olvera SC, Arciniega-González JA, García-Ponce B, Sánchez MDLP, Álvarez-Buylla ER (2024) The MADS-box genes *SOC1* and *AGL24* antagonize *XAL2* functions in *Arabidopsis thaliana* root development. Front Plant Sci 15:1331269

Chaiwanon J, Wang ZY (2015) Spatiotemporal brassinosteroid signaling and antagonism with auxin pattern stem cell dynamics in *Arabidopsis* roots. Curr Biol 25(8):1031–1042. https://doi.org/10.1016/j.cub.2015.02.046

Cheng NH, Liu JZ, Liu X, Wu Q, Thompson SM, Lin J, Chang J, Whitham SA, Park S, Cohen JD, Hirschi KD (2011) *Arabidopsis* monothiol glutaredoxin, *At*GRXS17, is critical for temperature-dependent postembryonic growth and development via modulating auxin response. J Biol Chem 286(23):20398–20406

Clouse SD (2011) Brassinosteroids. Arabidopsis Book 9:e0151. https://doi.org/10.1199/tab.0151

Clouse SD, Sasse JM (1998) Brassinosteroids: essential regulators of plant growth and development. Annu Rev Plant Biol 49(1):427–451

Cruz-Ramírez A, Diaz-Trivino S, Wachsman G, Du Y, Arteaga-Vazquez M, Zhang H, Benjamins R, Blilou I, Neef AB, Chandler V, Scheres B (2013) A SCARECROW-RETINOBLASTOMA protein network controls protective quiescence in the *Arabidopsis* root stem cell organizer. PLoS Biol 11(11):e1001724

Cutler SR, Rodriguez PL, Finkelstein RR, Abrams SR (2010) Abscisic acid: emergence of a core signaling network. Annu Rev Plant Biol 61(1):651–679

Dard A, Weiss A, Bariat L, Auverlot J, Fontaine V, Picault N, Pontvianne F, Riondet C, Reichheld JP (2023) Glutathione-mediated thermomorphogenesis and heat stress responses in *Arabidopsis thaliana*. J Exp Bot 74(8):2707–2725

Davis LOMM, Ogita N, Inagaki S, Takahashi N, Umeda M (2016) DNA damage inhibits lateral root formation by up-regulating cytokinin biosynthesis genes in *Arabidopsis thaliana*. Genes Cells 21(11):1195–1208

Davletova S, Schlauch K, Coutu J, Mittler R (2005) The zinc-finger protein Zat12 plays a central role in reactive oxygen and abiotic stress signaling in *Arabidopsis*. Plant Physiol 139(2):847–856

de Ollas C, Arbona V, GóMez-Cadenas A (2015) Jasmonoyl isoleucine accumulation is needed for abscisic acid build-up in roots of *Arabidopsis* under water stress conditions. Plant Cell Environ 38(10):2157–2170

De Smet I, Zhang H, Inzé D, Beeckman T (2006) A novel role for abscisic acid emerges from underground. Trends Plant Sci 11(9):434–439

De Smet I, Lau S, Voß U, Vanneste S, Benjamins R, Rademacher EH, Schlereth A, De Rybel B, Vassileva V, Grunewald W, Naudts M (2010) Bimodular auxin response controls organogenesis in *Arabidopsis*. Proc Natl Acad Sci 107(6):2705–2710

De Smet S, Cuypers A, Vangronsveld J, Remans T (2015) Gene networks involved in hormonal control of root development in *Arabidopsis thaliana*: a framework for studying its disturbance by metal stress. Int J Mol Sci 16(8):19195–19224

Deak KI, Malamy J (2005) Osmotic regulation of root system architecture. Plant J 43(1):17–28

Devaiah BN, Karthikeyan AS, Raghothama KG (2007) WRKY75 transcription factor is a modulator of phosphate acquisition and root development in *Arabidopsis*. Plant Physiol 143(4):1789–1801

Dietrich D (2018) Hydrotropism: how roots search for water. J Exp Bot 69(11):2759–2771

Ding ZJ, Yan JY, Li CX, Li GX, Wu YR, Zheng SJ (2015) Transcription factor WRKY 46 modulates the development of *Arabidopsis* lateral roots in osmotic/salt stress conditions via regulation of ABA signaling and auxin homeostasis. Plant J 84(1):56–69

Dinneny JR (2019) Developmental responses to water and salinity in root systems. Ann Rev Cell Dev 35(1):239–257

Dixit AR, Dhankher OP (2011) A novel stress-associated protein 'AtSAP10' from *Arabidopsis thaliana* confers tolerance to nickel, manganese, zinc, and high temperature stress. PLoS One 6:e20921

Dolgikh VA, Pukhovaya EM, Zemlyanskaya EV (2019) Shaping ethylene response: the role of EIN3/EIL1 transcription factors. Front Plant Sci 10:1030

Du M, Wang Y, Chen H, Han R (2020) Actin filaments mediated root growth inhibition by changing their distribution under UV-B and hydrogen peroxide exposure in *Arabidopsis*. Biol Res 53(1):54

Ebstrup E, Ansbøl J, Paez-Garcia A, Culp H, Chevalier J, Clemmens P, Coll NS, Risueno MAM, Rodriguez E (2024) NBR1-mediated selective autophagy of ARF7 modulates root branching. EMBO Rep 25(6):2571–2591

Fei Q, Wei S, Zhou Z, Gao H, Li X (2017) Adaptation of root growth to increased ambient temperature requires auxin and ethylene coordination in *Arabidopsis*. Plant Cell Rep 36(9):1507–1518

Feng Y, Xu P, Li B, Li P, Wen X, An F, Gong Y, Xin Y, Zhu Z, Wang Y, Guo H (2017) Ethylene promotes root hair growth through coordinated EIN3/EIL1 and RHD6/RSL1 activity in *Arabidopsis*. Proc Natl Acad Sci 114(52):13834–13839

Fonseca S, Chini A, Hamberg M, Adie B, Porzel A, Kramell R, Miersch O, Wasternack C, Solano R (2009) (+)-7-iso-Jasmonoyl-L-isoleucine is the endogenous bioactive jasmonate. Nat Chem Biol 5(5):344–350

Frugier F, Kosuta S, Murray JD, Crespi M, Szczyglowski K (2008) Cytokinin: secret agent of symbiosis. Trends Plant Sci 13(3):115–120

Fu Q, Li S, Yu D (2010) Identification of an *Arabidopsis* nodulin-related protein in heat stress. Mol Cell 29:77–84

Fu Y, Yang Y, Chen S, Ning N, Hu H (2019) *Arabidopsis* IAR4 modulates primary root growth under salt stress through ROS-mediated modulation of auxin distribution. Front Plant Sci 10:522

Fukaki H, Tasaka M (2009) Hormone interactions during lateral root formation. Plant Mol Biol 69:437–449

Fulcher N, Sablowski R (2009) Hypersensitivity to DNA damage in plant stem cell niches. Proc Natl Acad Sci 106(49):20984–20988

Galinha C, Hofhuis H, Luijten M, Willemsen V, Blilou I, Heidstra R, Scheres B (2007) PLETHORA proteins as dose-dependent master regulators of *Arabidopsis* root development. Nature 449(7165):1053–1057. https://doi.org/10.1038/nature06206

Gao J, Zhao Y, Zhao Z, Liu W, Jiang C, Li J, Zhang Z, Zhang H, Zhang Y, Wang X, Sun X (2023) RRS1 shapes robust root system to enhance drought resistance in rice. New Phytol 238(3):1146–1162

Garrido-Vargas F, Godoy T, Tejos R, O'Brien JA (2020) Overexpression of the auxin receptor *AFB3* in *Arabidopsis* results in salt stress resistance and the modulation of *NAC4* and *SZF1*. Int J Mol Sci 21(24):9528

Germon A, Laclau JP, Robin A, Jourdan C (2020) Tamm review: deep fine roots in forest ecosystems: why dig deeper? For Ecol Manag 466:118135

Goel P, Bhuria M, Sinha R, Sharma TR, Singh AK (2019) Promising transcription factors for salt and drought tolerance in plants. In: Molecular approaches in plant biology and environmental challenges. Springer Singapore, pp 7–50

Goldfarb T, Lichten M (2010) Frequent and efficient use of the sister chromatid for DNA double-strand break repair during budding yeast meiosis. PLoS Biol 8(10):e1000520

González-García MP, Vilarrasa-Blasi J, Zhiponova M, Divol F, Mora-García S, Russinova E, Caño-Delgado AI (2011) Brassinosteroids control meristem size by promoting cell cycle progression in *Arabidopsis* roots. Development 138(5):849–859. https://doi.org/10.1242/dev.057331

González-García MP, Conesa CM, Lozano-Enguita A, Baca-González V, Simancas B, Navarro-Neila S, Sánchez-Bermúdez M, Salas-González I, Caro E, Castrillo G, Del Pozo JC (2023) Temperature changes in the root ecosystem affect plant functionality. Plant Commun 4(3):100514

Grosjean N, Blaudez D, Chalot M, Flayac J, Gross EM, Le Jean M (2024) Rare earth elements perturb root architecture and ion homeostasis in *Arabidopsis thaliana*. J Hazard Mater 468:133701

Guo P, Chong L, Jiao Z, Xu R, Niu Q, Zhu Y (2025) Salt stress activates the CDK8-AHL10-SUVH2/9 module to dynamically regulate salt tolerance in *Arabidopsis*. Nat Commun 16(1):2454

Gupta A, Rico-Medina A, Caño-Delgado AI (2020) The physiology of plant responses to drought. Science 368(6488):266–269

Han S, Wang Y, Li Y, Zhu R, Gu Y, Li J, Guo H, Ye W, Nabi HG, Yang T, Wang Y (2024) The *Os*NAC41-RoLe1-*Os*AGAP module promotes root development and drought resistance in upland rice. Mol Plant 17(10):1573–1593

Hanzawa T, Shibasaki K, Numata T, Kawamura Y, Gaude T, Rahman A (2013) Cellular auxin homeostasis under high temperature is regulated through a sorting NEXIN1-dependent endosomal trafficking pathway. Plant Cell 25(9):3424–3433

He XJ, Mu RL, Cao WH, Zhang ZG, Zhang JS, Chen SY (2005) *At*NAC2, a transcription factor downstream of ethylene and auxin signaling pathways, is involved in salt stress response and lateral root development. Plant J 44(6):903–916

Hirakawa T, Matsunaga S (2019) Characterization of DNA repair foci in root cells of *Arabidopsis* in response to DNA damage. Front Plant Sci 10:990

Hiya HJ, Nakashima Y, Takeuchi A, Nakamura T, Nakamura Y, Murata Y, Munemasa S (2024) Outward-rectifying potassium channels GORK and SKOR function in regulation of root growth under salt stress in *Arabidopsis thaliana*. J Plant Physiol 302:154322

Hu H, Xiong L (2014) Genetic engineering and breeding of drought-resistant crops. Annu Rev Plant Biol 65(1):715–741. https://doi.org/10.1146/annurev-arplant-050213-040000

Hu YF, Zhou G, Na XF, Yang L, Nan WB, Liu X, Zhang YQ, Li JL, Bi YR (2013) Cadmium interferes with maintenance of auxin homeostasis in *Arabidopsis* seedlings. J Plant Physiol 170(11):965–975

Hwang JH, Seo DH, Kang BG, Kwak JM, Kim WT (2015) Suppression of *Arabidopsis At* PUB30 resulted in increased tolerance to salt stress during germination. Plant Cell Rep 34:277–289

Iglesias MJ, Terrile MC, Windels D, Lombardo MC, Bartoli CG, Vazquez F, Estelle M, Casalongué CA (2014) MiR393 regulation of auxin signaling and redox-related components during acclimation to salinity in Arabidopsis. PLoS One 9:e107678

Ishka MR, Sussman H, Hu Y, Alqahtani MD, Craft E, Sicat R, Wang M, Yu LA, Ait-Haddou R, Li B, Drakakaki G (2025) Natural variation in salt-induced changes in root: shoot ratio reveals SR3G as a negative regulator of root suberization and salt resilience in *Arabidopsis*. elife 13:RP98896

Ito H, Kim J-M, Matsunaga W, Saze H, Matsui A, Endo TA, Harukawa Y, Takagi H, Yaegashi H, Masuta Y, Masuda S (2016) A stress-activated transposon in *Arabidopsis* induces transgenerational abscisic acid insensitivity. Sci Rep 6(1):23181

Iwata S, Miyazawa Y, Fujii N, Takahashi H (2013) MIZ1-regulated hydrotropism functions in the growth and survival of *Arabidopsis thaliana* under natural conditions. Ann Bot 112(1):103–114

Jalmi SK, Bhagat PK, Verma D, Noryang S, Tayyeba S, Singh K, Sharma D, Sinha AK (2018) Traversing the links between heavy metal stress and plant signaling. Front Plant Sci 9:12

Jeon J, Cho C, Lee MR, Van Binh N, Kim J (2016) *CYTOKININ RESPONSE FACTOR2* (*CRF2*) and *CRF3* regulate lateral root development in response to cold stress in *Arabidopsis*. Plant Cell 28(8):1828–1843. https://doi.org/10.1105/tpc.15.00909

Julkowska MM, Koevoets IT, Mol S, Hoefsloot H, Feron R, Tester MA, Keurentjes JJ, Korte A, Haring MA, de Boer GJ, Testerink C (2017) Genetic components of root architecture remodeling in response to salt stress. Plant Cell 29(12):3198–3213

Kaldis A, Tsementzi D, Tanriverdi O, Vlachonasios KE (2011) *Arabidopsis thaliana* transcriptional co-activators ADA2b and SGF29a are implicated in salt stress responses. Planta 233:749–762

Karaba A, Dixit S, Greco R, Aharoni A, Trijatmiko KR, Marsch-Martinez N, Krishnan A, Natraja K, Udaykumar M, Pereira A (2007) Improvement of water use efficiency in rice by expression of *HARDY*, an *Arabidopsis* drought and salt tolerance gene. Proc Natl Acad Sci 104(39):15270–15275

Karlova R, Boer D, Hayes S, Testerink C (2021) Root plasticity under abiotic stress. Plant Physiol 187(3):1057–1070

Keerthana K, Ramakrishnan M, Ahmad Z, Amali P, Vijayakanth V, Wei Q (2025) Root-derived small peptides: key regulators of plant development, stress resilience, and nutrient acquisition. Plant Sci 354:112433

Khan MIR, Khan NA (2014) Ethylene reverses photosynthetic inhibition by nickel and zinc in mustard through changes in PS II activity, photosynthetic nitrogen use efficiency, and antioxidant metabolism. Protoplasma 251:1007–1019

Khan MA, Gemenet DC, Villordon A (2016) Root system architecture and abiotic stress tolerance: current knowledge in root and tuber crops. Front Plant Sci 7:1584

Kim E-J, Kim Y-J, Hong W-J, Lee C, Jeon J-S, Jung K-H (2019) Genome-wide analysis of root hair preferred RBOH genes suggests that three *RBOH* genes are associated with auxin-mediated root hair development in rice. J Plant Biol 62:229–238

Kim JH, Dubey SK, Ryu TH, Lee SS, Chung BY (2025) Argonaute 2 regulates nuclear DNA damage, repair, and phenotypes in *Arabidopsis* under genotoxic stress. Plant Physiol Biochem 109528:0981–9428

Krishnamurthy A, Rathinasabapathi B (2013) Auxin and its transport play a role in plant tolerance to arsenite-induced oxidative stress in *Arabidopsis thaliana*. Plant Cell Environ 36(10):1838–1849

Kutashev K, Meschichi A, Reeck S, Fonseca A, Sartori K, White CI, Sicard A, Rosa S (2024) Differences in *RAD51* transcriptional response and cell cycle dynamics reveal varying sensitivity to DNA damage among *Arabidopsis thaliana* root cell types. New Phytol 243(3):966–980

Laplaze L, Benkova E, Casimiro I, Maes L, Vanneste S, Swarup R, Weijers D, Calvo V, Parizot B, Rodriguez MBH, Offringa R, Graham N, Doumas P, Friml J, Bogusz D, Beeckman T, Bennett M (2007) Cytokinins act directly on lateral root founder cells to inhibit root initiation. Plant Cell 19(12):3889–3900

Leal AR, Belo J, Beeckman T, Barros PM, Oliveira MM (2022) The combined effect of heat and osmotic stress on suberization of *Arabidopsis* roots. Cells 11(15):2341

Lee S, Wang W, Huq E (2021) Spatial regulation of thermomorphogenesis by HY5 and PIF4 in *Arabidopsis*. Nat Commun 12(1):3656

Lequeux H, Hermans C, Lutts S, Verbruggen N (2010) Response to copper excess in *Arabidopsis thaliana*: impact on the root system architecture, hormone distribution, lignin accumulation and mineral profile. Plant Physiol Biochem 48:673–682

Li J, Qian W (2023) Translational control of SOG1 expression in response to replication stress in *Arabidopsis*. Stress Biol 3(1):28

Li X, Mo X, Shou H, Wu P (2006) Cytokinin-mediated cell cycling arrest of pericycle founder cells in lateral root initiation of *Arabidopsis*. Plant Cell Physiol 47:1112–1123

Li Y, Yuan W, Li L, Dai H, Dang X, Miao R, Baluška F, Kronzucker HJ, Lu C, Zhang J, Xu W (2020a) Comparative analysis reveals gravity is involved in the *MIZ1*-regulated root hydrotropism. J Exp Bot 71(22):7316–7330

Li Y, Yuan W, Li L, Miao R, Dai H, Zhang J, Xu W (2020b) Light-dark modulates root hydrotropism associated with gravitropism by involving amyloplast response in *Arabidopsis*. Cell Rep 32(13):108198

Li C, Liu G, Geng X, He C, Quan T, Hayashi K, De Smet I, Robert HS, Ding Z, Yang ZB (2021a) Local regulation of auxin transport in root-apex transition zone mediates aluminium-induced *Arabidopsis* root-growth inhibition. Plant J 108(1):55–66. https://doi.org/10.1111/TPJ.15424

Li SM, Zheng HX, Zhang XS, Sui N (2021b) Cytokinins as central regulators during plant growth and stress response. Plant Cell Rep 40:271–282

Liu W, Li RJ, Han TT, Cai W, Fu ZW, Lu YT (2015) Salt stress reduces root meristem size by nitric oxide-mediated modulation of auxin accumulation and signaling in *Arabidopsis*. Plant Physiol 168(1):343–356

Liu Q, Ma W, Chen R, Li ST, Wang Q, Wei C, Hong Y, Sun HX, Cheng Q, Zhao J, Kang J (2024) Multiome in the same cell reveals the impact of osmotic stress on *Arabidopsis* root tip development at single-cell level. Adv Sci 11(24):2308384

López-Bucio J, Cruz-Ramırez A, Herrera-Estrella L (2003) The role of nutrient availability in regulating root architecture. Curr Opin Plant Biol 6(3):280–287

Lorts CM, Lasky JR (2020) Competition × drought interactions change phenotypic plasticity and the direction of selection on *Arabidopsis* traits. New Phytol 227:1060–1072. https://doi.org/10.1111/nph.16593

Lucob-Agustin N, Kawai T, Kano-Nakata M, Suralta RR, Niones JM, Hasegawa T, Hasegawa T, Inari-Ikeda M, Yamauchi A, Inukai Y (2021) Morpho-physiological and molecular mechanisms of phenotypic root plasticity for rice adaptation to water stress conditions. Breed Sci 71(1):20–29

Lynch JP (2013) Steep, cheap and deep: an ideotype to optimize water and N acquisition by maize root systems. Ann Bot 112(2):347–357

Mähönen AP, Ten Tusscher K, Siligato R, Smetana O, Díaz-Triviño S, Salojärvi J, Wachsman G, Prasad K, Heidstra R, Scheres B (2014) *PLETHORA* gradient formation mechanism separates auxin responses. Nature 515:125–129. https://doi.org/10.1038/nature13663

Martins S, Montiel-Jorda A, Cayrel A, Huguet S, Roux CP, Ljung K, Vert G (2017) Brassinosteroid signaling-dependent root responses to prolonged elevated ambient temperature. Nat Commun 8:309

Meng Y, Huang J, Jing H, Wu Q, Shen R, Zhu X (2022) Exogenous abscisic acid alleviates Cd toxicity in *Arabidopsis thaliana* by inhibiting Cd uptake, translocation and accumulation, and promoting Cd chelation and efflux. Plant Sci 325:111464

Miyawaki K, Tarkowski P, Matsumoto-Kitano M, Kato T, Sato S, Tarkowska D, Tabata S, Sandberg G, Kakimoto T (2006) Roles of *Arabidopsis* ATP/ADP isopentenyltransferases and tRNA isopentenyltransferases in cytokinin biosynthesis. Proc Natl Acad Sci 103(44):16598–16603

Msanne J, Lin J, Stone JM, Awada T (2011) Characterization of abiotic stress-responsive *Arabidopsis thaliana RD29A* and *RD29B* genes and evaluation of transgenes. Planta 234(1):97–107

Müssig C, Shin GH, Altmann T (2003) Brassinosteroids promote root growth in *Arabidopsis*. Plant Physiol 133:1261–1271. https://doi.org/10.1104/pp.103.028662c

Nestler J, Liu S, Wen TJ, Paschold A, Marcon C, Tang HM, Li D, Li L, Meelay RB, Sakai H, Bruce W, Schnable PS, Hochholdinger F (2014) Root hairless 5, which functions in maize (*Zea mays* L.) root hair initiation and elongation encodes a monocot-specific NADPH oxidase. Plant J 79:729–740

Nolan TM, Brennan B, Yang M, Chen J, Zhang M, Li Z, Wang X, Bassham DC, Walley J, Yin Y (2017) Selective autophagy of BES1 mediated by DSK2 balances plant growth and survival. Dev Cell 41:33–46. https://doi.org/10.1016/j.devcel.2017.03.013

Ogura T, Goeschl C, Filiault D, Mirea M, Slovak R, Wolhrab B, Satbhai SB, Busch W (2019) Root system depth in *Arabidopsis* is shaped by *EXOCYST70A3* via the dynamic modulation of auxin transport. Cell 178(2):400–412

Oliver C, Santos JL, Pradillo M (2014) On the role of some ARGONAUTE proteins in meiosis and DNA repair in *Arabidopsis thaliana*. Front Plant Sci 5:177

Orosa-Puente B, Leftley N, Von Wangenheim D, Banda J, Srivastava AK, Hill K, Truskina J, Bhosale R, Morris E, Srivastava M, Kümpers B, Tatsuaki C, Fukaki H, Vermeer JEM, Vernoux T, Dinneny JRPA, Bishopp A, Sadanandom A, Bennett MJ (2018) Root branching toward water involves posttranslational modification of transcription factor ARF7. Science 362(6421):1407–1410

Osakabe Y, Yamaguchi-Shinozaki K, Shinozaki K, Tran LSP (2014) ABA control of plant macroelement membrane transport systems in response to water deficit and high salinity. New Phytol 202(1):35–49

Pagano P, Pagano A, Paternolli S, Balestrazzi A, Macovei A (2023) Integrative transcriptomics data mining to explore the functions of *TDP1α* and *TDP1β* genes in the *Arabidopsis thaliana* model plant. Gene 14(4):884

Pagano P, Bertoncini A, Pagano A, Nisa MU, Raynaud C, Balestrazzi A, Macovei A (2024) Exposure of *Arabidopsis thaliana* mutants to genotoxic stress provides new insights for the involvement of *TDP1α* and *TDP1β* genes in DNA-damage response. Plant Cell Environ 47(12):5483–5497

Pan W, You Y, Shentu JL, Weng YN, Wang ST, Xu QR, Liu HJ, Du ST (2020) Abscisic acid (ABA)-importing transporter 1 (AIT1) contributes to the inhibition of Cd accumulation via exogenous ABA application in *Arabidopsis*. J Hazard Mater 391:122189

Peng J, Li Z, Wen X, Li W, Shi H, Yang L, Zhu H, Guo H (2014) Salt-induced stabilization of EIN3/EIL1 confers salinity tolerance by deterring ROS accumulation in *Arabidopsis*. PLoS Genet 10(10):e1004664

Perilli S, Moubayidin L, Sabatini S (2010) The molecular basis of cytokinin function. Curr Opin Plant Biol 13(1):21–26

Pierik R, Testerink C (2014) The art of being flexible: how to escape from shade, salt, and drought. Plant Physiol 166(1):5–22

Planas-Riverola A, Gupta A, Betegón-Putze I, Bosch N, Ibañes M, Caño-Delgado AI (2019) Brassinosteroid signaling in plant development and adaptation to stress. Development 146(5):dev151894

Przybyla-Toscano J, Chetouhi C, Pennera L, Boursiac Y, Galeone A, Devime F, Balliau T, Santoni V, Bourguignon J, Alban C, Ravanel S (2025) New insights into uranium stress responses of *Arabidopsis* roots through membrane-and cell wall-associated proteome analysis. Chemosphere 370:143873

Rajappa S, Krishnamurthy P, Huang H, Yu D, Friml J, Xu J, Kumar PP (2024) The translocation of a chloride channel from the Golgi to the plasma membrane helps plants adapt to salt stress. Nat Commun 15(1):3978

Rellán-Álvarez R, Lobet G, Lindner H, Pradier PL, Sebastian J, Yee MC, Geng Y, Trontin C, LaRue T, Schrager-Lavelle A, Haney CH (2015) GLO-Roots: an imaging platform enabling multidimensional characterization of soil-grown root systems. elife 4:e07597

Ren C, Bodendorf J, Moreno-Risueno MA, Kleine-Vehn J (2024a) Two distinct oscillatory auxin signals define the plasticity of lateral rooting in *Arabidopsis thaliana*. bioRxiv:2024-08

Ren Y, Jiang M, Zhu JK, Zhou W, Zhao C (2024b) Simultaneous mutations in *ITPK4* and *MRP5* genes result in a low phytic acid level without compromising salt tolerance in *Arabidopsis*. J Integr Plant Biol 66(10):2109–2125

Renault H, Roussel V, El Amrani A, Arzel M, Renault D, Bouchereau A, Deleu C (2010) The *Arabidopsis pop2-1* mutant reveals the involvement of GABA transaminase in salt stress tolerance. BMC Plant Biol 10:1–16

Revenkova E, Masson J, Koncz C, Afsar K, Jakovleva L, Paszkowski J (1999) Involvement of *Arabidopsis thaliana* ribosomal protein S27 in mRNA degradation triggered by genotoxic stress. EMBO J 18:490–499

Richter S, Anders N, Wolters H, Beckmann H, Thomann A, Heinrich R, Schrader J, Singh MK, Geldner N, Mayer U, Jürgens G (2010) Role of the *GNOM* gene in *Arabidopsis* apical-basal patterning–from mutant phenotype to cellular mechanism of protein action. Eur J Cell Biol 89(2–3):138–144

Rosales MA, Maurel C, Nacry P (2019) Abscisic acid coordinates dose-dependent developmental and hydraulic responses of roots to water deficit. Plant Physiol 180(4):2198–2211

Rowe JH, Topping JF, Liu J, Lindsey K (2016) Abscisic acid regulates root growth under osmotic stress conditions via an interacting hormonal network with cytokinin, ethylene and auxin. New Phytol 211(1):225–239

Ruiz-Aguilar B, Raya-González J, López-Bucio JS, Reyes de la Cruz H, Herrera-Estrella L, Ruiz-Herrera LF, Martínez-Trujillo M, López-Bucio J (2020) Mutation of *MEDIATOR 18* and chromate trigger twinning of the primary root meristem in *Arabidopsis*. Plant Cell Environ 43(8):1989–1999

Sabagh AE, Hossain A, Islam MS, Iqbal MA, Amanet K, Mubeen M, Nasim W, Wasaya A, Llanes A, Ratnasekera D, Singhal RK (2021) Prospective role of plant growth regulators for tolerance to abiotic stresses. J Plant Growth Regul 2021:1–38

Sabagh A, Islam MS, Hossain A, Iqbal MA, Mubeen M, Waleed M, Reginato M, Battaglia M, Ahmed S, Rehman A, Arif M (2022) Phytohormones as growth regulators during abiotic stress tolerance in plants. Front Agron 4:765068

Schellingen K, Van Der Straeten D, Vandenbussche F, Prinsen E, Remans T, Vangronsveld J, Cuypers A (2014) Cadmium-induced ethylene production and responses in *Arabidopsis thaliana* rely on *ACS2* and *ACS6* gene expression. BMC Plant Biol 14:1–14

Schellingen K, Van Der Straeten D, Remans T, Loix C, Vangronsveld J, Cuypers A (2015) Ethylene biosynthesis is involved in the early oxidative challenge induced by moderate Cd exposure in *Arabidopsis thaliana*. Environ Exp Bot 117:1–11

Seifert GJ, Xue H, Acet T (2014) The *Arabidopsis thaliana FASCICLIN LIKE ARABINOGALACTAN PROTEIN 4* gene acts synergistically with abscisic acid signalling to control root growth. Ann Bot 114(6):1125–1133

Shahid M, Dumat C, Khalid S, Schreck E, Xiong T, Niazi NK (2017) Foliar heavy metal uptake, toxicity and detoxification in plants: a comparison of foliar and root metal uptake. J Hazard Mater 325:36–58

Sharma V, Goel P, Kumar S, Singh AK (2019) An apple transcription factor, MdDREB76, confers salt and drought tolerance in transgenic tobacco by activating the expression of stress-responsive genes. Plant Cell Rep 38:221–241

Sharp RE, LeNoble ME (2002) ABA, ethylene and the control of shoot and root growth under water stress. J Exp Bot 53(366):33–37

Sharp RE, Wu Y, Voetberg GS, Saab IN, LeNoble ME (1994) Confirmation that abscisic acid accumulation is required for maize primary root elongation at low water potentials. J Exp Bot 45(Special_Issue):1743–1751

Sharp RE, Poroyko V, Hejlek LG, Spollen WG, Springer GK, Bohnert HJ, Nguyen HT (2004) Root growth maintenance during water deficits: physiology to functional genomics. J Exp Bot 55(407):2343–2351

Shen H, Hou N, Schlicht M, Wan Y, Mancuso S, Baluska F (2008) Aluminium toxicity targets PIN2 in *Arabidopsis* root apices: effects on PIN2 endocytosis, vesicular recycling, and polar auxin transport. Chin Sci Bull 53(16):2480–2487

Sheridan ML, Simonelli L, Giustozzi M, Casati P (2022) Ultraviolet-B radiation represses primary root elongation by inhibiting cell proliferation in the meristematic zone of *Arabidopsis* seedlings. Front Plant Sci 13:829336

Shibasaki K, Uemura M, Tsurumi S, Rahman A (2009) Auxin response in *Arabidopsis* under cold stress: underlying molecular mechanisms. Plant Cell 21(12):3823–3838

Shibata M, Sugimoto K (2019) A gene regulatory network for root hair development. J Plant Res 132(3):301–309

Shinozaki K, Yamaguchi-Shinozaki K (2007) Gene networks involved in drought stress response and tolerance. J Exp Bot 58(2):221–227

Siddiqui MN, Léon J, Naz AA, Ballvora A (2021) Genetics and genomics of root system variation in adaptation to drought stress in cereal crops. J Exp Bot 72(4):1007–1019

Singh AK, Sharma V, Pal AK, Acharya V, Ahuja PS (2013) Genome-wide organization and expression profiling of the *NAC* transcription factor family in potato (*Solanum tuberosum* L.). DNA Res 20(4):403–423

Smolko A, Bauer N, Pavlović I, Pěnčík A, Novák O, Salopek-Sondi B (2021) Altered root growth, auxin metabolism and distribution in *Arabidopsis thaliana* exposed to salt and osmotic stress. Int J Mol Sci 22(15):7993

Soltabayeva A, Ongaltay A, Omondi JO, Srivastava S (2021) Morphological, physiological and molecular markers for salt-stressed plants. Plants 10:243. https://doi.org/10.3390/plants10020243

Soma F, Takahashi F, Yamaguchi-Shinozaki K, Shinozaki K (2021) Cellular phosphorylation signaling and gene expression in drought stress responses: ABA-dependent and ABA-independent regulatory systems. Plants 10(4):756

Staňová A, Ďurišová E, Banásová V, Gurinová E, Nadubinská M, Kenderešová L, Ovečka M, Čiamporová M (2012) Root system morphology and primary root anatomy in natural non-metallicolous and metallicolous populations of three *Arabidopsis* species differing in heavy metal tolerance. Biol 67:505–516

Stohs SJ, Bagchi D (1995) Oxidative mechanisms in the toxicity of metal ions. Free Radic Biol Med 18(2):321–336

Sun P, Tian QY, Chen J, Zhang WH (2010) Aluminium-induced inhibition of root elongation in *Arabidopsis* is mediated by ethylene and auxin. J Exp Bot 61(2):347–356

Sun LR, Wang YB, He SB, Hao FS (2018) Mechanisms for abscisic acid inhibition of primary root growth. Plant Signal Behav 13(9):e1500069

Sun L, Feraru E, Feraru MI, Waidmann S, Wang W, Passaia G, Wang ZY, Wabnik K, Kleine-Vehn J (2020) PIN-LIKES coordinate brassinosteroid signaling with nuclear auxin input in *Arabidopsis thaliana*. Curr Biol 30(9):1579–1588.e6

Sun C, Zhao C, Wang G, Han R (2025) Cerium oxide nanoparticles ameliorate *Arabidopsis thaliana* root damage under UV-B stress by modulating the cell cycle and auxin pathways. Protoplasma 262(4):1–19

Takahashi N, Goto N, Okada K, Takahashi H (2002) Hydrotropism in abscisic acid, wavy, and gravitropic mutants of *Arabidopsis thaliana*. Planta 216:203–211

Takeda S, Gapper C, Kaya H, Bell E, Kuchitsu K, Dolan L (2008) Local positive feedback regulation determines cell shape in root hair cells. Science 319(5867):1241–1244

Tchounwou PB, Yedjou CG, Patlolla AK, Sutton DJ (2012) Heavy metal toxicity and the environment. Mol Clin Environ Toxicol 101:133–164

Thiébaut N, Sarthou M, Richtmann L, Pergament Persson D, Ranjan A, Schloesser M, Boutet S, Rezende L, Clemens S, Verbruggen N, Hanikenne M (2025) Specific redox and iron homeostasis responses in the root tip of *Arabidopsis* upon zinc excess. New Phytol 246(4):1796–1815

Timilsina R, Kim JH, Nam HG, Woo HR (2019) Temporal changes in cell division rate and genotoxic stress tolerance in quiescent center cells of *Arabidopsis* primary root apical meristem. Sci Rep 9(1):3599

Tiryaki İ (2007) The role of auxin-signaling gene *axr1* in salt stress and jasmonic acid inducible gene expression in *Arabidopsis thaliana*. J Mol Cell Biol 6:189–195

Tiwari M, Kumar R, Subramanian S, Doherty CJ, Jagadish SK (2023) Auxin–cytokinin interplay shapes root functionality under low-temperature stress. Trends Plant Sci 28(4):447–459

Tuteja N (2007) Abscisic acid and abiotic stress signaling. Plant Signal Behav 2(3):135–138

Uga Y, Sugimoto K, Ogawa S, Rane J, Ishitani M, Hara N, Kitomi Y, Inukai Y, Ono K, Kanno N, Inoue H (2013) Control of root system architecture by *DEEPER ROOTING 1* increases rice yield under drought conditions. Nat Genet 45(9):1097–1102

Ullah A, Manghwar H, Shaban M, Khan AH, Akbar A, Ali U, Ali E, Fahad S (2018) Phytohormones enhanced drought tolerance in plants: a coping strategy. Environ Sci Pollut Res 25:33103–33118

Vanstraelen M, Benková E (2012) Hormonal interactions in the regulation of plant development. Annu Rev Cell Dev Biol 28(1):463–487

Vaseva II, Qudeimat E, Potuschak T, Du Y, Genschik P, Vandenbussche F, Van Der Straeten D (2018) The plant hormone ethylene restricts *Arabidopsis* growth via the epidermis. Proc Natl Acad Sci 115(17):E4130–E4139

Vilarrasa-Blasi J, González-García MP, Frigola D, Fabregas N, Alexiou KG, López-Bigas N, Rivas S, Jauneau A, Lohmann JU, Benfey PN, Ibanes M (2014) Regulation of plant stem cell quiescence by a brassinosteroid signaling module. Dev Cell 30:36–47. https://doi.org/10.1016/j.devcel.2014.05.020

Vishal B, Krishnamurthy P, Kumar PP (2024) *Arabidopsis* class II TPS controls root development and confers salt stress tolerance through enhanced hydrophobic barrier deposition. Plant Cell Rep 43(5):115

Vitti A, Nuzzaci M, Scopa A, Tataranni G, Remans T, Vangronsveld J, Sofo A (2013) Auxin and cytokinin metabolism and root morphological modifications in *Arabidopsis thaliana* seedlings infected with Cucumber mosaic virus (CMV) or exposed to cadmium. Int J Mol Sci 14:6889–6902

Wan J, Zhang P, Wang R, Sun L, Wang W, Zhou H, Xu J (2018) UV-B radiation induces root bending through the flavonoid-mediated auxin pathway in *Arabidopsis*. Front Plant Sci 9:618

Wang J, Yang S (2021) Dose-dependent responses of *Arabidopsis thaliana* to zinc are mediated by auxin homeostasis and transport. Environ Exp Bot 189:104554

Wang Y, Zhang W, Li K, Sun F, Han C, Wang Y, Li X (2008) Salt-induced plasticity of root hair development is caused by ion disequilibrium in *Arabidopsis thaliana*. J Plant Res 121:87–96

Wang R, Zhang Y, Kieffer M, Yu H, Kepinski S, Estelle M (2016) HSP90 regulates temperature-dependent seedling growth in *Arabidopsis* by stabilizing the auxin co-receptor F-box protein TIR1. Nat Commun 7:10269

Wasson AP, Richards RA, Chatrath R, Misra SC, Prasad SS, Rebetzke GJ, Kirkegaard JA, Christopher J, Watt M (2012) Traits and selection strategies to improve root systems and water uptake in water-limited wheat crops. J Exp Bot 63(9):3485–3498

Werner T, Motyka V, Laucou V, Smets R, Van Onckelen H, Schmülling T (2003) Cytokinin-deficient transgenic *Arabidopsis* plants show multiple developmental alterations indicating opposite functions of cytokinins in the regulation of shoot and root meristem activity. Plant Cell 15(11):2532–2550

Werner T, Nehnevajova E, Köllmer I, Novák O, Strnad M, Krämer U, Schmülling T (2010) Root-specific reduction of cytokinin causes enhanced root growth, drought tolerance, and leaf mineral enrichment in *Arabidopsis* and tobacco. Plant Cell 22(12):3905–3920

Wu X, Xu J, Meng X, Fang X, Xia M, Zhang J, Cao S, Fan T (2022) Linker histone variant *HIS1-3* and *WRKY1* oppositely regulate salt stress tolerance in *Arabidopsis*. Plant Physiol 189(3):1833–1847

Xing L, Zhao Y, Gao J, Xiang C, Zhu JK (2016) The ABA receptor PYL9 together with PYL8 plays an important role in regulating lateral root growth. Sci Rep 6(1):27177

Xu P, Zhao PX, Cai XT, Mao JL, Miao ZQ, Xiang CB (2020) Integration of jasmonic acid and ethylene into auxin signaling in root development. Front Plant Sci 11:271

Yabe N, Takahashi T, Komeda Y (1994) Analysis of tissue-specific expression of *Arabidopsis thaliana* HSP90-family gene HSP81. Plant Cell Physiol 35:1207–1219

Yang CJ, Zhang C, Lu YN, Jin JQ, Wang XL (2011) The mechanisms of brassinosteroids' action, from signal transduction to plant development. Mol Plant 4:588–600

Yang X, Dong G, Palaniappan K, Mi G, Baskin TI (2017) Temperature-compensated cell production rate and elongation zone length in the root of *Arabidopsis thaliana*. Plant Cell Environ 40:264–276

Yang L, Yang J, Hou C, Shi P, Zhang Y, Gu X, Liu W (2023) Hydrogen sulfide alleviates salt stress through auxin signaling in *Arabidopsis*. Environ Exp Bot 211:105354

Yoon EK, Yang JH, Lee WS (2010) Auxin and abscisic acid responses of Auxin Response Factor 3 in *Arabidopsis* lateral root development. J Plant Biol 53:150–154

Yu H, Chen X, Hong YY, Wang Y, Xu P, Ke SD, Liu HY, Zhu JK, Oliver DJ, Xiang CB (2008) Activated expression of an *Arabidopsis* HD-START protein confers drought tolerance with improved root system and reduced stomatal density. Plant Cell 20(4):1134–1151

Yuan HM, Huang X (2016) Inhibition of root meristem growth by cadmium involves nitric oxide-mediated repression of auxin accumulation and signalling in *Arabidopsis*. Plant Cell Environ 39:120–135

Yuan H-M, Xu H-H, Liu W-C, Lu Y-T (2013) Copper regulates primary root elongation through *PIN1*-mediated auxin redistribution. Plant Cell Physiol 54(5):766–778

Zhang L, Li G, Wang M, Di D, Sun L, Kronzucker HJ, Shi W (2018) Excess iron stress reduces root tip zone growth through nitric oxide-mediated repression of potassium homeostasis in *Arabidopsis*. New Phytol 219(1):259–274

Zhang P, Wang R, Wang Y, Xu J (2020) Ultraviolet-B radiation induces cell death in root tips and reprograms metabolism in *Arabidopsis*. Biol Plant 64(1):764–772

Zhang C, Zeng J, Xie W, Liu C, Niu L, Wang Y, Wang Y, Shi M, Shao J, Wang W, Schiefelbein J (2024) SPIRRIG is required for BRICK1 stability and salt stress induced root hair developmental plasticity in *Arabidopsis*. Stress Biol 4(1):48

Zhang Y, Bao Z, Smoljan A, Liu Y, Wang H, Friml J (2025) Foraging for water by MIZ1-mediated antagonism between root gravitropism and hydrotropism. Proc Natl Acad Sci 122(20):e2427315122

Zhao L, Wang P, Hou H, Zhang H, Wang Y, Yan S, Huang Y, Li H, Tan J, Hu A, Gao F (2014) Transcriptional regulation of cell cycle genes in response to abiotic stresses correlates with dynamic changes in histone modifications in maize. PLoS One 9(8):e106070

Zhao WT, Feng SJ, Li H, Faust F, Kleine T, Li LN, Yang ZM (2017) Salt stress-induced *FERROCHELATASE 1* improves resistance to salt stress by limiting sodium accumulation in *Arabidopsis thaliana*. Sci Rep 7(1):14737

Zhao Z, Wu S, Gao H, Tang W, Wu X, Zhang B (2023) The BR signaling pathway regulates primary root development and drought stress response by suppressing the expression of *PLT1* and *PLT2* in *Arabidopsis thaliana*. Front Plant Sci 14:1187605

Zheng L, Hu Y, Yang T, Wang Z, Wang D, Jia L, Xie Y, Luo L, Qi W, Lv Y, Beeckman T (2024) A root cap-localized NAC transcription factor controls root halotropic response to salt stress in *Arabidopsis*. Nat Commun 15(1):2061

Zhou X, Sun Z, Huang Y, He D, Lu L, Wei M, Lin S, Luo W, Liao X, Jin S, Guo M (2025) WRKY45 positively regulates salinity and osmotic stress responses in *Arabidopsis*. Plant Physiol Biochem 219:109408

Zhu XF, Jiang T, Wang ZW, Lei GJ, Shi YZ, Li GX, Zheng SJ (2012) Gibberellic acid alleviates cadmium toxicity by reducing nitric oxide accumulation and expression of *IRT1* in *Arabidopsis thaliana*. J Hazard Mater 239–240:302–307

Zhu J, Zhang KX, Wang WS, Gong W, Liu WC, Chen HG, Xu HH, Lu YT (2015) Low temperature inhibits root growth by reducing auxin accumulation via *ARR1/12*. Plant Cell Physiol 56(4):727–736

Zhu Y, You Y, Zheng S, Li J, Wang Y, Wu R, Fang Z, Liu H, Du S (2024) ABA-importing transporter (AIT1) synergies enhances exogenous ABA minimize heavy metals accumulations in *Arabidopsis*. J Hazard Mater 473:134718

Zolla G, Heimer YM, Barak S (2010) Mild salinity stimulates a stress-induced morphogenic response in *Arabidopsis thaliana* roots. J Exp Bot 61(1):211–224

Chapter 4
Understanding Intricacies of Stamen Development Pathways Holds Key to Trait Manipulation for Hybrid Seed Production

Ekta Pokhriyal, Richa Shukla, and Sandip Das

Abstract Food and nutritional security will remain a major challenge in coming years because of climate change and growing population. Dual objectives to achieve zero-hunger and sustainable food production practices will require innovative crop improvement strategies and agricultural practices. Heterosis breeding and hybrid seed production programmes rely on male sterile phenotype which forces allogamy or outcrossing and prevents self-pollination. Male sterile phenotype manifests due to a range of genetic, environmental, and developmental factors. Until now, the most widely used male sterile system is based on nuclear-cytoplasmic incompatibility (CMS system) that has been successfully deployed in crops for heterosis breeding. The cytoplasmic male sterile system has limitations in wide applicability and has been deployed only in a handful of crops because of the following major reasons (i) poor translation of knowledge from model plants to crop species, (ii) variability in stamen structures across species, and, (iii) diversity in mating models and out--crossing strategies in plants. A detailed understanding of stamen development pathway including ontogeny and regulatory network across a variety of crops is critical to devise innovative strategies for promoting outcrossing for heterosis breeding. For example, molecular genetic regulation of spatio-temporal control and crosstalks in tapetum development, endothecial wall thickening, pollen wall patterning, filament development and elongation, water and nutrient transport need to be understood. The knowledge and information gained can be deployed not only to induce sterility but also develop male fertile restorer lines. The power of such approaches, either in isolation, or in combination, has not been fully explored in crop species. This lack of information and knowledge opens a rich field of investigative pursuits. The knowledge gained can be deployed to devise novel heterosis breeding and crop improvement strategies by harnessing power of technologies such as precise genome editing, single-cell genomics, and molecular breeding.

Keywords Stamen · Anther · Filament · Tapetum · Male sterility · Hybrid seed

E. Pokhriyal · R. Shukla · S. Das (✉)
Department of Botany, University of Delhi, Delhi, India
e-mail: sdas@botany.du.ac.in

A. Singh et al. (eds.), *Sustainable Crop Production*, Sustainability Sciences in Asia and Africa, https://doi.org/10.1007/978-981-95-7844-3_4

Editor's Notes

In this chapter, the authors urge research scholars to explore novel and unconventional strategies for producing hybrid seeds. Whereas breeders develop hybrid seeds by exploiting hybrid vigor, seed industry and corporations are driven by commercial interests to develop improved varieties that can no longer be propagated. Generation and use of male sterile lines is a popular approach to favor out-crossing, necessary for hybrid seed production. Hitherto, there has been an overemphasis on exploiting nucleo-cytoplasmic incompatibility to engineer male sterility. The authors posit deployment of alternative strategies that lead to remodeled stamen morphology. Gross or subtle modifications in stamen phenotype can manifest as male sterility. Altered stamen phenotypes can also obviate the requirement to generate male sterile lines altogether. To this end, authors highlight the urgent need to decipher molecular mechanisms governing stamen development, and to identify candidate genes that can be targeted to generate useful stamen phenotypes. Attention of upcoming researchers is being drawn toward examining the potential of following strategies.

Actionable Points

1. Manipulation of filament development and elongation is a novel strategy to favor outcrossing and hybrid seed production while obviating the need to generate male sterile lines.
2. Manipulating genes responsible for tapetum-derived deposits will hinder exine--intine patterning and pollen wall development. This will result in aberrant pollen grains leading to male sterility, a prerequisite to hybrid seed production.
3. Anther dehiscence and pollen release is dependent on differential hydration of anther wall. Genes regulating water transport are required to be characterized to impair anther dehiscence. This will naturally promote outcrossing and hybrid seed production while obviating the requirement to generate male sterile lines.

4.1 Introduction

The United Nations estimates that the world population will reach from 8.2 billion in 2024 to about 10.3 billion by the 2080s. Goals of food and nutrition security with zero-hunger will thus remain a serious challenge (https://population.un.org/wpp/Publications/; https://desapublications.un.org/file/20622/download). A World Food Programme (WFP) Hunger-Map analysis reveals large areas around the world prone to experiencing high to very levels of food and nutrition deficiency (https://hungermap.wfp.org/; https://www.fao.org/interactive/state-of-food-security-nutrition/2--1-1/en/). An additional yet overarching problem facing humanity is that of climate change threatening the delicate environmental balance. Rapidly increasing population along with changing environmental conditions pose a serious threat to sustainable developmental goals (SDG) of meeting food production targets in a sustainable manner (Lobell and Gourdji 2012). This unmet demand for food, vegetable, and horticultural crops coupled with either stagnant or decrease in area under cultivation

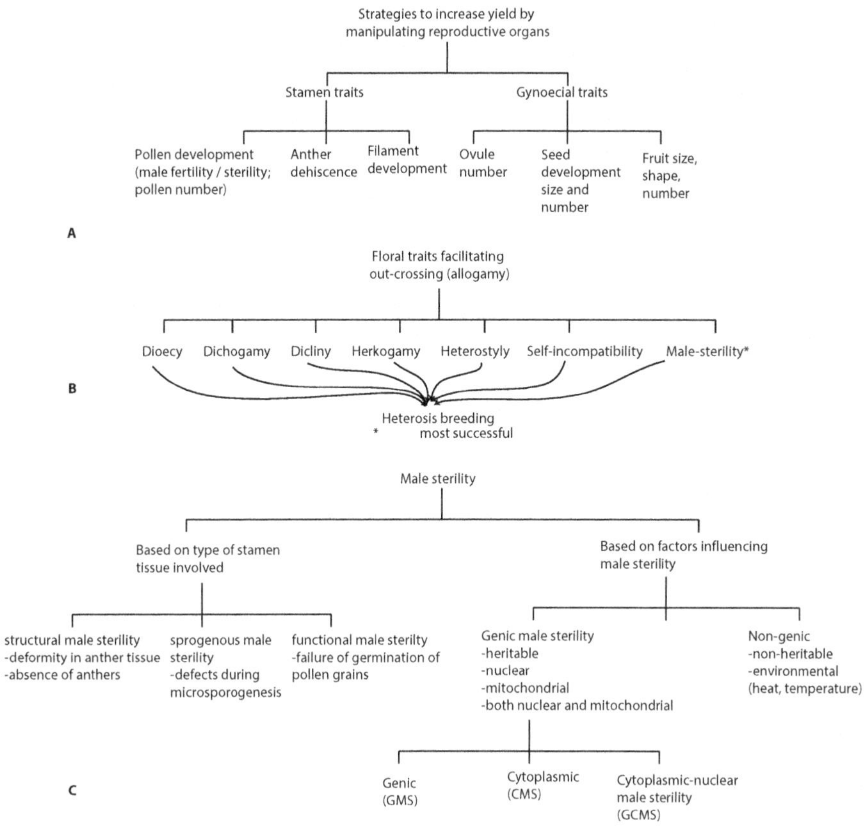

Fig. 4.1 Flowcharts depicting (**a**) strategies to increase yield by manipulating reproductive organs; (**b**) floral traits facilitating allogamy that can be exploited for hybrid seed production; and (**c**) classification of male sterility

requires systematic efforts to develop crop varieties with increased yield and improved traits.

A number of developmental and adaptive features are targeted to improve overall crop yield and performance; for instance, productivity can be improved by targeting reproductive as well as vegetative traits such as plant architecture, leaf area, plant vasculature, root system architecture (RSA), flowering time, fruit and seed size and number, biotic and abiotic stress tolerant and climate resilient varieties, microbiome, rhizobiome engineering, to name a few (Kissoudis et al. 2016). One such example of improvement in crop productivity via modification of plant architecture is development of semi-dwarf varieties in rice and wheat. Reduction of plant height was achieved by targeting GA biosynthesis and perception, resulting in decreased chances of lodging thereby improving overall crop productivity (Mathan et al. 2016). Studies have also suggested that increase in leaf area is associated with improvement in the photosynthetic efficiency of plants and thus targeting genes that improve the leaf area can also lead to increase in overall plant productivity (Meyer and Purugganan 2013). Factors such as plant vasculature also influence plant productivity by improving overall mechanical strength of plant and by serving as channels for transport of photosynthates and mineral across the plant (Mathan et al. 2016).

Besides targeting vegetative traits, crop improvement strategies include manipulating and modifying reproductive characters such as stamen and gynoecium development. Prior knowledge and understanding pollen development and male sterility, self-compatibility, ovule/seed number, seed size, seed developmental programmes, and fruit development, to name a few, are key to their successful manipulation (Fig. 4.1a). The challenge is to understand the multitude of reproductive strategies and developmental programs that exists in nature. This chapter aims to provide an overview of developmental events of stamen which allow identifying key regulators that can be or have been utilised to induce male sterility or promote outcrossing for heterosis breeding toward crop improvement and yield increase.

4.2 Heterosis

Nearly 150 years ago, Charles Darwin recorded an observation in the book *The Effects of Cross- and Self-Fertilization in the Vegetable Kingdom* that progenies obtained from cross-pollination or out-breeding exhibit higher vigour in a variety of traits than those obtained through self-pollination or inbreeding in a range of plant species, with a specific example of plant height in maize (Darwin 1876; also see Kolreuter 1766 cited in Jones 1917). Forgotten for more than a century, hybrid vigour was rediscovered and reported concurrently by East (1908) and Shull (1908); the concept of "heterosis" was proposed by Shull in 1908–1909, and the term "heterosis" was coined in 1914 (Shull 1908, 1909a, b, 1914); a closely related term "heterozygosis" meaning "stimulus of hybridity" was coined by Dr. Spillman (see Shull 1914). Heterosis thus refers to the increase in/elevated/superior performance in trait parameters of hybrids that are heterozygous in nature when compared to their parents which are maintained as homozygous lines (Hochholdinger and Hoecker 2007).

The concept of heterosis breeding depends on reproductive strategies which favour allogamy (cross-pollination, outcrossing; Fig. 4.1b). Continuous inbreeding or selfing for several generations in plants leads to inbreeding depression and yield plateau (Charlesworth and Willis 2009). Various reproductive strategies exist naturally in the flowering plants to favour and achieve cross-pollination or allogamy. These include dichogamy, dicliny, herkogamy, heterostyly, male sterility, and self--incompatibility (Boucher et al. 2024). Breeders have attempted to exploit one or a combination of pollination control systems to induce or enforce reproductive strategies listed above for heterosis. Among these, male sterility has proved to the most powerful and successful strategy for heterosis breeding in crop plants as was demonstrated for hybrid seed production in *Sorghum* (Stephens 1937).

4.2.1 Basis of Heterosis

Heterosis or hybrid vigour has been observed at multiple stages of life cycle and is not limited only to yield-related trait. Effects of hybrid vigour have been observed in maize embryos, as early as 6-days post-fertilisation that is associated with differential gene expression pattern of regulatory genes involved in signal transduction, transcription, and cellular transport and transport mechanism, to name a few categories (Meyer et al. 2007). Similarly, positive effects of heterosis have also been observed in several seed-related traits such as weight, volume, length, plumule length and dry weight, and radicle dry weight in maize (Gutiérrez-Hernández et al. 2023); during root development (Hoecker et al. 2006); enhanced photosynthetic efficiency (Fujimoto et al. 2012; Yamauchi and Yoshida 1985); and yield (Longin et al. 2012).

Heterosis is evaluated using two metrices: mid-parent heterosis (MPH) and best-parent heterosis (BPH). As the terms suggest, MPH refers to the trait value of the hybrid (F1) compared to the average trait value of the inbred lines of the two parents (P1, P2); BPH, on the other hand, refers to the trait value of F1 when compared to the trait value of the best parent (P_b). Three hypotheses – dominance, over-dominance, and epistasis hypothesis which deal with allelic and non-allelic interactions at multiple loci derived from the two parents – have been used to explain genetic basis of heterosis (Bruce 1910; Jones 1917; Powers 1945; Birchler et al. 2006; Hochholdinger and Hoecker 2007). However, none of these three hypotheses can explain the genetic basis of heterosis when applied in a stand-alone manner. A contemporary challenge is to integrate the genetic basis with the underlying molecular basis of heterosis. The fact that heterosis can be observed at nearly all stages of life cycle indicates its highly dynamic and complex nature.

4.2.2 Strategies of Heterosis Breeding

The primary strategy toward successful heterosis is to promote outbreeding while ensuring seed set. Ensuring seed set excludes the possibility of manipulating or modifying gynoecial traits or processes such as female-meiosis, whereas no such barrier exists for androecium. Up until now, inducing male sterility has been the primary strategy toward ensuring out-breeding or "failure to self-fertilize" leading to hybrid vigour.

4.2.3 Male Sterility

Male sterility in plants is the inability to produce viable and functional male gametes and forms an important strategy to promote allogamy. Any defect/s in gene regulatory network and pathways starting from establishment of stamen primordia till the differentiation and release of the functional viable pollen grains can result in male sterility.

Structurally, the male gametophyte or the pollen grain is relatively superficially positioned, enclosed with the outer anther wall, and an inner tapetal layer. This is in contrast to the female gametophyte which is deep-seated and buried inside several protective cell layers such as ovary wall, outer and inner integument of ovule; the female gamete-egg cell itself is localised inside the embryo sac. This offers the female gametophyte and the gamete additional layers of protection from environmental perturbations and this difference could account for common occurrence of male sterility as compared to rare events of female sterility across plant kingdom.

Male sterility can be classified into (Fig. 4.1c) the following:

A. Three types depending on the type of stamen tissue involved and responsible for sterility:

   (i) Structural male sterility that arises either due to complete absence of anthers or due to deformity in anther tissue
   (ii) Sporogenous male sterility that arises because of defects arising during microporogenesis
   (iii) Functional male sterility that is characterised by failure of germination of mature pollen grains

B. Two types based on factors influencing the male sterile phenotype:

   (i) Genic male sterility that arises either due to defects in nuclear gene or mitochondrial gene or both. Male sterility can occur spontaneously or can be induced by mutations, protoplasmic fusion, or genetic engineering. These can be further divided into the following:

      (a) Nuclear (genic) male sterility (GMS)
      (b) Cytoplasmic male sterility (CMS)
      (c) Cytoplasmic-nuclear male sterility (GCMS)

   (ii) Non-genic male sterility such as those caused by environmental factors (heat, temperature), and are non-heritable in nature

Natural male sterility occurs due to multiple factors; the primary means of inducing male sterility for crop breeding and hybrid vigour has been through manipulating the nuclear-cytoplasmic incompatibility (i.e. GMS, CMS, and GCMS). In order to deploy male sterility trait for hybrid breeding and seed production, female fertility must remain intact.

The following section provides brief summaries in the advances in understanding and deploying the three types-GMS, CMS, and GCMS toward hybrid seed production.

4.2.3.1 Nuclear (Genic) Male Sterility (GMS)

GMS is governed by nuclear genes and is not dependent on cytoplasmic factors. Normally GMS is controlled by recessive nuclear genes, designated as *ms*, and they follow Mendelian inheritance pattern. Any given species can have multiple, non-allelic *ms* genes that are responsible for male sterility, and these are nomenclatured as *ms1*, *ms2*, *ms3*, and so on. For example, 13 nuclear loci were identified in maize associated with male sterility and were annotated as *ms1-ms2, ms5-ms14*, and *ms17* (Albertsen and Philips 1981); similarly, 18 GMS loci were identified in *O. sativa* (*ms46-ms63;* Hu and Rutger 1992). In contrast to control of GMS by recessive loci, GMS due to dominant alleles (DGMS) has been reported in *Brassica oleracea* var. capitata (cabbage; Han et al. 2023), *B. oleracea* var. botrytis (Ruffio-Chable et al. 1993), *B. napus* (Li et al. 2022), and *Gossypium barbadense* (cotton; Cheng et al. 2020), to cite a few examples. Maintenance of nuclear male sterility involves crossing male sterile plants with homozygous recessive (*ms/ms*) or dominant (*Ms/Ms*) alleles with their heterozygous (*Ms/ms*) counterparts. However, maintenance of male sterility governed by the dominant alleles poses a challenge (Gautam et al. 2023).

4.2.3.2 Cytoplasmic Male Sterility

CMS is governed by defective mitochondrial DNA leading to nuclear-cytoplasmic incompatibility. The lack of fertile pollen grains in CMS plants is due to the presence of non-restoring recessive (*msms*) alleles in the nucleus. CMS can be maintained by crossing the CMS lines having defective cytoplasm denoted by "S" with the maintainer lines carrying fertile or normal cytoplasm designated as "F" or "N" and recessive nuclear alleles. This type of male sterility is widely used in propagation of horticultural crops where seeds are not of economic importance. CMS is maternally inherited as the mitochondrial genome is contributed by the female parent (Bohra et al. 2016).

4.2.3.3 Cytoplasmic Nuclear/Genic Male Sterility (CGMS)

Cytoplasmic nuclear male sterility is also governed by the interaction of cytoplasmic and nuclear genes. However, the CGMS differs from CMS in their fertility restoration mechanism. CGMS is maintained by dominant nuclear fertility restoring genes of the maintainer lines. This type of male sterility is commercially exploited in hybrid breeding programmes for a number of field crops. CGMS is often influenced by a number of environmental factors like photoperiod, temperature, etc. The defective mitochondrial genome results in reduction of respiration or production of toxic proteins due to chimeric mitochondrial ORFs in the cytoplasm thereby causing male sterility in CGMS system (Bohra et al. 2016).

The maintenance of CGMS involves the use of three distinct genotypes:

- The male sterile line with defective cytoplasm (S) and recessive nuclear alleles (*frfr*) is used as a female parent. This line is referred to as "A".
- The maintainer line has fertile cytoplasm (F) and recessive nuclear genes (*frfr*). This line is the B line and when crossed with A line gives male sterile progeny.
- The restorer line has dominant fertility restoring alleles in the nucleus (*FrFr*).

PPR proteins are an important class of plant proteins which are involved in restoration of fertility in plants. PPRs are mostly localised in the organelles like mitochondria and chloroplast and are involved in post-transcriptional organellar gene expression by binding to RNA in sequence-specific manner (Delannoy et al. 2007; Barkan and Small 2014). P-type PPR genes act as Rf genes and are involved in fertility restoration of male gametophyte in CMS plants (Yasumoto et al. 2009). Various studies conducted on Rf genes have identified a number of proteins involved in fertility restoration including aldehyde dehydrogenase (Rf2 in maize) and glycine-rich protein (GRP) (Rf2 in rice) (Fujii and Toriyama 2009; Matsuhira et al. 2012).

4.2.3.4 Environment-Sensitive Male Sterility

The expression of male sterility and fertility is also controlled by various environmental conditions. Kaul (1988) reported that environmental conditions like temperature, humidity, and photoperiod can regulate expression of male sterility genes. This type of sterility system eliminates the need of maintainer line in the breeding programme. Shi (1981) reported the presence of environment-sensitive male sterile lines in rice which was later on exploited by Yuan (1987) in hybrid breeding programme.

Hybrid breeding programmes until now have relied heavily on exploiting the CMS mechanism based on cytoplasmic-nuclear incompatibilities. The enormous potential of understanding and leveraging stamen development pathway has not yet been fully tapped.

4.3 Why Stamen Development?

Stamen is the site for male-gametogenesis and thus crucial for sexual reproduction (Wiese et al. 2024). Stamen development involves several discrete yet overlapping events regulated at multiple levels and governed by complex gene regulatory networks (GRNs). Each of the GRNs are also interconnected via regulatory proteins, regulatory RNAs, epigenetic modifiers, hormones, lipids, sugars, and other biomolecules. Stamen developmental events involve diploid somatic cells and haploid gametic cells, with exchange of genetic and biochemical information. For instance, the stages of male meiosis and production of functional mature haploid pollen grains are dependent on information and biochemical resources being contributed by the surrounding diploid, somatic cells. Accurate completion of developmental events thus is key to plant survival, and knowledge gained can be exploited to manipulate plant fertilisation as a strategy for crop improvement programs (see Table 4.1).

4.3.1 Establishment of Primordia

ABCE model of flower development defines the genetic framework of floral organ primordia identity in *Arabidopsis thaliana* and *Antirrhinum majus* (Coen and Meyerowitz 1991; Bowman et al. 1989, 1991; Bowman and Moyroud 2024). Based on this model, stamen identity in *Arabidopsis thaliana* is established by products of B-class (*APETALA3-AP3* and *PISTILLATA-PI*) and C-class (*AGAMOUS-AG)*, through interaction with E-class gene (*SEPALLATA-SEP*). In *Antirrhinum majus*, B-class genes *DEFICIENS* (*DEF*) and *GLOBOSA* (*GLO*), and C-class gene (PLENA) interact to establish stamen identity (Scott et al. 2004; Fig. 4.2). In maize, *STERILE TASSEL SILKY EAR1* and *MALE-STERILE SILKY* belong to B-class and *ZEA AGAMOUS1*, and *ZEA MAYS MADS2* belongs to C-class which regulates the establishment of stamen primordia (Schmidt et al. 1993; Mena et al. 1996; Ambrose et al. 2000; Bartlett et al. 2015; Marchant and Walbot 2022). The *BEARDED EAR* gene in maize is similar to *SEP* gene and thus is a putative E-class gene (Thompson et al. 2009).

The molecular basis of B-class (*AP3, PI* in *A. thaliana*; *DEF, GLO* in *A. majus*), C- (*AG, PLENA*), and E-class (*SEP*) genes for stamen primordia was established when protein-protein interactions were analysed. Comparative analysis revealed that both in *Arabidopsis* and in *Antirrhinum*, formation of heterodimers of AP3/PI, and DEF/GLO are essential for interacting with the CArG Box motif present in downstream target DNA—CC(A/T)$_6$CG. The heterodimers of AP3/PI also interact with the homodimer of AG/AG only in the presence of SEP3 protein. Thus, AP3/PI-AG/AG-SEP (BCE) proteins form a quaternary complex which is essential for establishment of stamen primordia. The genetic basis of role of B- and C-class genes was established through analysis of loss-of-function mutants where *ap3* and *pi* mutants form carpel instead of stamen and *ag* mutant forms petals instead of stamen. Loss of both Class-B and Class-C genes results in sepal; floral organs in *sep1sep2sep3* triple mutant also resemble sepal. Thus, mutations in any one or all of the class B-, C,- or E-class genes result in the loss of establishment of stamen primordia (Theißen 2001; Riechmann et al. 1996; Honma and Goto 2001). Auxin also plays significant roles not only in establishment of stamen primordia but during the entire developmental programme (Staldal et al. 2012; Cardarelli and Cecchetti 2014).

Table 4.1 Use of CRISPR-Cas9 technology-based genome editing to generate male sterile lines in different crop species

Plant system	Target gene	Function	Mutant	References
O. sativa	Os*CSA* (*carbon starved anther*)	*CSA* encodes R2R3 MYB transcription regulator which is involved in mediating sugar partitioning	Photosensitive male-sterile line, in the two-line hybrid rice system	Li et al. (2016)
O. sativa	Os*TMS5*	Temperature-sensitive male sterile line	Temperature--sensitive male sterile line	Barman et al. (2019)
T. aestivum	Ta*MS1* (*male sterile1*)	*MS1* encodes a glycosyl phosphatidyl inositol--anchored lipid transfer protein (LTP) and plays an important role in pollen exine development	Complete male sterile lines	Okada et al. (2019)
S. lycopersicum	*SlSTR1*	STR1 is involved in pollen development and proper exine formation	Complete male sterile lines	Du et al. (2020)
Ryegrass	*LpDMC1* (*disrupted meiotic cDNA1*)	*DMC1, a meiotic recombinase, mutations of which results in* disordered meiosis with univalents and multivalents appearing at diakinesis	Complete male sterile lines	Zhang et al. (2020)
O. sativa	*OsROS1* (*repressor of silencing 1*)	*ROS1* encodes a bi-functional DNA demethylase that removes 5-methylcytosine and nicks double-stranded DNA and plays an important role in development of gametophyte	Os*ROS1* knockout mutants have defective and aborted pollens and defects in embryo sac.	Xu et al. (2020)
Z. mays	Zm*MS8* (*male sterility 8*)	Zm*MS8* encodes a putative b-1,3,-galactosyltransferase, which plays an important role in meiotic stages of anther development	Transgene-free genome-edited *Z. mays*	Chen et al. (2018)
Cotton	*GhA12G0337* (*GhAMS11*)	*GhAMS11* is a key transcription factor, expressed during S4-S6 stages of anther development. It plays an important role in tapetum function and aids in pollen formation	*GhAMS11* knockout lines result in pollen inviability, due to defective pollen exine formation	Hao et al. (2025)

(continued)

Table 4.1 (continued)

Plant system	Target gene	Function	Mutant	References
Z. mays	Zm*DRP1*	*DRP1* is a desiccation--related gene which encodes a pcC13–62 protein. DRP1 belongs to a desiccation-related protein family, which is involved in providing dessication tolerance	Complete male sterility is observed in homozygous mutant lines	Han et al. (2022)
G. max	*GmAMS1* (Aborted microspores 1)	*GmAMS1* regulates degradation of tapetum and helps in formation of pollens	Complete male sterile lines	Chen et al. (2021)
S. lycopersicum	Sl*MS10* (*Solyc02g079810_ms1035*)	*SlMS10* is a homologue to *AtDYT1* and *OsUDT1* and is involved in regulation of tapetum PCD and meiosis	Complete male sterile lines	Jung et al. (2020)
S. lycopersicum	*SlAMS* (*aborted microspores*)	*SlAMS* encodes for a bHLH TF which is involved in pollen development	Complete male sterile lines	Bao et al. (2022)
S. lycopersicum	Sl*PHD_ms1* (*Solyc04g008420*)	*SlPHD_ms1*, a PHD-type TF, plays an important role in tapetum development and pollen formation	Complete male sterile lines	Gökdemir et al. (2022)
Cucumis melo	eIF4E	eIF4E is a cap-binding protein	Complete male sterile lines	Pechar et al. (2022)
Citrullus lanatus L. (watermelon)	*ClATM1* (*Abnormal Tapetum 1*)	Cl*ATM1* encodes a basic helix-loop-helix (bHLH) transcription factor which expresses in tapetum and in microsporocytes during stages 6–8 of anther development	Complete male sterile lines	Zhang et al. (2021a, b)
Z. mays	Zm*MS8* (*male sterile8*)	ZmMS8 encodes a putative β-1,3--galactosyltransferase. It regulates cell division, expansion, and differentiation during early *Z. mays* anther development	Complete male sterile lines	Chen et al. (2018)
O. sativa	Os*HXK5* (*hexokinase 5*)	Os*HXK5* is a hexokinase which plays an important role in starch biosynthesis and utilisation during pollen maturation	Complete male sterile lines	Lee et al. (2020)

(continued)

Table 4.1 (continued)

Plant system	Target gene	Function	Mutant	References
Z. mays	ZmABCG26	ZmABCG26, encodes a key transporter protein which is involved in pollen exine formation by transferring lipid precursors and polyketides	Complete Male Sterile lines	Jiang et al. (2021)
Z. mays	ZmFAR1	ZmFAR1 is involved in lipid metabolism	Complete Male Sterile lines	Jiang et al. (2021)
T. aestivum	*TaNP1*	*TaNP1* encodes a putative glucose-methanol-choline oxidoreductase, which is involved in tapetum degeneration and pollen exine formation	Triple homozygous mutations in *TaNP1* genes result in complete male sterility	Li et al. (2020)
Z. mays	Zm*Ms45* (*male sterile45*)	Ms*45* encodes a strictosidine synthase-like enzyme, which plays a critical role in pollen wall formation resulting and microspore development.	Tams45-abd results in complete male sterility due to defects at uninucleate stage in triple homozygous mutants	Singh et al. (2018)
Cotton	Gh*GPAT12* and Gh*GPAT25*	Gh*GPAT12/25* are sn-2 glycerol-3-phosphate acyltransferases involved in transfer of the acyl group of palmitoyl-CoA to glycerol-3-phosphate (*G3P*), which plays a key role in regulation of degradation of tapetum, anther cuticle formation, and pollen exine development	Gh*GPAT12/25* knockout results in complete male sterile phenotype	Zhang et al. (2021a, b)
Z. mays	*ZmGAMYB* (*Zm00001d012544*) and *ZmGAMYB2* (*Zm00001d043131*)	Zm*GAMYB* plays an important role in pollen grain development as mutations of Zmgamyb results in complete male sterility with shrunken and brown anthers and no pollen grain	*ZmGAMYB/2 knockout mutant results in complete male sterility*	Li et al. (2021)

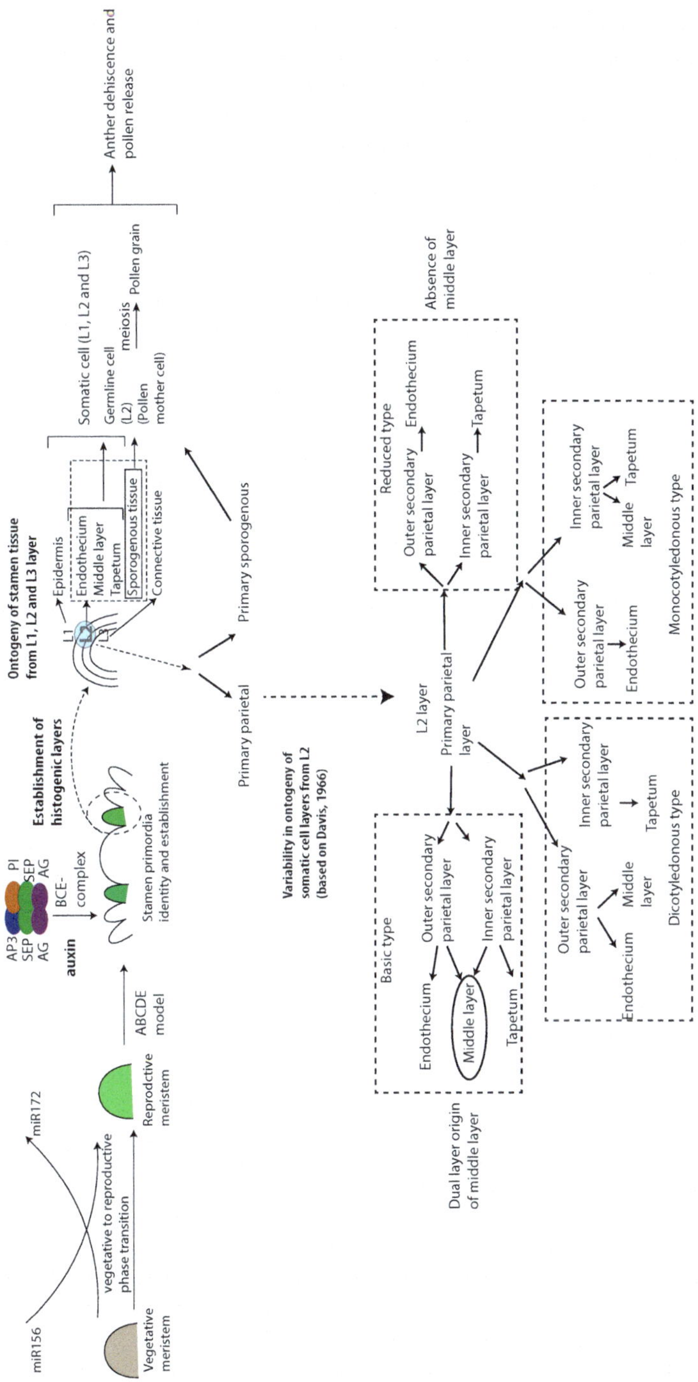

Fig. 4.2 A schematic representation of establishment of primordia, histogenic layers, and ontogeny of stamen tissue

4.3.2 Diversity in Stamen Structure

Structurally, the stamen consists of two distinct components – the male gametophyte bearing anther, and the stalk or filament. The anther which is the site of microsporogenesis typically contains two (bi-) or four (tetra-) sporangia, also termed as pollen sac. A constricted region between the two sporangia is termed as stomium and acts as dehiscence zone facilitating pollen release. A pair of the sporangia is located within a theca or lobe. The tetra-sporangiate condition is most commonly encountered in angiosperms whereas bi-sporangiate condition being less common (Davis 1966). A pair of sporangia within a theca usually merge or fuse during late stage of anther development and dehiscence to form a locule. Structurally, each sporangium comprises of outermost layer of epidermis, followed by endothecium, middle layer, tapetum, and the innermost group of archesporial cells that eventually give rise to pollen grains. The anther wall layers are of sporophytic origin (Gomez et al. 2015).

The filament (i) serves as means of attachment, (ii) elongates to bring anther to level of stigmatic surface (albeit see didynamous and tetradynamous condition), (iii) contains vascular system for nutrient and water transport, and (iv) provides support to anthers (Wilson et al. 2011). The filament consists of outer epidermal cell layer and vasculature at the core.

Stamen can be variously classified based on the following:

(a) Orientation of dehiscence zone relative to the gynoecium as (i) introse (facing the gynoecium or centre of the flower) and (ii) extrose (facing away from gynoecium or facing toward the petals).
(b) Attachment of anther lobe to filament (laminar or sessile type and filamentous type). The laminar or sessile type bears sporangia typically on the adaxial surface. The filamentous type can be further divided into (i) basifixed, (ii) adnate, (iii) dorsifixed, and (iv) versatile. The filament can also extend as connective and stamens are further classified based on connective tissue as (i) discrete, (ii) diverticate or divergent, (iii) distractile, and (iv) appendiculate type.
(c) Fusion properties of filament (i) free or polyandrous stamen and (ii) fused filaments. Filament fusion has been further classified as (i) monadelphous, (ii) diadelphous, and (iii) polyaldelphous.
(d) Fusion properties of anther (i) syngenesious (fused anther; filaments are free) and (ii) synandrous (anther and filaments are both fused).
(e) Stamen attachment and origin properties such as (i) epiphyllous, (ii) epipetalous, and (iii) gynandrous (gynostegium and gynostemium types).
(f) Stamen arrangement: Classified based on (i) number of whorls (one whorl or haplostemonous; two whorls or diplostemonous; multi- or polystemanous); (ii) arrangement in relation to sepals and petals (antisepalous, opposite to sepal, and antipetalous, opposite to petals); (iii) combination of number of whorls and arrangement as obdiplostemanous when stamens are in two whorls with each whorl in antisepalous (inner) antipetalous (outer) configuration.
(g) Functionality: fertile and sterile (staminodes).

(h) Anther dehiscence patterns as (i) longitudinal, (ii) transverse, (iii) apical or poricidal, and (iv) valvular.

4.3.3 Stamen Development

The male reproductive organ developmental programme is a tightly coordinated event, divided into several distinct pre-meiotic and post-meiotic steps. Developmental pathway of stamen, which consists of the gamete-bearing anther borne on the stalk or filament is relatively well studied in model plant *A. thaliana* and rice. In *A. thaliana* flower development process beginning with visible appearance of floral buttress and ends until stigma extends beyond the long stamens. This has been divided into 15 stages (Smyth et al. 1990). The development of stamen primordia initiates at floral stage 5 which has been further divided into 14 stages (Smyth et al. 1990; Sanders et al. 1999). Stamen development has been divided into 15 stages in *Triticum aestivum* (Browne et al. 2018), 10 stages in *Brachypodium* (Sharma et al. 2015), and 9 stages in rice (Raghavan 1988; Zhang and Wilson 2009).

Events in development of microsporangia include formation of archesporial cell, establishment of adaxial-abaxial polarity where in the two adaxial locules (facing away from the petals) are smaller than the two abaxial locules, the specification of cell types, and the formation of the radially symmetrical microsporangia.

Ontogeny of four layers of anther wall and the sporogenous tissue: In *Arabidopsis*, the floral meristem histologically consists of three cell layers—L1, L2, and L3. Ontogenetically, cells from L1 give rise to epidermal layer of anther wall. Periclinal division of cells in the L2 layer is involved in the establishment of stamen primordia; most of the subepidermal cell layer including the sporogenous layer originates from L2 layer. The L3 layer gives rise to vascular tissue and connective tissues which are located at the core region of the stamen (Jenik and Irish 2000; Scott et al. 2004). The process of somatic and germline differentiation and patterning involves communication across multiple cell layers. Much of cellular and tissue layers undergo programmed cell death (PCD) during or after meiosis (post-meiotic stages; Taylor et al. 1998). The cell layers of anther at post-meiotic dehiscence stage therefore vary from that of the pre-meiotic stage.

Ontogeny of the five distinct cell or tissue layers at the pre-meiotic stage and the connective tissue are (Fig. 4.2):

(a) Epidermis (arising from L1 layer)
(b) Endothecium (arising from L2 layer)
(c) Middle layer (arising from L2 layer)
(d) Tapetum (arising from L2 layer)
(e) Sporogenous tissue (arising from L2 layer)
(f) Connective tissue (arising from L3 layer)

A single cell termed as archesporial cell in the L2 layer undergoes periclinal division to give rise to primary parietal cell (PPC) (close to epidermis) and a primary sporogenous cell (PSC). PSC divides to form the sporophtyic cell layer. PPC undergoes further periclinal divisions to give rise to endothecial cell (toward epidermis) and secondary parietal cell (SPC). The two outermost cell layers were previously termed as exothecium and endothecium by Purkinje (1830; cited in Keijzer 1999). The SPC undergoes another round of periclinal division and gives rise to middle layer and tapetum. The PSC gives rise to microspore mother cell (MMC) or PMC which undergoes meiosis to form microspores (Canales et al. 2002; Scott et al. 2004). The archesporial cells in L2 layer thus are responsible for origin of somatic cells (endothecium, middle layer, tapetum) as well as germ line cells (pollen mother cells). In maize, development of different anther cell layers arises due to asymmetric division of cells derived from L2 layer. The first asymmetric division leads to generation of endothecium and secondary parietal cell. Another round of asymmetric division of SPC gives rise to middle layer and tapetum. The sporogenous cell layer is derived from the centralised L2 cells (Fig. 4.2; Kelliher and Walbot 2011; Marchant and Walbot 2022).

Davis (1966) described four types of anther wall development based on the role of secondary parietal cell layers (Fig. 4.2). Basic (type I) in which the one SPC layer gives rise to endothecium, and middle layer, and the second SPC layer gives rise to middle layer and tapetum. In type II or dicotyledonous type, periclinal division of the outer SPC layer gives rise to endothecium and middle layer, while the inner SPC gives rise only to tapetum (Carrizo-Garcia 2002). In type III or monocotyledonous type, the outer SPC layer gives rise to only endothecium, whereas the inner SPC layer forms middle layer and tapetum; this type of anther cell wall is characteristic of members of Poaceae (Teng et al. 2005). In the type IV or reduced type of development, the primary parietal layer (PPC) directly gives rise to endothecium and tapetum (Davis 1966). A few plant families have more than one type of anther wall, e.g. both type I and II are present in Solanaceae; type I and III class of anther wall is seen in Commelinaceae (Hardy and Stevenson 2000).

In *Arabidopsis thaliana*, the TCP-class II transcription factor *NOZZLE/SPOROCYTELESS (NZZ/SPL)* is induced by the MADS box transcription factor *AGAMOUS* which leads to specification of archesporial cell thus initiating anther development (Schiefthaler et al. 1999; Yang et al. 1999; Ito et al. 2004; Fig. 4.3). In *nzz/spl* mutants, the primary parietal layer is not established because of the absence of periclinal cell division in the archesporial cell layer. The lack of primary parietal layer leads to the absence of PMC and pollen sacs, and only homogenous mass of subepidermal cell is observed. *SPL/NZZ* is considered a key regulator of male (as well as female) sporogenesis and controls stamen identity as *spl/nzl* mutants form archesporial cells, but these cells fail to undergo meiosis (Schiefthaler et al. 1999; Yang et al. 1999; Liu et al. 2009). *SPL/NZZ* has been shown to encode a novel nuclear localised, adaptor-like transcriptional repressor and interact with TOPLESS (TPL) via the EAR-motif (ethylene-responsive element-binding factor-associated amphiphilic repression) present at the C-terminus, and with TCP proteins (Chen et al. 2014). SPL/NZZ regulates and promotes stamen identity by interacting with AP2 (B-class), AG (C-class), and SEP3 (E-class) products (Liu et al. 2009). The transcription of *SPL/NZZ* is directly regulated by AG, epigenetically by SET

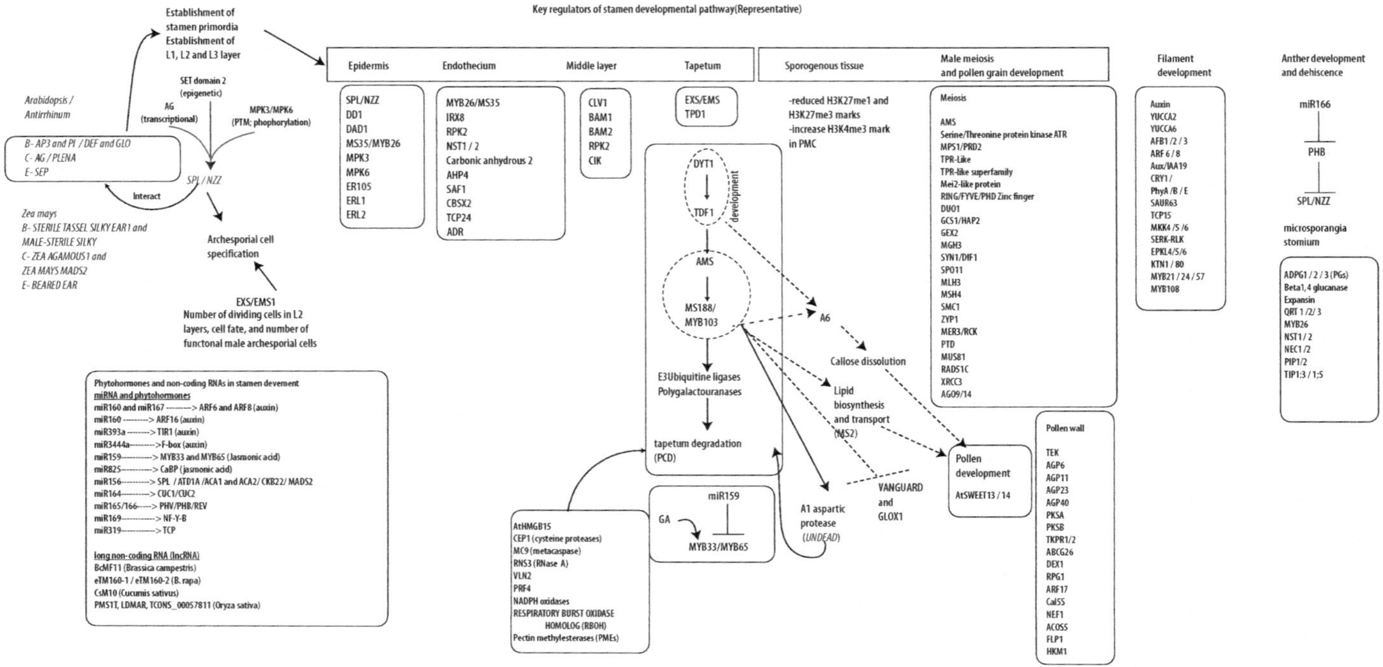

Fig. 4.3 Brief summary of molecular regulators controlling aspects of stamen development pathways

Domain group 2, and that of SPL/NZZ proteins by MAPK3/MAPK6-mediated phosphorylation (Berr et al. 2010; Zhao et al. 2017).

AG expression occurs from early in floral initiation until late in flower development. Early expression of *AG* helps in specification of stamen and carpels (Ito et al. 2004; Yanofsky et al. 1990); however, the expression at later stages is involved in anther morphogenesis and dehiscence, and filament formation and elongation (Ito et al. 2007a, b). AG regulates stamen development by controlling the induction of *Defective in Anther Dehiscence 1*, which encodes a chloroplastic phospholipase A1 that catalyses the initial step in jasmonic acid (JA) biosynthesis (Ito et al. 2007a, b).

Whereas NZZ/SPL decides the archesporial cell specification, *EXS (EXTRA SPOROGENOUS CELLS)/EMS1 (EXCESS MICROSPOROCYTES1)*, which encodes a plasma-membrane bound putative serine/threonine Leu-rich repeat receptor-like kinase (LRR-RLK), was identified as a regulator that determines cell fate and controls the number of dividing cells in the L2 layer which in turn controls number of functional male archesporial cell (Canales et al. 2002; Zhao et al. 2002). In the four-lobed anther of wild-type *A. thaliana*, four archesporial initial cells give rise to primary sporogenous cells (PSC) and anther wall layer through series of periclinal division. The sporogenous cells then divide meiotically to form microspores. The L2 layer in *ems1/exs* mutant undergoes increased number of cell division forming large number of sporogenous cells which also undergo meiosis to form tetrad which subsequently degenerates resulting in the absence of pollen grains; this suggests that EXS/EMS is not a meiotic regulator (Canales et al. 2002). The mutants do not form tapetum and middle layer, because the secondary parietal cells are not allowed to develop normally by the excess sporogenous mass of cells and endothecium and degenerate (Canales et al. 2002). The degeneration of pollen grain is thus an outcome of lack of tapetal layer. A combinatorial interaction between *SPOROCYTELESS/NOZZLE*, miR166, and *PHABULOSA* (*PHB*) controls the inner boundary between inner and outer microsporangia in *A. thaliana* which is important to ensure male fertility (Li et al. 2019b).

4.3.3.1 Anther Wall Epidermis Development

The outermost cell layer of the anther termed epidermis originates from the L1 layer, plays a protective role, and contains specialised cells in the stomium region that facilitate anther dehiscence and pollen release. Being the outermost cell layer with protective function, no mutant or genes have yet been identified that specifically regulates epidermis development in anther wall. It is most likely that such defects lead to complete abortion of the developmental pathway. Indirect evidences however link *DD1* (*DELAYED DEHISCENCE1*; Sanders et al. 2000), *DAD1* (*DEFECTIVE IN ANTHER DEHISCENCE1*; Ishiguro et al. 2001), *MS35* (*MYB26/ MALE STERILE 35*; Yang et al. 2007a, b), *mitogen-activated protein kinases* (*MPK3, MPK6*) and *ERECTA and related receptor-like kinases* (*ER105, ERL1. ERL2*; Hord et al. 2008), and *SPL/NZZ* (Wijeratne et al. 2007) to be involved in epidermal wall development and function. In maize, a number of mutants with defects in anther wall development have been reported including *ms23* and *ms32* in which tapetum does not form (Chaubal et al. 2003).

4.3.3.2 Endothecium

The cellular layer immediately anterior to the anther epidermis is primarily responsible for dehiscence and release of pollen grains. Ontogenetically, endothecium arises out of the secondary (outer) parietal layer in basic, dicotyledonous, and monocotyledonous type, except in the reduced type of anther wall where it arises from primary (inner) parietal layer (Davis 1966). In *A. thaliana*, endothecium is established between stage 4 and 5 and persists until anther dehiscence between stages 13 and 14. The endothecium develops secondary thickening (secondary cell wall-SCW) at stage 11 of anther development (Wilson et al. 2011).

As mentioned earlier, Purkinje (1830) coined the terms exothecium (now known as epidermis) and endothecium and correlated "evaporation" in the endothecium layer with anther dehiscence (Keijzer 1999). Previously, "fibrillar structures" were observed in hypodermal layer (corresponding to endothecium) by Meyen (1828) and Von Mohl (1830) (Cited in Keijzer 1999).

Anther dehiscence involves lytic opening of stomium and retraction of the anther wall which causes release of pollens (Dawson et al. 1999). Deposition of the cellulose and lignin in the endothecium during microspore maturation is critical for successful release of mature pollen grains; failure of lignification of endothecial layer results in occurrence of male sterility.

Molecular regulators affecting endothecium development often exhibit defects in the characteristic secondary thickening and anther dehiscence. For instance, *myb26*, *nst1*, and *nst2* loss-of-function mutants exhibit defects in secondary cell wall thickening of ligno-cellulose and as a result affect anther dehiscence (Yang et al. 2007a, b; Mitsuda et al. 2005). In *ms35/myb26* mutant, the lack of secondary thickening in the endothecium results in the failure of opening of anther. The absence of wall thickening results in flattening and distortion of the endothecial cells, thereby preventing the cells to generate the force which is required for outward bending of anther wall (Dawson et al. 1999; Yang et al. 2007a, b). NST1 and NST2 play a key role in endothecium wall thickening by regulating *MYB26* (Mitsuda et al. 2005). Similarly, defective secondary cell wall thickening and anther dehiscence are also observed when genes such as *IRREGULAR XYLEM (IRX; GAUT12/IRX8)* and *receptor-like protein kinase 2 (RPK2)* are mis-expressed (Brown et al. 2005; Mizuno et al. 2007; Hao et al. 2014). A TCP class of transcription factor TCP24 has been shown to be a negative regulator of SCW thickening as downregulation of *TCP24* led to an increase in lignification (Wang et al. 2015). Another gene-*ANTHER DEHISCENCE REPRESSOR* (*ADR*) has been shown to regulate *NST1, NST2* expression, and ROS accumulation to control endothecial wall thickening, and overexpression of *ADR* alters the fine balance of NST1/NST2, ROS levels. The study also showed post-translational modification of ADR (N-myristoylation), and targeting to peroxisome is necessary for its role in SCW thickening, endothecium formation, and stomium rupture (Dai et al. 2019).

Detailed analysis of male-sterile phenotype of *eceriferum* (*cer*) mutants –*cer2* and *cer2-like2* (*cer2l2*) revealed that the endothecial cell layer is the source of very long-chain fatty acid (VLCFA) that deposited on the pollen coat and essential for hydration in *A. thaliana* (Zhan et al. 2018). Functional analysis of *ZmMs33/ZmGPAT6* (*Zea mays male sterile 33/glycerol-3-phosphate acyltransferase 6*) demonstrated that the endothecium chloroplasts are maintained by lipids synthesised in tapetal cell layer which in turn is responsible for proper pollen development in *Z. mays* (Zhu et al. 2020).

Endothecium is also considered to play a dominant site for photosynthesis and starch synthesis in the anther (Marchant and Walbot 2022). In maize, the sub-epidermal endothecial layer, viz. the layer of endothecium closer to epidermis is the primary site of photosynthesis due to occurrence of more number of chloroplasts as compared to inter-endothecium, the layer adjacent to the connective tissue (Murphy et al. 2015; Zhu et al. 2020). Studies conducted on tapetum-specific *ms23* mutant in maize showed involvement of tapetum in controlling the production of chloroplasts in endothecium and in controlling the function of endothecium (Nan et al. 2022).

4.3.3.3 Middle Layer

The middle layer, as the name suggests, is located between the endothecium on the outer side and tapetum on the inner side. This cell layer is short-lived and exists between stage 4—when it is first established, and by stage 7, only remnants are present which coincide with termination of male meiosis and formation of pollen tetrads (Goldberg et al. 1993; Sanders et al. 1999; Walbot and Egger 2016). However, independent reports using transmission electron microscopy have also suggested that middle layer persists until uninucleate microspore, late in development (Owen and Makaroff 1995; Quilichini et al. 2014a, b). Ontogenetically, middle layer has been observed to initiate from inner secondary parietal cells (Feng and Dickinson 2010) as well as outer secondary parietal cells (Zhao et al. 2002), as in *ems1* (*excess microsporocytes 1*) mutant. The middle layer, irrespective of the layer from which it originates, degenerates by stage 6–7 of anther development.

As the middle layer is formed early and is extremely short-lived, molecular regulators controlling middle layer development are expressed during early, pre-meiotic stages. For instance, mutants of *CLAVATA1-related Leu-rich repeat receptor-like kinases* (*CLV1*-related *LRR-RLKs*)—*BAM1 and BAM2* (*BARELY ANY MERISTEM 1/2*) lack endothecium, middle layer, and tapetal cell layers. The *bam1bam2* mutants are defective in cell-to-cell communication and cell-fate decision. As a result, in *bam1bam2* double mutant, cells that are present on the inner side of epidermis transition into pollen mother-like cells (PMC), instead of endothecium, middle layer, and tapetum. However, these pollen mother-like cells (PMC) also degenerate without completing the meiotic process signifying the importance of somatic cells and cell-to-cell communication (Hord et al. 2006). Another LRR-RLK,

RECEPTOR-LIKE PROTEIN KINASE2 (RPK2), has also been shown to be involved in early anther development with the *rpk2* mutant completely lacking the middle layer due to lack of differentiation from inner secondary parietal cells (Mizuno et al. 2007). Similarly, genetic and molecular analyses of *bam1bam2* and *rpk2* mutants revealed that *CLAVATA3 INSENSITIVE RECEPTOR KINASEs (CIK)* acts as a co-receptor for both *BAM1/BAM2* and *RPK2* (Cui et al. 2018).

4.3.3.4 Tapetum

Development of functional pollen grain is dependent on formation of the tapetum—the innermost somatic cell layer (Wei and Ma 2023; Zhang and Yang 2014). Tapetum surrounds the reproductive tissue and acts as source of (i) nutrients for pollen development; (ii) components (lipids, sugars, enzymes) for pollen wall, exine/intine patterning; and (iii) enzymes for dissolution of callose wall. Analysis of mutants in *A. thaliana* allowed identification of regulators that control differentiation and development of tapetal cell layer. *EMS/EXS1* controls cell fate in association with *TAPETAL DETERMINANT 1 (TPD1)*. TPD1, a small protein of 176 amino acids in *A. thaliana*, has been shown to interact with EMS1/EXS under in vivo and in vitro conditions and also induces phosphorylation of EMS1/EXS. In *tpd1* mutants, instead of tapetal cells, microsporocytes are differentiated from tapetal cell initials (Yang et al. 2003). Segregation analysis of sterility and fertility using reciprocal crosses using wild-type and *tpd1* mutant established their clear role in tapetum development and not in either male or female gametophyte development. This also outlines the independent developmental processes that govern microsporocytes and tapetum. The molecular regulators that form part of this bifurcation can be explored for manipulating anther developmental pathway for crop improvement (Yang et al. 2003).

The formation of functional pollen grain depends not only on tapetum development but also on apoptosis/programmed cell death leading to degradation of tapetal cells. This balancing act of tapetal cell development, tapetal degradation, and pollen grain development is mediated via a network consisting of *DYSFUNCTIONAL TAPETUM1 (DYT1)*, *DEFECTIVE in TAPETAL DEVELOPMENT and FUNCTION1 (TDF1)*, *ABORTED MICROSPORES (AMS)*, and *AtMYB103/MS188* in *A. thaliana*. Two of these encode basic helix-loop-helix (bHLH; *DYT1, AMS1*), and two encode R2R3-MYBs (TDF1; *AtMYB103/MS188*) proteins. Studies conducted on the mutants of suggest that *EMS1/EXS*, *SERK1/SERK2*, and *TPD1* genes act at earlier stages than *DYT1*. DYT1 and TDF1 are responsible for early stages in tapetal development, controlling cell differentiation and endomitosis, whereas AMS and MS1 are responsible for late-stage controlling apoptosis/PCD and degeneration process (Zhu et al. 2008; Li et al. 2017). The gene encoding ARID-HMG DNA-binding protein, *AtHMGB15*, has also been shown to regulate tapetal apoptosis; additionally, the AtHMGB15 protein also controls cytoskeleton dynamics of actin distribution and f-actin concentration. AtHMGB15 controls tapetal PCD via *CEP1 (cysteine protease)*, *MC9 (metacaspase)*, and *RNS3* (*ribonuclease A family member 3*) genes and affects pollen germination through modulating actin dynamics via *VLN2* (*villin protein 2*) and *PRF4* (*Profilin 4*) genes (Zhang et al. 2014; Biswas and Choudhary 2024).

In a reconstructed genetic pathway of tapetum development, MS188 acts downstream of AMS which in turn acts downstream of TDF1 which is downstream of DYT1. Comparative analysis of laser-capture microdissection of tapetal cells from *dyt1*, *ams1*, *tdf1*, and *myb103/ms1* mutants allowed identification of downstream targets specific to DYT1 (238), TDF1 (226), and AMS (289); a total of 536 genes that were downregulated in all the four mutants are considered as common targets of *DYT1*, *AMS1*, *TDF1*, and *MYB103/MS1* (Li et al. 2017). DYT1-specific downstream genes include *bHLH089* and *bHLH010* that interact with DYT1 (itself encoding a bHLH protein), regulate anther development, and act as positive feedback regulatory mechanism for DYT1 (Zhu et al. 2015; Cui et al. 2016; Li et al. 2017). This interaction between bHLH089, bHLH010, bHLH091, and DYT1 has been mapped to the conserved C-terminal localised BIF domain (bHLH protein interaction and function) of DYT1 and thus is a critical determinant of dimerisation and nuclear localisation to regulate anther development at stage 6 (Cui et al. 2016).

Genes that are enriched in *tdf1* mutant include glutaredoxins (GRX), NADPH oxidases RESPIRATORY BURST OXIDASE HOMOLOG (RBOH), peroxidases, metacaspases, cysteine proteases, and pectin methylesterases (PMEs) that are likely to be involved in degradation and PCD of tapetal cells (Xing and Zachgo 2008; Xing et al. 2011; Xie et al. 2014). Genes enriched in *ams* mutants include type III lipid transfer proteins (LTPs) and indicate initiation of pollen wall development (Li et al. 2017). A total of 536 genes, such as pectate lyases, polygalacturonases (PGs), xyloglucan endo-transglucosylases, arabinogalactan proteins (AGPs), fasciclin-like AGPs (FLAs), cytochrome P450, members of zinc finger and LOB domain-containing protein (LBD) family transcription factors, receptor kinases, calcium-related proteins, G-proteins, rapid alkalisation factor (RALF), etc., were identified as common downstream targets of *DYT1*, *AMS1*, *TDF1*, and *MYB103/MS1* as these were downregulated in all the four mutants, namely, *dyt1*, *ams1*, *tdf1*, as well as *myb103/ms1*. Comparative analysis of differentially expressed genes among *dyt1*, *tdf1*, *ams*, and *ms188* also revealed their independent roles in modulating tapetum development (Li et al. 2017).

Leucine-rich repeat receptor-like protein kinases (*BARELY ANY MERISTEM*-BAM1, BAM2), somatic embryogenesis receptor-like kinase 1 (SERK1, SERK2), ER family of LRR-RLKs (ER, ERL1, ERL2), and MPK3-6 play important roles in specification of anther cell layers (Hord et al. 2006). Male sterility due to the absence of tapetum and increased sporocytes has been reported in *serk1-2* double mutant, *ems1/exs*, and *tpd1* mutants. BAM1 and BAM2 act redundantly and play important roles in maintenance of cell number in sporogenous layer and formation of endothecium, middle layer, and tapetum (Hord et al. 2006). In *bam1bam2* double mutant, occurrence of PMC-like cells is observed which soon degenerate resulting in male sterility. Comparative genomic studies have led to identification of orthologs of *BAM1/BAM2* in *Oryza sativa* and poplar (*Populus trichocarpa*) thereby suggesting a conservation in *BAM1/BAM2* function (Gomez et al. 2015).

Identification of orthologs of *EMS1* and *TPD1* in rice and maize has suggested a difference in the signalling mechanisms in various species. *In rice, multiple sporocyte1 (MSP1)*, a homolog of *EXS/EMS1*, is involved in regulation of female fertility in addition to male fertility. In rice, *OsTDL1/MIL2*, an ortholog *of AtTPD1*, plays a key role in differentiation of PPC as is indicated by the presence of two anther wall layers in *mil2* mutant (Hong et al. 2012). In maize *MULTIPLE ARCHESPORIAL CELLS1 (MAC1)* promotes periclinal division in sub-epidermal cells and suppresses the activity of archesporial cells. *MAC1* is a homolog of *AtTPD1* and *OsMIL2*, but it differs in its expression and localisation (Marchant and Walbot 2022).

After development of tapetal layer, a number of genes such as *DYSFUNCTIONAL TAPETUM1 (DYT1)*, *ABORTED MICROSPORE (AMS)*, and *MALE STERILITY 1 (MS1)* act together to regulate the function of tapetum and pollen development. *AtDYT1* is essential for normal tapetum development after the establishment of tapetal layer. The expression of At*DYT1* is first observed in the tapetum at stage 5 of anther development. *AtDYT1* belongs to basic helix loop helix family of transcription factor which is the second largest family of transcription factors (Toledo-Ortiz et al. 2003). In *dyt1* mutants though the pollen mother cells undergo meiosis, formation of thin callose walls after meiosis and defects in cytokinesis leads to male sterility. In stage 7–8 of wild-type anther, the end of meiosis leads to formation of tetrad and ultimately the microspores; however, in the *dyt1* mutant, defects in tapetum development were observed due to formation of highly vacuolated tapetum leading to degeneration of the meiocytes (Zhang et al. 2006).

Similar defects in tapetum development are observed in *OsUDT1*, an ortholog of *AtDYT1*. In *udt1* mutants, tapetum vacuoles becomes enlarged and microspores degenerate (Wilson and Zhang 2009). Although *OsUDT1* and *AtDYT1* belong to basic helix loop helix family, studies show that the expression pattern of *OsUDT1* is different than that of *AtDYT1* (Jung et al. 2005).

AMS belongs to MYC sub-family of transcription factors and contains basic helix-loop-helix domain. *AMS* gene expression starts from pre-meiotic stage of anther development but shows higher expression during post-meiotic stages (Sorensen et al. 2003). In anthers of *ams* mutants, meiotic division is normal with the formation of normal microspores; however, the microspores degenerate after formation, thereby causing male sterility. In mutant of rice *OsTDR*, an ortholog of *AtAMS1*, male sterility is caused by degeneration of microspores due to delayed tapetal degradation (Wilson and Zhang 2009). *DYT1* acts during the meiotic stages of pollen development, earlier to *AMS* and *MS1*. Further, interaction of many genes like *ATA20*, *ASHR3* (*Arabidopsis* SET domain) with *AMS* has been reported during tapetum development. Overexpression of *ASHR3* in *ams* resulted in reduction of male sterility, and *ASHR3* regulates stamen development by targeting AMS to chromatin (Thorstensen et al. 2008).

In *Arabidopsis*, *AMS* and *MS1* genes act during the post-meiotic stages of the tapetum development. Failure to produce viable pollen grains due to defects in tapetal secretion and tapetum programmed cell death (PCD) leads to male sterility in *ms1* (Vizcay-Barrena and Wilson 2006). Expression of *MS1* begins during callose degradation stage till free microspore stage. *AtMS1 (A. thaliana)*, and its homologs in rice *(OsPTC)*, and barley (*HvMS1*) regulate PCD in tapetum (Li et al. 2011; Fernandez Gomez and Wilson 2014). In *ms1*, instead of PCD, the tapetum breakdown follows necrotic pathway thereby leading to male sterility (Vizcay-Barrena and Wilson 2006). A number of genes of late tapetum development pathway and pollen wall deposition are targets of *MS1* as these genes are downregulated in *ms1* (Alves-Ferreira et al. 2007; Ito et al. 2007a; Yang et al. 2007a). One such target is *MYB99*; in *myb99*, the tapetum thickness decreases along with decrease in fertility. However, *myb99* mutant plants do not show complete sterility thereby indicating that *myb99* acts redundantly with other MYB transcription factors (Alves-Ferreira et al. 2007).

Transcription factors such as *MYB33* and *MYB65* also regulate the process of male fertility in temperature-dependent manner (Millar and Gubler 2005). Studies conducted on *myb33myb65* double mutants show defects in tapetum leading to pre-meiotic abortion of pollen thereby causing male sterility, and the sterility could be reversed under high light and low temperature conditions (Millar and Gubler 2005). *AtMYB33* and *AtMYB65* are GA-regulated MYBs with a close similarity to *HvGAMYB* gene. The *HvGAMYB* gene is involved in gibberellic acid-mediated starch mobilisation from aleurone layer of seed. Similar to *HvGAMYB*, *AtMYB33* and *AtMYB65* are involved in starch mobilisation through the tapetum layer. Similar to *Atmyb33myb65* double mutants, rice *gamyb* mutants exhibit failure of meiosis and tapetal hypertrophy (Kaneko et al. 2004). Such studies suggest a conserved expression and functional pattern of GA-MYBs in cereals and *Arabidopsis* (Millar and Gubler 2005).

MYB103/MS188 regulates tapetum development and function by causing downregulation of *VANGUARD1/VGD1* (pectin methylesterase) and *GLOX1* (glyoxal oxidase) and upregulation of A1 aspartic protease (*UNDEAD*). Mutation of *myb103* causes premature tapetum degradation leading to defects in exine deposition due to failure of release of oil bodies, vesicles, and plastids resulting in abortion of pollen grains proving the fact that *MYB103 (MYB80)* is important for development of tapetum (Phan et al. 2011; Higginson et al. 2003; Li et al. 2007). A1 aspartate protease exhibits a phenotype similar to *myb80* mutant. *OsMYB103*, the ortholog of *AtMYB103* in rice exhibits similar expression pattern (Zhang et al. 2010; Phan et al. 2011). *AtMYB103* regulates expression of MS2, a fatty acyl carrier protein reductase, which is also seen in rice where *OsMYB103* interacts with *OsDPW*, which regulates exine formation.

4.3.3.5 Gametogenesis

The most critical step in gametogenesis is timing and control of reductional division or meiosis. Male meiosis co-occurs with other events of overall stamen development. Successful male meiosis is essential to formation of fertile and viable pollen grain, eventual sexual reproduction, and crop yield (Masoomi-Aladizgeh et al. 2025). The process of gametogenesis involves origin of sporogenous tissue, specification of microspore mother cell (MMC), meiosis, mitosis, formation of pollen dyad/triad/tetrad, dissolution of the callose wall, and release of pollen grains coinciding with anther wall dehiscence.

As mentioned earlier, *NOZZLE/SPOROCYTELESS* is responsible for archesporial cell specification, and *EXS (EXTRA SPOROGENOUS CELLS)/EMS1 (EXCESS MICROSPOROCYTES1)* determines cell fate and controls the number of dividing cells in the L2 layer which in turn controls number of functional male archesporial cell in *A. thaliana* (Canales et al. 2002; Zhao et al. 2002).

The first step is transition of sporogenous somatic cells into pollen mother cells (PMC) that are committed to undergo meiosis. This transitory stage has been shown to be marked by chromatin reorganisation in *A. thaliana* including the presence of histone variant H2A.Z and eviction of linker histones in the PMC to mark chromatin decondensation, increase in nuclear volume, and decreased heterochromatin levels (Hou et al. 2022; She and Baroux 2015). The authors observed reduced H3K27me1 and H3K27me3 marks and increase in H3K4me3 in PMC compared to somatic cells; H3K4me2 marks remain unchanged. A similar fate was also observed in the megaspore mother cell suggesting a common mechanism of somatic to reproductive fate transition (She and Baroux 2015).

The tapetum-derived product *ABORTED MICROSPORES (AMS;* bHLH class of transcription factor) is a major regulator of meiosis progression; in *ams* mutants, even though meiosis initiates normally, progression is delayed after pachytene and eventually fails during late meiosis stages (Tidy et al. 2022). The *ams* mutants also exhibit defective radial microtubule array (RMA) at late stage which in turn disrupts tetrad formation and improper callose deposition and cell wall. Immunoprecipitation experiments showed that AMS protein binds to promoter regions of *Ataxia telangiectasia-mutated* and *RAD3-like protein (serine/threonine-protein kinase ATR; ATR), multipolar spindle 1/putative recombination initiation defects 2 (MPS1/PRD2), tetratricopeptide repeat-like (TPR-like), tetratricopeptide repeat (TPR)-like superfamily protein (T21.H19), MEI2-like protein 1 (ML1),* and *RING/FYVE/PHD zinc finger superfamily protein (At1g33420)*—genes that are known to be involved in meiosis (Tidy et al. 2022).

The meiotic process is highly complex and is conserved across organisms (Mercier et al. 2015). The complex and dynamic nature was uncovered when comparative transcriptome analysis identified nearly 20,000 genes, 800 of which were preferentially found in the male meiocytes. The conserved nature of meiosis was also revealed when orthologs of nearly 1000 *A. thaliana* transcripts were found in fission yeast (*Schizosaccharomyces pombe*) and mouse (Yang et al. 2011). The authors also identified transcripts arising from intronic regions and transposons, suggesting generation of intronic derived non-coding RNAs and/or mis-annotated exons (Yang et al. 2011). A similar study identified nearly 1000 genes that were deemed to be specific to meiosis; some of these genes include *SYN1/DIF1, AtSPO11-1, AtDMC1, AtRAD51, AtRAD51c, AtXRCC3*, ASY1, *AtMSH4*, SMC1, ZYP1, *AtMER3/RCK, SDS, AtDMC1, ASY1, AtMLH3, AtRAD51C, AtXRCC3, AtMSH4, AtMSH5, AtMER3/RCK, PTD*, and *AtMUS81*. The study also identified 55 mitochondrial origin genes and several transcripts from LTR-GYPSY, LTR-Copia, En-Spm, LINE, and MuDR transposons (Chen et al. 2010).

Several transcription factors regulate the functioning of germ cell; for example, *DUO1*, a MYB transcription factor, is involved in germ cell differentiation and fertilisation and regulates expression of 63 germ line-specific genes (Borg et al. 2011). *DUO1* directly regulates expression of three genes, namely, *GCS1/HAP2*, *GEX2*, and *MGH3*. In rice, three homologs of *Arabidopsis H3*, *TRT704*, *HRT11*, and *HRT12* have been identified, thereby establishing the conserved nature of the pathway in regulation of germ cell differentiation (Russell et al. 2012).

Genes involved in meiosis in *A. thaliana* include *AGO9/AGO14*, *MEI1*, *XRI1*, *SPO11-1* and *SPO11-2*, *MRE11*, *SYN1/DIF1/REC8*, *JASON*, and *ASK1,* to cite handful of examples (Mercier et al. 2015). Translation of JASON (JAS) protein is initiated using alternate initiation signal, and the shorter form of protein is produced during meiosis (Cabout et al. 2017). Several non-coding RNAs such as phasi-RNAs are also involved in meiotic regulation such as population of 24-nt phasi-RNA regulated by a cascade of bHLH transcription factors (MS23, MS32, bHLH122, and bHLH51) (Nan et al. 2022). Several studies have now shown that epigenetic reprogramming plays a vital and critical role in plant reproduction including in germline specification (Kuo et al. 2025; Hou et al. 2022; Ono and Kinoshita 2021; Lei and Liu 2020; Gehring 2019; Han et al. 2019; Mercier et al. 2015; Gutierrez-Marcos and Dickinson 2012).

Pollen development is characterised by unequal mitotic division in the microspore to produce one each of generative cell and vegetative cells. Subsequent division and developmental events lead formation of a tricellular pollen grain with two sperm cells and one vegetative cell. The absence of unequal division, as was observed in the *SIDECAR POLLEN* (*SDP*) mutant, causes abnormal pollen development including production of extra vegetative cells (Chen and McCormick 1996). Other regulators include microtubule-associated protein MAP215 (GEMINI POLLEN 1; GEM1), and TWO-IN-ONE are also know to control unequal cell division and pollen grain development (Twell et al. 2002).

4.3.3.6 Pollen Wall

In *A. thaliana*, the pollen wall starts to develop at stage 5 and on completion coincides with the tricellular pollen at stage 12 stage. The A typical pollen wall is composed of outer exine and inner intine, and components of the pollen wall are largely contributed by the tapetal layer.

The exine layer has been subdivided into outer sculptured sexine having tectum and baculae, and inner nexine. The nexine layer is composed primarily of glycoproteins; the *TRANSPOSABLE ELEMENT SILENCING VIA AT-HOOK (TEK)* gene which is part of the *AHL (AT-HOOK NUCLEAR LOCALIZED)* gene family is a major regulator of nexine formation (Edlund et al. 2004). The TEK protein is found to be expressed in the tapetum (Lou et al. 2014), and TEK protein has been shown to directly regulate expression of four Arabinogalactan protein genes *AGP6, AGP11, AGP23*, and *AGP40* most likely through interaction and binding to promoters of AGP to contribute to nexine development (Jia et al. 2015).

The principal component of exine, sporopollenin, is a complex polymer made of phenyl propanoids, oxygenated aromatic rings, and long-chain fatty acids which are synthesised in tapetum (Li et al. 2019a; Ma et al. 2021). Deposition of tryphine and pollenkit helps in interaction of pollen with the stigmatic surface and is also thought to be involved in self-incompatibility reactions. A number of genes involved exine patterning and sporopollenin biosynthesis have been identified which regulates male fertility (Wang et al. 2018; Ma et al. 2021). The biosynthesis of sporopollenin involves the polyketide pathway and involves the enzymes polyketide synthase A (PKSA), polyketide synthase B (PKSB), tetraketide α-pyrone reductase1 (TKPR1), and tetraketide α-pyrone reductase2 (TKPR2). These enzymes have been localised to the endoplasmic reticulum of tapetal cells and shown to use fatty acid derivatives as precursors for the polyketide pathway (Lallemand et al. 2013). Sporopollenin biosynthesis and transport is directly regulated by the action of AMS via MS188 (biosynthesis) and ABCG26 (transport).

Below the nexine layer is the intine layer composed of pectin, hemicellulose, and cellulose. Intine is divided into outer exintine and inner endintine (Ellis et al. 2010). The intine layer is composed of cellulose and is similar to "normal plant cell". The development of the intine is primarily regulated by the pollen grain itself (Xu et al. 2020).

The microspore plasma membrane plays a critical role in exine patterning as reported in *dex1* mutants (Paxson-Sowders et al. 2001). DEX1 expresses at stage 9, is a plasma membrane-localised protein which is essential for sporopollenin secretion and pollen wall formation, and is highly conserved in plant species such as barley, *Brachypodium*, and rice (Ma et al. 2013;Yu et al. 2016). In *defective in exine patterning (dex1)* mutant, male sterility due to pollen degeneration is reported. The pollen degeneration takes place due to aberrant primexine deposition because of defective plasma membrane undulations which mark the sites for sporopollenin deposition. As a result of defective plasma membrane undulations, sporopollenin deposition and anchoring fails. In *dex1* mutant, reduced expression of CalS5 (callose synthase) gene is observed (Ma et al. 2013).

The *RUPTURED POLLEN GRAIN1 (RPG1)* is a plasma membrane-localised protein belonging to MtN3/saliva family proteins. *Rpg1* mutant shows defects in exine pattern formation due to defects in tapetal and microspore plasma membrane development. Aberrant sporopollenin deposition resulting in microspore degeneration results in reduced fertility in *rpg1* mutant. The presence of phosphorylation sites in RPG1 protein suggests interaction of RPG1 protein with other membrane proteins thereby influencing primexine development (Guan et al. 2008).

The deposition of callose wall by microsporocytes begins during the initiation of meiosis. The microspores surrounded by the callose wall begin to develop primexine layer which later on helps in deposition of sporopollenin and exine formation. Tapetum is actively involved in the synthesis and breakdown of callose layer and development of pollen wall. In *A. thaliana*, 12 genes encoding for callose synthase have been so far identified. In *cals5*, the male fertility is affected because of failure of deposition of callose wall around the microspores during microsporogenesis (Dong et al. 2005). Callose wall deposition also affects the formation of sexine on the pollen surface thereby affecting fertility. Both premature or delays in callose deposition and breakdown result in male sterility. Expression of callase secretion is critical for maintenance of pollen fertility, and maximum activity of callase is observed during microspore release.

The callose wall acts as a regulator to movement of substances between the sporophytic tissue and the microspores. An extracellular matrix (ECM) that develops between the callose wall and the microspore plasma membrane is termed as primexine. This layer is believed to be a crucial regulator of exine formation as primexine is deposited during the tetrad stage of pollen development after the callose wall is deposited and disappears along with callose breakdown (Wang et al. 2021). Primexine is composed of acidic polysaccharides, cellulosic components such as microfibrils and proteins. Primexine deposition also controls pollen viability (Ma et al. 2021). *ARF17* is involved in primexine deposition, and positively regulates expression of Callose synthase (CalS5; Dong et al. 2005; Yang et al. 2013b); *arf17* exhibits male sterility due to defective callose layer, absence of primexine layer, and pollen degeneration (Yang et al. 2013b; Dong et al. 2005).

Lipid transfer proteins also play an important role in maintenance of male fertility. For instance, defects in transfer or polymerisation of sporopollenin and excessive tryphine deposition in *flp1* (*faceless pollen-1*) mutant cause male sterility. In *flp1* mutant, the pollen wall is smooth and lacks reticulate pattern. *FLP1* shows conditional male sterility as the mutants can be rescued by increasing humidity (Ariizumi et al. 2003).

NO EXINE FORMATION1 (NEF1) is a critical gene which plays an important role during early exine formation. Pollen degeneration in *nef1* mutant is due to formation of sporopollenin aggregates resulting in failure of sporopollenin deposition onto plasma membrane leading to failure of exine formation. However, normal primexine layer develops in the microspores (Ariizumi et al. 2004). *MS2, DEX1,* and *FLP1* act downstream of NEF1 during exine formation. NEF1 encodes for a membrane protein which helps in maintaining integrity of membrane in plastids (Ariizumi et al. 2003, 2004, 2005; Arizumi and Toriyama 2011).

Several genes such as *ACOS5*, *PKSA*, *PKSB*, *CYP703A2*, and *CYP704B1* are involved in sporopollenin biosynthesis pathway (Wang et al. 2003; de Azevedo Souza et al. 2009; Kim et al. 2010; Wallace et al. 2011; Yang et al. 2014). Phylogenetic analyses have shown the presence of homologs in other plants suggesting conservation in sporopollenin structure and biosynthesis pathways. For instance, the identification of homologs of *Arabidopsis PKSA* gene in *Physcomitrella patens* (*PpASCL*) suggests a conservation of pathway of sporopollenin in land plants (Colpitts et al. 2011). Similarly, *AtMS2* and its ortholog in rice, *OsDPW* gene, are involved in exine formation, and restoration of fertility in *ms2* mutant by *Osdpw* gene has been reported, indicating a conservation in sporopollenin synthesis across monocots and dicots (Shi et al. 2011). Identification of ortholog of *Arabidopsis CYP704B1* in rice, *OsCYP704B2*, also establishes conservation in fatty acid metabolism pathway (Li et al. 2010).

The mature pollen is released at the bicellular or tricellular stage of pollen. In *Arabidopsis*, Singh et al. (2008) have reported that the sperm cells are metabolically active cells and transcriptome analysis of sperm cells reveal the presence of genes involved in cell signalling, cell wall metabolism, cytoskeleton development, and genes involved in pollen-pistil interaction. Although pollen development in higher plants seems to be a highly conserved process, however reports suggest occurrence of some species-specific genes is also involved in the pollen development pathway. In rice and *Arabidopsis*, many shared transcription factors have been identified. In *Arabidopsis* five pollen-specific MIKC-MADS box proteins involved in pollen maturation have been identified (Liu et al. 2013). Similarly in rice S-(*OsMADS62, OsMADS63*) and P-(*OsMADS68*), MIKC-MADS box proteins have been reported (Verelst et al. 2007a, b).

4.3.3.7 Late Stages of Pollen Development

The breakdown of tapetum after the release of microspores marks the initiation of the late stages of pollen development. The disintegration of the tapetum follows the mechanisms of programmed cell death which includes chromosome condensation, swelling of endoplasmic reticulum, shrinkage of cytoplasm, and separation of cytoplasm from the cell wall (Rogers et al. 2005). The breakdown of tapetum is essential for release of pollen wall material and anther dehiscence. During the process of degradation of tapetum, sporopollenin is released into the anther locules which are used for formation of pollen wall. The tryphine and pollenkit produced by the tapetal cells which coats the mature pollen grains are also released at late stages. The process of development and degradation of tapetum is closely linked with the different stages of pollen development, and any process which disrupts this process leads to reduced fertility (Kaul 1988).

The timing of tapetal breakdown coincides with meiosis in the MMC; however, the timing of PCD varies between different plant species (Pacini 2010). Zhang et al. (2011) have reported that tapetal PCD begins at tetrad stage in Rice and *Brachypodium* and completes by the bicellular stage of pollen. In barley and wheat, the tapetum degradation begins during the vacuolated microspore stage (Mizelle et al. 1989; Fernandez Gomez and Wilson 2014).

In grasses, tapetum is associated with production of Ubisch bodies, which are electron dense structures and are commonly found in the inner tangential walls of tapetum cells (Heslop-Harrison 1968; Quilichini et al. 2014a, b). Studies conducted on the rice tapetum suggest the presence of Ubisch bodies that are associated with export of sporopollenin precursors to the locule; however, wheat tapetum is involved in secretion of pre-Ubisch bodies. Wang et al. (2003) have identified an anther specific protein, RAFTIN, in rice and wheat which is involved in late pollen development and is associated with accumulation of Ubisch bodies. In *Arabidopsis*, no homolog of *RAFTIN* gene is so far reported.

4.3.3.8 Anther Dehiscence and Pollen Release

Anther dehiscence is the final step in the developmental event leading to release of mature pollen grains. The dehiscence process involves the dissolution of the septum between the anther locules to form a single sporangium (in bilocular anthers) and the stomium or the dehiscence zone that is present as a constricted region in the epidermis between the sporangia. The process is facilitated primarily by endothecium, the sub-epidermal layer, and the filament which conducts the water necessary for differential contraction and hydration. Endothecium also interacts with both the middle layers and the tapetum in the final stages of anther and pollen development and dehiscence. The stomium or dehiscence zone has been classified based on the pattern and are termed as "longitudinal", "transverse", "poricidal", "poricidal-longitunidal", and "valvate"; the dehiscence process can be introse (toward the centre) or extrose (away from the centre) (Åstrand et al. 2021). Out of the four types, longitudinal dehiscence is the most common, and valvate type is rare.

The primary regulator of anther dehiscence is secondary thickening of endothecial wall comprising of lignin, cellulose, and callose. The secondary wall thickening is present in entire endothecial layer except in the stomium region that separates two adjacent locules within an anther lobe. The cells of endothecium are radially elongated and are characterised by presence of incomplete "fibrous bands", starting from inner/lower edge and ending at the outer edge, along the tangential wall. Species-specific variations have been observed in types of endothecial thickening, and four distinct classes have been recorded: unconnected and parallelly arranged radial rings (annular type); helical thickening that is formed along the periclinal axis (U-shaped or helical type); irregular, branched, and connected (reticulate type); and solid plate on inner periclinal wall (palmate type).

Cell wall-degrading enzymes catalysing hydrolysis of cell wall include members of polygalacturonase (PGs), β-1,4-glucanase, and expansin family of proteins which participate in septum dissolution and anther dehiscence. Among the polygalactouronases, *ARABIDOPSIS DEHISCENCE ZONE POLYGALACTURONASE1 (ADPG1), ADPG2* and *ADPG3* are involved (Ogawa et al. 2009); *QUARTET1 (QRT1)* which encodes a pectin methylesterase, and *QRT2 and QRT3 (encodes polygalacturonases) are required for separation of microspores from tetrads (Rhee and Somerville* 1998; Rhee et al. 2003). Three transcription factors, one MYB class (*MYB26*) and two NAC domain classes, *NAC Secondary Wall Promoting Factor 1 (NST1)* and *NST2,* are the major regulators of secondary wall thickening in endothecium; MYB26 is known to control expression of *NST1* and *NST2*. Together, these three regulate lignin biosynthesis and deposition in endothecial cell layers, and the process is mediated via auxin (Mitsuda et al. 2005; Yang et al. 2007a, b, 2017; Cecchetti et al. 2013).

A combination of factors that include enzymatic lysis of pollen tetrads and release of pollen grains, pollen hydration, ligno-cellulosic deposition in endothecial cells leading to expansion and dehydration, and enzymatic breakdown of the septum between the anther locules leads to differential or localised hydration/dehydration of anther locule and endothecial wall. The mechanical constraints because of the ligno-cellulosic deposition in the endothecial cell layer along the stomium/dehiscence zone are responsible for dehiscence (Bonner and Dickinson 1990; Wilson et al. 2011; Nelson et al. 2012). Molecular regulators that control differential, localised hydration and dehydration processes include elevated levels of (i) sucrose transporter in the connective tissue (AtSUC1; Stadler et al. 1999); (ii) *NEC1* and *NEC2* in filament and stomium (*Petunia NECTARY 1/2*; Ge et al. 2001); and (iii) aquaporins in anther (*PIP1, PIP2*; Bots et al. 2005a, b; *MIPa*; Ruiter et al. 1997; AtTIP1;3 and AtTIP5;1; Soto et al. 2008), to cite few examples.

Lignin-based endothecial thickening is not the only mechanism that is responsible for anther dehiscence. For instance, formation of fibrous resorption tissue (Hermann and Palser 2000); hypodermal and sub-hypodermal cells with thick walls (Marazzi et al. 2007); and differential dehydration of the dehiscence pore, and hydration of the surrounding cells because of cuticle (Cortez et al. 2014); and presence of specialised circular cell cluster intersporangial or hypodermal septum or oxalate package forming a "notch" region under the stomium (Iwano et al. 2004; Sanders et al. 2005) are also known.

4.3.3.9 Filament Development and Elongation

Filament consists of the epidermal layer and the vasculature essential for conducting water and nutrient for the anther tissue. The filament not only serves as means of attachment of the anther locule to the floral disc but is also instrumental in positioning the dehisced anthers to facilitate self-pollination or cross-pollination (Wilson et al. 2011). Rapid filament elongation is a characteristic feature just prior to anthesis and is observed between flower development stages 12–13 in *A. thaliana* because of cell elongation.

Auxin is the primary regulator of filament development and elongation. Analysis of members of YUCCA gene family – *TRYPTOPHAN AMINOTRANSFERASE OF ARABIDOPSIS (TAA) / YUCCA – (YUC2, YUC6)* and auxin receptor genes—auxin-signalling F-box (AFB) and transport inhibitor response 1 (TIR1)—AFB1, AFB2, AFB3, and TIR1 mutants, showed that biosynthesis of auxin in the anther tissue is necessary for pollen meiosis, development, and anther dehiscence, and transport of auxin is critical to filament elongation (Cecchetti et al. 2008). The study postulated that auxin biosynthesis genes (*YUC*) are transcriptionally active during stages 8–10 and decrease by stage 11 and are turned off at stage 13—the anthesis stage; in contrast, the auxin receptor genes *TIR1* and *AFB* are activated at stage 10 which coincides with beginning of pre-anthesis filament elongation. This study also demonstrates common genetic mechanisms in controlling filament elongation.

Several auxin response factors such as ARF8 control filament elongation by activating the expression of *Aux/IAA19* (*auxin/indole-3-acetic acid inducible 19*) which acts as repressor of auxin (Marzi et al. 2020). This ARF8-mediated filament elongation involves cryptochromes (CRY1 and CRY2) and phytochromes (PhyA, PhyB, and PhyE) mediated via COP1, HYH, and HY5 (Marzi et al. 2020). The blue and red lights are perceived in a sequential manner; during the early stages (10–11) when the bud is enclosed by sepals and petals, CRY1 and CRY2 regulate filament elongation, whereas at stages 12 onward when the sepals and petals start to open the red—and far-red light receptors—PhyA, PhyB, and PhyE act to control filament elongation (Marzi et al. 2020).

The developmental programme of filament elongation is tightly coordinated not only with pollen meiosis and mitosis but also with gynoecium and other floral organs. For instance, blocking the flow of auxin through *A. thaliana* and *N. tabacum* filaments not only leads to shorter cell size and stunted filaments but also defective pollen grains because of fault in mitosis; meiosis was found to proceed normally (Feng et al. 2006).

Filament development and elongation is an outcome of crosstalk between auxin and GA pathways. One of the most important regulators of filament development and elongation is members of the *small auxin-upregulated RNA (SAUR)* gene family members of the SAUR63 sub-clade. TCP15, a TCP class I member, directly interacts with promoter of *SAUR63* to regulate GA-dependent filament elongation (Gastaldi et al. 2020). An intricate network of cascade of map kinases (MKK4/MKK5 and MPK6), SERK-RLK (SOMATIC EMBRYOGENESIS RECEPTOR KINASE family RLK), and three peptides- EPFL4, EPFL5, and EPFL6 (EPIDERMAL PATTERNING FACTOR-LIKE) have been shown to control cell proliferation and filament elongation (He et al. 2023). Similar crosstalk between auxin and ethylene controlling filament elongation was suggested in *Gaillardia grandiflora* (Koning 1983).

Filament elongation is also regulated by the microtubule-severing enzyme KTN1 (katanin p60 subunit) which is negatively regulated by MYB21 (Wang et al. 2023b). The microtubule-severing by KTN1-KTN80 heterodimer complex has been linked to Brassinosteroid signalling (Wang et al. 2023a).

Other molecular regulators of filament development and elongation include MYC2, MYC3, MYC4, and MYC5 as the quadruple mutant myc2myc3myc4myc5 triple mutants exhibit delayed and slower filament elongation (Qi et al. 2015); TCP15 which is negatively regulated by KNAT1/BREVIPEDICELLUS to control elongation at late stage of stamen development (Gastaldi et al. 2023); interaction of DELLA proteins to repress the transcriptional activity of MYB21 and MYB24 (Huang et al. 2020); and network of *MYB21, MYB24,* and *MYB57* activity regulated by GA-mediated jasmonic acid biosynthesis (Cheng et al. 2009).

Figure 4.3 provides a broad summary of some of the regulators that regulate multiple aspects of stamen development.

4.3.4 Hormonal Control of Stamen Development

Stamen development is a tightly regulated process which involves interactions of various hormones, transcription factors, and miRNA. Various plant hormones regulate male fertility by controlling the timing of gene expression at various stages of stamen development.

Auxin plays an important role in regulation of expression of early and late anther and pollen development genes. Auxin biosynthesis in stamen is controlled by redundantly functional *flavin-containing monooxygenases* (*YUC2 and YUC6*) and *yuc2yuc6* double mutants are male sterile due to failure of stamen filament elongation and the absence of pollen grains (Cheng et al. 2006; Cecchetti et al. 2008). Loss of male fertility is also reported in *tir1afb1afb2afb3* quadruple mutant, defective in auxin signalling. Auxin-mediated response requires auxin response factors (ARFs) and AUX/IAA which represses the activity of ARF proteins. It has been reported that *ARF1* and *ARF2* are also involved in anther dehiscence as *arf1arf2* double mutant exhibit delayed anther dehiscence (Ellis et al. 2005). *AFR6* and *ARF8* act redundantly and are negative regulators of class-1 KNOX genes; altered expression of class 1 KNOX genes is observed *arf6arf8* double mutant. *ARF6* and *ARF8* also regulate JA biosynthesis by targeting DAD1 proteins and act by influencing JA biosynthesis or by inhibiting the breakdown of JA. The downstream targets of ARF6 and ARF8 are several small auxin up-regulated genes (*SAUR*) (Wu et al. 2006; Reed et al. 2018). The *ARF17-miR160* module is involved in endothecium lignification by regulating the expression of *MYB108*. *ARF17* overexpression mutants exhibit defects in development of tapetum leading to male sterility. *CALS5* plays an important role in pollen wall formation which is regulated by ARF17 (Yang et al. 2013b; Xu et al. 2019). Auxin also influences the process of anther dehiscence by controlling the process of endothecium lignification (Yang et al. 2007a, b; Cecchetti et al. 2008). Auxin negatively regulates the activity of *MYB26/MS35* via ARF8.4 which in turn controls endothecium lignification and anther dehiscence (Yang et al. 2007a, b; Cecchetti et al. 2008).

Gibberellic acid (GA) plays several important roles in the event of stamen development, maintenance of tapetum, and pollen functioning (Hu et al. 2008). GA regulates cell elongation, maintenance of pollen viability, and pollen tube elongation. Male sterility due to defects in stamen filament elongation has been reported in *ga1-3* mutant which can be reversed by exogenous application of GA. It has been reported that GA-DELLA proteins and AGAMOUS (AG) proteins regulate the expression of *DAD*1 gene which is involved in JA biosynthesis (Ito et al. 2007a, b; Peng 2009), indicating that GA regulates JA-mediated stamen development.

A complex and coordinated interplay of three phytohormones—auxin, JA, and GA—regulate late stamen development via MYB21, MYB24, and MYB57 (Song et al. 2013). Auxin bound to its receptors SCF-TIR1 is required for marking Aux/IAA for 26S proteasome-mediated degradation. In the absence of auxin-SCF-TIR1 complex, Aux/IAA interacts with ARF6/ARF8 and represses JA biosynthesis. The release of ARF6/ARF from Aux/IAA activates JA biosynthesis and late stamen development including filament elongation. In a similar manner, JA is perceived by COI1/SCF and together recruit JAZ for 26S proteasome-mediated degradation. In the absence of JA-COI1-SCF, JAZ interacts with MYB21/24/57 and represses transcription of downstream regulators. A parallel pathway of interaction between GA-GID1-DELLA-SCF is necessary to activate transcription of MYB21, MYB24, and MYB57 (Song et al. 2013). MYB21, MYB24, and MYB57 are functionally redundant proteins which regulate stamen filament elongation together with MYB108 (Mandaokar and Browse 2009).

The absence of meiosis and impaired anther development has been reported in mutants of rice soluble GA receptor—*GID1* (Aya et al. 2009). GID1-GA complex is known to interact with a DELLA repressor protein—*SLENDER RICE1 (SLR1)*—and degrades DELLA proteins which are essential for GA-mediated stamen development (Cao et al. 2006; Ueguchi-Tanaka et al. 2007).

JA regulates various developmental responses including role in the late stages of stamen development. Many JA biosynthetic and signalling genes regulate the process of stamen maturation. The first step in JA biosynthesis is catalysed by a phospholipase A1 encoded by *DEFECTIVE IN ANTHER DEHISCENCE1 (DAD1)* gene *A. thaliana*; *dad1* mutants exhibit male sterile phenotype due to failure of dehiscence of anthers and pollen maturation, delay in opening of floral bud (Ishiguro et al. 2001). JA biosynthesis mutants including *opr3*, *fad3fad7 fad8* triple mutant, and *lox3lox4* double mutant exhibit male sterility due to defects in JA biosynthesis. In *opr3* mutants, male sterility is attributed to failure of filament elongation, anther dehiscence, and reduced pollen viability, whereas endothecium lignification is not affected. Male fertility in JA biosynthesis mutant can be restored by application of methyl jasmonates on the flower buds. However, restoration of male fertility is a stage-specific event, and fertility can only be restored when exogenous JA is supplied to flower buds in the middle of stage 12 (Stintzi and Browse 2000; Park et al. 2002); application of JA fails to restore fertility if JA is applied at any stage earlier or later than stage 12 (Stintzi and Browse 2000).

MYC2 transcription factor belonging to the bHLH family plays an important role in jasmonate-regulated responses. In the process of stamen maturation, *MYC 2* works redundantly with its close paralogs *MYC 3*, *MYC 4*, and *MYC 5* (Qi et al. 2015). Delayed anther dehiscence and low pollen viability are seen in the triple mutant *myc3myc4myc5*. Analysis of quadruple MYC mutant—*myc2myc3myc4myc5*—exhibits reduced pollen viability and slow filament elongation (Qi et al. 2015).

CORONATINE INSENSITIVE 1 (COI1) is an F-box protein and plays an important role in JA signalling by causing degradation of JAZ proteins which are JA repressors. In *coi-1* mutants, defects in stamen maturation due to defects in stamen elongation, reduced pollen viability, and failure of anther dehiscence are observed. Unlike in JA-biosynthetic mutants, in JA signalling mutants, fertility cannot be restored by external application of Me-JA implying involvement of yet unidentified factors in these complex GRNs (Katsir et al. 2008; Acosta and Przybyl 2019).

Reports have suggested that JA mediates stamen development by inducing transcription of MYB21 which in turn regulates MYB24, and this subsequently regulates MYB108 (Preston et al. 2004; Tatematsu et al. 2004; Mandaokar and Browse 2009). It has been reported that in JA mutants, water transport in anther is defective which is influenced by *SUC1* (Ishiguro et al. 2001; Acosta and Przybyl 2019). *MYB32* and *MYB4* also regulate the events of stamen development as in *myb32myb4* double mutants pollen grains have cytoplasmic defects resulting in male sterility (Preston et al. 2004). The mutants also exhibit alterations in components of phenylpropanoid pathway which in turn results in defects in pollen wall formation. *MYB32* and *IAA19* also play an important role in stamen or pollen development as the mutation in these genes leads to reduction in male fertility (Acosta and Przybyl 2019).

Male sterility has been reported in Brassinosteroids (BR) synthesis and signalling mutants like *cpd, brassinosteroid-insensitive mutant bin2*, and *BR receptor mutant bri1-201* (Bouquin et al. 2001). Reports have suggested that the key regulator of brassinosteroid signalling, BRI1-EMS-SUPPRESSOR 1 (BES1) transcription factor, binds to promoter of *SPL/NZZ, TDF1, AMS, MS1*, and *MS2* which are known to regulate multiple aspects of stamen development (Ye et al. 2010).

4.3.5 Non-coding RNAs and Stamen Development

Non-coding RNAs are classified as long non-coding RNA and small ncRNAs depending on their length. Non-coding RNA regulates gene expression at the level of transcription (TGS), post-transcriptional (PTGS), and translational level. Small ncRNA molecules include miRNA, si-RNA, hc-siRNAs, phasiRNAs, and NAT-siRNAs on the basis of their biogenesis (Cao et al. 2024; Wierzbicki et al. 2021; Axtell 2013).

MiRNA regulates gene expression at post-transcriptional level by binding with their target mRNA and causes target cleavage. Several plant miRNAs such as *miR156, miR159, miR160, miR162, miR171*, and *miR172* that are conserved, as well as species-specific miRNAs such as *miR415, miR416, miR417*, and *miR418* have also been identified (Floyd and Bowman 2004). For example, the presence of *miR1885* and *miR5718* has only been reported in *Brassica campestris* during pollen development (Jiang et al. 2014).

miR158 regulates expression of bra027656 protein, a PPR gene which controls pollen intine deposition, in *B. campestris* spp. chinensis (Ma et al. 2017). Overexpression of miR158 led to down-regulation of bra027656 influencing pollen viability, fertility, and germination. miR159 in *Arabidopsis* regulates male fertility by targeting MYB-TFs—*MYB33, MYB65*, and *MYB101*. Defects in male fertility due to premature degradation of tapetum, aberration of pollen wall, and defect in endothecium lignification are reported in miR159 overexpression lines (Anand et al. 2023; Zhang et al. 2007). In cotton, *mir396* is known to regulate expression of ACC oxidase 3, an ethylene biosynthetic gene involved in anther development (Wei et al. 2013). In radish, miR396 controls anther development by regulating AGRONAUTE protein *AGO5* (Zhang et al. 2016).

In maize, the ncRNA, *zma-miR601* regulates flower morphogenesis and pollen development by targeting MADs-box TF family proteins. *zma-miR601* also regulates expression of *flavin-containing monooxygenase* (*FMO*) and *enoyl-CoA hydratase* (*ECH*) gene. ECH regulates fatty acid metabolism in tapetum and plays an important role in tapetum functioning (Shen et al. 2011). Studies have revealed low expression of ECH genes in CMS anthers due to up-regulation of *zma-miR601* in tapetal cells. In *B. juncea*, low expression of miR395 and higher expression of *ATP Sulfurylase 1 (APS1)* were observed in CMS lines as compared to male fertile line (Yang et al. 2013a).

lncRNAs are known to control several plant biological processes like control of male fertility and flowering. In maize, the lncRNA, *zm401*, regulates expression of *MZm3-3*, *ZmMADS2*, and *ZmC5*, pollen development-specific genes. A high expression level of *zm401* is reported in mature pollen grains and male gametophyte (Dai et al. 2007; Ma et al. 2008; Wang et al. 2009).

The precise role of lncRNAs in anther development is now beginning to be discovered. In a recently published report in *A. thaliana*, a total of 1283 lncRNAs were identified from various stages of anther and pollen development (Zhou et al. 2025). 524 out these 1283 lncRNAs were differentially expressed. The expression of several lncRNAs was correlated with those of protein coding genes located adjacent to the lcnRNA loci; these protein coding genes were previously known to be associated with pollen wall formation, pollen germination, and tube growth. Similarly, several lncRNAs were identified as endogenous target mimics (eTMs) of miRNAs demonstrating the interactome of lncRNA-miRNA-protein coding RNAs and exemplified by functional analysis of aHG-miR167-ARF6/8 module (Zhou et al. 2025). Similar developmentally modulated landscape of lncRNA and ncRNA has been identified in tomato (Lamin-Samu et al. 2022), maize (Yan et al. 2025), and rice (Babaei et al. 2022).

4.4 Concluding Remarks and Future Perspectives

Stamen development involves intricate and complex pathways involving interaction of various genetic, epigenetic, phyto-hormones, and environmental signals (Fig. 4.3). The advent of new knowledge and newer genomic strategies means that modulating stamen development trait is challenging but achievable. The following are a summary of some of the challenges and opportunities that need to be addressed in the future:

(a) *Beyond model plants*: Significant advances—although incomplete, have been made toward unravelling the components and participants of stamen development, albeit in only a handful of species such as *Arabidopsis*, maize, rice, tomato, etc. It is quite evident that even in these species, the entire compendium of regulatory network is yet to be revealed. The challenge and opportunity therefore lie in exploring stamen developmental pathways in crops and plants exhibiting structural diversity in stamens to investigate the molecular basis of natural variation.

(b) *Variations and GRNs*: Comparative analysis of genetic regulatory networks across crops will allow identifying common and unique targets for trait manipulation. For instances, molecular regulators unique to the male reproductive system could be targeted, whereas regulators common to male and female reproductive systems (such as meiotic processes, archesporial cell specifying regulators, etc.) are to be avoided to induce sterility and favour allogamy. Future investigations must be able to identify GRNs and their components in the wide diversity of stamen classes that are known in angiosperms. Apart from intellectual curiosity, such investigations also allow understanding the common, divergent, and unique developmental programmes existing across diverse stamen types.

(c) *Molecular basis of ontogeny*: Even though ontogeny of the anther wall layers is relatively well understood, molecular basis such as genetic reprogramming and cell fate, epigenetic control, etc. still remains incomplete.

(d) *Genomes, developmental, and adaptive programmes*: Crops being polyploids are endowed with complex genome structures. Extrapolation of information from simple model systems such as *Arabidopsis* is therefore fraught with challenges of identification of orthologs, paralogs, and homeologs, and isoforms that form part of GRNs. Additionally, studies on functional and regulatory diversification as part of stamen developmental programme require special attention. Developmental plasticity as manifested during adaptive processes (e.g. stamen developmental programme under abiotic and biotic stresses) also needs to be thoroughly investigated.

(e) *Cell-to-cell communication and newer frontiers in investigation*: Communications across boundaries of the various somatic cell layers, as well as between somatic (2n) and gametic (n) cells. Additional investigative frontiers include identifying tissue and developmental stage-specific transcripts (including isoforms), promoters and cis-regulators, and epigenetic marks. Such studies will allow identification of specific targets for trait modulation. A combination of these, for example, can be used to design and deliver mRNA or ncRNA or lncRNAs in specific tissue/s or developmental stage/s using developmentally regulated or inducible promoters for inducing male sterility or fertility, as the case may be.

(f) *Newer targets*: Apart from modifying well-known genetic programmes such as those in tapetum and pollen grains, newer targets such as filament development, water transport, modulated anther dehiscence, etc. can be evaluated. The potential of non-coding RNAs such as small RNAs and lncRNAs for modulating gene expression and induce allogamy requires to be explored.

(g) *Emerging technologies and strategies*: Extensive application of precise genome and gene editing tools using CRISPR/Cas9 by targeting genes involved in stamen development needs to be harnessed for generating variants in stamen developmental pathway (as has been demonstrated by Barman et al. 2019; Okada et al. 2019; Jung et al. 2020; Jeong et al. 2014; Bao et al. 2022; Gökdemir et al. 2022; Pechar et al. 2022; Zhang et al. 2021a, b; Chen et al. 2018; Lee et al. 2020).

Acknowledgements SD would like to acknowledge DBT, Govt. of India for EMR grants, Institute of Eminence, and University of Delhi for infrastructure and financial support. The award of JRF/SRF from CSIR to RS and from UGC to EP is gratefully acknowledged.

Glossary

Autogamy and Allogamy The process of fusion of male and female gametes that are derived from the same (autogamy) or different (allogamy) plants. Hybrid seed production or heterosis breeding exploits allogamy

Chasmogamy and Cleistogamy Chasmogamy refers to reproductive strategy in which the male and female gametophyte-bearing organs, i.e. stamens and gynoecium, are exposed prior to pollination to facilitate cross-pollination. Cleistogamy refers to the strategy when flowers remain closed, i.e. stamens and gynoecium are not exposed prior to pollination; cleistogamy therefore promotes self-pollination

Dichogamy Temporal separation of maturation of male and female reproductive organs to prevent self-pollination

Dicliny The presence of either male or female reproductive organ in a flower. This is manifested as being monoecious (unisexual flowers borne on same plant) or dioecious (unisexual flowers borne on separate plants, i.e. separate male plant and female plant)

Filament The stalk-like structure on which the anthers are borne. All filaments in a flower could be of equal length, or didynamous (two long and two short), tetradynamous (four long and two short)

Herkogamy Spatial separation of male and female reproductive organs in the same flower

Heterantherous Flowers with multiple anther types differing in function (fertile and sterile) and structure (differing in overall length or filament length; shape and colour of anther)

Heterosis Refers to the phenomena when the progeny arising out of cross between two parents outperforms or has superior performance in trait characteristics than the two parents

Heterostyly Polymorphism in size of reproductive organ such as distylous (two size classes) and tristylous (three size classes) condition (e.g. "pin" and "thrum" morphs of heterostyly)

Heterozygosity The presence of multiple allelic forms at any locus

Male Sterility Failure of development of functional anther or production of viable and fertile pollen grains/male gametes

Self-Incompatibility Failure of production of viable zygote after successful self-pollination due to pre- and post-fertilisation barriers

References

Acosta IF, Przybyl M (2019) Jasmonate signaling during *Arabidopsis* stamen maturation. Plant Cell Physiol 60(12):2648–2659

Albertsen MS, Phillips RL (1981) Developmental cytology of 13 genetic male sterile loci in maize. Can J Genet Cytol 23:195–208

Alves-Ferreira M, Wellmer F, Banhara A, Kumar V, Riechmann JL, Meyerowitz EM (2007) Global expression profiling applied to the analysis of *Arabidopsis* stamen development. Plant Physiol 145:747–762

Ambrose BA, Lerner DR, Ciceri P, Padilla CM, Yanofsky MF, Schmidt RJ (2000) Molecular and genetic analyses of the *Silky1* gene reveal conservation in floral organ specification between eudicots and monocots. Mol Cell 5:569–579

Anand S, Lal M, Bhardwaj E, Shukla R, Pokhriyal E, Jain A, Sri T, Srivastava PS, Singh A, Das S (2023) *MIR159* regulates multiple aspects of stamen and carpel development and requires dissection and delimitation of differential downstream regulatory network for manipulating fertility traits. Physiol Mol Biol Plants 29(10):1437–1456

Ariizumi T, Toriyama K (2011) Genetic regulation of sporopollenin synthesis and pollen exine development. Annu Rev Plant Biol 62:437–460

Ariizumi T, Hatakeyama K, Hinata K, Sato S, Kato T, Tabata S, Toriyama K (2003) A novel male-sterile mutant of *Arabidopsis thaliana*, *faceless pollen-1*, produces pollen with a smooth surface and an acetolysis-sensitive exine. Plant Mol Biol 53:107–116

Ariizumi T, Hatakeyama K, Hinata K, Inatsugi R, Nishida I, Sato S, Kato T, Tabata S, Toriyama K (2004) Disruption of the novel plant protein NEF1 affects lipid accumulation in the plastids of the tapetum and exine formation of pollen, resulting in male sterility in *Arabidopsis thaliana*. Plant J 39:170–181

Ariizumi T, Hatakeyama K, Hinata K, Sato S, Kato T et al (2005) The *HKM* gene, which is identical to the *MS1* gene of *Arabidopsis thaliana*, is essential for primexine formation and exine pattern formation. Sex Plant Reprod 18:1–7

Åstrand J, Knight C, Robson J, Talle B, Wilson ZA (2021) Evolution and diversity of the angiosperm anther: trends in function and development. Plant Reprod 34(4):307–319. Erratum in: Plant Reprod. 2021 34(4):385

Axtell MJ (2013) Classification and comparison of small RNAs from plants. Annu Rev Plant Biol 64:137–159

Aya K, Ueguchi-Tanaka M, Kondo M, Hamada K, Yano K, Nishimura M, Matsuoka M (2009) Gibberellin modulates anther development in rice via the transcriptional regulation of *GAMYB*. Plant Cell 21(5):1453–1472

Babaei S, Singh MB, Bhalla PL (2022) Role of long non-coding RNAs in rice reproductive development. Front Plant Sci 13:1040366

Bao H, Ding Y, Yang F, Zhang J, Xie J, Zhao C et al (2022) Gene silencing, knockout and over-expression of a transcription factor *ABORTED MICROSPORES (SlAMS)* strongly affects pollen viability in tomato (*Solanum lycopersicum*). BMC Genomics 23:346

Barkan A, Small I (2014) Pentatricopeptide repeat proteins in plants. Annu Rev Plant Biol 65:415–442

Barman HN, Sheng Z, Fiaz S, Zhong M, Wu Y, Cai Y et al (2019) Generation of a new thermo-sensitive genic male sterile rice line by targeted mutagenesis of *TMS5* gene through CRISPR/Cas9 system. BMC Plant Biol 19:109

Bartlett ME, Williams SK, Taylor Z, DeBlasio S, Goldshmidt A, Hall DH, Schmidt RJ, Jackson DP, Whipple CJ (2015) The maize *PI/GLO* ortholog *Zmm16/sterile tassel silky ear1* interacts with the zygomorphy and sex determination pathways in flower development. Plant Cell 27:3081–3098

Berr A, McCallum EJ, Ménard R, Meyer D, Fuchs J, Dong A, Shen WH (2010) *Arabidopsis* SET DOMAIN GROUP2 is required for H3K4 trimethylation and is crucial for both sporophyte and gametophyte development. Plant Cell 22:3232–3248

Birchler JA, Yao H, Chudalayandi S (2006) Unraveling the genetic basis of hybrid vigor. Proc Natl Acad Sci USA 103(35):12957–12958

Biswas R, Chaudhuri S (2024) AtHMGB15 regulates tapetal apoptosis in pollen development and actin dynamics during pollen germination in *Arabidopsis*. Plant Reprod 37(4):469–478

Bohra A, Jha UC, Adhimoolam P, Bisht D, Singh NP (2016) Cytoplasmic male sterility (CMS) in hybrid breeding in field crops. Plant Cell Rep 35:967–993

Bonner LJ, Dickinson HG (1990) Anther dehiscence in *Lycopersicon esculentum*: II. Water relations. New Phytol 115(2):367–375
Borg M, Brownfield L, Khatab H, Sidorova A, Lingaya M, Twell D (2011) The R2R3 MYB transcription factor DUO1 activates a male germline-specific regulon essential for sperm cell differentiation in *Arabidopsis*. Plant Cell 23:534–549
Bots M, Feron R, Uehlein N, Weterings K, Kaldenhoff R, Mariani T (2005a) PIP1 and PIP2 aquaporins are differentially expressed during tobacco anther and stigma development. J Exp Bot 56:113–121
Bots M, Vergeldt F, Wolters-Arts M, Weterings K, van As H, Mariani C (2005b) Aquaporins of the PIP2 class are required for efficient anther dehiscence in tobacco. Plant Physiol 137:1049–1056
Boucher JJ, Ireland HS, Wang R, David KM, Schaffer RJ (2024) The genetic control of herkogamy. Funct Plant Biol 51:FP23315
Bouquin T, Meier C, Foster R, Nielsen M, Mundy J (2001) Control of specific gene expression by gibberellin and brassinosteroid. Plant Physiol 127:450–458
Bowman JL, Moyroud E (2024) Reflections on the ABC model of flower development. Plant Cell 36(5):1334–1357
Bowman JL, Smyth DR, Meyerowitz EM (1989) Genes directing flower development in *Arabidopsis*. Plant Cell 1:37–52
Bowman JL, Drews GN, Meyerowitz EM (1991) Expression of the *Arabidopsis* floral homeotic gene *AGAMOUS* is restricted to specific cell types late in flower development. Plant Cell 3:749–758
Brown DM, Zeef LA, Ellis J, Goodacre R, Turner SR (2005) Identification of novel genes in *Arabidopsis* involved in secondary cell wall formation using expression profiling and reverse genetics. Plant Cell 17:2281–2295
Browne RG, Iacuone S, Li SF, Dolferus R, Parish RW (2018) Anther morphological development and stage determination in *Triticum aestivum*. Front Plant Sci 9:228
Bruce AB (1910) The Mendelian theory of heredity and the augmentation of vigor. Science 32:627–628
Cabout S, Leask MP, Varghese S, Yi J, Peters B, Conze LL, Köhler C, Brownfield L (2017) The meiotic regulator JASON utilizes alternative translation initiation sites to produce differentially localized forms. J Exp Bot 68(15):4205–4217
Canales C, Bhatt AM, Scott R, Dickinson H (2002) EXS, a putative LRR receptor kinase, regulates male germline cell number and tapetal identity and promotes seed development in *Arabidopsis*. Curr Biol 12:1718–1727
Cao D, Cheng H, Wu W, Soo HM, Peng J (2006) Gibberellin mobilizes distinct DELLA-dependent transcriptomes to regulate seed germination and floral development in *Arabidopsis*. Plant Physiol 142(2):509–525
Cao W, Yang L, Zhuang M, Lv H, Wang Y, Zhang Y, Ji J (2024) Plant non-coding RNAs: the new frontier for the regulation of plant development and adaptation to stress. Plant Physiol Biochem 208:108435
Cardarelli M, Cecchetti V (2014) Auxin polar transport in stamen formation and development: how many actors? Front Plant Sci 5:333
Carrizo-Garcia C (2002) Anther wall formation in solanaceae species. Ann Bot 90:701–706
Cecchetti V, Altamura MM, Falasca G, Costantino P, Cardarelli M (2008) Auxin regulates *Arabidopsis* anther dehiscence, pollen maturation, and filament elongation. Plant Cell 20(7):1760–1774
Cecchetti V, Altamura MM, Brunetti P, Petrocelli V, Falasca G, Ljung K, Costantino P, Cardarelli M (2013) Auxin controls *Arabidopsis* anther dehiscence by regulating endothecium lignification and jasmonic acid biosynthesis. Plant J 74(3):411–422
Charlesworth D, Willis JH (2009) The genetics of inbreeding depression. Nat Rev Genet 10(11):783–796
Chaubal R, Anderson JR, Trimnell MR, Fox TW, Albertsen MC, Bedinger P (2003) The transformation of anthers in the *msca1* mutant of maize. Planta 216:778–788

Chen YC, McCormick S (1996) *sidecar pollen*, an *Arabidopsis thaliana* male gametophytic mutant with aberrant cell divisions during pollen development. Development 122(10):3243–3253

Chen C, Farmer AD, Langley RJ, Mudge J, Crow JA, May GD, Huntley J, Smith AG, Retzel EF (2010) Meiosis-specific gene discovery in plants: RNA-Seq applied to isolated *Arabidopsis* male meiocytes. BMC Plant Biol 10:280

Chen GH, Sun JY, Liu M, Liu J, Yang WC (2014) SPOROCYTELESS is a novel embryophyte-specific transcription repressor that interacts with TPL and TCP proteins in *Arabidopsis*. J Genet Genomics 41(12):617–625

Chen R, Xu Q, Liu Y, Zhang J, Ren D, Wang G, Liu Y (2018) Generation of transgene-free maize male sterile lines using the CRISPR/Cas9 system. Front Plant Sci 9:1180

Chen X, Yang S, Zhang Y, Zhu X, Yang X, Zhang C et al (2021) Generation of male-sterile soybean lines with the CRISPR/Cas9 system. Crop J 9(6):1270–1277

Cheng Y, Dai X, Zhao Y (2006) Auxin biosynthesis by the YUCCA flavin monooxygenases controls the formation of floral organs and vascular tissues in *Arabidopsis*. Genes Dev 20(13):1790–1799

Cheng H, Song S, Xiao L, Soo HM, Cheng Z, Xie D, Peng J (2009) Gibberellin acts through jasmonate to control the expression of *MYB21, MYB24*, and *MYB57* to promote stamen filament growth in *Arabidopsis*. PLoS Genet 5(3):e1000440

Cheng XQ, Zhang XY, Xue F, Zhu SH, Li YJ, Zhu QH, Liu F, Sun J (2020) Characterization and transcriptome analysis of a dominant genic male sterile cotton mutant. BMC Plant Biol 20(1):312

Coen ES, Meyerowitz EM (1991) The war of the whorls: genetic interactions controlling flower development. Nature 353:31–37

Colpitts CC, Kim SS, Posehn SE, Jepson C, Kim SY, Wiedemann G, Reski R, Wee AGH, Douglas CJ, Suh D-Y (2011) PpASCL, a moss ortholog of anther-specific chalcone synthase-like enzymes, is a hydroxyalkylpyrone synthase involved in an evolutionarily conserved sporopollenin biosynthesis pathway. New Phytol 192:855–868

Cortez PA, Caetano AP et al (2014) Elucidating the mechanism of poricidal anther dehiscence in *Miconia* species (Melastomataceae). Flora 209(10):571–579

Cui J, You C, Zhu E, Huang Q, Ma H, Chang F (2016) Feedback regulation of DYT1 by interactions with downstream bHLH factors promotes DYT1 nuclear localization and anther development. Plant Cell 28(5):1078–1093

Cui Y, Hu C, Zhu Y, Cheng K, Li X, Wei Z, Xue L, Lin F, Shi H, Yi J, Hou S, He K, Li J, Gou X (2018) CIK receptor kinases determine cell fate specification during early anther development in *Arabidopsis*. Plant Cell 30(10):2383–2401

Dai X, Yu J, Ma J, Ao G, Zhao Q (2007) Overexpression of Zm401, an mRNA-like RNA, has distinct effects on pollen development in maize. Plant Growth Regul 52:229–239

Dai SY, Hsu WH, Yang CH (2019) The gene *ANTHER DEHISCENCE REPRESSOR (ADR)* controls male fertility by suppressing the ROS accumulation and anther cell wall thickening in *Arabidopsis*. Sci Rep 9(1):5112

Darwin CR (ed) (1876) The effects of cross- and self-fertilization in the vegetable kingdom, 1st edn. John Murray. https://darwin-online.org.uk/converted/published/1876_CrossandSelfFertilisation_F1249/1876_CrossandSelfFertilisation_F1249.html; https://darwin-online.org.uk/manuscripts.html

Davis G (1966) Systematic embryology of the angiosperms. Wiley, New York

Dawson J, Sözen E, Vizir I, Van Waeyenberge S, Wilson ZA, Mulligan BJ (1999) Characterization and genetic mapping of a mutation (*ms35*) which prevents anther dehiscence in *Arabidopsis thaliana* by affecting secondary wall thickening in the endothecium. New Phytol 144:213–222

de Azevedo Souza C, Kim SS, Koch S, Kienow L, Schneider K, McKim SM, Haughn GW, Kombrink E, Douglas CJ (2009) A novel fatty Acyl-CoA Synthetase is required for pollen development and sporopollenin biosynthesis in *Arabidopsis*. Plant Cell 21:507–525

Delannoy E, Stanley WA, Bond CS, Small ID (2007) Pentatricopeptide repeat (PPR) proteins as sequence-specificity factors in post-transcriptional processes in organelles. Biochem Soc Trans 35:1643–1647

Dong X, Hong Z, Sivaramakrishnan M, Mahfouz M, Verma DP (2005) Callose synthase (CalS5) is required for exine formation during microgametogenesis and for pollen viability in *Arabidopsis*. Plant J 42(3):315–328

Du M, Zhou KE, Liu Y, Deng L, Zhang X, Lin L et al (2020) A biotechnology-based male-sterility system for hybrid seed production in tomato. Plant J 102(5):1090–1100

East EM (1908) Inbreeding in corn. Conn Agric Exp Stat Rep 1907:419–428

Edlund AF, Swanson R, Preuss D (2004) Pollen and stigma structure and function: the role of diversity in pollination. Plant Cell 16(Suppl):S84–S97

Ellis CM, Nagpal P, Young JC, Hagen G, Guilfoyle TJ, Reed JW (2005) *AUXIN RESPONSE FACTOR1* and *AUXIN RESPONSE FACTOR2* regulate senescence and floral organ abscission in *Arabidopsis thaliana*. Development 132(20):4563–4574

Ellis M, Egelund J, Schultz CJ, Bacic A (2010) Arabinogalactan-proteins: key regulators at the cell surface? Plant Physiol 153:403–419

Feng X, Dickinson HG (2010) Tapetal cell fate, lineage and proliferation in the *Arabidopsis* anther. Development. 137(14):2409–2416

Feng XL, Ni WM, Elge S, Mueller-Roeber B, Xu ZH, Xue HW (2006) Auxin flow in anther filaments is critical for pollen grain development through regulating pollen mitosis. Plant Mol Biol 61(1–2):215–226

Fernandez Gomez J, Wilson ZA (2014) A barley PHD finger transcription factor that confers male sterility by affecting tapetal development. Plant Biotechnol J 12:765–777

Floyd SK, Bowman JL (2004) Gene regulation: ancient microRNA target sequences in plants. Nature 428:485–486

Fujii S, Toriyama K (2009) Suppressed expression of *RETROGRADE-REGULATED MALE STERILITY* restores pollen fertility in cytoplasmic male sterile rice plants. Proc Natl Acad Sci USA 106:9513–9518

Fujimoto R, Taylor JM, Shirasawa S, Peacock WJ, Dennis ES (2012) Heterosis of *Arabidopsis* hybrids between C24 and Col is associated with increased photosynthesis capacity. Proc Natl Acad Sci USA 109(18):7109–7114. https://doi.org/10.1073/pnas.1204464109

Gastaldi V, Lucero LE, Ferrero LV, Ariel FD, Gonzalez DH (2020) Class-I TCP transcription factors activate the *SAUR63* gene subfamily in gibberellin-dependent stamen filament elongation. Plant Physiol 182(4):2096–2110

Gastaldi V, Alem AL, Mansilla N, Ariel FD, Viola IL, Lucero LE, Gonzalez DH (2023) BREVIPEDICELLUS/KNAT1 targets TCP15 to modulate filament elongation during *Arabidopsis* late stamen development. Plant Physiol 191(1):29–34

Gautam R, Shukla P, Kirti PB (2023) Male sterility in plants: an overview of advancements from natural CMS to genetically manipulated systems for hybrid seed production. Theor Appl Genet 136(9):195

Ge YX, Angenent GC, Dahlhaus E, Franken J, Peters J, Wullems GJ, Creemers-Molenaar J (2001) Partial silencing of the NEC1 gene results in early opening of anthers in *Petunia hybrida*. Mol Gen Genomics 265:414–423

Gehring M (2019) Epigenetic dynamics during flowering plant reproduction: evidence for reprogramming? New Phytol 224(1):91–96

Gökdemir G, Seçgin Z, Uluisik S, Kavas M (2022) CRISPR/Cas9 knock-out of *SlPHD_MS1 (Solyc04g008420)* gene results in complete male sterility in tomato. Plant Growth Regul 98:329–341

Goldberg RB, Beals TP, Sanders PM (1993) Anther development: basic principles and practical applications. Plant Cell 5(10):1217–1229

Gómez JF, Talle B, Wilson ZA (2015) Anther and pollen development: a conserved developmental pathway. J Integr Plant Biol 57(11):876–891

Guan YF, Huang XY, Zhu J, Gao JF, Zhang HX, Yang ZN (2008) *RUPTURED POLLEN GRAIN1*, a member of the MtN3/Saliva gene family, is crucial for exine pattern formation and cell integrity of microspores in *Arabidopsis*. Plant Physiol 147:852–863

Gutiérrez-Hernández GF, Arellano-Vázquez JL, Ceja-Torres LF, García-Ramírez E, Quiroz-Figueroa FR (2023) Heterosis in the germination process and seed characteristics of the maize hybrid (*Zea mays* L.) HAZUL10E. Agro Prod 16(11):119–125

Gutierrez-Marcos JF, Dickinson HG (2012) Epigenetic reprogramming in plant reproductive lineages. Plant Cell Physiol 53(5):817–823

Han Q, Bartels A, Cheng X, Meyer A, An YC, Hsieh TF, Xiao W (2019) Epigenetics regulates reproductive development in plants. Plants (Basel) 8(12):564

Han Y, Hu M, Ma X, Yan G, Wang C, Jiang S et al (2022) Exploring key developmental phases and phase-specific genes across the entirety of anther development in maize. J Integr Plant Biol 64(7):1394–1410

Han F, Yuan K, Sun W et al (2023) A natural mutation in the promoter of Ms-cd1 causes dominant male sterility in *Brassica oleracea*. Nat Commun 14:6212

Hao Z, Avci U, Tan L, Zhu X, Glushka J, Pattathil S et al (2014) Loss of *Arabidopsis GAUT12/IRX8* causes anther indehiscence and leads to reduced G lignin associated with altered matrix polysaccharide deposition. Front Plant Sci 5:357

Hao J, Chen L, Zhao C, Qiao K, Wang N, Wang J et al (2025) CRISPR/Cas9-mediated knockout of *GhAMS11* and *GhMS188* reveals key roles in tapetal development and pollen exine formation in upland cotton. Int J Biol Macromol 293:139362

Hardy CR, Stevenson DW (2000) Development of the gametophytes, flower, and floral vasculature in *Cochliostema odoratissimum* (Commelinaceae). Bot J Linn Soc 134:131–157

He Y, He X, Wang X, Hao M, Gao J, Wang Y, Yang ZN, Meng X (2023) An EPFL peptide signaling pathway promotes stamen elongation via enhancing filament cell proliferation to ensure successful self-pollination in *Arabidopsis thaliana*. New Phytol 238(3):1045–1058

Hermann PM, Palser BF (2000) Stamen development in the Ericaceae. I. Anther wall, microsporogenesis, inversion, and appendages. Am J Bot 87(7):934–957

Heslop-Harrison J (1968) Tapetal origin of pollen-coat substances in *Lilium*. New Phytol 67:779–786

Higginson T, Li SF, Parish RW (2003) *AtMYB103* regulates tapetum and trichome development in *Arabidopsis thaliana*. Plant J 35:177–192

Hochholdinger F, Hoecker N (2007) Towards the molecular basis of heterosis. Trends Plant Sci 12(9):427–432

Hoecker N, Keller B, Piepho HP, Hochholdinger F (2006) Manifestation of heterosis during early maize (*Zea mays* L.) root development. Theor Appl Genet 112(3):421–429. https://doi.org/10.1007/s00122-005-0139-4

Hong L, Tang D, Zhu K, Wang K, Li M, Cheng Z (2012) Somatic and reproductive cell development in rice anther is regulated by a putative glutaredoxin. Plant Cell 24:577–588

Honma T, Goto K (2001) Complexes of MADS-box proteins are sufficient to convert leaves into floral organs. Nature 409:525–529

Hord CL, Chen C, Deyoung BJ, Clark SE, Ma H (2006) The BAM1/BAM2 receptor-like kinases are important regulators of *Arabidopsis* early anther development. Plant Cell 18(7):1667–1680

Hord CL, Sun YJ, Pillitteri LJ, Torii KU, Wang H, Zhang S, Ma H (2008) Regulation of *Arabidopsis* early anther development by the mitogen-activated protein kinases, MPK3 and MPK6, and the ERECTA and related receptor-like kinases. Mol Plant 1(4):645–658

Hou Q, Zhang T, Qi Y, Dong Z, Wan X (2022) Epigenetic dynamics and regulation of plant male reproduction. Int J Mol Sci 23(18):10420

Hu J, Rutger JN (1992) Pollen characteristics and genetics of induced and spontaneous genetic male-sterile mutants in rice. Plant Breed 109:97–107

Hu J, Mitchum MG, Barnaby N, Ayele BT, Ogawa M, Nam E et al (2008) Potential sites of bioactive gibberellin production during reproductive growth in *Arabidopsis*. Plant Cell 20(2):320–336

Huang H, Gong Y, Liu B, Wu D, Zhang M, Xie D, Song S (2020) The DELLA proteins interact with MYB21 and MYB24 to regulate filament elongation in *Arabidopsis*. BMC Plant Biol 20(1):64

Ishiguro S, Kawai-Oda A, Ueda J, Nishida I, Okada K (2001) The *DEFECTIVE IN ANTHER DEHISCIENCE1* gene encodes a novel phospholipase A1 catalyzing the initial step of jasmonic acid biosynthesis, which synchronizes pollen maturation, anther dehiscence, and flower opening in *Arabidopsis*. Plant Cell 13(10):2191–2209

Ito T, Wellmer F, Yu H, Das P, Ito N, Alves-Ferreira M, Riechmann JL, Meyerowitz EM (2004) The homeotic protein AGAMOUS controls microsporogenesis by regulation of SPOROCYTELESS. Nature 430:356–360

Ito T, Nagata N, Yoshiba Y, Ohme-Takagi M, Ma H, Shinozaki K (2007a) *Arabidopsis MALE STERILITY1* encodes a PHD-type transcription factor and regulates pollen and tapetum development. Plant Cell 19:3549–3562

Ito T, Ng K-H, Lim T-S, Yu H, Meyerowitz EM (2007b) The homeotic protein AGAMOUS controls late stamen development by regulating a jasmonate biosynthetic gene in *Arabidopsis*. Plant Cell 19:3516–3529

Iwano M, Entani T, Shiba H, Takayama S, Isogai A (2004) Calcium crystals in the anther of petunia: the existence and biological significance in the pollination process. Plant Cell Physiol 45(1):40–47

Jenik PD, Irish VF (2000) Regulation of cell proliferation patterns by homeotic genes during *Arabidopsis* floral development. Development 127:1267–1276

Jeong HJ, Kang JH, Zhao M, Kwon JK, Choi HS, Bae JH et al (2014) Tomato *Male sterile 10^{35}* is essential for pollen development and meiosis in anthers. J Exp Bot 65:6693–6709. https://doi.org/10.1093/jxb/eru389

Jia QS, Zhu J, Xu XF, Lou Y, Zhang ZL, Zhang ZP, Yang ZN (2015) *Arabidopsis* AT-hook protein TEK positively regulates the expression of arabinogalactan proteins for Nexine formation. Mol Plant 8:251–260

Jiang J, Lv M, Liang Y, Ma Z, Cao J (2014) Identification of novel and conserved miRNAs involved in pollen development in *Brassica campestris* ssp. chinensis by high-throughput sequencing and degradome analysis. BMC Genomics 15:146

Jiang Y, Li Z, Liu X, Zhu T, Xie K, Hou Q et al (2021) *ZmFAR1* and *ZmABCG26* regulated by microRNA are essential for lipid metabolism in maize anther. Int J Mol Sci 22(15):7916

Jones DF (1917) Dominance of linked factors as a means of accounting for heterosis. Genetics 2:466–475. https://doi.org/10.1093/genetics/2.5.466

Jung KH, Han MJ, Lee YS, Kim YW, Hwang I, Kim MJ, Kim YK, Nahm BH, An G (2005) Rice undeveloped tapetum1 is a major regulator of early tapetum development. Plant Cell 17:2705–2722

Jung YJ, Kim DH, Lee HJ, Nam KH, Bae S, Nou IS et al (2020) Knockout of *SlMS10* gene (*Solyc02g079810*) encoding bHLH transcription factor using CRISPR/Cas9 system confers Male sterility phenotype in tomato. Plants 9(9):1198. https://doi.org/10.3390/plants9091189

Kaneko M, Inukai Y, Ueguchi-Tanaka M et al (2004) Loss-of-function mutations of the rice *GAMYB* gene impair alpha-amylase expression in aleurone and flower development. Plant Cell 16:33–44

Katsir L, Schilmiller AL, Staswick PE, He SY, Howe GA (2008) COI1 is a critical component of a receptor for jasmonate and the bacterial virulence factor coronatine. Proc Natl Acad Sci USA 105(19):7100–7105

Kaul MLH (1988) Male sterility in higher plants. Springer, Berlin

Keijzer CJ (1999) Mechanisms of Angiosperm anther dehiscence, a historical review. In: Clément C et al (eds) Anther and pollen. Springer, Berlin Heidelberg, pp 55–67

Kelliher T, Walbot V (2011) Emergence and patterning of the five cell types of the *Zea mays* anther locule. Dev Biol 350:32–49

Kim SS, Grienenberger E, Lallemand B, Colpitts CC, Kim SY, Souza CA, Geoffroy P, Heintz D, Krahn D, Kaiser M, Kombrink E, Heitz T, Suh DY, Legrand M, Douglas CJ (2010) *LAP6/*

POLYKETIDE SYNTHASE A and *LAP5/POLYKETIDE SYNTHASE B* encode hydroxyalkyl a-pyrone synthases required for pollen development and sporopollenin biosynthesis in *Arabidopsis thaliana*. Plant Cell 22:4045–4066

Kissoudis C, van de Wiel C, Visser RG, van der Linden G (2016) Future-proof crops: challenges and strategies for climate resilience improvement. Curr Opin Plant Biol 30:47–56

Kolreuter JG (1766) Dritte Fortsetzung der Vorlaufigen Nachricht von einigen das Geschlecht der Pflanzen betreffenden Versuchen und Beobachstungen. Gleditschen Handlung, Leipzig. Reprinted 1893 in Ostwald's Klassiker der exakteii Wissenschaften, No. 41. Leipzig

Koning RE (1983) The roles of auxin, ethylene, and acid growth in filament elongation in *Gaillardia grandiflora* (Asteraceae). Am J Bot 70(4):602–610

Kuo P, Tock AJ, Liu X, Topp SD, Zhong Z, Henderson IR, Lambing C (2025) Histone variant H2A.W7 represses meiotic crossover formation in *Arabidopsis* heterochromatin. Proc Natl Acad Sci USA 122(22):e2414166122

Lallemand B, Erhardt M, Heitz T, Legrand M (2013) Sporopollenin biosynthetic enzymes interact and constitute a metabolon localized to the endoplasmic reticulum of tapetum cells. Plant Physiol 162(2):616–625

Lamin-Samu AT, Zhuo S, Ali M, Lu G (2022) Long non-coding RNA transcriptome landscape of anthers at different developmental stages in response to drought stress in tomato. Genomics 114(4):110383

Lee SK, Kim H, Cho JI, Nguyen CD, Moon S, Park JE et al (2020) Deficiency of rice hexokinase HXK5 impairs synthesis and utilization of starch in pollen grains and causes male sterility. J Exp Bot 71:116–125. https://doi.org/10.1093/jxb/erz436

Lei X, Liu B (2020) Tapetum-dependent male meiosis progression in plants: increasing evidence emerges. Front Plant Sci 16(10):1667

Li SF, Iacuone S, Parish RW (2007) Suppression and restoration of male fertility using a transcription factor. Plant Biotechnol J 5:297–312

Li H, Pinot F, Sauveplane V, Werck-Reichhart D, Diehl P, Schreiber L, Franke R, Zhang P, Chen L, Gao Y, Liang W, Zhang D (2010) *Cytochrome P450* family member *CYP704B2* catalyzes the v-hydroxylation of fatty acids and is required for anther cutin biosynthesis and pollen exine formation in rice. Plant Cell 22:173–190

Li H, Yuan Z, Vizcay-Barrena G, Yang C, Liang W, Zong J, Wilson ZA, Zhang D (2011) *PERSISTENT TAPETAL CELL1* Encodes a PHD-Finger protein that is required for tapetal cell death and pollen development in rice. Plant Physiol 156:615–630

Li Q, Zhang D, Chen M, Liang W, Wei J, Qi Y, Yuan Z (2016) Development of japonica photosensitive genic male sterile rice lines by editing carbon starved anther using CRISPR/Cas9. J Genet Genomics/Yi chuan xue bao 43(6):415–419

Li DD, Xue JS, Zhu J, Yang ZN (2017) Gene regulatory network for tapetum developm0ent in *Arabidopsis thaliana*. Front Plant Sci 8. https://doi.org/10.3389/fpls.2017.01559

Li FS, Phyo P, Jacobowitz J, Hong M, Weng JK (2019a) The molecular structure of plant sporopollenin. Nat Plants 5(1):41–46

Li X, Lian H, Zhao Q, He Y (2019b) MicroRNA166 monitors *SPOROCYTELESS/NOZZLE* for building of the anther internal boundary. Plant Physiol 181(1):208–220

Li J, Wang Z, He G, Ma L, Deng XW (2020) CRISPR/Cas9-mediated disruption of *TaNP1* genes results in complete male sterility in bread wheat. J Genet Genomics 47(5):263–272

Li Z, Zhu T, Liu S, Jiang Y, Liu H, Zhang Y et al (2021) Genome-wide analyses on transcription factors and their potential microRNA regulators involved in maize male fertility. Crop J 9(6):1248–1262

Li Z, Yuan R, Wang M, Hong M, Zhu L, Li X, Guo R, Wu G, Zeng X (2022) Development of the PARMS marker of the dominant genic male sterility (DGMS) line and its utilization in rapeseed (*Brassica napus* L.) breeding. Plants (Basel) 11(3):421

Liu X, Huang J, Parameswaran S, Ito T, Seubert B, Auer M, Rymaszewski A, Jia G, Owen HA, Zhao D (2009) The *SPOROCYTELESS/NOZZLE* gene is involved in controlling stamen identity in *Arabidopsis*. Plant Physiol 151(3):1401–1411

Liu Y, Cui S, Wu F, Yan S, Lin X, Du X, Chong K, Schilling S, Theissen G, Meng Z (2013) Functional conservation of *MIKC-Type MADS box* genes in *Arabidopsis* and rice pollen maturation. Plant Cell 25:1288–1303

Lobell DB, Gourdji SM (2012) The influence of climate change on global crop productivity. Plant Physiol 160(4):1686–1697

Longin CF, Mühleisen J, Maurer HP, Zhang H, Gowda M, Reif JC (2012) Hybrid breeding in autogamous cereals. Theor Appl Genet 125(6):1087–1096

Lou Y, Xu XF, Zhu J, Gu JN, Blackmore S, Yang ZN (2014) The tapetal AHL family protein TEK determines nexine formation in the pollen wall. Nat Commun 5:3855

Ma J, Yan B, Qu Y, Qin F, Yang Y, Hao X et al (2008) *Zm401*, a short-open reading-frame mRNA or noncoding RNA, is essential for tapetum and microspore development and can regulate the floret formation in maize. J Cell Biochem 105(1):136–146

Ma LJ, Yang ZN, Zhang S (2013) DEX1, a plasma membrane-localized protein, functions in microspore development by affecting *CalS5* expression in *Arabidopsis thaliana*. Chin Sci Bull 58:2855–2861

Ma Z, Jiang J, Hu Z, Lyu T, Yang Y, Jiang J et al (2017) Over-expression of miR158 causes pollen abortion in *Brassica campestris* ssp. chinensis. Plant Mol Biol 93:313–326

Ma X, Wu Y, Zhang G (2021) Formation pattern and regulatory mechanisms of pollen wall in Arabidopsis. J Plant Physiol 260:153388

Mandaokar A, Browse J (2009) *MYB108* acts together with *MYB24* to regulate jasmonate-mediated stamen maturation in *Arabidopsis*. Plant Physiol 149:851–862

Marazzi B, Conti E, Endress PK (2007) Diversity in anthers and stigmas in the buzz-pollinated genus *Senna* (Leguminosae, Cassiinae). Int J Plant Sci 168(4):371–391

Marchant DB, Walbot V (2022) Anther development—the long road to making pollen. Plant Cell 34(12):4677–4695

Marzi D, Brunetti P, Mele G, Napoli N, Calò L, Spaziani E, Matsui M, De Panfilis S, Costantino P, Serino G, Cardarelli M (2020) Light controls stamen elongation via cryptochromes, phytochromes and COP1 through HY5 and HYH. Plant J 103(1):379–394

Masoomi-Aladizgeh F, Atwell BJ, Bokshi AI, Thistlethwaite RJ, Khoddami A, Trethowan R, Tan DKY, Roberts TH (2025) Pinpointing the timing of meiosis: a critical factor in evaluating the impact of abiotic stresses on the fertility of cereal crops. New Phytol 245(4):1341–1354

Mathan J, Bhattacharya J, Ranjan A (2016) Enhancing crop yield by optimizing plant developmental features. Development 143(18):3283–3294

Matsuhira H, Kagami H, Kurata M et al (2012) Unusual and typical features of a novel restorer-of-fertility gene of sugar beet (*Beta vulgaris* L.). Genetics 192:1347–1358

Mena M, Ambrose BA, Meeley RB, Briggs SP, Yanofsky MF, Schmidt RJ (1996) Diversification of C-function activity in maize flower development. Science (80-) 274:1537–1540

Mercier R, Mézard C, Jenczewski E, Macaisne N, Grelon M (2015) The molecular biology of meiosis in plants. Annu Rev Plant Biol 66:297–327

Meyen FJF (1828) Anatomische und Physiologische Untersuchungen tiber den Inhalt der PflanzenzelIen. Joseephy, Berlin

Meyer RS, Purugganan MD (2013) Evolution of crop species: genetics of domestication and diversification. Nat Rev Genet 14(12):840–852

Meyer S, Pospisil H, Scholten S (2007) Heterosis associated gene expression in maize embryos 6-days after fertilization exhibits additive, dominant and overdominant pattern. Plant Mol Biol 63(3):381–391. https://doi.org/10.1007/s11103-006-9095-x

Millar AA, Gubler F (2005) The *Arabidopsis* GAMYB-Like genes, *MYB33* and *MYB65*, are microRNA-regulated genes that redundantly facilitate anther development. Plant Cell 17:705–721

Mitsuda N, Seki M, Shinozaki K, Ohme-Takagi M (2005) The NAC transcription factors *NST1* and *NST2* of *Arabidopsis* regulate secondary wall thickenings and are required for anther dehiscence. Plant Cell 17(11):2993–3006

Mizelle MB, Sethi R, Ashton ME, Jemen WA (1989) Development of the pollen grain and tapetum of wheat (*Triticum aestivum*) in untreated plants and plants treated with chemical hybridizing agent RH0007. Sex Plant Reprod 2:231–253

Mizuno S, Osakabe Y, Maruyama K, Ito T, Osakabe K, Sato T, Shinozaki K, Yamaguchi-Shinozaki K (2007) Receptor-like protein kinase 2 (RPK 2) is a novel factor controlling anther development in *Arabidopsis thaliana*. Plant J 50(5):751–766

Murphy KM, Egger RL, Walbot V (2015) Chloroplasts in anther endothecium of *Zea mays* (Poaceae). Am J Bot 102(11):1931–1937

Nan GL, Teng C, Fernandes J, O'Connor L, Meyers BC, Walbot V (2022) A cascade of bHLH-regulated pathways programs maize anther development. Plant Cell 34(4):1207–1225

Nelson MR, Band LR, Dyson RJ, Lessinnes T, Wells DM, Yang C, Everitt NM, Jensen OE, Wilson ZA (2012) A biomechanical model of anther opening reveals the roles of dehydration and secondary thickening. New Phytol 196(4):1030–1037

Ogawa M, Kay P, Wilson S, Swain SM (2009) *ARABIDOPSIS DEHISCENCE ZONE POLYGALACTURONASE1 (ADPG1)*, *ADPG2*, and *QUARTET2* are polygalacturonases required for cell separation during reproductive development in *Arabidopsis*. Plant Cell 21:216–233

Okada A, Arndell T, Borisjuk N, Sharma N, Watson-Haigh NS, Tucker EJ et al (2019) CRISPR/Cas9-mediated knockout of Ms1 enables the rapid generation of male-sterile hexaploid wheat lines for use in hybrid seed production. Plant Biotechnol J 17:1905–1913

Ono A, Kinoshita T (2021) Epigenetics and plant reproduction: multiple steps for responsibly handling succession. Curr Opin Plant Biol 61:102032

Owen HA, Makaroff CA (1995) Ultrastructure of microsporogenesis and microgametogenesis in *Arabidopsis thaliana* (L.) Heynh. ecotype Wassilewskija (Brassicaceae). Protoplasma 185:7–21

Pacini E (2010) Relationships between tapetum, loculus and pollen during development. Int J Plant Sci 171:1–11

Park JH, Halitschke R, Kim HB, Baldwin IT, Feldmann KA, Feyereisen R (2002) A knock-out mutation in allene oxide synthase results in male sterility and defective wound signal transduction in *Arabidopsis* due to a block in jasmonic acid biosynthesis. Plant J 31(1):1–12

Paxson-Sowders DM, Dodrill CH, Owen HA, Makaroff CA (2001) DEX1, a novel plant protein, is required for exine pattern formation during pollen development in *Arabidopsis*. Plant Physiol 127:1739–1749

Pechar GS, Donaire L, Gosalvez B, Garcia-Almodovar C, Sanchez-Pina MA, Truniger V et al (2022) Editing melon eIF4E associates with virus resistance and male sterility. Plant Biotechnol J 20:2006–2022. https://doi.org/10.1111/pbi.13885

Peng J (2009) Gibberellin and jasmonate crosstalk during stamen development. J Integr Plant Biol 51(12):1064–1070

Phan HA, Iacuone S, Li SF, Parish RW (2011) The MYB80 transcription factor is required for pollen development and the regulation of tapetal programmed cell death in *Arabidopsis thaliana*. Plant Cell 23:2209–2224

Powers L (1945) An expansion of Jones's theory for the explanation of heterosis. Am Nat 78:275–280. https://www.journals.uchicago.edu/doi/epdf/10.1086/281199

Preston J, Wheeler J, Heazlewood J, Li SF, Parish RW (2004) *AtMYB32* is required for normal pollen development in *Arabidopsis thaliana*. Plant J 40(6):979–995

Purkinje JE (1830) De cellulis antherarum fibrosis nec non de granorum pollinarium formis. Bratislava

Qi T, Huang H, Song S, Xie D (2015) Regulation of jasmonate-mediated stamen development and seed production by a bHLH-MYB complex in *Arabidopsis*. Plant Cell 27:1620–1633

Quilichini TD, Douglas CJ, Samuels AL (2014a) New views of tapetum ultrastructure and pollen exine development in *Arabidopsis thaliana*. Ann Bot 114(6):1189–1201

Quilichini TD, Grienenberger E, Douglas CJ (2014b) The biosynthesis, composition and assembly of the outer pollen wall: a tough case to crack. Phytochemistry 113:170–182

Raghavan V (1988) Anther and pollen development in rice (*Oryza sativa*). Am J Bot 75:183–196

Reed JW, Wu MF, Reeves PH, Hodgens C, Yadav V, Hayes S, Pierik R (2018) Three auxin response factors promote hypocotyl elongation. Plant Physiol 178(2):864–875

Rhee SY, Somerville CR (1998) Tetrad pollen formation in quartet mutants of *Arabidopsis thaliana* is associated with persistence of pectic polysaccharides of the pollen mother cell wall. Plant J 15:79–88

Rhee SY, Osborne E, Poindexter PD, Somerville CR (2003) Microspore separation in the *quartet 3* mutants of *Arabidopsis* is impaired by a defect in a developmentally regulated polygalacturonase required for pollen mother cell wall degradation. Plant Physiol 133:1170–1180

Riechmann JL, Wang MQ, Meyerowitz EM (1996) DNA-binding properties of *Arabidopsis* MADS domain homeotic proteins APETALA1, APETALA3, PISTILLATA and AGAMOUS. Nucleic Acids Res 24:3134–3141

Rogers LA, Dubos C, Surman C, Willment J, Cullis IF, Mansfield SD, Campbell MM (2005) Comparison of lignin deposition in three ectopic lignification mutants. New Phytol 168:123–140

Ruffio-Chable V, Bellis H, Herve Y (1993) A dominant gene for male sterility in cauliflower (*Brassica oleracea* var. botrytis): phenotype expression, inheritance, and use in FI hybrid production. Euphytica 67:9–17, 199

Ruiter RK, van Eldik GJ, van Herpen MM, Schrauwen JA, Wullems GJ (1997) Expression in anthers of two genes encoding *Brassica oleracea* transmembrane channel proteins. Plant Mol Biol 34(1):163–168

Russell SD, Gou X, Wong CE, Wang X, Yuan T, Wei X, Bhalla PL, Singh MB (2012) Genomic profiling of rice sperm cell transcripts reveals conserved and distinct elements in the flowering plant male germ lineage. New Phytol 195:560–573

Sanders P, Bui A, Weterings K, McIntire KN, Hsu Y-C, Lee PY, Truong MT, Beals TP, Goldberg RB (1999) Anther developmental defects in *Arabidopsis thaliana* male-sterile mutants. Sex Plant Reprod 11:297–322

Sanders PM, Lee PY, Biesgen C, Boone JD, Beals TP, Weiler EW, Goldberg RB (2000) The *Arabidopsis DELAYED DEHISCENCE1* gene encodes an enzyme in the jasmonic acid synthesis pathway. Plant Cell 12(7):1041–1061

Sanders PM, Bui AQ, Le BH, Goldberg RB (2005) Differentiation and degeneration of cells that play a major role in tobacco anther dehiscence. Sex Plant Reprod 17:219–241

Schiefthaler U, Balasubramanian S, Sieber P, Chevalier D, Wisman E, Schneitz K (1999) Molecular analysis of *NOZZLE*, a gene involved in pattern formation and early sporogenesis during sex organ development in *Arabidopsis thaliana*. Proc Natl Acad Sci USA 96:11664–11669

Schmidt RJ, Veit B, Mandel MA, Mena M, Hake S, Yanofsky MF (1993) Identification and molecular characterization of *ZAG1*, the maize homolog of the *Arabidopsis* floral homeotic gene *AGAMOUS*. Plant Cell 5:729–737

Scott RJ, Spielman M, Dickinson HG (2004) Stamen structure and function. Plant Cell 16(Suppl):S46–S60

Sharma A, Singh MB, Bhalla PL (2015) Anther ontogeny in *Brachypodium distachyon*. Protoplasma 252:439–450

She W, Baroux C (2015) Chromatin dynamics in pollen mother cells underpin a common scenario at the somatic-to-reproductive fate transition of both the male and female lineages in *Arabidopsis*. Front Plant Sci 6:294

Shen Y, Zhang Z, Lin H et al (2011) Cytoplasmic male sterility regulated novel microRNAs from maize. Funct Integr Genom 11:179–191

Shi M (1981) Preliminary report on breeding and application of late japonica natural dual-use lines. J Hubei Agric Sci 7:1–3

Shi J, Tan H, Yu XH, Liu Y, Liang W, Ranathunge K, Franke RB, Schreiber L, Wang Y, Kai G, Shanklin J, Ma H, Zhang D (2011) Defective pollen wall is required for anther and microspore development in rice and encodes a fatty acyl carrier protein reductase. Plant Cell 23:2225–2246

Shull GH (1908) The composition of a field of maize. Am Breed Assoc Rep IV:296–301. https://doi.org/10.1093/jhered/os-4.1.296

Shull GH (1909a) A pure-line method in corn breeding. Am Breed Assoc Rep V:51–59
Shull GH (1909b) Hybridization methods in corn breeding. Am Breed Assoc Rep VI:63–72
Shull GH (1914) Duplicate genes for capsule form in *Bursa bursa-pastoris*. Z.Ver-erbungslehre (Now MGAG) 12:97–149
Singh M, Bhalla P, Russell S (2008) Molecular repertoire of flowering plant male germ cells. Sex Plant Reprod 21:27–36
Singh M, Kumar M, Albertsen MC, Young JK, Cigan AM (2018) Concurrent modifications in the three homeologs of *Ms45* gene with CRISPR-Cas9 lead to rapid generation of male sterile bread wheat (*Triticum aestivum* L.). Plant Mol Biol 97:371–383
Smyth DR, Bowman JL, Meyerowitz EM (1990) Early flower development in *Arabidopsis*. Plant Cell 2:755–767
Song S, Qi T, Huang H, Xie D (2013) Regulation of stamen development by coordinated actions of jasmonate, auxin, and gibberellin in *Arabidopsis*. Mol Plant 6(4):1065–1073
Sorensen AM, Kröber S, Unte US, Huijser P, Dekker K, Saedler H (2003) The *Arabidopsis ABORTED MICROSPORES (AMS)* gene encodes a MYC class transcription factor. Plant J 33(2):413–423
Soto G, Alleva K, Mazzella MA, Amodeo G, Muschietti JP (2008) *AtTIP1;3* and *AtTIP5;1*, the only highly expressed *Arabidopsis* pollen-specific aquaporins, transport water and urea. FEBS Lett 582:4077–4082
Stadler R, Truernit E, Gahrtz M, Sauer N (1999) The *AtSUC1* sucrose carrier may represent the osmotic driving force for anther dehiscence and pollen tube growth in *Arabidopsis*. Plant J 19:269–278
Ståldal V, Cierlik I, Chen S, Landberg K, Baylis T, Myrenås M, Sundström JF, Eklund DM, Ljung K, Sundberg E (2012) The *Arabidopsis thaliana* transcriptional activator *STYLISH1* regulates genes affecting stamen development, cell expansion and timing of flowering. Plant Mol Biol 78(6):545–559
Stephens JC (1937) Male sterility in sorghum: its possible utilization in production of hybrid seed. J Am Soc Agron 29:690–696
Stintzi A, Browse J (2000) The *Arabidopsis* male-sterile mutant, *opr3*, lacks the 12-oxophytodienoic acid reductase required for jasmonate synthesis. Plant Biol 97(19):10625–10630
Tatematsu K, Kumagai S, Muto H, Sato A, Watahiki MK, Harper RM et al (2004) *MASSUGU2* encodes *Aux/IAA19*, an auxin-regulated protein that functions together with the transcriptional activator *NPH4/ARF7* to regulate differential growth responses of hypocotyl and formation of lateral roots in *Arabidopsis thaliana*. Plant Cell 16(2):379–393
Taylor PE, Glover JA, Lavithis M, Craig S, Singh MB, Knox RB, Dennis ES, Chaudhury AM (1998) Genetic control of male fertility in *Arabidopsis thaliana*: structural analyses of postmeiotic developmental mutants. Planta 205(4):492–505
Teng N, Huang Z, Mu X, Jin B, Hu Y, Lin J (2005) Microsporogenesis and pollen development in *Leymus chinensis* with emphasis on dynamic changes in callose deposition. Flora 200:256–263
Theißen G (2001) Development of floral organ identity: stories from the MADS house. Curr Opin Plant Biol 4:75–85
Thompson BE, Bartling L, Whipple C, Hall DH, Sakai H, Schmidt R, Hake S (2009) Bearded-ear encodes a MADS Box transcription factor critical for maize floral development. Plant Cell 21:2578–2590
Thorstensen T, Grini P, Mercy I, Alm V, Erdal S, Aasland R, Aalen R (2008) The *Arabidopsis* SET-domain protein ASHR3 is involved in stamen development and interacts with the bHLH transcription factor ABORTED MICROSPORES (AMS). Plant Mol Biol 66:47–59
Tidy AC, Ferjentsikova I, Vizcay-Barrena G, Liu B, Yin W, Higgins JD, Xu J, Zhang D, Geelen D, Wilson ZA (2022) Sporophytic control of pollen meiotic progression is mediated by tapetum expression of *ABORTED MICROSPORES*. J Exp Bot 73(16):5543–5558
Toledo-Ortiz G, Huq E, Quail PH (2003) The *Arabidopsis* basic/helix-loop-helix transcription factor family. Plant Cell 15(8):1749–1770

Twell D, Park SK, Hawkins TJ, Schubert D, Schmidt R, Smertenko A et al (2002) *Mor1/Gem1* has an essential role in the plant-specific cytokinetic phragmoplast. Nat Cell Biol 4:711–714
Ueguchi-Tanaka M, Nakajima M, Katoh E, Ohmiya H, Asano K, Saji S et al (2007) Molecular interactions of a soluble gibberellin receptor, GID1, with a rice DELLA protein, SLR1, and gibberellin. Plant Cell 19(7):2140–2155
Verelst W, Saedler H, Munster T (2007a) MIKC MADS-protein complexes bind motifs enriched in the proximal region of late pollen-specific *Arabidopsis* promoters. Plant Physiol 143:447–460
Verelst W, Twell D, de Folter S, Immink R, Saedler H, Munster T (2007b) MADS-complexes regulate transcriptome dynamics during pollen maturation. Genome Biol 8:R249
Vizcay-Barrena G, Wilson ZA (2006) Altered tapetal PCD and pollen wall development in the *Arabidopsis ms1* mutant. J Exp Bot 57:2709–2717
Von Mohl H (1830) Uber die fibrosen Zellen der Antheren. Flora
Walbot V, Egger RL (2016) Pre-meiotic anther development: cell fate specification and differentiation. Annu Rev Plant Biol 67:365–395
Wallace S, Fleming A, Wellman CH, Beerling DJ (2011) Evolutionary development of the plant and spore wall. AoB Plants 2011:plr027
Wang A, Xia Q, Xie W, Datla R, Selvaraj G (2003) The classical Ubisch bodies carry a sporophytically produced structural protein (RAFTIN) that is essential for pollen development. Proc Natl Acad Sci USA 100:14487–14492
Wang D, Li C, Zhao Q, Zhao L, Wang M, Zhu D et al (2009) *Zm401p10*, encoded by an anther-specific gene with short open reading frames, is essential for tapetum degeneration and anther development in maize. Funct Plant Biol 36(1):73–85
Wang H, Mao Y, Yang J, He Y (2015) *TCP24* modulates secondary cell wall thickening and anther endothecium development. Front Plant Sci 6:436
Wang K, Guo ZL, Zhou WT, Zhang C, Zhang ZY, Lou Y, Xiong SX, Yao XZ, Fan JJ, Zhu J, Yang ZN (2018) The regulation of sporopollenin biosynthesis genes for rapid pollen wall formation. Plant Physiol 178(1):283–294
Wang R, Owen HA, Dobritsa AA (2021) Dynamic changes in primexine during the tetrad stage of pollen development. Plant Physiol 187(4):2393–2404
Wang J, Wang G, Liu W, Yang H, Wang C, Chen W, Zhang X, Tian J, Yu Y, Li J, Xue Y, Kong Z (2023a) Brassinosteroid signals cooperate with katanin-mediated microtubule severing to control stamen filament elongation. EMBO J 42(4):e111883
Wang Y, Shi L, Feng W, Fu Y, Li C (2023b) *Arabidopsis MYB21* negatively regulates KTN1 to fine-tune the filament elongation. Plants (Basel) 12(22):3884
Wei S, Ma L (2023) Comprehensive insight into tapetum-mediated pollen development in *Arabidopsis thaliana*. Cells 12(2):247
Wei M, Wei H, Wu M et al (2013) Comparative expression profiling of miRNA during anther development in genetic male sterile and wild type cotton. BMC Plant Biol 13:66
Wierzbicki AT, Blevins T, Swiezewski S (2021) Long noncoding RNAs in plants. Annu Rev Plant Biol 72:245–271
Wiese AJ, Torutaeva E, Honys D (2024) The transcription factors and pathways underpinning male reproductive development in *Arabidopsis*. Front Plant Sci 15:1354418
Wijeratne AJ, Zhang W, Sun Y, Liu W, Albert R, Zheng Z, Oppenheimer DG, Zhao D, Ma H (2007) Differential gene expression in *Arabidopsis* wild-type and mutant anthers: insights into anther cell differentiation and regulatory networks. Plant J 52(1):14–29
Wilson ZA, Zhang DB (2009) From *Arabidopsis* to rice: pathways in pollen development. J Exp Bot 60:1479–1492
Wilson ZA, Song J, Taylor B, Yang C (2011) The final split: the regulation of anther dehiscence. J Exp Bot 62(5):1633–1649
Wu MF, Tian Q, Reed JW (2006) *Arabidopsis* microRNA167 controls patterns of *ARF6* and *ARF8* expression, and regulates both female and male reproduction. Development 133(21):4211–4218
Xie HT, Zhang YW, Li S, Zhang Y (2014) Spatiotemporal production of reactive oxygen species by NADPH oxidase is critical for tapetal programmed cell death and pollen development in *Arabidopsis*. Plant Cell 26:193–201

Xing SP, Zachgo S (2008) *ROXY1* and *ROXY2*, two *Arabidopsis* glutaredoxin genes, are required for anther development. Plant J 53:790–801

Xing S, Salinas M, Huijser P (2011) New players unveiled in early anther development. Plant Signal Behav 6(7):934–938

Xu XF, Wang B, Feng YF, Xue JS, Qian XX, Liu SQ, Zhou J, Yu YH, Yang NY, Xu P, Yang ZN (2019) AUXIN RESPONSE FACTOR17 directly regulates *MYB108* for anther dehiscence. Plant Physiol 181(2):645–655

Xu Y, Wang F, Chen Z, Wang J, Li W, Fan F et al (2020) CRISPR/Cas9-targeted mutagenesis of the *OsROS1* gene induces pollen and embryo sac defects in rice. Plant Biotechnol J 18(10):1999

Yamauchi M, Yoshida S (1985) Heteroses in net photosynthetic rate, leaf area, tillering, and some physiological characters of 35 F1 rice hybrids. J Exp Bot 36(163):274–280

Yan G, Ma X, Huang W, Wang C, Han Y, Wang S, Liu H, Zhang M (2025) Decoding the complexity of coding and non-coding RNAs across maize anther development at the isoform level. J Genet Genomics 16:S1673-8527(25)00149-3

Yang WC, Ye D, Xu J, Sundaresan V (1999) The *SPOROCYTELESS* gene of *Arabidopsis* is required for initiation of sporogenesis and encodes a novel nuclear protein. Genes Dev 13:2108–2117

Yang SL, Xie LF, Mao HZ, Puah CS, Yang WC, Jiang L, Sundaresan V, De Y (2003) *TAPETUM DETERMINANT1* is required for cell specialization in the *Arabidopsis* anther. Plant Cell 15:2792–2804

Yang C, Vizcay-Barrena G, Conner K, Wilson ZA (2007a) MALE STERILITY1 is required for tapetal development and pollen wall biosynthesis. Plant Cell 19(11):3530–3548

Yang C, Xu Z, Song J, Conner K, Vizcay Barrena G, Wilson ZA (2007b) *Arabidopsis MYB26/ MALE STERILE35* regulates secondary thickening in the endothecium and is essential for anther dehiscence. Plant Cell 19(2):534–548

Yang H, Lu P, Wang Y, Ma H (2011) The transcriptome landscape of *Arabidopsis* male meiocytes from high-throughput sequencing: the complexity and evolution of the meiotic process. Plant J 65(4):503–516

Yang J, Liu X, Xu B, Zhao N, Yang X, Zhang M (2013a) Identification of miRNAs and their targets using high-throughput sequencing and degradome analysis in cytoplasmic male-sterile and its maintainer fertile lines of *Brassica juncea*. BMC Genomics 14:9

Yang J, Tian L, Sun MX, Huang XY, Zhu J, Guan YF, Jia QS, Yang ZN (2013b) AUXIN RESPONSE FACTOR17 is essential for pollen wall pattern formation in Arabidopsis. Plant Physiol 162(2):720–731

Yang XJ, Wu D, Shi JX, He Y, Pinot F, Grausem B, Yin CS, Zhu L, Chen MJ, Luo ZJ, Liang WQ, Zhang DB (2014) Rice *CYP703A3*, a cytochrome P450 hydroxylase, is essential for development of anther cuticle and pollen exine. J Integr Plant Biol 56:979–994

Yang C, Song J, Ferguson AC, Klisch D, Simpson K, Mo R, Taylor B, Mitsuda N, Wilson ZA (2017) Transcription factor *MYB26* is key to spatial specificity in anther secondary thickening formation. Plant Physiol 175(1):333–350

Yanofsky MF, Ma H, Bowman JL, Drews GN, Feldmann KA, Meyerowitz EM (1990) The protein encoded by the *Arabidopsis* homeotic gene *agamous* resembles transcription factors. Nature 346:35–39

Yasumoto K, Terachi T, Yamagishi H (2009) A novel Rf gene controlling fertility restoration of Ogura male sterility by RNA processing of *orf138* found in Japanese wild radish and its STS markers. Genome 52:495–504

Ye Q, Zhu W, Li L, Zhang S, Yin Y, Ma H et al (2010) Brassinosteroids control male fertility by regulating the expression of key genes involved in *Arabidopsis* anther and pollen development. Proc Natl Acad Sci USA 107:6100–6105

Yu J, Meng Z, Liang W, Behera S, Kudla J, Tucker MR, Luo Z, Chen M, Xu D, Zhao G, Wang J, Zhang S, Kim YJ, Zhang D (2016) A rice Ca2+ binding protein is required for tapetum function and pollen formation. Plant Physiol 172(3):1772–1786

Yuan LP (1987) Strategy concepts of hybrid rice breeding. Hybrid Rice 1:1–4

Zhang D, Wilson ZA (2009) Stamen specification and anther development in rice. Chin Sci Bull 54:2342–2353

Zhang D, Yang L (2014) Specification of tapetum and microsporocyte cells within the anther. Curr Opin Plant Biol 17:49–55

Zhang W, Sun Y, Timofejeva L, Chen C, Grossniklaus U, Ma H (2006) Regulation of *Arabidopsis* tapetum development and function by *DYSFUNCTIONAL TAPETUM1 (DYT1)* encoding a putative bHLH transcription factor. Development 133(16):3085–3095

Zhang ZB, Zhu J, Gao JF et al (2007) Transcription factor *AtMYB103* is required for anther development by regulating tapetum development, callose dissolution and exine formation in *Arabidopsis*. Plant J 52(3):528–538

Zhang S, Fang Z, Zhu J, Gao J, Yang Z (2010) *OsMYB103* is required for rice anther development by regulating tapetum development and exine formation. Chin Sci Bull 55:3288–3297

Zhang D, Luo X, Zhu L (2011) Cytological analysis and genetic control of rice anther development. J Genet Genomics 38:379–390

Zhang D, Liu D, Lv X, Wang Y, Xun Z, Liu Z, Li F, Lu H (2014) The cysteine protease CEP1, a key executor involved in tapetal programmed cell death, regulates pollen development in *Arabidopsis*. Plant Cell 26(7):2939–2961

Zhang W, Xie Y, Xu L et al (2016) Identification of microRNAs and their target genes explores miRNA mediated regulatory network of cytoplasmic male sterility occurrence during anther development in Radish (*Raphanus sativus* L.). Front Plant Sci 7:105

Zhang Y, Ran Y, Nagy I, Lenk I, Qiu JL, Asp T et al (2020) Targeted mutagenesis in ryegrass (*Lolium* spp.) using the CRISPR/Cas9 system. Plant Biotechnol J 18(9):1854

Zhang M, Wei H, Hao P, Wu A, Ma Q, Zhang J et al (2021a) GhGPAT12/25 are essential for the formation of anther cuticle and pollen exine in cotton (*Gossypium hirsutum* L.). Front Plant Sci 12:667739

Zhang R, Chang J, Li J, Lan G, Xuan C, Li H, Wei C (2021b) Disruption of the bHLH transcription factor abnormal tapetum 1 causes male sterility in watermelon. Hortic Res 8:258

Zhan H, Xiong H, Wang S, Yang ZN (2018) Anther endothecium-derived very-long-chain fatty acids facilitate pollen hydration in *Arabidopsis*. Mol Plant 11(8):1101–1104

Zhao DZ, Wang GF, Speal B, Ma H (2002) The *EXCESS MICROSPOROCYTES1* gene encodes a putative leucine-rich repeat receptor protein kinase that controls somatic and reproductive cell fates in the *Arabidopsis* anther. Genes Dev 16:2021–2031

Zhao F, Zheng YF, Zeng T, Sun R, Yang JY, Li Y, Ren DT, Ma H, Xu ZH, Bai SN (2017) Phosphorylation of *SPOROCYTELESS/NOZZLE* by the MPK3/6 kinase is required for anther development. Plant Physiol 173(4):2265–2277

Zhou D, Song R, Fang Y, Liu R, You C, Wang Y, Huang L (2025) Global identification and regulatory network analysis reveal the significant roles of lncRNAs during anther and pollen development in *Arabidopsis*. Plant Cell Rep 44(2):44

Zhu J, Chen H, Li H, Gao JF, Jiang H, Wang C, Guan YF, Yang ZN (2008) *Defective in Tapetal development and function 1* is essential for anther development and tapetal function for microspore maturation in *Arabidopsis*. Plant J 55(2):266–277

Zhu E, You C, Wang S, Cui J, Niu B, Wang Y, Qi J, Ma H, Chang F (2015) The DYT1-interacting proteins bHLH010, bHLH089 and bHLH091 are redundantly required for *Arabidopsis* anther development and transcriptome. Plant J 83(6):976–990

Zhu T, Li Z, An X, Long Y, Xue X, Xie K, Ma B, Zhang D, Guan Y, Niu C, Dong Z, Hou Q, Zhao L, Wu S, Li J, Jin W, Wan X (2020) Normal structure and function of endothecium chloroplasts maintained by ZmMs33-mediated lipid biosynthesis in tapetal cells are critical for anther development in maize. Mol Plant 13(11):1624–1643

Chapter 5
Bioengineering Photosynthesis to Drive Crop Improvement: Engineering Carbon-Concentrating Mechanisms, Kranz Anatomy, Chloroplast Mimics, and New Cellular Compartments

Mouesanao K. Kandjoze and Pallavi Singh

Abstract Global food security is increasingly threatened by climate change and population growth, necessitating enhanced agricultural productivity. Photosynthesis, the foundation of life on Earth, represents a crucial target for bioengineering. This chapter explores innovative strategies for enhancing photosynthetic efficiency, focusing on approaches that optimise natural carbon fixation mechanisms, with a particular emphasis on three key research areas: (1) leveraging biophysical carbon--concentrating mechanisms to integrate functional pyrenoids and carboxysomes into higher plant chloroplasts; (2) introducing CAM, Kranz anatomy, and C_4 photosynthetic traits into staple crops; and (3) engineering chloroplast genomes. Moreover, the chapter highlights breakthroughs in gene editing, cellular reprogramming, and synthetic biology. Emerging tools, such as spatial-omics and synthetic gene circuits, which are transforming conceptual approaches into commercial applications, are also discussed. This exploration of cutting-edge bioengineering approaches offers promising avenues for developing next-generation crops with improved photosynthetic performance, contributing to a more sustainable and food-secure future.

Keywords Photosynthesis · Bioengineering · Chloroplast · Yield · Resilience

Editor's Notes
Crop productivity fundamentally hinges on efficiency of photosynthesis, a natural process powered by solar energy, resulting in fixation of atmospheric CO_2 into chemically available forms. However, photosynthetic efficiency is primarily limited by the sub-par catalytic rate of Rubisco posing a major bottleneck in boosting crop

M. K. Kandjoze · P. Singh (✉)
School of Life Sciences, University of Essex, Colchester, UK
e-mail: pallavi.singh@essex.ac.uk

A. Singh et al. (eds.), *Sustainable Crop Production*, Sustainability Sciences in Asia and Africa, https://doi.org/10.1007/978-981-95-7844-3_5

productivity. Enhancing photosynthetic efficiency in crop species has therefore been a long-standing goal in agbiotech research.

In this chapter, Kandjoze and Singh discuss the prospects of bioengineering photosynthesis to enhance its efficiency and reduce wasteful CO_2 emissions during photorespiration. To innovate mechanisms for efficient biological fixation of CO_2, the authors propose scholars to pursue three key areas of research: "*(1) leveraging biophysical carbon-concentrating mechanisms (CCMs) to integrate functional pyrenoids and carboxysomes into higher plant chloroplasts, (2) introducing CAM, Kranz anatomy, and C4 photosynthetic traits into staple crops, and (3) engineering chloroplast genomes.*" Appearing as impossible even in recent past, the authors inform how mutational studies, and computational modeling can support re--engineering of functional pyrenoid-based and carboxysome-based CCMs for achieving enhanced photosynthetic efficiency. Bioengineering of thylakoid membranes and stroma, and sequestering Rubisco by way of creating a CO_2 diffusion barrier are ingenious approaches that are being tested currently. The authors also propose integration of several other promising strategies such as introducing CAM, Kranz anatomy, and C4 photosynthetic traits into crops species. The authors present an elegant account on the potential of modern technologies in circumventing bottlenecks in each of these approaches which currently are at the proof-of-concept stage. The chapter is thus a starting point for researchers willing to explore this exciting field.

Actionable Points

1. **Enhancing photosynthetic efficiency:** Discusses current strategies for optimizing photosynthetic carbon fixation through biophysical carbon concentrating mechanisms integrating pyrenoids and carboxysomes.
2. **Engineering approaches of C_4 and CAM photosynthesis:** Explores approaches to introduce anatomical and biochemical traits from C_4 and CAM species into staple crops for improved yield resilience.
3. **Technological advancements:** Highlights experimental innovations and strategies that can pave the way to climate-smart crops.
4. **Future directions:** Provides insights into the technology-readiness levels of different photosynthetic improvement strategies, focusing on a path toward accelerating translational solutions.

5.1 Introduction

The global challenge of food security is intensifying, driven by the combined pressures of climate change and a rapidly growing population. As the global demand for food continues to escalate, traditional agricultural methods are struggling to keep pace. To meet this growing need, enhancing agricultural productivity through innovative approaches has become a priority. At the core of addressing this challenge is photosynthesis, the fundamental process that sustains life on Earth and drives plant productivity.

5.2 Background

5.2.1 Brief Evolution of Carbon Concentrating Mechanisms

Rubisco is responsible for 90% of inorganic carbon fixation (Erb and Zarzycki 2018). Globally, Rubisco is considered an abundant enzyme, with terrestrial Rubisco estimated to weigh 0.7 gigatons, about 14 times that of global estimates of collagen (Bar-On and Milo 2019). Despite having evolved 3.5 billion years ago, Rubisco is eight to ten times slower than other central metabolic enzymes (Barrett et al. 2021). Evolutionary biologists suggest that Rubisco evolved when atmospheric oxygen levels were low and conferred adaptive advantages due to its carbon-fixing activity (Sage 2004). As O_2 levels rose following the Great Oxygenation Event, Rubisco evolved the capacity to discriminate between CO_2 and O_2, which came at the cost of its catalytic speed (Erb and Zarzycki 2018; Giessen and Silver 2017). Additionally, the chemical similarity between CO_2 and O_2, under O_2-rich atmospheric conditions further impede Rubisco's selectivity for CO_2 (Whitney et al. 2010). Consequently, Rubisco, which has both oxygenase and carboxylase capabilities, still fixes an average of two molecules of O_2 for every five of CO_2 (Giessen and Silver 2017). The promiscuity of Rubisco is not only inefficient, but it also produces toxic 2-phosphoglycolate (2PG), which requires energy to metabolise and leads to the release of CO_2 during photorespiration (Bowes et al. 1971). To overcome this energetically wasteful process, which reduces photosynthetic efficiency by an estimated 30%, photosynthetic organisms have evolved mechanisms to reduce photorespiration by simply producing more Rubisco. In contrast, other systems evolved physical or chemical carbon-concentrating mechanisms (CCMs) in chloroplasts to increase CO_2 concentrations around the site of Rubisco activity (Giessen and Silver 2017; Price et al. 2012; Walsh et al. 2023).

Cyanobacteria, most algae, and most hornwort species developed biophysical CCMs that passively or actively concentrate inorganic bicarbonate (HCO_3^-), convert it to CO_2 via carbonic anhydrases (CA), and release it into a microcompartment that contains Rubisco (Hennacy and Jonikas 2020). Carboxysomes and pyrenoids are the two main types of compartments, both of which evolved physical barriers in the form of a protein shell or starch sheath, respectively, to limit CO_2 leakage (Rae et al. 2017). In cyanobacterial CCMs, the HCO_3^--CO_2 conversion and Rubisco--fixation occurs in the carboxysome matrix (Blikstad et al. 2023). Conversely, algal and hornwort CCM's convert HCO_3^- into CO_2 in specialised thylakoid membrane tubules that traverse thy pyrenoid, prior to releasing the converted CO_2 into the pyrenoid matrix for Rubisco fixation (He et al. 2020).

In contrast, biochemical CCMs, which include C_2, C_4, and crassulacean acid metabolism (CAM), concentrate carbon via organic intermediates such as oxaloacetate (OAA) and malate before delivering it to Rubisco (He et al. 2023). C_4 plants also developed a distinctive Kranz anatomy that further facilitates efficient carbon concentration (Lundgren et al. 2014). Biochemical CCMs have been demonstrated to increase photosynthesis in model species such as *Arabidopsis thaliana* (Arabidopsis) and *Nicotiana tabacum* (tobacco) but also in staple crops including rice and wheat (Sedelnikova et al. 2018).

5.2.2 Agricultural Benefits of CCMs

CCMs offer additional benefits beyond increasing photosynthesis and yield. The more efficient Rubisco activity induced by CCMs reduces the total Rubisco content, thereby improving nitrogen use efficiency (Adler et al. 2022). Furthermore, increased internal CO_2 for Rubisco would decrease stomatal conductance (g_s) and reduce water loss through transpiration in C_3 plants (Adler et al. 2022; Hennacy and Jonikas 2020), thereby enhancing water use efficiency (WUE). Therefore, investing in mechanisms to increase internal carbon concentrations are promising strategies to enhance photosynthetic carbon assimilation, resource use efficiency, and yield.

5.3 Pyrenoid-Based Carbon-Concentrating Mechanisms

Carbon fixation via pyrenoid-based CCMs accounts for about 28–44% of global carbon fixation (Field et al. 1998). *Chlamydomonas reinhardtii* (Chlamydomonas) is the most well-characterised pyrenoid-based CCM and will be used a model to describe how algal CCM systems operate (Adler et al. 2022). Several HCO_3^- transporters located at the plasma, chloroplast, and thylakoid membranes facilitate the accumulation of carbon in the thylakoid lumen (Rottet et al. 2021). HCO_3^- acidifies the lumen, which drives the decarboxylation reaction by carbonic anhydrase 3 (CA3). Finally, released CO_2 diffuses into the pyrenoid matrix for fixation by Rubisco (Barrett et al. 2021; Fei et al. 2022; He et al. 2023).

Pyrenoids have recently attracted research interest as a potential tool for improving photosynthesis in higher plants because of their carbon concentrating efficiency. Pyrenoid-based CCMs have been reported to facilitate an up to 40-fold increase in inorganic carbon (C_i) compared to chloroplasts without pyrenoids (Adler et al. 2022). In addition, the sub-cellular mechanism does not require changes to existing leaf anatomy, such as engineering efforts associated with introduction of C_4 traits (He et al. 2023). Moreover, the genetic components required to express and assemble pyrenoids already exist in eukaryotic genomes (He et al. 2023). Both these factors could streamline the transition of algal carbon-concentrating systems into global crop staples such as wheat or rice.

5.3.1 *Understanding Pyrenoid-Based CCM and Strategies for Engineering in Land Plants*

5.3.1.1 The Optimal Pyrenoid-Based CCM

Both mutational studies and computational modelling approaches detail the key steps required for functional pyrenoid assembly. Fei et al. (2022) modelled a pyrenoid-based CCM that includes three key modules; first, low-inducible B (LCIB) should be localised to the chloroplast stroma to convert passively imported CO_2 into HCO_3^-; second, a thylakoid membrane channel such as BCT1 (bicarbonate transporter 1) is required to deliver HCO_3^- to luminal CA3 for conversion back into CO_2 near Rubisco; and third, phase-separated Rubisco must be enclosed by a CO_2 diffusion barrier. Notably, the model suggests that a passive CO_2 uptake mechanism at ambient CO_2 levels is possible, while active HCO_3^- uptake is beneficial at low atmospheric CO_2 levels. To achieve this three-module pyrenoid-based CCM, land plants would need to be engineered to condense or phase-separate Rubisco, localise CA-rich thylakoids to the Rubisco condensate, integrate HCO_3^- transporters into the thylakoid membrane, and add a diffusion barrier to limit CO_2 diffusion out of the pyrenoid-based structure (Fig. 5.1).

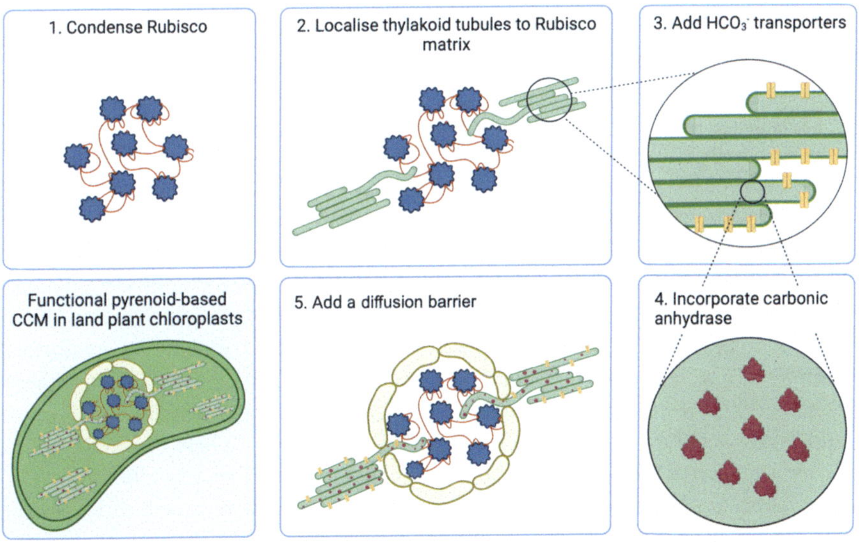

Fig. 5.1 Strategies for engineering a pyrenoid-based CCM in land plants. (1) Rubisco phase separates into a condensate via linker proteins (shown in red) such as EPYC1. (2) Thylakoid membrane tubules are recruited to the matrix core to form pyrenoid tubules, which deliver HCO_3^- to the Rubisco matrix, and supply ATP and NADPH for Rubisco-mediated carbon fixation. (3) HCO_3^- transporters are targeted to the thylakoid membrane to facilitate HCO_3^- accumulation. (4) Targeting carbonic anhydrase to the thylakoid lumen, necessary for HCO_3^- decarboxylation and CO_2 release near the Rubisco matrix. (5) Finally, a diffusion barrier comprised of either starch or stacked thylakoids encloses the pyrenoid to reduce CO_2 leakage. (Created with BioRender.com)

Approach 1: Understanding Pyrenoid Phase Separation and Rubisco Condensation

Understanding phase separation of pyrenoid-associated proteins is a key first step toward pyrenoid-based CCM assembly in higher plants. *Chlamydomonas* phase separation occurs partly due to the interaction between the essential pyrenoid component 1 (EPYC1) linker protein and the small subunit (SSU) of Rubisco (Hennacy and Jonikas 2020). Atkinson et al. (2017) investigated these SSUs by generating *Arabidopsis* mutants lacking a functional SSU and supplemented them with a partial *SSU* sequence encoding the *Chlamydomonas* alpha helix motif, or the complete *Chlamydomonas SSU* sequence. The hybrid Rubisco comprised of the partial *Chlamydomonas* sequence had similar catalytic properties to WT controls, reporting comparable maximum Rubisco carboxylation ($V_{c,\max}$) and max photosynthetic electron transport ($J_{\max}$) rates. Atkinson et al. (2020) then introduced the *EPYC1* linker gene in *Arabidopsis* that featured the *Chlamydomonas* Rubisco hybrid, but it failed to phase separate, partly due to EPYC1 being targeted for degradation. This was remedied by adding a native *Arabidopsis* chloroplast signal peptide to GFP-tagged *EPYC1* constructs (Atkinson et al. 2020). Confocal and transmission electron microscopy showed that EPYC1 was able to condense both the partial and mature *Chlamydomonas* Rubisco hybrids to form phase-separated pyrenoid-like compartments in *Arabidopsis*.

The evidence above indicates that modifying the Rubisco *SSU* is a potential avenue through which to initiate pyrenoid assembly in land plants. Rubisco *SSU* sequences are encoded in the nuclear genome rather than the chloroplast genome (He et al. 2023). Modifying nuclear-encoded Rubisco circumvents the challenges associated with editing chloroplast-encoded genes. This allows for a more standardised approach to CCM engineering, which can be translated to various C_3 plants.

Furthermore, research into under-explored algal pyrenoid systems is further increasing our understanding of how pyrenoids assemble. Oh et al. (2023) identified PYCO1 as a key linker protein that mediates *Phaeodactylum tricornutum* red algae pyrenoid phase separation. In *Chlorella sorokiniana*, sequence-independent bioinformatic analysis identified CsLinker proteins as EPYC1 functional analogues (Barrett et al. 2024). Most importantly, Barrett et al. (2024) demonstrated that these CsLinkers bind to highly conserved regions of Rubisco large subunit (LSU) and could phase separate native *Chlamydomonas* and tomato Rubisco in vivo, while tobacco Rubisco phase separated both in vitro and in vivo. This is the first report of native Rubisco phase separation in land plants and presents a critical milestone in pyrenoid-based CCM engineering efforts in C_3 plants.

The present strategies outlined highlight the potential of easily modifiable Rubisco SSUs or conserved LSUs as targets for Rubisco condensation. Additionally, the identification of a diverse set of linker proteins would further expand our repertoire of tools that can be matched to compatible C_3 pyrenoid-based CCMs.

Approach 2: Localising the Thylakoid to the Rubisco Matrix

For functional pyrenoid assembly, thylakoids must traverse the pyrenoid Rubisco matrix. Through cryo-electron transmission microscopy, it has been shown that thylakoid membranes merge as they traverse the starch sheath to form a pyrenoid tubule made up of several smaller, fenestrated tubules (He et al. 2023). The fenestrations connect the thylakoid lumen to the chloroplast stroma and are used to deliver concentrated CO_2 to the Rubisco matrix (Hennacy et al. 2024). Pyrenoid tubules are recruited to the pyrenoid by a starch-sheath-associated protein called Starch Granule Abnormal1 (SAGA1). This is evidenced by *Chlamydomonas* mutations in *SAGA1*, which produced pyrenoids without tubules, or with significantly fewer tubules compared to the WT (Itakura et al. 2019).

Other proteins further implicate the starch sheath in pyrenoid tubule formation. Proximity labelling identified a protein encoded by *Cre09.g394510*, which localised to the starch sheath periphery (Lau et al. 2023). The C-terminal domain of the protein-mediated vesicle fusion and was hypothesised to rearrange the thylakoid grana into a pyrenoid tubule as it traversed the sheath and entered the pyrenoid matrix. Additionally, the *mutation missing thylakoids 1 (MITH1)* in *Chlamydomonas* was associated with tubule-lacking pyrenoids (Hennacy et al. 2024). When expressed together with *SAGA1* in *Arabidopsis* previously engineered to form Rubisco condensates, thylakoid sheets extended into the pyrenoid-like assemblies (Hennacy et al. 2024). Researchers proposed that *MITH1* was required to form an MITH1-SAGA1 interaction that replaced existing Rubisco-EPYC1 interactions, which allowed the thylakoid to traverse the pyrenoid matrix. This was the first time that thylakoid-traversing structures in proto-pyrenoids were observed in planta, bringing this CCM approach one step closer to being fully assembled in land plants.

Approach 3: Expressing HCO_3^- Transporters in Thylakoid Membranes

Biophysical carbon concentration is mediated by several HCO_3^- transporters. Under atmospheric CO_2 levels, C_i passively diffuses across the plasma membrane via the low CO_2-inducible 1 (LCI1) channel (Adler et al. 2022; He et al. 2023). Cytoplasmic CO_2 then diffuses into the chloroplast stroma, where it is rapidly converted to HCO_3^- by the low-inducible B/C (LCIB/LCIC) complex (Rottet et al. 2021). However, at suboptimal CO_2 levels (below 200 ppm CO_2), pyrenoid-based CCMs switch to actively import HCO_3^- into the cytosol and stroma via HLA3 (high light activated 3) and LCIA, respectively (Adler et al. 2022). The stromal HCO_3^- is then shuttled into the thylakoid lumen via BCT1, 2, and 3, where it travels down into the pyrenoid membrane tubule that bisects the pyrenoid (Adler et al. 2022; Fei et al. 2022).

Research into LCI transporters is shedding light on their role in CO_2 accumulation. LCIA, which was suspected to be an active HCO_3^- transporter, was recently confirmed through a phenotypic rescue assay in *Arabidopsis* (Förster et al. 2023). *Arabidopsis* mutants lacking *bCA5*, which could not grow in ambient CO_2 levels, grew when rescued with LCIA, indicating that the channel was actively transporting HCO_3^- and facilitating sufficient carbon accumulation to support photoautotrophic growth. However, LCIA expression in WT tobacco did not improve photosynthetic performance (Förster et al. 2023). Similarly, LCIB expression was also able to restore *bCA5*-deficient *Arabidopsis* growth under atmospheric CO_2 levels (Kasili et al. 2023). Additionally, in low CO_2 conditions, LCIB transporters were recruited to the pyrenoid by the *CCM1* regulatory protein and to the surrounding pyrenoid area by starch-sheath-associated isoamylase1 (ISA1) (Toyokawa et al. 2020; Yamano et al. 2022). Furthermore, *Chlamydomonas* LCIB expression in tobacco, along with its target peptide sequence, increased the maximum Rubisco carboxylation rate (Vc_{max}) by 11–13%, fresh shoot biomass by 33% and root biomass by 57% compared to WT tobacco (Nölke et al. 2019). Based on these findings, LCIB is a potential promising target for photosynthetic improvement.

Approach 4: Targeting Functional CA to Pyrenoid Membrane Tubule Lumens

In many pyrenoid-based CCMs, the conversion of HCO_3^- back into CO_2 by CA in the thylakoid lumen is the final stage of carbon concentration in pyrenoids. Several CAs coordinate the accumulation of HCO_3^- in the pyrenoid matrix (Terentyev and Shukshina 2024). Of the various CAs, CA3 localises to the thylakoid lumen and mediates the delivery of CO_2 to the Rubisco matrix (Atkinson et al. 2016). In *Chlamydomonas*, mutations in CA3 impaired pyrenoid functioning and growth in atmospheric CO_2 conditions, despite demonstrating HCO_3^- accumulation in thylakoids and pyrenoid tubules (Mukherjee et al. 2019). CA is therefore crucial for the final step of CO_2 delivery to Rubisco.

In tobacco, although CA3 was correctly targeted to the chloroplasts, it was also largely mis-localised to the cytosol, despite expression with the *Chlamydomonas* transit peptide sequence (Atkinson et al. 2016). However, another study achieved preferential CA3 expression in tobacco thylakoid lumens, with only minor expression in the cytosol (Nölke et al. 2019). CA3 expression was also found together with a 16–17% increase in Vc_{max}, significantly increased fPSII (which indicates photosynthetic capacity), leaf number, and a 28% increase in shoot biomass (Nölke et al. 2019). Enhanced photosynthesis may be due to CA, which has previously been reported to be involved in PSII operation in thylakoids (Terentyev and Shukshina 2024). Therefore, although the findings suggest the potential of algal CA to improve photosynthetic performance, more research is required to determine whether improvement is due to enhanced PSII activity or increased CO_2 accumulation. Nonetheless, the findings confirm the benefits of introducing algal CA into land plants.

Additionally, the role of stromal CA in biophysical CCMs is debated. Some studies suggest stromal CA prevents HCO_3^- from accumulating in the cytosol by causing HCO_3^--CO_2 equilibration, which increases CO_2 leakage (Atkinson et al. 2016; Giessen and Silver 2017). This has motivated the proposal that native stromal CA should be eliminated and replaced with strategically localised CA isoforms.

However, reaction-diffusion modelling demonstrates that a pyrenoid-based CCM can still function with stromal CA, provided the system is otherwise well-optimised (Fei et al. 2022). Their findings emphasise the importance of targeting CA to the thylakoid lumen for efficient CO_2 delivery to the pyreoid matrix but do not support a strict requirement to remove stroml CA. Therefore, engineering efforts are advised to prioritise precise CA localisation rather than fully knocking out native CA—a task yet to be demonstrated in C_3 plants.

Approach 5: Creating a Diffusion Barrier

Experimental data implies that a diffusion barrier surrounding the pyrenoid-based CCM reduces CO_2 leakage (Toyokawa et al. 2020). In *Chlamydomonas*, the barrier is polysaccharide-based, comprising of curved starch granules that enclose the Rubisco condensate (He et al. 2023). Several proteins are also associated with the starch sheath, helping to assemble the individual plates, tether them together, and link the matrix to the sheath (Barrett et al. 2021; He et al. 2023). Of these starch proteins, SAGA1 and SAGA2 are well-characterised proteins that attach the Rubisco condensate to the sheath (Atkinson et al. 2024).

Under atmospheric (0.04% [v/v] CO_2) or very low (0.01% [v/v] CO_2) CO_2 conditions, a starch sheath is essential for pyrenoid-based carbon concentration. In *Chlamydomonas*, mutants lacking *SAGA1* observed increased pyrenoid numbers per chloroplast, thinner and uneven starch sheaths, decreased growth rates under low CO_2, and reduced photosynthetic rates under saturating C_i at both low and high CO_2 levels (Itakura et al. 2019). Additionally, SAGA1 was also demonstrated to recruit stromal HCO_3^- transporters such as HLA3 and LCIA to the surrounding pyrenoid area under CO_2-limiting conditions (Shimamura et al. 2023). The starch sheath may play a role in the proper morphological pyrenoid assembly, which contributes to optimal protein recruitment and functioning, particularly in atmospheric and suboptimal CO_2 conditions (Barrett et al. 2021; Hennacy and Jonikas 2020).

Recently, *Chlamydomonas SAGA 1* and *SAGA2* expression resulted in a partial starch sheath formation around the synthetic proto-pyrenoid in *Arabidopsis* (Atkinson et al. 2024). To form the partial sheath, *SAGA1* was hypothesised to primarily aggregate starch to form plate-like sheaths, while *SAGA2* facilitated the sheath-condensate interactions (Atkinson et al. 2024).

Altogether, starch-based diffusion barriers are essential for pyrenoid function, and its partial engineering into land plants is a promising step toward achieving biophysical carbon concentration in land plants.

Thylakoid-based diffusion barriers have also been observed and modelled to be effective in a pyrenoid-based CCM (Fei et al. 2022; Robison et al. 2025). Experimentally derived data about the gaps between thylakoid membranes and diffusivity were used to model realistic thylakoid geometry and membrane permeability (Fei et al. 2022). Simulation results suggest that tight spacing between thylakoid membranes ($\Delta_n = 5.6$ nm) and low diffusivity ($\approx 10^3$) substantially reduce CO_2 diffusion in the stroma, making them an efficient barrier against CO_2 leakage (Fei et al.

2022). Different thylakoid-based barriers have been observed in other algal species, such as in *Cyanophora paradoxa*, where the thylakoid stacks surround a naked Rubisco condensate, or in *Euglena carterae*, where thylakoid stacks both traverse and surround the Rubisco matrix (He et al. 2023). Additionally, anatomical research into Hornwort species, the only land plants known to have pyrenoid-based CCMs, were recently reported to contain multiple pyrenoids per chloroplast, which are collectively surrounded by thylakoid membrane stacks(Robison et al. 2025). Although the magnitude of CO_2 leakage of thylakoid-enclosed pyrenoids is yet to be experimentally determined, modelled estimates, along with experimental observations, support the feasibility of a pyrenoid-based CCM that encompasses a thylakoid membrane diffusion barrier. It should be noted that research on thylakoid-based pyrenoids are still in its preliminary discovery phase and no attempts at engineering them into higher land plants have been made. Overall, the evidence above confirms the importance of a diffusion barrier in effective pyrenoid-based CCMs.

5.4 Carboxysome-Based Carbon-Concentrating Mechanisms

5.4.1 Introduction to Carboxysomes

Like algae, cyanobacteria and most photosynthesising proteobacteria have also developed a physical component to compensate for the inefficiency of Rubisco, called the carboxysome (Blikstad et al. 2023). Through convergent evolution, two carboxysomes have evolved (Giessen and Silver 2017). α-carboxysomes, expressing Form 1A Rubisco, which is encoded by the *cso* operon, while β-carboxysomes, expressing 1B Rubisco, located on the *ccm* operon (Hennacy and Jonikas 2020). For carboxysome-based carbon concentration, CO_2 passively diffuses across the plasma membrane, while HCO_3^- is pumped into the cytosol, where it diffuses across the selectively permeable protein shell and into the carboxysome (Trettel et al. 2024). Inside the carboxysome, CA converts HCO_3^- back into CO_2 for fixation by Rubisco (Hennacy and Jonikas 2020). Cyanobacteria have been reported to have 10–40 times higher HCO_3^- concentrations in the cytosol compared to stromal levels in C_3 species, making carboxysomes an area of interest for researchers (Long et al. 2018).

5.4.2 Strategies for Carboxysome Engineering

Price et al. (2012) proposed a four-step plan to engineer carboxysomes into higher plants. This plan involves (1) transferring HCO_3^- transporters to chloroplast membranes, (2) removing endogenous stromal CA activity, (3) assembling a functional carboxysome in chloroplast stromal regions, and (4) incorporating an NADH1-CO_2

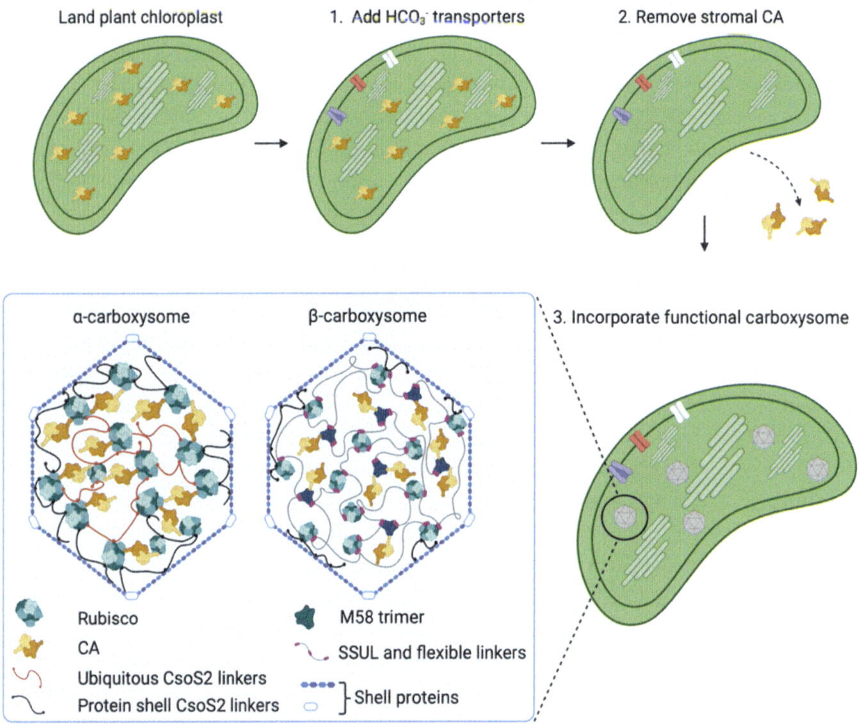

Fig. 5.2 Stepwise construction of a carboxysome-based CCM in land plant chloroplasts. (1) HCO_3^- transporters such as BTC1 (purple), BicA (pink), and SbtA (white) need to express in the inner-envelope membrane of chloroplasts. (2) Stromal CA is removed to allow for efficient HCO_3^- accumulation and to prevent CO_2 leakage. (3) Proteins that comprise the carboxysome shell need to self-assemble to form a carboxysome structure within chloroplasts. Transgenic α- or β-carboxysomes (blue diagram box) need to incorporate the appropriate linkers, condensation modules, and scaffold proteins to enable matrix condensation and protein shell encapsulation. (Created with BioRender.com)

uptake complex within the thylakoid membrane (Fig. 5.2). So far, extensive research has gone into the first three steps, which will be discussed further in this section.

5.4.2.1 Step 1: To Insert Cyanobacterial HCO_3^- Transporters in Chloroplast Membranes

Historically, an early strategy toward carboxysome-based CCMs in land plants involved engineering chloroplasts to express HCO_3^- transporters instead of incorporating whole carboxysomes (Hennacy and Jonikas 2020). This hypothesis is partly supported by reaction-diffusion modelling, which suggests that transforming non-CCM crops with cyanobacterial HCO_3^- transporters could increase daily biomass accumulation by 5–10%, which can have a significant cumulative effect on final yield (McGrath and Long 2014; Price et al. 2011). Initially, transporters such as bicarbonate transporter A (BicA) and sodium-dependent bicarbonate transporter A (SbtA) were transformed into higher plants, because they were encoded by a single gene (Rae et al. 2017). Single-gene proteins avoid the challenge of targeting multiple domains to the correct cellular location. However, the StbA failed to accumulate carbon in chloroplasts or to increase photosynthetic performance in higher plants, emphasising the need for further research into the mechanisms that regulate HCO_3^- transporter activity (Nguyen et al. 2024; Rae et al. 2017).

Recently, updated reaction-diffusion modelling is challenging the widely cited notion that expressing HCO_3^- transporters in the inner envelope membrane (IEM) of land plant chloroplasts would improve photosynthetic efficiency (Kaste et al. 2024). In this spatially resolved model of land plants with HCO_3^- transporters, Kaste et al. (2024) demonstrate that previous models underestimate lipid membrane permeability, which, when modelled, supports a higher photosynthetic efficiency. However, when the membrane porosity is adjusted to physiologically observed levels, more CO_2 leakage, reduced Rubisco fixation and lower light use efficiency are predicted. This results in energetically wasteful HCO_3^- pumping. The contrasting findings direct researchers and the wider scientific community to be cautious and through when simulating and interpreting modelled data. Additionally, neither simulation model has thus far been experimentally demonstrated.

Inorganic carbon transporter B (ictB) has attracted interest and debate regarding whether it contributes to carbon concentration as an active transporter or passive channel (Koester et al. 2021; Price et al. 2012). *IctB* expression in soybean and maize was associated with modestly increased photosynthesis and yield in greenhouse and field trials, representing a major milestone in the agricultural translation toward carboxysome-based crops (Hay et al. 2017; Koester et al. 2021). However, when expressed in tobacco, *ictB* failed to improve photosynthesis or yield in field trials (Ruiz-Vera et al. 2022). Thus, the role of IctB in photosynthesis and yield appears to be species and environmentally dependent.

There has been recent success in targeting the larger, multi-domain BCT1 to chloroplast membranes in tobacco (Rottet et al. 2024). Rational design and directed evolution were used to optimise *BCT1* expression, localisation, and functionality in *Escherichia coli* prior to transformation in *Arabidopsis*. Although BCT1 domains localised to the IEM, and in the correct orientation, transgenic *Arabidopsis* observed lower CO_2 assimilation rates compared to WT controls. Although the correct localisation and orientation have been achieved, HCO_3^-transporter functionalisation remains to be achieved in planta.

5.4.2.2 Step 2: To Eliminate Endogenous CA from the Chloroplast Stroma

Similar to pyrenoid-based chloroplasts, carboxysome-based CCMs are expected to accumulate carbon more efficiently if native CA is removed from the chloroplast stroma and replaced with carboxysome-associated CAs (Nguyen et al. 2024; Price et al. 2012). In carboxysome-engineered chloroplasts, the single step of removing CA from the stroma was estimated to increase HCO_3^- to 2–3 mM, which was enough to saturate CO_2 fixation (Price et al. 2012). This was achieved through CRISPR/Cas9 editing in tobacco or by transforming *Arabidopsis* with non-functional variant of CA, which expectedly hindered growth and fertility. Stromal CA was involved in metabolic regulation in C_3 crops. Although it was assumed that relocalised CA would not affect its role in anaplerotic pathway activity, this remains to be demonstrated as no study has concurrently reduced or eliminated stromal CA and expressed encapsulated CA (Nguyen et al. 2024).

5.4.2.3 Step 3: To Assemble a Functional Carboxysome in the Chloroplast Stroma

To assemble a functional carboxysome, cyanobacterial Rubisco must be stoichiometrically co-expressed with CA, scaffold, carboxysome shell, and linker proteins (Chen et al. 2023). In tobacco, a 'minimal carboxysome' was created by replacing endogenous Rubisco with *Cyanobium marinum* PCC7001 (*Cyanobium*) Rubisco, which was introduced together with the shell protein CsoS1A and a CsoS2 linker, which tethers Rubisco to the carboxysome shell (Long et al. 2018). At 2% CO_2, exogenous Rubisco in tobacco had similar catalytic activity to WT *Cyanobium* Rubisco. However, when enclosed in the minimal carboxysome, Rubisco's catalytic rate dropped drastically even at high CO_2 levels. Recently, Chen et al. (2023) demonstrated Rubisco catalytic activity in a carboxysome comprised from nine genes. Despite functionality, transformed tobacco failed to germinate under ambient CO_2 conditions, grew slowly, and had poor net CO_2 assimilation rates at 1% CO_2. Presently, although carboxysomes can be assembled in higher plants and demonstrate functional Rubisco, they can only support autophagic growth under elevated CO_2 (Chen et al. 2023; Long et al. 2018; Nguyen et al. 2023).

Advances in structural biology are paving the way for resolving the mechanisms behind carboxysome assembly, which will be important steps in the engineering of carboxysomal traits into chloroplasts. In α-carboxysomes from two distinct species, cryo-electron transmission resolved that the CsoS2 linker proteins either bind ubiquitously to Rubisco throughout the condensate or preferentially to Rubisco near the carboxysome shell in *Cyanobium* (Ni et al. 2022). Additionally, in cryo-electron microscopy, Blikstad et al. (2023) showed that CA is recruited to the a-carboxysome condensate by interacting directly with Rubisco. In β-carboxysomes, the CcmM protein, termed M58, has an N-terminus gCAL binding site, which organises the scaffold protein into a trimer and binds to the N-terminus of CA (Zang et al. 2021). The M58 trimer also recruits Rubisco via small subunit-like modules, which are strung together by linker proteins. Understanding the interactions between carboxysome enzymes and their associated scaffolds and linkers could optimise carboxysome assembly in planta.

5.5 Biochemical Carbon-Concentrating Mechanisms

In contrast to biophysical CCMs, biochemical CCMs follow a two-step process to metabolically fix and concentrate carbon into organic intermediates, which are then delivered to Rubisco (Rottet et al. 2021). These two processes are achieved either spatially, across two different cells as with C_4 plants, or temporally in CAM plants (Yuan et al. 2020). In C_4 species, the relevant biochemical and anatomical adaptations result in increased photosynthesis, water, and nitrogen use efficiency compared to C_3 species (Zhao et al. 2023). In addition to a tenfold increase in carbon concentration near Rubisco, C_4 photosynthesis also confers adaptive advantages to plants in hot and dry climates (Schlüter and Weber 2016; Simpson et al. 2021). As such, biochemical CCMs present an equally attractive approach for engineering plants to achieve enhanced photosynthetic capacity. This section will discuss the approaches, successes, and challenges currently underway to engineer biochemical CCMs, focusing primarily on engineering C_4 traits into C_3 plants (Fig. 5.3). The C_3 to C_4 transition could equip global crop staples like rice, wheat, and soybean with C_4 characteristics to facilitate faster growth and improved yield. Wherever possible, a perspective on the feasibility of current approaches as well as recommendations for future research will also be discussed.

5.6 Background and Brief Evolution C_4 Photosynthesis

C_4 photosynthesis occurs across two cell types, mesophyll (M) and bundle sheath (BS) (Schlüter and Weber 2016). HCO_3^- produced by CA in M cells is fixed into a four-carbon OAA by phospho*enol*pyruvate carboxylase (PEPC) (Schlüter and Weber 2016). OAA is either converted to malate or aspartate and subsequently shuttled across the BS-M membrane interface and into BS cells by transporters (Singh et al. 2022). C_4 photosynthesis is divided into three subtypes, based on the decarboxylating enzyme (Lin et al. 2020). In the first subtype, NADP-dependent malate dehydrogenase (NADP-ME) converts OAA to malate, which is transported to BS chloroplasts and decarboxylated by NADP-ME (Simpson et al. 2021). The second subtype involves aspartate transport to the BS mitochondria, where it is converted back into malate and decarboxylated by NAD-ME. The third subtype incorporates the NAD-ME pathway alongside a second pathway, in which aspartate is converted back into OAA and decarboxylated by phosphoenolpyruvate carboxykinase (PEPCK). Through these mechanisms, C_4 photosynthesis can dramatically reduce photorespiration, thus enhancing photosynthetic carbon fixation.

C_4 evolution is widely accepted as a notable example of convergent evolution event, having evolved independently over 65 times from the C_3 ancestral state (Sage and Coleman 2001). Understanding how plants arrived at a C_4 system could provide foundational knowledge that could be used to recapitulate the C_3 to C_4 transition. Increased leaf venation and subsequent BS enlargement are believed to have been

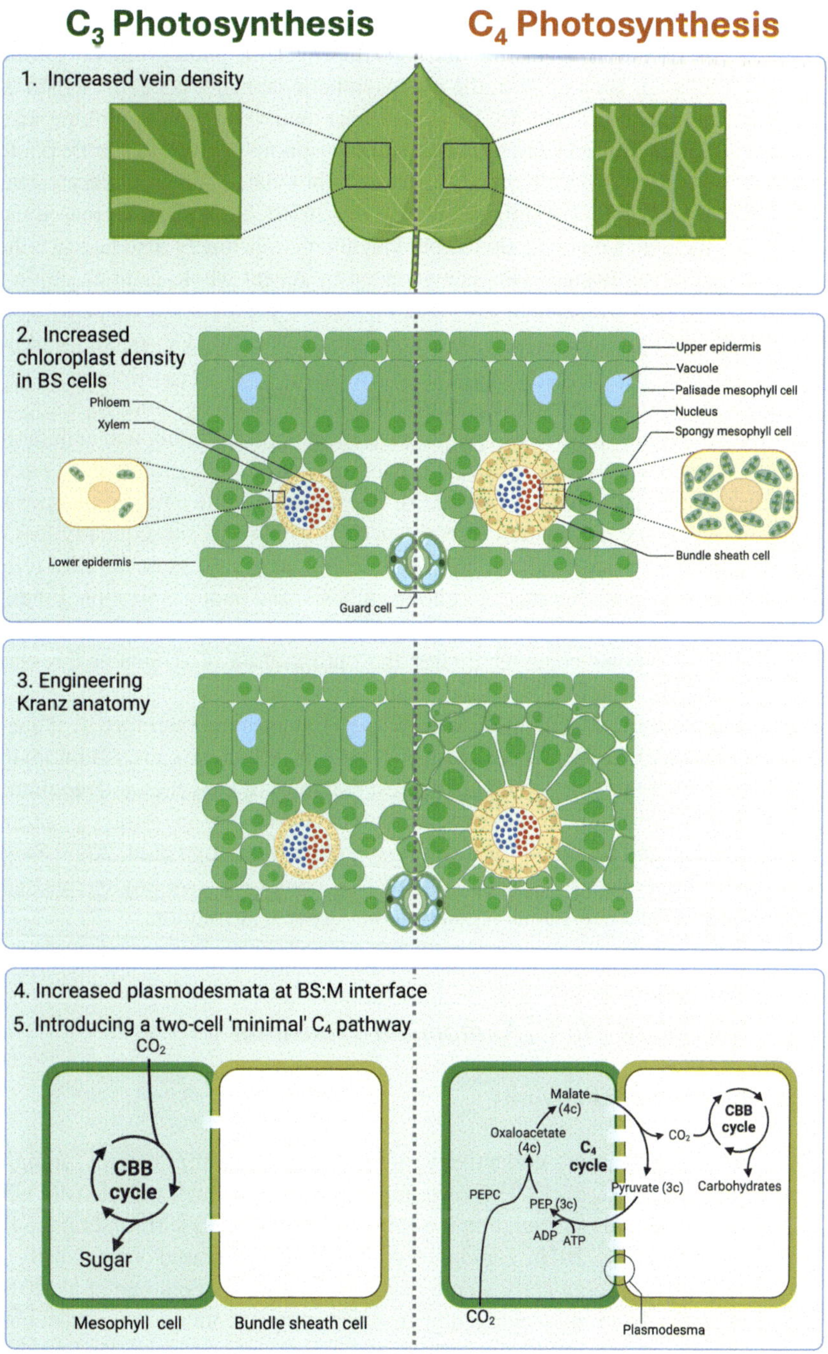

Fig. 5.3 Anatomical and biochemical differences between C_3 and C_4 photosynthesis. To achieve a C_4 anatomy in C_3 species, (1) leaf vein density is increased to improve the supply of water and nutrients to the photosynthesising cells. (2) BS cells increase in size and incorporate more chloroplasts for optimised Calvin-Benson-Bassham cycle operation. (3) M cells are concentrically oriented around the BS cells to supply C_4 intermediates to the BS. (4) Finally, plasmodesmata junctions frequency increases at the BS:M interface, which ensures efficient delivery and regeneration of C_4 intermediates. (Created with BioRender.com)

early steps in the C_3-C_4 transition (Bräutigam and Gowik 2016; Simpson et al. 2021). Increased access to water and nutrients improved the photosynthetic abilities of proximal cells. To keep up with the photosynthetic demand, BS cells, which are closest to the veins, evolved to become larger and house more chloroplasts. Additionally, under limited CO_2, a photorespiratory pump evolved to shuttle photorespiratory glycine to BS cells, where it is decarboxylated for secondary carbon concentration in BS cells (Bräutigam and Gowik 2016). To support increased carbon fixation, the genes that encode the photosynthetic machinery also had to either be upregulated or duplicated. This is evidenced by recent whole-genome duplication of existing C_3 genes which provided additional copies for a C_4 pathway to develop in the background, without compromising the existing C_3 system (Schlüter and Weber 2016).

In addition, whole-genome sequencing efforts are shedding more light on the evolution of C_4 photosynthesis by enabling the end-to-end assembly and annotation of C_4 *Gynandropsis gynandra* (Gynandropsis) and its close C_3 relative *Tarenaya hassleriana* (Tarenaya) (Hoang et al. 2023). The findings led to a model that postulates a whole-genome duplication in both species, which was subsequently lost in the C_3 species. In *Gynandropsis*, genome duplication expanded gene families relating to vein density, heat tolerance, C_4 photosynthesis, and photorespiration. Finally, a recent genome expansion by 50% integrated long terminal repeat retrotransposons (LTR-RT) into promoter genes facilitated the optimisation of C_4 gene expression (Zhao et al. 2023).

High-throughput next-generation sequencing strategies are therefore a valuable tool that evolutionary biologists can use to fill in the phylogenetic gaps. Additionally, research into gene regulatory elements suggest that most of the genes and regulatory elements that facilitated the C_4 pathway were derived from the ancestral C_3 genome and were rewired and recruited for C_4 photosynthesis (Amy Lyu et al. 2023; Singh et al. 2023). The next steps include constructing comprehensive phylogenies and understanding *cis* and *trans* gene regulation that drove C_4 evolution.

5.6.1 Toward a C_3 to C_4 Anatomical Transition

5.6.1.1 Increasing Vein Density

Notable anatomical features, collectively referred to as the Kranz anatomy, complement the biochemical C_4 cycle. This anatomy is defined by a high vein density with chloroplast-rich BS cells surrounding the veins and then M cells surrounding the BS (Bräutigam and Gowik 2016; Lundgren et al. 2014). Increased vein density is widely considered to be an early step toward a C_4 anatomy (Feldman et al. 2017; Walsh et al. 2023). Not only would a higher vein density encourage the correct ratio of BS to M cells, but it would also facilitate the rapid transport of metabolites (Liu and Cheng 2024; Lundgren et al. 2019). Additionally, if appropriate precursor structures exist in a plant, increasing leaf venation could be sufficient to induce a C_3 to

C_4 transition (Lundgren et al. 2019). However, engineering increased venation is not an easy task. The temporal and spatial regulation of venation is a complex process influenced by several known and unknown mechanisms, including auxin and other hormone signalling, protein-protein interactions, and cis-regulatory element signalling, among other factors (Liu and Cheng 2024).

Mutagenic screening of genetically diverse insertion and deletion libraries could yield candidate C_3 lines with increased vein density, thereby circumventing engineering constraints related to venation (Feldman et al. 2014, 2017; Lo et al. 2022). Screening mutant rice libraries identified eight lines with significantly increased venation compared to WT lines Feldman et al. (2014). Phenotypically, there was a strong correlation between vein density and leaf width, which could provide breeders with a quick phenotypic trait marker to accelerate vein density screening. Feldman et al. (2017) then analysed the photosynthetic performance of five of the eight lines and found significantly increased Vc_{max}, J_{max}, net photosynthesis rate, and g_s in mutants, compared to WT lines. Additionally, a mutagenic screening of the same insertion mutant rice line identified a leucine-rich repeat protein as a key regulator of venation (Lo et al. 2022). However, when overexpressed in T_3 rice lines, the initial 12% increase in vein density compared to WT rice dwindled to 4% in later developmental stages. Furthermore, no changes in photosynthetic parameters were reported.

The above findings highlight the need for more detailed transcriptomic and gene-regulatory assays that can help identify candidate genes and molecular mechanisms linked to increased venation in mutant lines. Nonetheless, candidate C_3 mutant lines could provide researchers with the necessary preconditions in which C_4 cycle modifications could be integrated.

5.6.1.2 Increasing Chloroplast Density in BS Cells

One of the four characteristics of the Kranz anatomy deemed essential for the proper functioning of a C_4 system is that the cells where the Calvin-Benson-Bassham cycle occurs (predominantly in the BS cells) should be chloroplast rich (Lundgren et al. 2014). Therefore, engineering the C_4 anatomy in C_3 species involves increasing chloroplast density in cells surrounding leaf veins. *GOLDEN2-LIKE* (GLK) genes are transcription factors that are known to promote chloroplast development and have emerged as promising targets for chloroplast enrichment in BS cells (Susila et al. 2023). When overexpressed under either under constitutive and cell-specific maize promoters, *GLK2*, which is preferentially expressed in BS cells, significantly increased chloroplast size and/or number in a tissue specific manner (Wang et al. 2017; Yeh et al. 2021). When looking at yield data, maize-specific promoters outperformed constitutive promoters, evidenced by an 80% increase in growth and 70% increase in grain yield in a double *GLK1/2* overexpression rice line (Yeh et al. 2021). Identifying candidate C_4 genes and combining them with appropriate promoter sequences are a feasible strategy that is realising the goal of improving photosynthesis in C_3 crops.

Mutagenic screening, sequencing, and gene editing are key driving forces in understanding BS development and manipulation. Transgenic *Arabidopsis* lines that expressed luciferase or chloroplast-tagged GFP under a C_4 BS-specific promoter were chemically mutagenized (Döring et al. 2019). Screening via confocal microscopy lines revealed nine M_2 lines with increased or diminished reporter signalling. In lines with increased reporter signalling, BS cell counts were higher and more radially arranged compared to the reference line. Mutants showing increased reporter signalling were sequenced and mapped to identify candidate genes, which were then investigated further using CRISPR/Cas9-knockout assays (Bano and Westhoff 2024). Knocking out the mutated *At2g25970* gene was shown to increase the GFP reporter signal and as well as leaf surface area. WT *At2g25970* is likely a multifunctional gene that negatively regulates the cell cycle. Although the identified KH domain gene may not be the most ideal candidate for enhancing BS proliferation in the C_3 plants given its pleiotropic functions, the study indicates that reporter-based forward genetics could elucidate BS regulation.

Through modifications of CRISPR/Cas9 technology, Lee et al. (2021) demonstrated a proof of concept for multiple gene regulation in a tissue-specific manner using a dead Cas9 (dCas9). The team generated transgenic rice lines that overexpressed a native chloroplast biosynthesis-gene *OsCGA1*. Transgenic lines were reported to have significantly increased BS chloroplast numbers. By fusing Cas9 to activation modules, dCas9 acquired a transcriptional activation function. In addition, a series of guide RNAs (gRNAs), which were complementary to different promoter regions, targeted the dCas9 module to upregulate both endogenous and exogenous CGA1. As the dCas9 and transgenic *OsCGA1* constructs were both placed under a C_4 *Flaveria trinervia* promoter, which is highly active in BS cells, both *CGA1* genes were expressed exclusively in BS cells. These findings introduce the possibility of regulating multiple Kranz anatomy or C_4 cycle genes using a streamlined system using gRNAs.

5.6.2 The Road to Kranz: Engineering Enhanced Photosynthetic Leaf Anatomy

The Kranz anatomy formation is delineated by three steps: (1) the innermost leaf cells differentiate into procambium, which then develop into vasculature systems; (2) M and BS cells are radially arranged in a 'wreath-like' structure, around the vasculature; and (3) sheath and mesophyll cells acquire increased photosynthetic capacity (Sedelnikova et al. 2018). Engineering the first two steps presents an incredible anatomical feat that requires correct spatial and temporal regulation (Sedelnikova et al. 2018). Additionally, the complex interactions of the gene-regulatory pathways involved are yet to be elucidated (Sedelnikova et al. 2018).

Despite these challenges, progress is evident. Wang et al. (2017) observed features of a Kranz-like structure in rice via BS-specific GLK2 overexpression, which included enlarged mitochondria in sheath cells and anatomical ratios comparable to a proto-Kranz structure. However, no improvements in photosynthesis or yield were found when compared to WT controls, suggesting that the anatomical changes were insufficient to confer photosynthetic advantages.

Singh and Reeves (2020) highlight the relevance of epigenetics and gene regulatory networks that regulate leaf development and anatomy by expanding on research done by van Rooijen et al. (2019). A post-transcriptional repressor NAC052 was found to regulate leaf development and arrest it before it progresses into a Kranz-like anatomy in *Arabidopsis*. However, when ectopically expressed, NAC052 assumed transcriptional activator function, which was noted by increased BS cells among other changes. Further studies into gene regulatory networks and *trans* elements in C_4 species are therefore crucial in unravelling the intricacies that define C_4 regulation (Singh and Reeves 2020). Given the complexity of leaf development and Kranz anatomy, characterising the multitude of genes and regulatory elements will take years (Sedelnikova et al. 2018). Consequently, modifying C_3 leaf anatomy will require more fundamental research and understanding before being considered as a feasible photosynthetic improvement strategy.

5.6.3 Intercellular Communication Through Plasmodesmata

For a dual-cell photosynthetic system to be efficient, robust intercellular transport of metabolites is crucial. Intercellular junctions called plasmodesmata (PD) facilitate this flux and are more frequent and closely clustered in C_4 species compared to C_3 counterparts (Danila et al. 2016; Simpson et al. 2021). It is estimated that if the C_3 to C_4 biochemical transition is established, current C_3 plasmodesmata structures would need to achieve a flux 12 times that of current C_4 rates (Danila et al. 2016). Therefore, optimising the BS-M plasmodesmata frequency is necessary to support the biochemical changes that could accompany the C_3 to C_4 transition.

Electron and confocal microscopy, as well as immunological assays, are common techniques used to visualise and quantify plasmodesmata frequency (Danila et al. 2016; Liu and Cheng 2024). Using these approaches, Schreier et al. (2024) found that C_4 *Gynandropsis* had an eightfold increase in PD frequency at the M-BS junction compared to C_3 *Tarenaya* and *Arabidopsis*. This M-BS PD frequency is partly mediated by photosynthetic metabolite formation during the early developmental stages following de-etiolation. These findings may explain why increased plasmodesmata frequency between mestome and BS cells was observed upon *GLK2* overexpression in BS cells discussed above (Wang et al. 2017). Increased photosynthetic activity and metabolite synthesis may necessitate intercellular connections between sheath cells in the form of increased PD frequency.

Previous work with understanding plasmodesmata clustering and pore size has been hindered by the inaccessibility to penetrate the cell walls and challenges in isolating the associated proteins that localise to plasmodesmata. Recent advancements such as Plant-Enzyme-Assisted (PEA) CLARITY renders clear plant tissue for high-resolution confocal or electron microscopy, allowing for more accurate PD visualisation and quantification (Liu and Cheng 2024; Tsang et al. 2024). Other approaches include GF tagging of known PD proteins for quantification using confocal microscopy, together with mRNA sequencing and gene co-expression network analysis (Tsang et al. 2024). By applying these techniques at different developmental stages of rice and millet development, a spatiotemporal atlas of PD density was created, along with the identification of up to 57 genes that mediate PD frequency (Tsang et al. 2024). With these advances, researchers can look forward to more streamlined quantification of PD-associated proteins and can investigate the identified candidate genes for functional analysis.

5.6.4 *Toward a C_3 to C_4 Biochemical Transition*

Recapitulating the key biochemical events that facilitated C_4 evolution one at a time will help fill in the gaps in our understanding of C_4 photosynthesis. For example, the re-localisation of CA from the mesophyll chloroplast to the mesophyll cytosol is considered an important step that facilitated and sustained the development of a C_4 biochemical pathway (Singh et al. 2022). This was successfully achieved in rice by genetically mutating or truncating the chloroplast transit peptide sequence that targets CA to chloroplasts (Jethva et al. 2024). However, cytosolic CA showed no differences in terms of yield, growth rates, photosynthetic rates, or CO_2 concentrations near Rubisco. Despite this, successful re-localisation indicates that the current C_3 genome can be edited to direct its evolution toward a C_4 system.

Advances are being made to develop a functional 'minimal' C_4 cycle, which consists of genes from the NADP-ME subtype. Initially, C_4 genes were inserted individually and then crossed and verified repeatedly to achieve trait stacking and was a time-consuming process (Lin et al. 2020). The advent of Golden Gate cloning has fast-tracked this process, allowing a suite of C_4 genes to be integrated into host genomes in one transformation event (Ermakova et al. 2021). In addition, incorporating cell-specific promoters in the constructs also enable tissue-specific gene expression, which partly circumvents localisation challenges (Ermakova et al. 2021). Both traditional and Golden Gate trait stacking were combined with $^{13}CO_2$ labelling to measure carbon flux through the modified pathway and demonstrated a partially functional C_4 pathway (Ermakova et al. 2021).

Similarly, single-celled omics is quickly becoming a key player in uncovering C_4 photosynthetic regulation. Single-celled indexed assay for transpose accessible chromatin sequencing (sciATACS-Seq) of different cell types from four different C_4 grasses successfully produced a C_4 cell-specific gene regulatory map (Mendieta et al. 2024). This map suggests that cell-specific regulatory networks are a necessary regulatory element for a functional C_4 pathway. Additionally, the rapid sequence diversification of central regulatory elements across C_4 species is a strong indicator of ongoing evolution (Zhang and Zhang 2022). More specifically, transposable elements (TEs), which modulate proximally located genes, are found to be instrumental components of accessible chromatin regions (ACRs) in rice (Zhang and Zhang 2022). TE-derived ACRs were found to be highly genetically variable, enabling the selection of advantageous agronomic traits. These traits, in turn, drove the selective fixation of the TE-derived ACRs during rice domestication. The C_3 to C_4 transition will therefore need to identify strongly conserved *cis* regulatory elements as well as consider suitable methods to recapitulate key regulators in a cell-specific manner.

Despite these advancements, the road toward C_3 to C_4 photosynthesis bioengineering comes with its challenges. It has been evolutionarily shown that the C_4 elements already exist in C_3 genomes (Singh et al. 2023). It has also been demonstrated that in part, rewiring these elements to mimic a C_4 pathway is possible. However, it seems that C_4 ancestral genes are primarily involved in stress responses in C_3 genes (Chen et al. 2022; Singh et al. 2022). Developing mechanisms that will coopt current C_3 isoforms to perform photosynthetic functions in addition to their current stress-related functions remains a challenge. Another stumbling block is that C_4 gene expression is significantly elevated compared to C_3 pathway enzymes (Furbank et al. 2023). RNAi knockdown studies suggest that reducing any of the C_4 enzymatic pool by up to 50% contributes to a maximal 21% reduction in photosynthesis (Furbank et al. 2023). This means that the full C_4 enzyme levels are not required to achieve a functional system in C_3 species. However, current C_3 to C_4 expression levels do not meet the threshold for functional C_4 photosynthesis (Ermakova et al. 2020). Furthermore, although concurrent gene regulation of two genes was recently demonstrated, this technology would need to be developed further to regulate the five 'minimal' C_4 enzymes for optimal gene expression and modulation. Overall, engineering the anatomical components of the biochemical C_4 transition in C_3 crops is forging ahead in a non-sequential order with successes and challenges being experienced simultaneously.

5.7 CAM Photosynthesis

High temperatures and arid conditions created a unique selection pressure for CAM species, which evolved the ability to fix CO_2 at night (Perron et al. 2024). CAM photosynthesis emerged through alterations in the circadian regulation of ancestral C_3 photosynthesis genes and in stomatal opening and closing patterns (Perron et al. 2024). Due to nocturnal stomatal opening, CAM plants have a sixfold increase in WUE compared to C_3 crops (Yang et al. 2015). CAM-engineered C_3 crops present a unique opportunity to grow crops using less water while also expanding the use of marginal land for crop farming, both of which are pertinent in meeting rising food demands amidst dwindling fresh water supplies (Yuan et al. 2020).

CAM photosynthesis is broadly divided into four phases. The first phase begins at night, when CO_2 diffuses into leaves through open stomata and is fixed into OAA by PEPC (Perron et al. 2024). The four-carbon compound is then converted to malate and is stored in the vacuole as malic acid until daybreak (Yuan et al. 2020).

Phase two is initiated in the early morning hours and is a transitional phase during which carbon is fixed by both PEPC and Rubisco (Burgos et al. 2022). Malic acid is decarboxylated back into PEP by malate dehydrogenase (MDH) during phase three, which releases CO_2 for fixation by rubisco (Guan et al. 2020). The stomata remain closed during this phase. Finally, during phase four, stomatal reopening is partly triggered by the depletion of low malate levels toward the end of the day, which restarts the cycle (Tan and Chen 2023).

CAM photosynthesis is another example of convergent evolution and is found in at least 400 genera (Yuan et al. 2020). Facultative CAM species have evolved to operate the C_3 pathway when conditions are optimal but switch to CAM photosynthesis under abiotic stress (Winter and Holtum 2024). Harnessing this facultative CAM system in C_3 would confer crops with the dual benefits of high yield through C_3 photosynthesis and CAM-mediated drought tolerance (Yuan et al. 2020). Yang et al. (2024) proposed a simplified strategy to engineer facultative CAM in C_3 species (Fig. 5.4). The first step proposed reprogramming the timing of gene expression

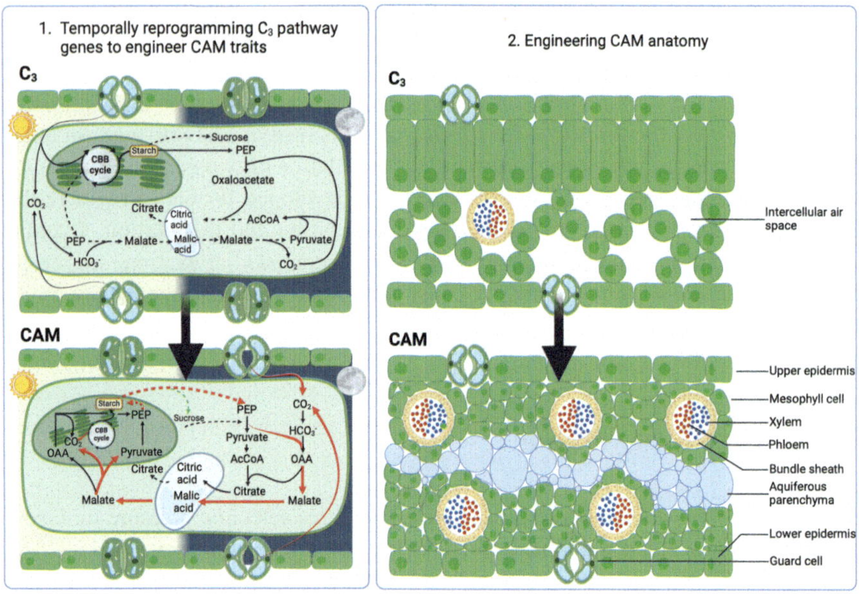

Fig. 5.4 Metabolic and anatomical modifications that enable C_3-CAM transition. (1) Transitioning from C_3 to CAM photosynthesis involves modifying the circadian expression of C_3 genes. This includes engineering stomatal opening at night to facilitate CO_2 uptake. Additional nocturnal reprogramming upregulates starch digestion for glycolysis, PEP conversion to malate, and malic acid storage in the enlarged vacuole. During the day, malate decarboxylation is prioritised to supply CO_2 to the CBB cycle for starch production. Starch accumulates throughout the day, partly due to the downregulation of starch hydrolysis into sucrose. (2) CAM anatomy features densely packed mesophyll cells and specialised water storage cells called aquiferous parenchymal cells, the latter of which ensures diurnal photosynthesis when stomata are closed. Dotted arrows denote a multistep process. Upregulated metabolic pathways are shown in red, and downregulated pathways are illustrated with green arrows. (Created with BioRender.com)

for genes common to both C_3 and CAM pathways. Secondly, C_3 leaf anatomy must be modified to a CAM-like anatomy, which would sustain the temporal shift in metabolic activity.

5.7.1 Reprogramming C_3 Circadian Regulation to Induce CAM

Reshuffling the circadian cycle of the genes that mediate carboxylation, decarboxylation, and stomatal activity is the foremost factor for engineering CAM photosynthesis. A comparative genomics study identified 54 CAM genes that showed altered diurnal cycle patterning (Yuan et al. 2020). A key challenge will be to determine the minimal number of genes that need to be temporally rewired to achieve CAM-like photosynthesis in C_3 species. Among these genes and regulatory changes, the malate pathway regulation is believed to be crucial in inducing CAM biochemistry (Burgos et al. 2022). Stomatal closure is regulated by high C_i through negative feedback (Schiller and Bräutigam 2021). It will therefore be important to increase malate levels to a high enough concentration such that decarboxylation increases CO_2 release and C_i, thus triggering stomatal closure (Burgos et al. 2022). Presently, circadian reprogramming of any of the CAM metabolic pathways has not been demonstrated in model or C_3 species.

Recent advances in omics studies are uncovering the mechanisms of CAM photosynthesis and regulation. Following 6 days of salinity treatment, the facultative CAM species *Mesembryanthemum crystallinum* (ice plant) upregulated V-type H^+-ATPase expression, which pumps H^+ into the vacuole and mediates salt and drought tolerance (Guan et al. 2020). Additionally, significant changes in abscisic acid (ABA) regulation initially induced drought-mediated stomatal closure, which was followed by the nocturnal upregulation of ABA inhibitors. Other omics assays are continuing to shed light of CAM induction, particularly with stomatal regulation, which also undergo a C_3-CAM transition when under salt stress (Kong et al. 2020). Although CAM engineering in C_3 is still in the preliminary research stages, advancements in single-cell omics are rapidly filling the gaps in our understanding of temporally regulated CAM.

5.7.2 Engineering CAM Anatomy

Increased succulence is a characteristic feature of most CAM species. Succulence is mediated by an extensive parenchymal network (Yuan et al. 2020). CAM plants have also densely packaged mesophyll cells, which reduce both intercellular air spaces (IAS) and the mesophyll surface area that is exposed to IAS (Yuan et al. 2018). These anatomical features reduce the conductance of CO_2, which allows for efficient CAM CO_2 fixation (Yuan et al. 2020). Overexpressing succulence-associated genes, such as *xyloglucan endotransglucosylase/hydrolase (XTH)* and *cell expansion basic helix-loop-helix (CEB1)* has shown moderate success in reducing IAS, increasing cellular packing, higher salt tolerance, and increased WUE (Yuan et al. 2020). Additionally, C_3 vacuoles need to expand to accommodate large malate pool that supports CAM photosynthesis (Burgos et al. 2022). Modulating V-type H^+-ATPase mentioned earlier alongside circadian modification may facilitate vacuole enlargement. Further research is needed to fully understand the CAM anatomy before optimisation into C_3 species can proceed.

Insights into C_4 species that feature CAM photosynthesis demonstrate the plasticity of CAM photosynthesis. *Portulaca oleracea*, a C_4 species known to operate CAM under stress conditions, was initially thought to operate both systems in different cells through unknown mechanisms (Guralnick et al. 2002). However, recent spatial transcriptomics study has shown that, under drought conditions, C_4 and CAM pathways both operate in the same BS and M cells, but at different times (Moreno-Villena et al. 2022). C_4 gene homologues evolved to be repurposed for CAM, which are upregulated at night to avoid futile cycling between the two pathways. Additionally, malate produced during nocturnally active CAM is integrated into the C_4 pathway during the day. Furthermore, the success of C_4 + CAM operation is partly due to mild succulence, combined with sufficient vein density to support a high BS:M ratio (Moreno-Villena et al. 2022). This finding demonstrates that CAM is a flexible system that integrates seamlessly with C_4 photosynthesis, provided the appropriate temporal rearrangements are made. Spatial transcriptomics is resolving long-unsolved questions while expanding the possibilities regarding integrating different systems for improved photosynthesis and climate resilience in crops.

5.8 Engineering Chloroplast Genomes and New Cellular Compartments

5.8.1 Chloroplast Genome Engineering

Concurrently with nuclear photosynthetic improvement strategies, chloroplast genome editing is an active area of exploration and could serve as a foundational step toward engineering chloroplast mimics. Their small size, maternal transmission, and lack of epigenetic silencing make chloroplast genomes, or plastomes, an attractive system to introduce suites of transgenes that can be expressed at high levels (Bock 2022). Plastomes are characterised by multiple operon-mediated gene expression and several steps of post-transcriptional mRNA processing (Zhang et al. 2023). This operon style of gene expression can easily be replicated through synthetic operons introduced by Golden Gate cloning. Additionally, integrating the operons with the appropriate post-transcriptional machinery is also important for successful plastid gene expression. For example, intercistronic elements (IEEs) are essential recognition sites which mediate the cleavage of polysictronic mRNA transcripts into individual transcripts and are associated with mRNA stability and translational efficiency (Bock 2022). In transplastomic tobacco, IEE incorporation into synthetic operons carrying carotenoid-related enzymes were shown to redirect the carotenoid pathway by increasing zeaxanthin production relative to astaxanthin and other carotenoids in tobacco (Tanwar et al. 2023).

Additionally, more efficient transformation strategies are encouraging the progression from modifying endogenous plastomes to synthetically engineering entirely new ones. Homologous recombination has been the standard plastid transformation protocol for the last 30 years and often involves applying sustained selective pressure across several rounds of positive selection to ensure the polyploidal plastomes all acquire the transgene (Jakubiec et al. 2021; Occhialini and Lenaghan 2023). Episomal DNA, which can replicate independently from plastid or nuclear DNA, circumvents the challenge of homoplastic integration (Occhialini and Lenaghan 2023). These independent replication vectors were either incorporated into the geminivirus replication system in tobacco or into a chloroplast specific origin of replication into potatoes (Occhialini and Lenaghan 2023). The resulting ‘minichromosomes’ or ‘mini-synthetic plastome’ replicate independently and are transmitted to subsequent generations (Jakubiec et al. 2021; Occhialini et al. 2023). It is easy to envisage the evolution of plastome engineering from individual gene transformations toward the bespoke design and integration of whole, independently replicating and heritable plastid genomes that deliver the photosynthetic code directly to the source.

5.9 New Cellular Compartments

As new technologies or advancements in current methods develop, so does the discovery of potentially groundbreaking science. Recently, reports of the nitrogen-fixing Candidatus *Atelocyanobacterium thalassa* (UCYN-A) symbiont and its algal host *Braarudosphaera bigelowii* (Braarudosphaera) suggest that UCYN-A is likely in the early stages of evolving into an organelle, aptly termed a nitroplast (Coale et al. 2024). By combining proteomics and high-resolution 3D X-ray technology, nitroplasts were observed to have undergone a genome contraction and recruited proteins expressed by nuclear *Braarudosphaera*. Additionally, the UCYN-A to *Braarudosphaera* size ratio is also consistent with those found in endosymbionts that evolved to become organelles (Cornejo-Castillo et al. 2024). Together, the speculative nitroplast provides a novel model system through which to study the evolution of an exceptionally rare event, perhaps with the hope of eventually replicating primary endosymbiosis. Although still in its infancy, possibilities of engineering a nitrogen-fixing organelle into land plants would undoubtedly revolutionise current agricultural practices, not to mention the benefits to plant productivity and yield.

5.10 Conclusions and Future Perspectives

Advancing photosynthesis remains a cornerstone of the global agricultural strategy to combat the challenges posed by climate change and an increasing global population. Technological advancements are accelerating the transition from concept development toward the commercial deployment of CCM-engineered crops. This process is structured into Technology Readiness Levels (TRLs), which outline the pathway from fundamental research and conceptual development to the full commercialisation of crops with enhanced photosynthetic efficiency (Fig. 5.5) (CAPITALISE 2024).

Researchers have achieved several key milestones which advance the progression of different photosynthetic improvement strategies toward commercially viable CCM-engineered crops. Partial pyrenoid assembly in planta places this strategy on the border between concept development and proof-of-concept testing (TRL 2–3). Meanwhile, an assembled and lowly functional carboxysome in tobacco together with modestly successful field trials advance carboxysome-based CCMs to TRL 4. Similarly, moderate success with C_3 BS expansion in improving photosynthesis and yield in rice also infer proof of principle and a translational readiness of level four. In contrast to other strategies, CAM photosynthesis research is preferentially focused on understanding how it works, thus positioning it at TRL 2–3. Technical advancements are also enabling potentially revolutionary concepts for photosynthetic improvement through synthetic chloroplast-like compartments and the discovery of entirely new organelles.

Progression along the TRLs is often hindered by common bottlenecks, including gaps in understanding of regulatory elements, achieving correct protein localisation, and overcoming challenges in anatomical reprogramming. However, emerging techniques like spatial transcriptomics and spatial proteomics, directed evolution, and synthetic gene circuit engineering are unlocking new pathways to decipher the molecular mechanisms underpinning various CCMs. These breakthroughs hold transformative potential to revolutionise agricultural productivity by increasing yield, enhancing WUE, improving climate adaptability, and reducing fertilizer dependency, thereby providing innovative solutions for crop improvement. Ultimately, enhancing photosynthesis will be central to ensuring sustainable, resilient agricultural systems in the face of environmental pressures.FundingMKK and PS were supported by the UKRI-Future Leaders Fellowship (MR/Y016882/1), *REVOLUTON*, awarded to PS. The authors declare that they have no competing interests.

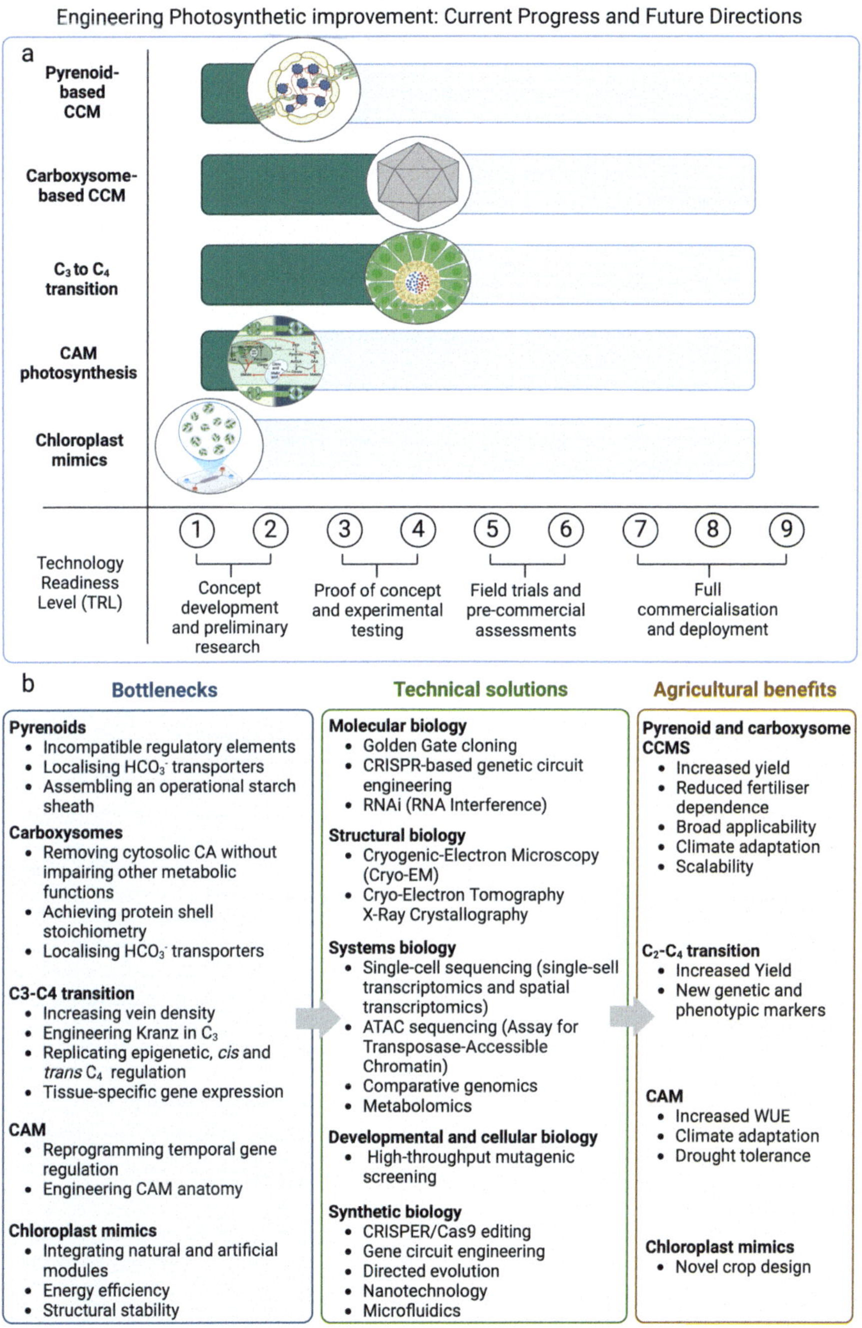

Fig. 5.5 Mapping photosynthetic pathways for crop improvement. (**a**) Technology readiness level (TRL) of photosynthetic improvement strategies. The timeline positions pyrenoid-based CCMs and CAM photosynthesis at TRL 2–3, carboxysome-based CCMs at TRL 4, and chloroplast mimics at TRL 1. (**b**) Identified bottlenecks of current improvement strategies, proposed technical solutions to accelerate these strategies, and an outline of the potential agricultural benefits. (Created with BioRender.com)

References

Adler L, Díaz-Ramos A, Mao Y, Pukacz KR, Fei C, McCormick AJ (2022) New horizons for building pyrenoid-based CO_2-concentrating mechanisms in plants to improve yields. Plant Physiol 190(3):1609–1627. https://doi.org/10.1093/plphys/kiac373

Amy Lyu M-J, Tang Q, Wang Y, Essemine J, Chen F, Ni X, Chen G, Zhu X-G (2023) Evolution of gene regulatory network of C4 photosynthesis in the genus *Flaveria* reveals the evolutionary status of C3-C4 intermediate species. Plant Commun 4(1). https://doi.org/10.1016/j.xplc.2022.100426

Atkinson N, Feike D, Mackinder LCM, Meyer MT, Griffiths H, Jonikas MC, Smith AM, McCormick AJ (2016) Introducing an algal carbon-concentrating mechanism into higher plants: location and incorporation of key components. Plant Biotechnol J 14(5):1302–1315. https://doi.org/10.1111/pbi.12497

Atkinson N, Leitão N, Orr DJ, Meyer MT, Carmo-Silva E, Griffiths H, Smith AM, McCormick AJ (2017) Rubisco small subunits from the unicellular green alga Chlamydomonas complement Rubisco-deficient mutants of *Arabidopsis*. New Phytol 214(2):655–667. https://doi.org/10.1111/nph.14414

Atkinson N, Mao Y, Chan KX, McCormick AJ (2020) Condensation of Rubisco into a proto-pyrenoid in higher plant chloroplasts. Nat Commun 11(1):6303. https://doi.org/10.1038/s41467-020-20132-0

Atkinson N, Stringer R, Mitchell SR, Seung D, McCormick AJ (2024) SAGA1 and SAGA2 promote starch formation around proto-pyrenoids in *Arabidopsis* chloroplasts. Proc Natl Acad Sci USA 121(4):e2311013121. https://doi.org/10.1073/pnas.2311013121

Bano Z, Westhoff P (2024) A K homology (KH) domain protein identified by a forward genetic screen affects bundle sheath anatomy in *Arabidopsis thaliana*. Plant Direct 8(4):e577. https://doi.org/10.1002/pld3.577

Bar-On YM, Milo R (2019) The global mass and average rate of rubisco. Proc Natl Acad Sci 116(10):4738–4743. https://doi.org/10.1073/pnas.1816654116

Barrett J, Girr P, Mackinder LCM (2021) Pyrenoids: CO(2)-fixing phase separated liquid organelles. Biochim Biophys Acta, Mol Cell Res 1868(5):118949. https://doi.org/10.1016/j.bbamcr.2021.118949

Barrett J, Naduthodi MIS, Mao Y, Dégut C, Musiał S, Salter A, Leake MC, Plevin MJ, McCormick AJ, Blaza JN, Mackinder LCM (2024) A promiscuous mechanism to phase separate eukaryotic carbon fixation in the green lineage. Nat Plants. https://doi.org/10.1038/s41477-024-01812-x

Blikstad C, Dugan EJ, Laughlin TG, Turnšek JB, Liu MD, Shoemaker SR, Vogiatzi N, Remis JP, Savage DF (2023) Identification of a carbonic anhydrase-Rubisco complex within the alpha-carboxysome. Proc Natl Acad Sci USA 120(43):e2308600120. https://doi.org/10.1073/pnas.2308600120

Bock R (2022) Transplastomic approaches for metabolic engineering. Curr Opin Plant Biol 66:102185. https://doi.org/10.1016/j.pbi.2022.102185

Bowes G, Ogren WL, Hageman RH (1971) Phosphoglycolate production catalyzed by ribulose diphosphate carboxylase. Biochem Biophys Res Commun 45(3):716–722. https://doi.org/10.1016/0006-291x(71)90475-x

Bräutigam A, Gowik U (2016) Photorespiration connects C3 and C4 photosynthesis. J Exp Bot 67(10):2953–2962. https://doi.org/10.1093/jxb/erw056

Burgos A, Miranda E, Vilaprinyo E, Meza-Canales ID, Alves R (2022) CAM models: lessons and implications for CAM evolution. Front Plant Sci 13:893095. https://doi.org/10.3389/fpls.2022.893095

CAPITALISE (2024, November 29) Time to translate—a roadmap for photosynthesis to drive crop improvement. https://www.capitalise.eu/wp-content/uploads/Photosynthesis-Roadmap.pdf

Chen S, Peng W, Ansah EO, Xiong F, Wu Y (2022) Encoded C_4 homologue enzymes genes function under abiotic stresses in C3 plant. Plant Signal Behav 17(1):2115634. https://doi.org/10.1080/15592324.2022.2115634

Chen T, Hojka M, Davey P, Sun Y, Dykes GF, Zhou F, Lawson T, Nixon PJ, Lin Y, Liu LN (2023) Engineering α-carboxysomes into plant chloroplasts to support autotrophic photosynthesis. Nat Commun 14(1):2118. https://doi.org/10.1038/s41467-023-37490-0

Coale TH, Loconte V, Turk-Kubo KA, Vanslembrouck B, Mak WKE, Cheung S, Ekman A, Chen JH, Hagino K, Takano Y, Nishimura T, Adachi M, Le Gros M, Larabell C, Zehr JP (2024) Nitrogen-fixing organelle in a marine alga. Science 384(6692):217–222. https://doi.org/10.1126/science.adk1075

Cornejo-Castillo FM, Inomura K, Zehr JP, Follows MJ (2024) Metabolic trade-offs constrain the cell size ratio in a nitrogen-fixing symbiosis. Cell 187(7):1762–1768.e1769. https://doi.org/10.1016/j.cell.2024.02.016

Danila FR, Quick WP, White RG, Furbank RT, von Caemmerer S (2016) The metabolite pathway between bundle sheath and mesophyll: quantification of plasmodesmata in leaves of C3 and C4 monocots. Plant Cell 28(6):1461–1471. https://doi.org/10.1105/tpc.16.00155

Döring F, Billakurthi K, Gowik U, Sultmanis S, Khoshravesh R, Das Gupta S, Sage TL, Westhoff P (2019) Reporter-based forward genetic screen to identify bundle sheath anatomy mutants in *A. thaliana*. Plant J 97(5):984–995. https://doi.org/10.1111/tpj.14165

Erb TJ, Zarzycki J (2018) A short history of RubisCO: the rise and fall (?) of nature's predominant CO_2 fixing enzyme. Curr Opin Biotechnol 49:100–107. https://doi.org/10.1016/j.copbio.2017.07.017

Ermakova M, Danila FR, Furbank RT, von Caemmerer S (2020) On the road to C_4 rice: advances and perspectives. Plant J 101(4):940–950. https://doi.org/10.1111/tpj.14562

Ermakova M, Arrivault S, Giuliani R, Danila F, Alonso-Cantabrana H, Vlad D, Ishihara H, Feil R, Guenther M, Borghi GL, Covshoff S, Ludwig M, Cousins AB, Langdale JA, Kelly S, Lunn JE, Stitt M, von Caemmerer S, Furbank RT (2021) Installation of C(4) photosynthetic pathway enzymes in rice using a single construct. Plant Biotechnol J 19(3):575–588. https://doi.org/10.1111/pbi.13487

Fei C, Wilson AT, Mangan NM, Wingreen NS, Jonikas MC (2022) Modelling the pyrenoid-based CO_2-concentrating mechanism provides insights into its operating principles and a roadmap for its engineering into crops. Nat Plants 8(5):583–595. https://doi.org/10.1038/s41477-022-01153-7

Feldman AB, Murchie EH, Leung H, Baraoidan M, Coe R, Yu SM, Lo SF, Quick WP (2014) Increasing leaf vein density by mutagenesis: laying the foundations for C4 rice. PLoS One 9(4):e94947. https://doi.org/10.1371/journal.pone.0094947

Feldman AB, Leung H, Baraoidan M, Elmido-Mabilangan A, Canicosa I, Quick WP, Sheehy J, Murchie EH (2017) Increasing leaf vein density via mutagenesis in rice results in an enhanced rate of photosynthesis, smaller cell sizes and can reduce interveinal mesophyll cell number. Front Plant Sci 8:1883. https://doi.org/10.3389/fpls.2017.01883

Field CB, Behrenfeld MJ, Randerson JT, Falkowski P (1998) Primary production of the biosphere: integrating terrestrial and oceanic components. Science 281(5374):237–240. https://doi.org/10.1126/science.281.5374.237

Förster B, Rourke LM, Weerasooriya HN, Pabuayon ICM, Rolland V, Au EK, Bala S, Bajsa-Hirschel J, Kaines S, Kasili RW, LaPlace LM, Machingura MC, Massey B, Rosati VC, Stuart-Williams H, Badger MR, Price GD, Moroney JV (2023) The *Chlamydomonas reinhardtii* chloroplast envelope protein LCIA transports bicarbonate in planta. J Exp Bot 74(12):3651–3666. https://doi.org/10.1093/jxb/erad116

Furbank R, Kelly S, von Caemmerer S (2023) Photosynthesis and food security: the evolving story of C4 rice. Photosynth Res 158(2):121–130. https://doi.org/10.1007/s11120-023-01014-0

Giessen TW, Silver PA (2017) Engineering carbon fixation with artificial protein organelles. Curr Opin Biotechnol 46:42–50. https://doi.org/10.1016/j.copbio.2017.01.004

Guan Q, Tan B, Kelley TM, Tian J, Chen S (2020) Physiological changes in *Mesembryanthemum crystallinum* during the C_3 to CAM transition induced by salt stress. Front Plant Sci 11:283. https://doi.org/10.3389/fpls.2020.00283

Guralnick LJ, Edwards G, Ku MSB, Hockema B, Franceschi V (2002) Photosynthetic and anatomical characteristics in the C4 crassulacean acid metabolism-cycling plant *Portulaca grandiflora*. Funct Plant Biol 29(6):763–773. https://doi.org/10.1071/pp01176

Hay WT, Bihmidine S, Mutlu N, Hoang KL, Awada T, Weeks DP, Clemente TE, Long SP (2017) Enhancing soybean photosynthetic CO_2 assimilation using a cyanobacterial membrane protein, ictB. J Plant Physiol 212:58–68. https://doi.org/10.1016/j.jplph.2017.02.003

He S, Chou HT, Matthies D, Wunder T, Meyer MT, Atkinson N, Martinez-Sanchez A, Jeffrey PD, Port SA, Patena W, He G, Chen VK, Hughson FM, McCormick AJ, Mueller-Cajar O, Engel BD, Yu Z, Jonikas MC (2020) The structural basis of Rubisco phase separation in the pyrenoid. Nat Plants 6(12):1480–1490. https://doi.org/10.1038/s41477-020-00811-y

He S, Crans VL, Jonikas MC (2023) The pyrenoid: the eukaryotic CO_2-concentrating organelle. Plant Cell 35(9):3236–3259. https://doi.org/10.1093/plcell/koad157

Hennacy JH, Jonikas MC (2020) Prospects for engineering biophysical CO_2 concentrating mechanisms into land plants to enhance yields. Annu Rev Plant Biol 71:461–485. https://doi.org/10.1146/annurev-arplant-081519-040100

Hennacy JH, Atkinson N, Kayser-Browne A, Ergun SL, Franklin E, Wang L, Eicke S, Kazachkova Y, Kafri M, Fauser F, Vilarrasa-Blasi J, Jinkerson RE, Zeeman SC, McCormick AJ, Jonikas MC (2024) SAGA1 and MITH1 produce matrix-traversing membranes in the CO_2-fixing pyrenoid. Nat Plants. https://doi.org/10.1038/s41477-024-01847-0

Hoang NV, Sogbohossou EOD, Xiong W, Simpson CJC, Singh P, Walden N, van den Bergh E, Becker FFM, Li Z, Zhu XG, Brautigam A, Weber APM, van Haarst JC, Schijlen E, Hendre PS, Van Deynze A, Achigan-Dako EG, Hibberd JM, Schranz ME (2023) The *Gynandropsis gynandra* genome provides insights into whole-genome duplications and the evolution of C4 photosynthesis in Cleomaceae. Plant Cell 35(5):1334–1359. https://doi.org/10.1093/plcell/koad018

Itakura AK, Chan KX, Atkinson N, Pallesen L, Wang L, Reeves G, Patena W, Caspari O, Roth R, Goodenough U, McCormick AJ, Griffiths H, Jonikas MC (2019) A Rubisco-binding protein is required for normal pyrenoid number and starch sheath morphology in *Chlamydomonas reinhardtii*. Proc Natl Acad Sci USA 116(37):18445–18454. https://doi.org/10.1073/pnas.1904587116

Jakubiec A, Sarokina A, Choinard S, Vlad F, Malcuit I, Sorokin AP (2021) Replicating minichromosomes as a new tool for plastid genome engineering. Nat Plants 7(7):932–941. https://doi.org/10.1038/s41477-021-00940-y

Jethva J, Hahn F, Giuliani R, Peeters N, Cousins A, Kelly S (2024) Realisation of a key step in the evolution of C4 photosynthesis in rice by genome editing. bioRxiv:2024.2005.2021.595093. https://doi.org/10.1101/2024.05.21.595093

Kasili RW, Rai AK, Moroney JV (2023) LCIB functions as a carbonic anhydrase: evidence from yeast and *Arabidopsis* carbonic anhydrase knockout mutants. Photosynth Res 156(2):193–204. https://doi.org/10.1007/s11120-023-01005-1

Kaste JAM, Walker BJ, Shachar-Hill Y (2024) Reaction-diffusion modeling provides insights into biophysical carbon-concentrating mechanisms in land plants. Plant Physiol 196(2):1374–1390. https://doi.org/10.1093/plphys/kiae324

Koester RP, Pignon CP, Kesler DC, Willison RS, Kang M, Shen Y, Priest HD, Begemann MB, Cook KA, Bannon GA, Oufattole M (2021) Transgenic insertion of the cyanobacterial membrane protein ictB increases grain yield in *Zea mays* through increased photosynthesis and carbohydrate production. PLoS One 16(2):e0246359. https://doi.org/10.1371/journal.pone.0246359

Kong W, Yoo MJ, Zhu D, Noble JD, Kelley TM, Li J, Kirst M, Assmann SM, Chen S (2020) Molecular changes in *Mesembryanthemum crystallinum* guard cells underlying the C_3 to CAM transition. Plant Mol Biol 103(6):653–667. https://doi.org/10.1007/s11103-020-01016-9

Lau CS, Dowle A, Thomas GH, Girr P, Mackinder LCM (2023) A phase-separated CO_2-fixing pyrenoid proteome determined by TurboID in *Chlamydomonas reinhardtii*. Plant Cell 35(9):3260–3279. https://doi.org/10.1093/plcell/koad131

Lee DY, Hua L, Khoshravesh R, Giuliani R, Kumar I, Cousins A, Sage TL, Hibberd JM, Brutnell TP (2021) Engineering chloroplast development in rice through cell-specific control of endogenous genetic circuits. Plant Biotechnol J 19(11):2291–2303. https://doi.org/10.1111/pbi.13660

Lin H, Arrivault S, Coe RA, Karki S, Covshoff S, Bagunu E, Lunn JE, Stitt M, Furbank RT, Hibberd JM, Quick WP (2020) A partial C_4 photosynthetic biochemical pathway in rice. Front Plant Sci 11:564463. https://doi.org/10.3389/fpls.2020.564463

Liu Z, Cheng J (2024) C_4 rice engineering, beyond installing a C_4 cycle. Plant Physiol Biochem 206:108256. https://doi.org/10.1016/j.plaphy.2023.108256

Lo SF, Chatterjee J, Biswal AK, Liu IL, Chang YP, Chen PJ, Wanchana S, Elmido-Mabilangan A, Nepomuceno RA, Bandyopadhyay A, Hsing YI, Quick WP (2022) Closer vein spacing by ectopic expression of nucleotide-binding and leucine-rich repeat proteins in rice leaves. Plant Cell Rep 41(2):319–335. https://doi.org/10.1007/s00299-021-02810-5

Long BM, Hee WY, Sharwood RE, Rae BD, Kaines S, Lim YL, Nguyen ND, Massey B, Bala S, von Caemmerer S, Badger MR, Price GD (2018) Carboxysome encapsulation of the CO_2-fixing enzyme Rubisco in tobacco chloroplasts. Nat Commun 9(1):3570. https://doi.org/10.1038/s41467-018-06044-0

Lundgren MR, Osborne CP, Christin PA (2014) Deconstructing Kranz anatomy to understand C4 evolution. J Exp Bot 65(13):3357–3369. https://doi.org/10.1093/jxb/eru186

Lundgren MR, Dunning LT, Olofsson JK, Moreno-Villena JJ, Bouvier JW, Sage TL, Khoshravesh R, Sultmanis S, Stata M, Ripley BS, Vorontsova MS, Besnard G, Adams C, Cuff N, Mapaura A, Bianconi ME, Long CM, Christin P-A, Osborne CP (2019) C4 anatomy can evolve via a single developmental change. Ecol Lett 22(2):302–312. https://doi.org/10.1111/ele.13191

McGrath JM, Long SP (2014) Can the cyanobacterial carbon-concentrating mechanism increase photosynthesis in crop species? A theoretical analysis. Plant Physiol 164(4):2247–2261. https://doi.org/10.1104/pp.113.232611

Mendieta JP, Tu X, Jiang D, Yan H, Zhang X, Marand AP, Zhong S, Schmitz RJ (2024) Investigating the *cis*-regulatory basis of C_3 and C_4 photosynthesis in grasses at single-cell resolution. Proc Natl Acad Sci 121(40):e2402781121. https://doi.org/10.1073/pnas.2402781121

Moreno-Villena JJ, Zhou H, Gilman IS, Tausta SL, Cheung CYM, Edwards EJ (2022) Spatial resolution of an integrated C_4+CAM photosynthetic metabolism. Sci Adv 8(31):eabn2349. https://doi.org/10.1126/sciadv.abn2349

Mukherjee A, Lau CS, Walker CE, Rai AK, Prejean CI, Yates G, Emrich-Mills T, Lemoine SG, Vinyard DJ, Mackinder LCM, Moroney JV (2019) Thylakoid localized bestrophin-like proteins are essential for the CO_2 concentrating mechanism of *Chlamydomonas reinhardtii*. Proc Natl Acad Sci USA 116(34):16915–16920. https://doi.org/10.1073/pnas.1909706116

Nguyen ND, Pulsford SB, Hee WY, Rae BD, Rourke LM, Price GD, Long BM (2023) Towards engineering a hybrid carboxysome. Photosynth Res 156(2):265–277. https://doi.org/10.1007/s11120-023-01009-x

Nguyen ND, Pulsford SB, Förster B, Rottet S, Rourke L, Long BM, Price GD (2024) A carboxysome-based CO_2 concentrating mechanism for C_3 crop chloroplasts: advances and the road ahead. Plant J 118(4):940–952. https://doi.org/10.1111/tpj.16667

Ni T, Sun Y, Burn W, Al-Hazeem MMJ, Zhu Y, Yu X, Liu LN, Zhang P (2022) Structure and assembly of cargo Rubisco in two native α-carboxysomes. Nat Commun 13(1):4299. https://doi.org/10.1038/s41467-022-32004-w

Nölke G, Barsoum M, Houdelet M, Arcalís E, Kreuzaler F, Fischer R, Schillberg S (2019) The integration of algal carbon concentration mechanism components into tobacco chloroplasts increases photosynthetic efficiency and biomass. Biotechnol J 14(3):e1800170. https://doi.org/10.1002/biot.201800170

Occhialini A, Lenaghan SC (2023) Plastid engineering using episomal DNA. Plant Cell Rep 42(7):1125–1132. https://doi.org/10.1007/s00299-023-03020-x

Occhialini A, Pfotenhauer AC, Daniell H, Neal Stewart C Jr., Lenaghan SC (2023) Genetic engineering of potato (*Solanum tuberosum*) chloroplasts using the small synthetic plastome "mini-synplastome". Methods Mol Biol 2653:73–92. https://doi.org/10.1007/978-1-0716-3131-7_5

Oh ZG, Ang WSL, Poh CW, Lai SK, Sze SK, Li HY, Bhushan S, Wunder T, Mueller-Cajar O (2023) A linker protein from a red-type pyrenoid phase separates with Rubisco via oligomerizing sticker motifs. Proc Natl Acad Sci USA 120(25):e2304833120. https://doi.org/10.1073/pnas.2304833120

Perron N, Kirst M, Chen S (2024) Bringing CAM photosynthesis to the table: paving the way for resilient and productive agricultural systems in a changing climate. Plant Commun 5(3). https://doi.org/10.1016/j.xplc.2023.100772

Price GD, Badger MR, von Caemmerer S (2011) The prospect of using cyanobacterial bicarbonate transporters to improve leaf photosynthesis in C3 crop plants. Plant Physiol 155(1):20–26. https://doi.org/10.1104/pp.110.164681

Price GD, Pengelly JJL, Forster B, Du J, Whitney SM, von Caemmerer S, Badger MR, Howitt SM, Evans JR (2012) The cyanobacterial CCM as a source of genes for improving photosynthetic CO_2 fixation in crop species. J Exp Bot 64(3):753–768. https://doi.org/10.1093/jxb/ers257

Rae BD, Long BM, Förster B, Nguyen ND, Velanis CN, Atkinson N, Hee WY, Mukherjee B, Price GD, McCormick AJ (2017) Progress and challenges of engineering a biophysical CO_2-concentrating mechanism into higher plants. J Exp Bot 68(14):3717–3737. https://doi.org/10.1093/jxb/erx133

Robison TA, Oh ZG, Lafferty D, Xu X, Villarreal JCA, Gunn LH, Li F-W (2025) Hornworts reveal a spatial model for pyrenoid-based CO_2-concentrating mechanisms in land plants. Nat Plants 11(1):63–73. https://doi.org/10.1038/s41477-024-01871-0

Rottet S, Förster B, Hee WY, Rourke LM, Price GD, Long BM (2021) Engineered accumulation of bicarbonate in plant chloroplasts: known knowns and known unknowns. Front Plant Sci 12:727118. https://doi.org/10.3389/fpls.2021.727118

Rottet S, Rourke LM, Pabuayon ICM, Phua SY, Yee S, Weerasooriya HN, Wang X, Mehra HS, Nguyen ND, Long BM, Moroney JV, Price GD (2024) Engineering the cyanobacterial ATP-driven BCT1 bicarbonate transporter for functional targeting to C3 plant chloroplasts. J Exp Bot 75(16):4926–4943. https://doi.org/10.1093/jxb/erae234

Ruiz-Vera UM, Acevedo-Siaca LG, Brown KL, Afamefule C, Gherli H, Simkin AJ, Long SP, Lawson T, Raines CA (2022) Field-grown *ictB* tobacco transformants show no difference in photosynthetic efficiency for biomass relative to the wild type. J Exp Bot 73(14):4897–4907. https://doi.org/10.1093/jxb/erac193

Sage RF (2004) The evolution of C4 photosynthesis. New Phytol 161(2):341–370. https://doi.org/10.1111/j.1469-8137.2004.00974.x

Sage RF, Coleman JR (2001) Effects of low atmospheric CO_2 on plants: more than a thing of the past. Trends Plant Sci 6(1):18–24. https://doi.org/10.1016/s1360-1385(00)01813-6

Schiller K, Bräutigam A (2021) Engineering of crassulacean acid metabolism. Annu Rev Plant Biol 72:77–103. https://doi.org/10.1146/annurev-arplant-071720-104814

Schlüter U, Weber APM (2016) The road to C4 photosynthesis: evolution of a complex trait via intermediary states. Plant Cell Physiol 57(5):881–889. https://doi.org/10.1093/pcp/pcw009

Schreier TB, Müller KH, Eicke S, Faulkner C, Zeeman SC, Hibberd JM (2024) Plasmodesmal connectivity in C_4 *Gynandropsis gynandra* is induced by light and dependent on photosynthesis. New Phytol 241(1):298–313. https://doi.org/10.1111/nph.19343

Sedelnikova OV, Hughes TE, Langdale JA (2018) Understanding the genetic basis of C4 Kranz anatomy with a view to engineering C3 crops. Annu Rev Genet 52:249–270. https://doi.org/10.1146/annurev-genet-120417-031217

Shimamura D, Yamano T, Niikawa Y, Hu D, Fukuzawa H (2023) A pyrenoid-localized protein SAGA1 is necessary for Ca(2+)-binding protein CAS-dependent expression of nuclear genes encoding inorganic carbon transporters in *Chlamydomonas reinhardtii*. Photosynth Res 156(2):181–192. https://doi.org/10.1007/s11120-022-00996-7

Simpson CJC, Reeves G, Tripathi A, Singh P, Hibberd JM (2021) Using breeding and quantitative genetics to understand the C4 pathway. J Exp Bot 73(10):3072–3084. https://doi.org/10.1093/jxb/erab486

Singh P, Reeves G (2020) Constructing the bundle sheath towards enhanced photosynthesis. J Exp Bot 71(4):1206–1209. https://doi.org/10.1093/jxb/erz537

Singh J, Garai S, Das S, Thakur JK, Tripathy BC (2022) Role of C4 photosynthetic enzyme isoforms in C3 plants and their potential applications in improving agronomic traits in crops. Photosynth Res 154(3):233–258. https://doi.org/10.1007/s11120-022-00978-9

Singh P, Stevenson SR, Dickinson PJ, Reyna-Llorens I, Tripathi A, Reeves G, Schreier TB, Hibberd JM (2023) C_4 gene induction during de-etiolation evolved through changes in cis to allow integration with ancestral C_3 gene regulatory networks. Sci Adv 9(13):eade9756. https://doi.org/10.1126/sciadv.ade9756

Susila H, Nasim Z, Gawarecka K, Jung JY, Jin S, Youn G, Ahn JH (2023) Chloroplasts prevent precocious flowering through a *GOLDEN2-LIKE-B-BOX DOMAIN PROTEIN* module. Plant Commun 4(3):100515. https://doi.org/10.1016/j.xplc.2023.100515

Tan B, Chen S (2023) Defining mechanisms of C3 to CAM photosynthesis transition toward enhancing crop stress resilience. Int J Mol Sci 24(17):13072. https://www.mdpi.com/1422-0067/24/17/13072

Tanwar N, Rookes JE, Cahill DM, Lenka SK (2023) Carotenoid pathway engineering in tobacco chloroplast using a synthetic operon. Mol Biotechnol 65(11):1923–1934. https://doi.org/10.1007/s12033-023-00693-3

Terentyev VV, Shukshina AK (2024) CAH3 from *Chlamydomonas reinhardtii*: unique carbonic anhydrase of the thylakoid lumen. Cells 13(2). https://doi.org/10.3390/cells13020109

Toyokawa C, Yamano T, Fukuzawa H (2020) Pyrenoid starch sheath is required for LCIB localization and the CO_2-concentrating mechanism in green algae. Plant Physiol 182(4):1883–1893. https://doi.org/10.1104/pp.19.01587

Trettel DS, Pacheco SL, Laskie AK, Gonzalez-Esquer CR (2024) Modeling bacterial microcompartment architectures for enhanced cyanobacterial carbon fixation. Front Plant Sci 15:1346759. https://doi.org/10.3389/fpls.2024.1346759

Tsang HT, Ganguly DR, Furbank RT, von Caemmerer S, Danila FR (2024) Novel resources to investigate leaf plasmodesmata formation in C and C monocots. Plant J. https://doi.org/10.1111/tpj.17113

van Rooijen R, Schulze S, Petzsch P, Westhoff P (2019) Targeted misexpression of *NAC052*, acting in H3K4 demethylation, alters leaf morphological and anatomical traits in *Arabidopsis thaliana*. J Exp Bot 71(4):1434–1448. https://doi.org/10.1093/jxb/erz509

Walsh CA, Bräutigam A, Roberts MR, Lundgren MR (2023) Evolutionary implications of C_2 photosynthesis: how complex biochemical trade-offs may limit C_4 evolution. J Exp Bot 74(3):707–722. https://doi.org/10.1093/jxb/erac465

Wang P, Khoshravesh R, Karki S, Tapia R, Balahadia CP, Bandyopadhyay A, Quick WP, Furbank R, Sage TL, Langdale JA (2017) Re-creation of a key step in the evolutionary switch from C_3 to C_4 leaf anatomy. Curr Biol 27(21):3278–3287.e3276. https://doi.org/10.1016/j.cub.2017.09.040

Whitney SM, Houtz RL, Alonso H (2010) Advancing our understanding and capacity to engineer nature's CO_2-sequestering enzyme, Rubisco. Plant Physiol 155(1):27–35. https://doi.org/10.1104/pp.110.164814

Winter K, Holtum JAM (2024) Shifting photosynthesis between the fast and slow lane: facultative CAM and water-deficit stress. J Plant Physiol 294:154185. https://doi.org/10.1016/j.jplph.2024.154185

Yamano T, Toyokawa C, Shimamura D, Matsuoka T, Fukuzawa H (2022) CO2-dependent migration and relocation of LCIB, a pyrenoid-peripheral protein in *Chlamydomonas reinhardtii*. Plant Physiol 188(2):1081–1094. https://doi.org/10.1093/plphys/kiab528

Yang X, Cushman JC, Borland AM, Edwards EJ, Wullschleger SD, Tuskan GA, Owen NA, Griffiths H, Smith JAC, De Paoli HC, Weston DJ, Cottingham R, Hartwell J, Davis SC, Silvera K, Ming R, Schlauch K, Abraham P, Stewart JR et al (2015) A roadmap for research on crassulacean acid metabolism (CAM) to enhance sustainable food and bioenergy production in a hotter, drier world [Note]. New Phytol 207(3):491–504. https://doi.org/10.1111/nph.13393

Yang X, Liu Y, Yuan G, Weston DJ, Tuskan GA (2024) Engineering crassulacean acid metabolism in C_3 and C_4 plants. Cold Spring Harb Perspect Biol 16(4). https://doi.org/10.1101/cshperspect.a041674

Yeh S-Y, Lin H-H, Chang Y-M, Chang Y-L, Chang C-K, Huang Y-C, Ho Y-W, Lin C-Y, Zheng J-Z, Jane W-N, Ng C-Y, Lu M-Y, Lai I-L, To K-Y, Li W-H, Ku MSB (2021) Maize *Golden2-like* transcription factors boost rice chloroplast development, photosynthesis, and grain yield. Plant Physiol 188(1):442–459. https://doi.org/10.1093/plphys/kiab511

Yuan Y, Mei L, Wu M, Wei W, Shan W, Gong Z, Zhang Q, Yang F, Yan F, Zhang Q, Luo Y, Xu X, Zhang W, Miao M, Lu W, Li Z, Deng W (2018) *SlARF10*, an auxin response factor, is involved in chlorophyll and sugar accumulation during tomato fruit development. J Exp Bot 69(22):5507–5518. https://doi.org/10.1093/jxb/ery328

Yuan G, Hassan M, Liu D, Lim SD, Yim WC, Cushman JC, Markel K, Shih PM, Lu H, Weston DJ, Chen JG, Tschaplinski TJ, Tuskan GA, Yang X (2020) Biosystems design to accelerate C3-to-CAM progression [Review]. BioDesign Res 2020:Article 3686791. https://doi.org/10.34133/2020/3686791

Zang K, Wang H, Hartl FU, Hayer-Hartl M (2021) Scaffolding protein CcmM directs multiprotein phase separation in β-carboxysome biogenesis. Nat Struct Mol Biol 28(11):909–922. https://doi.org/10.1038/s41594-021-00676-5

Zhang A, Zhang W (2022) Characterization of transposon-derived accessible chromatin regions in rice (*Oryza sativa*). Int J Mol Sci 23(16). https://doi.org/10.3390/ijms23168947

Zhang Y, Tian L, Lu C (2023) Chloroplast gene expression: recent advances and perspectives. Plant Commun 4(5):100611. https://doi.org/10.1016/j.xplc.2023.100611

Zhao W, Li J, Sun X, Zheng Q, Liu J, Hua W, Liu J (2023) Integrated global analysis in spider flowers illuminates features underlying the evolution and maintenance of C4 photosynthesis. Hortic Res 10(8). https://doi.org/10.1093/hr/uhad129

Chapter 6
Advancements in Carbon Concentrating Mechanisms and Synthetic Pathways to Enhance Photosynthetic Efficiency in C_3 Plants

Gurbir Kaur Sidhu and Sangram K. Lenka

Abstract To meet global food security demands, improving the photosynthetic efficiency of C3 plants like rice and wheat is essential. Traditional C3 photosynthesis is inherently limited by Rubisco, an enzyme with suboptimal CO_2/O_2 specificity, resulting in significant photorespiration and decreased carbon fixation. Carbon-concentrating mechanisms (CCMs) naturally evolved in C4, and CAM plants, cyanobacteria, and algae offer promising avenues for improving C3 crop productivity. These mechanisms work by concentrating CO_2 around Rubisco, thereby enhancing carbon assimilation and reducing photorespiration losses. Advanced synthetic pathways, such as the synthetic photorespiratory bypass and the CETCH (crotonyl-CoA/ethylmalonyl-CoA/hydroxybutyryl-CoA) cycle, represent novel methods for improving photosynthetic efficiency. These methods aim to create bypass pathways for photorespiration or introduce entirely new carbon fixation routes in chloroplasts. Integrating elements of algal and cyanobacterial CCMs, such as pyrenoids and carboxysomes, respectively, into C3 plants further enhances this potential by creating microenvironments conducive to efficient carbon capture. Together, these approaches signify substantial progress toward sustainable crop productivity, with applications extending to climate resilience and resource-efficient agriculture.

Keywords CO_2-concentrating mechanism (CCM) · Carboxysomes · Pyrenoid · CETCH cycle · Photorespiratory bypass

G. K. Sidhu
TERI School of Advanced Studies, New Delhi, India

S. K. Lenka (✉)
Gujarat Biotechnology University, Gandhinagar, Gujarat, India

A. Singh et al. (eds.), *Sustainable Crop Production*, Sustainability Sciences in Asia and Africa, https://doi.org/10.1007/978-981-95-7844-3_6

Editor's Notes

Rubisco, the most abundant protein on Earth, is central to photosynthesis wherein atmospheric CO_2 is biologically fixed into chemical energy of sugars. However, the catalytic activity of Rubisco is relatively low mainly because of its dual activity, characterized as competitive binding between CO_2 and O_2. In other words, Rubisco can act both as a carboxylase and oxygenase, depending on relative concentrations of CO_2 and O_2. Under high light intensity and temperature conditions, Rubisco drives a wasteful metabolic process of photorespiration, considered as a major scourge by agriculturists. For these reasons, protein engineering of Rubisco has remained a major strategy to improve photosynthetic efficiency though with limited success.

This chapter by Kaur and Lenka explores the potential of unconventional approaches that work around Rubisco. In one of the strategies, the authors discuss the prospects of bioengineering mimics of CO_2 capturing mechanisms (CCMs), carboxysomes, and pyrenoids in C3 plants. These are naturally occurring systems in algae and cyanobacteria. The authors rationalize that establishing CCMs will result in higher concentrations of CO_2 necessary for favoring carboxylase activity of Rubisco. In addition, they explore the potential of engineering biosynthetic pathways in C3 plants to achieve photorespiratory bypasses, recycling of glycolate, and engineering resource-efficient CAM and C4 pathways in C3 crops. The chapter also explores exciting possibility of expressing enoyl-CoA carboxylases/reductases (ECRs) as an alternate to Rubisco. The authors urge researchers to consider following research targets in order to achieve sustainable food production.

Actionable Points

1. For enhancing CO_2 fixation in C3 plants, it is critical to focus on mitigating Rubisco's biocatalytic limitations through synthetic and naturally evolved CCMs.
2. Engineered pathways can recycle glycolate within chloroplasts, reducing photorespiration losses and potentially increasing plant biomass and yield.
3. Translating water-efficient CAM and high-efficiency C4 pathways into C3 crops could improve photosynthetic output under varying climates.
4. Bioengineering carboxysomes and pyrenoids from cyanobacteria and algae into C3 plants holds promise for Rubisco compartmentalization, leading to enhanced CO_2 assimilation.
5. The CETCH cycle demonstrates a feasible synthetic CO_2 pathway in vitro, providing an alternative to Rubisco with greater CO_2 specificity and reduced oxygenation.

6.1 Introduction

The world population is projected to reach 9.8 billion in 2050 from 7.9 billion in 2021. To ensure the food security of the increasing population, the global crop production should double by 2050 (Tilman et al. 2011). To achieve this target, yield of major cereal crops like rice and wheat need to be boosted by about 2.4% per year. Conversely, this increase in demand will be unlikely to be achieved. Factors that corroborate this slowdown are limited availability of land, water scarcity, salinization, and conversion of cultivation areas for other allied activities. With the advent of Green Revolution (GR) in the mid-1960s in Asia, yield in rice increased rapidly between 1970 and 1985 and then slowed down and attained stagnancy from 1991 to present (Janaiah et al. 2006; Pingali 2012). The yield of plants cannot increase beyond a certain point solely depending on the fertilizers, irrigation, or high yielding varieties that were introduced during GR. This is because the ultimate limiting factor that reduces photosynthesis in C3 plants is the enzyme Rubisco, which has poor discrimination between CO_2 and O_2. The lower carboxylation rate in vivo in C3 plants is due to the ability of Rubisco to catalyze both the carboxylation and oxygenation of RuBP. When Rubisco catalyzes the oxygenation of RuBP, O_2 is fixed, and a wasteful process known as photorespiration takes place. Photorespiration leads to the loss of CO_2 that was fixed during Calvin cycle, reducing photosynthetic output by 25% in C3 plants.

In nature to circumvent or cope with the inefficiencies of Rubisco, algae (Mackinder 2018), cyanobacteria (Rae et al. 2013), and C4 and CAM (crassulacean acid metabolism) plants (Ludwig et al. 2024) have evolved CCM to increase the carbon concentration near Rubisco. In cyanobacteria, the carbon fixation is accomplished inside proteinaceous compartments called carboxysome, and in algae, these structures are called pyrenoids. As a result of which, the process of photorespiration could be either partially or fully suppressed. This in turn leads to higher photosynthetic output in these organisms compared to C3 plants. Other than the mentioned natural CCM, there are synthetically designed pathways which if transformed in C3 plants could augment in concentrating carbon in the chloroplasts. One such method is the synthetic photorespiratory bypass pathway (Xin et al. 2015). This pathway follows an alternate way (designed from genes of heterologous organisms) to bypass endogenous photorespiration cycle of C3 plants, reducing the loss of CO_2 from the stroma of the chloroplasts. The other method is to engineer the synthetic carbon cycle, CETCH (crotonyl-CoA/ethylmalonyl-CoA/hydroxybutyryl-CoA) in C3 plants (Naseem et al. 2020). This synthetic cycle has been successfully demonstrated in vitro to fix CO_2 using the enoyl-CoA carboxylases/reductases (ECR's), as an alternate to Rubisco. The genes responsible for catalyzing the reactions in this cycle has been incorporated from all the three forms of life (humans, plants, bacteria and archae). The methods which can be implemented to maximize the carbon fixations efficiency in C3 plants are depicted in Fig. 6.1.

Due to the scarcity of natural resources (land and water), it is necessary to implement sustainable methods to increase crop productivity to feed the growing population. Therefore, with the help of advance molecular biology techniques, photosynthetic efficiency in C3 plants can be improved by the genetic engineering of various carbon concentrating mechanisms (CCM). This chapter overviews components of diverse CCMs and the recent significant advancements toward engineering them in C3 plants to improve photosynthetic efficiency.

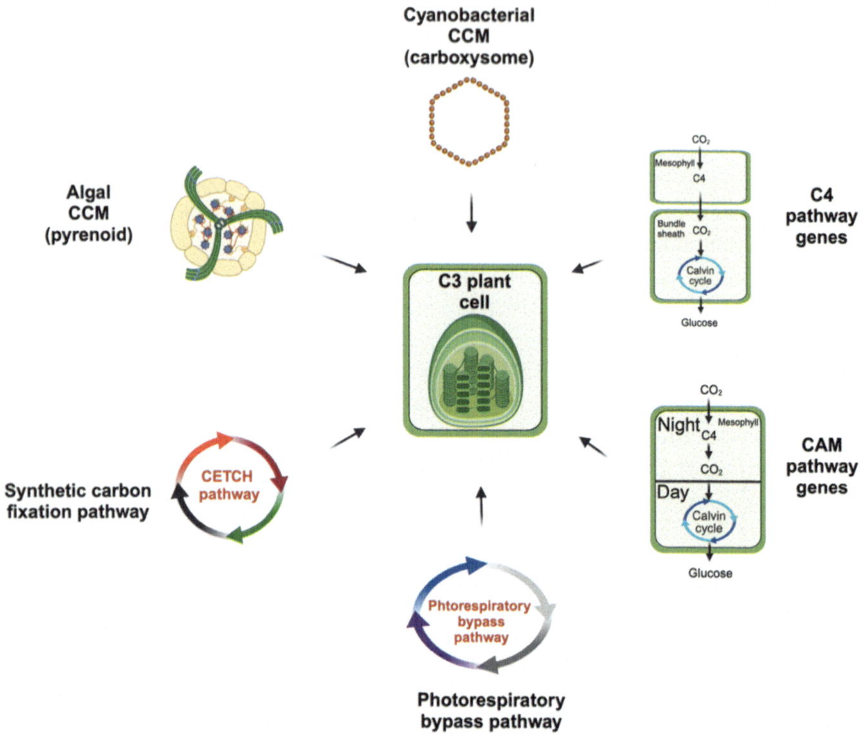

Fig. 6.1 Various mechanisms to maximize photosynthetic efficiency by concentrating carbon in C3 plants. The photosynthetic output in C3 plants can be enhanced by targeting various naturally evolved components to chloroplasts such as the carboxysomes of cyanobacteria, pyrenoids of algae, and C4 and CAM pathway. Additionally, the artificial CCMs such as the photorespiratory bypass pathway that could recycle glycolate produced during photorespiration and the recently designed in vitro synthetic CO_2 pathway, the CETCH cycle (crotonyl-CoA/ethylmalonyl-CoA/hydroxybutyryl-CoA) which provides an alternative way of carbon fixation, can be expressed in C3 plants

6.2 Naturally Occurring CO_2-Concentrating Mechanisms

Many plants, algae, and cyanobacteria have evolved mechanisms to reduce or suppress the oxygenation of Rubisco and inhibit photorespiration either completely or to an extent, thereby decreasing the loss of CO_2. Approximately 50% of the carbon fixed in a year is contributed by CCMs which is at least 53 $PgCyear^{-1}$ of the global photosynthetic productivity of the total 105 $PgCyear^{-1}$ (Raven et al. 2017). The Rubisco and photosynthetic carbon reduction cycle of oxygen evolving organisms cannot be replaced with other inorganic carbon fixation pathways of autotrophs. Thus, the CCMs have been evolved to make the conditions more conducive for carbon fixation when there is decreased availability of CO_2. The earliest CCM might have evolved about 2.4 billion years (Gyr) ago in the oxygenic cyanobacteria in stromatolites despite the levels of CO_2 being high in the atmosphere. It was stated that these cyanobacteria had low intercellular CO_2 and high O_2 concentrations because of the presence of diffusion barriers around the cells (Giordano et al. 2005; Raven et al. 2008). However, the functional CCM might have originated much later around 350 million years ago (Badger and Price 2003). Subsequently, plastids of algae originated from ancestral cyanobacteria by endosymbiosis. Later, algal lineages gave rise to C3 photosynthesis. CAM photosynthesis originated from C3 plants approximately 20 million years ago when atmospheric CO_2 decreased, which is an adaptive mechanism to grow well in hot and arid conditions (Heyduk 2022). Since then, it has evolved multiple times across at least 35 plant families (Silvera et al. 2010). C4 photosynthesis has evolved 65 times independently from its C3 ancestors (Heyduk et al. 2019).

6.2.1 C4 Photosynthetic Carbon Fixation (C4) Pathway

This cycle is called C4 cycle as the first metabolite formed after the carbon fixation is a four-carbon compound (malate and/or aspartate). Plants with C4 cycle are often found in hot climates. Maize, sugarcane, and tropical grasses are some of the plants that have C4 type cycle. C4 plants have evolved over the past 40 million years when the CO_2 concentrations in the atmosphere declined increasing the rates of photorespiration. Over this period of time, the transition from C3 to C4 photosynthesis followed in two stages, formation of proto-Kranz anatomy in close C3 relatives of C4 species and C2 photosynthesis in C3-C4 intermediate species (Sage et al. 2012). There are two salient features that have been acquired in C4 plants that act as CCM to enhance photosynthetic efficiency (Ludwig et al. 2024). First is the CO_2 insensitive enzyme phosphoenolpyruvate carboxylase (PEPCase) whose substrate is HCO_3^- and not oxygen, so there is no competition for carbon dioxide fixation. Second is the spatial separation of photosynthetic process into two compartments. It involves two types of cells, the mesophyll where CO_2 fixation occurs by PEPCase and the bundle sheath cells for photosynthesis. They form specialized structures known as Kranz anatomy where mesophyll cells surround bundle sheath cells in a ringlike arrangement. These cells are merely two or three cells distant apart which are connected via plasmodesmata, and the shuttling of metabolites takes place along the diffusion gradients, and transportation into the cells occurs through the translocators. The process of C4 carbon concentration occurs through the four steps: (1) Fixation of CO_2 occurs in the mesophyll cells. CO_2 in the form of bicarbonate ions acts as the substrate for the PEPCase. The carboxylation of phosphoenolpyruvate results in the formation oxaloacetate, which is then catalyzed by malate dehydrogenase to form malate and/or aspartate (a C4 acid). (2) Transportation of malate and/or aspartate to the bundle sheath cells. (3) Decarboxylation of C4 acids occurs in the bundle sheath cells, into pyruvate, catalyzed by malic enzyme, and CO_2 is released which is then fixed by Rubisco and reduced to carbohydrate in the Calvin cycle. (4) Transportation of pyruvate to mesophyll cells, where it is regenerated to phosphoenolpyruvate (PEP), the CO_2 acceptor catalyzed by pyruvate phosphate dikinase (PPDK). The affinity of PEPCase for its substrate HCO_3^- is high, and therefore, it enables these plants to reduce stomatal aperture and conserve water. As the CO_2 is concentrated in the bundle sheath cells, photorespiration does not occur in these cells.

High photosynthetic capacity of C4 plants as compared to C3 plants has attracted the attention of researchers to introduce the features of C4 such as low photorespiration and increased water, nitrogen, and light use efficiency into C3 plants (Furbank 2017; Sedelnikova et al. 2018). The *Atriplex prostrata*, a C3-like species has been crossed with C4-like species, *Atriplex rosea* (Oakley et al. 2014). The F1 hybrids had reduced levels of net CO_2 assimilation compared to *A. prostrata*, and PEPCase, PPDK, and NAD-malic enzyme had low levels of activity when compared to C4 system and leaves did not develop Kranz anatomy. F2 hybrid lines had an impaired C4 photosynthesis; C3 pathway was also improperly assembled. The expression of incomplete photorespiratory pathway and mismatch in compartmentalization of C3 components resulted in C2-like photosynthesis in these lines. For the efficient working of C4 system and photosynthesis, it is necessary that all the components of C4 system are in place.

Prime limitation of crossing is that it can be done only with C3 species that are closely related to C4 relatives. Most of the chief C3 crops does not have C4 relatives. Therefore, efforts were made to express genes coding for C4 enzymes in C3 plants to decrease photorespiration and increase photosynthesis and yield. Single-cell C4-like pathway genes from *Hydrilla verticillata* were introduced in rice (a C3 plant), but it did not lead to improvements in photosynthesis (Miyao et al. 2011). The phosphoenolpyruvate carboxylase (PEPCase) gene from maize has been introduced into rice plants. The transgenic plants showed high expression of the PEPCase gene in rice accounting for the 12% total soluble protein of leaf. The transgenic plants had lower O_2 inhibition of photosynthesis, while the photosynthetic rates were similar to WT plants (Ku et al. 1999). Introduction of maize pyruvate phosphate dikinase (PPDK) in rice leads to increase in photosynthetic rate of up to 35% which was partly due to stomatal conductance. The grain yield also increased to 30–35% in transgenic PPDK rice compared to WT (Ku et al. 2001). C4-specific enzymes from maize when overexpressed in rice resulted in stunted growth (Taniguchi et al. 2008). Five enzymes of C4 pathway from maize when expressed in rice were able to localize in appropriate subcellular compartments, and CO_2 was incorporated in the C4 acid, demonstrating that the C4 cycle was partially functional in rice (Ermakova et al. 2021). These results prove that it is possible to attain a fully functional C4 photosynthesis in plants with further optimizations.

Bioengineering of two-cell approach of attaining C4 Kranz anatomy into C3 plants would require the following biochemical and anatomical modifications (Gowik and Westhoff 2011): (1) Anatomical changes such as the vein density and ratio of bundle sheath to mesophyll cells should increase in leaves of C3 plants. (2) Moreover, development of Kranz-like anatomy is crucial where mesophyll cells surround the bundle sheath cells. Also, the adaptation of bundle sheath (BS) morphology is a prerequisite. (3) BS cells should be enlarged and have higher number of chloroplasts and well connected with mesophyll cells via plasmodesmata. (4) Compartmentalization of cell-type specific enzymes involved in photosynthesis between mesophyll and bundle sheath is also necessary (Schuler et al. 2016). It has been suggested that C4 photosynthesis gene expression shares similar gene regulatory mechanism like C3 plants (Furbank 2017). Our knowledge and studies related to conversion of C3 plant systems into C4-like plant system are very limited. Also, it is difficult to engineer Kranz-like anatomy in C3 plants as the genes involved in development of the anatomy are not known. And a fully functional C4 cycle requires a large number of genes to be engineered in C3 plants which poses the risk of genetic load and regulation (Von Caemmerer et al. 2012). Due to the complexity of the C4 system, it is very challenging to introduce it into C3 plants (Cui 2021; Pradhan et al. 2022). It is clear that it is not so complicated to express the genes of C4 pathway into C3 plants rather than introducing the C4 Kranz anatomy of leaf.

6.2.2 Crassulacean Acid Metabolism (CAM)

This CO_2 concentration mechanism is found in the crassulacean family. Plants such as cacti and pineapple are some of the examples which are found mostly in the desert areas. It has evolved from C3 plants via the convergent or parallel evolution. In the course of evolution, the facultative CAM species have also originated. They exhibit plasticity, which means they generally follow C3 cycle for carbon fixation under non-stressed conditions but change their mode from C3 to CAM under stress condition such as drought (Borland and Griffiths 1990) or salinity (Kong et al. 2020). CAM is an efficient pathway for photosynthetic CO_2 fixation in semiarid or desert plants as an adaption mechanism to hot climates where water availability is scarce. In these plants, carbon fixation is temporally separated (Ludwig et al. 2024). The stomata close during daytime when the temperatures are high and open at night for CO_2 fixation which leads to low levels of transpiration and use water efficiently. During night, CO_2 is taken up when stomata remain open and converted into bicarbonate (HCO_3^-) ions in the mesophyll cells. In the cytosol, HCO_3^- acts as the substrate for phosphoenolpyruvate and is converted to oxaloacetate catalyzed by PEP carboxylase. Oxaloacetate is then converted to malic acid by the enzyme NAD^+ malic dehydrogenase, which is stored in the vacuole. During the daytime, when stomata are closed, malic acid moves to the chloroplast and NADP+ malic enzyme converts it to pyruvate, and CO_2 is released. CO_2 is then reduced to carbohydrates through the Calvin cycle. During the daytime, when stomata are closed, oxygen cannot enter the cells; therefore, CO_2 levels increase near Rubisco. This results in suppression of photorespiration.

High water use efficiency (WUE) of CAM plants is a sustainable approach for biomass production in high temperature areas (Yang et al. 2015). There are numerous reasons to engineer the CAM in C3 crops: (1) C3 plants are the agriculturally important crops that feed the world population but require adequate water supply; (2) climate change and unexpected weather conditions had led to high temperatures and drought in most areas; and (3) world's 40% of the land accounts for arid, semi-arid, or dry area.

Tissue succulence is one of the characteristics of CAM plants which help them thrive under drought conditions by storing water within the tissues. This property of CAM plants was mimicked in *Arabidopsis* plants by overexpressing the helix-loop-helix transcription factor responsible for cell expansion from grape wine (Lim et al. 2020). The resulting *Arabidopsis* plants had increased cell size and succulence, along with reduced stomatal aperture and density. All these factors alleviated water-deficit stress. C4 metabolism genes of CAM pathway has also been expressed in *Arabidopsis* to analyze the relative contribution of each of the components, which is critical for the CAM bio-design (Lim et al. 2019).

Not much research has been done in this area of utilizing or incorporating the CCM machinery of CAM in C3 plants. However, modelling strategies have suggested that it is possible to engineer CAM into C3 plants. CAM requires single cell for concentrating CO_2 unlike C4 which requires specialized bundle sheath and mesophyll cells. To convert C3 system into CAM like, manipulation of CAM genes at different levels is required: (1) CO_2 fixation at night and its accumulation in vacuole in the form of malic acid, (2) malic acid decarboxylation leading to CO_2 formation, (3) stomatal opening during night and closing during daytime, and (4) anatomical changes at the leaf level to acquire higher degree of succulence (Tuskan et al. 2015; Yang et al. 2015). It has been suggested that to convert C3 into CAM, not only the starch-malate cycle of CAM is necessary but metabolic fluxes from other cycles such tricarboxylic acid and electron transport chain are also needed to be optimized in C3 plants (Tay et al. 2021). To engineer CAM machinery in C3 plants, there are different levels at which studies are yet to be accomplished: the anatomical traits of leaf surface, temporal reprogramming of the genes, silencing endogenous C3 genes and expressing CAM pathway genes. It is not clear what the minimum number of genes from large pool of CAM pathway machinery are required for transfer into C3 plants.

6.2.3 CCM in Algae

One third of the global carbon fixation is done by the algae. These photosynthetic eukaryotes have developed a carbon-concentrating mechanism (CCM), which sequesters CO_2 around Rubisco, to grow better at the ambient carbon dioxide concentrations (which are generally low) (Mackinder 2018). The CCM machinery of algae includes the following components: (1) the inorganic carbon transporters present on the plasma membrane and chloroplast envelope that increase the inorganic carbon concentration inside the cytosol and chloroplast, respectively, (2) the carbonic anhydrases, and (3) the Rubisco-encapsulated microcompartment called pyrenoid. Pyrenoid containing algae fix up to 44% of global carbon (Mackinder et al. 2016). However, not all algae possess pyrenoid-based CCM, such as

Nannochloropsis oceanica and *Chloromonas rosae*. Alternately, these species solely have carbonic anhydrases that are attributed to CCM.

Pyrenoid-based CCM are well studied in the model organism, *Chlamydomonas reinhardtii*. They belong to the genus *Chlamydomonas* in which all the species contain pyrenoid. The pyrenoids are spherical, chloroplast stroma localized, proteinaceous bodies. They are composed of a matrix covered by sheaths of starch and thylakoid tubules (known as pyrenoid tubules) that traverse through them (Fig. 6.2). Pyrenoid tubules are enriched in photosystem 1 (PSI) and carbonic anhydrase. Matrix of pyrenoid is phase separated, liquid like condensate that dissolves and condenses during cell division. Pyrenoid matrixes densely packed with Rubisco, Rubisco activase (the chaperone), and the linker protein, EPYC1 (essential pyrenoid component 1). Additionally, matrix is composed of ten proteins that are recently identified (He et al. 2023). Pyrenoid matrix has Form I Rubisco which is composed of four large subunit (LSU) dimers which have catalytic activity, surrounded by four small subunit (SSU) dimers. EPYC1 has five Rubisco binding regions that interact with the α-helices of SSU via salt bridges and hydrophobic interactions (He et al. 2020). This leads to aggregation and localization of Rubisco inside the matrix resulting in the assembly of a functional pyrenoid (Mackinder et al. 2016; Meyer et al. 2012). EPCY1 is not only essential for the functionality of a CCM but also important in maintaining pyrenoid size, number, and matrix density.

There are two models of pyrenoid-based CCM operation in *C. reinhardtii* based on computational analysis done using all available experimental evidence (Fig. 6.2)

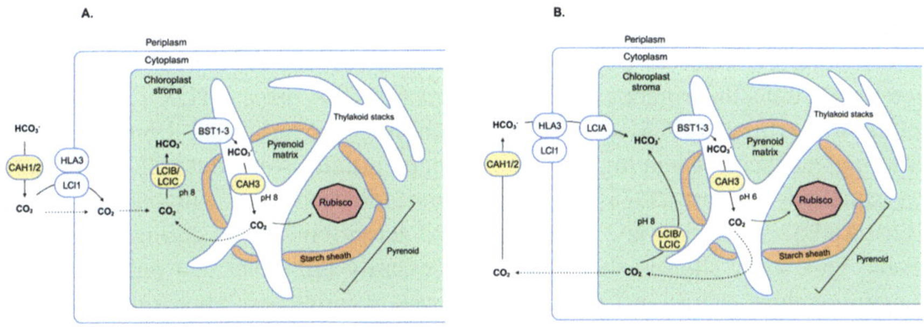

Fig. 6.2 Cartoon diagrams depicting pyrenoid-based CO_2 concentrating pathway model with its various components in *Chlamydomonas reinhardtii*. (**a**) Passive CO_2 uptake mode in the ambient CO_2 levels which are approximately 400 ppm. CO_2 diffuses passively through the cell membrane and chloroplast envelope into the stroma CO_2, which is then transported into thylakoid lumen by BST-1. (**b**) Active CO_2 uptake mode when the external CO_2 concentrations are very low (<200 ppm). HLA3 present in the cell membrane actively pumps HCO_3^- in the cell, which is transported to the chloroplast stroma by LCIA and eventually to the thylakoid lumen by BST1-3. The reactions occurring inside the thylakoid lumen are similar in both modes. The carbonic anhydrase (CAH3) residing inside the acidic thylakoid lumen (where thylakoids traverse through pyrenoid matrix) converts HCO_3^- to CO_2, which then diffuses into pyrenoid matrix creating high CO_2 levels near Rubisco. The LCIB/LCIC complex recaptures the CO_2 that diffuses out from the pyrenoid matrix (via thylakoid tubules) into the stroma, by converting it into HCO_3^-, which is transported back into the lumen by BST1-3. (Adapted from Fei et al. 2022; He et al. 2023)

(Fei et al. 2022; He et al. 2023). The first mode is the passive CO_2 uptake which works under ambient air conditions, with CO_2 concentration of approximately 400 ppm. CO_2 diffuses inside the cytoplasm from the periplasmic space through the cell membrane or the LCI1 (low CO_2 inducible 1) channel and then across the chloroplast envelope into stroma, while the second mode is the active HCO_3^- uptake that is active under very low inorganic carbon concentrations which are generally less than 200 ppm (Wang and Spalding 2014). HCO_3^- is converted to CO_2 by the carbonic anhydrase (CAH1 and CAH2) in the periplasmic space. The plasma membrane transporter, HLA3 (high light activated 3), transports HCO_3^- into the cytoplasm, which then travels to stroma via the putative active pump in chloroplast envelope, LCIA (low CO_2 inducible channel 1A). Hereafter, passive and active CO_2 models share a common pathway for transportation of inorganic carbon from stroma to thylakoid lumen and eventually to pyrenoid matrix. HCO_3^- from stroma is likely to be transported through bestrophin-like channels (BST1, BST2, and BST3) into the thylakoid lumen. However, the functions of all the three bestrophin-like channels are still unknown. The part of the thylakoid lumen which traverses through the pyrenoid is the site where the conversion of HCO_3^- to CO_2 takes place. This reaction is catalyzed by carbonic anhydrase, CAH3, and driven by low pH of the thylakoid lumen. CO_2 then diffuses into pyrenoid matrix containing Rubisco where carbon fixation takes place. CO_2 can freely diffuse or leak out from the thylakoid lumen into the chloroplast stroma. To concentrate this CO_2 back into the thylakoid lumen, the carbonic anhydrase complex (LCIB and LCIC) localized to the periphery of pyrenoid converts it back to HCO_3^-. It is suggested the CO_2 diffusion from the pyrenoid matrix is prevented by the starch sheaths that act as barriers to prevent carbon leak.

Pyrenoid-less algae could also manifest the presence of CCM. Two strains of free living, unicellular genus of *Chloromonas* (*Cr*), *Cr. rosae* UTEX 1337 and *Cr. serbinowii*, exhibited CCM at low CO_2 concentrations (Morita et al. 1998). Both the *Chloromonas* species had higher photosynthetic affinity when the culture medium was supplied with low CO_2 as compared to the culture medium with 5% CO_2 (high CO_2). And also, the carbonic anhydrase (CA) activity increased during lower CO_2 levels. This study showed that it is not necessarily important that only pyrenoid containing algae can act as CCM. Another such example is of photosynthetic stramenopile, *Nannochloropsis oceanica*, a unicellular picoplanktonic alga. In *Nannochloropsis* spp., a simple (leaky) CCM model has emerged (Gee and Niyogi 2017). The model demonstrated that the α-type carbonic anhydrase1 (CAH1) localized in the lumen of periplastid endoplasmic reticulum acts as a component of the CCM and inorganic carbon (DIC) concentrations play very important role in the expression of CAH1.

In terms of Rubisco, non-green algae are different from other algae and higher plants. Their Rubisco has relatively higher CO_2 specificity compared to O_2. Therefore, the possibility that transferring algal Rubisco into tobacco plants could enhance photosynthesis was explored. Transplastomic tobacco lines were developed using Form1 rubisco (large and small subunit) operon from non-green algae, rhodophyte *Galdieria sulphuraria* and the diatom *Phaeodactylum tricornutum* into plastid genome without replacing native Rubisco large subunit (Whitney et al. 2001). However, it was discovered that the algal Rubisco large and small subunit protein did not fold properly and were recovered in the insoluble fractions of leaf extract from tobacco transplastomic lines. It was stipulated that for the folding and assembly of red-algal Rubisco in higher plants, chaperone coding genes of algae, *dna*K and *gro*EL, are possibly required.

For the assembly of a functional algal CCM in C3 plants, it is important that the proteins are targeted to their appropriate subcellular locations in plants. To check this feasibility, ten potentially essential CCM components of *Chlamydomonas* have been transiently expressed in tobacco leaves by agroinfiltration technique (Atkinson et al. 2016). The CCM components included putative plasma membrane (LCI1 and HLA3) and chloroplast envelope (LCIA, CCP1, and CCP2) inorganic carbon (Ci) transporters that import Ci from the external environment into the cell along with carbonic anhydrases (CAH1, CAH3, CAH6) and proteins (LCIB and LCIC) of unknown functions that surround the pyrenoid under CO_2 limiting conditions. It was found that out of ten proteins, eight of them could effectively localize to locations same as *Chlamydomonas*. Additionally, in this study, the Ci transporters (LCIA and HLA3) were stably expressed in Arabidopsis, but the transgenic lines were unable to enhance growth and photosynthesis compared to WT plants. This reveals that it is important to stack up other CCM components in C3 plants for improved photosynthesis.

Arabidopsis double mutants deficient in Rubisco small subunit coding genes, *rbcS1A* and *rbcS3B* (called *1a3b*), could be complemented by the Rubisco small subunit (SSU) or the surface exposed two α-helices of SSU from *C. reinhardtii* (Atkinson et al. 2017). It was found that these *Arabidopsis* plants can assemble a plant-algal hybrid Rubisco using native Rubisco large subunit (LSU) and *Chlamydomonas* SSU with kinetic properties similar to native Rubisco. It proves to be an advancement toward achieving a functional CCM in plants. In *Chlamydomonas*, EPYC1 is known to interact with two α-helices of SSU to form aggregates of Rubisco, and SSU is essential for liquid-like aggregation in an algal pyrenoid (Wunder et al. 2018). Consequently, *Arabidopsis* plants developed earlier with plant-algal hybrid Rubisco (Atkinson et al. 2017) were tested for Rubisco condensation using *Chlamydomonas* full coding sequence of EPYC1 (Atkinson et al. 2019). It was observed that though EPYC1 could not lead to aggregation of Rubisco in planta. However, the plant-algal hybrid rubisco could phase-separate in vitro. Earlier studies in *Chlamydomonas* suggest that stoichiometric ratios of EPYC1 and Rubisco play an important role in phase separation (Wunder et al. 2018, 2019). Recent study in *Arabidopsis* has demonstrated the condensation of Rubisco by the expression of a mature EPYC1 along with plant-algal hybrid Rubisco (Atkinson et al. 2020). The Rubisco aggregation was able to phase-separate into a compartment in chloroplasts which had liquid-like properties similar to pyrenoid matrix of algae. This work proves to be an important initial step toward enhancing photosynthesis in C3 plants by engineering algal-based CCM.

Computational modelling of various components of pyrenoid-based *Chlamydomonas* CCM has extended the understanding of the fundamentals of its operation and engineering in higher plants (Fei et al. 2022). Based on this study, four steps were proposed for the assembly of a functional pyrenoid-based CCM in plants. The first step is the localization of plant Rubisco in pyrenoid matrix, and the second is the localization of carbonic anhydrases to thylakoids tubules that traverse through the pyrenoid matrix. The third step is to incorporate the bicarbonate channels to the thylakoid membrane to increase inorganic carbon flux. The last step is to introduce starch sheaths around the pyrenoid matrix which could prevent leakage of CO_2 from the matrix. Engineering of this pathway using minimal algal CCM components has been suggested to increase the CO_2 fixation in plants by threefold and could prove to be useful for future studies.

6.2.4 CCM in Cyanobacteria

All photosynthetically competent cyanobacteria and some chemoautotrophic proteobacteria have evolved CCM which allows these bacteria to grow competently under low CO_2 environment. CCM machinery in cyanobacteria comprises two major components, the inorganic carbon transporters systems and the carboxysomes (Fig. 6.3a) (Rae et al. 2013). The first part of a CCM involves accumulation of inorganic carbon in the cell facilitated by membrane bicarbonate transporters. And the second part involves the fixation of carbon inside the specialized proteinaceous bacterial microcompartments known as carboxysomes. Carboxysomes are composed of polyhedral protein shells which encompass CO_2 fixing enzyme, D-ribulose 1,5-bisphosphate carboxylase/oxygenase (Rubisco), and CO_2 converting enzyme, carbonic anhydrase (CA) which dehydrates $HCO3^-$ ions to CO_2 thereby increasing the carbon dioxide concentrations in the vicinity of Rubisco and suppressing oxygen fixation (Fig. 6.3). CA increases the levels of CO_2 from the steady state to above the K_M of Rubisco which results in saturation of all the enzyme active sites. Carboxysomes along with the active CO_2 and bicarbonate membrane transporters which accumulate CO_2 in the cell comprise the CCM in photosynthetic bacteria.

6.2.4.1 Ci Uptake and Transporter Systems

There are two CO_2 uptake systems present in the thylakoid membrane which convert cytosolic CO_2 to HCO_3^- and three HCO_3^- transporters in the plasma membrane of cyanobacteria which transports $HCO3^-$ into the cell of cyanobacteria.

The CO_2 uptake systems are dependent on NAD(P)H dehydrogenase type1 (NDH-1) complexes (Hualing 2022). In the genome of *Synechocystis* PCC6803, NDH-1 genes are present in single copies such as *ndhB*, *ndhH*, and *ndhL*, whereas some genes are in multiple copies such as *ndhD* has six homologues and *ndhF* has three homologues. There are two types of NDH-1 complexes, NDH-1_3 and NDH-1_4. NDH-1_3 complex is encoded by *ndhD3*, *ndhF3*, *chpY* (*cupA*) genes which is a low-CO_2 inducible high-affinity CO_2 uptake system, whereas NDH-1_4 complex is

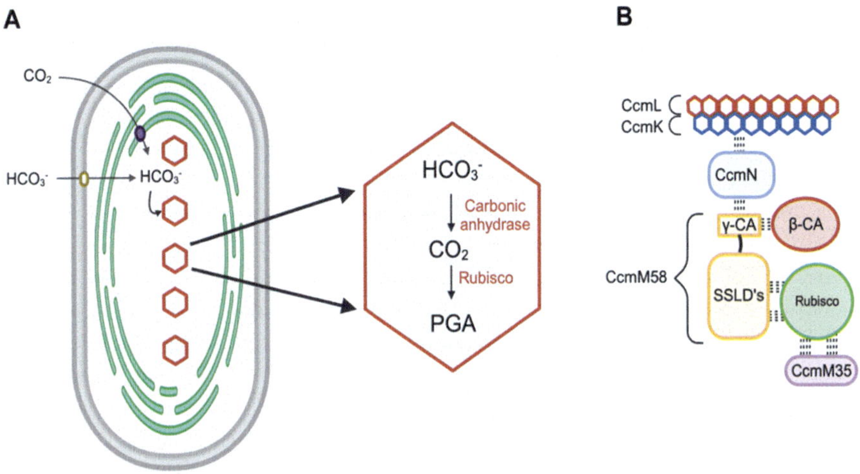

Fig. 6.3 Schematic representation of (**a**) a cyanobacterial cell consisting of plasma membrane bound bicarbonate transporters which transfers the HCO_3^- ions into the cytosol and thylakoid-bound CO_2 pumps that convert CO_2 to HCO_3^-, leading to an increase in CO_2 concentration in the cell. The red-colored hexagonal structures are carboxysomes. The bicarbonate ions enter into these structures by diffusion. HCO_3^- ions are then converted to CO_2 by the carbonic anhydrase enzyme, resulting in an increase in carbon concentration near the Rubisco. The cyanobacterial Rubisco then catalyzes the carbon fixation step, converting CO_2 into 3-phosphoglycerate (PGA). (**b**) The organization of a β-carboxysome (Syn7942) structural and lumen proteins. The shell of a carboxysome consists of outer and inner layer of hexagonal proteins, CcmL and CcmK, respectively. The lumen is comprised of CcmN, CcmM58, CcmM35, β-CA (carbonic anhydrase), and Rubisco. The shell is connected to the lumen proteins by CcmN which acts as a bridge between the two interfaces. CcmM58 has a γ-class of carbonic anhydrase and small subunits like domains (SSLD's)

encoded by *ndhD4*, *ndhF4*, *chpX* (*cupB*) genes and is a low-affinity CO_2 uptake system. ndhD and ndhF protein subunits of NDH-1 complex are involved in CO_2 uptake (Shibata et al. 2001), while chpY (cupA) and chpX (cupB) subunits are responsible for the hydration of CO_2 to HCO_3^- (Maeda et al. 2002).

The transportation of HCO_3^- ions into the cytosol is facilitated by three different types of transporters located at the plasma membrane, namely BicA, SbtA, BCT1 (Price et al. 2008). BCT1 is the first transporter to have been identified and characterized. This transporter is a member of the subfamily of bacterial ATP-binding cassette (ABC) transporters where hydrolysis of ATP leads to the uptake of HCO_3^-. In *Synechococcus* PCC7942 (hereafter, Syn7942), it is coded by the *cmp*ABCD operon which is expressed under limited Ci conditions. *cmp*ABCD operon encodes four proteins. The plasma membrane anchored 42-kDa cmpA protein has substrate binding site which binds HCO_3^- with moderate affinity, cmpB protein forms a dimer in membrane, and both cmpC and cmpD are extrinsic membrane protein with an ATP binding site (Omata et al. 1999). BCT1 has moderate HCO_3^- affinity but low photosynthetic flux rate. BicA transporter belongs to the family of prokaryotic and eukaryotic anion transporters also annotated as sulfate transporters in bacteria (SulP family). It is a SulP-type sodium-dependent low-affinity HCO_3^- transporter, which has high photosynthetic flux rate of HCO_3^- uptake (Price et al. 2004). The third transporter is SbtA, a Na^+-dependent HCO_3^- transporter which was first identified in *Synechococcus* PCC7002 (hereafter, Syn7002). The transporter is most likely driven by Na^+ electrochemical gradient across the plasma membrane (Shibata et al. 2002). SbtA transporter has high affinity for bicarbonate but low flux rate of HCO_3^- uptake. BicA and SbtA are highly upregulated in Syn7002 under Ci limiting conditions, and both transporters are known to transport most of the HCO_3^- into the cytosol. The uptake of Ci (inorganic carbon) via the transporters leads to high concentrations of Ci in the cytosol which then diffuses into carboxysomes. The cyanobacterium Syn7942 also encodes another type of highly conserved transmembrane ictB (inorganic carbon transporter) protein, encoded by *ictB* gene which leads to inorganic carbon accumulation in their cells, and a mutant of this gene shows HCR (high CO_2 requiring) phenotype (Bonfil et al. 1998).

Efforts have been made to transform the bicarbonate transporters in C3 plants. *Arabidopsis thaliana* (Arabidopsis) and *Nicotiana tabacum* (tobacco) transgenic plants transformed with cyanobacterial *ictB* gene had resulted in lower CO_2 compensation points compared to WT plants (Lieman-Hurwitz et al. 2003). This could be due to an increase in concentrations of CO_2 in the vicinity of Rubisco facilitated by ictB transporter which led to enhanced photosynthesis and growth in transgenic plants. Syn7942 *ictB* gene when transformed in rice plants induced higher photosynthesis, growth, and grain yield (Yang et al. 2008). Recently, the cyanobacterial transmembrane bicarbonate transporters, BicA and SbtA, were efficiently targeted to chloroplast envelope of tobacco (Rolland et al. 2016) and *Arabidopsis* (Uehara et al. 2020). These studies reveal that the bicarbonate transporters could translocate to C3 chloroplast envelope and in future could be essential in increasing C3 crop productivity along with other CCM components.

6.2.4.2 Carboxysomes

The carboxysomes were initially recognized as polyhedral inclusions present in the cells of cyanobacteria and proteobacteria, while their functions were not known (Shively et al. 1970). The polyhedral inclusions isolated from *Thiobacillus neapolitanus* were for the first called the carboxysomes (Ridge et al. 1973). It was demonstrated that carboxysomes contain Rubisco enzyme within them which is responsible for carbon fixation. Based on Rubisco phylogeny, there are two distinct types of carboxysomes, α-type carboxysomes that encapsulate form 1A Rubisco found in oceanic cyanobacteria, while β-type carboxysomes encapsulate form 1B Rubisco which are found in freshwater cyanobacteria (Kerfeld and Melnicki 2016). α- and β-carboxysomes differ from each other in genetic architecture, shell composition, and type of Rubisco they encapsulate.

α-Carboxysomes

They are found in marine cyanobacteria, *Bradyrhizobium* (alphaproteobacteria), *Nitrobacter*, *Nitrosomonas* (betaproteobacteria), *Nitrococcus* (gammaproteobacteria), and *Acidimicrobium* (actinobacteria). Gammaproteobacterium, *Halothiobacillus neapolitanus, is the model cyanobacterium having* α-carboxysomes *that has been well studied.* α-carboxysomes of *H. neapolitanus* are self-assembling structures that are icosahedral in shape, and their size varies from 94–95 nm to 97–100 nm (Schmid et al. 2006). The α-carboxysome proteins are encoded by the conserved *cso* operon in *H. neapolitanus* genome which are arranged in the transcriptional order as, *cbbL, cbbS, csoS2, csoS3, csoS4, csoS1C, csoS1A, csoS1B* (Shively et al. 1998; Tanaka et al. 2008). The *cbbL* and *cbbS* genes code for Rubisco large and small subunit proteins that forms the core of α-carboxysomes. Rubisco is arranged in concentric layers inside carboxysomes. *The shell is composed of proteins, namely,* CsoS1 paralogs (CsoS1A, CsoS1B, CsoS1C, CsoS1D) and CsoS4 paralogs (CsoS4A, CsoS4B). CsoS1A, CsoS1B, and CsoS1C are pentameric proteins which form the facets of the icosahedron, while the pentameric CsoS4A, CsoS4B proteins are present at the vertices of the shell. Six monomeric subunits of CsoS1A form a cyclic hexagonal structure that binds to the complementary edges of other hexagons tightly packing into a flat molecular layer. The average thickness of CsoS1A layer is approximately 18.6 Å, perforated with narrow pores which are presumed to be the passage for bicarbonate ions (Tsai et al. 2007). CsoS1D, the pseudohexamer protein, has a pore that is thought to allow the passage of metabolites, 3-PGA and RuBP (Roberts et al. 2012). CsoS2 is a linker protein that has two isoforms, the truncated CsoS2A and full-length, CsoS2B that binds Rubisco to induce its condensation (Oltrogge et al. 2020). Additionally, CsoS2 interacts with surface proteins of the shell and assists phase separation with Rubisco to form a structured carboxysome. *csoS3* gene encodes the β-class of CA which is tightly associated with the shell proteins and converts bicarbonate ions into CO_2.

Components of α-carboxysomes have been successfully assembled in heterologous organisms. An autotrophic strain of *E. coli* was developed which could photosynthesize using CCM components of α-carboxysome (Flamholz et al. 2020). Seven CCM structural genes along with Rubisco large and small subunits, their carbonic anhydrase, an inorganic transporter, three Rubisco chaperones, and four genes of unknown function of *H.* neapolitanus were transformed in *E. coli*. This was done to test whether CCM functioning could be established in a non-native host *E. coli* and, also, that the current knowledge is sufficient to establish a fully functional CCM in heterologous systems. Eventually, it was found in this study that carboxysome structures assembled in the bacterial cell and growth of *E. coli* were found to be dependent on 20 genes of CCM for biomass production from CO_2.

CCM genes of *Cyanobium* have been successfully transformed in tobacco chloroplasts using biolistic bombardment (Long et al. 2018). The endogenous Rubisco large subunit gene in tobacco was replaced with Rubisco (large and small subunit) of *Cyanobium* alone or in conjugation with the two key α-carboxysome structural genes, *CsoS1A* and *CsoS2*. This study followed a bottom-up approach for the construction of simplified carboxysomes successfully from minimal set of proteins, CsoS1A and CsoS2. Carboxysomes isometrically similar to that of the *Cyanobium* were observed in tobacco chloroplast, which were able to encapsulate Rubisco. Isolated Rubisco from these lines also had catalysis similar to that of *Cyanobium* Rubisco. The transgenic tobacco showed autotrophic growth, fully reliant on cyanobacterial Rubisco but at elevated levels of CO_2. It was suggested that the absence of specific components of a carboxysome such as the carbonic anhydrase (CsoSCA), vertex proteins (CsoS4A and CsoS4B), and shell components (CsoS1D and CsoS1E) could have hindered the formation of a fully functional carboxysome.

Recently, the full set of components of α-carboxysomes encoding nine genes in the *cso* operon of *H. neapolitanus* were incorporated in the tobacco chloroplast genome replacing the endogenous Rubisco large subunit gene (Chen et al. 2023). The targeted α-carboxysomes genes of H. *neapolitanus* were the form1-A Rubisco (large subunit, *cbbL* and small subunit, *cbbS*), proteinaceous shell (*csoS4A*, *csoS4B*, *csoS1A*, *csoS1B*, *csoS1B*), carbonic anhydrase (*csoSCA*), and the linker protein (CsoS2) coding genes. The study successfully demonstrated the assembly of a α-carboxysome in transplastomic tobacco plants which displayed structural and functional similarities comparable to that of a *H. neapolitanus* carboxysomes. These transgenic plants were also capable of driving photosynthesis fully dependent on cyanobacterial Rubisco. However, plants were unable to grow at ambient CO_2 concentrations; they required high CO_2 for autotrophic growth. This study revealed that for the assembly of a fully functional CCM in plants, carboxysome components along with the cyanobacterial bicarbonate transporters should also be incorporated in the plants.

β-Carboxysomes

They are mainly found in the freshwater strains of cyanobacteria. Much of the information regarding the genetic composition, structure, and physiological aspects of these types of carboxysomes has been derived from the research done in the model strains, *Synechococcus elongatus* PCC 7942 (*Syn* 7942) and PCC7002, and *Synechocystis* PCC 6803 as their complete genome has been sequenced.

Based on the current models of carboxysome organization and assembly, in case of β-carboxysome of Syn7942, the protein coding genes are present in the *ccmKLMN* operon which is closely associated with Rubisco coding genes (*RbcL* for large subunit and *RbcS* for small subunit) (Cannon et al. 2001). While *ccaA* gene coding for CcaA, a CA is present at a different place in the genome (Yu et al. 1992). These proteins make up the shell and lumen of the carboxysomes. The shell of carboxysome is composed of CcmL, CcmO, and CcmK proteins, while the lumen comprises of CcmN, CcmM, CcaA, RbcL, and RbcS proteins (Fig. 6.3b). The protein encoded by *RbcX* gene present between *RbcL* and *RbcS* operon might act as a chaperone but is not so important for the folding and assembly of Rubisco in Syn7942 (Emlyn-Jones et al. 2006; Saschenbrecker et al. 2007).

CcmK is a hexameric protein and its homologues are CcmK1, CcmK2, CcmK3, and CcmK4 encoded by *ccmK* gene (Rae et al. 2012). The hexameric units have flat edges which are packed tightly forming sheets of solid molecular layers in the carboxysome shell. The pores within the hexamers possibly act as conduit for the metabolites such as bicarbonate ions, 3PGA, RuBP. CcmL is a pentameric protein coded by *ccmL* gene that is present at the vertices of the icosahedral structure of carboxysomes where it seals the edges of the flat molecular sheets of ccmK hexamers (Tanaka et al. 2008). CcmO is a hexameric shell protein encoded by *ccmO* gene. It acts as a zipper protein interfacing the edges of carboxysome facets. Recently, a new conserved pseudo-hexameric protein, ccmP, is recognized which forms a bi-layered building block of carboxysomal shell with the central pore for allowing the passage of RuBP (Larsson et al. 2017).

In Syn7942 species, *ccmM* gene expresses two protein isoforms, namely, ccmM-35 and ccmM-58 of 35 KDa and 58 KDa, respectively (Fig 6.3b) (Long et al. 2005). CcmM-58 isoform is the full-length protein which has a γ-class of carbonic anhydrase domain present at the N-terminus, and three tandem repeats of protein domains of Rubisco small subunit (SSU-like domains) of about 85 amino acids which has 30% homology with their native RbcS are present at the C-terminus, while the other truncated isoform CcmM-35 has five repeats of small subunit like domains which is present in the lumen of carboxysomes where it links and organizes the adjacent Rubisco enzyme. CcmM-58 present in the lumen has an important function in the assembly of carboxysomes; it interlinks the Rubisco molecules and recruits CcaA, a carbonic anhydrase which in turn forms a scaffold for recruiting outer shell proteins (Fig 6.3b) (Long et al. 2011). CcmN is a protein encoded by *ccmN* gene which is mostly conserved in all β-carboxysome containing cyanobacteria. The protein has a N-terminal region that interacts with ccmM, and its C-terminal peptide interacts with the carboxysome shell proteins; both terminals are connected via a linker. These proteins are important in the assembly of microcompartments by interacting with shell proteins (Kinney et al. 2012). The β-carboxysome biogenesis starts with CcmM assisted Rubisco aggregation forming a nucleation around which the shell proteins assemble (Chen et al. 2013).

There are two types of carbonic anhydrases in β-carboxysomes which are a γ-CA present at N terminal of ccmM and a specific carboxymal β-CA, ccaA (Sommer et al. 2017). N-terminal γ-CA domain of M58 is essential for binding the carboxysomal β-CA and recruiting it to compartment during carboxysome biogenesis in Syn7942 (Long et al. 2007; Cot et al. 2008). In most cyanobacterium species such as Syn7942 and Syn6803, CcaA is the main CA involved in CO_2 formation, the absence, or mutation which leads to high CO_2 requiring strain (Yu et al. 1992).

Cyanobacterial Rubisco has relatively higher catalytic activity compared to plant Rubisco. Attempts have been made to replace the Rubisco of C3 plants with cyanobacterial Rubisco, to determine whether the cyanobacterial Rubisco could surpass the CO_2 fixing abilities of plant Rubisco. Transplastomic tobacco lines were created by the insertion of *RbcL* gene from cyanobacteria (*Synechococcus* PCC6301) for the formation of hybrid Rubisco (Kanevski et al. 1999). The cyanobacterial *RbcL* gene was transcribed but not translated to form protein in the transgenic lines, and no Rubisco activity was found. It was suggested that cyanobacterial Rubisco small subunit is to be incorporated along with the large subunit in plastids to maintain the activity of heterologous Rubisco.

To investigate whether the cyanobacterial large and small subunits could assemble to form active Rubisco in plants, large subunit of Rubisco in chloroplasts of tobacco was replaced with the cyanobacterial (Syn7942) large and small subunit of Rubisco (Lin et al. 2014b). Two transplastomic lines of tobacco were created in which the native Rubisco large subunit was knocked out and replaced with Syn7942 Rubisco large and small subunit along with either *SeRbcX* or with *ccmM35* genes. Both the transgenic lines grew autotrophically fully dependent on cyanobacterial Rubisco for carbon fixation. This reveals that the cyanobacterial Rubisco of Syn7942 has actively been assembled in tobacco plants. However, these plants were not able to grow in ambient air but required supplementation of external CO_2. This is because of the higher CO_2 requirements of the cyanobacterial Rubisco for photosynthesis (Whitney et al. 2011). Thus, it becomes necessary to transplant the additional cyanobacterial CCM machinery and bicarbonate transporters (to increase the carbon flux) inside the plant cells. In another study, it was found that the Syn7942 proteins, RbcX or CcmM35, are not essential for the assembly of active cyanobacterial Rubisco in tobacco plants (Occhialini et al. 2016). It was inferred that cyanobacterial Rubisco could be assembled with minimal engineering, by the assembly systems present in chloroplast of plants. This probability is due to the fact that the cyanobacteria and plant have similar type of Rubisco, i.e., Form1B (Cheng et al. 2008; Whitehead et al. 2014).

Attempts have been made to transiently express β-carboxysome proteins (CcmL, CcmM58, CcmN, CcmK2, and CcmO) of Syn7942 in tobacco leaves by agroinfiltration (Lin et al. 2014a). This study successfully demonstrated the formation of empty Syn7942 carboxysome compartments without Rubisco expression in the tobacco chloroplasts. Transplastomic tobacco plants have been created that expressed hybrid Rubisco in chloroplasts consisting of cyanobacterial Rubisco large (Se LSU) and tobacco small subunit (SSU) with and without the CcmM35 linker protein (Orr et al. 2020). The cyanobacterial genes were transformed in chloroplasts of tobacco by replacing the large subunit of Rubisco. Inside the chloroplasts of transgenic tobacco plants, large aggregations of Rubisco occurred resulting in formation of pro-carboxysome microcompartments. The aggregations also incorporated small amounts of native Rubisco small subunits (SSU). However, these transgenic lines had delayed growth, lower photosynthesis, and Rubisco activity compared to WT plants. These results substantiated that functional Rubisco assembly could take place in tobacco with hybrid Rubisco without cyanobacterial small subunit but resulting in slower pace of plant growth at high CO_2.

If we are to adopt the idea of bioengineering carboxysome machinery in plants, a comprehensive study and knowledge of the proteins involved in structure formation, their proper folding, and assembly are needed to be understood. Apart from the cyanobacterial Rubisco, a minimum of six other proteins such as CcmL, CcmK, CcmO, CcmM, CcmN, and CcaA are required for carboxysome biogenesis. Expressing these many genes would pose the complications of genetic load on the plants. On these grounds, a simplified design of carboxysome was illustrated in which the domains of four proteins of Syn7942 carboxysomes were incorporated into a single chimeric protein, 67-KDa ccmC (where C stands for chimeric) (Gonzalez-Esquer et al. 2015). Chimeric gene, *ccmC*, constituted domains from CcmM, CcaA, and CcmN proteins. *ccmC* gene could effectively complement the triple mutant cyanobacterial strain (where *ccmM*, *ccmN*, and *ccaA* native genes were knocked out), displaying growth in ambient CO_2. This chimeric gene reduced the genetic load by 18% in Syn7942. It can be inferred from these results that expressing the chimeric gene, *ccmC*, which replaces full length genes of Syn7942 would result in a functional carboxysome with reduced cyanobacterial genetic load in C3 plants. Recently, transgenic rice lines were generated using the chimeric ccmC gene from Syn7942, along with other carboxysome genes (ccmL, ccmO, ccmK) and their Rubisco. The CCM proteins were targeted to the rice chloroplast using various transit peptides. Additionally, the endogenous Rubisco and stromal carbonic anhydrase genes of rice were suppressed using RNAi. It was demonstrated that carboxysome assembly is achievable in the rice chloroplast; however, the RNAi plants exhibited reduced growth and yield (Sidhu et al. 2025).

Studies done so far in plants by incorporating cyanobacterial Rubisco with either few (Lin et al. 2014b; Occhialini et al. 2016; Long et al. 2018; Orr et al. 2020) or full set (Chen et al. 2023) of α- or β-carboxysomes CCM components replacing the tobacco large subunit of Rubisco have not resulted in autotrophic growth in ambient CO_2 conditions. The transplastomic lines required supplementation of external CO_2 for growth. This is understood from the fact that cyanobacterial Rubisco requires CO_2 higher than that found in the ambient air. Therefore, the first step for the assembly of a fully functional CCM in C3 plants to enhance crop photosynthesis requires the incorporation of the full sets of genes coding for structural and lumen proteins of a carboxysome. Secondly, to create higher influx of inorganic carbon, bicarbonate transporters (SbtA/B and BicA) should be targeted to the chloroplast inner envelope which could raise the chloroplastic levels of HCO_3^- concentration to 5 mM (Dean Price et al. 2011; McGrath and Long 2014). Additionally, inhibition of plant stromal carbonic anhydrase (CA) is crucial as it is speculated that this could lead to increase in chloroplast levels of bicarbonate ions to 2–3 mM. Otherwise, the active stromal CA could dissipate the inorganic carbon concentrations by converting bicarbonate to CO_2. Thus, incorporating CCM components alongside bicarbonate transporters in C3 plants could result in photosynthetic capacity and yields up to 60% (McGrath and Long 2014).

6.3 Synthetic CCM Pathway

6.3.1 Manipulation of Photorespiratory Pathway

Photorespiration occurs when oxygenation of Rubisco takes place which leads to production of toxic by-products such as 2-phosphoglycolate, glycolate. Detoxification of these by products and recycling them to phosphoglycerate is an energy-consuming process during which fixed carbon is released as CO_2. Furthermore, ammonia is also lost which is fixed via an energy-driven reaction. Eventually, this process leads to lowering of photosynthetic efficiency and biomass production in C3 plants.

Past efforts that have been made to reduce the photorespiration have failed to improve photosynthesis or carbon gain (Campbell and Ogren 1990; Peterhanse and Maurino 2011; Hagemann and Bauwe 2016). This is because photorespiration is necessary as it recovers 75% of the lost CO_2 in the form of 2-phosphoglycolate and remove inhibitors of photosynthesis (Givan and Kleczkowski 1992). Therefore, ideal choice would be to introduce a photorespiratory bypass pathway in C3 plants by directing away part of the 2-phosphoglycolate away from photorespiratory pathway in a less energy-intensive process and minimizing carbon and nitrogen losses (Xin et al. 2015). Besides, it should also avoid or detoxify the photorespiratory intermediates. It is known that bacteria such as *E. coli* can metabolize glycolate by glycolate catabolic pathway (Lord 1972). Transferring of the E. coli glycolate pathway in *Arabidopsis thaliana* has resulted in reduced photorespiration and increased biomass and photosynthesis (Kebeish et al. 2007; Maier et al. 2012). Transgenic tobacco plants harboring alternative photorespiratory pathway have been tested under photorespiratory stress (where light intensity is high and CO_2 is low) (Cavanagh et al. 2022). They grew trasformed tobacco lines to overexpress genes for malate synthesis (MS) and *Chlamydomonas* glycolate dehydrogenase (CrGDH) in chloroplasts. Additionally, in these plants RNAi was targeted to reduce the expression of a plastidic glycolate-glycerate transporter (PLGG1). It was found that transgenic tobacco plants had higher resilience, photosynthetic assimilation, and biomass compared to untransformed plants at higher temperatures. Modelling studies have also revealed that expressing cyanobacterial glycolate decarboxylation pathway in C3 plants could lead to 10% increase in photosynthetic assimilation (Khurshid et al. 2020). Photorespiratory bypass pathway in rice was introduced using the *glycolate oxidase*1 and *ascorbate peroxidase* genes from rice and *malate synthase* gene from *Cucurbita maxima* (Xu et al. 2023). These genes were driven by light-induced Rubisco small subunit promoter, and transgenic plants exhibited higher photosynthesis and grain yield under both greenhouse and field conditions. A photorespiratory bypass pathway was established in the chloroplasts of rice using its endogenous genes, glycolate oxidase, oxalate oxidase, and catalase for the oxidation of glycolate (Shen et al. 2019). Under greenhouse and field conditions, transgenic rice plants carrying these genes had higher photosynthetic assimilation, biomass, and nitrogen content compared to non-transformed plants.

Three alternative synthetic pathways of photorespiration were expressed in *Nicotiana tabacum*, a C3 plant to recycle glycolate and investigate whether this could enhance the photosynthetic efficiency (Fig. 6.4) (South et al. 2019). First alternate pathway (AP1) had five *E. coli* genes of glycolate oxidation pathway which converts glycolate to glycerate. Second pathway (AP2) had genes encoding glycolate oxidase and malate synthase from plants, and catalase from *E. coli*. Third pathway, AP3 had gene encoding glycolate dehydrogenase from *Chlamydomonas reinhardtii* and malate synthase from *Cucurbita maxima*. All the genes were targeted to chloroplasts with or without RNAi construct having *PLGG1* gene, which encodes the plastid glycolate-glycerate transporter. The purpose of downregulation of native *PLGG1* was to reduce the metabolite flux, i.e., glycolate export from chloroplasts, which could lead to detoxification of glycolate by synthetic alternative pathways in the chloroplast and save energy and increase

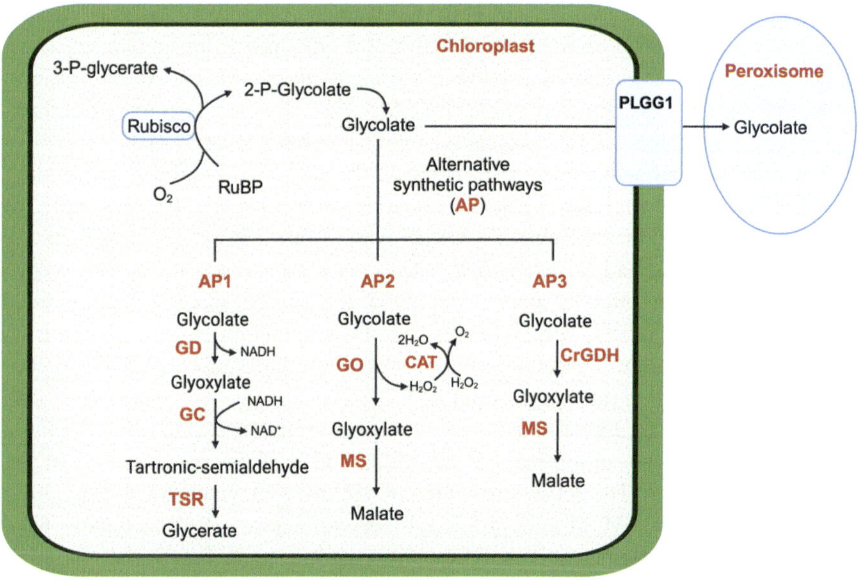

Fig. 6.4 Demonstration of three alternative pathways to bypass photorespiration in tobacco plants. Three alternative pathways (AP) are represented as AP1 (alternative pathway 1), AP2 (alternative pathway 2), and AP3 (alternative pathway 3). The enzymes encoded by the genes of the pathway were targeted to the chloroplasts. AP1 has five genes of the glycolate oxidation pathway from *E. coli* coding for three enzymes which are glycolate dehydrogenase (GD), glyoxylate carboligase (GC), and tartrate semi-aldehyde reductase (TSR). These enzymes convert glycolate to glycerate. AP2 relies on three genes encoding the enzymes, glycolate oxidase (GO) from *Arabidopsis*, malate synthase (MS) from *Cucurbita maxima*, and catalase (CAT) from *E. coli* for converting glycolate to malate. AP3 expressed two genes which encoded the enzymes, glycolate dehydrogenase (CrGDH) from *Chlamydomonas reinhardtii* and malate synthase (MS) from *Cucurbita maxima* for converting glycolate to malate. The targeted enzymes of the pathway are denoted in red color. *PLGG1* is the plastid membrane, glycolate-glycerate transporter that exports glycolate from the chloroplast

photosynthetic efficiency. AP1 lines showed improved productivity, which was lost when *PLGG1* RNAi gene was also expressed in these plants. This implies that AP1 lines were not able to manage the flux of glycolate under high rates of Rubisco oxygenation. Productivity of AP2 lines had hardly any enhancement. AP3 lines with or without *PLGG1* RNAi gene showed enhanced biomass accumulation and photosynthetic efficiency in field trials. Dry mass of AP3 lines expressing *PLGG1* RNAi increased post-harvesting when compared to AP3 lines without RNAi. These results demonstrated that alternate pathways could recycle glycolate and result in an increase in biomass production and crop productivity in the field trials.

6.3.2 Synthetic CO_2 Pathway

Apart from the CO_2 fixation mechanisms found in nature, synthetic CO_2 pathway has been successfully established in vitro (Schwander et al. 2018). They followed a bottom-up approach by engineering 17 different enzymes from archae, bacteria, plants, and humans to construct a synthetic carbon fixation cycle called the crotonyl-CoA/ethylmalonyl-CoA/hydroxybutyryl-CoA (CETCH). Identification of carboxylases for CO_2 fixation was done by analyzing the biochemical and kinetic properties of coenzyme A (CoA)-dependent carboxylases, the enoyl-CoA carboxylases/reductases (ECRs). These enzymes are not involved in autotrophic CO_2 fixation cycle but rather participate in secondary metabolism and carbon metabolism in *Alphaproteobacteria* and *Streptomycetes*. They do not have specificity for oxygen and possess a higher catalytic activity compared to Rubisco, making them an ideal choice for carbon fixation. The cycle consists of three major steps: (1) carbon fixation with ECR reaction, (2) regeneration of the carboxylated substrate, and (3) conversion of fixed carbon into the product.

Out of 17 enzymes, 13 formed the core reactions of the pathway, while the other enzymes were used to minimize side reactions and perform proofreading of the dead-end metabolites. The active sites of the two of these enzymes, methyl succinyl-CoA dehydrogenase (Mcd) and propionyl-CoA reductase (Pcd) enzymes, were engineered to alter their activities. These were converted to functional enzymes, succinyl-CoA oxidase (Mco) and propionyl-CoA oxidase (Pco), respectively. So, Mco could act as an electron acceptor.

The continuous cycle starts by adding propionyl-CoA to the $NaH^{13}CO_3$ buffer containing cofactors, enzymes, and artificial electron acceptors such as ferrocenium. $NaH^{13}CO_3$ enriched with ^{13}C-isotope used as carbon source which could be tracked through the flux of intermediates in the cycle via HPLC-MS analysis. After optimization of the reaction sequence in the cycle, the final version was called CETCH 5.4, which resulted in formation of one molecule of glyoxylate from two molecules of CO_2 in one cycle (Fig. 6.5). Glyoxylate concentration was measured by converting it into malate-by-malate synthase (isolated from *E. coli*). Enzymes, polyphosphate kinase, and formate dehydrogenase were added for the regeneration of ATP and NADPH, respectively. Cycle was also supplemented with the enzyme,

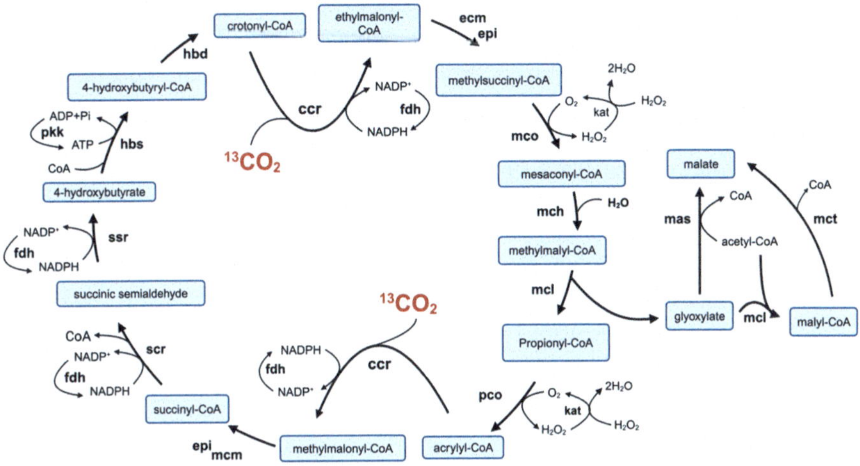

Fig. 6.5 A schematic representation of the synthetic carbon fixation pathway, the CETCH cycle (5.4). The cycle starts by adding propionyl-CoA to the $NaH^{13}CO_3$ buffer, where $NaH^{13}CO_3$ enriched with ^{13}C- isotope is used as a carbon source. The enzymes in the cycle are abbreviated, pco as propionyl Co-A oxidase, ccr as crotonyl-CoA carboxylase/reductase, mcm as methylmalonyl-CoA mutase, epi as isomerase, scr as Succinyl-CoA reductase, ssr as succinic semialdehyde reductase, hbs as 4-hydroxybutyrate-CoA synthetase, hbd as 4-hydroxybutyryl-CoA dehydrogenase, mco as methylsuccinyl-CoA oxidase, mch as mesaconyl-CoA hydratase, mcl1 as β-methylmalyl-CoA lyase, mas as malate synthase, and mcl2 as malyl-CoA thioesterase. (Adapted from Schwander et al. 2018)

catalase, to prevent the oxidative damage by H_2O_2 generated from the reactions. The CETCH cycle reactions are described as follows: (1) Propionyl-CoA is oxidized to acrylyl-Co-A catalyzed by propionyl Co-A oxidase in the presence of molecular oxygen, (2) crotonyl-CoA carboxylase/reductase (ccr) catalyzes the reductive carboxylation of acrylyl-CoA to methylmalonyl-CoA with reduced NADPH as reductant, (3) isomerization of methylmalonyl-CoA to succinyl-CoA by methylmalonyl-CoA mutase (mcm) and isomerase (epi), (4) succinyl-CoA reductase (scr) catalyzes the NADPH-dependent reduction of succinyl-CoA to succinic semialdehyde and releasing Co-A, (5) succinic semialdehyde reductase (ssr) reduces succinic semialdehyde with NADPH to 4-hydroxy butyrate, (6) 4-hydroxybutyrate-CoA synthetase (hbs) catalyzes the conversion of 4-hydroxy butyrate and CoA to hydroxybutyryl-CoA through the dephosphorylation of ATP to ADP, (7) 4-hydroxybutyryl-CoA dehydrogenase (hbd) catalyzes the dehydration of 4-hydroxy butyryl-CoA to crotonyl-CoA, (8) crotonyl-CoA carboxylase/reductase (ccr) catalyzes the reductive carboxylation of crotonyl-CoA to ethylmalonyl-CoA using NADPH as reductant, (9) ethylmalonyl-CoA mutase (ecm) and epimerase (epi) catalyze the transformation of ethylmalonyl-CoA to methylsuccinyl Co-A, (10) methylsuccinyl-CoA oxidase (mco) catalyzes the oxidation of methylsuccinyl Co-A in presence of molecular oxygen into trans-2-enoyl-CoA thioester, mesaconyl-CoA, (11) mesaconyl-CoA is converted to methylmalonyl CoA by the addition of a water molecule, which is catalyzed by mesaconyl-CoA hydratase (mch), (12) methylmalyl-CoA is cleaved into propionyl-CoA and glyoxylate by β-methylmalyl-CoA lyase (mcl1), (13) malate synthase (mas) catalyzes the condensation of glyoxylate with acetyl-CoA leading to formation of malate, and (14) β-methylmalyl-CoA lyase (mcl1) catalyzes the condensation of glyoxylate and acetyl-CoA to yield malyl-CoA.β-methylmalyl-CoA lyase (mcl1) is a promiscuous enzyme which catalyzes both the cleavage (as in step 12) and condensation reactions, thereby leading to unwanted side reactions. Consequently, malyl-CoA thioesterase (mcl2) was introduced in the cycle to hydrolyze malyl-CoA into malate.

The CETCH cycle fixes CO_2 at a rate of 5 nmol min^{-1} mg^{-1} compared to the Calvin cycle which fixes with the rate of 1–3 nmol min^{-1} mg^{-1} in cell extracts. There are six naturally occurring carbon fixation pathways that are known so far. With the successful demonstration of CETCH pathway, this synthetic cycle becomes the seventh known pathway to incorporate inorganic carbon. The limitation of this cycle is the formation of glyoxylate, which is a less active metabolite which is not so much connected to other pathways. Nevertheless, glyoxylate is converted to malate via another reaction catalyzed by malate synthase. But then again, this cycle incorporates the crotonyl-CoA carboxylase/reductase which fixes 80 CO_2 molecules per second in vitro compared to plant Rubisco, which only sequesters 2–5 CO_2 molecule per second.

Recently in 2020, artificial chloroplasts have been created in vitro, by encapsulating photosynthetic membranes in cell-sized droplets coupled with the synthetic CO_2 fixation cycle, CETCH (Miller et al. 2020). These semisynthetic photosynthesizing structures could execute light-driven carbon fixation, replicating the function of chloroplasts found in nature. These artificial cell-like structures are known as chloroplast-mimics. In this study, they followed the bottom-up approach using microfluidics to construct the thylakoid membranes (isolated from young spinach leaves) encapsulated in miniaturized droplets. The structures provided light-driven chemical energy in the form of energy-rich cofactors, ATP and reduced NADPH. These chemical cofactors then could power the reactions of the integrated CETCH cycle, which leads to the formation of a self-sufficient system for fixing inorganic carbon into multicarbon compound, i.e., gylcolate. This study demonstrates and forms an interface between the natural and synthetic biological modules which can be coupled to create life-like structures in vitro to achieve specific functions. With the rising CO_2 levels in the atmosphere through carbon emissions, these studies could be helpful to mitigate the problem in the future.

CETCH cycle could act as a new synthetic CO_2-concentrating mechanism to achieve higher photosynthetic efficiency if replicated in C3 plants. To do so, at least 17 heterologous enzymes are required to be expressed in C3 plants. This could pose a threat of genetic load on the plant system. Therefore, to test the feasibility of the synthetic cycle in the living system, it is best to start standardizing and reconstituting in the biological systems such as *E. coli* (Naseem et al. 2020). Later, the pathway genes could be translated in the model plant systems such as *Arabidopsis* and *Nicotiana* using Calvin cycle promoters. Alternatively, CETCH cycle could be expressed in the plants while concomitantly inhibiting the native Calvin cycle genes either suppressed by RNAi or knocked out by CRISPR-Cas9. However, inhibiting or completely shutting down of the Calvin cycle enzymes could result in detrimental effects on plants. Therefore, to operate a fully function CETCH cycle in plants would require the steady optimization of carbon fixation efficiency, regulation of non-native genes, and adjustment of nutrient availability in the soil to meet the requirements of the novel metabolic pathway.

6.4 Conclusion

This chapter has underscored the potential of engineering advanced carbon-concentrating mechanisms (CCMs) and synthetic pathways to enhance photosynthetic efficiency in C3 plants. By harnessing the traits found in C4 and CAM plants, as well as cyanobacteria and algae, we can create promising pathways for reducing photorespiration and increasing carbon fixation. It would also be beneficial to incorporate combination of C4 pathway and cyanobacterial CCM for improved crop productivity in C3 plants (Yin and Struik 2017). Key advancements, such as the integration of carboxysomes and pyrenoids, synthetic photorespiratory bypasses, and the innovative CETCH cycle, present powerful strategies to overcome the limitations of C3 photosynthesis. Integration of photorespiratory bypass and CETCH pathway as major CCM together at the same time in plants could lead the way to future endeavors in research (Naseem et al. 2020). These approaches not only hold promise for improved crop productivity and resilience but also align with sustainable agricultural practices that address global food security in the face of climate change. Future studies should focus on optimizing the introduced pathways and understanding the regulatory mechanisms within these complex systems, enabling the translation of these bioengineering solutions into field-ready applications.

References

Atkinson N, Feike D, Mackinder LCM, Meyer MT, Griffiths H, Jonikas MC, Smith AM, Mccormick AJ (2016) Introducing an algal carbon-concentrating mechanism into higher plants: location and incorporation of key components. Plant Biotechnol J 14:1302–1315. https://doi.org/10.1111/pbi.12497

Atkinson N, Leitão N, Orr DJ, Meyer MT, Carmo-Silva E, Griffiths H, Smith AM, McCormick AJ (2017) Rubisco small subunits from the unicellular green alga *Chlamydomonas* complement Rubisco-deficient mutants of *Arabidopsis*. New Phytol 214:655–667. https://doi.org/10.1111/nph.14414

Atkinson N, Velanis CN, Wunder T, Clarke DJ, Mueller-Cajar O, Mccormick AJ (2019) The pyrenoidal linker protein EPYC1 phase separates with hybrid *Arabidopsis-Chlamydomonas* Rubisco through interactions with the algal Rubisco small subunit. J Exp Bot 70:5271–5285. https://doi.org/10.1093/jxb/erz275

Atkinson N, Mao Y, Chan KX, McCormick AJ (2020) Condensation of Rubisco into a proto-pyrenoid in higher plant chloroplasts. Nat Commun 11. https://doi.org/10.1038/s41467-020-20132-0

Badger MR, Price GD (2003) CO2 concentrating mechanisms in cyanobacteria: molecular components, their diversity and evolution. J Exp Bot 54:609–622. https://doi.org/10.1093/jxb/erg076

Bonfil DJ, Ronen-Tarazi M, Sültemeyer D, Lieman-Hurwitz J, Schatz D, Kaplan A (1998) A putative HCO3/− transporter in the cyanobacterium *Synechococcus* sp. strain PCC 7942. FEBS Lett 430:236–240. https://doi.org/10.1016/S0014-5793(98)00662-0

Borland AM, Griffiths H (1990) The regulation of CAM and respiratory recycling by water supply and light regime in the C3-CAM intermediate *Sedum telephium*. Funct Ecol 4:33. https://doi.org/10.2307/2389649

Campbell WJ, Ogren WL (1990) Glyoxylate inhibition of ribulosebisphosphate carboxylase/oxygenase activation in intact, lysed, and reconstituted chloroplasts. Photosynth Res 23:257–268. https://doi.org/10.1007/BF00034856

Cannon GC, Bradburne CE, Aldrich HC, Baker SH, Heinhorst S, Shively JM (2001) Microcompartments in prokaryotes: carboxysomes and related polyhedra. Appl Environ Microbiol 67:5351–5361. https://doi.org/10.1128/AEM.67.12.5351-5361.2001

Cavanagh AP, South PF, Bernacchi CJ, Ort DR (2022) Alternative pathway to photorespiration protects growth and productivity at elevated temperatures in a model crop. Plant Biotechnol J 20:711–721. https://doi.org/10.1111/pbi.13750

Chen AH, Robinson-Mosher A, Savage DF, Silver PA, Polka JK (2013) The bacterial carbon-fixing organelle is formed by shell envelopment of preassembled cargo. PLoS One 8:1–13. https://doi.org/10.1371/journal.pone.0076127

Chen T, Hojka M, Davey P, Sun Y, Dykes GF, Zhou F, Lawson T, Nixon PJ, Lin Y, Liu LN (2023) Engineering α-carboxysomes into plant chloroplasts to support autotrophic photosynthesis. Plant J 14:1–12. https://doi.org/10.1038/s41467-023-37490-0

Cheng S, Liu Y, Crowley CS, Yeates TO, Bobik TA (2008) Bacterial microcompartments: their properties and paradoxes. BioEssays 30:1084–1095. https://doi.org/10.1002/bies.20830

Cot SSW, So AKC, Espie GS (2008) A multiprotein bicarbonate dehydration complex essential to carboxysome function in cyanobacteria. J Bacteriol 190:936–945. https://doi.org/10.1128/JB.01283-07

Cui H (2021) Challenges and approaches to crop improvement through C3-to-C4 engineering. Front Plant Sci 12. https://doi.org/10.3389/fpls.2021.715391

Dean Price G, Badger MR, von Caemmerer S (2011) The prospect of using cyanobacterial bicarbonate transporters to improve leaf photosynthesis in C3 crop plants. Plant Physiol 155:20–26. https://doi.org/10.1104/pp.110.164681

Ermakova M, Arrivault S, Giuliani R, Danila F, Alonso-Cantabrana H, Vlad D, Ishihara H, Feil R, Guenther M, Borghi GL, Covshoff S, Ludwig M, Cousins AB, Langdale JA, Kelly S, Lunn JE, Stitt M, von Caemmerer S, Furbank RT (2021) Installation of C4 photosynthetic pathway enzymes in rice using a single construct. Plant Biotechnol J 19:575–588. https://doi.org/10.1111/pbi.13487

Fei C, Wilson AT, Mangan NM, Wingreen NS, Jonikas MC (2022) Modelling the pyrenoid-based CO2-concentrating mechanism provides insights into its operating principles and a roadmap for its engineering into crops. Nat Plants 8:583–595. https://doi.org/10.1038/s41477-022-01153-7

Flamholz AI, Dugan E, Blikstad C, Gleizer S, Ben-Nissan R, Amram S, Antonovsky N, Ravishankar S, Noor E, Bar-Even A, Milo R, Savage DF (2020) Functional reconstitution of a bacterial CO_2 concentrating mechanism in *E. coli*. eLife 9:1–57. https://doi.org/10.7554/eLife.59882

Furbank RT (2017) Walking the C4 pathway: past, present, and future. J Exp Bot 68:e1–e10. https://doi.org/10.1093/jxb/erx006

Gee CW, Niyogi KK (2017) The carbonic anhydrase CAH1 is an essential component of the carbon-concentrating mechanism in *Nannochloropsis oceanica*. Proc Natl Acad Sci USA 114:4537–4542. https://doi.org/10.1073/pnas.1700139114

Giordano M, Beardall J, Raven JA (2005) CO_2 concentrating mechanisms in algae: mechanisms, environmental modulation, and evolution. Annu Rev Plant Biol 56:99–131. https://doi.org/10.1146/annurev.arplant.56.032604.144052

Givan CV, Kleczkowski LA (1992) The enzymic reduction of glyoxylate and hydroxypyruvate in leaves of higher plants. Plant Physiol 100:552–556. https://doi.org/10.1104/pp.100.2.552

Gonzalez-Esquer CR, Shubitowski TB, Kerfeld CA (2015) Streamlined construction of the cyanobacterial CO_2-fixing organelle via protein domain fusions for use in plant synthetic biology. Plant Cell 27:2637–2644. https://doi.org/10.1105/tpc.15.00329

Gowik U, Westhoff P (2011) The path from C3 to C4 photosynthesis. Plant Physiol 155:56–63. https://doi.org/10.1104/pp.110.165308

Hagemann M, Bauwe H (2016) Photorespiration. Encycl Appl Plant Sci 1:86–89. https://doi.org/10.1016/B978-0-12-394807-6.00094-0

He S, Chou HT, Matthies D, Wunder T, Meyer MT, Atkinson N, Martinez-Sanchez A, Jeffrey PD, Port SA, Patena W, He G, Chen VK, Hughson FM, McCormick AJ, Mueller-Cajar O, Engel BD, Yu Z, Jonikas MC (2020) The structural basis of Rubisco phase separation in the pyrenoid. Nat Plants 6:1480–1490. https://doi.org/10.1038/s41477-020-00811-y

He S, Crans VL, Jonikas MC (2023) The pyrenoid: the eukaryotic CO_2-concentrating organelle. Plant Cell 35:3236–3259. https://doi.org/10.1093/plcell/koad157

Heyduk K (2022) Evolution of Crassulacean acid metabolism in response to the environment: past, present, and future. Plant Physiol 190:19–30. https://doi.org/10.1093/plphys/kiac303

Heyduk K, Moreno-Villena JJ, Gilman IS, Christin PA, Edwards EJ (2019) The genetics of convergent evolution: insights from plant photosynthesis. Nat Rev Genet 20:485–493. https://doi.org/10.1038/s41576-019-0107-5

Hualing M (2022) Cyanobacterial NDH-1 complexes. Front Microbiol 13:1–7. https://doi.org/10.3389/fmicb.2022.933160

Janaiah A, Hossain M, Otsuka K (2006) Productivity impact of the modern varieties of rice in India. Dev Econ 44:190–207. https://doi.org/10.1111/j.1746-1049.2006.00013.x

Kanevski I, Maliga P, Rhoades DF, Gutteridge S (1999) Plastome engineering of ribulose-1,5-bisphosphate carboxylase/oxygenase in tobacco to form a sunflower large subunit and tobacco small subunit hybrid. Plant Physiol 119:133–141. https://doi.org/10.1104/pp.119.1.133

Kebeish R, Niessen M, Thiruveedhi K, Bari R, Hirsch HJ, Rosenkranz R, Stäbler N, Schönfeld B, Kreuzaler F, Peterhänsel C (2007) Chloroplastic photorespiratory bypass increases photosynthesis and biomass production in *Arabidopsis thaliana*. Nat Biotechnol 25:593–599. https://doi.org/10.1038/nbt1299

Kerfeld CA, Melnicki MR (2016) Assembly, function and evolution of cyanobacterial carboxysomes. Curr Opin Plant Biol 31:66–75. https://doi.org/10.1016/j.pbi.2016.03.009

Khurshid G, Abbassi AZ, Khalid MF, Gondal MN, Naqvi TA, Shah MM, Chaudhary SU, Ahmad R (2020) A cyanobacterial photorespiratory bypass model to enhance photosynthesis by rerouting photorespiratory pathway in C3 plants. Sci Rep 10:1–13. https://doi.org/10.1038/s41598-020-77894-2

Kinney JN, Salmeen A, Cai F, Kerfeld CA (2012) Elucidating essential role of conserved carboxysomal protein CcmN reveals common feature of bacterial microcompartment assembly. J Biol Chem 287:17729–17736. https://doi.org/10.1074/jbc.M112.355305

Kong W, Yoo MJ, Zhu D, Noble JD, Kelley TM, Li J, Kirst M, Assmann SM, Chen S (2020) Molecular changes in *Mesembryanthemum crystallinum* guard cells underlying the C3 to CAM transition. Plant Mol Biol 103:653–667. https://doi.org/10.1007/s11103-020-01016-9

Ku MSB, Agarie S, Nomura M, Fukayama H, Tsuchida H, Ono K, Hirose S, Toki S, Miyao M, Matsuoka M (1999) High-level expression of maize phosphoenolpyruvate carboxylase in transgenic rice plants. Nat Biotechnol 17:76–81. https://doi.org/10.1038/5256

Ku MSB, Cho D, Li X, Jiao DM, Pinto M, Miyao M, Matsuoka M (2001) Introduction of genes encoding C4 photosynthesis enzymes into rice plants: physiological consequences. Novartis Found Symp 236:100–116. https://doi.org/10.1002/9780470515778.ch8

Larsson AM, Hasse D, Valegård K, Andersson I (2017) Crystal structures of β-carboxysome shell protein CcmP: ligand binding correlates with the closed or open central pore. J Exp Bot 68:3857–3867. https://doi.org/10.1093/jxb/erx070

Lieman-Hurwitz J, Rachmilevitch S, Mittler R, Marcus Y, Kaplan A (2003) Enhanced photosynthesis and growth of transgenic plants that express ictB, a gene involved in HCO^{3-} accumulation in cyanobacteria. Plant Biotechnol J 1:43–50. https://doi.org/10.1046/j.1467-7652.2003.00003.x

Lim SD, Lee S, Choi WG, Yim WC, Cushman JC (2019) Laying the foundation for crassulacean acid metabolism (CAM) biodesign: expression of the C4 metabolism cycle genes of CAM in *Arabidopsis*. Front Plant Sci 10:1–20. https://doi.org/10.3389/fpls.2019.00101

Lim SD, Mayer JA, Yim WC, Cushman JC (2020) Plant tissue succulence engineering improves water-use efficiency, water-deficit stress attenuation and salinity tolerance in *Arabidopsis*. Plant J 103:1049–1072. https://doi.org/10.1111/tpj.14783

Lin MT, Occhialini A, Andralojc PJ, Devonshire J, Hines KM, Parry MAJ, Hanson MR (2014a) β-Carboxysomal proteins assemble into highly organized structures in *Nicotiana* chloroplasts. Plant J 79:1–12. https://doi.org/10.1111/tpj.12536

Lin MT, Occhialini A, Andralojc PJ, Parry MAJ, Hanson MR (2014b) A faster Rubisco with potential to increase photosynthesis in crops. Plant J 513:547–550. https://doi.org/10.1038/nature13776

Long BM, Price GD, Badger MR (2005) Proteomic assessment of an established technique for carboxysome enrichment from *Synechococcus* PCC7942. Can J Bot 83:746–757. https://doi.org/10.1139/b05-058

Long BM, Badger MR, Whitney SM, Price GD (2007) Analysis of carboxysomes from *Synechococcus* PCC7942 reveals multiple Rubisco complexes with carboxysomal proteins CcmM and CcaA. Plant J 282:29323–29335. https://doi.org/10.1074/jbc.M703896200

Long BM, Rae BD, Badger MR, Price GD (2011) Over-expression of the β-carboxysomal CcmM protein in *Synechococcus* PCC7942 reveals a tight co-regulation of carboxysomal carbonic anhydrase (CcaA) and M58 content. Photosynth Res 109:33–45. https://doi.org/10.1007/s11120-011-9659-8

Long BM, Hee WY, Sharwood RE, Rae BD, Kaines S, Lim YL, Nguyen ND, Massey B, Bala S, von Caemmerer S, Badger MR, Price GD (2018) Carboxysome encapsulation of the CO2-fixing enzyme Rubisco in tobacco chloroplasts. Nat Commun 9. https://doi.org/10.1038/s41467-018-06044-0

Lord JM (1972) Glycolate oxidoreductase in *Escherichia coli*. BBA-Bioenergetics 267:227–237. https://doi.org/10.1016/0005-2728(72)90111-9

Ludwig M, Hartwell J, Raines CA, Simkin AJ (2024) The Calvin-Benson-Bassham cycle in C4 and Crassulacean acid metabolism species. Semin Cell Dev Biol 155:10–22. https://doi.org/10.1016/j.semcdb.2023.07.013

Mackinder LCM (2018) The *Chlamydomonas* CO2-concentrating mechanism and its potential for engineering photosynthesis in plants. New Phytol 217:54–61. https://doi.org/10.1111/nph.14749

Mackinder LCM, Meyer MT, Mettler-altmann T, Chen V, Madeline C (2016) A repeat protein links Rubisco to form the eukaryotic carbon concentrating organelle. Proc Natl Acad Sci USA 113(21):5958–5963

Maeda SI, Badger MR, Price GD (2002) Novel gene products associated with NdhD3/D4-containing NDH-1 complexes are involved in photosynthetic CO_2 hydration in the cyanobacterium, *Synechococcus* sp. PCC7942. Mol Microbiol 43:425–435. https://doi.org/10.1046/j.1365-2958.2002.02753.x

Maier A, Fahnenstich H, von Caemmerer S, Engqvist MKM, Weber APM, Flügge UI, Maurino VG (2012) Transgenic introduction of a glycolate oxidative cycle into *A. thaliana* chloroplasts leads to growth improvement. Front Plant Sci 3:1–12. https://doi.org/10.3389/fpls.2012.00038

McGrath JM, Long SP (2014) Can the cyanobacterial carbon-concentrating mechanism increase photosynthesis in crop species? A theoretical analysis. Plant Physiol 164:2247–2261. https://doi.org/10.1104/pp.113.232611

Meyer MT, Genkov T, Skepper JN, Jouhet J, Mitchell MC, Spreitzer RJ, Griffiths H (2012) Rubisco small-subunit α-helices control pyrenoid formation in *Chlamydomonas*. Proc Natl Acad Sci USA 109:19474–19479. https://doi.org/10.1073/pnas.1210993109

Miller TE, Beneyton T, Schwander T, Diehl C, Girault M, Mclean R, Chotel T, Claus P, Cortina NS, Baret J-C, Erb TJ (2020) Bottom-up construction of a chloroplast mimic capable of light-driven synthetic CO_2 fixation. Science (80-)368:649–654. https://doi.org/10.1126/science.aaz6802.Bottom-up

Miyao M, Masumoto C, Miyazawa SI, Fukayama H (2011) Lessons from engineering a single-cell C4 photosynthetic pathway into rice. J Exp Bot 62:3021–3029. https://doi.org/10.1093/jxb/err023

Morita E, Abe T, Tsuzuki M, Fujiwara S, Sato N, Hirata A, Sonoike K, Nozaki H (1998) Presence of the CO2-concentrating mechanism in some species of the pyrenoid-less free-living algal genus *Chloromonas* (Volvocales, Chlorophyta). Planta 204:269–276. https://doi.org/10.1007/s004250050256

Naseem M, Osmanoglu Ö, Dandekar T (2020) Synthetic rewiring of plant CO2 sequestration galvanizes plant biomass production. Trends Biotechnol 38:354–359. https://doi.org/10.1016/j.tibtech.2019.12.019

Oakley JC, Sultmanis S, Stinson CR, Sage TL, Sage RF (2014) Comparative studies of C3 and C4 Atriplex hybrids in the genomics era: physiological assessments. J Exp Bot 65:3637–3647. https://doi.org/10.1093/jxb/eru106

Occhialini A, Lin MT, Andralojc PJ, Hanson MR, Parry MAJ (2016) Transgenic tobacco plants with improved cyanobacterial Rubisco expression but no extra assembly factors grow at near wild-type rates if provided with elevated CO2. Plant J 85:148–160. https://doi.org/10.1111/tpj.13098

Oltrogge LM, Chaijarasphong T, Chen AW, Bolin ER, Marqusee S, Savage DF (2020) Multivalent interactions between CsoS2 and Rubisco mediate α-carboxysome formation. Nat Struct Mol Biol 27:281–287. https://doi.org/10.1038/s41594-020-0387-7

Omata T, Price GD, Badger MR, Okamura M, Gohta S, Ogawa T (1999) Identification of an ATP-binding cassette transporter involved in bicarbonate uptake in the cyanobacterium *Synechococcus* sp. strain PCC 7942. Proc Natl Acad Sci USA 96:13571–13576. https://doi.org/10.1073/pnas.96.23.13571

Orr DJ, Worrall D, Lin MT, Carmo-Silva E, Hanson MR, Parry MAJ (2020) Hybrid cyanobacterial-tobacco Rubisco supports autotrophic growth and procarboxysomal aggregation1[CC-By]. Plant Physiol 182:807–818. https://doi.org/10.1104/pp.19.01193

Peterhanse C, Maurino VG (2011) Photorespiration redesigned. Plant Physiol 155:49–55. https://doi.org/10.1104/pp.110.165019

Pingali PL (2012) Green revolution: impacts, limits, andthe path ahead. Proc Natl Acad Sci USA 109:12302–12308. https://doi.org/10.1073/pnas.0912953109

Pradhan B, Panda D, Bishi SK, Chakraborty K, Muthusamy SK, Lenka SK (2022) Progress and prospects of C4 trait engineering in plants. Plant Biol 24:920–931. https://doi.org/10.1111/plb.13446

Price GD, Woodger FJ, Badger MR, Howitt SM, Tucker L (2004) Identification of a SulP-type bicarbonate transporter in marine cyanobacteria. Proc Natl Acad Sci USA 101:18228–18233. https://doi.org/10.1073/pnas.0405211101

Price GD, Badger MR, Woodger FJ, Long BM (2008) Advances in understanding the cyanobacterial CO2-concentrating- mechanism (CCM): functional components, Ci transporters, diversity, genetic regulation and prospects for engineering into plants. J Exp Bot 59:1441–1461. https://doi.org/10.1093/jxb/erm112

Rae BD, Long BM, Badger MR, Price GD (2012) Structural determinants of the outer shell of β-carboxysomes in *Synechococcus elongatus* PCC 7942: roles for CcmK2, K3-K4, CcmO, and CcmL. PLoS One 7:14–19. https://doi.org/10.1371/journal.pone.0043871

Rae BD, Long BM, Badger MR, Price GD (2013) Functions, compositions, and evolution of the two types of carboxysomes: polyhedral microcompartments that facilitate CO 2 fixation in cyanobacteria and some proteobacteria. Microbiol Mol Biol Rev 77:357–379. https://doi.org/10.1128/mmbr.00061-12

Raven JA, Cockell CS, De La Rocha CL (2008) The evolution of inorganic carbon concentrating mechanisms in photosynthesis. Philos Trans R Soc B Biol Sci 363:2641–2650. https://doi.org/10.1098/rstb.2008.0020

Raven JA, Beardall J, Sánchez-Baracaldo P (2017) The possible evolution and future of CO2-concentrating mechanisms. J Exp Bot 68:3701–3716. https://doi.org/10.1093/jxb/erx110

Ridge O, Laboratories N, Elmis S, Johnson E, Or N (1973) Electron microscopy of the carboxysomes (polyhedral bodies) of *Thiobacillus neapolitanus*. J Bacteriol 116:1405–1411

Roberts EW, Cai F, Kerfeld CA, Cannon GC, Heinhorst S (2012) Isolation and characterization of the *Prochlorococcus* carboxysome reveal the presence of the novel shell protein CsoS1D. J Bacteriol 194:787–795. https://doi.org/10.1128/JB.06444-11

Rolland V, Badger MR, Price GD (2016) Redirecting the cyanobacterial bicarbonate transporters BicA and SbtA to the chloroplast envelope: soluble and membrane cargos need different chloroplast targeting signals in plants. Plant J 7:1–19. https://doi.org/10.3389/fpls.2016.00185

Sage RF, Sage TL, Kocacinar F (2012) Photorespiration and the evolution of C4 photosynthesis. Annu Rev Plant Biol 63:19–47. https://doi.org/10.1146/annurev-arplant-042811-105511

Saschenbrecker S, Bracher A, Rao KV, Rao BV, Hartl FU, Hayer-Hartl M (2007) Structure and function of RbcX, an assembly chaperone for hexadecameric Rubisco. Cell 129:1189–1200. https://doi.org/10.1016/j.cell.2007.04.025

Schmid MF, Paredes AM, Khant HA, Soyer F, Aldrich HC, Chiu W, Shively JM (2006) Structure of *Halothiobacillus neapolitanus* carboxysomes by cryo-electron tomography. J Mol Biol 364:526–535. https://doi.org/10.1016/j.jmb.2006.09.024

Schuler ML, Mantegazza O, Weber APM (2016) Engineering C4 photosynthesis into C3 chassis in the synthetic biology age. Plant J 87:51–65. https://doi.org/10.1111/tpj.13155

Schwander T, Borzyskowski LS Von, Burgener S (2018) A synthetic pathway for the fixation of carbon dioxide in vitro. Science 354:900–904. https://doi.org/10.1126/science.aah5237.A

Sedelnikova OV, Hughes TE, Langdale JA (2018) Understanding the genetic basis of C4 Kranz anatomy with a view to engineering C3 crops. Annu Rev Genet 52:249–270. https://doi.org/10.1146/annurev-genet-120417-031217

Shen BR, Wang LM, Lin XL, Yao Z, Xu HW, Zhu CH, Teng HY, Cui LL, Liu EE, Zhang JJ, He ZH, Peng XX (2019) Engineering a new chloroplastic photorespiratory bypass to increase photosynthetic efficiency and productivity in rice. Mol Plant 12:199–214. https://doi.org/10.1016/j.molp.2018.11.013

Shibata M, Ohkawa H, Kaneko T, Fukuzawa H, Tabata S, Kaplan A, Ogawa T (2001) Distinct constitutive and low-CO_2-induced CO_2 uptake systems in cyanobacteria: genes involved and their phylogenetic relationship with homologous genes in other organisms. Proc Natl Acad Sci USA 98:11789–11794. https://doi.org/10.1073/pnas.191258298

Shibata M, Katoh H, Sonoda M, Ohkawa H, Shimoyama M, Fukuzawa H, Kaplan A, Ogawa T (2002) Genes essential to sodium-dependent bicarbonate transport in cyanobacteria: function and phylogenetic analysis. J Biol Chem 277:18658–18664. https://doi.org/10.1074/jbc.M112468200

Shively JM, Decker GL, Greenawalt JW (1970) Comparative ultrastructure of the *Thiobacilli*. J Bacteriol 101:618–627. https://doi.org/10.1128/jb.101.2.618-627.1970

Shively JM, Bradburne CE, Aldrich HC, Bobik TA, Mehlman JL, Jin S, Baker SH (1998) Sequence homologs of the carboxysomal polypeptide CsoS1 of the Thiobacilli are present in cyanobacteria and enteric bacteria that form carboxysomes—polyhedral bodies. Can J Bot 76:906–916. https://doi.org/10.1139/b98-088

Sidhu GK, Pandey R, Kaur G, Singh A, Lenka SK, Reddy PM (2025) Towards assembling functional cyanobacterial β-carboxysomes in *Oryza sativa* chloroplasts. Funct Integr Genomics 25:1–17. https://doi.org/10.1007/s10142-024-01518-5

Silvera K, Neubig KM, Whitten WM, Williams NH, Winter K, Cushman JC (2010) Evolution along the crassulacean acid metabolism continuum. Funct Plant Biol 37:995–1010. https://doi.org/10.1071/FP10084

Sommer M, Cai F, Melnicki M, Kerfeld CA (2017) β-Carboxysome bioinformatics: identification and evolution of new bacterial microcompartment protein gene classes and core locus constraints. J Exp Bot 68:3841–3855. https://doi.org/10.1093/jxb/erx115

South PF, Cavanagh AP, Liu HW, Ort DR (2019) Synthetic glycolate metabolism pathways stimulate crop growth and productivity in the field. Science (80-) 363. https://doi.org/10.1126/science.aat9077

Tanaka S, Kerfeld CA, Sawaya MR, Cai F, Heinhorst S, Cannon GC, Yeates TO (2008) Atomic-level models of the bacterial carboxysome shell. Science (80-) 319:1083–1086. https://doi.org/10.1126/science.1151458

Taniguchi Y, Ohkawa H, Masumoto C, Fukuda T, Tamai T, Lee K, Sudoh S, Tsuchida H, Sasaki H, Fukayama H, Miyao M (2008) Overproduction of C4 photosynthetic enzymes in transgenic rice plants: an approach to introduce the C4-like photosynthetic pathway into rice. J Exp Bot 59:1799–1809. https://doi.org/10.1093/jxb/ern016

Tay IYY, Odang KB, Cheung CYM (2021) Metabolic modeling of the C3-CAM continuum revealed the establishment of a starch/sugar-malate cycle in CAM evolution. Front Plant Sci 11:1–10. https://doi.org/10.3389/fpls.2020.573197

Tilman D, Balzer C, Hill J, Befort BL (2011) Global food demand and the sustainable intensification of agriculture. Proc Natl Acad Sci USA 108:20260–20264. https://doi.org/10.1073/pnas.1116437108

Tsai Y, Sawaya MR, Cannon GC, Cai F, Williams EB, Heinhorst S, Kerfeld CA, Yeates TO (2007) Structural analysis of CsoS1A and the protein shell of the *Halothiobacillus neapolitanus* carboxysome. PLoS Biol 5:1345–1354. https://doi.org/10.1371/journal.pbio.0050144

Tuskan GA, Yang X, Cushman JC (2015) Engineering crassulacean acid metabolism to improve water-use efficiency. Trends Plant Sci 19:327–338. https://doi.org/10.1016/j.tplants.2014.01.006.Engineering

Uehara S, Sei A, Sada M, Ito-Inaba Y, Inaba T (2020) Installation of authentic BicA and SbtA proteins to the chloroplast envelope membrane is achieved by the proteolytic cleavage of chimeric proteins in Arabidopsis. Sci Rep 10:1–10. https://doi.org/10.1038/s41598-020-59190-1

Von Caemmerer S, Quick WP, Furbank RT (2012) The development of C4 rice: current progress and future challenges. Science (80-) 336:1671–1672. https://doi.org/10.1126/science.1220177

Wang Y, Spalding MH (2014) Acclimation to very low CO_2: contribution of limiting CO_2 inducible proteins, LCIB and LCIA, to inorganic carbon uptake in *Chlamydomonas reinhardtii*. Plant Physiol 166:2040–2050. https://doi.org/10.1104/pp.114.248294

Whitehead L, Long BM, Dean Price G, Badger MR (2014) Comparing the in vivo function of α-carboxysomes and β-carboxysomes in two model cyanobacteria. Plant Physiol 165:398–411. https://doi.org/10.1104/pp.114.237941

Whitney SM, Baldet P, Hudson GS, John Andrews T (2001) Form I Rubiscos from non-green algae are expressed abundantly but not assembled in tobacco chloroplasts. Plant J 26:535–547. https://doi.org/10.1046/j.1365-313X.2001.01056.x

Whitney SM, Houtz RL, Alonso H (2011) Advancing our understanding and capacity to engineer nature's CO_2-sequestering enzyme, Rubisco. Plant J 155:27–35. https://doi.org/10.1104/pp.110.164814

Wunder T, Cheng SLH, Lai SK, Li HY, Mueller-Cajar O (2018) The phase separation underlying the pyrenoid-based microalgal Rubisco supercharger. Nat Commun 9:1–10. https://doi.org/10.1038/s41467-018-07624-w

Wunder T, Oh ZG, Mueller-Cajar O (2019) CO_2-fixing liquid droplets: towards a dissection of the microalgal pyrenoid. Traffic 20:380–389. https://doi.org/10.1111/tra.12650

Xin CP, Tholen D, Devloo V, Zhu XG (2015) The benefits of photorespiratory bypasses: how can they work? Plant Physiol 167:574–585. https://doi.org/10.1104/pp.114.248013

Xu H, Wang H, Zhang Y, Yang X, Lv S, Hou D, Mo C, Wassie M, Yu B, Hu T (2023) A synthetic light-inducible photorespiratory bypass enhances photosynthesis to improve rice growth and grain yield. Plant Commun 4:1–11. https://doi.org/10.1016/j.xplc.2023.100641

Yang S-M, Chang C-Y, Yanagisawa M, Park I, Tseng T-H, Ku MSB (2008) Transgenic rice expressing cyanobacterial bicarbonate transporter exhibited enhanced photosynthesis, growth and grain yield. Plant J:1243–1246. https://doi.org/10.1007/978-1-4020-6709-9_268

Yang X, Cushman JC, Borland AM, Edwards EJ, Wullschleger SD, Tuskan GA, Owen NA, Griffiths H, Smith JAC, De Paoli HC, Weston DJ, Cottingham R, Hartwell J, Davis SC, Silvera K, Ming R, Schlauch K, Abraham P, Stewart JR, Guo HB, Albion R, Ha J, Lim SD, Wone BWM, Yim WC, Garcia T, Mayer JA, Petereit J, Nair SS, Casey E, Hettich RL, Ceusters J, Ranjan P, Palla KJ, Yin H, Reyes-García C, Andrade JL, Freschi L, Beltrán JD, Dever LV, Boxall SF, Waller J, Davies J, Bupphada P, Kadu N, Winter K, Sage RF, Aguilar CN, Schmutz J, Jenkins J, Holtum JAM (2015) A roadmap for research on crassulacean acid metabolism (CAM) to enhance sustainable food and bioenergy production in a hotter, drier world. New Phytol 207:491–504. https://doi.org/10.1111/nph.13393

Yin X, Struik PC (2017) Can increased leaf photosynthesis be converted into higher crop mass production? A simulation study for rice using the crop model GECROS. J Exp Bot 68:2345–2360. https://doi.org/10.1093/jxb/erx085

Yu JW, Price GD, Song L, Badger MR (1992) Isolation of a putative carboxysomal carbonic anhydrase gene from the cyanobacterium *Synechococcus* PCC7942. Plant Physiol 100:794–800. https://doi.org/10.1104/pp.100.2.794

Chapter 7
Quantum Dots and Photosynthesis

Özge İbiş and Caner Ünlü

Abstract Quantum dots (QDs), with carbon dots (CDs) emerging as a prominent subclass, have become transformative nanomaterials for enhancing photosynthesis and agricultural productivity. Their tunable photophysical properties enable the modification of the light spectrum by converting ultraviolet light into photosynthetically active radiation, which optimizes chloroplast light absorption and energy utilization. Early research using traditional QDs such as CdTe and CdSe demonstrated improved electron transport, enhanced chlorophyll fluorescence, and increased energy conversion in photosystems, though heavy metal toxicity remains a concern. In contrast, carbon dots offer a safer and more biocompatible alternative, exhibiting excellent light absorption, customizable luminescence, and versatile surface chemistry. Studies have shown that CDs boost key photosynthetic processes by increasing RuBisCO activity, stimulating chlorophyll synthesis, and mitigating abiotic stresses through effective reactive oxygen species scavenging and enhanced nutrient uptake. Integration of these QDs into agricultural practices, ranging from greenhouse covering materials and LED lighting systems to foliar applications and nanocapsules, has led to significant improvements in crop yield and plant resilience. Future research should focus on optimizing QD formulations and delivery methods to maximize photosynthetic enhancement while ensuring environmental sustainability, thereby paving the way for revolutionary advances in modern agriculture and artificial photosynthesis.

Keywords Quantum dots · Photosynthesis · Plant growth

Ö. İbiş · C. Ünlü (✉)
Department of Chemistry, İstanbul Technical University, Istanbul, Turkey
e-mail: canerunlu@itu.edu.tr

A. Singh et al. (eds.), *Sustainable Crop Production*, Sustainability Sciences in Asia and Africa, https://doi.org/10.1007/978-981-95-7844-3_7

Editor's Notes

Nanomaterials such as quantum-dot (Q-dots; QDs) are semiconductor particles of less than 10 nm dimension. These particles hold promise in agriculture, particularly in improving photosynthetic efficiency of crop species. This stems from the unique size-dependent tunability of Q-dots. In other words, the dimensional shift in Q-dots modifies their electrophysical properties. Thus Q-dots can be optimized for photophysical properties such as absorption, fluorescence, and emission spectra. These properties qualify QDs as new agents for increasing photosynthetic properties and transforming agriculture practices and productivity.

This chapter by Ibis and Unlu introduces the readers to the types, photophysical properties, and biocompatible nature of Q-dots. They illustrate their customizable luminescence and emission properties, which have been deployed in various applications of agriculture. Carbon Q-dots are now being widely used as these are considered safer than other heavy metal-based forms such as those based on CdS, CdSe, CdTe, PbS, and PbSe. Although nascent, there is adequate evidence that such innovative approaches can play a significant role in not only enhancing photosynthetic efficiency but a wide array of applications in agriculture. This chapter shows how quantum dots can be integrated in sustainable crop production.

Actionable Points

1. Quantum dots have emerged as transformative tools in enhancing natural photosynthesis and promoting plant growth.
2. Manipulation of the photophysical characteristics of quantum dots can render them highly effective for manipulating the light quality and in turn photosynthesis. Quantum dots can be used to make agriculture more efficient by modulating natural photosynthesis.
3. Future research should focus on the development of highly efficient, biocompatible quantum dots; on optimizing QD formulations and delivery methods; to maximize photosynthetic enhancement while ensuring environmental sustainability and minimize environmental risks.

7.1 Introduction

Solids that exhibit significant diversity in optical and electronic features depending on the change in their particle size below 100 nm are called nanocrystals and can be categorized as two-dimensional (thin films or quantum wells), one-dimensional (quantum wires), and zero-dimensional (quantum dots) (Bera et al. 2010).

Quantum wells are formed by epitaxial semiconductor crystal growth techniques, that is, sandwiching of a narrow-band-gap layer between two layers of wide-band-gap material. Electrons are confined within only one axis but are free to move in the other two directions. *Quantum wires* are semiconductor materials that are ultrasmall rectangular shaped, surrounded by another semiconductor material of wider band gap. The motion of electrons is restricted to two axes, and they are allowed to move only in the third direction. *Quantum dots* are zero-dimensional semiconductor materials in which electrons are restricted along three axes (Fig. 7.1).

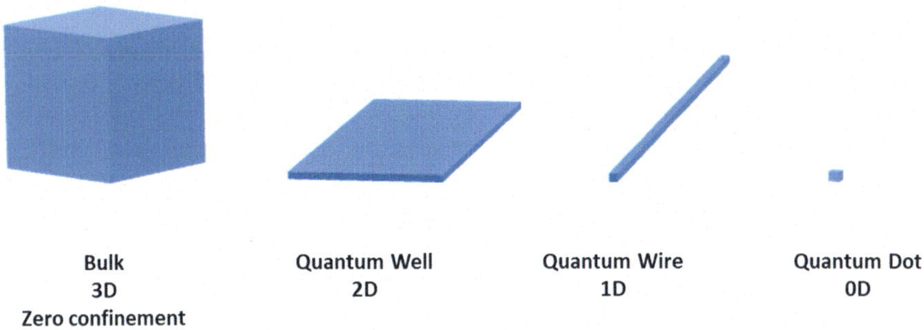

Fig. 7.1 Schematic illustration of the bulk material, quantum well, quantum wire, and quantum dot

7.1.1 Quantum Dots

Quantum dots are semiconductor nanoparticles, ranging from 1.5 to 10 nm in size, capable of trapping electrons in tiny, specific regions. Often referred to as "zero-dimensional electronic structures," these quantum dots are distinctive because their semiconductor energy levels can be customized by changing their size, shape, and charge potential because of the quantum confinement effect (Fargi et al. 2023). This tunability leads to unique color signatures for quantum dots of various sizes. Quantum dots have different properties from their bulk counterparts; this ensures remarkably distinct electronic and optical properties to use in a wide area. Different types of quantum dots have been studied after the pioneer work of Aleksey I. Ekimov in 1981. His research on small semiconductor materials that exhibit quantum confinement was the first experimental observation. Therefore, it is no wonder that the Royal Swedish Academy of Sciences awarded the 2023 Nobel Prize in Chemistry to Moungi G. Bawendi, Louis E. Brus, and Aleksey I. Ekimov for their groundbreaking contributions to the discovery and synthesis of quantum dots.

7.1.2 Properties of Quantum Dots

Semiconductors function by exciting an electron from its ground state to an elevated energy level. The gap between these two states is known as the energy band gap. When an electron is excited and leaves the ground state (valence band), it creates a corresponding hole (positive charge) (Sumanth Kumar et al. 2018). This separation between the excited electron (negative charge) and hole, also referred to as an exciton, generates a potential difference (Maxwell et al. 2019). Once a circuit is completed, the electron starts to flow through the conduction band of the semiconductor to return to the ground state. During this process, electron and hole recombine, releasing photons across a broad and continuous wavelength range. The most common radiative relaxation processes are band-edge and near-band-edge emissions. The recombination of an excited electron in the conduction band with a hole in the valence band is called band-edge emission. However, this electron and hole may recombine from conduction band minima to the valence band maxima of the semiconductors. Therefore, radiative recombination of an exciton leads to near band-edge emission at energies slightly lower than the band gap. This high-density electron-hole recombination increases the rate of radiative recombination, resulting in intense photoluminescence (Hassan et al. 2019).

For particles that have an electron-hole separation as small as the Bohr radius size, this particular photon is very large compared to the size of the particle. Thus, the energy required to build an exciton increases. This energy and particle size relation is expressed as "quantum confinement effect," which is observed in semiconductor nanocrystals that are as small as 10 nm, e.g., quantum dots. This distinctive confinement property results in the transformation of the continuous "band" of energies into discrete energy levels within quantum dots (QDs). Therefore, QDs are also called "artificial atoms" because of these discrete electronic states available for the atoms (Kastner 1993).

The relation between the size of the quantum dots and their band gap energy is inverse; namely, as the size of the quantum dot increases, the band gap energy decreases, so it will emit shorter-wavelength photons and vice versa. This phenomenon can be observed as colors, such that small-band gap energy/large-sized quantum dots emit red photons, while higher-band gap energy/small-sized quantum dots emit blue/purple photons (Fig. 7.2). The quantum confinement effect provides the ability to tune the optical and electrical properties of quantum dots to be used in certain areas (Bera et al. 2010).

As mentioned above, the excited electron recombines with the corresponding hole to a lower ground state upon relaxation while releasing energy from recombination and relaxation stages, which is emitted as radiative or nonradiative (Bera et al. 2010). The radiative and nonradiative transitions are described with a Jablonski diagram (Fig. 7.3), which is an energy level diagram (Omary and Patterson 2016).

An excited electron relaxes to the ground electronic state that has the same multiplicity as the excited electronic state by emitting a photon. This phenomenon is called fluorescence, which is shown in Fig. 7.3 of the Jablonski diagram as solid lines through the transition from S_1 to S_0. Fluorescence consists of only spin-allowed transitions faster than phosphorescence relaxation (Hassan et al. 2019). Phosphorescence occurs when an excited electron moves to a metastable excited electronic state (intersystem crossing), which is also called the triplet state (T_1). After intersystem crossing, the electron further relaxes to the ground state (S_0) by phosphorescence (T_1 to S_0 transition in Fig. 7.3). The triplet-to-singlet transition is less feasible than singlet-to-singlet transitions, which results in a longer average lifetime (Hassan et al. 2019). This transition is a spin-forbidden transition because the excited triplet state has a different spin multiplicity from the ground state.

Radiative transitions can be divided into band-edge, near-band-edge, and defect emissions. Band-edge and near-band-edge emissions are explained above. Defect emissions from QDs arise from impurities or intrinsic structural anomalies that occur during QD fabrication. According to the type of these defects, the state can act as a donor or acceptor, which can attract a hole or an electron due to Coulombic forces. The emissions that come from these defects can be shallow or deep. Because

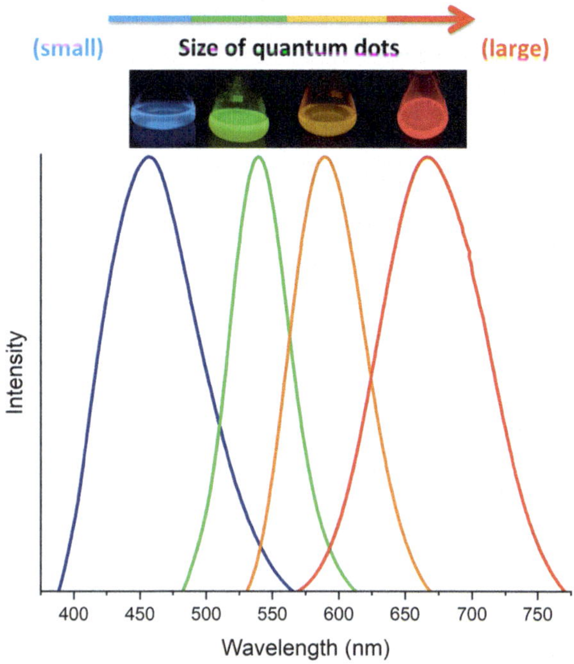

Fig. 7.2 Photoluminescence spectra of QDs excited at 350 nm and their emission color under 366 nm UV-light irradiation

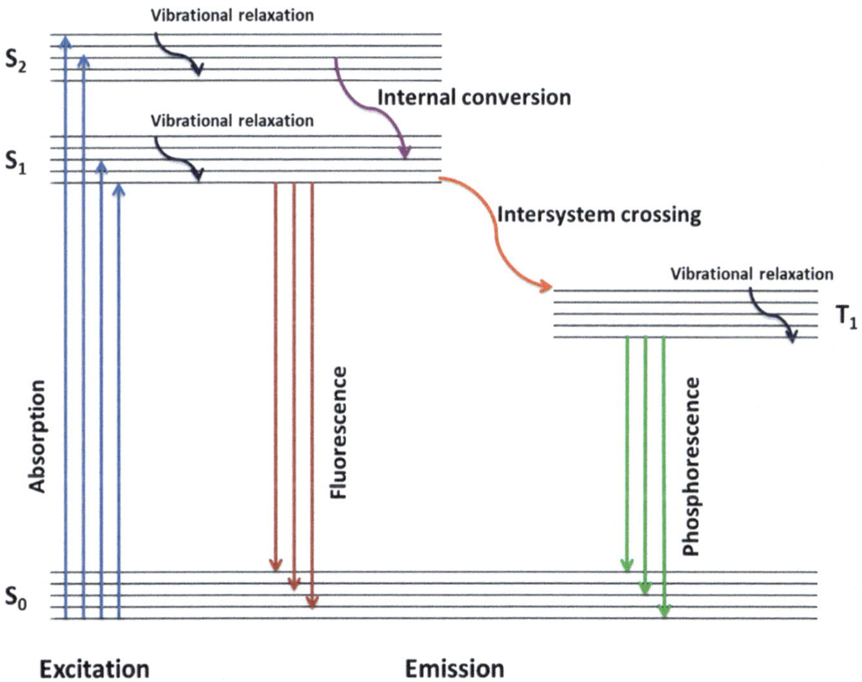

Fig. 7.3 The Jablonski diagram. Radiative and nonradiative processes are depicted as solid and dashed lines, respectively

shallow defect states are close to the conduction or valence band, transitions are counted as radiative transitions. However, deep defect states in the band gap are referred to as "traps"; since they do not emit light, they experience nonradiative recombination (Bera et al. 2010).

Nonradiative transitions from the excited state to the ground state do not emit light; however, they are in competition with radiative transitions. The energy from nonradiative transitions emerges as changes in thermodynamic states. This type of relaxation occurs rapidly through internal conversion, external conversion, and vibrational relaxation (Omary and Patterson 2016). Quantum yield is a property of luminescence that is related to lifetimes. Luminescence quantum yield (ϕ) is the ratio of the number of photons emitted to the number of photons absorbed. Therefore, the maximum value of ϕ is 1. Quantum yield is less than unity for almost all luminescent materials because of the large number of nonradiative processes that lead to a decrease in the number of photons emitted.

7.2 Types of Quantum Dots

Quantum dots with different structures and chemical compositions from the groups of I-VI (e.g., Ag_2Se, Ag_2S, Ag_2Te), II-VI (e.g., CdSe, CdS, ZnO, ZnS), III-V (e.g., GaN, GaP, InP), IV-VI (e.g., PbSe, PbS), and I-III-VI (e.g., CuInS), which are called metal-based QDs (MBQDs), from perovskite materials that have general formula of ABX_3, which are called perovskite QDs (PQDs), and from carbon-based materials that are called as carbon-based dots (carbon dots, CDs, and graphene quantum dots, GQDs) have been synthesized and studied (Piao et al. 2022).

7.2.1 Metal Based QDs

Metal chalcogenide quantum dots (MCQDs) have been mostly studied with the aim of reducing the toxicity of QDs derived from heavy metal compounds such as CdS, CdSe, CdTe, PbS, and PbSe. Nevertheless, their use in medical applications is unsuitable due to their toxicity. To reduce the toxicity of newly developed quantum dots (QDs), various approaches, mainly the production of entirely heavy-metal-free QDs, such as those based on In (InP) (Almeida et al. 2023), Zn (ZnSe (Gao et al. 2021), ZnSSe (İbiş et al. 2025), ZnTe (Gumaste and Kulkarni 2021), ZnS (Khenthaphak et al. 2023)), Ag (Ag_2S (Xue et al. 2019), Ag_2Se (An et al. 2023), and Ag_2Te (Wang et al. 2024)) and the reduction of the heavy metal content by substituting heavy metals with safer metals or encapsulating the heavy metal core with a protective shell (Granados-Oliveros et al. 2022), are employed. Together with their tunable optical properties, high photostability, and good affinity of metal ions for biomolecules, MCQDs have attracted great interest in biomedical applications such as biological imaging, biosensing, and cancer therapy (Piao et al. 2022).

Among the colloidal QDs, perovskite quantum dots (PQDs) have gained interest due to their superior electronic and optical properties. PQDs show high photoluminescence quantum yields of tunable narrow emission wavelength with various emission colors depending on size and content. Considering these utilities, they are well-suited for light-emitting diodes (LEDs). Perovskite quantum dots involve either organic or metal cations categorized as hybrid organic-inorganic PQDs (e.g., $CH_3NH_3PbBr_3$ (Zhang et al. 2015)) or all-inorganic PQDs (e.g., $CsPbI_3$ (Yoshida et al. 2024)), respectively (Wang et al. 2018a).

7.2.2 *Carbon-Based QDs*

Carbon-based QDs exhibit broad, tunable photoluminescence wavelengths ranging from deep ultraviolet to near infrared (NIR), tuned by their size, shape, surface functional groups, and heteroatom doping (e.g., nitrogen, boron, or phosphorus). With their biocompatibility and water-solubility properties, carbon-based QDs form the basis for biological applications. According to the synthesis method, carbon-based quantum dots can be categorized as carbon quantum dots (CDs) and graphene quantum dots (GQDs). Carbon-based QDs are synthesized via top-down methods that break down larger carbon structures, i.e., graphite or its derivatives, into smaller ones. During synthesis, functionality for solubility and applicability can be acquired, as well as surface defects. To obtain carbon-based QDs, a bottom-up approach can also be applied via carbonization of the precursor, cage-opening of fullerene, and chemical vapor deposition (Facure et al. 2020; Yuan et al. 2019; Zheng et al. 2015).

CDs and GQDs show different structural properties such that CDs are quasi-spherical nanoparticles around 5 nm in diameter (Sun et al. 2006). CDs can be amorphous or nanocrystalline with sp^2-hybridized carbon clusters (Ray et al. 2009) and have diamond-like structure (Hu et al. 2009). GQDs resemble the crystalline structure of a single or a few layered graphenes (Pan et al. 2010).

7.3 Synthesis of Quantum Dots

Quantum dot synthesis can be mainly divided into two methods. One is the top-down synthesis method in which bulk materials are divided to produce nanostructured materials by molecular beam epitaxy (MBE), mechanical milling, laser ablation, etching, sputtering, electro-explosion, and electron beam lithography. These methods have limitations, such as the introduction of impurities and structural defects into quantum dots during synthesis. The other is the bottom-up method, which involves the assembly of atoms or molecules into nanosized materials via attractive forces. This approach can be carried out through reactions between materials that are in solid, liquid, or gas phases. The bottom-up process relies on the fabrication of QDs via various self-assembly techniques of wet-chemical methods and vapor-phase methods (Bera et al. 2010).

Small-scale (laboratory-scale) quantum dot syntheses have succeeded and been reported widely in the literature. However, the struggle to scale up QD production remains significantly challenging in achieving low-cost production processes, high reproducibility, and user- and environmentally friendly synthesis methods. With applications of QDs in electronic areas such as solar cells, photovoltaic cells, image sensors, panel displays, and light-emitting devices, the demand for large-scale production methods has emerged. Colloidal synthesis methods have been very practical processes for the synthesis of binary, ternary, or core/shell cadmium (Tao et al. 2023), lead (Zhou et al. 2019), and mercury-based (Rogach et al. 1999) and perovskite quantum dots (Kirmani et al. 2022) for large-scale production.

QDs have played a significant role in promoting stem and root growth, increasing photosynthesis efficiency, and improving nutrient uptake for applications in agriculture (Gupta et al. 2022). The utility of QDs in agricultural studies depends on their low toxicity and good biocompatibility properties. In addition to QDs' chemical composition, the production methods are crucial details; green synthesis methods and low-toxicity precursors are preferred for their production. Hydrothermal methods are commonly preferred for the production of quantum dots for agricultural applications. For example, carbon dots synthesized from cysteine and citric acid have been shown to enhance the growth of mung bean plants (Zhang et al. 2018). In the study, water-soluble carbon dots that were prepared from citric acid and urea via a simple, low-cost, one-step microwave synthesis route showed non-toxicity to the bean sprout plant and therefore potential application in agriculture (Qu et al. 2012).

7.3.1 Colloidal Synthesis Methods

A colloid is a mixture that contains nanometer size particles, yet remains evenly distributed throughout the solution. Colloidal synthesis can be performed in aqueous and nonaqueous media. The reactions usually proceed in an inert atmosphere to prevent oxidation. By altering the temperature, the pH of the aqueous medium, and reaction time, the quantum yield of the reaction can be controlled. CdSe QDs are mostly produced using this method; however, the method is also applicable to produce other metal quantum dots, such as ZnSe. As an example, in 1998, highly luminescent ZnSe nanocrystals from organometallic precursors were synthesized in hexadecylamine (Hines and Guyot-Sionnest 1998). Subsequently, to improve the photoluminescence properties of the ZnSe, the colloidal solutions were irradiated with the "white light" of a 100 W xenon lamp. As a result, up to 30% photoluminescence quantum yield increase was observed (Shavel et al. 2004). Hot injection method is the most commonly used colloidal synthesis method. This involves rapidly injecting precursors into a hot solvent, where they react quickly to form QDs.

7.3.2 *Hydrothermal Synthesis Methods*

The procedures for hydrothermal synthesis are simple, cheap, and mostly environmentally friendly, requiring simple equipment such as an autoclave and an oven. Reactant solubility and reactivity under high temperature and pressure are enhanced and can form particles with high crystallinity and high purity. Changing the parameters of the reactions, such as reaction time, temperature, and concentration of the reactants, provides control over size, morphology, and shape. Two-step hydrothermal synthesis of blue fluorescent carbon dots with surface modification using EGTA has provided crosslinking of carboxyl and amino groups so that CDs exhibit high sensitivity and selectivity to Ca^{2+} ions (Yue et al. 2019). Despite the various benefits hydrothermal synthesis, high temperature and pressure conditions can be challenging to obtain or maintain.

7.3.3 *Microwave-Assisted Synthesis Methods*

This method enables the rapid synthesis of uniformly sized quantum structures as heat is provided by the electromagnetic signal to quickly reach high-reaction temperatures. Since there is no delay to reach and maintain high temperature during the reaction, i.e., no fluctuation in heating rate, the reaction starts immediately and the formation of QDs with narrow size distribution (Nuryadin et al. 2014). High crystallinity is achieved in a shorter time providing reduction of energy consumption and reproducibility compared with a traditional heating method (Wu et al. 2010).

7.4 Characterization of Quantum Dots

Structural and chemical characterizations play an important role in providing the data required to calculate the electronic and optical properties of quantum dots. Characterization methods such as X-ray diffraction analysis, transmission electron microscopy (TEM), photoluminescence spectroscopy, and X-ray photoelectron spectroscopy and their complementary use provide comprehensive information on the chemical, structural, and optical properties of quantum dots.

X-ray diffraction analysis (XRD) shows detailed information about the crystallographic structure, chemical composition, and physical properties of a material. It is related to constructive interference of monochromatic X-rays and a crystalline sample. When monochromatic incident X-rays interact with a crystalline sample, they produce diffracted rays. These diffracted rays are subsequently detected, processed, and counted. The intensity of the diffracted rays scattered at different angles is plotted to display a diffraction pattern. Depending on the arrangement of atoms, every material produces a distinctive diffraction pattern (Raja et al. 2021). The width of peaks on an XRD spectrum reflects the crystalline size of a sample, where sharp peaks indicate a large size of crystallites, whereas broad peaks indicate smaller crystallites. As the crystallite size decreases, the number of diffracting planes available for constructive interference reduces. This leads to a decrease in the sharpness and intensity of the diffraction peaks, resulting in peak broadening.

Yang and Holloway produced Mn^{2+}-doped ZnS-passivated CdS:Mn nanocrystals and using XRD patterns reduced with Debye-Scherrer, Eq. (7.1), which concluded that the particles were 2.3 nm in size (Yang and Holloway 2004). Ramanery et al. produced ZnS quantum dots in different pH media. When the pH of the solution is increased, the width of the XRD peaks is observed to increase, which indicates the formation of small crystals (Ramanery et al. 2013).

The Debye-Scherrer equation

$$D = \frac{K * \lambda}{\beta * \cos\theta} \tag{7.1}$$

where D is the nanoparticles' crystalline size, K represents the Scherrer constant (0.98), λ denotes the wavelength (1.54), and β denotes the full width at half maximum (FWHM) (Sivagami and Asharani 2022).

Emission and excitation spectra are collected from spectrofluorometers, which provide the simplest information about the size of quantum dots relative to the quantum confinement effect. To obtain emission spectra, the excitation wavelength is fixed and the emission monochromator is scanned. Excitation is usually fixed at a wavelength at which the sample has significant absorbance. To obtain excitation spectra, the emission wavelength is fixed and the excitation monochromator is scanned that theoretically expected to mimic the absorption spectra of photoluminescent materials; however, because the lamp output is wavelength dependent, a correction is needed. Emission is normally fixed at a wavelength corresponding to the emission peak of the sample (Omary and Patterson 2016). Far-red carbon dots (FR-CDs) produced by Li et al. have absorption bands at 263, 340–440, and 570–720 nm that show the electron promotions, which are related to the type of bonding of atoms (Li et al. 2020c). They observed that the excitation spectrum overlapped with the absorption bands for these FR-CDs, which proves that the same kind of transitions lead to emission. When excited at 415 nm, FR-CDs gave emission peaks at 683 and 750 nm, and the QD solution appeared red under UV light, evidence of quantum confinement.

Transmission electron microscopy provides high-resolution images using high-energy electron beams that provide information about the size, shape, order, and crystallinity of nanomaterials. Electronic lenses allow the electron beams to be focused onto the material, and when the electron beam strikes the nanomaterial, the electrons pass through the sample, providing a magnified image of the material. Mercaptosuccinic acid (MSA)-capped silver selenide (Ag_2Se) quantum dots were synthesized by Pereira et al.; these were spherical shaped and 5.7 nm in size, as obtained by TEM (Pereira et al. 2020). High-resolution transmission electron microscopy (HRTEM) is a special type of transmission electron microscopy used to study the surface structure, elemental distribution, and morphology of the sample, as it produces interference images using both transmitted and scattered beams (Mitić et al. 2017).

7.5 Applications of Quantum Dots

QDs are receiving significant interest because they have applications in a variety of technological innovations including solar cells (Coşkun et al. 2022; Nadalini et al. 2023; Ding et al. 2019; Chen et al. 2022), light-emitting diodes (Efros and Brus 2021; Zhu et al. 2015; Yin et al. 2019), biological imaging (Huang et al. 2018; Cai et al. 2006), drug delivery (Baker and Baker 2010), and agriculture (Ünlü et al. 2022). This is attributed to their tunable electrical and optical properties, which can be controlled by size, resulting in wide absorption and sharp emission bands.

There is increasing interest in QDs in agricultural sciences. They can be used in plant bioimaging, increasing plant yield (Li et al. 2020a, b, c, d), eliminating plant diseases (Chakravarty et al. 2015), and so on.

Quantum dots have many merits along with some drawbacks. Due to their distinctive electrical and optical properties, which can be customized by changing their size, they find application in a wide range of technological innovations. However, their usage may have limits because of the toxicity of precursor reagents that are used in the fabrication. Traditional QDs have emerged with a Cd-based precursor, which is a very toxic heavy metal in the periodic table. Recent advances in decreasing the toxicity of QDs include the development of core/shell structures, such as CdSe/ZnS (Granados-Oliveros et al. 2022), surface encapsulation encapsulating of the QDs with surfactant ligands, the use of less toxic reagents in the production of QDs, or choosing greener and safer precursors (e.g., Zn, In, C). Quantum dots synthesized by more traditional organic methods often have hydrophobic surface properties, causing poor aqueous solubility and stability for in vivo applications. The surface needs to be modified post-synthesis through encapsulation (Jańczewski et al. 2011), ligand exchange (Zylstra et al. 2011), etc.

7.6 Quantum Dots & Plants and Natural Photosynthesis

Quantum dots have emerged as pivotal tools in enhancing photosynthetic efficiency due to their unique photophysical properties and tunable surface chemistry (Budak et al. 2021; Rakovich et al. 2014). Early research demonstrated that QDs such as CdTe and CdSe, when appropriately surface modified, can interact with key photosynthetic proteins like bacteriorhodopsin and light-harvesting complexes, facilitating efficient energy transfer and filling spectral absorption gaps (Erker et al. 2010; Nabiev et al. 2010; Rakovich et al. 2009). Subsequent studies have shown that these interactions enhance chlorophyll fluorescence, improve electron transfer, and ultimately boost the overall energy conversion processes in photosystems (Jung et al.

2010; Maksimov et al. 2012). Moreover, variations in QD surface coatings, including PEG and metallic formulations, allow fine-tuning of energy transfer rates and even increase photophosphorylation efficiency (Krivenkov et al. 2014; Olejnik et al. 2013; Xu et al. 2018). The development of hybrid bio-nanostructures that integrate QDs with photosynthetic reaction centers further highlights their potential to mimic or even enhance natural photosynthesis, offering promising applications in artificial photosynthesis and solar energy conversion (Liu et al. 2020; Lukashev et al. 2016).

7.6.1 Cadmium-Based Quantum Dots

Cadmium-based quantum dots have been extensively studied for their impact on photosynthetic organisms. Exposure to Cd^{2+} ions reduces chlorophyll content and photosynthetic activity (Haider et al. 2021). Early work with CdSe/ZnS QDs on *Chlamydomonas* sp. demonstrated that although these QDs are adsorbed onto the cell surface—prevented from entering the cell by a thick cell wall—they still inhibit photosynthetic activity (Lin et al. 2009). Comparative studies revealed that the toxicity of Cd-based QDs is largely due to the release of Cd^{2+} ions upon quantum dot decomposition; core/shell structures like CdSe/ZnS exhibit lower toxicity because the protective shell minimizes ion release (Domingos et al. 2011; Lin et al. 2009). However, core-type Cd-based QDs have been shown to be even more toxic than Cd^{2+} ions in green algae such as *Chlorella pyrenoidosa* and *Scenedesmus obliquus*, with oxidative stress markedly affecting cell morphology and facilitating quantum dot uptake despite the cell wall barrier (Yan et al. 2019). Additionally, the surface properties of these QDs play a critical role, as different organic coatings can lead to the formation of an "ecological corona" that reduces toxicity (Chakraborty et al. 2021). In higher plants, studies on *Arabidopsis thaliana* have indicated that while polymer-coated CdSe/ZnS QDs tend to remain localized around the roots, they still induce oxidative stress and negatively affect metabolism (Koo et al. 2015; Navarro et al. 2012). Research on other plant species such as onion, maize, and soybean further supports the conclusion that Cd-based QDs are toxic, prompting suggestions to either develop tolerant mutants or employ safer core/shell QDs (Banerjee et al. 2021; Majumdar et al. 2019; Sun et al. 2021; Wroblewska-Wolna et al. 2020).

7.6.2 Carbon Dots

Due to toxicity concerns, attention has shifted to carbon dots. Comparative studies on *Microcystis aeruginosa* revealed that carbon QDs exert much lower toxicity than CdTe QDs (Yan et al. 2015). Further comparisons involving various doped and undoped carbon QDs on microalgae such as *Chlorella pyrenoidosa* and *Scenedesmus obliquus* have consistently shown that carbon-based QDs are less harmful (Xiao et al. 2016; Yao et al. 2018). Nevertheless, certain degradable carbon QDs can still trigger oxidative stress and toxicity, as observed in *Chlorella vulgaris* (Zhang et al. 2019), and graphene quantum dots have been found to inhibit the growth of tomato and mung bean plants (Feng et al. 2019). High GQD concentrations have been observed to reduce chlorophyll *a* and chlorophyll *b* levels and cause cell damage by affecting membrane permeability and damaging the nucleic acid or proteins due to an increase in the content of reactive oxygen species (ROS) and H_2O_2 levels.

Recent advances in the application of carbon dots (CDs) in agriculture have revealed their transformative potential for enhancing photosynthesis and promoting plant growth. CDs not only expand the usable light spectrum by converting ultraviolet light into photosynthetically active radiation and acting as optical amplifiers, but they also boost the efficiency of the photosynthetic machinery by enhancing electron transport, increasing RuBisCO activity, and improving key enzymatic functions. Furthermore, various functionalized CDs have been shown to facilitate nutrient uptake and mitigate abiotic stresses, ultimately leading to improved crop productivity and resilience. These multifaceted benefits show the significant role that CDs play as sustainable nanomaterials in modern agricultural practices. Photosynthetic pigments: chlorophyll *a*, chlorophyll *b*, and carotenoids are packed together to build photosystems (Lao and Plaza 2006). Photosystems are divided into two groups and designated as photosystem I (PSI) and photosystem II (PSII). The pigment chlorophyll *a* is in the heart of the photosynthetic reaction center and is referred to as P700 for PSI and P680 for PSI in cyanobacteria, algae, or plants. Here, 700 and 680 indicate the wavelength of the light that is most efficiently absorbed by the chlorophyll *a*. Other pigments absorb different wavelengths of light and channel this energy from the absorbed light to chlorophyll *a*. The amount of photosynthesis depends on the duration and intensity of sunlight, and the ability of the green parts of a plant to capture it. The visible spectrum of the 400–700 nm waveband of the electromagnetic spectrum empowers photosynthesis and referred to as photosynthetically active radiation (PAR) (Parrish et al. 2021). Therefore, diverse actions have been taken to mimic photosynthetic processes or create hybrid biological systems. Typically, less than 10% of sunlight fulfills the requirements of the photosynthetic apparatus (Zhu et al. 2010). Photosynthesis directly utilizes longer wavelengths more efficiently than UV light. Moreover, wavelengths between 575 and 675 nm (orange-red waveband) are approximately 30% more efficient for photosynthesis than those between 400 and 500 nm (blue waveband) (Parrish et al. 2021). Some carbon dots (CDs) exhibit enhanced absorption in these longer wavelengths, making them promising nanomaterials for improving solar energy capture in plants. Absorbing light in the UV and blue bands minimizes PAR self-absorption and shifts the spectrum toward wavelengths that are more efficient for photosynthetic pigments (Parrish et al. 2021). CDs have often been used to enhance photosynthesis by converting and expanding the light spectrum available for energy capture. In a recent study, the ability of CDs to convert ultraviolet light into photosynthetically active radiation, increasing the quantum yield effect in rice, was shown in the literature. By enhancing electron transport in photosystem II, CDs improved photosynthetic efficiency by nearly 30% and stimulated RuBisCO activity (Hallaji et al. 2022). Also, the light conversion properties of CDs and their role as nanosensors and delivery tools were emphasized as critical for sustainable agriculture. These features enabled precise intervention to improve photosynthetic efficiency and crop resilience (Maholiya et al. 2023). In a different study, N-doped graphene quantum dots (N-GQDs) exhibited broadband emission from UV to near-infrared, making them highly effective for light manipulation in photosynthesis. As a consequence, N-GQDs acted as bridge gaps in the light spectrum, providing chloroplasts with optimized light conditions for improved photosynthetic efficiency (Tang et al. 2014). Studies have shown that the combination of blue and red light is highly effective for plant photosynthesis and growth (Lin et al. 2013). These wavelengths serve as the primary energy sources for CO_2 assimilation during photosynthesis (Lee et al. 2007). The effect of dual-emissive CDs on photosynthesis was explored by applying them to chloroplasts; it was observed that photosynthesis was enhanced by increasing ATP production by 2.8 times in vitro and improving electron transport in vivo. During photosynthetic light reactions, generated electrons from light absorption are transported through the electron transport chain to photosystem II (PS II) and photosystem I (PS I). In PS I, the excited electrons are transferred to ferredoxin; thereafter, they can return to the electron transport chain of PS II, creating a cyclic pathway for ATP generation. Artificial electron acceptors and carbon dots enhance electron transport rates compared to pure chloroplasts. This process enables the estimation of reduction and electron transfer rates, ultimately increasing ATP production. The dual emission, matching the absorption spectra of chlorophyll, expanded the usable light spectrum, especially in the UV and red regions. This study highlighted the transformative role of CDs in engineering hybrid biological systems to optimize solar energy capture (Li et al. 2018b). Furthermore, it was shown that far-red-emitting CDs expanded the photosynthetically active radiation spectrum by converting UV-A light to far-red emission, improving photosynthetic activity in lettuce. These CDs enhanced electron transfer between photosystems and increased ATP synthesis, boosting fresh and dry weights by over 50%. This demonstrated the role of CDs in enhancing light utilization beyond the visible spectrum (Li et al. 2020c). It should be noted that CDs may act as optical amplifiers, converting UV to visible light, enhancing chloroplast activity, and boosting photosynthetic rates. The effectiveness of CDs in various delivery methods, including foliar feeding and hydroponics, proved that CDs could be versatile tools for improving light absorption in modern agriculture (Li et al. 2020d; Nabiev et al. 2010). In addition, fluorescent carbon dot nanocapsules can effectively adhere to leaves and emit visible light, which is then absorbed by the chloroplasts. This enables the chloroplasts to capture a broader spectrum of light, thereby accelerating electron transport. The adhesive property of polyvinyl alcohol (PVA) allowed the nanocapsules to adhere to leaf surfaces, prolonging their effect and significantly boosting light absorption and energy transfer (Xu et al. 2020).

Carbon dots enhance photosynthetic performance by improving key enzymatic activities, facilitating energy transfer, and optimizing the overall photosynthetic apparatus. In a recent study, biomass-derived carbon dots (B-CDs) showed enhanced efficiency in light conversion and nutrient provisioning. As the activity of RuBisCO and ferredoxin improved, B-CDs increased net photosynthetic rates by up to 300%

and biomass production by up to 150%. Their ability to release nutrients directly addressed deficiencies while simultaneously augmenting photosynthesis. Engineered nanomaterials (ENMs) are transforming agriculture by reducing nutrient loss and enhancing environmental sustainability. B-CDs improve photosynthesis by facilitating light conversion and nutrient supply. Synthesized from organic waste such as soybeans, walnuts, and dried laver, these CDs provide essential nutrients that regulate plant physiological processes. By attaching to chloroplasts, they increase Mg, Cu, and Fe levels—critical for chlorophyll synthesis, electron transport, and ferritin accumulation. Their small size and hydrophilicity enable them to penetrate leaves and enhance photosynthetic activity based on their raw material composition. B-CDs effectively release Mg (2.5%), Cu (0.45%), and Fe (6.0%), addressing nutrient deficiencies that cause chlorosis, poor growth, and leaf deformation. Foliar application of B-CDs significantly boosts lettuce growth under nutrient-deficient conditions, increasing biomass, dry weight, and carbohydrate content. Additionally, B-CDs enhance chlorophyll content, electron transport rate, and ferredoxin levels, further promoting photosynthesis. These findings highlight the potential of B-CDs as nanofertilizers, supporting crop growth in nutrient-limited environments, which shows sustainability of using biomass-derived materials for large-scale applications (Cheng et al. 2023). Furthermore, CDs can enhance photosystem activity in mung beans, boosting electron transfer and RuBisCO function. This resulted in a 21.9% increase in carbohydrate production, directly linked to improved chlorophyll content and photosystem I activity (Wang et al. 2018b). Also, degradable CDs improved RuBisCO enzyme activity and promoted growth in dicotyledons like soybeans and tomatoes. The CDs accelerated seed germination, enhanced carbohydrate production, and contributed to nutrient absorption. The first step in plant growth is seed germination and root development. *Arabidopsis thaliana* seeds treated with CDs showed increased moisture content, which enhanced seed germination. The hydrophilic surface groups of the CDs (–OH and –COOH) help attract water molecules into the seeds. The germination rate and root length were significantly improved compared to control seeds, with CDs yielding the longest roots. The metal ion adsorptive capacity of CDs, particularly for calcium ions, was tested, and it was found that the CDs could deliver these ions into the plant, enhancing nutrient uptake. This nutrient delivery is critical for plant growth. Additionally, CD treatment increased RuBisCO activity in *Arabidopsis thaliana*, which is important for carbohydrate production in the Calvin cycle. The increased enzyme activity led to a boost in carbohydrate levels in the plants. CDs were also found to degrade in plants, releasing CO_2, which could further support photosynthesis by contributing to higher intercellular CO_2 concentrations. The degradation products of CDs were found to act as plant hormone analogs, promoting plant growth. Experiments with 13C-labeled CDs confirmed that the degradation products of the CDs were incorporated into plant carbohydrates, indicating their role in supporting plant metabolism.

Overall, CDs enhance seed germination, root development, nutrient uptake, and photosynthetic activity, improving plant growth and yield. They demonstrate the potential to act as a multifunctional tool in modern agriculture presenting a promising approach to reducing fertilizer use in agriculture (Li et al. 2019a). In addition, CDs facilitated electron transfer from light-harvesting complexes to chloroplasts, by boosting photosynthetic efficiency. It was demonstrated that ATP production and oxygen evolution increase by threefold, highlighting transformative impact of CDs on the photosynthetic electron transport chain. Biocompatible, amine-functionalized fluorescent carbon dots (CQDs) enhance photosynthesis by promoting efficient electron transfer in chloroplasts (CLPs). When CQDs are conjugated to CLPs, they facilitate a substantial increase in electron transfer, as demonstrated by a 74.3% increase in ferricyanide reduction and a 6.7-fold higher oxygen uptake. CQDs also enhance the reduction of electron acceptors like DCPIP and $NADP^+$, leading to increased ATP and NADPH production, essential for the "dark biochemistry" of photosynthesis. In vivo and in vitro toxicity studies confirmed CQDs' low toxicity, showing no significant alterations in organ morphology or blood parameters in animal tests. These findings underscore CQDs' potential as a biocompatible nanomaterial for boosting photosynthetic efficiency, with promising applications in agriculture and renewable energy (Chandra et al. 2014). Also, magnesium and nitrogen co-doped CDs significantly improved photosynthetic activity in rice plants by increasing chlorophyll *a* and *b* content by 14.39% and 26.54%, respectively. They enhanced RuBisCO activity and photosynthetic electron transport by over 100%, demonstrating the ability of such doped CDs to enhance photosynthesis at both molecular and physiological levels. Mg,N-CDs promoted plant photosynthesis by influencing key processes like chlorophyll biosynthesis, light absorption, electron transport, and CO_2 assimilation. The Mg^{2+} ions play a crucial role in chlorophyll synthesis, where they facilitate the chelation into protoporphyrin IX, a key step in chlorophyll production. Nitrogen and functional groups on Mg,N-CDs significantly affect chlorophyll synthesis by enhancing gene expression for chlorophyll synthetase and Mg-chelatase subunits, as well as chlorophyllase, involved in chlorophyll degradation. The surface groups, especially nitrogen-containing ones, are essential for regulating chlorophyll metabolism and maintaining high levels of chlorophyll in plants, which is vital for light harvesting and energy conversion. Additionally, Mg,N-CDs absorb UV light in chloroplasts and emit blue fluorescence, protecting the chloroplasts from UV damage while enhancing photosynthesis (Li et al. 2021). Furthermore, chiral CDs synthesized via hydrothermal methods from citric acid and L- or D-cysteine showed enantioselective effects on mung bean plants. Chirality plays a key role in the effects of CDs on plant metabolism, particularly in enzyme activity. While achiral CDs had no significant impact on root length, stem length, or chlorophyll content compared to chiral CDs, the activity of key metabolic enzymes was strongly influenced by chirality. D-CDs notably increased succinodehydrogenase activity more than L-CDs, demonstrating their stronger effect on root vigor. Additionally, RuBisCO enzyme activity and carbohydrate content were significantly higher in plants treated with D-CDs compared to both L-CDs and achiral CDs. These findings confirm that the chirality of CDs directly affects enzyme activity, with D-CDs exhibiting superior properties in promoting photosynthesis and plant metabolism. Thus, the D-enantiomer enhanced RuBisCO activity by 67.5%, improving photosynthetic efficiency and carbohydrate accumulation, which revealed the potential of chirality in CDs for optimizing photosynthetic pathways (Zhang et al. 2018).

In a recent study (Zheng et al. 2017), the dual role of pollen-derived carbon dots (CDs) in bioimaging and growth promotion was explored in detail. Carbon dots (CDs) synthesized from rapeseed pollen were found to enhance lettuce growth and significantly reduce nitrate (NO_3^-) content by facilitating nutrient transport through the transpiration stream. At 20 mg/L, nitrate content decreased by 33%, while the highest production yield was achieved at 30 mg/L. High nitrate levels in vegetables are concerning as they can be converted into harmful nitrites in the human body, posing health risks such as methemoglobinemia and potential carcinogenic effects. CDs acted as ion carriers, improving stomatal conductance and transpiration, which promoted leaf expansion, increased biomass, and enhanced root growth. Their ability to trap and gradually release essential ions, including potassium (K^+) and calcium (Ca^{2+}), contributed to better nutrient absorption, supporting overall plant development. Additionally, CDs exhibited excellent biocompatibility and water solubility, making them a promising tool for improving crop yield and quality in sustainable agriculture while reducing health risks associated with excessive nitrate accumulation.

Another study focused on the role of CDs in alleviating abiotic stresses such as salinity and UV radiation in rice (Li et al. 2020a). By scavenging reactive oxygen species (ROS) and enhancing antioxidant enzyme activity, CDs improved chloroplast functionality and nutrient assimilation under stress conditions. This protective role demonstrated the adaptability of CDs in enhancing photosynthetic performance under adverse environmental conditions. Proline-functionalized CDs mitigated salinity stress in grapevine plants by enhancing antioxidant activity, reducing oxidative stress, and protecting photosynthetic machinery. Chlorophyll fluorescence and electron transport rates were significantly improved as well (Gohari et al. 2021b). In the study, the impact of graphene quantum dots (GQDs) on coriander and garlic plants resulted in enhanced root and shoot elongation, leaf expansion, and increased biomass (Chakravarty et al. 2015). GQDs improved nutrient absorption and regulated hormonal balance, leading to more efficient photosynthesis. Their ability to optimize light absorption suggests their potential as biostimulants, particularly for crops where maximizing photosynthetic efficiency is crucial for growth and productivity. Also, GQDs enhanced chlorophyll content in mung bean and tomato plants, boosting photosynthetic efficiency under optimal concentrations (Feng et al. 2019). They affected mung bean and tomato seedlings differently. While germination rates remained unchanged, high GQD concentrations ($\geq$1250 mg L^{-1}) severely inhibited mung bean growth, reducing stem/root lengths by over 75% and increasing oxidative stress markers like proline and H_2O_2. Tomato seedlings were less sensitive, showing growth reductions only at 1500 mg L^{-1}. Moderate GQD levels (250–1250 mg L^{-1}) enhanced chlorophyll content, but excessive exposure caused declines. Antioxidant enzyme activities (CAT, GR) increased as a defense response, and fluorescence analysis confirmed GQD accumulation, mainly in roots and stems. High concentrations caused oxidative stress and disrupted physiological processes, highlighting the importance of dosing when using QDs to optimize photosynthesis and prevent adverse effects.

In addition, CDs derived from *Spirulina* showed significant potential in promoting plant growth (Agnol et al. 2021). Their nitrogen-rich composition boosted photosynthetic pigments and facilitated the electron transport chain in photosynthesis. These findings demonstrated the dual role of CDs as light converters and nutrient providers, enhancing the sustainability of agricultural practices. Furthermore, CDs improved rice growth and disease resistance by enhancing RuBisCO activity, carbohydrate production, and gene expression for defense proteins (Li et al. 2018a). The degradation of CDs into plant hormone analogs also stimulated growth, emphasizing their multifaceted role in agricultural applications. Also, it was shown that CDs enhanced wheat growth by increasing chlorophyll content and optimizing photosynthetic pathways and improved drought resistance, making them valuable for crops in water-limited environments (Gong and Dong 2021). Also, functionalized CDs with carboxyl and ester groups improved water uptake, nutrient assimilation, and overall plant metabolism (Chen et al. 2020). Their enhanced dispersibility facilitated interaction with photosynthetic machinery, directly impacting photosynthetic efficiency. Eco-friendly synthesized CDs using peanut bran as raw material enhanced the growth of Italian lettuce by 24.3% and improved nutrient utilization efficiency; the potential of CDs as sustainable nanofertilizers for improving crop productivity through enhanced photosynthesis and nutrient uptake was demonstrated successfully (An et al. 2022). The orange CDs have shown improved light absorption and photosynthetic pigment content in maize (Milenković et al. 2021). By reducing oxidative stress and enhancing energy transfer to photosystems, these CDs increased photosynthetic rates and biomass production. Furthermore, N-doped CDs derived from *Piper betle* leaves enhanced strawberry plant growth by increasing chlorophyll content by 24.7% and carbohydrate content by 48.61% and promoted secondary metabolite production, demonstrating their dual role in enhancing photosynthesis and plant health (Abu Salha et al. 2023). Also, putrescine-functionalized CDs improved photosynthesis in grapevines under salinity stress by increasing chlorophyll fluorescence, antioxidant enzyme activity, and membrane stability, which demonstrated the combined potential of functionalized CQDs and polyamines in stress resilience and photosynthesis enhancement (Gohari et al. 2021a).

7.6.3 Metal Quantum Dots

Other than CDs, it was shown that silicon quantum dots (SiQDs) could enhance light harvesting, boost chlorophyll synthesis, promote plant growth, and enable real-time stress monitoring through their unique optical properties. In a recent study, the role of silicon quantum dots (SiQDs) as artificial antennas to amplify light harvesting was discussed in detail. SiQDs could convert ultraviolet (UV) light into blue light, matching with chloroplast absorption spectra, enhancing photosystem II activity, and increasing biomass by over 50%. SiQDs improved chlorophyll *a* and *b* content significantly and demonstrated their effectiveness even under high concentrations. Also, in this study, the importance of SiQDs in optimizing the photosynthesis process by making UV radiation bioavailable for plant growth was mentioned (Li et al. 2020b). The biomimetic synthesis of SiQDs demonstrated their ability to enhance cucumber seedling growth by 51% through improved water uptake and aquaporin gene expression. This study showed that the multifunctionality of SiQDs in promoting physiological processes is critical to photosynthesis and water management in plants (Li et al. 2019b). Also, it was shown that SiQDs could act as fluorescent probes to detect H_2O_2in plant tissues. By mitigating oxidative stress and promoting metabolic activity, SiQDs have the potential to support photosynthetic processes under stress conditions. The fluorescence properties allow for real-time monitoring of plant health (Li et al. 2020d).

Also, various quantum dots—including ZnO, CdS Mn/ZnS, CdSe/CdZnS, and sulfur- and Zn-doped QDs—could be used for enhancing photosynthetic efficiency, promoting plant growth, and facilitating the tracking of QD movement within plant systems. In a recent study, ZnO quantum dots demonstrated dose-dependent effects, improving photosynthetic performance and growth in lettuce. At optimal concentrations, ZnO QDs enhanced chlorophyll content and increased photosynthetic efficiency by optimizing light absorption and reducing oxidative stress (Liang et al. 2021). It was shown that N-acetyl cysteine (NAC)-coated CdS Mn/ZnS QDs improved seed germination and seedling growth of snow pea (*Pisum sativum L.*). These QDs enhanced chlorophyll content and overall plant vigor, acting as a nutrient medium. The fluorescence properties enabled efficient tracking within plant tissues. This work highlights the dual role of QDs as both growth promoters and imaging tools, showcasing their potential in agricultural nanotechnology (Das et al. 2015). Furthermore, CdSe/CdZnS QDs were studied for their uptake and translocation in *Arabidopsis*. The QDs moved through the plant vascular system and showed fluorescence in leaves and insect herbivores feeding on treated plants, which showed the potential for QDs to influence photosynthesis while raising questions about environmental and ecological implications (Koo et al. 2015). The uptake and localization of CdSe/ZnS QDs in plants were examined. While QDs remained localized in roots, their oxidative effects were minimal, suggesting their potential for controlled applications in enhancing photosynthesis without significant toxicity (Navarro et al. 2012). Furthermore, Zn-doped QDs combined with biochar and proline enhanced photosynthetic activity in chili plants under drought conditions. This combination mitigated oxidative stress, improved chlorophyll fluorescence parameters, and boosted yield, emphasizing the role of QDs in supporting photosynthesis under environmental stress (Alotaibi et al. 2024). In addition, the potential of sulfur nanoparticles in alleviating manganese-induced stress in sunflowers was explored. While not directly addressing QDs, the implications for nanoscale interventions in photosynthetic pathways are clear. The findings suggest similar benefits for QDs in mitigating abiotic stresses and enhancing photosynthetic resilience (Ragab and Saad-Allah 2020).

7.6.4 *Quantum Dots in Greenhouses*

Quantum dots have recently been employed as additives in greenhouse covering materials and lighting systems (Pavlov et al. 2017; Liu et al. 2021; Parrish et al. 2021). Initial investigations into their use in greenhouse settings focused on Cd-based quantum dots (Pavlov et al. 2017; Liu et al. 2021; Parrish et al. 2021). For example, studies demonstrated that by modifying the sunlight spectrum with coverings infused with Cd-based quantum dots, the yield of vegetables—such as cucumber (*Cucumis sativus L.*), tomato, and cabbage (*Brassica oleracea L. var. capitata*)—could be boosted by up to 70% (Pavlov et al. 2017). However, the commercial application of Cd-based quantum dots is limited by their inherent toxicity (Pavlov et al. 2017). As the effectiveness of quantum dot-enriched greenhouse coverings became evident, researchers began exploring Cd-free alternatives. For instance, $CuInS_2/ZnS$ quantum dot films, which emit between 600 and 660 nm, were used to adjust the solar spectrum into the photosynthetically active region (PAR) for red romaine lettuce (*Lactuca sativa L.*), resulting in a 10% increase in both fresh and dry edible mass (Parrish et al. 2021). Additionally, quantum dots have been proposed as light sources in the form of light-emitting diodes (LEDs), where CdZnS-based quantum dots are utilized to maximize photosynthetic performance (Liu et al. 2021).

In summary, the exceptional photophysical characteristics of quantum dots—such as their narrow emission bands and high absorption capacities—render them highly effective for manipulating the light quality experienced by plants in greenhouse systems. Carbon dots (CDs) and quantum dots (QDs) have shown significant potential in enhancing plant growth and photosynthesis. CDs, derived from various sources like *Spirulina*, peanut bran, and *Piper betle*, promote photosynthesis, increase chlorophyll content, and improve nutrient uptake, leading to better plant growth, stress resilience, and disease resistance. Functionalized CDs and SiQDs, which can amplify light absorption and reduce oxidative stress, also enhance photosynthetic efficiency and biomass production. Various QDs, including ZnO, CdS, and $CuInS_2/ZnS$, optimize light absorption, mitigate stress, and promote growth in crops like lettuce, maize, and grapevines. Additionally, QDs are being explored for use in greenhouse coverings and light systems, where they adjust the solar spectrum to boost vegetable yield and improve photosynthesis, with Cd-free alternatives being developed to address toxicity concerns.

7.7 Conclusion and Future Perspective

Quantum dots have emerged as transformative tools in enhancing natural photosynthesis and promoting plant growth. Their unique photophysical properties—such as narrow emission bands, high absorption capacity, and tunable surface chemistry—enable them to effectively modify the light spectrum, improve energy transfer, and boost photosynthetic activity. Studies have demonstrated that both cadmium-based and carbon-based quantum dots can interact with key photosynthetic proteins, enhance chlorophyll fluorescence, and increase biomass production. However, the use of Cd-based quantum dots is limited by toxicity concerns, which has led to a growing interest in safer alternatives such as carbon dots and Cd-free formulations. These alternatives offer a promising path forward, mitigating the risks associated with heavy metals while maintaining the beneficial properties of quantum dots. Overall, these advancements underscore the significant potential of quantum dots as sustainable nanomaterials in modern agricultural practices and solar energy conversion.

Future research should focus on the development of highly efficient, biocompatible quantum dots that minimize environmental risks while maximizing photosynthetic enhancement. Emphasis must be placed on optimizing core/shell structures, surface modifications, and doping strategies to reduce toxicity and improve stability in various agricultural settings. Additionally, integrating quantum dots into greenhouse covering materials, into LED lighting systems, and as additives in field crops holds great promise for increasing crop yield and resilience. Long-term studies assessing environmental impacts and exploring genetic adaptations in plants to quantum dot exposure will be crucial. Ultimately, these innovations could revolutionize sustainable agriculture by enabling precise control over light quality and energy conversion in photosynthetic systems.

References

Abu Salha B, Saravanan A, Maruthapandi M, Perelshtein I, Gedanken A (2023) Plant-derived nitrogen-doped carbon dots as an effective fertilizer for enhanced strawberry growth and yield. ACS ES T Eng 3(8):1165–1175. https://doi.org/10.1021/acsestengg.3c00046

Agnol LD, Neves RM, Maraschin M, Moura S, Ornaghi HL, Dias FTG, Bianchi O (2021) Green synthesis of *Spirulina*-based carbon dots for stimulating agricultural plant growth. Sustain Mater Technol 30. https://doi.org/10.1016/j.susmat.2021.e00347

Almeida G, Ubbink RF, Stam M, du Fossé I, Houtepen AJ (2023) InP colloidal quantum dots for visible and near-infrared photonics. Nat Rev Mater 8(11). https://doi.org/10.1038/s41578-023-00596-4

Alotaibi NM, Alotibi MM, Younis U, Hussain GS, Dawar K, Hareem M, Ghoneim AM, Fahad S, Danish S (2024) Zn-quantum dot biochar regulates antioxidants and nutrient uptake to improve rapeseed growth and yield in drought stress. Plant Stress 11. https://doi.org/10.1016/j.stress.2023.100286

An Z, Zhang H, Li W, Yang X, Kang Y, Zheng M, Lei B (2022) Large-scale preparation of peanut-bran-derived carbon dots and their promoting effect on Italian lettuce. ACS Agric Sci Technol 2(2). https://doi.org/10.1021/acsagscitech.1c00177

An MN, Eom SY, Lee JH, Song H, Cho M, Jeong KS (2023) Room temperature synthesis of self-doped silver selenide quantum dots sensitive to mid-infrared light. ACS Appl Nano Mater 6(23). https://doi.org/10.1021/acsanm.3c05043

Baker SN, Baker GA (2010) Luminescent carbon nanodots: emergent nanolights. Angew Chem Int Ed 49(38):6726–6744. https://doi.org/10.1002/anie.200906623

Banerjee R, Goswami P, Chakrabarti M, Chakraborty D, Mukherjee A, Mukherjee A (2021) Cadmium selenide (CdSe) quantum dots cause genotoxicity and oxidative stress in *Allium cepa* plants. Mutat Res Genet Toxicol Environ Mutagen 865. https://doi.org/10.1016/j.mrgentox.2021.503338

Bera D, Qian L, Tseng TK, Holloway PH (2010) Quantum dots and their multimodal applications: a review. Materials 3(4):2260–2345. https://doi.org/10.3390/ma3042260

Budak E, Erdoğan D, Ünlü C (2021) Enhanced fluorescence of photosynthetic pigments through conjugation with carbon quantum dots. Photosynth Res 147(1):1–10. https://doi.org/10.1007/s11120-020-00786-z

Cai W, Shin DW, Chen K, Gheysens O, Cao Q, Wang SX, Gambhir SS, Chen X (2006) Peptide-labeled near-infrared quantum dots for imaging tumor vasculature in living subjects. Nano Lett 6(4). https://doi.org/10.1021/nl052405t

Chakraborty D, Ethiraj KR, Chandrasekaran N, Mukherjee A (2021) Mitigating the toxic effects of CdSe quantum dots towards freshwater alga *Scenedesmus obliquus*: role of eco-corona. Environ Pollut 270. https://doi.org/10.1016/j.envpol.2020.116049

Chakravarty D, Erande MB, Late DJ (2015) Graphene quantum dots as enhanced plant growth regulators: effects on coriander and garlic plants. J Sci Food Agric 95(13):2772–2778. https://doi.org/10.1002/jsfa.7106

Chandra S, Pradhan S, Mitra S, Patra P, Bhattacharya A, Pramanik P, Goswami A (2014) High throughput electron transfer from carbon dots to chloroplast: a rationale of enhanced photosynthesis. Nanoscale 6(7):3647–3655. https://doi.org/10.1039/c3nr06079a

Chen Q, Chen L, Nie X, Man H, Guo Z, Wang X, Tu J, Jin G, Ci L (2020) Impacts of surface chemistry of functional carbon nanodots on the plant growth. Ecotoxicol Environ Saf 206. https://doi.org/10.1016/j.ecoenv.2020.111220

Chen J, Jia D, Zhuang R, Hua Y, Zhang X (2022) Highly orientated perovskite quantum dot solids for efficient solar cells. Adv Mater 34(37). https://doi.org/10.1002/adma.202204259

Cheng B, Yang Z, Chen F, Yue L, Cao X, Li J, Qian HL, Yan XP, Wang C, Wang Z (2023) Biomass-derived carbon dots with light conversion and nutrient provisioning capabilities facilitate plant photosynthesis. Sci Total Environ 901. https://doi.org/10.1016/j.scitotenv.2023.165973

Coşkun Y, Ünlü FY, Yllmaz T, Türker Y, Aydogan A, Kuş M, Ünlü C (2022) Development of highly luminescent water-insoluble carbon dots by using Calix[4]pyrrole as the carbon precursor and their potential application in organic solar cells. ACS Omega 7(22):18840–18851. https://doi.org/10.1021/acsomega.2c01795

Das S, Wolfson BP, Tetard L, Tharkur J, Bazata J, Santra S (2015) Effect of N-acetyl cysteine coated CdS:Mn/ZnS quantum dots on seed germination and seedling growth of snow pea (*Pisum sativum* L.): imaging and spectroscopic studies. Environ Sci Nano 2(2). https://doi.org/10.1039/c4en00198b

Ding X, Cai M, Liu X, Ding Y, Liu X, Wu Y, Hayat T, Alsaedi A, Dai S (2019) Lead sulfide quantum dots as a bifunctional layer for efficient and stable all-inorganic cesium lead iodide perovskite solar cells. ChemistrySelect 4(45). https://doi.org/10.1002/slct.201904141

Domingos RF, Simon DF, Hauser C, Wilkinson KJ (2011) Bioaccumulation and effects of CdTe/CdS quantum dots on *Chlamydomonas reinhardtii*—nanoparticles or the free ions? Environ Sci Technol 45(18):7664–7669. https://doi.org/10.1021/es201193s

Efros AL, Brus LE (2021) Nanocrystal quantum dots: from discovery to modern development. ACS Nano 15(4):6192–6210. https://doi.org/10.1021/acsnano.1c01399

Erker W, Boggasch S, Xie R, Grundmann G, Paulsen H, Basch T (2010) Assemblies of semiconductor quantum dots and light-harvesting-complex II. J Lumin 130(9):1624–1627. https://doi.org/10.1016/j.jlumin.2009.12.002

Facure MHM, Schneider R, Mercante LA, Correa DS (2020) A review on graphene quantum dots and their nanocomposites: from laboratory synthesis towards agricultural and environmental applications. Environ Sci Nano 7(12). https://doi.org/10.1039/d0en00787k

Fargi A, Ghedira S, Kalboussi A (2023) Capacitance spectroscopy of InAs quantum dots inserted in an AlGaAs/GaAs HEMT for photodetector applications. AIP Adv 13(10). https://doi.org/10.1063/5.0167563

Feng P, Geng B, Cheng Z, Liao X, Pan D, Huang J (2019) Graphene quantum dots-induced physiological and biochemical responses in mung bean and tomato seedlings. Rev Bras Bot 42(1):29–41. https://doi.org/10.1007/s40415-019-00519-0
Gao M, Yang H, Shen H, Zeng Z, Fan F, Tang B, Min J, Zhang Y, Hua Q, Li LS, Ji B, Du Z (2021) Bulk-like ZnSe quantum dots enabling efficient ultranarrow blue light-emitting diodes. Nano Lett 21(17). https://doi.org/10.1021/acs.nanolett.1c02284
Gohari G, Panahirad S, Sadeghi M, Akbari A, Zareei E, Zahedi SM, Bahrami MK, Fotopoulos V (2021a) Putrescine-functionalized carbon quantum dot (put-CQD) nanoparticles effectively prime grapevine (*Vitis vinifera* cv. 'Sultana') against salt stress. BMC Plant Biol 21(1). https://doi.org/10.1186/s12870-021-02901-1
Gohari G, Panahirad S, Sepehri N, Akbari A, Zahedi SM, Jafari H, Dadpour MR, Fotopoulos V (2021b) Enhanced tolerance to salinity stress in grapevine plants through application of carbon quantum dots functionalized by proline. Environ Sci Pollut Res 28(31). https://doi.org/10.1007/s11356-021-13794-w
Gong Y, Dong Z (2021) Transfer, transportation, and accumulation of cerium-doped carbon quantum dots: promoting growth and development in wheat. Ecotoxicol Environ Saf 226. https://doi.org/10.1016/j.ecoenv.2021.112852
Granados-Oliveros G, Pineros BSG, Calderon FGO (2022) CdSe/ZnS quantum dots capped with oleic acid and L-glutathione: structural properties and application in detection of Hg2+. J Mol Struct 1254. https://doi.org/10.1016/j.molstruc.2021.132293
Gumaste MR, Kulkarni GA (2021) Dominant UV emission and n-type conductivity in manganese doped Zinc Telluride quantum dots. Phys B Condens Matter 615. https://doi.org/10.1016/j.physb.2021.413076
Gupta A, Mehta SK, Kunal K, Mukhopadhyay K, Singh S (2022) Quantum dots as promising nanomaterials in agriculture. In: Agricultural nanobiotechnology: biogenic nanoparticles, nanofertilizers and nanoscale biocontrol agents, pp 243–296. https://doi.org/10.1016/B978-0-323-91908-1.00016-X
Haider FU, Liqun C, Coulter JA, Cheema SA, Wu J, Zhang R, Wenjun M, Farooq M (2021) Cadmium toxicity in plants: Impacts and remediation strategies. Ecotoxicol Environ Saf 211. https://doi.org/10.1016/j.ecoenv.2020.111887
Hallaji Z, Bagheri Z, Tavassoli Z, Ranjbar B (2022) Fluorescent carbon dot as an optical amplifier in modern agriculture. Sustain Mater Technol 34. https://doi.org/10.1016/j.susmat.2022.e00493
Hassan A, Iqbal MF, Maksym S, Tetiana M, Azam M, Kanwal Z, Irfan I, Li B, Jiang Y (2019) Near-band-edge emission enhancement and suppression of the deep levels in Ga-doped ZnO via surface plasmon-exciton coupling without a dielectric spacer. J Mater Sci Mater Electron 30(23). https://doi.org/10.1007/s10854-019-02418-2
Hines MA, Guyot-Sionnest P (1998) Bright UV-blue luminescent colloidal ZnSe nanocrystals. J Phys Chem B 102(19). https://doi.org/10.1021/jp9810217
Hu SL, Niu KY, Sun J, Yang J, Zhao NQ, Du XW (2009) One-step synthesis of fluorescent carbon nanoparticles by laser irradiation. J Mater Chem 19(4). https://doi.org/10.1039/b812943f
Huang S, Yang E, Yao J, Liu Y, Xiao Q (2018) Carbon dots doped with nitrogen and boron as ultrasensitive fluorescent probes for determination of α-glucosidase activity and its inhibitors in water samples and living cells. Microchim Acta 185(8). https://doi.org/10.1007/s00604-018-2932-1
İbiş Ö, Ük N, Nar I, Ünlü C (2025) Manipulation of defect state emission in Zn chalcogenide quantum dots and their effects on chlorophyll spectral response. Spectrochim Acta A Mol Biomol Spectrosc 327:125348. https://doi.org/10.1016/j.saa.2024.125348
Jańczewski D, Tomczak N, Han MY, Vancso GJ (2011) Synthesis of functionalized amphiphilic polymers for coating quantum dots. Nat Protoc 6(10). https://doi.org/10.1038/nprot.2011.381
Jung H, Gulis G, Gupta S, Redding K, Gosztola DJ, Wiederrecht GP, Stroscio MA, Dutta M (2010) Optical and electrical measurement of energy transfer between nanocrystalline quantum dots and photosystem I. J Phys Chem B 114(45):14544–14549. https://doi.org/10.1021/jp102291e
Kastner MA (1993) Artificial Atoms. Phys Today 46(1). https://doi.org/10.1063/1.881393

Khenthaphak S, Juleang P, Wonsawat W, Chailapakul O, Punrat E (2023) Development of inexpensive device for methanol detection in hand sanitizer gel using manganese-doped zinc sulfide quantum dots modified by N-methylpolypyrrole. J Met Mater Miner 33(1). https://doi.org/10.55713/jmmm.v33i1.1597

Kirmani AR, Woodhouse M, Luther JM (2022) Technoeconomic model suggests scaling-up perovskite quantum dots for optoelectronics warrants improved synthesis yield, solvent recycling, and automation. ACS Energy Lett 7(4):1255–1259. https://doi.org/10.1021/acsenergylett.2c00250

Koo Y, Wang J, Zhang Q, Zhu H, Chehab EW, Colvin VL, Alvarez PJJ, Braam J (2015) Fluorescence reports intact quantum dot uptake into roots and translocation to leaves of *Arabidopsis thaliana* and subsequent ingestion by insect herbivores. Environ Sci Technol 49(1):626–632. https://doi.org/10.1021/es5050562

Krivenkov VA, Solovyeva DO, Samokhvalov PS, Brazhnik KI, Kotkovskiy GE, Chistyakov AA, Lukashev EP, Nabiev IR (2014) Photoinduced modification of quantum dot optical properties affects bacteriorhodopsin photocycle in a (quantum dot)—bacteriorhodopsin hybrid material. J Phys Conf Ser 541(1). https://doi.org/10.1088/1742-6596/541/1/012045

Lao MT, Plaza BM (2006) Water management in ornamental crops. In: Floriculture, ornamental and plant biotechnology: advances and topical issues, vol III. Global Science Book

Lee S-H, Tewari RK, Hahn E-J, Paek K-Y (2007) Photon flux density and light quality induce changes in growth, stomatal development, photosynthesis and transpiration of *Withania somnifera* (L.) Dunal. plantlets. Plant Cell Tissue Organ Cult 90(2):141–151. https://doi.org/10.1007/s11240-006-9191-2

Li H, Huang J, Lu F, Liu Y, Song Y, Sun Y, Zhong J, Huang H, Wang Y, Li S, Lifshitz Y, Lee ST, Kang Z (2018a) Impacts of carbon dots on rice plants: boosting the growth and improving the disease resistance. ACS Appl Bio Mater 1(3):663–672. https://doi.org/10.1021/acsabm.8b00345

Li W, Wu S, Zhang H, Zhang X, Zhuang J, Hu C, Liu Y, Lei B, Ma L, Wang X (2018b) Enhanced biological photosynthetic efficiency using light-harvesting engineering with dual-emissive carbon dots. Adv Funct Mater 28(44). https://doi.org/10.1002/adfm.201804004

Li H, Huang J, Liu Y, Lu F, Zhong J, Wang Y, Li S, Lifshitz Y, Lee ST, Kang Z (2019a) Enhanced RuBisCO activity and promoted dicotyledons growth with degradable carbon dots. Nano Res 12(7). https://doi.org/10.1007/s12274-019-2397-5

Li Y, Li W, Zhang H, Dong R, Li D, Liu Y, Huang L, Lei B (2019b) Biomimetic preparation of silicon quantum dots and their phytophysiology effect on cucumber seedlings. J Mater Chem B 7(7):1107–1115. https://doi.org/10.1039/c8tb02981d

Li Y, Gao J, Xu X, Wu Y, Zhuang J, Zhang X, Zhang H, Lei B, Zheng M, Liu Y, Hu C (2020a) Carbon Dots as a protective agent alleviating abiotic stress on rice (*Oryza sativa* L.) through promoting nutrition assimilation and the defense system. ACS Appl Mater Interfaces 12(30). https://doi.org/10.1021/acsami.0c11724

Li Y, Li W, Zhang H, Liu Y, Ma L, Lei B (2020b) Amplified light harvesting for enhancing Italian lettuce photosynthesis using water soluble silicon quantum dots as artificial antennas. Nanoscale 12(1):155–166. https://doi.org/10.1039/c9nr08187a

Li D, Li W, Zhang H, Zhang X, Zhuang J, Liu Y, Hu C, Lei B (2020c) Far-red carbon dots as efficient light-harvesting agents for enhanced photosynthesis. ACS Appl Mater Interfaces 12(18):21009–21019. https://doi.org/10.1021/acsami.9b21576

Li Y, Zhang H, Yao Y, Gong T, Dong R, Li D, Liu Y, Lei B (2020d) Promoted off-on recognition of H_2O_2 based on the fluorescence of silicon quantum dots assembled two-dimensional PEG-MnO_2 nanosheets hybrid nanoprobe. Microchim Acta 187(6). https://doi.org/10.1007/s00604-020-04276-w

Li Y, Xu X, Lei B, Zhuang J, Zhang X, Hu C, Cui J, Liu Y (2021) Magnesium-nitrogen co-doped carbon dots enhance plant growth through multifunctional regulation in photosynthesis. Chem Eng J 422. https://doi.org/10.1016/j.cej.2021.130114

Liang Z, Pan X, Li W, Kou E, Kang Y, Lei B, Song S (2021) Dose-dependent effect of ZnO quantum dots for lettuce growth. ACS Omega 6(15):10141–10149. https://doi.org/10.1021/acsomega.1c00205

Lin S, Bhattacharya P, Rajapakse NC, Brune DE, Ke PC (2009) Effects of quantum dots adsorption on algal photosynthesis. J Phys Chem C 113(25):10962–10966. https://doi.org/10.1021/jp904343s

Lin K-H, Huang M-Y, Huang W-D, Hsu M-H, Yang Z-W, Yang C-M (2013) The effects of red, blue, and white light-emitting diodes on the growth, development, and edible quality of hydroponically grown lettuce (*Lactuca sativa* L. var. capitata). Sci Hortic 150:86–91. https://doi.org/10.1016/j.scienta.2012.10.002

Liu J, Mantell J, Jones MR (2020) Minding the gap between plant and bacterial photosynthesis within a self-assembling biohybrid photosystem. ACS Nano 14(4):4536–4549. https://doi.org/10.1021/acsnano.0c00058

Liu Z (2021) Spectral design of light-emitting diodes for plant photosynthesis based on quantum dots. IEEE Access 9:156229–156238. https://doi.org/10.1109/ACCESS.2021.3054929

Lukashev EP, Knox PP, Oleinikov IP, Seifullina NK, Grishanova NP (2016) Investigation of stability of photosynthetic reaction center and quantum dot hybrid films. Biochem Mosc 81(1):58–63. https://doi.org/10.1134/S0006297916010065

Maholiya A, Ranjan P, Khan R, Murali S, Nainwal RC, Chauhan PS, Sathish N, Chaurasia JP, Srivastava AK (2023) An insight into the role of carbon dots in the agriculture system: a review. Environ Sci Nano 10(4):959–995. https://doi.org/10.1039/d2en00954d

Majumdar S, Pagano L, Wohlschlegel JA, Villani M, Zappettini A, White JC, Keller AA (2019) Proteomic, gene and metabolite characterization reveal the uptake and toxicity mechanisms of cadmium sulfide quantum dots in soybean plants. Environ Sci Nano 6(10):3010–3026. https://doi.org/10.1039/c9en00599d

Maksimov EG, Kurashov VN, Mamedov MD, Paschenko VZ (2012) Hybrid system based on quantum dots and photosystem 2 core complex. Biochem Mosc 77(6):624–630. https://doi.org/10.1134/S0006297912060090

Maxwell T, Nogueira Campos MG, Smith S, Doomra M, Thwin Z, Santra S (2019) Quantum dots. In: Nanoparticles for biomedical applications: fundamental concepts, biological interactions and clinical applications, pp 243–265. https://doi.org/10.1016/B978-0-12-816662-8.00015-1

Milenković I, Borišev M, Zhou Y, Spasić SZ, Leblanc RM, Radotić K (2021) Photosynthesis enhancement in maize via nontoxic orange carbon dots. J Agric Food Chem 69(19):5446–5451. https://doi.org/10.1021/acs.jafc.1c01094

Mitić Ž, Stolić A, Stojanović S, Najman S, Ignjatović N, Nikolić G, Trajanović M (2017) Instrumental methods and techniques for structural and physicochemical characterization of biomaterials and bone tissue: a review. Mater Sci Eng C 79. https://doi.org/10.1016/j.msec.2017.05.127

Nabiev I, Rakovich A, Sukhanova A, Lukashev E, Zagidullin V, Pachenko V, Rakovich YP, Donegan JF, Rubin AB, Govorov AO (2010) Fluorescent quantum dots as artificial antennas for enhanced light harvesting and energy transfer to photosynthetic reaction centers. Angew Chem Int Ed 49(40):7217–7221. https://doi.org/10.1002/anie.201003067

Nadalini G, Borghi F, Piseri P, Di Vece M (2023) Persistent quantum confinement in a Germanium quantum dot solid. Physica E Low Dimens Syst Nanostruct 151. https://doi.org/10.1016/j.physe.2023.115708

Navarro DA, Bisson MA, Aga DS (2012) Investigating uptake of water-dispersible CdSe/ZnS quantum dot nanoparticles by *Arabidopsis thaliana* plants. J Hazard Mater 211–212:427–435. https://doi.org/10.1016/j.jhazmat.2011.12.012

Nuryadin BW, Septia EC, Iskandar F, Ogi T, Okuyama K, Abdullah M, Khairurrijal. (2014) Microwave-assisted solid state synthesis of red-emitting BCNO phosphor and its characteristics. Adv Mater Res 896:464–467. https://doi.org/10.4028/www.scientific.net/AMR.896.464

Olejnik M, Krajnik B, Kowalska D, Lin G, Mackowski S (2013) Spectroscopic studies of plasmon coupling between photosynthetic complexes and metallic quantum dots. J Phys Condens Matter 25(19). https://doi.org/10.1088/0953-8984/25/19/194103

Omary MA, Patterson HH (2016) Luminescence, theory. In: Encyclopedia of spectroscopy and spectrometry. https://doi.org/10.1016/B978-0-12-803224-4.00193-X

Pan D, Zhang J, Li Z, Wu M (2010) Hydrothermal route for cutting graphene sheets into blue-luminescent graphene quantum dots. Adv Mater 22(6). https://doi.org/10.1002/adma.200902825

Pavlov SA, Sherstneva NE, Koryakin SL, Maksimova EY, Makovetskiy VV, Krikushenko VV & Antipov EM (2017) Features of light conversion process with covering materials containing quantum dots and their application in agriculture. Nano Hybrids Compos 13:162–175. https://doi.org/10.4028/www.scientific.net/nhc.13.162

Parrish CH, Hebert D, Jackson A et al (2021) Optimizing spectral quality with quantum dots to enhance crop yield in controlled environments. Commun Biol **4**:124. https://doi.org/10.1038/s42003-020-01646-1

Pereira CF, Viegas IMA, Souza Sobrinha IG, Pereira G, Pereira GAL, Krebs P, Mizaikoff B (2020) Surface-enhanced infrared absorption spectroscopy using silver selenide quantum dots. J Mater Chem C 8(30). https://doi.org/10.1039/d0tc02653k

Piao Z, Yang D, Cui Z, He H, Mei S, Lu H, Fu Z, Wang L, Zhang W, Guo R (2022) Recent advances in metal chalcogenide quantum dots: from material design to biomedical applications. Adv Funct Mater 32(44). https://doi.org/10.1002/adfm.202207662

Qu S, Wang X, Lu Q, Liu X, Wang L, Qu S, Lu Q, Liu X, Wang L, Wang X (2012) A biocompatible fluorescent ink based on water-soluble luminescent carbon nanodots. Angew Chem Int Ed 51(49):12215–12218. https://doi.org/10.1002/ANIE.201206791

Ragab GA, Saad-Allah KM (2020) Green synthesis of sulfur nanoparticles using *Ocimum basilicum* leaves and its prospective effect on manganese-stressed *Helianthus annuus* (L.) seedlings. Ecotoxicol Environ Saf 191. https://doi.org/10.1016/j.ecoenv.2020.110242

Raja PB, Munusamy KR, Perumal V, Ibrahim MNM (2021) Characterization of nanomaterial used in nanobioremediation. In: Nano-bioremediation: fundamentals and applications. https://doi.org/10.1016/B978-0-12-823962-9.00037-4

Rakovich A, Sukhanova A, Bouchonville N, Molinari M, Troyon M, Cohen JHM, Rakovich Y, Donegan JF, Nabiev I (2009) Energy transfer processes in semiconductor quantum dots: bacteriorhodopsin hybrid system. Photon Mater Dev Appl III 7366:736620. https://doi.org/10.1117/12.821731

Rakovich A, Donegan JF, Oleinikov V, Molinari M, Sukhanova A, Nabiev I, Rakovich YP (2014) Linear and nonlinear optical effects induced by energy transfer from semiconductor nanoparticles to photosynthetic biological systems. J Photochem Photobiol C: Photochem Rev 20(1):17–32. https://doi.org/10.1016/j.jphotochemrev.2014.04.001

Ramanery FP, Mansur AAP, Mansur HS (2013) One-step colloidal synthesis of biocompatible water-soluble ZnS quantum dot/chitosan nanoconjugates. Nanoscale Res Lett 8(1). https://doi.org/10.1186/1556-276X-8-512

Ray SC, Saha A, Jana NR, Sarkar R (2009) Fluorescent carbon nanoparticles: synthesis, characterization, and bioimaging application. J Phys Chem C 113(43). https://doi.org/10.1021/jp905912n

Rogach A, Kershaw SV, Burt M, Harrison MT, Kornowski A, Eychmüller A, Weller H (1999) Colloidally prepared HgTe nanocrystals with strong room-temperature infrared luminescence. Adv Mater 11(7):552–555. https://doi.org/10.1002/(sici)1521-4095(199905)11:7<552::aid-adma552>3.0.co;2-q

Shavel A, Gaponik N, Eychmüller A (2004) Efficient UV-blue photoluminescing thiol-stabilized water-soluble alloyed ZnSe(S) nanocrystals. J Phys Chem B 108(19). https://doi.org/10.1021/jp037941t

Sivagami M, Asharani IV (2022) Phyto-mediated Ni/NiO NPs and their catalytic applications—a short review. Inorg Chem Commun 145. https://doi.org/10.1016/j.inoche.2022.110054

Sumanth Kumar D, Jai Kumar B, Mahesh HM (2018) Quantum nanostructures (QDs): an overview. In: Synthesis of inorganic nanomaterials: advances and key technologies. https://doi.org/10.1016/B978-0-08-101975-7.00003-8

Sun YP, Zhou B, Lin Y, Wang W, Fernando KAS, Pathak P, Meziani MJ, Harruff BA, Wang X, Wang H, Luo PG, Yang H, Kose ME, Chen B, Veca LM, Xie SY (2006) Quantum-sized carbon dots for bright and colorful photoluminescence. J Am Chem Soc 128(24):7756–7757. https://doi.org/10.1021/ja062677d

Sun H, Wang M, Lei C, Li R (2021) Cell wall: an important medium regulating the aggregation of quantum dots in maize (*Zea mays* L.) seedlings. J Hazard Mater 403. https://doi.org/10.1016/j.jhazmat.2020.123960

Tang L, Ji R, Li X, Bai G, Liu CP, Hao J, Lin J, Jiang H, Teng KS, Yang Z, Lau SP (2014) Deep ultraviolet to near-infrared emission and photoresponse in layered n-doped graphene quantum dots. ACS Nano 8(6). https://doi.org/10.1021/nn501796r

Tao CL, Ma J, Wei C, Xu D, Xie Z, Jiang Z, Ge F, Zhang H, Xie M, Ye Z, Cheng F, Xu B, Tian Y, Wu XJ (2023) Scalable synthesis of high-quality core/shell quantum dots with suppressed blinking. Adv Opt Mater 11(17). https://doi.org/10.1002/ADOM.202300533

Ünlü C, Budak E, Kestir SM (2022) Altering natural photosynthesis through quantum dots: effect of quantum dots on viability, light harvesting capacity and growth of photosynthetic organisms. Funct Plant Biol 49(6):444–451. https://doi.org/10.1071/FP21136

Wang HC, Bao Z, Tsai HY, Tang AC, Liu RS (2018a) Perovskite quantum dots and their application in light-emitting diodes. Small 14(1). https://doi.org/10.1002/smll.201702433

Wang H, Zhang M, Song Y, Li H, Huang H, Shao M, Liu Y, Kang Z (2018b) Carbon dots promote the growth and photosynthesis of mung bean sprouts. Carbon 136:94–102. https://doi.org/10.1016/j.carbon.2018.04.051

Wang Y, Peng L, Schreier J, Bi Y, Black A, Malla A, Goossens S, Konstantatos G (2024) Silver telluride colloidal quantum dot infrared photodetectors and image sensors. Nat Photonics 18(3). https://doi.org/10.1038/s41566-023-01345-3

Wroblewska-Wolna AM, Harvie AJ, Rowe SF, Critchley K, Butt JN, Jeuken LJC (2020) Quantum dot interactions with and toxicity to *Shewanella oneidensis* MR-1. Nanotechnology 31(13). https://doi.org/10.1088/1361-6528/ab5f78

Wu CC, Shiau CY, Ayele DW, Su WN, Cheng MY, Chiu CY, Hwang BJ (2010) Rapid microwave-enhanced solvothermal process for synthesis of CuInSe 2 particles and its morphologic manipulation. Chem Mater 22(14). https://doi.org/10.1021/cm1006263

Xiao A, Wang C, Chen J, Guo R, Yan Z, Chen J (2016) Carbon and metal quantum dots toxicity on the microalgae *Chlorella pyrenoidosa*. Ecotoxicol Environ Saf 133:211–217. https://doi.org/10.1016/j.ecoenv.2016.07.026

Xu Y, Wang X, Zhang WL, Lv F, Guo S (2018) Recent progress in two-dimensional inorganic quantum dots. Chem Soc Rev 47(2):586–625. https://doi.org/10.1039/c7cs00500h

Xu X, Mao X, Zhuang J, Lei B, Li Y, Li W, Zhang X, Hu C, Fang Y, Liu Y (2020) PVA-coated fluorescent carbon dot nanocapsules as an optical amplifier for enhanced photosynthesis of lettuce. ACS Sustain Chem Eng 8(9):3938–3949. https://doi.org/10.1021/acssuschemeng.9b07706

Xue J, Li H, Liu J, Wang Y, Liu Y, Sun D, Wang W, Huang L, Tang J (2019) Facile synthesis of silver sulfide quantum dots by one pot reverse microemulsion under ambient temperature. Mater Lett 242. https://doi.org/10.1016/j.matlet.2019.01.121

Yan Z, Chen J, Xiao A, Shu J, Chen J (2015) Effects of representative quantum dots on microorganisms and phytoplankton: a comparative study. RSC Adv 5(129):106406–106412. https://doi.org/10.1039/c5ra23730k

Yan K, Liu Y, Yang Q, Liu W, Guo R, Sui J, Yan Z, Chen J (2019) Evaluation of the novel nanoparticle material—CdSe quantum dots on *Chlorella pyrenoidosa* and *Scenedesmus obliquus*: concentration-time-dependent responses. Ecotoxicol Environ Saf 171:728–736. https://doi.org/10.1016/j.ecoenv.2019.01.018

Yang H, Holloway PH (2004) Efficient and photostable ZnS-passivated CdS:Mn luminescent nanocrystals. Adv Funct Mater 14(2). https://doi.org/10.1002/adfm.200305011

Yao K, Lv X, Zheng G, Chen Z, Jiang Y, Zhu X, Wang Z, Cai Z (2018) Effects of carbon quantum dots on aquatic environments: comparison of toxicity to organisms at different trophic levels. Environ Sci Technol 52(24):14445–14451. https://doi.org/10.1021/acs.est.8b04235

Yin L, Zhou J, Li W, Zhang J, Wang L (2019) Yellow fluorescent graphene quantum dots as a phosphor for white tunable light-emitting diodes. RSC Adv 9(16). https://doi.org/10.1039/c8ra10353d

Yoshida K, Oshita N, Kimura T, Goto M, Oikawa R, Uwano M, Asakura S, Chiba T, Masuhara A (2024) Synthetic method of pure-red emissive CsPbI3 under ambient conditions for quantum-dot light-emitting diode application. Appl Phys Express 17(2). https://doi.org/10.35848/1882-0786/ad2025

Yuan T, Meng T, He P, Shi Y, Li Y, Li X, Fan L, Yang S (2019) Carbon quantum dots: an emerging material for optoelectronic applications. J Mater Chem C 7(23). https://doi.org/10.1039/c9tc01730e

Yue J, Li L, Cao L, Zan M, Yang D, Wang Z, Chang Z, Mei Q, Miao P, Dong WF (2019) Two-step hydrothermal preparation of carbon dots for calcium ion detection. ACS Appl Mater Interfaces 11(47). https://doi.org/10.1021/acsami.9b13737

Zhang F, Zhong H, Chen C, Wu XG, Hu X, Huang H, Han J, Zou B, Dong Y (2015) Brightly luminescent and color-tunable colloidal CH3NH3PbX3 (X = Br, I, Cl) quantum dots: potential alternatives for display technology. ACS Nano 9(4). https://doi.org/10.1021/acsnano.5b01154

Zhang M, Hu L, Wang H, Song Y, Liu Y, Li H, Shao M, Huang H, Kang Z (2018) One-step hydrothermal synthesis of chiral carbon dots and their effects on mung bean plant growth. Nanoscale 10(26):12734–12742. https://doi.org/10.1039/c8nr01644e

Zhang M, Wang H, Liu P, Song Y, Huang H, Shao M, Liu Y, Li H, Kang Z (2019) Biotoxicity of degradable carbon dots towards microalgae: *Chlorella vulgaris*. Environ Sci Nano 6(11):3316–3323. https://doi.org/10.1039/c9en00829b

Zheng XT, Ananthanarayanan A, Luo KQ, Chen P (2015) Glowing graphene quantum dots and carbon dots: properties, syntheses, and biological applications. Small 11(14):1620–1636. https://doi.org/10.1002/smll.201402648

Zheng Y, Xie G, Zhang X, Chen Z, Cai Y, Yu W, Liu H, Shan J, Li R, Liu Y, Lei B (2017) Bioimaging application and growth-promoting behavior of carbon dots from pollen on hydroponically cultivated rome lettuce. ACS Omega 2(7):3958–3965. https://doi.org/10.1021/acsomega.7b00657

Zhou S, Liu Z, Wang Y, Lu K, Yang F, Gu M, Xu Y, Chen S, Ling X, Zhang Y, Li F, Yuan J, Ma W (2019) Towards scalable synthesis of high-quality PbS colloidal quantum dots for photovoltaic applications. J Mater Chem C 7(6):1575–1583. https://doi.org/10.1039/C8TC05353G

Zhu X-G, Long SP, Ort DR (2010) Improving photosynthetic efficiency for greater yield. Annu Rev Plant Biol 61(1):235–261. https://doi.org/10.1146/annurev-arplant-042809-112206

Zhu R, Luo Z, Chen H, Dong Y, Wu S-T (2015) Realizing Rec 2020 color gamut with quantum dot displays. Opt Express 23(18). https://doi.org/10.1364/oe.23.023680

Zylstra J, Amey J, Miska NJ, Pang L, Hine CR, Langer J, Doyle RP, Maye MM (2011) A modular phase transfer and ligand exchange protocol for quantum dots. Langmuir 27(8). https://doi.org/10.1021/la104542n

Chapter 8
Molecular Cues of Phytohormone-Mediated Modulation of Plant Abiotic Stress Responses

Yogesh Negi, Shiva Sai Ravva, Shivam Singh, Susithra Priyadarshni Mugunthan, Kundan Kumar, and Penna Suprasanna

Abstract Plants are sessile organisms and are often challenged with multitude of stress conditions, of both natural and anthropogenic origin. Abiotic stress conditions such as drought, extreme temperature regimes, and salinity pose a serious threat to crop production and are the primary cause of crop loss and reduction in yield across the globe. Phytohormones are key players involved in various stages of plant growth and development and also confer tolerance to a wide array of abiotic stressors. Therefore, understanding the molecular networks in conjunction with phytohormones would offer great advantages toward divulging the complex signaling networks that underlie improvement and development of stress-resilient crop plants. Further, cross-talks between different plant hormones plays critical roles and helps plants to respond to abiotic stress. The elucidation of molecular mechanisms that regulate plant hormone biosynthesis, signaling, and functions is crucial for fine-tuning the regulation of the hormonal biosynthesis pathways and for the generation of genetically modified crop plants with improved stress tolerance responses toward multiple abiotic stress conditions. Advances in genomics-accelerated breeding and gene editing techniques offer promise to speed up improvement of traits of stress tolerance. In this chapter, we have summarized different abiotic stress factors that severely affect growth and development of plants and involvement of different class of phytohormones that play imperative roles in delivering tolerance responses to counteract abiotic stress conditions.

Keywords Climate change · Plant abiotic stress · Phytohormones · Auxins · Cytokinins · Cross-talk · Abscisic acid · Brassinosteroids · Gibberellins

Y. Negi · S. S. Ravva · S. Singh · K. Kumar (✉)
Department of Biological Sciences, Birla Institute of Technology & Science Pilani, K. K. Birla Goa Campus, Sancoale, Goa, India

S. P. Mugunthan · P. Suprasanna (✉)
Amity Institute of Biotechnology, Amity University of Maharashtra (AUM), Mumbai, Maharashtra, India

A. Singh et al. (eds.), *Sustainable Crop Production*, Sustainability Sciences in Asia and Africa, https://doi.org/10.1007/978-981-95-7844-3_8

Editor's Notes

Crops grow and survive in complex field environments, sensing and responding to multiple kinds of environmental stresses. Stress response, being an adaptive trait, manifests as both phenotypic modifications and metabolic shifts. While developmental regulatory genes confer resilience directly by establishment of suitable phenotypes, the role of phytohormones is currently being unveiled in influencing morphogenetic changes in response to stress. A deep understanding of the role of phytohormones in ameliorating abiotic stress needs to be explored at molecular level to realize the goal of sustainable crop production.

The chapter by Negi et al. summarizes the detrimental effects of abiotic stresses on plant growth and development and phytohormone-mediated stress mitigation. Multiple abiotic stresses are a major cause of worldwide yield losses across crops. The complex and quantitative nature of stress perception and response have rendered conventional breeding-based approaches ineffective. The authors draw readers' attention to the role played by phytohormones in ameliorating stresses by regulating development for conferring resilience. A detailed account of molecular regulators participating in abiotic stress response mediated by auxin, cytokinin, ABA, SA, brassinosteroids, and GA is provided for researchers, scientists, and breeders. A greater understanding and dissection of the phytohormone signaling, and cross-talk with stress-related pathways in crops is, however, required. This is an opportune moment for stakeholders to investigate and identify key genes and phytohormones for engineering abiotic stress tolerance.

Actionable Points

1. *Molecular breeding technologies*: Molecular breeding offers the scope of accelerating the selection of stress-tolerant crops with increased precision, and phytohormone interactions involving ABA-ethylene and ABA-cytokinin relative abundances represent targets for yield resilience under a wide range of stress conditions.
2. *Marker-Assisted Selection (MAS) for abiotic stress tolerance*: Segment RM 201 on rice chromosome 9, QLt.ksu.1D, QFv/fm.ksu-3B, QChl.ksu-3B, QGyp.ksu-4A in wheat, Root-ABA1 related to root development along with abscisic acid levels in maize can be implicated in developing an efficient strategy of MAS for drought tolerance. Also, direct selection for yield under drought conditions, along with a selection of secondary traits like root thickness, depth, and penetration ability, will fasten the development of cultivars with improved performance under water deficit environments.
3. *Gene pyramiding*: Pyramiding of genes involved in hormonal cross-talks offers tremendous potential for improvement of crop yield.
4. *CRISPR-Cas9 editing*: Breeders have successfully used knock-out strategy to generate pyl1/4/6 triple mutants of OsPYL abscisic acid receptor, which conferred increased tolerance to heat stress and improved yield. The employment of CRISPR technology to modify stress-related genes is essential to precise plant breeding in agriculture."

Chapter Snippets

- Phytohormones regulate various stages of plant growth and development.
- Adaptation to stress conditions involves complex interplay of phytohormonal networks.
- Cross talk between different phytohormones occurs to deliver effective tolerance response under abiotic stress conditions.
- Unravelling the phytohormonal pathways and their involvement in abiotic stress conditions is the key toward development of stress-resilient crops.

Hormonal Regulation in Plants

Plant hormones serve as critical signaling molecules in regulating growth, development, and stress responses. Among these, gibberellins and cytokinins are pivotal in modulating plant adaptations to abiotic stresses. GAs primarily influence cell elongation and growth, while CKs mediate cell division, senescence delay, and stress tolerance. The interplay of these hormones is governed by complex genetic networks, including GA-responsive DELLA proteins and CK signaling components like ARR transcription factors. Understanding these pathways can shed light on how plants coordinate hormonal responses under adverse conditions.

Genetic Basis of Hormonal Interplay

The genetic regulation of GA and CK pathways involves several key genes. In the gibberellin pathway, genes such as *GA20ox* and *GA3ox* encode enzymes responsible for GA biosynthesis, while DELLA proteins act as growth repressors modulated by GAs. For cytokinins, *IPT* genes are crucial for CK biosynthesis, and *ARR* genes mediate CK signal transduction. Studies show that DELLA proteins interact with ARR transcription factors, integrating GA and CK signaling pathways to fine-tune plant responses to stress.

Hormonal Regulation of Gene Expression

GAs and CKs influence the expression of stress-responsive genes, such as *RD29A* and *DREB2A*, through direct or indirect pathways. DELLA proteins suppress GA-induced transcriptional changes by interacting with transcription factors like PIF4. Similarly, CKs regulate gene expression through type B ARR proteins that act as transcriptional activators. Emerging evidence suggests that stress-induced hormonal shifts modulate the expression of genes involved in ROS detoxification, including SOD, CAT, and APX, reinforcing the antioxidant capacity of plants under stress.

Molecular Mechanisms of GA-CK Interplay

The molecular basis of GA-CK interplay lies in their shared regulation of downstream targets, including cell cycle regulators and stress-responsive genes. DELLA proteins act as molecular hubs, physically interacting with CK signaling components to modulate their activity. For instance, DELLAs suppress GA-mediated growth by stabilizing type B ARRs, which are required for CK signaling. This balance ensures that resources are allocated efficiently between growth and stress adaptation, particularly under resource-limiting conditions.

Hormonal Cross-Talk Under Abiotic Stress
Abiotic stresses such as drought, salinity, and extreme temperatures disrupt homeostasis, triggering hormonal cross-talk to maintain growth and survival. Under salinity stress, the upregulation of *GA2ox* genes reduces bioactive GA levels, restricting growth and conserving energy. Concurrently, CK-responsive genes such as *ARR5* and *ARR7* are activated to enhance antioxidant defenses and modulate stomatal behavior. The interaction between DELLA genes and ARR highlights a synergistic mechanism, wherein CKs mitigate oxidative damage while GAs promotes recovery post-stress.

Applications in Crop Improvement
Harnessing the genetic and hormonal interplay between GAs and CKs offers promising avenues for crop improvement. Genetic engineering approaches, such as overexpressing *GA20ox* or *IPT* genes, have shown potential in enhancing growth and stress resilience. Additionally, manipulating DELLA-ARR interactions could optimize hormonal responses, enabling crops to better withstand combined stresses like drought and salinity. These insights are particularly relevant for developing sustainable solutions in agriculture.

8.1 Introduction

Climate change has become a serious threat to food security, and consequent to the surge in different biotic and abiotic stresses, crop productivity has shown a decline in many regions of the world demanding immediate measures for sustainable agriculture. It is estimated that in the coming years, water deficits and drought will be on the rise) with repeated occurrences and magnitude of drought impact (Eckardt et al. 2023). Abiotic stresses have shown a negative impact on crop productivity with estimated crop yields dropping significantly, and, in most of the cultivated crop plants, an estimated 42% yield loss has been observed. Abiotic stresses including drought, salinity, heat, cold, and heavy metals affect plant growth and development through several molecular, cellular, physiological, and morphological processes (Baidya et al. 2022; Shakespear et al. 2024).

Development of plant varieties with a high level of tolerance to abiotic stresses is crucial for establishing full yield potential and to stabilize production under environmental extremities (Purohit et al. 2026). Due to multitude of abiotic stresses and their complex genetic control, progress of breeding for tolerance to abiotic stresses using conventional approaches has not been very rewarding, and hence, there is a continued development of new techniques for mitigation (Suprasanna 2020; Elsisi et al. 2024). Abiotic stresses lead to dehydration or osmotic stress through reduced availability of water for vital cellular functions and maintenance of turgor pressure. Stomatal closure, reduced supply of CO_2, and slower rate of biochemical reactions during prolonged periods of dehydration, high light intensity, and high and low temperatures lead to high production of reactive oxygen species (ROS) causing cellular damage (Gu et al. 2022). In response to stress, a series of compatible osmolytes is accumulated for osmotic adjustment, water retention, and free radical scavenging. Similarly, upregulation of certain enzymes such as superoxide dismutase, ascorbate peroxidase, and glutathione reductase has been implicated in free radical detoxification and scavenging under oxidative stress. Plants have evolved mechanisms for avoidance and/or tolerance to one or more of the abiotic stresses.

Further understanding of the molecular mechanisms of stress perception, signal transduction, and response by plants and other organisms may help to engineer plants with high levels of tolerance to multiple stresses. The intervention of omics-based approaches such as genomics, transcriptomics, miRNAomics, proteomics, metabolomics, phenomics, and pangenomics has been established for molecular breeding and engineering of crop plants for trait improvement. Further, there have been multitude of other genomics-based techniques for unravelling genetic variation for developing superior cultivars (Sinha et al. 2021). Several transcription factors regulating water channels, proteases, key enzymes such as proline, sucrose, osmolytes, chaperons, detoxification of ROS, and regulatory genes such as abscisic acid (ABA) and ethylene regulations, phospholipid metabolism, protein kinases, and phosphatase genes have been functionally characterized uncovering different regulatory factors affecting traits associated with abiotic stress tolerance responses and which could be targeted to manipulate stress tolerance (Kumar et al. 2024; Singh et al. 2024).

Soil salinity is one of the primary environmental factors that restrict the yield of horticultural crops. For an expanding population, soil salinity is a primary focus of food security research (Mir and Somasundaram 2021). Over 20% of all farmed fields and 6% of the area that is properly irrigated are affected by salinity (Pirasteh-Anosheh et al. 2016). An imbalance in cellular ionic and osmotic balances has led to the loss of plant yield caused by salt stress (Khan et al. 2012a, b). The consequences of salinity stress include increased ion toxicity, osmotic stress, nutrient-acquisition, and homeostasis/deficiency, impaired stomatal conductance, increased cell turgor loss, decreased reduction in leaf water potential, altered physiological/biochemical processes, and elevated ROS-caused oxidative stress (Munns and Tester 2008; Khan et al. 2014). Significant water evaporation, brought on by rising temperatures, fuels global warming and, consequently, drought stress. Horticultural crops' growth and productivity are significantly hampered by drought stress (Toscano et al. 2019; Hanaka et al. 2021). The growth of plants and food security are seriously threatened by global warming. Heat stress results in sterility, decreased yield, delayed development, and disruption of plant cellular homeostasis (Saini et al. 2021). Heat stress negatively impacts the growth status of water *Ipomoea aquatica*, *Raphanus sativus*, *Cucumis sativus*, and *Lactuca sativa* (Zhang et al. 2013; Guo et al. 2020; Nijabat et al. 2020). Cold stress is one of the key environmental factors that reduces crop productivity. Cold stress poses a significant danger to the sustainability of crop production. Plants under cold stress undergo physiological and metabolic changes that ultimately lead to plant mortality (Atayee and Noori 2020). Environmental systems and sustainable agriculture are now very concerned about metal-/metalloid-induced stress. Metal contaminants can enter agricultural soils from various sources, such as residential rubbish dumps, sewage waste disposal, urban runoff, burning liquid and solid fuel, industrial effluents, and agricultural hazardous chemicals. Growth delay, leaf chlorosis, dying, and cell death are some obvious harmful signs that heavy metals can induce. They can also be readily absorbed by plant roots and transferred into shoots.

Plants evoke different plant hormones as chemical messengers for triggering plant defense mechanisms to survive under stress conditions (Raza et al. 2019; Malhi et al. 2021; Imran et al. 2021). Phytohormones promote plant growth and enable plants to perceive and respond to unfavorable environmental stimuli. Plant hormones are known to modulate physiological and molecular responses either at their site of production or transported to regulate stress responses elsewhere in plants (Arif et al. 2020; Kundu et al. 2022). Cytokinin (CKs), abscisic acid (ABA), ethylene (ET), jasmonic acid (JA), and salicylic acid (SA) have been documented as critical players for the regulation of abiotic stress responses (Tian and Li 2018; Zhao et al. 2021a; Mukherjee et al. 2022; EL Sabagh et al. 2022). In this review, we have presented an account of the involvement of phytohormones in abiotic stress responses and the molecular regulation of different stress-related pathways under the regulation of different phytohormones. Currently, several of the candidate genes associated with phytohormone biosynthesis are being used for improving tolerance to abiotic stress in crop plants.

8.2 Phytohormones as Crucial Regulators of Plant Growth and Defense

Phytohormones represent crucial regulators of plant growth and development and primary mediators of environmental stress responses (Sah et al. 2016). The phytohormone class first identified, auxin, represents low molecular weight compounds that exerts growth inducing effects. Auxin is synthesized in tissues where cells maintain high cell division activity and is known to be transported through specific transmembrane proteins (Gomes and Scortecci 2021). For instance, the early root development and growth is primarily regulated by auxin synthesized in shoot tips and young leaves and transported to root tips through the phloem, while in the later part, root growth is principally dependent on local auxin synthesis in roots (Yu et al. 2022). Even when present at trace levels, auxins potentially regulate the expression of genes through modulation of specific transcription factors and proteins and respond to environmental cues. These molecules are involved in several biological processes such as response to environmental stimuli, including cell signaling, cell cycle regulation, endocytosis, embryogenesis, organogenesis, and growth regulation (Saini et al. 2013; Korver et al. 2018; Gomes and Scortecci 2021). Auxin also regulates plant responses to environmental stresses by modulating processes such as biosynthesis, transport, signaling, conjugation, and degradation (Potters et al. 2007; Saini et al. 2017). Recognized as a key regulator among phytohormones, auxin governs all aspects of root development under both normal and stress conditions (Bücker-Neto et al. 2017; Saini et al. 2013). Auxin is crucial for plant responses to drought stress, with its biosynthesis, transport, and conjugation with amino acids or sugars being regulated during abiotic stress conditions (Shibasaki et al. 2009). Auxin can positively influence the activity of four antioxidant enzymes such as superoxide dismutase, catalase, peroxidase, and glutathione reductase under drought stress, facilitating effective detoxification of reactive oxygen species (ROS) and enhancing drought resistance. Auxin plays a positive role in enhancing root biomass and branching, potentially improving water uptake efficiency and contributing to drought stress resistance (Shi et al. 2014). The phytohormone abscisic acid (ABA) regulates various cellular processes such as regulation of seed development, dormancy, germination, and modulation of root architecture (Vishwakarma et al. 2017). Abscisic acid is also the most sought phytohormone known to be involved in plant's adaptive responses to a wide variety of abiotic stress, particularly salinity and drought (Wani and Kumar 2015). ABA levels are known to significantly increase under drought and salinity stress consequently bringing about stomatal closure, changes in expression of genes responsible for accumulation of compatible osmolytes, and synthesis of Late Embryogenesis Abundant (LEA) proteins, dehydrins, and other protective proteins beside antioxidant enzymes (Wani and Kumar 2015; Sah et al. 2016). The transcription factors (TFs) DREB2A/2B, AREB1, RD22BP1, and MYC/MYB regulate the ABA-responsive gene expression by interacting with their corresponding cis-acting elements, viz., DRE/CRT(A/GCCGAC), ABRE (PyACGTGGC), and MYCRS(CANNTG)/MYBRS (C/TAACNA/G), respectively. These cis-acting elements play an important role in inducing ABA-dependent gene expression in plants under stress conditions (Dar et al. 2017). Salicylic acid is an important endogenous growth regulator and a potent signaling molecule that plays important roles in plant tolerance responses to abiotic stressors such as cold, heat, drought, salinity, and heavy metals. SA is also known to be involved in developmental signaling pathways vital for normal plant growth, such as flowering and senescence. Studies have shown that SA can enhance tolerance to salt stress by regulating ionic homeostasis, maintenance of cell membrane integrity, and modulation of antioxidative enzyme system to efficiently counteract salinity stress. The phytohormone had also been reported to confer drought tolerance in plants by improving water-use efficiency, modulating root growth, membrane stabilization, and mitigation of oxidative stress (Munne-Bosch and Penuelas 2003; Emamverdian et al. 2020; Mohamed et al. 2020; Santisree et al. 2020; Prakash et al. 2021; Raza et al. 2021; Song et al. 2023). SA also finds its tremendous scope in horticulture industry as an important biostimulator (Chen et al. 2023). Brassinosteroids (BRs) are active plant steroid hormones, initially discovered in rapeseed pollen, and represent the sixth major category of naturally occurring plant hormones. They play crucial roles in growth and development by regulating processes like cell division and expansion. Brassinosteroids (BRs) are present in small quantities across all plant organs, with concentrations ranging from 1 to 100 ng g^{-1} of fresh weight in young tissues. In contrast, mature tissues exhibit significantly lower levels, ranging from 0.01 to 0.1 ng g^{-1} (Hayat et al. 2003). These hormones influence various metabolic processes, enhancing tolerance to biotic and abiotic stresses, promoting cell expansion and elongation, facilitating redox signaling, and supporting numerous other physiological functions (Silva et al. 2024). BR-related compounds have been identified, with 24-epibrassinolide (24-EBR) and 28-homobrassinolide (28-HBR) being the most active compounds, crucial for BR function and signaling pathways (Khripach et al. 2000; Nomura and Bishop 2006). BRs enhance plant resilience against various biotic and abiotic stresses, including heavy metals, salinity, drought, pathogen attacks, and extreme temperatures (Vardhini et al. 2010; Chen et al. 2021). In most plant cells, BRs bind to the membrane receptor-kinase BR-INSENSITIVE 1 (BRI1) and promotes growth and stress responses (Gupta et al. 2023). Additionally, BRs play a crucial role in regulating ion uptake (Khripach et al. 2000) and have been found to reduce the accumulation of heavy metals and radioactive elements (Khripach et al. 1995). Studies confirm that BRs enhance abiotic stress tolerance in plants. For example, BR application increased salinity tolerance in rice, tomato, and chickpea. Additionally, BR-induced stress tolerance in *Arabidopsis* was associated with the ABA, SA, and ethylene pathways (Prakash et al. 1999; Anuradha and Rao 2001; Ali et al. 2007; Divi and Krishna 2010). BRs also function synergistically with GA and IAA to promote hypocotyl elongation in *Arabidopsis* (Tanaka et al. 2003). The summary of the phytohormonal functions in abiotic stress responses has been summarized in Table 8.1.

8.3 Overview of Major Classes of Phytohormones

8.3.1 Auxins

Natural auxins are a heterogenous group of aromatic carboxylic acids (Batista-Silva et al. 2024). Auxin biosynthesis, primarily of indole-3-acetic acid (IAA), which is a major auxin in plants occurs through both tryptophan-dependent and tryptophan-independent mechanisms. The tryptophan-dependent pathway involves four routes: the indole-3-acetamide (IAM) pathway, indole-3-pyruvic acid (IPA) pathway, tryptamine (TAM) pathway, and indole-3-acetaldoxime (IAOX) pathway (Mano and Nemoto 2012). Due to structural similarity between IAA and tryptophan (Trp), tryptophan has been widely regarded as the precursor of IAA. Strong evidence confirms that in *Arabidopsis*, IAA is primarily synthesized from tryptophan via the tryptophan-dependent pathway (Cohen et al. 2003). A complete two-step auxin biosynthesis pathway, where tryptophan is converted into IAA in plants, was identified (Mashiguchi et al. 2011; Won et al. 2011). In this pathway, TRYPTOPHAN AMINOTRANSFERASE OF *ARABIDOPSIS* 1/WEAK ETHYLENE INSENSITIVE 8/SHADE AVOIDANCE 3/CYTOKININ INDUCED ROOT CURLING 1 (TAA1/SAV3) first converts tryptophan into indole-3-pyruvate (IPA). Then, IPA is converted into IAA by the YUCCA (YUC) family of flavin monooxygenases (Mashiguchi et al. 2011; Zhao 2012). Auxin concentration and distribution

Table 8.1 Effect of gene or exogenous phytohormones in mitigating abiotic stress in plants

Hormone	Stress	Plant species	Gene(s)/ treatment	Key findings	References
Auxin	Drought stress	*Arabidopsis thaliana*	*YUCCA7*	Increased endogenous auxin levels	Lee and Luan (2012)
	Drought stress	*Glycine max*	*GmARF33,50*	Responsive to water deficit by increasing auxin levels	Van Ha et al. (2013)
	Drought stress	*Sorghum bicolor*	*SbARF10,16,21*	Induced in leaf and root tissues, expressing auxin content	Wang et al. (2010)
	Drought stress	*Oryza sativa*	*OsPIN3t*	*OsPIN3t* overexpression in rice enhanced drought tolerance by increasing auxin levels	Zhang et al. (2012)
	Salinity stress	*Arabidopsis thaliana*	*IAA30*	Regulating auxin levels aids in maintaining ion balance during salt stress, as plants lower auxin concentration and signaling to adapt to salinity conditions	Jung and Park (2011)
	Salinity stress	*Sorghum bicolor*	*SbGH3*	Maintenance of auxin levels	Wang et al. (2010)
	Heat stress	*Arabidopsis thaliana*	*YUCCA 8; YUCCA 9*	Upregulated the expression of auxins	Hentrich et al. (2013)
	Heat stress	*Sorghum bicolor*	*SbARF17; SbARF24*	Increases auxin accumulation	Chen et al. (2019)
	Cold stress	*Arabidopsis thaliana*	*PIN2*	Extreme temperature stress decreases auxin levels by converting active IAA into its inactive conjugated form. However, applying exogenous auxin may mitigate plant damage caused by such stress	Shibasaki et al. (2009)
	Cold stress	*Oryza sativa*	*OsGH3–2*	Reduces free IAA, carotene, and ABA levels, enhancing cold tolerance by modulating IAA and ABA homeostasis	Du et al. (2012)
	Heavy metal stress	*Arabidopsis thaliana*	*YUCCA6*	Auxin accumulation and enhanced lateral and adventitious root density	Fattorini et al. (2017)
	Heavy metal stress	*Oryza sativa*	*OsAUX1*	Enhanced root development in response to cadmium stress	Yu et al. (2015)
	Heavy metal stress	*Oryza sativa*	Antioxidative-related genes	Increased Fe plague, which improved Mn, Cu, and Zn uptake in roots and restricted Cd and As root uptake and translocation	Xu et al. (2021)

(continued)

Table 8.1 (continued)

Hormone	Stress	Plant species	Gene(s)/ treatment	Key findings	References
Abscisic acid	Salinity stress	*Oryza sativa*	Exogenous ABA treatment	ABA treatment improved antioxidant enzyme levels and significantly reduced the Na^+ content of two varieties as well as increased the K^+, Mg^{2+}, and Ca^{2+} content in leaves and roots	Chen et al. (2022)
	Stimulated acid rain stress	*Oryza sativa*	Exogenous ABA treatment	Alleviated the SAR-induced inhibition of root growth by balancing endogenous phytohormones, promoted nutrient uptake, increased H^+-ATPase activity, and up-regulated expression levels of genes (*OSA2*, *OSA4*, *OSA9*, and *OSA10*)	Liu et al. (2018)
	Alkaline stress	*Oryza sativa*	Exogenous ABA treatment	Enhanced tolerance to alkaline stress by upregulation of the antioxidant defense system and stress tolerance-related genes	Liu et al. (2019)
	Salinity stress	*Oryza sativa*	Exogenous ABA treatment	Significant decrease in Na^+ content with higher accumulation of K^+ and Ca^{2+} in flag leaves	Gurmani et al. (2006)
	Salinity stress	*Oryza sativa*	Exogenous ABA treatment	Ionic homeostasis, increased proline accumulation, and higher grain yield	Gurmani et al. (2011)
	Salinity stress	*Oryza sativa*	Exogenous ABA treatment	Reduced Na^+ accumulation in shoots and lower Na^+/K^+ ratios	Gurmani et al. (2013)
	Heat stress	*Oryza sativa*	Exogenous ABA treatment	Increased vitality, significant increase in xylem sap exudation volume, and increased water uptake to cope with abiotic stresses	Chen et al. (2006)
	Drought stress	*Oryza sativa*	Exogenous ABA treatment	Improved shoot and root growth, physiological and biochemical contents	Ramachandran et al. (2021)
	Drought stress	*Arabidopsis thaliana*	*NCED3*	Overexpression of AtNCED3 in transgenic *Arabidopsis* caused an increase in endogenous ABA level and promoted transcription of drought- and ABA-inducible genes	Iuchi et al. (2001)
	Early drought stress	*Arabidopsis thaliana*	*NGTHA1* (*NGA1*)	Transcription factor, NGTHA1 (NGA1), positively regulates ABA accumulation during dehydration stress by activating the NCED3 gene	Sato et al. (2018)

	Water stress	*Arachis hypogaea L.* and *Arabidopsis thaliana*	*AhNCED1*	Ectopic expression of *AhNCED1* in 129B08/nced3 mutant *Arabidopsis* (with impaired AtNCED3 gene involved in ABA biosynthesis under water stress) driven by the AtNCED3 promoter restores its ability to accumulate ABA during drought stress and reverts its hypersensitivity to nonionic osmotic stress and soil drought	Wan and Li (2006)
	Salinity and water stress	*Oryza sativa* and *Arabidopsis thaliana*	*OsNCED5*	OsNCED5 could regulate plant endogenous ABA accumulation to confer high salt and water-stress tolerance and promote leaf senescence	Huang et al. (2019)
	Drought stress	*Zea mays* and *Arabidopsis thaliana*	*ZmPP2C15*	Overexpressing ZmPP2C15 in *Arabidopsis* under drought stress enhanced growth and survival compared with WT plants	Pang et al. (2024)
Brassinosteroids	Drought stress	*BZR1* and *BES1*	*Arabidopsis thaliana*	BZR1 and BES1 inhibited RD26 expression and drought tolerance	Ye et al. (2017)
	Drought stress	*CYP85A1*	*Nicotiana tabacum L*	Overexpression of the *CYP85A1* enhanced drought resistance	Duan et al. (2017)
	Drought stress	*GhBEE3*	*Gossypium hirsutum L.*	GhBZR1 inhibited GhBEE3-like gene expression and improved drought resistance	Chen et al. (2022)
	Drought stress	*TaBZR2* and *TaBES1*	*Triticum aestivum L*	Overexpression of *TaBZR2* and *TaBES1* increased drought resistance	Cui et al. (2019)
	Salinity stress	*BIN2*	*Arabidopsis thaliana*	The BIN2 negatively regulated salt stress	Zhu et al. (2023)
	Salinity stress	*StDWF4*	*Solanum tuberosum*	Overexpression of the *StDWF4* enhanced salt tolerance	Zhou et al. (2018)
	Heat stress	*TaBRI1*	*Triticum aestivum L*	The *TaBRI1* knockdown mutant reduced tolerance to high-temperature stress	Ibañez et al. (2018)
	Heat stress	*BES1* and *HSFA1*	*Arabidopsis thaliana*	BES1 and HSFA1 interacted to enhance heat resistance	Jacob et al. (2017)

(continued)

Table 8.1 (continued)

Hormone	Stress	Plant species	Gene(s)/ treatment	Key findings	References
	Cold stress	*CBF*	*Arabidopsis thaliana*	Overexpression improves cold tolerance	Ye et al. (2019)
	Cold stress	*DFW4*	*Arabidopsis*	Enhance freezing tolerance	Divi and Krishna (2010)
	Cold stress	*DWRF*	*Solanum lycopersicum L.*	DWRF enhanced the expression of GRX to improve cold tolerance	Xia et al. (2018)
	Heavy metal stress	Exogenous application of 24-epiBL	*Brassica juncea*	Enhanced the activity of antioxidant enzymes and ameliorated Ni stress	Kanwar et al. (2013)
	Heavy metal stress	*APX08*, *CATa*, *CATb*, *APX02*, *SOD-Fe2*, and *SOD-Cu-Zn*	*Vigna unguiculata*	Enhanced antioxidative defense-related genes expression to mitigate ROS species to ameliorate Cd-induced toxicity	Santos et al. (2018)
	Heavy metal stress	Antioxidative defense-related genes	*Oryza sativa* L.	Decreased ROS and increased carboxylation, CAT, SOD, and POD activities	Tadaiesky et al. (2021)
	Heavy metal stress	*BRZ1*	*Arabidopsis thaliana*	DELLA protein interacted with the BRZ1 to enhance plant resistance to heavy metals	Achard et al. (2008)

can be influenced by alterations in auxin transport in response to abiotic stresses (Shibasaki et al. 2009). Auxin, via the ARF gene family, activates stress-specific auxin response genes to alleviate the adverse effects of abiotic stresses in plants.

8.3.2 *Abscisic Acid (ABA)*

Abscisic acid (ABA) is a pivotal phytohormone which effectively ameliorates the adverse effects of abiotic stress in plants. The hormone is a sesquiterpene (C_{15}) with the biosynthesis occurring in chloroplast or plastids from a C_{40} carotenoid precursor derived through the 2-C-methyl-d-erythritol-4-phosphate (MEP) pathway (Dar et al. 2017). The distribution and transport of abscisic acid (ABA) are facilitated by various influx and efflux carrier proteins. The perception of this phytohormone occurs through ABA-binding receptor proteins that are present in the plasma membrane and cytosol (Lee and Luan 2012). The PYRABACTIN RESISTANCE 1 (PYR1)/PYR1-LIKE (PYL) protein family interacts with clade A type 2C protein phosphatases (PP2Cs). This interaction activates sucrose nonfermenting 1-related protein kinase 2 (SnRK2) (Cutler et al. 2010; Raghavendra et al. 2010). Through autophosphorylation, SnRK2s trigger downstream signaling pathways involving transcription factors such as NAC, basic leucine zipper (bZIP), ethylene-responsive factor/APETALA 2, and MYB (Antoni et al. 2013; Zhu 2016). In rice, the core ABA signaling pathway consists of 13 PYLs, 10 PP2Cs, and 10 SnRK2s, including stress/ABA-activated protein kinases (SAPKs) (Miao et al. 2018; Singh et al. 2015; Kobayashi et al. 2004). For instance, the rice ABA receptor OsPYL interacts with the PP2C family member OsPP2C30, thereby releasing the kinase OsSAPK2 to establish a fundamental ABA signaling cascade (Kim et al. 2012).

8.3.3 *Salicylic Acid (SA)*

Salicylic acid is a seven-carbon (C) containing, naturally occurring phenolic compound and endogenously synthesized signaling molecule in plants. Two different pathways, the phenylalanine ammonia-lyase (PAL) pathway and the isochorismate (ICS) pathway, are used by plants to biosynthesize salicylic acid (SA). The PAL route for SA biosynthesis occurs in the cytosol, and the ICS pathway occurs in the plastids. The shikimate pathway's end product, chorismate, is the primary metabolite necessary for both pathways (Hao et al. 2018; Mishra and Baek 2021; Zhang et al. 2021). Chorismite is produced from phosphoenolpyruvate (PEP) and erythrose-4-phosphate (E4P) in plastids specifically in the chloroplast via the seven-step shikimate pathway enzymatic activity (Ding and Ding 2020). While the PAL route is more specific in many other trees and crops, including rice, the ICS pathway is specific to the genome of the Brassicaceae family, including *Arabidopsis* (Yuan et al. 2009; Holland et al. 2019; Torrens-Spence et al. 2019; Wu et al. 2023). In

Glycine max, PAL and ICS pathway for SA biosynthesis have been reported (Lefevere et al. 2020).

8.3.4 Brassinosteroids (BRs)

Brassinosteroids (BRs) have been identified and are categorized based on the number of carbon atoms in their structure (Vardhini 2014). The brassinosteroid (BR) biosynthetic pathway begins with campesterol (CR). Based on whether campestanol (CN) is formed during the synthesis process, the pathway is classified into CN-dependent and CN-independent pathways (Wang et al. 2021; Zhao and Li 2012). The CN-dependent pathway includes two branches: the early and late C-6 oxidation pathways. In the early C-6 oxidation pathway, CN is oxidized at C-6 to form 6-oxocampestanol (6-oxoCN), hydroxylated at C-22 to produce cathasterone (CT), and converted through several steps into castasterone (CS) and then brassinolide (BL). In the late C-6 pathway, CN is first hydroxylated at C-22 to form 6-deoxocathasterone (6-deoxoCT) and follows similar steps, completing C-6 oxidation in the final stage to form CS and BL. The CN-independent pathway skips CN, converting CR directly to BL in eight catalytic steps involving enzymes like DWF4, CPD, DET2, ROT3/CYP90D1, and CYP85A1/2(Kim et al. 2005; Zhao et al. 2012). The identification of *Brassinosteroid Insensitive 1 (BRI1)* gene, supported by molecular and biochemical research, revealed that *BRI1* encodes a leucine-rich receptor kinase expressed throughout the plant. By phosphorylating serine and threonine residues, BRI1 plays a pivotal role in BR signaling (Friedrichsen et al. 2000). This receptor recognizes extracellular BRs and initiates an intracellular signaling cascade (Manghwar et al. 2022).

8.4 Mechanisms of Hormonal Action in Abiotic Stress Response

8.4.1 Role of Auxins in Abiotic Stress Response

Many auxin-responsive genes are regulated during both development and abiotic stress, highlighting their role in linking growth and stress responses (Jain and Khurana 2009). Evidence suggests that auxin transporters play a direct or indirect role in abiotic stress adaptation and tolerance, likely through the regulation of auxin transport and redistribution (Zhang et al. 2012). Salt stress in *Arabidopsis* triggers the expression of IAA biosynthesis genes, such as *nitrilase 1* and *nitrilase 2*, indicating a potential increase in IAA levels in response to salinity (Fang and Yang 2002). Additionally, salinity-induced expression of the *NAC* transcription factor *AtNAC2* influences lateral root (LR) development under salt stress via auxin signaling pathways (He et al. 2005). Furthermore, Jung and Park (2011) identified that the membrane-bound NAC transcription factor NTM2 mediates the interaction between auxin signaling and salt stress through regulation of the *IAA30* gene during seed germination in *Arabidopsis*. The activation of *YUCCA7* gene, involved in tryptophan-dependent auxin biosynthetic pathway in *yuc7-1D* mutant, led to increased endogenous auxin levels and improved drought resistance in *Arabidopsis* (Lee et al. 2012). In rice, drought stress led to a 72% reduction in IAA levels after 3 days (Du et al. 2013a). The rice gene *OsPIN3t* responded to drought stress, and its overexpression in rice enhanced drought tolerance (Zhang et al. 2012). Overexpression of *Aux/IAA6* in rice also showed enhanced drought tolerance (Jung et al. 2015). In *Sorghum bicolor*, the genes *SbARF10*, *SbARF16*, and *SbARF21* were significantly induced in leaf and root tissues under drought conditions (Wang et al. 2010). Expression analysis of the *GH3* genes in *Sorghum bicolor* shows that *SbGH3* is expressed at low levels under normal conditions, due to induction of salt stress (Wang et al. 2010). Similarly, in *Glycine max*, only *GmARF33* and *GmARF50* were found to be responsive to water deficit, making them promising candidates for drought stress response (Van Ha et al. 2013).

Morris (1979) demonstrated that temperature influences the rate of exogenous auxin transport in plants. It was found that high and low day/night temperatures (35/24 °C and 10/5 °C) reduced fruit set, pollen grain viability, and IAA levels in tomato (El-Abd et al. 1984). In *Sorghum bicolor*, high temperatures will be significantly upregulated by the expression of most *SbARF* genes, with *SbARF17* and *SbARF24* showing strong expression in vascular tissues (Chen et al. 2019). These findings highlight auxin's critical role in plant resistance to heat stress and thermal signal transduction. Exposure to mild high temperatures (28–29 °C) in *Arabidopsis* rapidly increases auxin accumulation and induces *YUCCA* (flavin-containing monooxygenase) gene expression (*YUCCA 8 and YUCCA 9*) (Hentrich et al. 2013). PIF4 regulates the expression of auxin biosynthetic genes and endogenous auxin levels, facilitating stem elongation under high temperatures (Franklin et al. 2011). Additionally, a family of *SMALL AUXIN UP RNA (SAUR)* genes is expressed in a PIF4-dependent manner, driving high-temperature-induced elongation responses. Wang et al. (2016a, b) demonstrated that HSP90, a key molecular chaperone during heat stress, also plays a role in auxin-mediated growth responses triggered by mild high-temperature increases in *Arabidopsis*. Expression analysis of the auxin-responsive marker IAA2-GUS and direct transport assays demonstrated that cold stress suppresses root basipetal (shootward) auxin transport. In *Arabidopsis*, it has been reported that cold stress significantly disrupted the trafficking of the auxin efflux carrier PIN2, which is critical for basipetal auxin transport. Besides, cold stress inhibited the lateral relocalization of PIN3, a protein implicated in the initial phase of root gravity response (Shibasaki et al. 2009). However, Du et al. (2013b) reported that under cold stress for 3 days, the IAA level increased by 1.6-fold, while it rose by 1.3-fold after 6 hours of heat stress, compared to the control. The *GH3* family gene *OsGH3–2*, which catalyzes IAA conjugation to amino acids, is induced by drought stress but suppressed by cold stress (Du et al. 2012). The carotenoid-deficient mutants with lower auxin levels also showed cold tolerance (Du et al.

2013b). A decline in auxin levels under cold and salinity stress was observed in *Malus sieversii* leaves and roots (Yuan et al. 2013). Certain *OsLAX* and *OsABCB* genes possess cold stress-responsive elements (ICEr2), suggesting their role in adaptation to cold stress (Chai and Subudhi 2016).

Since roots are the primary organs sensing heavy metal (HM) stress, understanding the interplay between HM stress and auxin homeostasis is crucial (Bücker-Neto et al. 2017; Ronzan et al. 2018). HMs affects auxin concentration and distribution, disrupting its homeostasis in plants (Fattorini et al. 2017; Krishnamurthy and Rathinasabapathi 2013). Studies show that HM stress alters endogenous auxin levels differentially in roots and shoots, demonstrating both positive and negative correlations between auxin levels and HM toxicity (Srivastava et al. 2013; Yuan and Huang 2016; Zelinová et al. 2015). Cadmium treatment significantly increased *YUCCA6* expression and auxin levels in *Arabidopsis* postembryonic roots (Fattorini et al. 2017). *OsAUX1* plays a key role in regulating various aspects of root development in rice and is involved in responses to cadmium (Cd) stress (Yu et al. 2015). Auxin transporter mutant's *aux1*, *pin1*, and *pin2* showed significantly greater sensitivity to As (III) compared to wild-type plants. However, the exogenous application of IAA enhanced As (III) tolerance in the *aux1* mutant through reactive oxygen species (ROS)-mediated signaling (Krishnamurthy and Rathinasabapathi 2013). The inhibition of root growth caused by aluminum is connected to the disruption of auxin transport and/or signaling in both *Zea mays* (Kollmeier et al. 2000) and *Arabidopsis* (Sun et al. 2010).

8.4.2 Role of Cytokinins in Abiotic Stress Response

Cytokinins, primarily recognized for promoting cell division, delay senescence, and maintain chloroplast function, are critical in plant responses to abiotic stress (Table 8.1). CK signaling operates through a two-component signaling system involving histidine kinases (AHKs), which detect CK levels and activate downstream response regulators (ARRs) and type B ARRs, which trigger specific gene expression (Inoue et al. 2001; Hai et al. 2020; Mandal et al. 2022). Under stress, CK levels are altered as plants attempt to optimize resource use, often saving energy by redirecting it away from growth-related processes. Cytokinins play a dual role in drought response. While CKs are generally downregulated under drought conditions to limit water usage, specific CK-related genes like *IPT* genes encoding isopentenyltransferase are upregulated in roots, effectively enhancing root development and water uptake. CK-deficient plants are often more tolerant to drought as they show reduced transpiration and enhanced ABA accumulation, favoring survival overgrowth (Seleiman et al. 2021).

During the early years of this century, significant advancements were made in identifying genes related to cytokinin (CK) metabolism and signaling in *Arabidopsis thaliana*. These findings, applicable to other plant species, have established a foundation for understanding CK action. Changes in endogenous CK levels in response to stress highlight the hormone's role in stress adaptation. For example, under drought conditions, the concentration and transport of trans-zeatin riboside within the plant significantly decrease, while abscisic acid (ABA) levels rise. Notably, when a partial root zone-drying technique was applied, CK levels dropped not only in roots but also in leaves, buds, and shoot tips, accompanied by increased ABA levels. These findings suggest that localized root stress can influence CK concentrations across plant organs, including shoots, thereby affecting developmental processes such as apical dominance (Schachtman and Goodger 2008). Typically, reduced CK levels strengthen apical dominance and, together with ABA-mediated stomatal regulation, enhance drought tolerance. The negative regulatory role of CKs in drought-stressed plants is further supported by genetic studies altering endogenous CK levels. For instance, plants with impaired CK biosynthesis, such as the *ipt1/ipt3/ipt5/ipt7* quadruple mutant or the *ipt8* single mutant, as well as those overexpressing cytokinin oxidase (CKX) genes, show reduced CK content and improved tolerance to both salt and drought stress (Nishiyama et al. 2011). Similarly, mutants with nonfunctional CK receptors exhibit enhanced stress resistance. Loss-of-function mutants of *ahk*, such as *ahk2/ahk3* and *ahk3/ahk4*, demonstrate greater resistance to freezing temperatures compared to the wild type (Jeon et al. 2010). Additionally, most single and multiple *ahk* mutants, except *ahk4*, display improved dehydration tolerance (Kang et al. 2012). These observations highlight the critical role of CK metabolism and signaling in modulating plant responses to abiotic stresses.

Under salt stress, CKs contribute to ionic balance and cellular homeostasis; they also upregulate antioxidant systems, which counteract the harmful reactive oxygen species produced by salt stress. Studies have shown that CKs can modulate Na^+/K^+ balance and reduce sodium accumulation, thereby alleviating ion toxicity and improving salt tolerance in crops like rice and tomato (Zhao et al. 2021). Genetic studies carried out using the disruption mutants of the *AtHKT1* gene, which encodes a sodium ion transporter, demonstrated its role in root/shoot Na^+ distribution and also mediate tolerance to salinity stress. The protein is localized on the plasma membrane of xylem parenchyma cells and unloads Na^+ from xylem vessels into xylem parenchyma cells. Then, via symplastic diffusion, the Na^+ moves through plasmodesmata from xylem parenchyma cells into the phloem, thus reducing Na^+ levels in xylem vessels and leaves (Maser et al. 2002; Sunarpi et al. 2005). This mechanism has been effectively utilized in breeding programs, especially in wheat, where the selection for *HKT1* alleles has resulted in higher yields under saline conditions (Munns et al. 2012). Various crops are being targeted for *HKT*-based marker-assisted selection, highlighting the broader applicability of this approach. *HKT* expression is mainly affected by hormonal regulation, which connects it to the complex interactions between salt stress responses and plant growth regulation. By including this perspective, we can better link the fundamental molecular mechanisms to practical breeding strategies to enhance crop salt tolerance. Cytokinins can also enhance plant tolerance to both high and low temperatures. CKs activate heat-shock proteins and other protective molecules that stabilize proteins and membranes, thus maintaining cellular integrity under extreme temperatures. Additionally, specific CKs can mitigate oxidative stress by activating antioxidant enzymes, such as superoxide dismutase and catalase, under heat stress conditions. Cytokinins play a pivotal role in nutrient signaling, especially for nitrogen. CKs promote root growth and enhance nutrient absorption capacity. Under nutrient-deficient conditions, CK signaling pathways adjust the plant's growth strategy by promoting lateral root formation and delaying senescence, which helps to maximize nutrient uptake and use efficiency (Hönig et al. 2018).

Cytokinins interact with other hormones, such as ABA and auxin, to fine-tune stress responses. Under drought conditions, CK levels often decline, allowing ABA to dominate and conserve resources. CK and auxin interactions are also significant, particularly in root growth regulation under stress, where CKs promote lateral root formation, while auxin influences root elongation and architecture. Research in CK biology has explored genetic and chemical modulation of CK pathways to enhance stress tolerance. For instance, transgenic approaches that upregulate IPT genes, leading to higher CK levels, have shown promise in improving drought resilience in crops like tobacco (Hönig et al. 2018). Additionally, CK analogues are being tested to optimize growth and stress tolerance in a controlled manner, making CKs a key area of interest for crop improvement strategies. The role of gibberellins and cytokinins in plant responses to abiotic stress represents a promising frontier in agricultural biotechnology. Through strategic manipulation of GA and CK pathways, scientists can effectively increase crop resilience to stressors such as drought, salinity, and extreme temperatures. With collaborative knowledge from molecular genetics, physiology, and biochemistry, we now understand how the phytohormone interactions enable plants to balance growth with stress responses, paving the way for crop improvement.

8.4.3 Role of Abscisic Acid in Abiotic Stress Response

Abscisic acid (ABA) is an important plant hormone that helps control how plants grow and develop, especially under stress conditions like drought or high salinity. The enzyme 9-cis-epoxy carotenoid dioxygenase (NCED) catalyzes the rate-limiting step in ABA biosynthesis. NCED cleaves 9-*cis* xanthophylls to xanthoxin that serves as a precursor of ABA (Tan et al. 2003). In *Arabidopsis*, nine *NCED genes* and five in *Oryza sativa* are likely involved in ABA biosynthesis (Tan et al. 2003; Ye et al. 2012). In *Arabidopsis*, the expression of *AtNCED3* plays a key role in ABA biosynthesis and under drought stress (Iuchi et al. 2001) The promoter region of *AtNCED3* contains an overlapping G-box recognition sequence (5′-CACGTG-3′) located at −2248 bp upstream of the transcriptional start site and serves as an important cis-acting element in the induction of the dehydration response (Behnam et al. 2013). The transcription factor NGA1 binds directly to the cis-element NGA-binding element (NBE) in the 5′ untranslated region (5′ UTR) of the *NCED3* promoter and was suggested to be transcriptional activators of *NCED3* (Sato et al. 2018). The expression of *AhNCED1* gene in peanut plants is significantly upregulated by dehydration and high salinity. The water-stress tolerance of *Arabidopsis* plants was shown to be improved by ectopic expression of the *AhNCED1* gene causing accumulation of endogenous ABA (Wan and Li 2006). Specific *NCED* genes in rice play critical roles in ABA biosynthesis and stress responses. For instance, *OsNCED3* regulates seed dormancy, stomatal aperture, stress tolerance, and senescence, with its expression induced under NaCl, PEG, and H_2O_2 stresses (Huang et al. 2018). *OsNCED5* is ubiquitously expressed in various tissues and was induced upon exposure to salt stress, water stress, and darkness. Mutational analysis revealed that *nced5* mutants had reduced ABA levels, increased sensitive to salinity and water stress, and displayed delayed leaf senescence. Conversely, overexpression of *OsNCED5* led to increased ABA levels, enhanced stress tolerance, and accelerated leaf senescence. *OsNCED5* also shows functional redundancy with *OsNCED3* and *OsNCED4* in regulating salt and water-stress tolerance (Huang et al. 2019). Additionally, *OsNCED1* overexpression enhances heat stress tolerance by improving pollen viability, seed setting rate, and antioxidant enzyme activities (Zhou et al. 2022). The *PP2C* gene family members have also been reported for their involvement in ABA signaling and confer stress tolerance in rice and maize (Singh et al. 2015; Zong et al. 2016; Bhatnagar et al. 2020; Miao et al. 2020; Pang et al. 2024). In maize, *ZmPP2C15* is known to improve drought tolerance. Heterologous expression of maize *ZmPP2C15* gene in *Arabidopsis* exhibited increased survival and growth under drought stress. The antioxidative enzyme system, superoxide dismutase (SOD), peroxidase (POD), ascorbate peroxidase (APX), and catalase (CAT) activities, was higher in shoots, with increased proline content and lesser amounts of malondialdehyde (MDA) (Pang et al. 2024). ABA is a central hormone and well documented for mitigating drought and salinity stress in plants and boosts antioxidant enzyme system (Zhu, 2002; Miao et al. 2020). In rice, utilization of differential patters of ABA signaling intensity determinants helps to cope with drought, salt, or cold stress (Ye et al. 2012). A study carried out by co-overexpression of the transcription factor *OsbZIP46CA1* and the protein kinase *SAPK6* significantly enhanced drought tolerance in rice overexpression lines (Chang et al. 2017). Exogenous application of ABA in rice, maize, barley, and grapes enhances tolerance to abiotic stress, including salinity, drought, cold, and simulated acid rain (SAR) (Wang et al. 2016a, b; Liu et al. 2018; Skowron and Trojak 2021; Habibpourmehraban et al. 2023). Under salinity stress, in studies carried out with two rice cultivars, "Chaoyouqianhao" and "Huanghuazhan," exogenous application of ABA was shown to mitigate salinity-induced toxicity by improving antioxidant enzyme levels (Liu et al. 2019; Chen et al. 2022). Priming of rice and sesame seeds with ABA had also been reported to ameliorate the salinity stress (Gurmani et al. 2006, 2011, 2013; Safari et al. 2018).

8.4.4 Role of Salicylic in Abiotic Stress Response

Studies in various plant species, including *Vicia faba*, *Medicago sativa*, Fragaria × ananassa, *Solanum lycopersicum*, *Spinacia oleracea*, *Glycine max*, *Cucumis sativus*, *Brassica juncea*, and *Vigna radiata*, have demonstrated the importance of SA in enhancing salt stress-tolerance mechanisms (Stevens et al. 2006; Azooz 2009; Karlidag et al. 2009; Nazar et al. 2011; Palma et al. 2013; Khan et al. 2014; Kim et al. 2017; Nigam et al. 2022). In strawberries (*Fragaria X ananassa*), under salt stress, SA significantly reduces sodium uptake and improves potassium uptake, along with proline level, total soluble solid content, and activity of the antioxidant enzymes (Roshdy et al. 2021). Naeem et al. (2020) found that SA positively influences the growth of tomato plants in unfavorable environmental conditions. It had also been reported that SA can potentially improve resistance of horticulture crops to abiotic stress and cause reduction in oxidative damage (Joseph et al. 2010; Altaf et al. 2023). Exogenously treatment with SA (0.5 mM) was found to enhance salt tolerance in *Triticum aestivum* by increasing the activity of ascorbate (AsA)-GSH pathway enzymes and the transcript level of antioxidant genes (Li et al. 2013). In *Hordeum vulgare*, *Phillyrea angustifolia*, and *Salvia officinalis*, the SA levels increased during drought stress reflecting the role of SA in plant's response to drought (Munne-Bosch and Penuelas 2003; Bandurska and Stroiski 2005; Abreu and Munné-Bosch 2008). SA pretreatment elevated proline content with its synthesis-related genes and eliminated the drought-induced superoxide radical with improved expression of redox-regulating genes (Lee et al. 2019). The application of SA also improved the growth and photosynthetic performance of *Citrullus lanatus*, *Cucumis melo*, *Ipomoea batatas*, *Solanum melongena*, *Capsicum annuum*, *Cucumis sativus*, and *Fragaria × ananassa* (Korkmaz et al. 2007; Baninasab 2010; Mozafari et al. 2018; Wakchaure et al. 2020; Kaya 2021; Huang et al. 2022; Silva et al. 2023).

In *Arabidopsis thaliana*, SA-induced expression of *PR* genes, namely, *PR1*, *PR2*, *and PR5* displayed stomatal closure and enhanced drought tolerance in SA-accumulating mutants *cpr5* and *acd6* (Liu et al. 2013). Studies have also shown that exogenous application of SA considerably improved the heat stress tolerance of horticultural crops including watermelon, strawberries, tomatoes, potatoes, and grape berries (Lopez-Delgado et al. 1998; Snyman and Cronjé 2008; Wen et al. 2008; Coltro et al. 2014; Ram et al. 2014). In rice, tomato, and *Arabidopsis*, exogenous SA treatment caused the production of heat shock proteins (HSPs) and conferred increased tolerance to heat stress (Clarke et al. 2004; Snyman and Cronjé 2008). Cold stress is known to cause physiological harm including poor seed germination, necrosis and chlorosis of the leaves, delayed growth of seedlings, changed enzyme activity, and decreased leaf expansion, ultimately leading to tissue death (Yadav 2010). The phytohormone SA is also known to exhibit the regulatory function in protecting plants against chilling stress and has been reported in *Zea mays*, *Secale montanum Guss*, *Citrullus lanatus*, *Phaseolus vulgaris L.*, *Triticum aestivum, and Hordeum vulgare* (Janda et al. 1999; Gharib and Hegazi 2010; Kosová et al. 2012; Ansari and Sharif-Zadeh 2012; Mutlu et al. 2013; Cheng et al. 2016;

Ignatenko et al. 2019). Application of SA topically to spinach under cold stress showed reduced oxidative damage and enhanced cold stress tolerance. SA treatment also increased proline content and displayed elevated antioxidant enzyme activity (Shin et al. 2018). Also, it was noted that exogenous SA (1 mM) increased chilling tolerance in cucumber plants by upregulating genes linked to SA metabolism (*PAL, ICS, and SABP2*) and cold signaling pathway (*ICE1, CBF1, and COR47*) (Fu et al. 2021). Zhao et al. (2021) reported that treatment with 1 μM SA reduced cold-induced injury in peach fruit and showed increased expression of cold-response genes (*DREB1A and DREB2A*) and associated genes (*SPS4, NINV2, SuSy2, and SUT1*). The role of SA had also been investigated in response to heavy metals stress. Under lead (Pb)-, copper (Cu)-, and nickel (Ni)-stressed conditions in *Oryza sativa*, *Phaseolus vulgaris*, and *Brassica juncea*, respectively, the administration of SA enhanced growth and photosynthetic capacities (Jing et al. 2007; Zengin 2014; Zaid et al. 2019). Increased endogenous SA levels have been shown to reduce H_2O_2 content in Cd-exposed *Oryza sativa*, by preventing cellular membrane damage (Chao et al. 2010).

8.4.5 Role of Brassinosteroids in Abiotic Stress Response

Brassinosteroids (BRs) are steroidal phytohormones known to regulate plant growth, development, and responses to environmental stresses, including heat stress (Divi and Krishna 2009). In *Brassica napus*, BRs enhance basal thermotolerance by increasing HSP synthesis and accumulation (Dhaubhadel et al. 2002). In tomatoes, exogenous BRs alleviate heat stress by boosting photosynthesis, stomatal conductance, RuBisCO activity, and electron transport linked to RuBP. They also enhance antioxidant enzyme activity and reduce lipid peroxidation under heat stress (Ogweno et al. 2008). Brassinosteroids (BRs) enhance photosynthetic efficiency in plants by boosting initial Rubisco activity, improving the quantum efficiency of PSII, and increasing chlorophyll fluorescence under high-temperature conditions (Bajguz and Hayat 2009; Hayat et al. 2010; Fariduddin et al. 2014). BR-induced heat tolerance is linked to the increased expression of *RBOH1*, *MPK1*, and *MPK2* transcripts, with *MPK2* playing a more prominent role than *MPK1* in BR-mediated H_2O_2 accumulation and subsequent stress resistance (Nie et al. 2013). The *TaBRI1* knockdown mutant reduced tolerance to high-temperature stress in *Triticum aestivum L.* (Ibañez et al. 2018). *BES1* and *HSFA1* interacted to enhance heat resistance in *Arabidopsis thaliana* (Jacob et al. 2017).

Expression of *CBF* genes has provided insights into how BRs enhance freezing tolerance in plants (Eremina et al. 2016; Li et al. 2017). Exogenous application of brassinosteroids (BRs) increases the expression of *C-repeat Binding Factor (CBF)* genes, which encode transcription factors essential for regulating *cold-responsive (COR)* genes and developing a proper cold acclimation response. Studies of various BR-signaling mutants revealed that BRs enhance freezing tolerance, at least partially, by activating the *CBF-COR* pathway (Eremina et al. 2016; Li et al. 2017). The application of BR enhances cold tolerance in various plant species, including chilling-sensitive crops like maize and cucumber (*Cucumis sativus*) (Xia et al. 2009; Singh et al. 2012; Jiang et al. 2013). When overexpressed, *DFW4* improved cold stress resistance in *Arabidopsis* and also elevated the mRNA levels of *COR15A* (Divi and Krishna 2010). DWRF enhanced the expression of *GRX* to improve cold tolerance in *Solanum lycopersicum* L. (Xia et al. 2018). These studies offer strong genetic evidence supporting BRs' role in *Arabidopsis* freezing tolerance and cold acclimation, an adaptive mechanism where plants boost their freezing resistance after exposure to low, nonfreezing temperatures.

Exogenous application of brassinosteroids (BRs) has been shown to mitigate the negative effects of salt stress on seed germination, root elongation, and overall plant growth (Anuradha and Rao 2001; Amzallag and Goloubinoff 2003). Under saline conditions, BRs improved seed germination and growth in rice, tomato, and chickpea (Prakash et al. 1999; Anuradha and Rao 2001; Ali et al. 2007). Sharma et al. (2013) demonstrated that EBL increased proline levels in salt-stressed rice seedlings. Zhu et al. (2023) showed that BIN2 negatively regulates salt stress in *Arabidopsis thaliana*. Overexpression of the *StDWF4* enhanced salt tolerance in *Solanum tuberosum* (Zhou et al. 2018). During drought stress, BRs regulate the expression of drought-resistance genes and related protein levels to improve plant tolerance. The BHLH protein plays a role in modulating the BR signaling pathway, with *BEE3*, a *bHLH* family gene, identified as an early BR response gene. In cotton, BRs activated *GhBZR1*, which suppressed the expression of *GhBEE3-Like*. Studies revealed that knocking out the *GhBEE3-Like* gene increased the expression of stress-related genes *GhERD10*, *GhCDPK1*, and *GhRD26*, thereby enhancing drought resistance (Chen et al. 2021). Additionally, under drought stress, *RD26* was upregulated to suppress BR-induced growth, contributing to improved drought tolerance (Tran et al. 2004). However, in *Arabidopsis thaliana*, BZR1 and BES1 inhibited *RD26* expression and drought tolerance (Ye et al. 2017). Overexpression of the *CYP85A1* enhanced drought resistance in *Nicotiana tabacum L* (Duan et al. 2017). Similarly, *TaBZR2* and *TaBES1* increased drought resistance in *Triticum aestivum* L. (Cui et al. 2019).

Brassinosteroids play a crucial role in regulating ion uptake (Khripach et al. 2000) and have been shown to reduce the accumulation of heavy metals and radioactive elements (Khripach et al. 1995). BRs also influence the accumulation of heavy metals such as Pb, Cu, Cr, Cd, and Zn, enabling plants to maintain developmental plasticity under HM stress (Anuradha and Rao 2007; Hayat et al. 2010; Hasan et al. 2011; Ramakrishna and Rao 2015). Brassinosteroids (BRs) also help in absorption of ions in plant cells by significantly reducing heavy metal accumulation. Exogenous application of 24-epiBL in *Brassica juncea* caused the enhancement activity in antioxidant enzymes and ameliorated Ni stress (Kanwar et al. 2013). Exogenous treatment of cucumber leaves with 24-epibrassinolide stimulated root and stem growth and enhanced resistance to cadmium stress (Shah et al. 2019). The upregulation of antioxidative defense-related genes, including *APX08*, *CATa*, *CATb*, *APX02*, *SOD-Fe2*, and *SOD-Cu-Zn*, helps mitigate ROS production, thereby alleviating cadmium-induced toxicity (Santos et al. 2018). Antioxidative defense-related genes in *Oryza sativa* L. decreased ROS levels and increased carboxylation (Tadaiesky et al. 2021). In *Arabidopsis thaliana*, DELLA protein interacted with the BRZ1 gene to enhance plant resistance to heavy metals (Achard et al. 2008).

8.4.6 Role of Gibberellins in Abiotic Stress Response

The GAs is one of the most well-known and longest classes of regulatory hormones that are involved in many plants' developmental processes, including induced flowering, fruit development, internodal elongation, and seed germination. Gibberellins (GAs) are a diverse class of tetracyclic diterpenoid carboxylic acids, though only a small subset, notably GA1 and GA4, function as key growth hormones in higher plants. These hormones play critical roles throughout the plant life cycle, promoting organ growth via cell division and elongation, as well as mediating developmental phase transitions such as seed dormancy to germination, juvenile to adult growth, and vegetative to reproductive development. While GAs are essential for normal plant development, seedlings unable to synthesize or perceive these hormones exhibit stunted growth and may fail to transition to flowering under specific light conditions (Koornneef and van der Veen 1980). The GA signaling process involves the biosynthesis of active forms, their detection, signal transduction, and inactivation, all of which are influenced by environmental factors, including abiotic stress.

Gibberellins are diterpenoid compounds that influence several developmental processes (Table 8.2). Using cutting-edge biochemical and genetic techniques, recent research has investigated the genetic foundation and genes encoding the manufacturing of GAs and deactivation enzymes (Islam et al. 2021). Although there are more than 250 members of the GAs group known to exist, only a small number of them are physiologically active and have several functions in plant development. It is claimed that the control of GAs at the cellular level is intricate. For instance, the steps of the GAs metabolic pathway are controlled by enzymes belonging to small multigenic families, each of which has a distinct pattern of expression. The GA signaling pathway is primarily mediated by the DELLA proteins, which are growth repressors (Degefu and Tesema 2020). Under stress conditions, GA signaling pathway is modulated leading to altered DELLA stability and subsequently altered growth and stress responses. DELLA proteins interact with other transcription factors and signaling molecules, making them critical in cross talk between GA and other phytohormones (Lantzouni et al. 2020).

GA biosynthesis begins with trans-geranylgeranyl diphosphate in plastids via the methylerythritol phosphate pathway (Kasahara et al. 2002). This pathway involves plastid-localized terpene cyclases, followed by oxidation processes mediated by cytochrome P450 monooxygenases in the endoplasmic reticulum and soluble 2-oxoglutarate-dependent dioxygenases (Hedden and Thomas 2012). Key enzymes in this pathway include GA 20-oxidase (GA20ox) and GA 3-oxidase (GA3ox), which synthesize bioactive GAs, and GA 2-oxidases (GA2ox), which inactivate

Table 8.2 Summary of gibberellins and cytokinins functions under abiotic stress conditions

S. no.	Abiotic stress	Role of gibberellins	Role of cytokinins	Hormone interaction	Benefits to plants
1.	Drought	Decreased growth rate	Increased root growth and senescence	Antagonistic with ABA	Improved water-use efficiency
2.	Salinity	Decreased NA^{+} toxicity	Decreases reactive oxygen species and balances ionic concentration	Synergistic with ethylene	Increased salt tolerance
3.	Cold	Maintains cell membrane stability	Increased antioxidant activity	Cross talk with auxin	Protects cells against freezing
4.	Nutrient deficiency	Regulates nutrient uptake	Promotes root system expansion	Synergistic with auxin	Enhances nutrient absorption
5.	High-temperature stress	Maintain cell elongation and photosynthesis	Enhance antioxidant enzyme activities	Cross talk with ABA	Sustained growth and photosynthetic efficiency
6.	Heavy metal stress	Reduce metal toxicity by improving ion transport and homeostasis	Enhance root system architecture for nutrient uptake	Enhances ROS scavenging in synergy with salicylic acid	Enhanced detoxification capacity
7.	High light stress	Modulate leaf expansion for better light absorption	Delay oxidative damage by reducing ROS	Synergizes with jasmonic acid for photoprotection and stress signaling	Sustained growth and productivity
8.	Flooding	Induce aerenchyma formation for oxygen transport	Delay senescence and promote regrowth post-submergence	Synergistic with ethylene for adaptive growth	Increased survival and recovery under submergence

them to regulate GA levels. The dioxygenase genes, particularly *GA2ox*, are responsive to abiotic stress and are key regulators of the GA biosynthetic pathway, with feedback mechanisms ensuring GA homeostasis (O'Neill et al. 2010). Recent advances have elucidated how GAs are perceived and how early signal transduction events occur, shedding light on their multifaceted roles in plant development (Ueguchi-Tanaka and Matsuoka 2010). GA signaling involves the degradation of DELLA proteins, which are transcriptional regulators belonging to the GRAS protein family. DELLA proteins, named after a conserved N-terminal domain, act as growth repressors and require GA-induced degradation for their function. GA binding to the nuclear receptor GID1 triggers a conformational change that allows DELLA interaction with SCF ubiquitin ligases, targeting DELLA for proteasomal degradation. Although DELLA proteins modulate gene expression by interacting with transcription factors, they do not directly bind to DNA (Hirano et al. 2012). DELLA proteins also integrate GA signaling with other hormonal pathways, highlighting their regulatory complexity (Gallego-Bartolomé et al. 2012). A pivotal discovery was the role of DELLA proteins in mediating growth restraint under abiotic stress, enhancing stress tolerance. For example, salt stress in *Arabidopsis thaliana* reduces endogenous GAs and increases DELLA protein levels, leading to growth inhibition and improved stress survival (Magome et al. 2008). Similarly, exposure to cold stress results in reduced bioactive GAs, DELLA accumulation, and associated growth restriction, which is linked to increased stress tolerance. DELLA proteins contribute to stress resilience by modulating the antioxidant system, reducing reactive oxygen species (ROS) levels, which are often elevated under stress conditions and can lead to cell death (Fukao et al. 2011). The dual role of DELLA proteins in restraining growth and enhancing stress tolerance suggests a shared regulatory mechanism. However, their contributions under unstressed versus stressed conditions remain to be fully understood, particularly regarding their effects on growth rate and stress adaptation (Claeys et al. 2012). The ability of DELLA proteins to regulate cell proliferation and mediate transitions during osmotic stress highlights their critical role in plant stress responses (Ubeda-Tomás et al. 2009). Understanding the precise mechanisms by which DELLA proteins influence growth and stress responses will be essential for advancing plant stress tolerance strategies.

Gibberellins are usually downregulated during drought conditions to reduce growth and conserve resources. Previous studies have shown that GA-deficient mutants exhibit higher drought tolerance due to enhanced water-use efficiency and ABA accumulation (Liao et al. 2023). Nevertheless, fine-tuning GA levels can regulate growth without significantly compromising drought tolerance. Like drought, high salinity triggers GA suppression, promoting DELLA accumulation, which reduces growth. However, certain GA-responsive genes may aid in salt tolerance by enhancing root architecture or ionic balance, thereby reducing ion toxicity and osmotic stress (Delgado et al. 2021). Low temperatures often prevent GA biosynthesis, which is critical in cold adaptation. Some species exhibit GA-induced cold tolerance via improved accumulation of specific proteins that protect cell membranes and maintain cellular homeostasis. GA application has been explored to improve cold resilience in sensitive crop species (Bouteraa et al. 2023). Nutrient stress, such as nitrogen and phosphate deficiency, can impact GA signaling, specifically by DELLA-regulated pathways. For instance, DELLA proteins help optimize nutrient usage under suboptimal conditions by adjusting growth patterns, potentially making plants more resilient to nutrient-poor soils (Cheng et al. 2024).

8.4.7 *Role of Gibberellins and Cytokinins*

As the global climate continues to evolve, abiotic stresses are expected to intensify, underscoring the urgency of developing crops that can thrive under suboptimal conditions. Abiotic stresses, including salinity, drought, extreme temperatures, and nutrient deficiencies, pose significant challenges to plant growth and productivity, directly impacting global agricultural sustainability. These stressors trigger complex molecular, cellular, and physiological responses in plants, orchestrated by an array of phytohormones that modulate growth, development, and survival mechanisms. Among these phytohormones, gibberellins (GAs) and cytokinins (CKs) have garnered considerable attention for their multifaceted roles in stress tolerance and yield optimization (Fig. 8.1). Gibberellins are widely recognized for their contributions to stem elongation, seed germination, and flowering, while cytokinin plays pivotal roles in cell division, chloroplast development, and the regulation of senescence. Together, these hormones are central to shaping plant adaptability under adverse environmental conditions. By deciphering the regulatory networks involving these hormones, researchers can identify key molecular targets for crop improvement. Furthermore, the development of hormone-based agronomic solutions, such as exogenous application of GA and CK analogues, offers promising avenues for mitigating stress-induced yield losses. This section explores current knowledge on the roles of GAs and CKs in abiotic stress tolerance, examines recent research breakthroughs, and discusses future directions in crop improvement and sustainable agriculture.

Recent advances in plant molecular biology have shed light on the intricate mechanisms through which GAs and CKs mediate abiotic stress responses. Both hormones act individually and in concert with other signaling molecules such as abscisic acid (ABA), auxin, and ethylene to fine-tune plant physiology. The cross

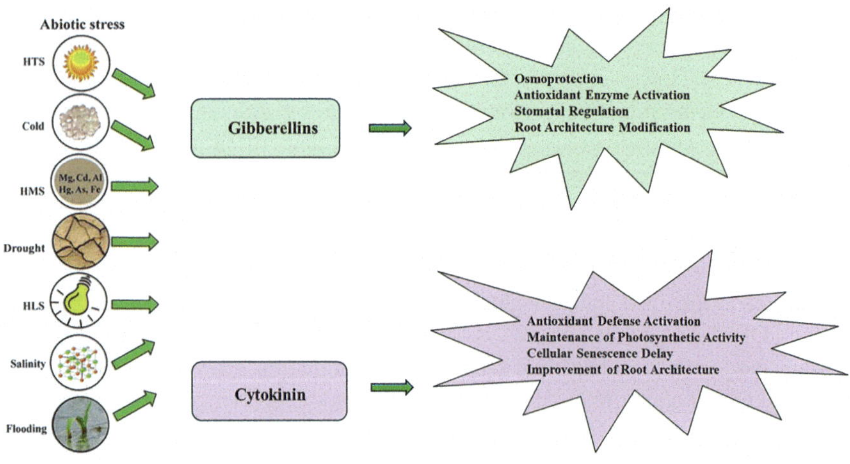

Fig. 8.1 Schematic representation of role of GA and CK in abiotic stress

talk between these hormonal pathways is a critical determinant of plant resilience, with ABA often antagonizing the actions of GAs and CKs to prioritize stress tolerance overgrowth. Emerging research highlights that the balance between these hormones is dynamically regulated, enabling plants to optimize resource allocation under stress conditions. For instance, studies by Napieraj et al. (2023) demonstrate that the modulation of GA and CK signaling pathways can enhance drought tolerance by improving water-use efficiency and maintaining photosynthetic activity. Understanding the roles of GAs and CKs in abiotic stress adaptation extends beyond their individual functions. Their synergistic interactions, as well as their interplay with stress-responsive transcription factors and downstream gene networks, present opportunities for targeted genetic and biotechnological interventions. Recent work by Rai et al. (2024) underscores the potential of manipulating GA and CK pathways to enhance crop performance in saline soils, a pressing concern in regions with high soil salinity. Additionally, advances in CRISPR-Cas9 genome editing and transgenic approaches have paved the way for engineering hormone biosynthesis and signaling components to create stress-resilient crop varieties.

This growing body of research underscores the dual role of GAs and CKs as growth promoters and stress mitigators, highlighting their potential in addressing the dual challenges of food security and environmental sustainability. However, significant gaps remain in our understanding of the precise molecular mechanisms underlying their functions in abiotic stress responses. For instance, the temporal dynamics of GA and CK signaling under fluctuating stress conditions and their interaction with nonhormonal stress pathways remain poorly elucidated. Moreover, the impact of environmental variables such as light and temperature on GA and CK homeostasis warrants further investigation to unravel the complexities of plant stress physiology.

8.5 Phytohormonal Cross Talk

Phytohormones regulate cellular activities in higher plants and play a crucial role in coordinating signal transduction pathways during abiotic stress responses (Pieterse et al. 2009). Abiotic stresses activate signal transduction cascades linked to baseline pathways mediated by plant hormones. It is also evident from recent research studies that several signaling pathways are coordinated to modulate hormonal functions, and hence, plant hormones act synergistically with each other to elicit stress responses (Fig. 8.2, Khan et al. 2023).

In addition to the classical phytohormones—auxins, cytokinins (CK), gibberellins (GA), abscisic acid (ABA), and ethylene (ET)—newly identified ones like salicylic acid (SA), brassinosteroids (BRs), jasmonates (JA), polyamines, and strigolactones also play key roles (Santner and Estelle 2009). ABA, SA, JA, and ET are particularly important for enhancing stress tolerance, and their cross talk with auxins, cytokinins, and gibberellins regulates plant defense responses (Pieterse et al. 2012; Nishiyama et al. 2013). This hormonal interplay orchestrates the

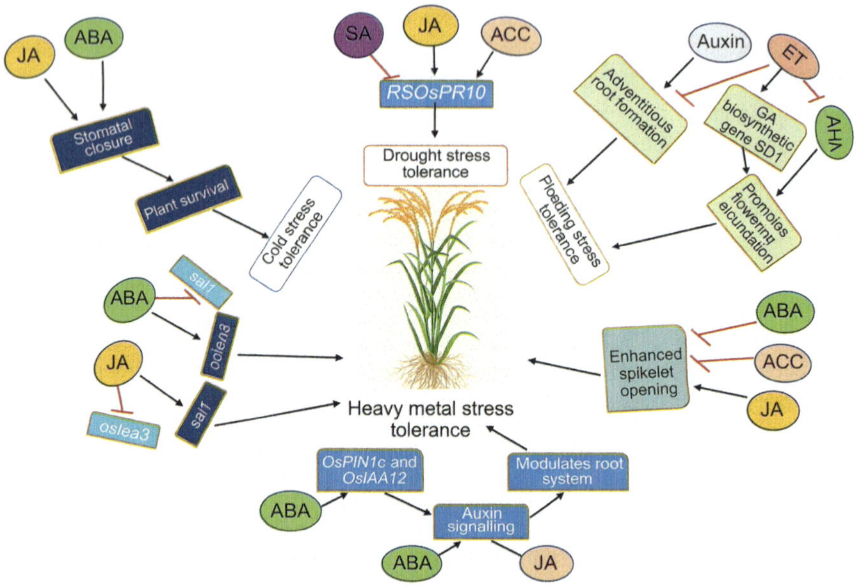

Fig. 8.2 Plant hormones modulate signaling mechanism in plant system. (Khan et al. 2023)

expression of numerous genes involved in stress mitigation, highlighting the need to understand the complex interactions among phytohormones (Pieterse et al. 2012). Cross talk among GA, auxin, ABA, ethylene, and cytokinin is essential for organ-specific and whole-plant adaptation under stress. Growth promoters like auxin, gibberellin, and cytokinin support growth during stress, while stress hormones ABA and ethylene inhibit growth by modulating these promoters in a tissue- and context-dependent manner (Achard et al. 2006; Wolters and Jürgens 2009; Peleg and Blumwald 2011). Further, Lackman et al. (2011) linked the ABA receptor NtPYL4, central to ABA signaling to JA responses in balancing growth and stress adaptation. Using a stress- or senescence-induced promoter to regulate cytokinin levels in rice led to a stay-green phenotype and increased yield under water stress (Peleg et al. 2011). Cytokinin-modulated transgenic rice showed altered levels of other phytohormones, indicating hormonal interplay by synergistic and antagonistic interactions among auxins, ethylene, cytokinins, GA, BR, and SA under stress (Peleg and Blumwald 2011). GA, along with auxins and BR, plays a central role in regulating plant growth, especially elongation (Golldack et al. 2013; Stamm and Kumar 2013). Cross talk between GA and ABA influences seedling development under salinity (Yuan et al. 2011), while GA and jasmonates interact to regulate growth and defense (Hou et al. 2013). Treating a transgenic *Arabidopsis* line overexpressing cytokinin oxidase/dehydrogenase with brassinolide significantly increased lateral root growth, suggesting alternative hormonal pathways to achieve similar morphologies (Vercruyssen et al. 2011). The Small Auxin Up-Regulated (SAUR) protein RAG1 affects hypocotyl elongation and fertility in *Arabidopsis*, indicating cross talk between auxin, GA, and possibly ABA through this protein (Stamm and Kumar 2013). Under submergence, ethylene increases GA sensitivity, promoting stem and leaf growth in rice (Xu et al. 2006). It also plays roles in drought and ozone stress with ABA, affecting stomatal opening (Wilkinson and Davies 2010), and in flooding and low-oxygen stress with ERFs, GA, and CIPK (Bailey-Serres and Voesenek 2010). Ethylene induces ERFs, which enhance endogenous ethylene biosynthesis (Zhang et al. 2009). ERFs include AP2, RAV, DREB (a cross talk point with ABA), and other transcription factors (TFs) based on DNA-binding domain similarity (Sakuma et al. 2002). These fluctuations in stress-responsive hormones adjust cellular dynamics, playing a key role in coordinating growth responses under stress conditions. Overexpression of *Arabidopsis YUC6* in potato and poplar resulted in drought-tolerant plants with narrower downward-curled leaves and increased height (Kim et al. 2013; Ke et al. 2015). The MYB96 transcription factor regulates lateral root meristem activation during drought, potentially through ABA-auxin signaling interactions. MYB96-knockout mutants exhibited increased susceptibility to drought stress (Seo et al. 2009). Overexpression of *OsGH3-2* in rice reduced free IAA, carotene, and ABA levels and conferred enhanced tolerance to cold stress by modulating IAA and ABA homeostasis. SA pretreatment reduces cell membrane damage and elevates ABA concentration in barley and maize leaves, indicating that SA and ABA interact during drought stress (Bandurska and Stroiski 2005; Tayyab et al. 2020). Thus, through these studies, the cross talk mechanism among phytohormones represents a sophisticated regulatory network that balances growth and stress responses in plants. This intricate interplay involves synergistic and antagonistic interactions between growth-promoting hormones (auxins, gibberellins, cytokinins, and brassinosteroids) and stress-related hormones (ABA, ethylene, jasmonates, and salicylic acid), enabling plants to adapt to diverse abiotic stresses. By modulating gene expression and physiological processes in a tissue- and context-specific manner, hormonal cross talk fine-tunes developmental and stress-adaptive pathways, highlighting its critical role in improving plant resilience and productivity under adverse environmental conditions (Fig. 8.3).

8.6 Conclusions and Future Perspectives

Abiotic stresses pose significant challenges to agriculture, particularly in the context of climate change and resource scarcity. Understanding the mechanisms underlying plant responses to these stresses is crucial for developing resilient crop varieties. Plant hormones play indispensable roles in modulating stress tolerance by regulating physiological, biochemical, and molecular processes. Quantitative trait loci (QTL) represent genomic regions that harbors genes governing traits of interest primarily concerned with crop productivity components. Abiotic stress conditions significantly reduce crop yield and are also a major factor for crop loss. As such, identification of genomic regions governing the yield-related traits particularly under abiotic stress conditions or involved in stress tolerances responses would ease the rapid discovery of genes whose introgression through plant breeding programs

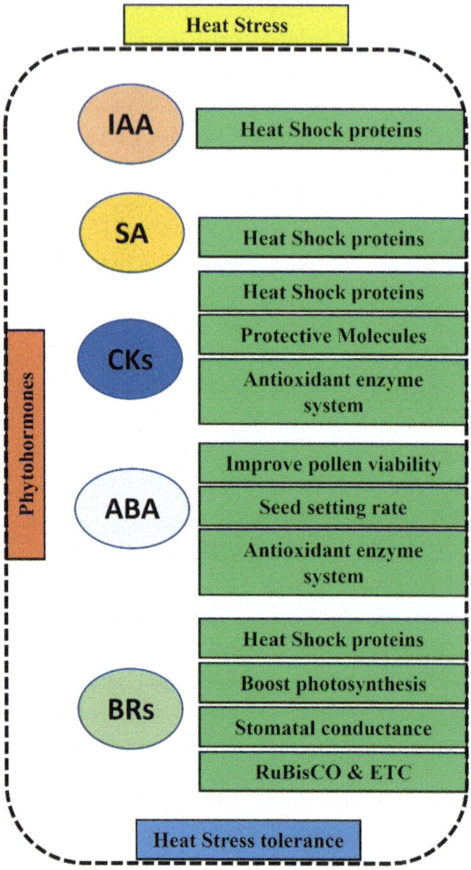

Fig. 8.3 Plant hormones modulate various physiological and biochemical responses to deliver an effective heat stress tolerance response

would immensely contribute toward development of stress-resilient crops with enhanced productivity (Devi et al. 2017). Considering the involvement of phytohormones in plant growth, development, and their pragmatic involvement in abiotic stress tolerance responses reflects their potentially employment in plant breeding programs to develop varieties with superior tolerance responses to a wide array of abiotic stress conditions that prevail in the environment. Advances in genetic engineering and biotechnology hold promise for harnessing these hormonal pathways to enhance crop productivity and ensure sustainable agriculture. By integrating knowledge of stress physiology, hormone signaling, and plant genetics, we can pave the way for a more resilient and productive agricultural system.

Future research on phytohormonal cross-talk offers exciting opportunities to enhance plant stress tolerance and productivity in the face of climate change and environmental challenges. Unravelling the molecular mechanisms underlying hormone interactions, particularly in species-specific and tissue-specific contexts, will provide insights into how plants integrate multiple signaling pathways. Advanced technologies such as CRISPR-Cas9 genome editing, multi-omics approaches, and single-cell transcriptomics can be employed to dissect the precise roles of hormone-regulated genes and their dynamic interactions during stress responses (Chen et al. 2019). Developing crops with tailored hormonal pathways through genetic engineering and synthetic biology could improve growth and stress adaptation. For example, manipulating key cross talk nodes, may optimize trade-offs between growth and defense (Wilkinson et al. 2012). Furthermore, leveraging natural allelic variation in hormone signaling components through breeding programs can help develop climate-resilient crop varieties.

References

Abreu ME, Munné-Bosch S (2008) Salicylic acid may be involved in the regulation of drought-induced leaf senescence in perennials: a case study in field-grown *Salvia officinalis* L. plants. Environ Exp Bot 64(2):105–112

Achard P, Cheng H, De Grauwe L, Decat J, Schoutteten H, Moritz T et al (2006) Integration of plant responses to environmentally activated phytohormonal signals. Science 311(5757):91–94

Achard P, Renou JP, Berthomé R, Harberd NP, Genschik P (2008) Plant DELLAs restrain growth and promote survival of adversity by reducing the levels of reactive oxygen species. Curr Biol 18(9):656–660

Ali B, Hayat S, Ahmad A (2007) 28-Homobrassinolide ameliorates the saline stress in chickpea (*Cicer arietinum* L.). Environ Exp Bot 59(2):217–223

Altaf MA, Shahid R, Kumar R, Altaf MM, Kumar A, Khan LU et al (2023) Phytohormones mediated modulation of abiotic stress tolerance and potential crosstalk in horticultural crops. J Plant Growth Regul 42(8):4724–4750

Amzallag GN, Goloubinoff P (2003) An Hsp90 inhibitor, geldanamycin, as a brassinosteroid antagonist: evidence from salt-exposed roots of *Vigna radiata*. Plant Biol 5(02):143–150

Ansari O, Sharif-Zadeh F (2012) Does Gibberelic acid (GA), Salicylic acid (SA) and Ascorbic acid (ASc) improve Mountain Rye (*Secale montanum*) seeds germination and seedlings growth under cold stress. Int Res J Appl Basic Sci 3(8):1651–1657

Antoni R, Gonzalez-Guzman M, Rodriguez L, Peirats-Llobet M, Pizzio GA, Fernandez MA et al (2013) Pyrabactin resistance1-like8 plays an important role for the regulation of abscisic acid signaling in root. Plant Physiol 161(2):931–941

Anuradha S, Rao SSR (2001) Effect of brassinosteroids on salinity stress induced inhibition of seed germination and seedling growth of rice (*Oryza sativa* L.). Plant Growth Regul 33:151–153

Anuradha S, Rao SSR (2007) Effect of 24-epibrassinolide on the growth and antioxidant enzyme activities in radish seedlings under lead toxicity toxicity. Indian J Plant Physiol 12(4):396–400

Arif Y, Sami F, Siddiqui H, Bajguz A, Hayat S (2020) Salicylic acid in relation to other phytohormones in plant: A study towards physiology and signal transduction under challenging environment. Environ Exp Bot 175:104040. https://doi.org/10.1016/j.envexpbot.2020.104040

Atayee AR, Noori MS (2020) Alleviation of cold stress in vegetable crops. J Sci Agric 4:38–44

Azooz MM (2009) Salt stress mitigation by seed priming with salicylic acid in two faba bean genotypes differing in salt tolerance. Int J Agric Biol 11(4):343–350

Baidya A, Anwar Ali M, Atta K (2022) Climate change and abiotic stresses in plants. APDM 351. https://doi.org/10.5772/intechopen.105575

Bailey-Serres J, Voesenek LA (2010) Life in the balance: a signaling network controlling survival of flooding. Curr Opin Plant Biol 13(5):489–494

Bajguz A, Hayat S (2009) Effects of brassinosteroids on the plant responses to environmental stresses. Plant Physiol Biochem 47(1):1–8

Bandurska H, Stroiski A (2005) The effect of salicylic acid on barley response to water deficit. Acta Physiol Plant 27:379–386

Baninasab B (2010) Induction of drought tolerance by salicylic acid in seedlings of cucumber (*Cucumis sativus* L.). J Hortic Sci Biotechnol 85(3):191–196

Batista-Silva W, de Paiva Gonçalves J, Siqueira JA, Martins AO, Ribeiro DM, Nunes-Nesi A et al (2024) Auxin metabolism and the modulation of plant growth. Environ Exp Bot 226:105917

Behnam B, Iuchi, Fujita M, Fujita Y, Takasaki H, Osakabe Y et al (2013) Characterization of the promoter region of an *Arabidopsis* gene for 9-cis-epoxycarotenoid dioxygenase involved in dehydration-inducible transcription. DNA Res 20(4):315–324

Bhatnagar N, Kim R, Han S, Song J, Lee GS, Lee S et al (2020) Ectopic expression of OsPYL/RCAR7, an ABA receptor having low signaling activity, improves drought tolerance without growth defects in rice. Int J Mol Sci 21(11):4163

Bouteraa MT, Ben Romdhane W, Baazaoui N, Alfaifi MY, Chouaibi Y, Ben Akacha B, Ben Saad R (2023) GASA proteins: review of their functions in plant environmental stress tolerance. Plants 12(10):2045

Bücker-Neto L, Paiva ALS, Machado RD, Arenhart RA, Margis-Pinheiro M (2017) Interactions between plant hormones and heavy metals responses. Genet Mol Biol 40(1 suppl 1):373–386

Chai C, Subudhi PK (2016) Comprehensive analysis and expression profiling of the OsLAX and OsABCB auxin transporter gene families in rice (*Oryza sativa*) under phytohormone stimuli and abiotic stresses. Front Plant Sci 7:593

Chang Y, Nguyen BH, Xie Y, Xiao B, Tang N, Zhu W et al (2017) Co-overexpression of the constitutively active form of OsbZIP46 and ABA-activated protein kinase SAPK6 improves drought and temperature stress resistance in rice. Front Plant Sci 8:1102

Chao YY, Chen CY, Huang WD, Kao CH (2010) Salicylic acid-mediated hydrogen peroxide accumulation and protection against Cd toxicity in rice leaves. Plant Soil 329:327–337. https://doi.org/10.1007/s11104-009-0161-4

Chen D, Wang W, Wu Y, Xie H, Zhao L, Zeng Q, Zhan Y (2019) Expression and distribution of the auxin response factors in *Sorghum bicolor* during development and temperature stress. Int J Mol Sci 20(19):4816

Chen K, Wang Y, Zhang R, Zhang H, Gao C (2019b) CRISPR/Cas genome editing and precision plant breeding in agriculture. Annu Rev Plant Biol 70(1):667–697

Chen L, Yang H, Fang Y, Guo W, Chen H, Zhang X et al (2021) Overexpression of *GmMYB14* improves high-density yield and drought tolerance of soybean through regulating plant architecture mediated by the brassinosteroid pathway. Plant Biotechnol J 19(4):702–716

Chen G, Zheng D, Feng N, Zhou H, Mu D, Zhao L et al (2022) Physiological mechanisms of ABA-induced salinity tolerance in leaves and roots of rice. Sci Rep 12(1):8228

Chen S, Zhao CB, Ren RM, Jiang JH (2023) Salicylic acid had the potential to enhance tolerance in horticultural crops against abiotic stress. Front Plant Sci 14:1141918

Cheng F, Lu J, Gao M, Shi K, Kong Q, Huang Y, Bie Z (2016) Redox signaling and CBF-responsive pathway are involved in salicylic acid-improved photosynthesis and growth under chilling stress in watermelon. Front Plant Sci 7:1519

Cheng S, Xue W, Gong X, Hu F, Yang Y, Liu M (2024) Reconciling plant and microbial ecological strategies to elucidate cover crop effects on soil carbon and nitrogen cycling. J Ecol 112:2901

Claeys H, Skirycz A, Maleux K, Inzé D (2012) DELLA signaling mediates stress-induced cell differentiation in *Arabidopsis* leaves through modulation of anaphase-promoting complex/cyclosome activity. Plant Physiol 159(2):739–747. https://doi.org/10.1104/pp.112.195032

Clarke SM, Mur LA, Wood JE, Scott IM (2004) Salicylic acid dependent signaling promotes basal thermotolerance but is not essential for acquired thermotolerance in *Arabidopsis thaliana*. Plant J 38(3):432–447

Cohen JD, Slovin JP, Hendrickson AM (2003) Two genetically discrete pathways convert tryptophan to auxin: more redundancy in auxin biosynthesis. Trends Plant Sci 8(5):197–199

Coltro S, Broetto L, Rotilli MCC, Moraes AJD, Barp FK, Braga GC (2014) Heat shock and salicylic acid on postharvest preservation of organic strawberries. Rev Ceres 61:306–312

Cui XY, Gao Y, Guo J, Yu TF, Zheng WJ, Liu YW et al (2019) BES/BZR transcription factor TaBZR2 positively regulates drought responses by activation of TaGST1. Plant Physiol 180(1):605–620
Cutler SR, Rodriguez PL, Finkelstein RR, Abrams SR (2010) Abscisic acid: emergence of a core signaling network. Annu Rev Plant Biol 61(1):651–679
Dar NA, Amin I, Wani W, Wani SA, Shikari AB, Wani SH, Masoodi KZ (2017) Abscisic acid: A key regulator of abiotic stress tolerance in plants. Plant Gene 11:106–111
Degefu MY, Tesema M (2020) Review of gibberellin signaling. Intl J Eng Appl Sci Technol 4(9):377–390
Delgado C, Mora-Poblete F, Ahmar S, Chen JT, Figueroa CR (2021) Jasmonates and plant salt stress: molecular players, physiological effects, and improving tolerance by using genome-associated tools. Int J Mol Sci 22(6):3082
Devi EL, Devi CP, Kumar S, Sharma SK, Beemrote A, Chongtham SK et al (2017) Marker assisted selection (MAS) towards generating stress tolerant crop plants. Plant Gene 11:205–218
Dhaubhadel S, Browning KS, Gallie DR, Krishna P (2002) Brassinosteroid functions to protect the translational machinery and heat-shock protein synthesis following thermal stress. Plant J 29(6):681–691
Ding P, Ding Y (2020) Stories of salicylic acid: a plant defense hormone. Trends Plant Sci 25(6):549–565
Divi UK, Krishna P (2009) Brassinosteroid: a biotechnological target for enhancing crop yield and stress tolerance. New Biotechnol 26(3–4):131–136
Divi UK, Krishna P (2010) Overexpression of the brassinosteroid biosynthetic gene *AtDWF4* in *Arabidopsis* seeds overcomes abscisic acid-induced inhibition of germination and increases cold tolerance in transgenic seedlings. J Plant Growth Regul 29:385–393
Du H, Wu N, Fu J, Wang S, Li X, Xiao J, Xiong L (2012) A GH3 family member, OsGH3-2, modulates auxin and abscisic acid levels and differentially affects drought and cold tolerance in rice. J Exp Bot 63(18):6467–6480
Du H, Liu H, Xiong L (2013a) Endogenous auxin and jasmonic acid levels are differentially modulated by abiotic stresses in rice. Front Plant Sci 4:397
Du H, Wu N, Chang Y, Li X, Xiao J, Xiong L (2013b) Carotenoid deficiency impairs ABA and IAA biosynthesis and differentially affects drought and cold tolerance in rice. Plant Mol Biol 83:475–488
Duan F, Ding J, Lee D, Lu X, Feng Y, Song W (2017) Overexpression of *SoCYP85A1*, a spinach cytochrome p450 gene in transgenic tobacco enhances root development and drought stress tolerance. Front Plant Sci 8:1909
Eckardt NA, Cutler S, Juenger TE et al (2023) Focus on climate change and plant abiotic stress biology. Plant Cell 35:1–3. https://doi.org/10.1093/PLCELL/KOAC329
El-Abd SO, El-Beltagy AS, Hall MA (1984) Physiological studies on flowering and fruit set in tomatoes. In: Symposium on tomato production on arid land, vol 190, pp 389–396
EL Sabagh A, Islam MS, Hossain A, Iqbal MA, Mubeen M, Waleed M, Reginato M, Battaglia M, Ahmed S, Rehman A, Arif M (2022) Phytohormones as growth regulators during abiotic stress tolerance in plants. Front Agron 4:765068
Elsisi M, Elshiekh M, Sabry N, Aziz M, Attia K, Islam F, Chen J, Abdelrahman M (2024) The genetic orchestra of salicylic acid in plant resilience to climate change induced abiotic stress: critical review. Stress Biol 4:31. https://doi.org/10.1007/s44154-024-00160-2
Emamverdian A, Ding Y, Mokhberdoran F (2020) The role of salicylic acid and gibberellin signaling in plant responses to abiotic stress with an emphasis on heavy metals. Plant Signal Behav 15(7):1777372
Eremina M, Unterholzner SJ, Rathnayake AI, Castellanos M, Khan M, Kugler KG, Poppenberger B (2016) Brassinosteroids participate in the control of basal and acquired freezing tolerance of plants. Proc Natl Acad Sci 113(40):E5982–E5991
Fang BAO, Yang LJ (2002) Evidence that the auxin signaling pathway interacts with plant stress response. J Integr Plant Biol 44(5):532

Fariduddin Q, Mir BA, Yusuf M, Ahmad A (2014) 24-epibrassinolide and/or putrescine trigger physiological and biochemical responses for the salt stress mitigation in *Cucumis sativus* L. Photosynthetica 52(3):464–474

Fattorini L, Ronzan M, Piacentini D, Della Rovere F, De Virgilio C, Sofo A et al (2017) Cadmium and arsenic affect quiescent centre formation and maintenance in *Arabidopsis thaliana* post-embryonic roots disrupting auxin biosynthesis and transport. Environ Exp Bot 144:37–48

Franklin KA, Lee SH, Patel D, Kumar SV, Spartz AK, Gu C et al (2011) Phytochrome-interacting factor 4 (PIF4) regulates auxin biosynthesis at high temperature. Proc Natl Acad Sci 108(50):20231–20235

Friedrichsen DM, Joazeiro CA, Li J, Hunter T, Chory J (2000) Brassinosteroid-insensitive-1 is a ubiquitously expressed leucine-rich repeat receptor serine/threonine kinase. Plant Physiol 123(4):1247–1256

Fu X, Feng YQ, Zhang XW, Zhang YY, Bi HG, Ai XZ (2021) Salicylic acid is involved in rootstock–scion communication in improving the chilling tolerance of grafted cucumber. Front Plant Sci 12:693344

Fukao T, Yeung E, Bailey-Serres J (2011) The submergence tolerance regulator SUB1A mediates crosstalk between submergence and drought tolerance in rice. Plant Cell 23(1):412–427. https://doi.org/10.1105/tpc.110.080325

Gallego-Bartolomé J, Minguet EG, Grau-Enguix F, Abbas M, Locascio A, Thomas SG, Alabadí D, Blázquez MA (2012) Molecular mechanism for the interaction between gibberellin and brassinosteroid signaling pathways in *Arabidopsis*. Proc Natl Acad Sci USA 109(33):13446–13451. https://doi.org/10.1073/pnas.1119992109

Gharib FA, Hegazi AZ (2010) Salicylic acid ameliorates germination, seedling growth, phytohormone and enzymes activity in bean (*Phaseolus vulgaris* L.) under cold stress. J Am Sci 6(10):675–683

Golldack D, Li C, Mohan H, Probst N (2013) Gibberellins and abscisic acid signal crosstalk: living and developing under unfavorable conditions. Plant Cell Rep 32:1007–1016

Gomes GLB, Scortecci KC (2021) Auxin and its role in plant development: structure, signalling, regulation and response mechanisms. Plant Biol 23(6):894–904

Gu Q, Xiao Q, Chen Z, Han Y (2022) Crosstalk between melatonin and reactive oxygen species in plant abiotic stress responses: an update. Int J Mol Sci 23(10):5666

Guo R, Wang X, Han X, Chen X, Wang-Pruski G (2020) Physiological and transcriptomic responses of water spinach (*Ipomoea aquatica*) to prolonged heat stress. BMC Genomics 21:1–15

Gupta A, Rico-Medina A, Lozano-Elena F, Marqués-Bueno M, Fontanet JB, Fàbregas N et al (2023) Brassinosteroid receptor BRL3 triggers systemic plant adaptation to elevated temperature from the phloem cells. bioRxiv:2023-03

Gurmani AR, Bano A, Salim M (2006) Effect of growth regulators on growth, yield and ions accumulation of rice (*Oryza sativa* L.) under salt stress. Pak J Bot 38(5):1415–1424

Gurmani AR, Bano A, Khan SU, Din J, Zhang JL (2011) Alleviation of salt stress by seed treatment with abscisic acid (ABA), 6-benzylaminopurine (BA) and chlormequat chloride (CCC) optimizes ion and organic matter accumulation and increases yield of rice (*Oryza sativa* L.). Aust J Crop Sci 5(10):1278–1285

Gurmani AR, Bano A, Ullah N, Khan H, Jahangir M, Flowers TJ (2013) Exogenous abscisic acid (ABA) and silicon (Si) promote salinity tolerance by reducing sodium (Na+) transport and bypass flow in rice (*Oryza sativa* indica). Aust J Crop Sci 7(9):1219–1226

Habibpourmehraban F, Wu Y, Masoomi-Aladizgeh F, Amirkhani A, Atwell BJ, Haynes PA (2023) Pre-treatment of rice plants with ABA makes them more tolerant to multiple abiotic stress. Int J Mol Sci 24(11):9628

Hai NN, Chuong NN, Tu NHC, Kisiala A, Hoang XLT, Thao NP (2020) Role and regulation of Cytokinins in plant response to drought stress. Plants (Basel, Switz) 9(4):422. https://doi.org/10.3390/plants9040422

Hanaka A, Ozimek E, Reszczyńska E, Jaroszuk-Ściseł J, Stolarz M (2021) Plant tolerance to drought stress in the presence of supporting bacteria and fungi: an efficient strategy in horticulture. Horticulturae 7(10):390

Hao Q, Wang W, Han X, Wu J, Lyu B, Chen F et al (2018) Isochorismate-based salicylic acid biosynthesis confers basal resistance to *Fusarium graminearum* in barley. Mol Plant Pathol 19(8):1995–2010

Hasan SA, Hayat S, Ahmad A (2011) Brassinosteroids protect photosynthetic machinery against the cadmium induced oxidative stress in two tomato cultivars. Chemosphere 84(10):1446–1451

Hayat S, Ahmad A, Fariduddin Q (2003) Brassinosteroids: a regulator of 21st century. In: Brassinosteroids: bioactivity and crop productivity. Springer Netherlands, Dordrecht, pp 231–246

Hayat SHAMSUL, Mori M, Fariduddin QAZI, Bajguz ANDRZEJ, Ahmad A (2010) Physiological role of brassinosteroids: an update. Indian J Plant Physiol 15:99–109

He XJ, Mu RL, Cao WH, Zhang ZG, Zhang JS, Chen SY (2005) *AtNAC2*, a transcription factor downstream of ethylene and auxin signaling pathways, is involved in salt stress response and lateral root development. Plant J 44(6):903–916

Hedden P, Thomas SG (2012) Gibberellin biosynthesis and its regulation. Biochem J 444(1):11–25. https://doi.org/10.1042/BJ20120245

Hentrich M, Böttcher C, Düchting P, Cheng Y, Zhao Y, Berkowitz O et al (2013) The jasmonic acid signaling pathway is linked to auxin homeostasis through the modulation of *YUCCA 8* and *YUCCA 9* gene expression. Plant J 74(4):626–637

Hirano K, Kouketu E, Katoh H, Aya K, Ueguchi-Tanaka M, Matsuoka M (2012) The suppressive function of the rice DELLA protein SLR1 is dependent on its transcriptional activation activity. Plant J Cell Mol Biol 71(3):443–453. https://doi.org/10.1111/j.1365-313X.2012.05000.x

Holland CK, Westfall CS, Schaffer JE, De Santiago A, Zubieta C, Alvarez S, Jez JM (2019) Brassicaceae-specific Gretchen Hagen 3 acyl acid amido synthetases conjugate amino acids to chorismate, a precursor of aromatic amino acids and salicylic acid. J Biol Chem 294(45):16855–16864

Hönig M, Plíhalová L, Husičková A, Nisler J, Doležal K (2018) Role of Cytokinins in senescence, antioxidant Defence and photosynthesis. Int J Mol Sci 19(12):4045. https://doi.org/10.3390/ijms19124045

Hou X, Ding L, Yu H (2013) Crosstalk between GA and JA signaling mediates plant growth and defense. Plant Cell Rep 32:1067–1074

Huang Y, Guo Y, Liu Y, Zhang F, Wang Z, Wang H et al (2018) 9-cis-epoxycarotenoid dioxygenase 3 regulates plant growth and enhances multi-abiotic stress tolerance in rice. Front Plant Sci 9:162

Huang Y, Jiao Y, Xie N, Guo Y, Zhang F, Xiang Z et al (2019) *OsNCED5*, a 9-cis-epoxycarotenoid dioxygenase gene, regulates salt and water stress tolerance and leaf senescence in rice. Plant Sci 287:110188

Huang C, Liao J, Huang W, Qin N (2022) Salicylic acid protects sweet potato seedlings from drought stress by mediating abscisic acid-related gene expression and enhancing the antioxidant defense system. Int J Mol Sci 23(23):14819

Ibañez C, Delker C, Martinez C, Bürstenbinder K, Janitza P, Lippmann R et al (2018) Brassinosteroids dominate hormonal regulation of plant thermomorphogenesis via BZR1. Curr Biol 28(2):303–310

Ignatenko A, Talanova V, Repkina N, Titov A (2019) Exogenous salicylic acid treatment induces cold tolerance in wheat through promotion of antioxidant enzyme activity and proline accumulation. Acta Physiol Plant 41(6):80

Imran QM, Falak N, Hussain A et al (2021) Abiotic stress in plants, stress perception to molecular response and role of biotechnological tools in stress resistance. Agronomy 11:1579. https://doi.org/10.3390/agronomy11081579

Inoue T, Higuchi M, Hashimoto Y, Seki M, Kobayashi M, Kato T et al (2001) Identification of *CRE1* as a cytokinin receptor from *Arabidopsis*. Nature 409(6823):1060–1063

Islam MS, Hasan MK, Islam B, Renu NA, Hakim MA, Islam MR et al (2021) Responses of water and pigments status, dry matter partitioning, seed production, and traits of yield and quality to foliar application of GA3 in mugnbean (*Vigna radiata* L.). Front Agron 2:596850. https://doi.org/10.3389/fagro.2020.596850

Iuchi S, Kobayashi M, Taji T, Naramoto M, Seki M, Kato T, Tabata S, Kakubari Y, Yamaguchi-Shinozaki K, Shinozaki K (2001) Regulation of drought tolerance by gene manipulation of 9-cis-epoxycarotenoid dioxygenase, a key enzyme in abscisic acid biosynthesis in *Arabidopsis*. Plant J 27(4):325–333

Jacob P, Hirt H, Bendahmane A (2017) The heat-shock protein/chaperone network and multiple stress resistance. Plant Biotechnol J 15(4):405–414

Jain M, Khurana JP (2009) Transcript profiling reveals diverse roles of auxin-responsive genes during reproductive development and abiotic stress in rice. FEBS J 276(11):3148–3162

Janda T, Szalai G, Tari I, Paldi E (1999) Hydroponic treatment with salicylic acid decreases the effects of chilling injury in maize (*Zea mays* L.) plants. Planta 208:175–180

Jeon J, Kim NY, Kim S, Kang NY, Novák O, Ku SJ, Cho C, Lee DJ, Lee EJ, Strnad M, Kim J (2010) A subset of cytokinin two-component signaling system plays a role in cold temperature stress response in *Arabidopsis*. J Biol Chem 285(30):23371–23386. https://doi.org/10.1074/jbc.M109.096644

Jiang YP, Huang LF, Cheng F, Zhou YH, Xia XJ, Mao WH et al (2013) Brassinosteroids accelerate recovery of photosynthetic apparatus from cold stress by balancing the electron partitioning, carboxylation and redox homeostasis in cucumber. Physiol Plant 148(1):133–145

Jing CH, Cheng Z, Li-ping L, Zhong-yang S, Xue-bo P (2007) Effects of exogenous salicylic acid on growth and H_2O_2-metabolizing enzymes in rice seedlings under lead stress. J Environ Sci 19:44–49

Joseph B, Jini D, Sujatha S (2010) Insight into the role of exogenous salicylic acid on. Asian J Crop Sci 2(4):226–235

Jung JH, Park CM (2011) Auxin modulation of salt stress signaling in *Arabidopsis* seed germination. Plant Signal Behav 6(8):1198–1200

Jung H, Lee DK, Do Choi Y, Kim JK (2015) *OsIAA6*, a member of the rice aux/IAA gene family, is involved in drought tolerance and tiller outgrowth. Plant Sci 236:304–312

Kang NY, Cho C, Kim NY, Kim J (2012) Cytokinin receptor-dependent and receptor-independent pathways in the dehydration response of *Arabidopsis thaliana*. J Plant Physiol 169(14):1382–1391. https://doi.org/10.1016/j.jplph.2012.05.007

Kanwar MK, Bhardwaj R, Chowdhary SP, Arora P, Sharma P, Kumar S (2013) Isolation and characterization of 24-Epibrassinolide from *Brassica juncea* L. and its effects on growth, Ni ion uptake, antioxidant defense of *Brassica* plants and in vitro cytotoxicity. Acta Physiol Plant 35:1351–1362

Karlidag H, Yildirim E, Turan M (2009) Salicylic acid ameliorates the adverse effect of salt stress on strawberry. Sci Agric 66:180–187

Kasahara H, Hanada A, Kuzuyama T, Takagi M, Kamiya Y, Yamaguchi S (2002) Contribution of the mevalonate and methylerythritol phosphate pathways to the biosynthesis of gibberellins in *Arabidopsis*. J Biol Chem 277(47):45188–45194. https://doi.org/10.1074/jbc.M208659200

Kaya C (2021) Nitrate reductase is required for salicylic acid-induced water stress tolerance of pepper by upraising the AsA-GSH pathway and glyoxalase system. Physiol Plant 172(2):351–370

Ke Q, Wang Z, Ji CY, Jeong JC, Lee HS, Li H et al (2015) Transgenic poplar expressing *Arabidopsis YUCCA6* exhibits auxin-overproduction phenotypes and increased tolerance to abiotic stress. Plant Physiol Biochem 94:19–27

Khan AL, Hamayun M, Waqas M, Kang SM, Kim YH, Kim DH, Lee IJ (2012a) *Exophiala* sp. LHL08 association gives heat stress tolerance by avoiding oxidative damage to cucumber plants. Biol Fertil Soils 48:519–529

Khan MIR, Syeed S, Nazar R, Anjum NA (2012b) An insight into the role of salicylic acid and jasmonic acid in salt stress tolerance. In: Phytohormones and abiotic stress tolerance in plants, pp 277–300

Khan MIR, Asgher M, Khan NA (2014) Alleviation of salt-induced photosynthesis and growth inhibition by salicylic acid involves glycinebetaine and ethylene in mungbean (*Vigna radiata* L.). Plant Physiol Biochem 80:67–74

Khan MIR, Kumari S, Nazir F, Khanna RR, Gupta R, Chhillar H (2023) Defensive role of plant hormones in advancing abiotic stress-resistant rice plants. Rice Sci 30(1):15–35

Khripach V, Zhabinskiĭ VN, Lakhvich FA (1995) Prospects for the practical use of brassinosteroids-a new class of phytohormones

Khripach V, Zhabinskii V, de Groot A (2000) Twenty years of brassinosteroids: steroidal plant hormones warrant better crops for the XXI century. Ann Bot 86(3):441–447

Kim TW, Hwang JY, Kim YS, Joo SH, Chang SC, Lee JS et al (2005) *Arabidopsis CYP85A2*, a cytochrome P450, mediates the Baeyer-Villiger oxidation of castasterone to brassinolide in brassinosteroid biosynthesis. Plant Cell 17(8):2397–2412

Kim H, Hwang H, Hong JW, Lee YN, Ahn IP, Yoon IS et al (2012) A rice orthologue of the ABA receptor, *OsPYL/RCAR5*, is a positive regulator of the ABA signal transduction pathway in seed germination and early seedling growth. J Exp Bot 63(2):1013–1024

Kim J, Baek D, Park HC, Chun HJ, Oh DH, Lee MK et al (2013) Overexpression of *Arabidopsis YUCCA6* in potato results in high-auxin developmental phenotypes and enhanced resistance to water deficit. Mol Plant 6(2):337–349

Kim Y, Kim S, Shim IS (2017) Exogenous salicylic acid alleviates salt-stress damage in cucumber under moderate nitrogen conditions by controlling endogenous salicylic acid levels. Hortic Environ Biotechnol 58:247–253

Kobayashi Y, Yamamoto S, Minami H, Kagaya Y, Hattori T (2004) Differential activation of the rice sucrose nonfermenting1–related protein kinase2 family by hyperosmotic stress and abscisic acid. Plant Cell 16(5):1163–1177

Kollmeier M, Felle HH, Horst WJ (2000) Genotypical differences in aluminum resistance of maize are expressed in the distal part of the transition zone. Is reduced basipetal auxin flow involved in inhibition of root elongation by aluminum? Plant Physiol 122(3):945–956

Koornneef M, van der Veen JH (1980) Induction and analysis of gibberellin sensitive mutants in *Arabidopsis thaliana* (L.) Heynh. Theor Appl Genet 58:257–263

Korkmaz A, Uzunlu M, Demirkiran AR (2007) Treatment with acetyl salicylic acid protects muskmelon seedlings against drought stress. Acta Physiol Plant 29:503–508

Korver RA, Koevoets IT, Testerink C (2018) Out of shape during stress: a key role for auxin. Trends Plant Sci 23(9):783–793

Kosová K, Prášil IT, Vítámvás P, Dobrev P, Motyka V, Floková K, Trávničková A (2012) Complex phytohormone responses during the cold acclimation of two wheat cultivars differing in cold tolerance, winter Samanta and spring Sandra. J Plant Physiol 169:567–576. https://doi.org/10.1016/j.jplph.2011.12.013

Krishnamurthy A, Rathinasabapathi B (2013) Auxin and its transport play a role in plant tolerance to arsenite-induced oxidative stress in *Arabidopsis thaliana*. Plant Cell Environ 36(10):1838–1849

Kumar R, Das SP, Choudhury BU (2024) Advances in genomic tools for plant breeding: harnessing DNA molecular markers, genomic selection, and genome editing. Biol Res 57:80. https://doi.org/10.1186/s40659-024-00562-6

Kundu P, Nehra A, Gill R et al (2022) Unraveling the importance of EF-hand mediated calcium signaling in plants. S Afr J Bot 148:615–633. https://doi.org/10.1016/J.SAJB.2022.04.045

Lackman P, González-Guzmán M, Tilleman S, Carqueijeiro I, Pérez AC, Moses T et al (2011) Jasmonate signaling involves the abscisic acid receptor PYL4 to regulate metabolic reprogramming in *Arabidopsis* and tobacco. Proc Natl Acad Sci 108(14):5891–5896

Lantzouni O, Alkofer A, Falter-Braun P, Schwechheimer C (2020) Growth-regulating factors interact with DELLAs and regulate growth in cold stress. Plant Cell 32(4):1018–1034. https://doi.org/10.1105/tpc.19.00784

Lee SC, Luan S (2012) ABA signal transduction at the crossroad of biotic and abiotic stress responses. Plant Cell Environ 35(1):53–60

Lee M, Jung JH, Han DY, Seo PJ, Park WJ, Park CM (2012) Activation of a flavin monooxygenase gene *YUCCA7* enhances drought resistance in Arabidopsis. Planta 235:923–938
Lee BR, Islam MT, Park SH, Jung HI, Bae DW, Kim TH (2019) Characterization of salicylic acid-mediated modulation of the drought stress responses: reactive oxygen species, proline, and redox state in *Brassica napus*. Environ Exp Bot 157:1–10
Lefevere H, Bauters L, Gheysen G (2020) Salicylic acid biosynthesis in plants. Front Plant Sci 11:338
Li G, Peng X, Wei L, Kang G (2013) Salicylic acid increases the contents of glutathione and ascorbate and temporally regulates the related gene expression in salt-stressed wheat seedlings. Gene 529(2):321–325
Li H, Ye K, Shi Y, Cheng J, Zhang X, Yang S (2017) BZR1 positively regulates freezing tolerance via CBF-dependent and CBF-independent pathways in *Arabidopsis*. Mol Plant 10(4):545–559
Liao Z, Zhang Y, Yu Q, Fang W, Chen M, Li T et al (2023) Coordination of growth and drought responses by GA-ABA signaling in rice. New Phytol 240(3):1149–1161
Liu P, Xu ZS, Pan-Pan L, Hu D, Chen M, Li LC, Ma YZ (2013) A wheat *PI4K* gene whose product possesses threonine autophophorylation activity confers tolerance to drought and salt in *Arabidopsis*. J Exp Bot 64(10):2915–2927
Liu H, Ren X, Zhu J, Wu X, Liang C (2018) Effect of exogenous abscisic acid on morphology, growth and nutrient uptake of rice (*Oryza sativa*) roots under simulated acid rain stress. Planta 248:647–659
Liu XL, Zhang H, Jin YY, Wang MM, Yang HY, Ma HY et al (2019) Abscisic acid primes rice seedlings for enhanced tolerance to alkaline stress by upregulating antioxidant defense and stress tolerance-related genes. Plant Soil 438:39–55
Lopez-Delgado H, Dat JF, Foyer CH, Scott IM (1998) Induction of thermotolerance in potato microplants by acetylsalicylic acid and H_2O_2. J Exp Bot 49(321):713–720
Magome H, Yamaguchi S, Hanada A, Kamiya Y, Oda K (2008) The DDF1 transcriptional activator upregulates expression of a gibberellin-deactivating gene, GA2ox7, under high-salinity stress in *Arabidopsis*. Plant J Cell Mol Biol 56(4):613–626. https://doi.org/10.1111/j.1365-313X.2008.03627.x
Malhi GS, Kaur M, Kaushik P (2021) Impact of climate change on agriculture and its mitigation strategies: a review. Sustainability 13:1–21. https://doi.org/10.3390/su13031318
Mandal S, Ghorai M, Anand U, Roy D, Kant N, Mishra T, Mane AB, Jha NK, Lal MK, Tiwari RK, Kumar M, Radha, Ghosh A, Bhattacharjee R, Proćków J, Dey A (2022) Cytokinins: A genetic target for increasing yield potential in the CRISPR era. Front Genet 13:883930. https://doi.org/10.3389/fgene.2022.883930
Manghwar H, Hussain A, Ali Q, Liu F (2022) Brassinosteroids (BRs) role in plant development and coping with different stresses. Int J Mol Sci 23(3):1012
Mano Y, Nemoto K (2012) The pathway of auxin biosynthesis in plants. J Exp Bot 63(8):2853–2872
Mäser P, Eckelman B, Vaidyanathan R, Horie T, Fairbairn DJ, Kubo M, Yamagami M, Yamaguchi K, Nishimura M, Uozumi N, Robertson W (2002) Altered shoot/root Na^+ distribution and bifurcating salt sensitivity in *Arabidopsis* by genetic disruption of the Na^+ transporter AtHKT1. FEBS Lett 531(2):157–161
Mashiguchi K, Tanaka K, Sakai T, Sugawara S, Kawaide H, Natsume M et al (2011) The main auxin biosynthesis pathway in *Arabidopsis*. Proc Natl Acad Sci 108(45):18512–18517
Miao C, Xiao L, Hua K, Zou C, Zhao Y, Bressan RA, Zhu JK (2018) Mutations in a subfamily of abscisic acid receptor genes promote rice growth and productivity. Proc Natl Acad Sci 115(23):6058–6063
Miao J, Li X, Li X, Tan W, You A, Wu S et al (2020) OsPP2C09, a negative regulatory factor in abscisic acid signalling, plays an essential role in balancing plant growth and drought tolerance in rice. New Phytol 227(5):1417–1433
Mir RA, Somasundaram R (2021) Salicylic acid and salt stress tolerance in plants: a review. J Stress Physiol Biochem 17(3):32–50

Mishra AK, Baek KH (2021) Salicylic acid biosynthesis and metabolism: a divergent pathway for plants and bacteria. Biomolecules 11(5):705
Mohamed HI, El-Shazly HH, Badr A (2020) Role of salicylic acid in biotic and abiotic stress tolerance in plants. In: Plant phenolics in sustainable agriculture, vol 1, pp 533–554
Morris DA (1979) The effect of temperature on the velocity of exogenous auxin transport in intact chilling-sensitive and chilling-resistant plants. Planta 146:603–605
Mozafari AA, Havas F, Ghaderi N (2018) Application of iron nanoparticles and salicylic acid in in vitro culture of strawberries (*Fragaria*× *ananassa* Duch.) to cope with drought stress. Plant Cell Tissue Organ Culture (PCTOC) 132:511–523
Mukherjee A, Gaurav AK, Singh S et al (2022) The bioactive potential of phytohormones: a review. Biotechnol Rep 35:e00748. https://doi.org/10.1016/j.btre.2022.e00748
Munne-Bosch S, Penuelas J (2003) Photo-and antioxidative protection, and a role for salicylic acid during drought and recovery in field-grown *Phillyrea angustifolia* plants. Planta 217:758–766
Munns R, Tester M (2008) Mechanisms of salinity tolerance. Annu Rev Plant Biol 59(1):651–681
Munns R, James RA, Xu B, Athman A, Conn SJ, Jordans C et al (2012) Wheat grain yield on saline soils is improved by an ancestral Na+ transporter gene. Nat Biotechnol 30(4):360–364
Mutlu S, Karadağoğlu Ö, Atici Ö, Nalbantoğlu B (2013) Protective role of salicylic acid applied before cold stress on antioxidative system and protein patterns in barley apoplast. Biol Plant 57(3):507–513
Naeem M, Basit A, Ahmad I, Mohamed HI, Wasila H (2020) Effect of salicylic acid and salinity stress on the performance of tomato plants. Gesunde Pflanzen 72(4):393
Napieraj N, Janicka M, Reda M (2023) Interactions of polyamines and phytohormones in plant response to abiotic stress. Plants (Basel, Switz) 12(5):1159. https://doi.org/10.3390/plants12051159
Nazar R, Iqbal N, Syeed S, Khan NA (2011) Salicylic acid alleviates decreases in photosynthesis under salt stress by enhancing nitrogen and sulfur assimilation and antioxidant metabolism differentially in two mungbean cultivars. J Plant Physiol 168(8):807–815
Nie WF, Wang MM, Xia XJ, Zhou YH, Shi K, Chen Z, Yu JQ (2013) Silencing of tomato *RBOH1* and *MPK2* abolishes brassinosteroid-induced H2O2 generation and stress tolerance. Plant Cell Environ 36(4):789–803
Nigam B, Dubey RS, Rathore D (2022) Protective role of exogenously supplied salicylic acid and PGPB (*Stenotrophomonas sp.*) on spinach and soybean cultivars grown under salt stress. Sci Hortic 293:110654
Nijabat A, Bolton A, Mahmood-ur-Rehman M, Shah AI, Hussain R, Naveed NH et al (2020) Cell membrane stability and relative cell injury in response to heat stress during early and late seedling stages of diverse carrot (*Daucus carota* L.) germplasm. HortScience 55(9):1446–1452
Nishiyama R, Watanabe Y, Fujita Y, Le DT, Kojima M, Werner T, Vankova R, Yamaguchi-Shinozaki K, Shinozaki K, Kakimoto T, Sakakibara H, Schmülling T, Tran LS (2011) Analysis of cytokinin mutants and regulation of cytokinin metabolic genes reveals important regulatory roles of cytokinins in drought, salt and abscisic acid responses, and abscisic acid biosynthesis. Plant Cell 23(6):2169–2183. https://doi.org/10.1105/tpc.111.087395
Nishiyama R, Watanabe Y, Leyva-Gonzalez MA, Van Ha C, Fujita Y, Tanaka M et al (2013) *Arabidopsis* AHP2, AHP3, and AHP5 histidine phosphotransfer proteins function as redundant negative regulators of drought stress response. Proc Natl Acad Sci 110(12):4840–4845
Nomura T, Bishop GJ (2006) Cytochrome P450s in plant steroid hormone synthesis and metabolism. Phytochem Rev 5(2):421–432
O'Neill DP, Davidson SE, Clarke VC, Yamauchi Y, Yamaguchi S, Kamiya Y, Reid JB, Ross JJ (2010) Regulation of the gibberellin pathway by auxin and DELLA proteins. Planta 232(5):1141–1149. https://doi.org/10.1007/s00425-010-1248-0
Ogweno JO, Song XS, Shi K, Hu WH, Mao WH, Zhou YH et al (2008) Brassinosteroids alleviate heat-induced inhibition of photosynthesis by increasing carboxylation efficiency and enhancing antioxidant systems in *Lycopersicon esculentum*. J Plant Growth Regul 27:49–57

Palma F, López-Gómez M, Tejera NA, Lluch C (2013) Salicylic acid improves the salinity tolerance of *Medicago sativa* in symbiosis with *Sinorhizobium meliloti* by preventing nitrogen fixation inhibition. Plant Sci 208:75–82

Pang Y, Cao L, Ye F, Ma C, Liang X, Song Y, Lu X (2024) Identification of the maize *PP2C* gene family and functional studies on the role of *ZmPP2C15* in drought tolerance. Plants 13(3):340

Peleg Z, Blumwald E (2011) Hormone balance and abiotic stress tolerance in crop plants. Curr Opin Plant Biol 14(3):290–295

Peleg Z, Reguera M, Tumimbang E, Walia H, Blumwald E (2011) Cytokinin-mediated source/sink modifications improve drought tolerance and increase grain yield in rice under water-stress. Plant Biotechnol J 9(7):747–758

Pieterse CM, Leon-Reyes A, Van der Ent S, Van Wees SC (2009) Networking by small-molecule hormones in plant immunity. Nat Chem Biol 5(5):308–316

Pieterse CM, Van der Does D, Zamioudis C, Leon-Reyes A, Van Wees SC (2012) Hormonal modulation of plant immunity. Annu Rev Cell Dev Biol 28(1):489–521

Pirasteh-Anosheh H, Ranjbar G, Pakniyat H, Emam Y (2016) Physiological mechanisms of salt stress tolerance in plants: an overview. In: Plant-environment interaction: responses and approaches to mitigate stress, pp 141–160

Potters G, Pasternak TP, Guisez Y, Palme KJ, Jansen MA (2007) Stress-induced morphogenic responses: growing out of trouble? Trends Plant Sci 12(3):98–105

Prakash M, Kannan K, Kumar JS, Veeramani BB, Ganeshan J (1999) Effect of brassinosteroids on germination and seedling behaviour of tomato. Ann Plant Physiol 13(2):178–180

Prakash V, Singh VP, Tripathi DK, Sharma S, Corpas FJ (2021) Nitric oxide (NO) and salicylic acid (SA): a framework for their relationship in plant development under abiotic stress. Plant Biol 23:39–49

Purohit A, Kounain S, Madala C, Chavan S, Guguloth N, Kethavathu S, Behera PP, Penna S, Thakur NR (2026) Sustainable crops for future food security: challenges, innovations, and solutions. In Penna S, Varshney RK (eds) Improving crops for future sustainability and climate resilience. Springer, Singapore. pp 1–40. https://doi.org/10.1007/978-981-95-2754-0_1

Raghavendra AS, Gonugunta VK, Christmann A, Grill E (2010) ABA perception and signalling. Trends Plant Sci 15(7):395–401

Rai GK, Khanday DM, Choudhary SM, Kumar P, Kumari S, Martínez-Andújar C, Martinez-Melgarejo PA, Rai PK, Pérez-Alfocea F (2024) Unlocking nature's stress buster: Abscisic acid's crucial role in defending plants against abiotic stress. Plant Stress 11:100359

Ram A, Verma P, Gadi BR (2014) Effect of fluoride and salicylic acid on seedling growth and biochemical parameters of watermelon (*Citrullus lanatus*). Fluoride 47(1):49–55

Ramachandran M, Arulbalachandran D, Dilipan E, Ramya S (2021) Comparative analysis of abscisic acid recovery on two varieties of rice (*Oryza sativa* L.) under drought condition. Biocatal Agric Biotechnol 33:102006

Ramakrishna B, Rao SSR (2015) Foliar application of brassinosteroids alleviates adverse effects of zinc toxicity in radish (*Raphanus sativus* L.) plants. Protoplasma 252:665–677

Raza A, Razzaq A, Mehmood SS et al (2019) Impact of climate change on crops adaptation and strategies to tackle its outcome: a review. Plants 8:34. https://doi.org/10.3390/plants8020034

Raza A, Charagh S, Zahid Z, Mubarik MS, Javed R, Siddiqui MH, Hasanuzzaman M (2021) Jasmonic acid: a key frontier in conferring abiotic stress tolerance in plants. Plant Cell Rep 40(8):1513–1541

Ronzan M, Piacentini D, Fattorini L, Della Rovere F, Eiche E, Riemann M et al (2018) Cadmium and arsenic affect root development in (*Oryza sativa* L.) negatively interacting with auxin. Environ Exp Bot 151:64–75

Roshdy AED, Alebidi A, Almutairi K, Al-Obeed R, Elsabagh A (2021) The effect of salicylic acid on the performances of salt stressed strawberry plants, enzymes activity, and salt tolerance index. Agronomy 11(4):775

Safari H, Hosseini SM, Azari A, Rafsanjani MH (2018) Effects of seed priming with ABA and SA on seed germination and seedling growth of sesame (*Sesamum indicum* L.) under saline condition. Aust J Crop Sci 12(9):1385–1392

Sah SK, Reddy KR, Li J (2016) Abscisic acid and abiotic stress tolerance in crop plants. Front Plant Sci 7:571

Saini S, Sharma I, Kaur N, Pati PK (2013) Auxin: a master regulator in plant root development. Plant Cell Rep 32:741–757

Saini K, AbdElgawad H, Markakis MN, Schoenaers S, Asard H, Prinsen E et al (2017) Perturbation of auxin homeostasis and signaling by *PINOID* overexpression induces stress responses in *Arabidopsis*. Front Plant Sci 8:1308

Saini N, Nikalje GC, Zargar SM, Suprasanna P (2021) Molecular insights into sensing, regulation and engineering of heat tolerance in plants. Plant Cell Rep 41(3):799–813

Sakuma Y, Liu Q, Dubouzet JG, Abe H, Shinozaki K, Yamaguchi-Shinozaki K (2002) DNA-binding specificity of the ERF/AP2 domain of *Arabidopsis* DREBs, transcription factors involved in dehydration-and cold-inducible gene expression. Biochem Biophys Res Commun 290(3):998–1009

Santisree P, Jalli LCL, Bhatnagar-Mathur P, Sharma KK (2020) Emerging roles of salicylic acid and jasmonates in plant abiotic stress responses. In: Protective chemical agents in the amelioration of plant abiotic stress: biochemical and molecular perspectives, pp 342–373

Santner A, Estelle M (2009) Recent advances and emerging trends in plant hormone signalling. Nature 459(7250):1071–1078

Santos LR, Batista BL, Lobato AKS (2018) Brassinosteroids mitigate cadmium toxicity in cowpea plants. Photosynthetica 56(2):591–605

Sato H, Takasaki H, Takahashi F, Suzuki T, Iuchi S, Mitsuda N et al (2018) *Arabidopsis thaliana* NGATHA1 transcription factor induces ABA biosynthesis by activating *NCED3* gene during dehydration stress. Proc Natl Acad Sci 115(47):E11178–E11187

Schachtman DP, Goodger JQ (2008) Chemical root to shoot signaling under drought. Trends Plant Sci 13(6):281–287. https://doi.org/10.1016/j.tplants.2008.04.003

Seleiman MF, Al-Suhaibani N, Ali N, Akmal M, Alotaibi M, Refay Y, Dindaroglu T, Abdul-Wajid HH, Battaglia ML (2021) Drought stress impacts on plants and different approaches to alleviate its adverse effects. Plants (Basel, Switz) 10(2):259. https://doi.org/10.3390/plants10020259

Seo PJ, Xiang F, Qiao M, Park JY, Lee YN, Kim SG et al (2009) The *MYB96* transcription factor mediates abscisic acid signaling during drought stress response in *Arabidopsis*. Plant Physiol 151(1):275–289

Shah AA, Ahmed S, Yasin NA (2019) 24-epibrassinolide triggers cadmium stress mitigation in *Cucumis sativus* through intonation of antioxidant system. S Afr J Bot 127:349–360

Shakespear S, Sivaji M, Kumar V, Pillai MA, Wani SH, Suprasanna P, Yasin JK (2024) Navigating through harsh conditions: coordinated networks of plant adaptation to abiotic stress. J Plant Growth Regul. https://doi.org/10.1007/s00344-023-11224-4

Sharma I, Ching E, Saini S, Bhardwaj R, Pati PK (2013) Exogenous application of brassinosteroid offers tolerance to salinity by altering stress responses in rice variety Pusa Basmati-1. Plant Physiol Biochem 69:17–26

Shi H, Chen L, Ye T, Liu X, Ding K, Chan Z (2014) Modulation of auxin content in *Arabidopsis* confers improved drought stress resistance. Plant Physiol Biochem 82:209–217

Shibasaki K, Uemura M, Tsurumi S, Rahman A (2009) Auxin response in *Arabidopsis* under cold stress: underlying molecular mechanisms. Plant Cell 21(12):3823–3838

Shin H, Min K, Arora R (2018) Exogenous salicylic acid improves freezing tolerance of spinach (*Spinacia oleracea* L.) leaves. Cryobiology 81:192–200

Silva JM, da Silva Júnior GB, Bonifácio A, Dutra AF, de Mello Prado R, de Alcântara Neto F et al (2023) Exogenous salicylic acid alleviates water stress in watermelon plants. Ann Appl Biol 182(1):121–130

Silva APSD, Alencar AADS, Sudré CP, Araújo MDSBD, Lobato AKDS (2024) Brassinosteroids: relevant evidence related to mitigation of abiotic and biotic stresses in plants. Agronomy 14(4):840

Singh I, Kumar U, Singh SK, Gupta C, Singh M, Kushwaha SR (2012) Physiological and biochemical effect of 24-epibrassinoslide on cold tolerance in maize seedlings. Physiol Mol Biol Plants 18:229–236

Singh A, Jha SK, Bagri J, Pandey GK (2015) ABA inducible rice protein phosphatase 2C confers ABA insensitivity and abiotic stress tolerance in *Arabidopsis*. PLoS One 10(4):e0125168

Singh J, Varshney V, Saddhe AA, Manuka R, Penna S, Jha S (2024) Molecular techniques for the improvement of abiotic stress tolerance in plants. In: Al-Khayri J, Alnaddaf LM, Jain SM, Penna S (eds) Innovative methods in horticultural crop improvement. advances in plant breeding strategies, vol 1. Springer, Cham, pp 51–96. https://doi.org/10.1007/978-3-031-61081-3_3

Sinha P, Singh VK, Bohra A, Kumar A, Reif JC, Varshney RK (2021) Genomics and breeding innovations for enhancing genetic gain for climate resilience and nutrition traits. Theor Appl Genet 134(6):1829–1843

Skowron E, Trojak M (2021) Effect of exogenously-applied abscisic acid, putrescine and hydrogen peroxide on drought tolerance of barley. Biologia 76(2):453–468

Snyman M, Cronjé MJ (2008) Modulation of heat shock factors accompanies salicylic acid-mediated potentiation of Hsp70 in tomato seedlings. J Exp Bot 59(8):2125–2132

Song W, Shao H, Zheng A, Zhao L, Xu Y (2023) Advances in roles of salicylic acid in plant tolerance responses to biotic and abiotic stresses. Plants 12(19):3475

Srivastava S, Srivastava AK, Suprasanna P, D'souza SF (2013) Identification and profiling of arsenic stress-induced microRNAs in *Brassica juncea*. J Exp Bot 64(1):303–315

Stamm P, Kumar PP (2013) Auxin and gibberellin responsive *Arabidopsis SMALL AUXIN UP RNA36* regulates hypocotyl elongation in the light. Plant Cell Rep 32:759–769

Stevens J, Senaratna T, Sivasithamparam K (2006) Salicylic acid induces salinity tolerance in tomato (*Lycopersicon esculentum* cv. Roma): associated changes in gas exchange, water relations and membrane stabilisation. Plant Growth Regul 49:77–83

Sun P, Tian QY, Chen J, Zhang WH (2010) Aluminium-induced inhibition of root elongation in *Arabidopsis* is mediated by ethylene and auxin. J Exp Bot 61(2):347–356

Sunarpi X, Horie T, Motoda J, Kubo M, Yang H, Yoda K et al (2005) Enhanced salt tolerance mediated by *AtHKT1* transporter-induced Na+ unloading from xylem vessels to xylem parenchyma cells. Plant J 44(6):928–938

Suprasanna P (2020) Plant abiotic stress tolerance: insights into resilience build-up. J Biosci 45:120. https://doi.org/10.1007/s12038-020-00088-5

Tadaiesky LB, da Silva BR, Batista BL, Lobato AKDS (2021) Brassinosteroids trigger tolerance to iron toxicity in rice. Physiol Plant 171(3):371–387

Tan BC, Joseph LM, Deng WT, Liu L, Li QB, Cline K, McCarty DR (2003) Molecular characterization of the *Arabidopsis* 9-cis epoxycarotenoid dioxygenase gene family. Plant J 35(1):44–56

Tanaka K, Nakamura Y, Asami T, Yoshida S, Matsuo T, Okamoto S (2003) Physiological roles of brassinosteroids in early growth of *Arabidopsis*: brassinosteroids have a synergistic relationship with gibberellin as well as auxin in light-grown hypocotyl elongation. J Plant Growth Regul 22:259–271

Tayyab N, Naz R, Yasmin H, Nosheen A, Keyani R, Sajjad M, Hassan MN, Roberts TH (2020) Combined seed and foliar pre-treatments with exogenous methyl jasmonate and salicylic acid mitigate drought-induced stress in maize. Plos One 15(5):e0232269

Tian LX, Li J (2018) The effects of exogenous ABA applied to maize (*Zea mays* L.) roots on plant responses to chilling stress. Acta Physiol Plant 40:1–13

Torrens-Spence MP, Bobokalonova A, Carballo V, Glinkerman CM, Pluskal T, Shen A, Weng JK (2019) PBS3 and EPS1 complete salicylic acid biosynthesis from isochorismate in *Arabidopsis*. Mol Plant 12(12):1577–1586

Toscano S, Ferrante A, Romano D (2019) Response of Mediterranean ornamental plants to drought stress. Horticulturae 5(1):6

Tran LSP, Nakashima K, Sakuma Y, Simpson SD, Fujita Y, Maruyama K et al (2004) Isolation and functional analysis of *Arabidopsis* stress-inducible NAC transcription factors that bind to a drought-responsive cis-element in the early *responsive to dehydration stress 1* promoter. Plant Cell 16(9):2481–2498

Ubeda-Tomás, Susana et al (2009) Gibberellin signaling in the endodermis controls *Arabidopsis* root meristem size. Curr Biol 19(14):1194–1199. https://doi.org/10.1016/j.cub.2009.06.023

Ueguchi-Tanaka M, Matsuoka M (2010) The perception of gibberellins: clues from receptor structure. Curr Opin Plant Biol 13(5):503–508. https://doi.org/10.1016/j.pbi.2010.08.004

Van Ha C, Le DT, Nishiyama RIE, Watanabe YASUKO, Sulieman S, Tran UT et al (2013) The auxin response factor transcription factor family in soybean: genome-wide identification and expression analyses during development and water stress. DNA Res 20(5):511–524

Vardhini BV (2014) Brassinosteroids' role for amino acids, peptides and amines modulation in stressed plants-a review. In: Plant adaptation to environmental change: significance of amino acids and their derivatives, pp 300–316

Vardhini BV, Anuradha S, Sujatha E, Rao SSR (2010) Role of brassinosteroids in alleviating various abiotic and biotic stresses-a review. Plant Stress 4(1):55–61

Vercruyssen L, Gonzalez N, Werner T, Schmülling T, Inzé D (2011) Combining enhanced root and shoot growth reveals cross talk between pathways that control plant organ size in *Arabidopsis*. Plant Physiol 155(3):1339–1352

Vishwakarma K, Upadhyay N, Kumar N, Yadav G, Singh J, Mishra RK et al (2017) Abscisic acid signaling and abiotic stress tolerance in plants: a review on current knowledge and future prospects. Front Plant Sci 8:161

Vishwakarma K, Upadhyay N, Kumar N, Yadav G, Singh J, Mishra RK, Kumar V, Verma R, Upadhyay RG, Pandey M, Sharma S (2017) Abscisic acid signaling and abiotic stress tolerance in plants: a review on current knowledge and future prospects. Front Plant Sci 8:161

Wakchaure GC, Minhas PS, Meena KK, Kumar S, Rane J (2020) Effect of plant growth regulators and deficit irrigation on canopy traits, yield, water productivity and fruit quality of eggplant (*Solanum melongena* L.) grown in the water scarce environment

Wan XR, Li L (2006) Regulation of ABA level and water-stress tolerance of *Arabidopsis* by ectopic expression of a peanut 9-cis-epoxycarotenoid dioxygenase gene. Biochem Biophys Res Commun 347(4):1030–1038

Wang S, Bai Y, Shen C, Wu Y, Zhang S, Jiang D et al (2010) Auxin-related gene families in abiotic stress response in *Sorghum bicolor*. Funct Integr Genomics 10:533–546

Wang J, Xia H, Lin LJ, He H, Liang D, Lv XL (2016a) Exogenous abscisic acid increases resistances against abiotic stress and improve fruit quality of grape. JAPS J Anim Plant Sci 26(5)

Wang R, Zhang Y, Kieffer M, Yu H, Kepinski S, Estelle M (2016b) HSP90 regulates temperature-dependent seedling growth in *Arabidopsis* by stabilizing the auxin co-receptor F-box protein TIR1. Nat Commun 7(1):10269

Wang L, Liu J, Shen Y, Pu R, Hou M, Wei Q et al (2021) Brassinosteroids synthesised by CYP85A/A1 but not CYP85A2 function via a BRI1-like receptor but not via BRI1 in *Picea abies*. J Exp Bot 72(5):1748–1763

Wani SH, Kumar V (2015) Plant stress tolerance: engineering ABA: a potent phytohormone. Transcriptomics 3(2):1000113

Wen PF, Chen JY, Wan SB, Kong WF, Zhang P, Wang W et al (2008) Salicylic acid activates phenylalanine ammonia-lyase in grape berry in response to high temperature stress. Plant Growth Regul 55:1–10

Wilkinson S, Davies WJ (2010) Drought, ozone, ABA and ethylene: new insights from cell to plant to community. Plant Cell Environ 33(4):510–525

Wilkinson S, Kudoyarova GR, Veselov DS, Arkhipova TN, Davies WJ (2012) Plant hormone interactions: innovative targets for crop breeding and management. J Exp Bot 63(9):3499–3509

Wolters H, and Jürgens G (2009) Survival of the flexible: hormonal growth control and adaptation in plant development. Nat Rev 10(5):305–317. https://doi.org/10.1038/nrg2558

Won C, Shen X, Mashiguchi K, Zheng Z, Dai X, Cheng Y et al (2011) Conversion of tryptophan to indole-3-acetic acid by TRYPTOPHAN AMINOTRANSFERASES OF *ARABIDOPSIS* and *YUCCAs* in *Arabidopsis*. Proc Natl Acad Sci 108(45):18518–18523

Wu J, Zhu W, Zhao Q (2023) Salicylic acid biosynthesis is not from phenylalanine in *Arabidopsis*. J Integr Plant Biol 65(4):881–887

Xia XJ, Wang YJ, Zhou YH, Tao Y, Mao WH, Shi K et al (2009) Reactive oxygen species are involved in brassinosteroid-induced stress tolerance in cucumber. Plant Physiol 150(2):801–814

Xia XJ, Fang PP, Guo X, Qian XJ, Zhou J, Shi K et al (2018) Brassinosteroid-mediated apoplastic H_2O_2-glutaredoxin 12/14 cascade regulates antioxidant capacity in response to chilling in tomato. Plant Cell Environ 41(5):1052–1064

Xu K, Xu X, Fukao T, Canlas P, Maghirang-Rodriguez R, Heuer S et al (2006) *Sub1A* is an ethylene-response-factor-like gene that confers submergence tolerance to rice. Nature 442(7103):705–708

Xu Y, Feng J, Li H (2021) Water management increased rhizosphere redox potential and decreased Cd uptake in a low-Cd rice cultivar but decreased redox potential and increased Cd uptake in a high-Cd rice cultivar under intercropping. Sci Total Env 751:141701

Yadav SK (2010) Cold stress tolerance mechanisms in plants. A review. Agron Sustain Dev 30(3):515–527

Ye N, Jia L, Zhang J (2012) ABA signal in rice under stress conditions. Rice 5:1–9

Ye H, Liu S, Tang B, Chen J, Xie Z, Nolan TM et al (2017) RD26 mediates crosstalk between drought and brassinosteroid signalling pathways. Nat Commun 8(1):14573

Ye K, Li H, Ding Y, Shi Y, Song C, Gong Z, Yang S (2019) BRASSINOSTEROID-INSENSITIVE2 negatively regulates the stability of transcription factor ICE1 in response to cold stress in *Arabidopsis*. PC 31(11):2682–2696

Yu C, Sun C, Shen C, Wang S, Liu F, Liu Y et al (2015) The auxin transporter, Os AUX 1, is involved in primary root and root hair elongation and in cd stress responses in rice (*Oryza sativa* L.). Plant J 83(5):818–830

Yu Z, Zhang F, Friml J, Ding Z (2022) Auxin signaling: research advances over the past 30 years. J Integr Plant Biol 64(2):371–392

Yuan HM, Huang X (2016) Inhibition of root meristem growth by cadmium involves nitric oxide-mediated repression of auxin accumulation and signalling in *Arabidopsis*. Plant Cell Environ 39(1):120–135

Yuan Y, Chung JD, Fu X, Johnson VE, Ranjan P, Booth SL et al (2009) Alternative splicing and gene duplication differentially shaped the regulation of isochorismate synthase in Populus and *Arabidopsis*. Proc Natl Acad Sci 106(51):22020–22025

Yuan K, Rashotte AM, Wysocka-Diller JW (2011) ABA and GA signaling pathways interact and regulate seed germination and seedling development under salt stress. Acta Physiol Plant 33:261–271

Yuan H, Zhao K, Lei H, Shen X, Liu Y, Liao X, Li T (2013) Genome-wide analysis of the GH3 family in apple (*Malus*× *domestica*). BMC Genomics 14:1–14

Zaid A, Mohammad F, Wani SH, Siddique KM (2019) Salicylic acid enhances nickel stress tolerance by up-regulating antioxidant defense and glyoxalase systems in mustard plants. EES 180:575–587

Zelinová V, Alemayehu A, Bočová B, Huttová J, Tamás L (2015) Cadmium-induced reactive oxygen species generation, changes in morphogenic responses and activity of some enzymes in barley root tip are regulated by auxin. Biologia 70(3):356–364

Zengin F. (2014) Exogenous treatment with salicylic acid alleviating copper toxicity in bean seedlings. PNAS, India Section B: Biological Sciences 84(3):749–755

Zhang Z, Zhang H, Quan R, Wang XC, Huang R (2009) Transcriptional regulation of the ethylene response factor LeERF2 in the expression of ethylene biosynthesis genes controls ethylene production in tomato and tobacco. Plant Physiol 150(1):365–377

Zhang Q, Li J, Zhang W, Yan S, Wang R, Zhao J et al (2012) The putative auxin efflux carrier *OsPIN3t* is involved in the drought stress response and drought tolerance. Plant J 72(5):805–816

Zhang Y, Xu L, Zhu X, Gong Y, Xiang F, Sun X, Liu L (2013) Proteomic analysis of heat stress response in leaves of radish (*Raphanus sativus* L.). Plant Mol Biol Report 31:195–203
Zhang H, Huang Q, Yi L, Song X, Li L, Deng G et al (2021) PAL-mediated SA biosynthesis pathway contributes to nematode resistance in wheat. Plant J 107(3):698–712
Zhao Y (2012) Auxin biosynthesis: a simple two-step pathway converts tryptophan to indole-3-acetic acid in plants. Mol Plant 5(2):334–338
Zhao B, Li J (2012) Regulation of brassinosteroid biosynthesis and inactivation. J Integr Plant Biol 54(10):746–759
Zhao S, Zhang Q, Liu M, Zhou H, Ma C, Wang P (2021) Regulation of plant responses to salt stress. Int J Mol Sci 22(9):4609. https://doi.org/10.3390/ijms22094609
Zhao B, Liu Q, Wang B, Yuan F (2021a) Roles of Phytohormones and their signaling pathways in leaf development and stress responses. J Agric Food Chem 69:3566–3584. https://doi.org/10.1021/acs.jafc.0c07908
Zhou X, Zhang N, Yang J, Tang X, Wen Y, Si H (2018) Functional analysis of *StDWF4* gene in response to salt stress in potato. Plant Physiol Biochem 125:63–73
Zhou H, Wang Y, Zhang Y, Xiao Y, Liu X, Deng H et al (2022) Comparative analysis of heat-tolerant and heat-susceptible rice highlights the role of *OsNCED1* gene in heat stress tolerance. Plants 11(8):1062
Zhu JK (2002) Salt and drought stress signal transduction in plants. Annu Rev Plant Biol 53(1):247–273
Zhu JK (2016) Abiotic stress signaling and responses in plants. Cell 167(2):313–324
Zhu T, Li B, Chen Y, Wang S, Li W, Gao N et al (2023) BRASSINOSTEROID-INSENSITIVE 2 regulates salt stress tolerance in *Arabidopsis* by promoting AGL16 activity. Biochem Biophys Res Commun 678:17–23
Zong W, Tang N, Yang J, Peng L, Ma S, Xu Y et al (2016) Feedback regulation of ABA signaling and biosynthesis by a bZIP transcription factor targets drought-resistance-related genes. Plant Physiol 171(4):2810–2825

Chapter 9
Plant Hormones and Abiotic Stress Response

Harsha Samtani, Aishwarye Sharma, Mukesh Kumar, Shital Sandhya, Sanjay Singh Rawat, Harshita Bharti Saksena, Sheeba Naaz, Anjali Pande, Aryaman, Prakhar Awasthi, Halidev Krishna Botta, and Ashverya Laxmi

Abstract Plant hormones act as essential signaling molecules that orchestrate vital processes like growth, development, and responses to environmental challenges. Abiotic stresses, such as drought, salinity, extreme temperatures, heavy metals, and flooding, significantly impact plant growth and survival. Plants must possess intricate systems for sensing, signaling, and responding to these stresses to adapt and endure. This chapter delves into various studies based on how plant hormones such as abscisic acid, auxin, brassinosteroids, cytokinins, ethylene, gibberellins, salicylic acid, jasmonic acid, and strigolactones regulate physiological responses to abiotic stress, with a special focus on crop species. The article further emphasizes the role of nitric oxide (NO) and reactive oxygen species (ROS), which act as crucial signaling molecules in stress mitigation. Furthermore, the intricate interplay of hormonal signaling pathways during stress conditions is also highlighted. This comprehensive overview lays a foundation for biotechnological innovations in phytohormone engineering, with the potential to improve crop resilience and nutrition, and substantially enhance agricultural productivity.

Keywords Abiotic stress · Crop · Phytohormones · Signaling · Tolerance

Abbreviations

ABA	Abscisic acid
APX	Ascorbate peroxidase
BR	Brassinosteroids
CAT	Catalase
CK	Cytokinins

H. Samtani · A. Sharma · M. Kumar · S. Sandhya · S. S. Rawat · H. B. Saksena · S. Naaz · A. Pande · Aryaman · P. Awasthi · H. K. Botta · A. Laxmi (✉)
National Institute of Plant Genome Research, New Delhi, India
e-mail: ashverya_laxmi@nipgr.ac.in

A. Singh et al. (eds.), *Sustainable Crop Production*, Sustainability Sciences in Asia and Africa, https://doi.org/10.1007/978-981-95-7844-3_9

GA	Gibberellic acid
HSP	Heat shock protein
JA	Jasmonic acid
MDA	Malondialdehyde
ROS	Reactive oxygen species
RWC	Relative water content
POD	Peroxidase
NO	Nitric oxide

Editor's Notes

The importance of phytohormones in mitigating abiotic stress responses is reiterated in this chapter (Samtani et al). Primarily associated with development, phytohormones modulate stress responses via stress-sensing, signaling, and cross-talk. The authors explain the roles played by the phytohormones in various abiotic stresses highlighting the specifics of key regulators, with added emphasis on the role of NO and ROS signaling. A poorly studied area that needs urgent attention is bioprospecting of useful genes and alleles from crop wild relatives. The chapter is a handy guide for researchers to investigate and elucidate mechanistic aspects of phytohormone action under stress in both model and crop species. A holistic approach driven by new technologies and experimental design is key to achieve progress in understanding the phytohormone-NO-ROS regulator cross-talks under stress conditions. This will permit devising strategies to impart phytohormone-mediated stress resilience in crops.

Actionable Points

1. *Identification of phytohormone-related genetic markers*: GWAS or QTL mapping can be conducted to pinpoint specific genetic variations (SNPs, indels) within phytohormone biosynthesis, signaling, or response genes that correlate with enhanced stress resilience. These could further be used as molecular markers for efficient selection.
2. *Precise editing of phytohormone genes with CRISPR-Cas9*: CRISPR-Cas9 can be utilized to make targeted changes (knock-outs, knock-ins, or base edits) in genes controlling phytohormone production or sensitivity. This will fine-tune hormone levels or alter pathway sensitivity for improved stress tolerance.
3. *Engineering phytohormone regulatory elements*: CRISPR-Cas9 can be employed to modify regulatory regions (promoters, enhancers) of phytohormone-related genes. This will enable conditional or tissue-specific activation of stress responses, thereby optimizing benefits and minimizing negative impacts on plant growth.
4. *Application of exogenous phytohormones for immediate benefit*: Field trials can be conducted to optimize the application of external phytohormones (e.g., brassinosteroids, auxins). This can provide a rapid way to alleviate stress symptoms, improve crop performance, and validate the potential of specific hormone pathways.

Key Points

- Phytohormones could function as critical stress reducer in plants, and thus, their engineering holds significant potential as a biotechnological strategy for enhancing crop nutrition and boosting economic yields.
- Comprehending phytohormone signaling pathways guides manipulation of hormone levels/signaling components for stress tolerance.
- Phytohormone gene knowledge could accelerate breeding by enabling marker-assisted selection of stress-tolerant plants.
- CRISPR-Cas9 technology offers a powerful tool for developing climate-resilient crops by precisely manipulating hormone pathways. This involves editing genes that control hormone production, signaling, and responses, enabling plants to better withstand stresses like drought, heat, and salinity.
- Proteomics represents a powerful tool for identifying key proteins involved in hormone signaling pathways that contribute stress tolerance. This knowledge could help breeding programs and provide novel strategies for enhancing crop resilience.

9.1 Introduction

Feeding a rapidly expanding global population demands a dramatic increase in agricultural productivity, presenting a critical challenge to global food security. Unfortunately, crop yields are severely hampered by a range of environmental stresses, both biotic and abiotic (Wani and Kumar Sah 2014). To adequately feed the projected 2.3 billion increase in the global population by 2050, agricultural productivity must increase by a remarkable 70% (Tilman et al. 2011). Unlike animals, plant responses to environmental stresses are intricate and multifaceted (Qin et al. 2011a). Therefore, a fundamental understanding of how plants perceive and react to various environmental challenges is paramount for plant scientists seeking to enhance crop resilience and ensure future food security.

A range of environmental factors, including drought, waterlogging, salinity, extreme temperatures, heavy metal contamination, and excessive solar radiation, significantly impede optimal plant growth and development (He et al. 2018). These stresses have a detrimental impact on global crop yields and biomass production (Wu et al. 2024). However, plants have evolved effective mechanisms to sense, signal, and respond to harmful stresses. A prominent example of plant stress responses involves phytohormones, which are plant hormones acting as chemical messengers at low concentrations. They play crucial roles in regulating plant responses to abiotic stresses (Davies 1995). Phytohormone engineering offers a promising platform for biotechnological advancements in crop nutrition and economic yield. Nine distinct types of plant hormones have been identified to date (Su et al. 2017) with auxins as the first phytohormone to be discovered (Enders and Strader 2015) and strigolactones, the most recently characterized class (Gomez-Roldan et al. 2008). Other important hormones encompass cytokinin, gibberellins, abscisic acid, ethylene, salicylates, brassinosteroids, and jasmonates. The first five (encompassing auxin, cytokinin, gibberellin, abscisic acid, and ethylene) are sometimes referred to as the "classical" phytohormones, while the other four are more recent additions to the growing phytohormonal family (Wang and Irving 2011). While nitric oxide and reactive oxygen species (ROS) are crucial signaling molecules in plants, their inorganic nature often precludes their classification as phytohormones (Su et al. 2017). This chapter explores the mechanisms by which phytohormones mediate plant responses to abiotic stresses, focusing on crops. This knowledge is crucial for developing climate-resilient crop varieties in our rapidly changing world.

9.2 Role of Auxins in Abiotic Stress Responses

Auxins, primarily produced in shoot tips, regulate various physiological processes such as cell elongation, apical dominance, root growth, and phototropism (Gomes and Scortecci 2021). It is biosynthesized in plants mainly through the tryptophan-dependent pathway. This pathway involves two key steps: conversion of tryptophan to indole-3-pyruvic acid by tryptophan aminotransferases and subsequent oxidative decarboxylation of indole-3-pyruvic acid to Indole-3-Acetic Acid (IAA) by YUCCA flavin monooxygenases. Auxin transport is a crucial process in plants, facilitating its directional movement within tissues. It primarily occurs through a mechanism called polar auxin transport (PAT), where auxin is actively transported from cell to cell in a specific direction. This directional movement is facilitated by a complex interplay of influx (AUX1/LAX family proteins) and efflux carrier (PIN-FORMED (PIN) proteins and ATP-binding cassette (ABC) B transporters) proteins located on the plasma membrane of plant cells (Casanova-Sáez et al. 2021). Auxin signaling begins with the binding of auxin to its receptors, the TIR1/AFB proteins. This binding triggers the degradation of Aux/IAA proteins, which act as repressors of auxin response factors (ARFs). Activated ARFs promote transcription of genes involved in various cellular responses such as cell elongation, division, and differentiation (Gomes and Scortecci 2021; Waadt et al. 2022).

9.2.1 *Heat Stress*

Elevated temperatures in *Arabidopsis* trigger *PIF4* activation, which stimulates YUCCA8-mediated auxin biosynthesis. Increased auxin levels are crucial for thermomorphogenesis, particularly hypocotyl elongation. Disrupting *YUCCA8* function inhibits this temperature-induced elongation, demonstrating auxin playing an essential role in plant growth responses to heat (Sun et al. 2012; Samtani et al. 2023; Sharma et al. 2023; Sharma et al. 2024a). Apart from this, heat stress also inhibits flower development in *Arabidopsis* by causing an imbalance in auxin signaling. Specifically, it activates mitogen-activated protein kinases (MPKs), which phosphorylate IAA8 (a repressor of auxin signaling), increasing its stability and leading to its accumulation, thereby perturbing the expression of auxin-regulated flower development genes (Kim et al. 2024). High temperatures in rice reduce spikelet fertility and grain yield. Heat and drought stress suppress auxin-related gene expression, lowering endogenous auxin levels and impairing male fertility. Exogenous IAA application mitigates these negative effects by improving chlorophyll content, restoring spikelet fertility across diverse rice varieties, and reducing lipid peroxidation. These findings suggest that manipulating auxin levels can enhance yield stability in rice under elevated temperatures (Sharma et al. 2018). Heat stress in cotton compromises male fertility, with auxin signaling playing a critical role. MicroRNAs like *miR156* and *miR160*, involved in auxin regulation, influence heat stress responses (Ding et al. 2017). Overexpression of *miR160* increases heat sensitivity, while elevated anther auxin levels exacerbate male sterility. Exogenous auxin application further worsens heat-induced male sterility, emphasizing the critical need for maintaining proper auxin homeostasis to prevent reproductive failure under heat stress (Ding et al. 2017). Auxin is also critical for root development, where fluctuations in temperature can impact the maintenance of meristems. Temperature stress further affects alternative splicing of pre-mRNA in roots, potentially altering the expression of vital meristematic regulators like WOX5. These modifications in root structure and auxin levels influence the plant's adaptability to temperature stress and growth optimization (Chen et al. 2024a). In maize, the transcription factor ZmHSF20 negatively impacts heat stress responses by repressing essential genes like *ZmHsf4* and *ZmCesA2*, which play a role in stress resilience. Seedlings with mutations in *ZmHSF20* demonstrate increased heat tolerance, suggesting that adjusting auxin pathways alongside heat shock factors may be a method for enhancing stress resilience (Li et al. 2024b). In case of rice, auxin signaling pathway also plays an important role in response to heat stress during grain filling stage. Under high-temperature conditions, *OsIAA29* plays a role in rice seed development by competing with *OsIAA21* for binding to *OsARF17*, thereby regulating the auxin signaling pathway (Chen et al. 2024b). A HEAT RESPONSIVE PROTEIN (GhHRP) in cotton plays a crucial role in plant adaptation to warm conditions by modulating phytohormone signaling, specifically fine-tuning the interplay between auxin and ethylene. *GhHRP* by promoting *GhPIF4* accumulation regulates auxin biosynthesis (via *AtYUC8*) and represses ethylene biosynthesis (via *AtACS8*), ultimately improving plant growth and protecting chloroplasts from heat-induced damage (Abdullah et al. 2023).

9.2.2 Drought Stress

Under drought stress, auxin levels and activity are modulated to enhance root system architecture, including increased depth for improved water uptake. OsPIN3t auxin transporters and OsIAA18 ARFs in rice regulate root architecture, enhancing drought tolerance by promoting the development of deeper and more extensive root systems for improved water uptake (Zhang et al. 2012). Auxin and ABA interact to regulate drought responses. Drought stress often leads to decreased auxin levels or altered distribution, while ABA levels increase. This interplay modulates stomatal closure and root system development to enhance water conservation (Du et al. 2013). In apple (*Malus domestica*), genes like *MdIAA24*, involved in auxin signaling, enhance drought resistance. They stimulate strigolactone production, promoting beneficial mycorrhizal associations that improve soil moisture retention. This demonstrates the crucial role of auxin signaling in facilitating drought tolerance through intricate interactions with other plant hormones (Huang et al. 2021).

9.2.3 Salinity Stress

ARFs regulate genes involved in stress adaptation, particularly root system development, essential for salt-stressed plants. *OsIAA18* overexpression in rice improves salt tolerance by increasing tissue ABA levels and activities associated with ROS detoxification, thus increasing proline content and aiding in enduring salinity stress (Wang et al. 2021a). The interaction between PIN1 auxin transporters and AZG1 cytokinin importers regulates lateral root density under salinity stress, influencing root system architecture. This interplay between auxin and cytokinin is crucial for optimizing root structure and overall plant performance in saline conditions (Tessi et al. 2023).

9.3 Role of Abscisic Acid in Abiotic Stress Responses

Abscisic acid (ABA) acts as a key integrator of various stress signals while orchestrating the responses of plants to such stresses (Tuteja 2007). ABA also plays crucial roles in various physiological processes, including seed dormancy and germination inhibition, seed development, embryo formation, synthesis of storage compounds, leaf senescence, and pathogen defense (Swamy and Smith 1999). ABA biosynthesis primarily occurs in plastids, originating from the cleavage of carotenoid precursors. This process begins with the synthesis of carotenoids, such as zeaxanthin. Subsequently, the enzyme 9-cis-epoxycarotenoid dioxygenase (NCED) cleaves zeaxanthin, initiating a series of conversions that ultimately lead to the formation of ABA. This pathway is meticulously regulated to ensure appropriate ABA levels in response to various environmental stimuli (Seo and Koshiba 2002). ABA triggers stomatal closure, minimizing water loss during drought; hence, it is designated as a "stress hormone" (Tuteja 2007). Under stress conditions, ABA accumulates inside the plant cells and binds to PYR/PYL/RCAR receptors. This inhibits PP2Cs leading to the activation of SnRK2s through autophosphorylation or phosphorylation by other kinases (Fujii et al. 2009; Miyazono et al. 2009). SnRK2 phosphorylates and regulates key targets, including ion channels (SLAC1, KAT1) in guard cells, leading to stomatal closure. Additionally, SnRK2s activate AREB1/2/3 transcription factors (ABFs), ABA insensitive 5 (ABI5), triggering the expression of stress-responsive genes, such as *LEA* and other stress-related TFs (such as *bZIP*, *MYB*, and *WRKY*) (Uno et al. 2000; Ng et al. 2014; Muhammad Aslam et al. 2022).

9.3.1 Drought Stress

ABA is best known for inducing stomatal closure during drought (Mittelheuser and Van Steveninck 1969). Also, ABA inhibits shoot growth (Yao and Finlayson 2015) while promoting primary root elongation (Yang et al. 2014) and suppressing lateral root formation (Koevoets et al. 2016), although these effects can vary with concentration and species (Rosales et al. 2019; Kavi Kishor et al. 2022). ABA downregulates leaf hydraulic conductance, and mesophyll conductance, and upregulates cuticular wax formation, all contributing to reduced water loss (Sorrentino et al. 2016; Coupel-Ledru et al. 2017). In sweet potato, ABA application enhanced leaf drought resistance by increasing relative water content, promoting ABA accumulation, and mitigating osmotic stress damage as evidenced by reduced malondialdehyde (MDA) content (Zhang et al. 2020). Drought stress response in rice is orchestrated by the ABA signaling pathway, which intricately involves multiple ABA receptors and a cascade of regulatory factors. PYR/PYL/RCAR proteins are recognized as crucial players in enhancing drought tolerance in rice, *Arabidopsis*, and tomato (Pizzio et al. 2013; Gonzalez-Guzman et al. 2014; Kim et al. 2014). OsPYL6 contributes to enhanced drought tolerance in rice by regulating ABA levels and the expression of drought-responsive genes, although its effects on yield under non-stress conditions suggest a potential trade-off between stress tolerance and optimal growth (Santosh Kumar et al. 2021). *OsPYL9-Knock-out* enhances both drought tolerance and yield in rice by regulating circadian rhythms and stress proteins (Zhao et al. 2016b; Usman et al. 2020). Overexpression of *OsPYL10* promotes drought stress tolerance through the regulation of ABA accumulation and the expression of drought-related genes (Verma et al. 2019). Furthermore, as a positive regulator of ABA signaling, OsPSKR15 enhances drought tolerance by interacting with OsPYL11 (Nagar et al. 2022). OsPP2C09 regulates the trade-off between plant growth and drought resistance in rice. The higher expression of *OsPP2C09* in roots compared to shoots suggests its role in promoting root growth, which may contribute to an increased root-to-shoot ratio under drought conditions (Miao et al. 2020). Additionally, OsPP2C09, by directly interacting with DRE-binding proteins, causes activation of genes with drought-responsive elements (DRE) in their promoters (Min et al. 2021).

OsSAPK2 activates the transcription factor *OsbZIP23*, leading to the upregulation of the ABA biosynthesis gene *OsNCED4* and several drought-responsive genes, including *OsLEA3-2*, *OsSRO1c*, and *OsETOL1*. This regulatory cascade facilitates seed dormancy, stomatal closure, and the activation of antioxidant defense mechanisms, ultimately contributing to improved drought tolerance in rice (Lou et al.

2017). OsSAPK4 activates OsbZIP42 (after protein modification), leading to the suppression of *OsPP2C* genes like *ABI2* and the activation of ABA-responsive genes like *LEA3* and *Rab16*, thereby improving drought resistance in rice (Joo et al. 2019). Additionally, OsSAPK9 enhances drought tolerance and yield by modulating cell osmotic potential, inducing stomatal closure, and regulating the expression of stress-responsive genes (Dey et al. 2016). Overexpression of *OsbZIP72* in rice enhances drought resistance by upregulating ABA-responsive genes, particularly LEA proteins (Lu et al. 2009). AP2/ERF transcription factors OsERF71 and OsERF101 are key regulators of drought stress through ABA signaling. *OsERF71* overexpression in rice enhances proline synthesis, downregulates proline degradation, and upregulates ABA- and drought-responsive genes, significantly improving drought tolerance (Jin et al. 2018). Interestingly, overexpression of HVA1, a *LEA3* gene from *H. vulgare*, promoted drought tolerance in doubled haploid wheat transgenics (Samtani et al. 2022b).

9.3.2 Heavy Metal Stress

Human activities on agricultural lands have led to elevated levels of certain heavy metals, including arsenic (As), chromium (Cr), lead (Pb), cadmium (Cd), mercury (Hg), nickel (Ni), and copper (Cu). Their exposure causes a serious threat to plants as their high reactivity disrupts crucial cellular processes such as energy production, senescence, and growth (Singh et al. 2016). Heavy metal stress impacts ABA biosynthesis and signaling in plants. Elevated ABA biosynthesis genes and ABA levels have been observed in various species, including rice, potato, oilseed rape, *Malus hupehensis*, and *Sedum alfredii*, upon Cd exposure (Hsu and Kao 2003; Stroïnski et al. 2013; Yan et al. 2016; Shi et al. 2019). In rice, *OsNCED3*, *OsNCED4*, and *OsNCED5* expression is upregulated by Cd (Tan et al. 2017). Furthermore, Cd-tolerant rice cultivars exhibit a more pronounced and rapid ABA production in response to Cd stress compared to sensitive cultivars, suggesting a positive correlation between endogenous ABA levels and Cd tolerance (Hsu and Kao 2003). Exposure to As in *Brassica juncea* led to increased ABA levels in both roots and shoots (Srivastava et al. 2013). In rice, exposure to As upregulated genes involved in ABA biosynthesis (*OsNCED1, OsNCED2, OsNCED3, OsABA4*) and signaling (*OsPP2Cs, OsbZIP10, OsbZIP12, OsbZIP66, OsbZIP72*) (Huang et al. 2012; Yu et al. 2012). Increased ABA levels were also observed in germinating chickpea and pea leaves exposed to Pb (Parys et al. 1998; Atici et al. 2005). In *Populus × canescens (syn. Populus tremula × P. alba)*, Pb exposure increased leaf ABA concentration by 107%, and exogenous ABA application mitigated Pb toxicity (Shi et al. 2019). Transcriptomic analysis in *Arabidopsis* and *Hirschfeldia incana* revealed upregulation of ABA biosynthesis genes in the roots and shoots of Pb-tolerant *H. incana* upon Pb exposure (Auguy et al. 2016). These findings suggest that ABA plays a beneficial role in mitigating heavy metal and metalloid accumulation and toxicity. Exogenous ABA application has been shown to alleviate stress in angiosperms exposed to Cd, As, or Pb. It can trigger three major detoxification mechanisms: inhibiting metal uptake, reducing root-to-shoot translocation, and promoting metal chelation (Hu et al. 2020). Manipulating ABA signaling in plants through the use of ABA-producing bacteria or synthetic ABA analogues holds promise for reducing the uptake and accumulation of toxic heavy metals and metalloids in crops. This approach could significantly enhance crop tolerance to these contaminants.

9.3.3 Salinity Stress

Soil salinity disrupts water and nutrient uptake by the roots, leading to growth retardation and even mortality. It induces ion toxicity, osmotic stress, and nutrient deficiencies. Furthermore, it causes membrane damage and stomatal closure, reducing CO_2 uptake and increasing oxidative stress. This oxidative stress, characterized by the accumulation of ROS such as O_2^-, H_2O_2, and OH^-, can damage cell membranes, proteins, DNA, and lipids, leading to the accumulation of MDA and disrupting cellular processes (Suzuki et al. 2012; Chen et al. 2022). ABA has been shown to have a protective role under salinity stress in crops such as wheat and rice (Sripinyowanich et al. 2013; Parveen et al. 2021). ABA also enhances alkaline stress tolerance in rice seedlings by activating antioxidant defense systems in roots and upregulating stress-related genes (Liu et al. 2019). ABA plays a key role in regulating ion homeostasis under salt stress. Increased ABA synthesis in response to salinity promotes ion efflux (K^+, Cl^-), leading to stomatal closure and thus providing a protective effect by reducing stomatal conductance, water loss, and wilting. Moreover, ABA helps to maintain low Na^+/K^+ ratios and enhances the activity of protective enzymes, ultimately preserving cell membrane integrity and minimizing salt-induced damage (Parveen et al. 2021; Zhao et al. 2021). Further, foliar ABA application mitigates salt stress in rice seedlings by improving growth parameters, enhancing photosynthesis, and reducing oxidative damage. ABA modulates endogenous hormone levels, including ACC, TZ, IPA, IAA, and ABA itself, suggesting a complex interplay of hormonal regulation in salt stress responses (Chen et al. 2022). Further, OsSAE1, an AP2-type transcription factor in rice, promotes seed germination and salt tolerance by directly repressing *OsABI5*, a key ABA signaling gene. *ossae1* mutants exhibit delayed germination, ABA sensitivity, and reduced salt tolerance, while *OsSAE1* overexpression lines show enhanced germination and salt tolerance. Genetic analyses demonstrated that OsABI5-mediated ABA signaling operates downstream of *OsSAE1* (Li et al. 2022). Interestingly, hormone profiling and transcriptional analysis demonstrate a significant role for ABA in tomato salt tolerance (Yang et al. 2014). Further, the impact of ABA deficiency in response to salt stress across different leaf ages was examined in tomato plants. The *notabilis* mutant, lacking functional stomatal closure due to an *NCED1* mutation, exhibited unexpected resilience to salt stress. Despite impaired stomatal closure, *notabilis* plants maintained the growth and water content better than the wild-type suggesting compensatory mechanisms involving upregulation of other *NCED* genes and altered osmoregulation (Holsteens et al. 2022). A recent study by Lamers et al. (2025) reveals a surprising and critical role for ABA in mitigating sodium-specific responses and cellular damage under salinity stress in *Arabidopsis thaliana* plants. Contrary to its general promoting role in abiotic stress responses, ABA represses a unique cluster of Na^+-specific genes and, when applied exogenously, completely abolishes sodium-induced responses. This suggests that the plant's initial response to the osmotic component of salinity stress triggers a delayed ABA response, which then acts to counteract the detrimental effects of sodium (Lamers et al. 2025).

9.3.4 Cold Stress

Cold stress significantly impacts plant growth and survival. Chilling temperatures (0–12 °C) inhibit vital biochemical processes, including water and nutrient uptake, leading to reduced growth (Lyons 1973). Chilling stress also triggers stomatal closure, inhibiting photosynthesis and transpiration, mimicking the effects of drought (Wilkinson et al. 2001). Freezing temperatures (below 0 °C) are more detrimental, causing ice crystal formation within cells, and damaging membranes and cell walls. Ice nucleation in intercellular spaces reduces extracellular water potential, drawing water out of cells and causing severe dehydration.

Plants enhance freezing tolerance through cold acclimation by triggering significant transcriptional and metabolic adjustments, including the accumulation of osmoprotectants and the activation of ROS scavenging pathways (Cook et al. 2004; Maruyama et al. 2009; Mittler et al. 2022). This overlap is reflected in the shared expression of certain cold- and drought-responsive genes (Maruyama et al. 2009). Furthermore, cold stress alters plasma membrane composition to maintain fluidity and prevents freezing injury (Miki et al. 2019). Additionally, exposure to non-lethal freezing temperatures (around −3 °C) induces modifications in cell walls and extracellular components, leading to an enhanced freezing tolerance known as subzero acclimation (Takahashi et al. 2019). While ABA is primarily recognized for its role in dehydration stress, evidence suggests it also contributes to cold acclimation by inducing specific cellular and molecular osmotic responses (Raza et al. 2023). Cold tolerance in plants is often associated with increased endogenous ABA levels (Mäntylä et al. 1995). Moreover, continuous ABA application has been shown to induce chilling tolerance in chilling-sensitive species like maize, rice, cucumber, and pepper. Additionally, exogenous ABA application in temperate plants such as poplar, barley, wheat, and *Arabidopsis* can partially mimic cold acclimation and improve freezing tolerance (Smoleńska-Sym et al. 1995; Thomashow 1999; Kadlecova et al. 2000; Zhu et al. 2000). Exogenous ABA treatment enhanced the activities of antioxidant enzymes, including catalase (CAT), superoxide dismutase (SOD), peroxidase (POD), ascorbate peroxidase (APX), and glutathione reductase (GR) and reduced H_2O_2 production in wheat plants subjected to chilling stress (0 °C to −24 °C) (Yu et al. 2020). While ABA is known to stimulate putrescine biosynthesis (Lee 1997), putrescine itself is a vital player in a plant's cold response. Crucial for freezing tolerance, putrescine mediates the cold-induced ABA response, boosting ABA production and the expression of ABA-dependent genes, which in turn ensures cold acclimation and survival (Cuevas et al. 2009). Abscisic acid and putrescine work together to enhance cold tolerance in melon seedlings. It has been found that in cold-tolerant melon varieties, ABA and putrescine levels rise in response to cold stress. These crucial signaling molecules mutually regulate each other's production and activate the plant's antioxidant defense system, thereby boosting the melon's ability to withstand cold (Li et al. 2021b). Exogenous ABA treatment at both 22 °C and 4 °C induced specific ultrastructural changes in wheat chloroplasts, including modifications to thylakoid membrane organization, thereby enhancing photosynthetic apparatus stability under chilling stress (Venzhik et al. 2016). *Fragaria daltoniana* (cold-tolerant) species accumulate higher levels of ABA as compared to cold *Fragaria vesca* (cold-sensitive) species under low temperature stress. *ABF2*, involved in glucose signaling, was upregulated in *F. daltoniana* but downregulated in *F. vesca* under cold stress. This differential expression likely contributed to enhanced glucose signaling in *F. daltoniana*, potentially explaining its higher peroxidase activity, reduced membrane damage, and, consequently, stronger cold tolerance (Shen et al. 2022). Similarly, in *Hevea brasiliensis*, exogenous application of ABA enhanced cold tolerance as expression of cold-regulated (*COR*) genes were upregulated. Also, HbSnRK2.6A and HbSnRK2.6B promoted the transcriptional activity of *HbICE2*, thus increasing the expression of *HbCBF1* (Wang et al. 2021c). Exogenous ABA also improved chilling stress tolerance in field-grown chickpea during the flowering stage. It alleviated stress-induced reductions in plant height, chlorophyll content, and relative water content while enhancing pollen viability, flower retention, and seed yield. This was accompanied by reduced oxidative stress, as indicated by increased antioxidant enzyme activities (SOD, CAT, APX) and higher levels of antioxidants (AsA, GSH, proline) (Kumar et al. 2008). Chilling stress in *Iris pseudacorus* decreased chlorophyll content, PSII efficiency, and cytokinin levels in leaves while increasing ABA levels and decreasing violaxanthin. In contrast, carotenoid and ABA levels increased in rhizomes. These changes suggest that ABA plays a crucial role in mitigating chilling stress by regulating leaf senescence (Caselles et al. 2021).

9.3.5 Heat Stress

Heat stress poses a significant threat to yield by disrupting plant development across different growth stages (Samtani et al. 2022a). Plant heat stress occurs when ambient temperatures exceed optimal growing conditions by 10–15 °C. The duration, intensity, and rate of temperature increase significantly influence HS severity and vary across climatic zones (Samtani et al. 2022a). During heat stress and low water status, ABA synthesized in leaf vascular cells is transported to guard cells, triggering stomatal closure. This reduces water loss through transpiration, limiting CO_2

uptake and photosynthesis. However, it helps maintain vital physiological functions, enhancing heat resistance and minimizing crop yield losses (Umezawa et al. 2006; Kuromori et al. 2018; Yaaran et al. 2019). Furthermore, ABA can boost glucose metabolism in response to heat stress, providing the necessary energy for heat resistance (Ruan et al. 2010). Interestingly, ABA leads to upregulation of *heat shock factors* (*Hsfs*) and *heat shock proteins* (*HSPs*), thereby contributing to thermotolerance in plants such as tall fescue and *Arabidopsis* (Wang et al. 2017b). Similarly, in case of *Pisum sativum*, ABA levels were found to be more in heat-tolerant varieties as compared to heat-sensitive varieties (Huang et al. 2023). In Arabidopsis, heat stress triggers a transient ABA surge (Larkindale et al. 2005). ABA enhances thermotolerance by inducing ROS production via RBOHs, strengthening antioxidant defenses (Suzuki et al. 2011). Exogenous ABA increases H_2O_2, improving heat tolerance in wild-type plants and rescuing it in ABA-deficient mutants (Larkindale and Knight 2002). Both ABA-deficient and signaling mutants display impaired heat tolerance (Larkindale et al. 2005).

In rice, high temperatures impair carbon metabolism, limiting sugar production and energy reserves, thus compromising pollen viability and leading to spikelet abortion (Ruan et al. 2010). Exogenous application of ABA alleviates this by increasing sugar levels, enhancing *HSP* expression, boosting antioxidant activity, and improving energy homeostasis, ultimately promoting heat tolerance (Rezaul et al. 2019). ABA also helps in sucrose transport and metabolism, supplying the energy needed for pollen thermotolerance (Chen et al. 2019).

9.4 Role of Cytokinin in Abiotic Stress Response

Cytokinin (CK), initially identified as a key regulator of organogenesis in plant tissue cultures (Mähönen et al. 2006; Skoog and Armstrong 1970), plays vital roles in diverse plant processes such as flowering transition and senescence, cell proliferation and differentiation, root nodulation, as well as response to abiotic and biotic stresses (Cortleven et al. 2019; Wybouw and De Rybel 2019). Natural CKs are primarily classified into two groups: isoprenoids and aromatics. Isoprenoids are the dominant type in plants and comprise four main groups: isopentenyladenine (iP), trans-zeatin (tZ), cis-zeatin (cZ), and dihydrozeatin (DZ) (Sakakibara 2006). Their presence has been documented in diverse organisms, ranging from primitive bacteria (Uniyal et al. 2022) and the moss *Physcomitrium patens* (von Schwartzenberg et al. 2007) to flowering plants like *Arabidopsis thaliana* (Ko et al. 2014) and rice (*Oryza sativa*) (Zhao et al. 2019). Cytokinin biosynthesis primarily involves the enzyme isopentenyltransferase (IPT), which catalyzes the addition of an isopentenyl group to adenine nucleotides, forming cytokinin precursors. This biosynthetic pathway is tightly regulated, ensuring appropriate cytokinin levels in different tissues and at various developmental stages. Cytokinins are perceived by membrane-bound receptors, which are histidine kinases. Upon cytokinin binding, the receptor undergoes autophosphorylation, initiating a phosphorelay system. This involves the transfer of phosphate groups between different proteins, ultimately leading to the activation of type-B response regulators (RRs). Activated type-B RRs act as transcription factors, regulating the expression of numerous genes involved in cell division, differentiation, and other developmental processes (Hwang and Sakakibara 2006).

Plants significantly alter their endogenous CK levels when subjected to prolonged abiotic stresses such as drought, salinity, and exposure to toxic metals (Zwack and Rashotte 2015; Mohan et al. 2016; Prerostova et al. 2018). Moreover, the proteins involved in perceiving and responding to CK signals are also regulated at various levels (transcription, translation, and posttranslation) by these abiotic stresses (Acheampong et al. 2020; Yan et al. 2021).

9.4.1 Drought Stress

In crop plants such as apple, peanut, and tomato, studies have elucidated the role of enzymes involved in CK synthesis and metabolism and CK signaling components in drought stress responses (Qin et al. 2011b; Farber et al. 2016; Feng et al. 2021). Plants employ various strategies to survive drought, including repressing CK levels and signaling. This downregulation of CK promotes physiological adaptations such as stomatal closure, increased anthocyanin production, and enhanced antioxidant activity. Concurrently, root system architecture (RSA) undergoes significant modifications, including alterations in root positioning, length, angle, and branching, and the formation and density of lateral roots (LRs), crown roots (CRs), and adventitious roots (ARs), to optimize water uptake and enhance drought tolerance (Cortleven et al. 2019; Karlova et al. 2021; Li et al. 2021a). Rice CR emergence is suppressed by OsRR2, a transcription factor directly activated by OsERF3 (Zhao et al. 2015). However, OsWOX11, a WUSCHEL-related homeobox protein, inhibits *OsRR2* transcription (Zhao et al. 2009b). CR initiation is facilitated by reduced CK signaling, a consequence of OsRR2 repression by OsERF3. Conversely, CR elongation is driven by increased CK signaling through repressed OsRR2 (Zhao et al. 2009b; Zhao et al. 2015). Enhanced CR formation and a more robust root system, observed in plants with increased *OsCKX4* overexpression, highlight its positive role in CR development. *OsCKX4* transcription is positively regulated by OsRR2 and OsRR3 (Gao et al. 2014), while OsNAC2 negatively regulates *OsCKX4* expression and root growth (Mao et al. 2020). This suggests a role for *OsNAC2* in modulating CK signaling. Consistent with these findings, overexpression of *OsNAC2* or *OsCKX4* knockdown/knockout lines exhibit reduced root growth, whereas root-specific *OsCKX4* expression and *OsNAC2* knockout lines display enhanced root growth (Gao et al. 2014; Mao et al. 2020). These results collectively underscore the critical role of CK signaling in shaping RSA plasticity in plants. Recent studies have highlighted significant regulatory interactions between CK and auxin pathways in controlling root development (Jing and Strader 2019). Notably, OsNAC2, a negative regulator of *OsCKX4*, also regulates the expression of genes involved in auxin metabolism and signaling, such as *OsGH3.6*, *OsGH3.8*, and *OsARF25* (Mao et al. 2020). These findings underscore the intricate interplay between hormonal pathways in shaping root architecture under stress conditions.

To mitigate the adverse effects of constitutively reduced CK levels on plant growth, researchers have employed tissue-specific or stress-inducible promoters to regulate CK synthesis in a spatiotemporal manner (Rivero et al. 2007; Reguera et al. 2013; Prerostova et al. 2018). This approach has been successfully implemented in various plant species, including tobacco, creeping bentgrass, cotton, wheat, and rice, by expressing IPT genes under the control of drought-inducible promoters, such as senescence-associated receptor protein kinase (SARK) (Rivero et al. 2007; Kuppu et al. 2013; Reguera et al. 2013; Beznec et al. 2021). Similarly, transgenic poplar lines expressing *IPT* under the control of the drought-inducible *PtRD26* promoter exhibited improved growth and drought tolerance (Wang et al. 2022a). These findings demonstrate the potential of strategically manipulating CK levels through inducible gene expression to enhance plant drought tolerance while minimizing negative impacts on overall plant growth.

9.4.2 Salt Stress

CKs play both positive and negative roles in regulation of salt tolerance in plants. Constitutive downregulation of CK levels through *IPT* gene inactivation or *CKX* overexpression can enhance salt tolerance in *Arabidopsis* and moss (Nishiyama et al. 2011; Hyoung et al. 2020). Conversely, overexpression of *SlIPT3* in tomato and *MsCKX* in *Arabidopsis* has been shown to improve salt tolerance (Žižková et al. 2015; Li et al. 2019). Additionally, suppressing the expression of *OsCKX2*, a rice CK oxidase, enhanced salt tolerance and grain yield (Joshi et al. 2018). This section emphasizes the pathways that link CK signaling to sodium ion homeostasis.

Knockdown of *OsAHP1* and *OsAHP2* genes attributed increased salt sensitivity by downregulating key genes like *OsSOS1*, *OsHKT1;1*, *OsHKT1;5*, and *OsNHX1* (Sun et al. 2014b). Conversely, loss-of-function mutations in *OsRR9* and *OsRR10* enhanced salt tolerance, likely due to upregulation of *OsHKT1;1*, *OsHKT1;3*, and *OsHKT2;1* (Wang et al. 2019). Apart from this, spatial distribution of CK plays a pivotal role in rice salt tolerance. Salt stress alters CK levels in rice shoots and roots, with isopentenyladenine decreasing in shoots and cis-zeatin increasing in roots. This spatial imbalance is partly due to the inhibited expression of *OsPUP4*, a transporter responsible for CK distribution between tissues (Xiao et al. 2019; Yin et al. 2020). ARGONAUTE 2 (AGO2) positively regulates *OsPUP4* expression by influencing epigenetic modifications, including increased H3K4me3 and decreased H3K27me3 levels (Yin et al. 2020). In summary, these studies elucidate how CK distribution and signaling influence sodium transport and salt resistance, contributing to strategies for enhancing global food security.

9.4.3 Heat Stress

CKs regulate numerous physiological, biochemical, and molecular mechanisms to enhance tolerance and sustain growth under heat stress (Calleja-Cabrera et al. 2020; Liu et al. 2020). They maintain cellular homeostasis, stabilize proteins, and influence stress-signaling pathways. CKs help mitigate the detrimental effects of heat stress by preserving chlorophyll content, photosynthetic activity, and cell membrane stability, thereby ensuring energy production and cellular integrity (Veerasamy et al. 2007; Kobayashi et al. 2009).

CKs also regulate the expression of *HSPs* and antioxidant enzymes, protecting cellular components from heat-induced damage and reducing oxidative stress (Srivastava and Pandey 2023; Mughal et al. 2024). They elevate enzymatic activities and upregulate genes involved in photosynthesis and carbohydrate metabolism, contributing to improved thermotolerance (Xing et al. 2009; Dobrá et al. 2015). Recent studies highlight the potential of manipulating CK levels to enhance heat stress tolerance. Exogenous CK application, inhibitors of CK degradation, or CK biosynthetic gene transformants have demonstrated significant improvements in plant thermotolerance and recovery after heat stress (Černý et al. 2014; Jespersen et al. 2015; Prerostova et al. 2020).

However, boosting stress tolerance via CKs requires careful consideration due to their dynamic regulation during stress responses. The exogenous application of CKs can mimic acclimation effects and mitigate crop losses caused by heat stress, as evidenced by studies demonstrating the upregulation of endogenous CK levels in rice plants during stress (Prerostova et al. 2023). Heat stress in rice often leads to reductions in panicle size and spikelet number, adversely affecting yield. High temperatures decrease CK levels in panicles, correlating with reduced spikelet numbers. Exogenous application of CK, such as 6-benzylaminopurine (BAP), has been found to mitigate these effects by maintaining panicle CK levels, thereby preserving spikelet formation under heat stress conditions (Wu et al. 2017). Additionally, foliar sprays of CK during flowering and grain-filling stages have been reported to improve physiological parameters like stomatal conductance and chlorophyll fluorescence (*Fv/Fm* ratio) in heat-stressed rice plants and favor the number of filled spikelets and decrease the percentage of panicle blanking in plants subjected to heat stress during the flowering stage (Liu et al. 2020; Pantoja-Benavides et al. 2022). Furthermore, exogenous kinetin treatment has been observed to curb ROS homeostasis in rice seedlings under heat stress, suggesting a role in enhancing antioxidant defense mechanisms. This modulation aids in preserving cellular integrity and function during heat stress (Mei et al. 2023).

Studies have demonstrated that applying CKs during the post-anthesis period can enhance the "stay-green" characteristics of wheat, thereby improving heat tolerance and grain yield (Prerostova et al. 2023). Similarly, foliar application of CKs during the grain-filling stage has been shown to improve grain development, increase yield, and mitigate the adverse effects of heat stress on wheat productivity (Abasi et al. 2024). In maize, heat stress significantly reduces yield and quality by impairing photosynthesis and increasing oxidative stress. Although specific studies on CKs in heat-stressed maize are limited, their potential role in alleviating these effects needs to be further explored (El-Sappah et al. 2022). In soybean, heat stress negatively affects yield and quality. Recent studies have explored the improvement of heat stress tolerance in soybean through various approaches, including genetic modifications and exogenous applications of growth regulators. However, specific studies on the role of CK in heat-stressed soybean are limited (Jianing et al. 2022). CKs show promise in mitigating heat stress in crops, enhancing resilience, and sustaining yield in a warming climate.

9.4.4 Cold and Freezing Stress

CKs are essential plant hormones involved in regulating numerous physiological and molecular processes, including responses to cold and freezing stress. CKs play a pivotal role in enhancing plant tolerance to low-temperature stress by influencing chlorophyll metabolism, delaying senescence, and maintaining photosynthetic activity. Low temperatures impose significant stress on plants, affecting membrane systems, energy production, and reproductive development. For example, exposure to cold during the flowering stage of rice leads to sterility and yield loss (Feng et al. 2014; Wang et al. 2017a). Freezing stress, when temperatures fall below 0 °C, causes ice crystal formation, leading to mechanical damage and metabolic dysfunction (Yang et al. 2013; Cheng et al. 2014). Cold stress disrupts cellular homeostasis by inducing ROS accumulation, reducing antioxidant enzyme activity, and causing membrane damage and ion leakage (Sui 2015; Liu et al. 2016). CKs counteract these effects by modulating ROS homeostasis and enhancing the antioxidant defense system. A stable cytokinin level is crucial for normal plant growth and development under low-temperature conditions, as rapid CK downregulation occurs during initial cold stress, followed by an increase during acclimation (Kosová et al. 2012).

In wheat, CKs improve freezing tolerance by maintaining cellular functions and delaying senescence during cold stress. Pretreatment with exogenous CKs enhances cold tolerance in wheat seedlings by increasing endogenous CK levels (Belintani et al. 2012). Cytokinin response factors (CRFs), such as CRF2 and CRF3, regulate lateral root development in *Arabidopsis* under low-temperature stress (Jeon et al. 2016). In maize, although direct studies on CKs in cold stress are limited, they aid in the recovery of photosynthetic performance after temperature fluctuations, suggesting a role in stress resilience. In rice, cold stress adversely affects growth and reproductive development, particularly during the seedling and flowering stages. CKs are hypothesized to play a role in modulating cold tolerance mechanisms in rice, though specific studies remain scarce. Earlier studies suggest that both endogenous and exogenous CKs improve cold tolerance by stabilizing cellular processes and enhancing stress-responsive pathways (Jeon and Kim 2013). Overall, CKs are integral to plant cold stress responses, regulating processes such as chloroplast function, ROS scavenging, and stress-responsive gene expression.

9.5 Role of Gibberellins in Abiotic Stress Responses

Gibberellins, also known as gibberellic acid (GAs), represent the class of diterpenoid phytohormones that are known for their critical role in plant growth and development. Their role extends beyond promoting growth to regulating plant responses to abiotic stresses. This regulation occurs by the intervention and modulation in GA biosynthesis, signaling pathways, and interaction with other hormones such as ABA, ethylene, and JA (Achard et al. 2006; Hedden and Sponsel 2015). GA biosynthesis in higher plants proceeds in three stages: initial synthesis of ent-kaurene within proplastids, its subsequent conversion to GA12 by microsomal cytochrome P450 monooxygenases, and finally, the formation of C20- and C19-GAs within the cytoplasm. The final stage of GA biosynthesis involves the conversion of GA12 and GA53 into bioactive GAs like GA1 and GA4. This crucial step is orchestrated by 2-oxoglutarate-dependent dioxygenases, notably GA 20-oxidases and GA 3-oxidases. GA signaling begins with the perception of the hormone by GID1 receptors. This interaction triggers the degradation of DELLA proteins, which act as negative regulators of GA signaling. The degradation of DELLA proteins relieves the repression of downstream transcription factors, leading to changes in gene expression that promote GA-responsive growth and development (Olszewski et al. 2002).

9.5.1 *Drought Stress*

Drought stress induces plant adaptations, including growth inhibition, root lengthening, and stomatal closure (Liao et al. 2023). GA-mediated growth is suppressed during drought to conserve resources (Litvin et al. 2016). The process of inhibition of GA synthesis in drought stress can lead to an increase in ABA levels, which then can lead to suppression of GA signaling. For instance, downregulation of GA20ox and upregulation of GA2ox in wheat under drought reduce GA availability to slow growth and conserve water (Liao et al. 2023). Study by Shohat et al. (2021) indicates that water deficiency in *Solanum lycopersicum* reduced the levels of the bioactive GA1 and GA3 in tomato guard cells, and this resulted in fast stomata closure, thereby preserving water loss (Shohat et al. 2021). Further, there occurs a major cross talk between DELLA and ABA under drought stress, as DELLA stabilization under drought induces transcription of ABA-responsive genes, including *RD29A* and *COR47*, enhancing stress tolerance mechanisms (Shuai et al. 2017). ABA strongly antagonizes GA signaling. DELLAs are stabilized under high ABA levels, which helps in promoting drought tolerance by enhancing stomatal closure and root elongation. *Arabidopsis* plants with stabilized DELLA proteins exhibited improved drought tolerance due to reduced transpiration and enhanced antioxidant responses (Achard et al. 2006; Shu et al. 2018). In case of rice under drought stress, reduced GA levels enhance drought tolerance. The lower GA increases SLENDER RICE 1

(SLR1), which stabilizes the ABA receptor PYL10 by competing for Tillering and Dwarf 1 (TAD1) binding, thereby boosting the ABA response and coordinating growth with survival (Liao et al. 2023).

9.5.2 Salinity Stress

Salinity stress can disrupt ionic and osmotic homeostasis, impairing plant growth. Similar to drought, salinity inhibits GA biosynthesis according to the rule of prioritizing resource allocation to root growth over the shoot elongation. In *Oryza sativa*, *OsGA2ox*, which is responsible for GA inactivation, is upregulated under salinity stress, reducing shoot elongation and mitigating ionic imbalance (Wang et al. 2021d). Salt stress upregulates genes like *OsGA2ox6* in rice, leading to GA inactivation and stunted growth, potentially an energy-saving adaptation (Magome et al. 2013). Genes like *OsGA2ox3* and *OsGA2ox6* suppress bioactive GA levels under saline conditions, resulting in stunted shoot growth but enhanced root-to-shoot ratios for improved ionic balance (Hedden and Thomas 2012; Liu et al. 2018). GAs particularly enhance the activity of antioxidant enzymes like SOD and CAT, which leads to the scavenging of ROS generated during salinity stress (Hedden and Thomas 2012). Shahzad et al. (2021) reported that GA3 application in salt-stressed maize increased antioxidant enzyme activity, mitigating ROS-induced oxidative stress, while also boosting chlorophyll content (Shahzad et al. 2021). Similarly, GA3 mitigates salinity stress in rice by rescuing seed germination and seedling growth, which are negatively impacted by salt. It achieves this by increasing bioactive GA content and upregulating α-amylase-related genes that are suppressed under saline conditions (Farooq et al. 2022). GA3 application significantly mitigates the negative impacts of salt stress on early seedling growth, emergence, and various biochemical parameters in case of *Sorghum* as well. It helps counteract the detrimental effects of salinity, improving plant health under stressful conditions (Ali et al. 2021).

9.5.3 Cold Stress

The preharvest application of GA3 to various fruits and vegetables directly affects their quality and postharvest durability (Erogul and Sen 2016). Exogenous GA3 treatment effectively regulates the activity of numerous enzymes within fruit cells, mitigating chilling injury and suppressing ethylene production, thereby enhancing the storage quality of fruits. These findings were corroborated by studies on tomato, green chili, and peach (Ding et al. 2015; Maurya et al. 2020; Sun et al. 2024).

9.5.4 Heat Stress

GA3 treatment alleviated heat stress in tomato plants by enhancing growth, photosynthesis, and antioxidant activity (Guo et al. 2022). In date palms, combined application of silicon (Si) and GA3 significantly alleviated heat stress by enhancing growth, reducing oxidative stress, and modulating hormone signaling pathways. This synergistic effect involved increased antioxidant enzyme activity, upregulation of stress-related genes, and reduced ABA levels, offering a promising strategy for improving date palm productivity under heat stress (Khan et al. 2020).

9.5.5 Flooding Stress

GA plays a crucial role in plant flood tolerance by influencing carbohydrate metabolism and regulating the growth of internodes and other plant structures (Jia et al. 2021). In flood-tolerant plants (particularly rice), submergence often triggers a reduction in growth, which typically involves the downregulation of GA signaling (Fukao and Bailey-Serres 2008; Jia et al. 2021). In lowland rice, submergence stress can lead to an increase in active GA (GA1), resulting in leaf sheath elongation (Dubois et al. 2011). In deepwater rice, submergence induces the expression of the GA biosynthesis gene *SD1*, which is regulated by the ethylene-responsive transcription factor OsEIL1a. This increase in GA levels then promotes significant internode elongation (Kuroha et al. 2018). Nagai et al. (2020) reported that *ACE1* and *DEC1* function antagonistically with GA to regulate internode elongation in deepwater rice. While the function of *ACE1* is currently unclear, it appears to influence internode elongation in conjunction with GA. Both genes contribute to the ability of deepwater rice to adapt to submergence stress (Nagai et al. 2020).

9.5.6 Nutrient Deficiency

Nutrient deficiencies, such as low nitrogen or phosphorus (Pi), disrupt GA biosynthesis. This reduction in GAs alters plant growth patterns, prioritizing root development over the shoot growth to enhance nutrient acquisition. In maize, Pi deficiency specifically decreases bioactive GA levels and increases DELLA protein levels. This hormonal imbalance inhibits shoot growth and promotes root hair elongation, facilitating improved phosphorus uptake (Jiang et al. 2007). GA signaling has also been implicated in plant responses to iron deficiency. Under iron-limiting conditions, DELLA proteins are excluded from the root epidermal cells of the differentiation zone, the primary site of iron uptake. This exclusion allows for the expression of iron uptake genes, which are activated by the FER-like IRON-DEFICIENCY INDUCED TRANSCRIPTION FACTOR (FIT) (Wild et al. 2016).

9.6 Role of Salicylic Acid in Abiotic Stress Responses

Salicylic acid (SA), also known as hydroxybenzoic acid, is a crucial plant growth regulator involved in mediating responses to various stresses and regulating plant growth and development. In plants, SA biosynthesis begins in the chloroplast from chorismate, following two primary metabolic pathways: the isochorismate synthase (ICS) pathway and the phenylalanine ammonia-lyase (PAL) pathway (Garcion et al. 2008; Chen et al. 2009; Zhang and Li 2019). Both pathways originate from chorismate, the final product of the shikimate pathway (Garcion et al. 2008; Chen et al. 2009; Lefevere et al. 2020; Peng et al. 2021). While SA can be modified into various forms, the biologically active molecule in plants is free SA. However, most SA in plants exists as a glucose-conjugated inactive form, which can undergo several modifications. These modifications often lead to the degradation, inactivation, or storage of SA in large quantities within the vacuole (Maris et al. 2011; Chen et al. 2013). SA signaling involves the perception of SA, which can trigger various cellular responses, including the activation of defense genes. One key component of SA signaling is NPR1 (Nonexpressor of PR genes 1), a protein that acts as a regulator of SA-mediated defense responses. Upon SA accumulation, NPR1 undergoes a conformational change, allowing it to interact with TGA transcription factors and regulate the expression of defense-related genes (Peng et al. 2021). SA is involved in abiotic stress responses, including drought, osmotic stress, extreme temperatures, salinity, and heavy metal toxicity (Khan et al. 2015).

9.6.1 Cold and Heat Stress

SA mitigates damage caused by cold injury (Taşgin et al. 2006). In plants like barley, winter wheat, and others, SA treatment enhances the accumulation of antioxidative enzymes such as CAT, SOD, and POX (Taşgin et al. 2006; Mutlu et al. 2013; Ignatenko et al. 2019; Zhang et al. 2021; Wang et al. 2022b; Golovatskaya et al. 2023). Additionally, pretreatment with SA significantly increases the grain-filling rate and weight in rice, ultimately contributing to higher grain yields (Chen et al. 2017). Applying SA as a foliar spray on grape leaves helped alleviate the negative impacts of chilling stress by boosting the activity of various antioxidant enzymes and increasing the levels of soluble sugars, proline, and chlorophyll (*Vitis vinifera* L.) (Li and Wang 2021). Additionally, SA significantly enhanced the expression of cold response genes (*CBF1*, *CBF2*, and *CBF3*), which are essential for enhancing the plant's tolerance to low temperatures. Furthermore, wheat plants that were primed with both SA and cold showed increased expression of the *WRKY* gene (*WRKY19*), heat shock transcription factor (*HSF3*), mitochondrial alternative oxidase (*AOX1a*), and heat shock protein (*HSP70*) when exposed to freezing stress (Wang et al. 2020). This dual priming led to enhanced antioxidant capacity, protection of the photosystem, reduced malondialdehyde levels, lower superoxide radical production, improved photosystem II efficiency, and better photosynthetic performance (Wang et al. 2020). Furthermore, free proline levels increased in SA-primed plants during freezing due to both enhanced synthesis and reduced degradation (Wang et al. 2022b). Transcriptome analysis revealed that the mitogen-activated protein kinase (MAPK) signaling pathway, involving genes like *MPK3*, *MPK9*, *WRKY22*, and *TGA1*, is critical in SA-mediated freezing stress tolerance (Wang et al. 2023b).

Besides cold stress, SA also regulates heat stress tolerance. In barley, exogenous SA protected root growth from the combined stresses of heat, drought, and salinity (Torun 2019). In wheat, SA treatment at anthesis improved photosynthetic performance in flag leaves by increasing leaf area, stomatal density, chlorophyll content, and other photosynthetic parameters, mitigating yield losses under high temperatures (Fan et al. 2022). Additionally, SA modulated antioxidant enzyme activities (CAT, SOD, and POD) under heat stress (Wassie et al. 2020). Combined NO and SA treatment enhanced heat stress tolerance in plants by modulating antioxidant systems (glutathione, ethylene), sulfur assimilation, and physiological traits (plant height, biomass, photosynthesis) while minimizing membrane damage (Rasheed et al. 2022a). SA alone also improved heat tolerance in alfalfa through similar mechanisms (Rasheed et al. 2022a).

9.6.2 *Heavy Metal Stress*

Several studies indicate the role of SA in enhancing plant resistance to various heavy metals. For example, SA pretreatment has been shown to alleviate cadmium toxicity. Presoaking dry barley seeds in 0.5 mM SA solution partially mitigated cadmium toxicity in seedlings (Metwally et al. 2003). SA treatment mitigated cadmium toxicity in both soybean and wheat plants. In soybean, SA reduced cadmium uptake, while in wheat, SA pretreatment decreased malondialdehyde accumulation and electrolyte leakage, indicating reduced oxidative stress and membrane damage (Noriega et al. 2012). Beyond cadmium, foliar application of 15 μM SA in alfalfa alleviated aluminum toxicity by reducing its accumulation, mitigating chloroplast damage, increasing biomass, and enhancing photosynthesis through upregulation of *LHCA3* and improved light capture efficiency (Cheng et al. 2020). Similarly, Bortolin et al. (2020) reported that seed priming with SA reduced aluminum toxicity in *Trifolium vesiculosum* and *T. repens* by boosting the expression of antioxidant enzymes (SOD and APX) and lowering MDA accumulation (Bortolin et al. 2020). In rice, SA restored root membrane integrity, reduced ROS levels, and oxidative damage under aluminum stress (Pandey et al. 2013).

SA application effectively mitigates zinc stress in various plants, as demonstrated in *Salvia officinalis*, by enhancing growth, nutrient uptake (Ca, P, K), and essential oil content (Es-sbihi et al. 2020). In wheat, seed priming with SA alleviated zinc and chromium stress by enhancing growth and biomass while increasing levels of non-enzymatic antioxidants like ascorbic acid and reduced glutathione (Mazumder et al.

2022). Exogenous application of 0.5 mM SA in chromium-stressed plants enhanced growth and photosynthesis, reduced chromium accumulation, and alleviated oxidative stress by bolstering antioxidant defenses (lower MDA and superoxide levels) in tomato plants (Islam et al. 2016; Gupta and Seth 2021). Under copper stress, SA enhanced antioxidant enzyme activities (SOD, APX, GR, glutathione peroxidase (GPX)) and restored diminished activities of CAT, DHAR, and GST (Mostofa and Fujita 2013). These studies underscore the potential of SA in alleviating heavy metal toxicity and enhancing plant resilience under these conditions.

9.6.3 Drought Stress

Research has demonstrated various morphological, physiological, and biochemical aspects associated with SA-mediated drought tolerance. SA pretreatment influences dehydrin accumulation in water-stressed barley. This response is bimodal, as the dehydrin protein levels increased with 0.2 mM SA treatment but decreased at 0.5 mM SA (Sun et al. 2009). In rice, seed soaking with SA effectively mitigated oxidative damage, with the regulation of proline levels and improved plant-water status (Sohag et al. 2020). Additionally, foliar application of SA significantly reduced membrane electrolyte leakage and leaf H_2O_2 and MDA contents while enhancing CAT activity (Farooq et al. 2009). In wheat, treatment with 0.5 mM SA resulted in enhanced biomass accumulation, improved physiological performance, reduced lipid peroxidation, increased antioxidant enzyme activities, and better yields (Sharma et al. 2017; Khalvandi et al. 2021). Proteomic analysis identified 76 proteins differentially regulated by SA pretreatment, associated with signal transduction, stress defense, photosynthesis, carbohydrate metabolism, protein metabolism, and energy production. Of these, 35 proteins were suggested to play critical roles in SA-induced drought tolerance (Kang et al. 2012). Similarly, plants grown from seeds imbibed in aqueous solutions of SA or acetyl salicylic acid (ASA) demonstrated improved tolerance to drought stress, as well as heat and chilling stresses (Senaratna et al. 2000). In maize, foliar application of SA combined with zinc and glycine betaine increased drought tolerance by reducing ROS accumulation, enhancing enzymatic antioxidants, and promoting osmolyte accumulation, particularly during reproductive growth stages (Shemi et al. 2021). In some cases, SA-induced drought tolerance was associated with adaptive mechanisms like leaf rolling, providing an alternative protective response to water deficit (Kadioglu et al. 2011).

In *Brassica rapa*, SA enhanced drought tolerance by maintaining redox homeostasis and promoting proline biosynthesis by upregulation of *P5CSA* and *P5CSB* (proline synthesis) and downregulation of *PDH* (proline degradation). These changes were not observed in plants without SA pretreatment, highlighting the protective role of SA under drought stress (La et al. 2019).

9.6.4 *Osmotic Stress*

In many crop species, SA has been shown to mitigate oxidative stress damage. In wheat, SA combined with NO pretreatment enhanced antioxidant activity, resulting in increased fresh weight, chlorophyll content, reduced MDA levels, and lower lipoxygenase activity (Mehdi et al. 2014). Furthermore, SA, in conjunction with ABA, alleviated PEG-induced growth inhibition by improving osmotic adjustment through elevated proline and carbohydrate content and enhanced antioxidant activity (Marcińska et al. 2013). In barley, SA pretreatment under osmotic stress enhanced stress tolerance by modulating ROS production and alleviating the decline in photosystem II proteins D1, D2, and LHC2 and their corresponding gene transcripts (Luo et al. 2009). In cowpeas, SA reduced the adverse effects of water stress by increasing germination rates, seedling height, chlorophyll a and b, carotenoid content, and proline levels while improving cellular electrolyte balance (De Araújo et al. 2018). Additionally, SA pretreatment mitigated salt stress-induced osmotic stress in maize seedlings, resulting in improved growth parameters, photosynthetic pigment levels, nitrate reductase activity, and reduced free proline and MDA levels (Gautam and Singh 2009). These findings highlight the diverse roles of SA in modulating osmotic stress tolerance across various plant species and conditions.

9.7 Role of Jasmonic Acid in Abiotic Stress Responses

Jasmonic acid (JA) is a cyclopentane fatty acid, along with its precursors and derivatives that are collectively known as jasmonates (JAs). JA is synthesized in the peroxisomes and subsequently transported to the cytosol, where it is conjugated with isoleucine by the enzyme JASMONATE-RESISTANT 1 (JAR1), a JA amino acid conjugate synthase. This process results in the formation of the bioactive compound JA-Ile, which activates the JA signaling pathway and promotes the expression of JA-responsive genes (Han 2017; Wasternack and Song 2017; Ruan et al. 2019; Wasternack and Hause 2019; Ghorbel et al. 2021). Key components of JA signaling include the F-box receptor, CORONANTINE INSENSITIVE1 (COI1), the signaling repressors JAZ proteins, and the transcription factor MYC2 along with its homologs (Wasternack and Hause 2019). JA metabolism and signaling are highly responsive to abiotic stress, with such conditions rapidly inducing the expression of genes involved in JA biosynthesis and signaling pathways. This dynamic response underscores the critical role of JA in helping plants adapt to abiotic stresses. Numerous studies have demonstrated the importance of JA in modulating plant responses to a variety of environmental challenges, as further discussed in the following sections.

9.7.1 Heat Stress

JA is a key regulator of plant growth under extreme environmental conditions such as heat and cold. JA in conjunction with SA confers basal thermotolerance in *Arabidopsis* as enhanced thermotolerance in the SA-overproducing *cpr5-1* mutant is lost in the *cpr5-1 jar1-1* double mutant (Clarke et al. 2009). Furthermore, JA is essential for acclimation to combined stresses like heat and light, as JA biosynthesis mutants, i.e., *aos* mutant plants, exhibit impaired acclimation due to reduced expression of JA-regulated *ZAT6*, ROS- and OPDA-regulated *ZAT10*, and redox-regulated *APX1* and *APX2* genes and elevated JA levels during stress (Balfagón et al. 2019). Exogenous application of methyl jasmonate (MeJA) improves heat tolerance in perennial ryegrass by enhancing osmotic adjustment, antioxidant defense, and the expression of JA-responsive genes. These effects included better maintenance of relative water content (RWC), reduced chlorophyll loss, decreased electrolyte leakage and MDA levels under heat stress (Su et al. 2021; Nie et al. 2022). Similarly, in soybean, JA priming enhanced plant biomass, photosynthetic efficiency, and RWC, thereby alleviating oxidative stress through increased activities of antioxidant enzymes such as SOD, CAT, glutathione peroxidase (GPX), and glutathione S-transferase (GST) under combined heat and drought conditions (Rahman et al. 2024). In eggplant (*Solanum melongena*), silencing *SmLOX4* and *SmLOX5*, key genes in JA biosynthesis, reduced heat tolerance, while JA influenced heat tolerance by regulating *DREB2A* expression (Liu et al. 2024).

9.7.2 Cold Stress

JA also plays a significant role in cold stress responses. Cold stress induces endogenous JA accumulation, which acts as an upstream signal for the ICE-CBF/DREB1 cascade, positively regulating cold tolerance in *Arabidopsis* (Hu et al. 2013). JA, together with ABA, mediates cold tolerance by upregulating the CBF pathway, with transcripts suppressed in JA biosynthetic mutant (Wang et al. 2016). Recent research in tomato (*Solanum lycopersicum*) has highlighted that JA promotes cold tolerance by influencing the arginine decarboxylase (ADC1) pathway involved in putrescine biosynthesis, a major polyamine that mitigates cold-induced oxidative stress. RNAi-mediated suppression of *MYC2*, a central regulator of JA signaling, reduced *ADC1* expression, diminished putrescine accumulation, and exacerbated cold-induced oxidative damage (Ding et al. 2021).

9.7.3 UV Radiation Stress

Exposure to UV-B radiation has been shown to elevate JA levels. UV-B radiation triggers oxidative stress by increasing ROS production. The simultaneous rise in JA levels and PDF1.2 transcripts upon UV-B exposure suggests JA's role in managing UV-B-induced stress (A. H. Mackerness et al. 1999). MeJA enhances barley seedling tolerance to UV-B stress. MeJA application increases free proline accumulation and activates antioxidant enzymes (SOD, CAT, POX), particularly inducing POX isoforms, suggesting a role in mitigating UV-B-induced oxidative stress (Fedina et al. 2009). Similarly, UV-C radiation increased total ascorbic acid and polyphenolic concentrations in apple plants (*Malus domestica*). This was accompanied by a rise in endogenous JA levels, suggesting a potential link between JA accumulation and UV-C-induced stress responses (Kondo et al. 2011).

9.7.4 Heavy Metal

JA plays a crucial role in alleviating heavy metal stress in various plants. In tomato, JA mitigates Cd-induced damage as Cd accumulation is higher in *spr2* (*suppressor of prosystemin-mediated responses2*) mutants, leading to increased oxidative stress markers (MDA, electrolyte leakage, H_2O_2). Wild-type plants, with higher endogenous JA, exhibit enhanced antioxidant enzyme activities (SOD, POD, CAT), suggesting JA's role in reducing Cd-induced oxidative stress (Zhao et al. 2016a). In *Glycine max*, pretreatment with JA before nickel exposure significantly reduced MDA and H_2O_2 levels, enhancing nickel stress tolerance compared to untreated controls (Sirhindi et al. 2015). Similarly, pre-sowing maize seeds with JA alleviated the adverse effects of oxidative stress caused by nickel, improving growth, biomass production, protein content, and antioxidant enzyme activity (Azeem 2018). In Pb-exposed tomato, JA treatment enhanced the accumulation of secondary metabolites (phenols, flavonoids, anthocyanins) by upregulating key biosynthetic genes like chalcone synthase and phenylalanine ammonia-lyase. JA also increases organic acids and metal-chelating compounds while enhancing the expression of genes involved in the tricarboxylic acid cycle, contributing to Pb stress tolerance (Bali et al. 2019). In alfalfa, JA mitigates Cu stress by reducing Cu accumulation, increasing biomass, and improving chlorophyll content. JA treatment enhanced antioxidant enzyme activities and reduced oxidative stress markers (MDA, H_2O_2), indicating improved stress tolerance (Dai et al. 2020). Similarly, in *Brassica parachinensis L.*, JA alleviated Cr stress by restoring mineral nutrient content and restricting Cr uptake and accumulation in roots and shoots compared to plants exposed to Cr alone (Kamran et al. 2021).

9.7.5 Drought Stress

JA plays a pivotal role in modulating plant responses to drought stress. In soybean, JA priming significantly improved plant biomass, photosynthetic efficiency, and leaf relative water content, which collectively reduced the accumulation of ROS and MDA, enhancing overall drought tolerance. The increased activities of GR, GPX, and glutathione S-transferase (GST) in JA-primed plants further highlight the activation of a glutathione-dependent peroxide scavenging system (Rahman et al. 2024). JA also interacts synergistically with other phytohormones and metabolites to confer drought resilience. In *Brassica rapa L.*, the combined application of BR and JA significantly alleviated drought stress by enhancing photosynthetic rate, pigment concentration, stomatal conductance, transpiration rate, and antioxidant defenses (Ahmad Lone et al. 2022). Moreover, JA accumulation during drought also increases ABA biosynthesis, suggesting a positive interaction between these hormones. In citrus roots, a rapid and transient accumulation of JA under stress conditions triggers ABA biosynthesis, emphasizing JA's role as an initiator of plant stress responses (De Ollas et al. 2013). Exogenous ABA and JA improve growth under PEG-simulated drought in pearl millet plants. This treatment enhanced chlorophyll content, relative water content, and antioxidative enzyme activities (APX, SOD), thereby strengthening the plant's defense system (Awan et al. 2021). In wheat, chemical inhibition of the JA pathway resulted in reduced endogenous ABA levels, particularly by downregulating the expression of *NCED1*. This finding underscores the interdependence of JA and ABA pathways in drought tolerance (Wang et al. 2021b). Additionally, JA-induced NO production enhances drought tolerance in wheat by upregulating the ascorbate-glutathione (AsA-GSH) cycle. This led to elevated activities of antioxidative enzymes and increased ratios of reduced to oxidized forms of ascorbate (AsA/DHA) and glutathione (GSH/GSSG), further supporting cellular redox homeostasis under drought stress (Shan et al. 2015). These findings demonstrate that JA, either alone or in combination with other signaling molecules, serves as a critical regulator of plant responses to drought stress.

9.7.6 Salinity Stress

JA has been widely studied for its potential to enhance salt stress tolerance in plants. Qiu et al. (2014) showed that foliar JA application to wheat seedlings decreased MDA content and ROS production. This treatment also increased the expression and activity of antioxidant enzymes (SOD, POD, CAT, APX) (Qiu et al. 2014). Similarly, JA priming in wheat enhanced vegetative growth and yield. This was associated with improved photosynthesis, increased stomatal conductance and CO_2 levels, and enhanced transpiration. JA reduced Na^+ accumulation while increasing K^+ content by inducing transporters. Additionally, JA increased α-tocopherol, phenolics, and flavonoids and enhanced SOD and APX activity, improving antioxidant defense (Sheteiwy et al. 2022). In barley, JA pretreatment significantly lowered Na^+ accumulation in shoot tissues under salt stress, alleviating the photosynthetic inhibition caused by salinity (Walia et al. 2007). Likewise, Zhang et al. (2017) observed that the JA signaling pathway was crucial in sweet potato (*Ipomoea batatas*) for salt stress responses. Salt-tolerant varieties exhibited higher JA content and increased expression of JA biosynthesis and signaling genes, including *LOX*, *AOS*, *OPR3*, *COI1*, and *PDF1.2*. JA also regulated stomatal closure and minimized water loss, further improving salt tolerance (Zhang et al. 2017). In rice, Kang et al. (2005) reported that JA application 24 hours after salt stress significantly improved leaf water potential, photosynthetic rate, and the maximum quantum yield of photosystem II compared to plants treated only with NaCl (Kang et al. 2005). These findings collectively highlight the multifaceted role of JA in enhancing plant resilience to salt stress through antioxidant defense, ion homeostasis, photosynthetic efficiency, and water regulation. A recent study in *Arabidopsis* has highlighted the molecular mechanism behind the role of JA signaling in combating salt stress. The study by Li and coworkers reveals that JA signaling disarms JAZ8, a repressor that normally inhibits the NUCLEAR FACTOR-Y (NF-Y) subunits NF-YA1, NF-YB2, and NF-YC9, which form a trimeric complex (NF-YA1-YB2-YC9), which is a positively regulator of salinity-responsive genes. Upon salt stress, JA-induced degradation of JAZ8 allows the complex to activate salinity-responsive genes, thereby boosting the plant's salt tolerance (Li et al. 2024a). Apart from this, JA helps safflower plants tolerate salinity by boosting catalase activity, which reduces harmful ROS and lipid peroxidation. It also improves photosynthetic efficiency by decreasing ROS production within photosystems and enhancing the light-dependent phase of photosynthesis (Lotfi et al. 2024).

9.8 Role of Ethylene in Abiotic Stress Responses

Ethylene, a gaseous hormone, has been well established in the context of its functions as a regulator in plant development as well as stress responses (Groen and Whiteman 2014). In ethylene biosynthesis, the initial step involves the production of AdoMet (SAM) from methionine. Further, the SAM is converted to the intermediate aminocyclopropane-1-carboxylic acid (ACC) catalyzed by ACC synthase. The final reaction culminates in the production of ethylene from ACC catalyzed by ACC oxidase (Kende 1993). In *Arabidopsis,* ethylene perception is carried out by a family of five receptors, including ETHYLENE RESPONSE1 (ETR1), ETR2, ETHYLENE RESPONSE SENSOR1 (ERS1), ERS2, and ETHYLENE INSENSITIVE4 (EIN4), which are similar to the bacterial two-component regulators (Guo and Ecker 2004; Kazan 2015). In the absence of ethylene, these ethylene receptors, which are localized in the endoplasmic reticulum membrane, activate CONSTITUTIVE TRIPLE RESPONSE1 (CTR1) kinase. CTR1 kinase phosphorylates EIN2, leading to its inhibition. The F-Box proteins EIN3-BINDING F BOX PROTEIN 1 (EBF1) and EBF2 target EIN3/EIL1 for proteasomal degradation. In contrast, the presence of ethylene shuts off the receptors, causing inactivation of CTR1 and dephosphorylation of EIN2. Consequently, the translation of *EBF*1 and *EBF2* is suppressed, thereby mitigating the proteasomal degradation of EIN3 and EIL1. Ethylene-induced stabilization of EIN3/EIL1 positively regulates the transcription of ethylene-responsive genes, including the *ETHYLENE RESPONSE FACTORs (ERFs)* (Guo and Ecker 2004; Chen et al. 2021). ERFs are AP2-containing transcription factors, which recognize Dehydration-Responsive or C-Repeat Element (DRE/CRT) present on the stress-responsive genes (Guo and Ecker 2004; Chen et al. 2021). The *eto1* mutant exhibits enhanced ethylene accumulation and has been utilized to investigate the effects of ethylene in diverse plant processes, including stress tolerance (Woeste et al. 1999; Kazan 2015). Accordingly, the presence of ethylene affects the tolerance of plants to diverse abiotic stresses through multiple layers of regulation (Chen et al. 2021).

9.8.1 Heat Stress

Ethylene plays a crucial role in regulating both high and low temperature responses in plants. Application of an ethylene mimic known as ethephon could induce the synthesis of a heat shock-induced protein APX1, which promoted heat stress tolerance in plants (Wu et al. 1994). Elevated temperatures caused ethylene accumulation in several plant species, including *Phaseolus vulgaris* (Sauter et al. 1990), *Mangifera indica* (Ketsa et al. 1999), *Actinidia deliciosa* (Antunes and Sfakiotakis 2000), *Triticum aestivum*, (Khan et al. 2013), *and Solanum lycopersicum* (Pan et al. 2019). However, ethylene synthesis was inhibited in *Lycopersicon esculentum* under heat stress (Biggs and Handa 1988). *Arabidopsis* and rice displayed distinguished expression of certain ethylene biosynthesis genes during heat stress. In Arabidopsis, *1-AMINOCYCLOPROPANE-1-CARBOXYLATE (ACS)* genes such as *ACS6, ACS7, ACS8, ACS10, ACS11, ACS12* and *1-AMINOCYCLOPROPANE-1-CARBOXYLIC ACID OXIDASE* (ACO) including *ACO2, ACO4* were exhibited enhanced expression under heat stress, whereas *ACS2, ACS4, ACS5, ACO1, ACO3*, and *ACO4* showed negative regulation. Contrastingly in rice, *OsACS2, OsACS6, OsACO5*, and *OsACO7* were upregulated in heat-tolerant and heat-sensitive lines at high temperatures, while *OsACO1* and *OsACO2* are repressed (Poór et al. 2021). Additionally, the ethylene signaling mutants *ein2* and *etr1* exhibited a reduction in basal thermotolerance. In contrast, the difference in acquired thermotolerance between these mutants and wild type was scarce (Larkindale et al. 2005). Interestingly, ethylene signaling helps rice seedlings withstand heat stress by improving their thermotolerance and regulating the expression of Hsfs. Ethylene treatment in rice seedlings reduces oxidative damage, maintains chlorophyll content, increases the levels of Hsfs (like HSFA1a and HSFA2a, c, d, e, and f) and ethylene-signaling-related genes (such as ethylene insensitive 2, ethylene insensitive-like 1, and ethylene insensitive-like 2). This indicates that ethylene orchestrates a complex genetic network to enhance thermotolerance in rice seedlings (Wu and Yang 2019).

9.8.2 Cold Stress

In *Arabidopsis*, the function of ethylene is likely controversial under low temperature stress, as application of ACC led to reduced freezing tolerance. Similarly, ethylene posed a negative effect on cold acclimation-dependent freezing tolerance in *Medicago truncatula* (Zhao et al. 2014). However, treatment with ACC biosynthetic inhibitor, aminoethoxyvinylglycine (AVG), enhanced the freezing tolerance in *Arabidopsis* (Kazan 2015). Likewise, an ethylene biosynthesis inhibitor, 1-methylcyclo propene (1-MCP), suppressed cold tolerance in tomato (*Lycopersicon esculentum*), thereby suggesting a beneficial role of ethylene in low temperature stress tolerance (Zhao et al. 2009a). Ethylene positively influences apple cold tolerance by inducing the expression of *MdERF1B*, which in turn upregulates cold-responsive genes like *MdCBF1*. The interaction between MdERF1B and MdCIbHLH1 forms a regulatory module that integrates cold and ethylene signaling pathways, impacting cold stress response and ethylene production in apples (Wang et al. 2021e). However, interestingly in soybean, the ethylene signaling pathway negatively regulates the CBF/DREB1 cold-responsive pathway, leading to poor cold tolerance. Modulating ethylene levels impacts GmDREB1A and GmDREB1B transcripts and downstream cold-responsive genes, suggesting ethylene transcriptionally inhibits the CBF/DREB1 pathway (Robison et al. 2019).

9.8.3 Drought Stress

Low-concentration foliar application of ACC escalated the shoot growth rate in drought-resistant and drought-sensitive wheat varieties due to the accumulation of carbohydrates (Valluru et al. 2016). However, *eto1* mutants in rice having high ethylene levels exhibited enhanced drought tolerance (Du et al. 2014). Elevated levels of ethylene were observed in various plant species under drought stress, including *Vicia faba*, orange, and French bean (Husain et al. 2020). Overexpression of *Triticum aestivum* transcription factor *TaERF1* had a crucial function in regulating the stomatal apertures and promoted tolerance toward drought stress (Xu et al. 2007). In case of squash, the *etr2b* mutant, with partial ethylene insensitivity, exhibits enhanced growth and improved drought adaptation, accumulating protective osmolytes, unsaturated fatty acids, and phenolics. This drought resilience in *etr2b* is not ABA-mediated, suggesting ethylene negatively regulates plant growth and drought response (Iglesias-Moya et al. 2023).

9.8.4 Salinity Stress

The ethylene biosynthesis pathway negatively regulates salt tolerance in plants. It has been documented that an ethylene biosynthesis mutant, *acs7*, showed enhanced salt tolerance and had better germination in the presence of salt stress (Dong et al. 2011). Similarly, *acs7* mutants in *Arabidopsis* consisting of reduced ethylene levels were resistant to salt stress. Furthermore, constitutive expression of an ACC oxidase gene from wheat, *TaACO1*, in *Arabidopsis* led to sensitivity to salt stress and suppressed expression of stress-responsive genes (Chen et al. 2014). Interestingly, even though ethylene biosynthesis had a contrasting function in regulating salt stress tolerance, ethylene signaling promoted resistance to salt stress. Mutants of *ETR1* and *EIN4* in *Arabidopsis* were sensitive to salt stress (Cao et al. 2007; Xu et al. 2007). Mutant of *CTR1*, a negative regulator of salinity stress, led to enhanced tolerance, thereby implying that *CTR1* suppresses salt stress tolerance (Xu et al. 2007; Jiang et al. 2013; Kazan 2015; Achard et al. 2006). EIN2, a positive regulator of ethylene signaling, promoted salt tolerance since the *ein2* mutants were sensitive to salt stress (Wang et al. 2007; Lei et al. 2011). EIN3, a positive regulator of ethylene signaling acting downstream of EIN2, was observed to function in providing resistance toward salt stress. Mutant of *EIN3* exhibited reduced germination under salt stress conditions (Achard et al. 2006; Cao et al. 2007; Wang et al. 2007; Yoo et al. 2008). EIN3 is known to target genes such as *SALT-INDUCED AND EIN3/EIL1-DEPENDENT* (*SIED1*) involved in the detoxification of reactive oxygen species (Zhang et al. 2011). Cebrián et al. (2021) established that gain-of-function mutations of ethylene receptors *CpETR1B, CpETR1A*, and *CpETR2B* in *Cucurbita pepo* were less receptive to ethylene; however, they displayed improved germination and vegetative growth during salt stress (Cebrián et al. 2021).

9.8.5 Submergence Stress

Plant tissues that have been submerged have less access to oxygen and carbon dioxide, which can seriously impair their metabolism. Furthermore, ethylene builds up in flooded plant tissues due to limited gas diffusion underwater (Waadt et al. 2022). Five transcription factors known as group VII ETHYLENE RESPONSE FACTORS (ERFs) are necessary for the transcription of hypoxia-responsive genes in *Arabidopsis* (Gasch et al. 2016). The presence of hypoxia as well as increased concentrations of ethylene leads to protein stabilization of group VII ERFs, thereby enhancing transcription of target genes (Gibbs et al. 2011; Licausi et al. 2011). In addition, a cluster of three related ERF genes is found in *SUBMERGENCE TOLERANCE 1*, a significant quantitative trait locus linked to increased flood tolerance in rice (Xu et al. 2006). *SNORKEL1* and *SNORKEL2*, two major quantitative trait loci, are expressed on the accumulation of ethylene in submerged tissues, thus causing internode elongation in deepwater rice (Hattori et al. 2009). In rice, overexpression of *RAP2.2*, a transcription factor belonging to the subfamily VII AP2/ERF related to *SUB1A*, led to improved survival hypoxia tolerance in rice via regulating ethylene biosynthesis, sugar, and fermentation metabolism (Hinz et al. 2010).

9.8.6 Heavy Metal Stress

Several evidences propose the metal-specific and concentration-dependent effect of ethylene in heavy metal stress tolerance in plants (Husain et al. 2020). In mustard, ethylene played a crucial role in sulfur-induced alleviation of cadmium stress and assisted in increasing photosynthesis through activating the antioxidant mechanism (Masood et al. 2012). Copper toxicity is also known to induce ethylene biosynthesis by upregulating the expression of *ACO1* and *ACO3* in *Nicotiana glutinosa* (Kim et al. 1998; Husain et al. 2020).

9.9 Role of Brassinosteroids in Abiotic Stress Responses

Brassinosteroids (BR) are a class of steroidal hormone, which play an important role in regulating plant growth and development. Discovered as a growth stimulator extracted from pollen called brassin, they are mostly found in Brassicaceae family crops (Mitchell et al. 1970). BRs bind to the BRI1/BAK1 receptor complex, triggering a signaling cascade that inactivates the negative regulator BIN2. This allows the activation of transcription factors like BEM/BZR1, which subsequently bind to DNA and activate the expression of BR-responsive genes. These genes control a wide range of cellular processes, including cell elongation, division, and differentiation, ultimately shaping plant architecture and influencing responses to environmental cues. BR biosynthesis, primarily occurring in the endoplasmic reticulum, begins with the conversion of campesterol, a plant sterol, through a series of enzymatic steps. Key enzymes involved include cytochrome P450 monooxygenases, which catalyze crucial oxidation reactions. This multistep pathway ultimately leads to the production of brassinolide, the most active form of BRs (Ye et al. 2011).

9.9.1 Drought Stress

BR plays a key role in helping plants tolerate drought by improving growth and water use efficiency. Applying DI-31 (a BR analogue) significantly boosts soybean growth and reduces drought-related yield losses by about 7% (Perez-Borroto et al. 2022). Overexpression of *GmBSK1* and *GmBSK1.5* in soybean has enhanced heat stress tolerance, which can be compounded by drought tolerance (Hou et al. 2024).

Interestingly, BR activates ROS scavenging system, through which they enhance drought tolerance, as overexpression of *IbBT4* in transgenic *Arabidopsis* activated ROS scavenging rates under drought stress (Zhou et al. 2020). In peanut plant, priming with BR helps plant to cope with drought stress (Huang et al. 2020b). Surina et al. (2024) demonstrated that BIL9 enhances plant growth and drought stress resistance by influencing brassinosteroid signaling. Through its interaction with the transcription factor HDG11, BIL9 upregulates the expression of genes involved in drought response, thereby improving the ability of plants to cope with water deficit (Surina et al. 2024).

9.9.2 Salinity Stress

Endogenous BR signaling plays a critical role in *Arabidopsis* salt stress responses, as evidenced by the hypersensitivity of the BR-deficient mutant *det2-1* to salinity. Exogenous EBR partially rescues the growth of *det2-1* under salt stress, highlighting the importance of BR-regulated proline accumulation and signaling for stress adaptation (Zeng et al. 2010). BRs alleviate salt stress in pepper by enhancing growth, photosynthesis, and antioxidant defenses. BR treatment increased osmolytes (sugars, proline, glycine betaine), reduced ROS/MDA levels, and upregulated stress-responsive genes, improving salt tolerance (Jin et al. 2024). In maize, a combined application of BR and SLs alleviates salt stress by enhancing antioxidant properties, maintaining growth, and reducing Na^+ accumulation. Transcriptome analyses reveal that BR and SL interactions modulate photosynthesis and hormone signaling while positively correlating gene expression with stress tolerance traits (Wang et al. 2024). BRs and ethylene signaling interact to enhance salt stress tolerance. BRs induce ethylene production via H_2O_2, and ethylene acts downstream in the BR-mediated salt stress response. Silencing ethylene signaling components reduces BR-induced salt tolerance in tomato plants, highlighting the interconnectedness of these pathways (Zhu et al. 2016).

9.9.3 Heavy Metal Stress

The role of BRs in enhancing plant resilience to heavy metal stress has been well documented across multiple studies. In tomato, exogenous application of BR mitigated cadmium toxicity by enhancing photosynthesis, antioxidant defenses, and alleviating oxidative stress (Hasan et al. 2011). Similarly, in *Solanum nigrum* L., pretreatment with 24-epibrassinolide (24-EBL) improves tolerance to nickel (Ni) stress by maintaining lipid peroxidation levels and enhancing antioxidant enzyme activities, such as SOD and ascorbate peroxidase (APX) (Soares et al. 2016). In grapevine cuttings, 24-epibrassinolide (EBR) mitigates copper (Cu) stress by reducing ROS levels, improving root growth, and increasing endogenous phytohormone levels, including JA, ABA, and SA (Zhou et al. 2018). Furthermore, EBR enhances the activities of antioxidant enzymes and regulates the ascorbate-glutathione (AsA-GSH) cycle in *Solanum lycopersicum* plants, highlighting its potential in alleviating oxidative stress and promoting detoxification of polycyclic aromatic hydrocarbons (PAHs) in contaminated environments (Ahammed et al. 2012).

9.9.4 Cold Stress

BRs are integral to enhancing cold tolerance in various plants, including tomatoes, *Arabidopsis*, mango, and apple. In *Arabidopsis*, BYPASS1-LIKE (B1L) interacts with TRANSTHYRETIN-LIKE (TTL), which is involved in BR-mediated plant growth. This interaction is critical in regulating growth and cold tolerance as mutation in B1L leads to increased growth and overexpression of TTL results in heightened freezing sensitivity (Chen et al. 2020). In tomatoes, BRs facilitate cold tolerance through the promotion of NBR1-dependent selective autophagy, activated by BZR1 under cold stress conditions. This mechanism effectiveness hinges on the stability of BZR1, as silencing of autophagy-related genes significantly reduces cold tolerance (Chi et al. 2020). Brassinolide application in mango fruit upregulates proteins involved in transport, defense, and stress response, thereby mitigating cold-induced damage and enhancing physiological properties (Li et al. 2012). In apple, MdBIM1, a key BR signaling component, enhances cold tolerance by activating MdCBF1 and forming a complex with MdCBF2. JAZ proteins (MdJAZ1/MdJAZ2) interact with MdBIM1, attenuating its activity and inhibiting cold tolerance. Furthermore, the E3 ubiquitin ligase MdATL73 targets MdBIM1 for degradation, further regulating BR and JA signaling in response to cold stress (An et al. 2023). BRs positively regulate photoprotection in tomato during chilling stress via a redox-PROTON GRADIENT REGULATION5 (PGR5)-dependent mechanism. BZR1, activated by cold and BR signaling, directly upregulates RBOH1, resulting in apoplastic H_2O_2 production. This triggers PGR5-dependent cyclic electron flow (CEF), non-photochemical quenching (NPQ), and accumulation of photoprotective proteins (Fang et al. 2019). Thus, BR signaling is crucial in cold stress response.

9.10 Role of Strigolactones in Abiotic Stress Responses

Strigolactones (SLs) are derived from naturally occurring carotenoids in plants and are known to regulate plant growth and development. SLs were discovered for their role in seed germination from parasitic plants (Tashjian et al. 1966). Later on, many studies reported the role of SLs as newly identified phytohormones, with their established role in biotic and abiotic stress responses. SLs are named after strigol, a compound, which is previously characterized to promote the germination of parasitic plant *Striga* (Cook et al. 1972). Biosynthesis of SLs in plants is a multistep process involving enzymes like carotenoid dioxygenases (CCDs), carotenoid isomerase (D27), carlactone synthase (CCD7), and cytochrome P450 monooxygenases (CYP711) (Sharma et al. 2024b). Biosynthesis of SLs is also dependent on the presence of other phytohormones. Auxin induces the expression of *CCD7/8*, while cytokinin negatively regulates the expression of D27/17/10 (Sharma et al. 2024b). Similar to other phytohormones (auxin, JA, GA), SL signaling is proteasome-dependent and comprises of receptor, α/β-fold hydrolase named D14, the F-box leucine-rich protein MAX2, and the repressor protein D53 (Saeed et al. 2017). In the presence of SLs, D14 and SL interact with MAX2 to ubiquitinate D53 and degrade the repressor protein to promote the expression of SL-responsive genes (Sharma et al. 2024b).

Using the synthetic analog (GR24) of SLs, the physiological and molecular role of SL is well studied. SLs are known to regulate light responses, leaf shape, senescence, drought, lateral roots, root hairs, nutrient assimilation, and microbial association with plants. SL acts in combination with other phytochromes to regulate plant growth response (Sharma et al. 2024b). SL with auxin regulates lateral root growth and seed dormancy. ABA and SL both regulate stomatal closure and germination. SL with JA and SA regulates plant defense responses (Sharma et al. 2024b). The role of SLs in other abiotic stress responses, such as drought, salinity, heat, and cold stress, is well studied (Tsuchiya et al. 2010; Saeed et al. 2017; Cooper et al. 2018; Wang et al. 2023a).

9.10.1 Cold and Heat Stress

Heat and cold stress stimulate SL production in tomatoes. Disruptions in SL biosynthesis or signaling pathways impair heat and cold tolerance. Conversely, application of the synthetic SL GR245DS enhances tolerance to both stresses. This protective effect is linked to increased ABA levels, accumulation of heat shock protein 70 (HSP70), upregulation of *C-REPEAT BINDING FACTOR 1* (*CBF1*) transcription, and enhanced antioxidant enzyme activity (Chi et al. 2021). SLs significantly enhance crown root elongation in tall fescue, both under normal temperatures and during heat stress, effectively mitigating the inhibitory effects of heat on root growth. GR24 increases cell numbers in root tips, likely through upregulating cell cycle genes and downregulating auxin transport. This demonstrates that SLs promote root growth in tall fescue, mitigating heat stress, by influencing key cellular processes (Hu et al. 2018). GR24 promotes heat stress tolerance in lupine seeds and seedlings as it enhanced the activity of antioxidant and glyoxalase enzymes. GR24 enhanced PSII thermotolerance by protecting the oxygen-evolving complex and preventing PSII reaction center inactivation (Omoarelojie et al. 2020).

9.10.2 Drought Stress

SLs play a crucial role in enhancing plant drought tolerance. In wheat and grapevine, SL application alleviates drought stress by reducing oxidative damage, improving water content and photosynthesis, and stabilizing membranes, likely through the activation of antioxidant defense systems (Min et al. 2019; Sedaghat et al. 2020). SLs also indirectly influence stomatal closure by modulating ABA signaling. Studies on rice and barley mutants deficient in SL biosynthesis or signaling pathways have demonstrated increased drought sensitivity, characterized by reduced SL and ABA levels, impaired water relations, disrupted photosynthesis, altered stomatal behavior, and impaired chloroplast structure (Haider et al. 2018; Marzec et al. 2020). Furthermore, SLs have been shown to enhance oil content in *Dracocephalum kotschyi* under drought conditions (Shirani Bidabadi and Sharifi 2021), highlighting their multifaceted role in improving plant resilience to drought stress.

9.10.3 Salinity Stress

SLs improved salt tolerance in cucumber and *Salvia nemorosa* by enhancing ROS scavenging mechanisms, leading to decreased H_2O_2 and MDA content and reduced cellular damage (Sharifi and Bidabadi 2020; Zhang et al. 2022). Arbuscular mycorrhizal (AM) symbiosis in lettuce plants alters SL production in response to salinity stress and enhances plant tolerance (Aroca et al. 2013).

9.10.4 Heavy Metal

SLs play a crucial role in mitigating heavy metal toxicity in plants. In *Panicum virgatum*, SLs effectively reverse the detrimental effects of Cd toxicity (Tai et al. 2017). In rice, SL-deficient mutants (*d10* and *d17*) exhibit increased sensitivity to As compared to wild-type plants. This enhanced sensitivity in mutants is associated with lower expression of antioxidant enzymes (SOD, APX) and higher expression of arsenic transporter genes in roots (Mostofa et al. 2021). These findings highlight the importance of SLs in enhancing plant tolerance to heavy metal stress by regulating antioxidant defense mechanisms and modulating heavy metal uptake and transport.

9.10.5 Nutrient Deficiency

In recent years, the role of SL in regulating the microbial association of plants has been uncovered especially, in the context of nutrient deficiency. Exudation of SLs in the soil promotes the colonization of AM with plant roots (Das et al. 2022). Interestingly, the biosynthesis and perception of SLs are also dependent on phosphate availability (Umehara et al. 2015; Saeed et al. 2017). In low phosphate conditions. SLs regulate root hair elongation, anthocyanin accumulation, and the expression of phosphate starvation marker genes. (Sun et al. 2014a; Ito et al. 2015). Phosphate starvation response 1 (PHR1), the principal transcription factor of phosphate signaling, is shown to promote the biosynthesis of SL by regulating the expression of the SL biosynthesis gene, *CCD7* of rice (Das et al. 2022).

9.11 Role of Nitric Oxide in Abiotic Stress Response

Nitric oxide (NO) is a gaseous messenger molecule that is involved in plant growth, development, senescence, and stress tolerance. NO can interact with protein kinases, alter gene expression, and release secondary messengers to carry out its biological activity. Also, NO has potential role in chromatin modifications, nitration, and S-nitrosylation-induced posttranslational modifications (PTM) in targeted proteins, biomolecules, and in nucleic acids. Transcriptomics studies on plants revealed how NO regulates gene expression, leading to the discovery of genes associated with histone or DNA methylation (Kumari et al. 2024). There are various reports that highlight the role of NO in different abiotic stress responses.

9.11.1 Heat Stress

Previous investigation has indicated that the external application of NO triggered the activation of genes linked to the development of thermotolerance and the heat shock response under stress (Aranda-Caño et al. 2019; Sun et al. 2021). Application of NO accelerated the rate of acclimatization for *Zea mays* and *Triticum aestivum* while suppressing the symptoms of high temperatures in *Oryza sativa* (Ren et al. 2023). Previous investigations confirmed that the NO overproducing lines, *hot5/atgsnor1* and *nox1/cue1*, are heat-sensitive proteins that regulate NO/SNO (SNO: S-nitrosothiols) to contribute to the thermotolerance process. Heat stress enhanced the NO/SNO concentration and accelerated the reactive nitrogen species (RNS) metabolism in *Brassica juncea* and *Pisum sativum* (Khator et al. 2024). A study on citrus leaf showed that a substantial rise in NO, SNO, and superoxide levels was directly linked to a decline in chlorophyll pigment concentrations and increased electrolyte leakage. This suggests adaptation to heat stress requires NO/RNS equilibrium (Khator et al. 2024). Also, in wheat, NO improved resistance to high temperatures by boosting the antioxidant system and osmolyte accumulation, which in turn improved photosynthesis (Sehar et al. 2023).

9.11.2 Cold Stress

Various research studies revealed that NO has the role in cold tolerance through increasing the activities of apoplastic ice nucleation proteins and apoplastic antioxidant enzymes. Moreover, SNP application reduces excessive ROS, MDA, and freeze damage and improves the damage caused by cold stress (Esim et al. 2014). When *Camellia sinensis* is exposed to cold stress, SNP treatment increases SOD and CAT activity, promotes the accumulation of proline, soluble protein, and sugar, and suppresses lipid peroxidation (Wang et al. 2022c). Sodium nitroprusside (SNP) reduces ROS accumulation, MDA content, and excessive intracellular ROS either directly or indirectly by increasing chlorophyll, soluble content, and antioxidant enzymes in cold stress in Cucumis (Diao et al. 2022).

9.11.3 Salinity Stress

It has been noted that NO signaling activation regulates proline biosynthesis in plants under salt stress by activating proline-producing enzymes like pyrroline-5-carboxylate synthase (P5CS) (Hayat et al. 2012). In rice, the expression of the stress-tolerant genes *OsHIPP38*, *OsGR1*, and *OsP5CS2* dramatically increased in seedlings treated with the NO donor, i.e., SNP, suggesting resistance to salt stress (Adamu et al. 2018). Tomato plants also showed the mitigating effect of NO in reducing the impacts of salinity, as evidenced by an increase in S-nitrosylated protein levels upon salt exposure (Wei et al. 2022). In addition to increasing the amount of nitrogen in rice seedlings, NO treatment also alleviates the negative effects of salt on biomass accumulation and plant height (Huang et al. 2020a). When NO was applied to tomato plants exposed to salinity, their photosynthetic parameters and machinery increased. Exogenous NO through SNP has also been shown in multiple studies to enhance intrinsic CO_2 concentration, chlorophyll levels, fluorescence, photosynthetic rate, stomatal conductance, pollen development, and transpiration in plants under salinity stress (Kumari et al. 2024). NO increased the photosynthetic efficiency by encouraging the formation of chlorophyll in salinized eggplant (Wu et al. 2013). In another study, NO produced by the foliar spray of SNP enhanced cytokinin production, which in turn helped cotton to delay senescence under salt stress (Kong et al. 2016). When SNP is given to *N. tangutorum* seedlings, the AsA-GSH cycle is triggered, and antioxidant metabolites are restored in soybean root nodules under salt stress by raising the AsA/DHAsA and GSH/GSSG ratios (Egbichi et al. 2013). Treatment with NO donors under salt stress elevate enzymatic activity of antioxidants (APX, ASA, CAT, GR, GSH, GPX, and SOD) in cotton, chickpea, sunflower, tomato, and *Aegiceras corniculatum* (Ahmad et al. 2016; Alkhatib et al. 2021; Kumari et al. 2024).

9.11.4 Drought Stress

Under drought stress, the application of SNP (150 μM) improved RWC and reduced ion leakage in turfgrass species (Hatamzadeh et al. 2015). RWC, proline concentration, and peroxidase activity were all significantly increased in drought-stressed sunflowers treated with SNP, while MDA levels were reduced. MDA content increased 231% in plants without the SNP treatment when there was a lack of water compared to control plants (Cechin et al. 2015). Apple rootstock drought tolerance was increased by exogenous SNP application, which also increased SOD activity and decreased ion leakage, ROS production, and MDA generated via lipid peroxidation (Pandey et al. 2023). It was discovered that NO donors, such as SNP, increase drought tolerance in *Antiaris toxicaria* seedlings by decreasing H_2O_2 accumulation and increasing antioxidant protein S-nitrosylation. These results suggested that NO has a role in the plant's ability to recover from stress in addition to influencing drought tolerance (Bai et al. 2011). By lowering the buildup of ROS in rice, the early-NO burst caused by NOS-like enzymes in ammonium-treated plants increases the antioxidant defense mechanism (Cao et al. 2019).

9.11.5 Heavy Metal Stress

Numerous studies have shown that NO can alter the expression of genes linked to defense, nutrition, and hormone metabolism in response to excess heavy metals (Wei et al. 2020). In plants under Cd stress, exogenous NO may affect a variety of physiological and biochemical functions. In order to mitigate Cd-induced toxicity, this effect can be attained by regulating the uptake of Cd and reducing its accumulation in plants (Rasheed et al. 2022b). NO is crucial for reducing As toxicity in *Pistia stratiotes* because it integrates key physiological systems, raising respiration and non-photochemical quenching rates in addition to lowering photorespiration levels (Talukdar 2013). In *Festuca arundinacea*, SNP administration improves Cd transport and detoxification by controlling the antioxidant system and the ABC transporters pathway (Luo et al. 2024). SNP administration decreases Pb uptake, safeguards photosynthetic pigments and proteins, and prevents ROS buildup by boosting antioxidant enzyme activity in *Solanum melongena* (Luo et al. 2024).

9.12 Conclusion

Climate change intensifies abiotic stresses like drought, salinity, heat, and flooding, adversely impacting plant growth and increasing susceptibility to diseases. Understanding how plants perceive and respond to these stresses, particularly through phytohormonal regulation, is crucial for the development of climate-resilient crops. While extensive research in *Arabidopsis thaliana* has elucidated the role of phytohormones in abiotic stress responses, this chapter focuses on conserved and divergent hormonal mechanisms operating in crop plants under stress conditions. Future research should focus on elucidating the molecular mechanisms of hormone action and their interactions in mitigating stress. New tools and technologies such as hormone-responsive reporters, protein interaction analysis, and single-cell sequencing will be instrumental in unraveling these intricate pathways. In parallel, advanced genetic tools, including genomics-assisted breeding and CRISPR-Cas9 genome editing, are revolutionizing our capacity to accelerate the development of stress-resilient traits. Integrating multi-omics approaches with high-throughput phenotyping will further enhance our ability to identify and predict crucial genes for stress tolerance, particularly from wild relatives. Knowledge-driven strategies for the de novo domestication of these climate-adapted wild relatives can serve as a valuable source for harnessing novel hormone-regulated stress tolerance pathways. Continued progress in understanding the complex interplay of plant hormones is crucial for engineering crops that can thrive under challenging environmental conditions.

Author Contribution Conceptualization: A.L., H.S., A.S.; Manuscript drafting and editing: H.S., A.S.; Reference addition and management: M.K.; Auxin and BR written by: M.K.; ABA written by: A.S.; CK written by: H.S., H.K.B; GA written by: A., SA written by: S.S.R.; JA written by: S.S.; Ethylene written by: H.B.S., SL written by: P.A.; NO written by: S.N. and A.P.

References

Abasi F, Raja NI, Mashwani Z, Ehsan M, Ali H, Shahbaz M (2024) Heat and wheat: adaptation strategies with respect to heat shock proteins and antioxidant potential; an era of climate change. Int J Biol Macromol. https://doi.org/10.1016/j.ijbiomac.2023.128379

Abdullah M, Ahmad F, Zang Y, Jin S, Ahmed S, Li J, Islam F, Ahmad M, Zhang Y, Hu Y et al (2023) HEAT-RESPONSIVE PROTEIN regulates heat stress via fine-tuning ethylene/auxin signaling pathways in cotton. Plant Physiol 191:772–788

Achard P, Cheng H, De Grauwe L, Decat J, Schoutteten H, Moritz T, Van Der Straeten D Peng J (2006) Integration of plant responses to environmentally activated phytohormonal signals. Science 311(5757):91–94

Acheampong AK, Shanks C, Cheng C-Y, Schaller GE, Dagdas Y, Kieber JJ (2020) EXO70D isoforms mediate selective autophagic degradation of type-A ARR proteins to regulate cytokinin sensitivity. Proc Natl Acad Sci 117:27034–27043

Adamu TA, Mun BG, Lee SU, Hussain A, Yun BW (2018) Exogenously applied nitric oxide enhances salt tolerance in rice (*Oryza sativa* L.) at seedling stage. Agronomy 8(8):276

Ahammed GJ, Gao CJ, Ogweno JO, Zhou YH, Xia XJ, Mao WH, Shi K, Yu JQ (2012) Brassinosteroids induce plant tolerance against phenanthrene by enhancing degradation and detoxification in *Solanum lycopersicum* L. Ecotoxicol Environ Saf 80:28–36

Ahmad Lone W, Majeed N, Yaqoob U, John R (2022) Exogenous brassinosteroid and jasmonic acid improve drought tolerance in *Brassica rapa* L. genotypes by modulating osmolytes, antioxidants and photosynthetic system. Plant Cell Rep. https://doi.org/10.1007/s00299-021-02763-9

Ahmad P, Latef AAA, Hashem A, Abd Allah EF, Gucel S, Tran LSP (2016) Nitric oxide mitigates salt stress by regulating levels of osmolytes and antioxidant enzymes in chickpea. Front Plant Sci 7:183720

Ali AYA, Ibrahim MEH, Zhou G, Nimir NEA, Elsiddig AMI, Jiao X, Zhu G, Salih EGI, Suliman MSES, Elradi SBM (2021) Gibberellic acid and nitrogen efficiently protect early seedlings growth stage from salt stress damage in *Sorghum*. Sci Rep 11:6672

Alkhatib R, Abdo N, Mheidat M (2021) Photosynthetic and ultrastructural properties of eggplant (*Solanum melongena*) under salinity stress. Horticulturae. https://doi.org/10.3390/horticulturae7070181

An JP, Liu ZY, Zhang XW, Wang DR, Zeng F, You CX, Han Y (2023) Brassinosteroid signaling regulator BIM1 integrates brassinolide and jasmonic acid signaling during cold tolerance in apple. Plant Physiol 193:1652–1674

Antunes MDC, Sfakiotakis EM (2000) Effect of high temperature stress on ethylene biosynthesis, respiration and ripening of 'Hayward' kiwifruit. Postharvest Biol Technol 20:251–259

Aranda-Caño L, Sánchez-Calvo B, Begara-Morales JC, Chaki M, Mata-Pérez C, Padilla MN, Valderrama R, Barroso JB (2019) Post-translational modification of proteins mediated by nitro-fatty acids in plants: nitroalkylation. Plants 8:82

Aroca R, Ruiz-Lozano JM, Zamarreño Ángel M, Paz JA, García-Mina JM, Pozo MJ, López-Ráez JA (2013) Arbuscular mycorrhizal symbiosis influences strigolactone production under salinity and alleviates salt stress in lettuce plants. J Plant Physiol. https://doi.org/10.1016/j.jplph.2012.08.020

Atici Ö, Ağar G, Battal P (2005) Changes in phytohormone contents in chickpea seeds germinating under lead or zinc stress. Biol Plant 49:215–222

Auguy F, Fahr M, Moulin P, El Mzibri M, Smouni A, Filali-Maltouf A, Béna G, Doumas P (2016) Transcriptome changes in *Hirschfeldia incana* in response to Lead exposure. Front Plant Sci 6:1231

Awan SA, Khan I, Rizwan M, Zhang X, Brestic M, Khan A, El-Sheikh MA, Alyemeni MN, Ali S, Huang L (2021) Exogenous abscisic acid and jasmonic acid restrain polyethylene glycol-induced drought by improving the growth and antioxidative enzyme activities in pearl millet. Physiol Plant. https://doi.org/10.1111/ppl.13247

Azeem U (2018) Ameliorating nickel stress by Jasmonic acid treatment in *Zea mays* L. Russ Agric Sci. https://doi.org/10.3103/s1068367418030035

Bai X, Yang L, Tian M, Chen J, Shi J, Yang Y, Hu X (2011) Nitric oxide enhances desiccation tolerance of recalcitrant *Antiaris toxicaria* seeds via protein S-Nitrosylation and Carbonylation. PLoS One 6:e20714

Balfagón D, Sengupta S, Gómez-Cadenas A, Fritschi FB, Azad RK, Mittler R, Zandalinasc SI (2019) Jasmonic acid is required for plant acclimation to a combination of high light and heat stress. Plant Physiol. https://doi.org/10.1104/pp.19.00956

Bali S, Jamwal VL, Kohli SK, Kaur P, Tejpal R, Bhalla V, Ohri P, Gandhi SG, Bhardwaj R, Al-Huqail AA et al (2019) Jasmonic acid application triggers detoxification of lead (Pb) toxicity in tomato through the modifications of secondary metabolites and gene expression. Chemosphere. https://doi.org/10.1016/j.chemosphere.2019.06.188

Belintani NG, Guerzoni JTS, Moreira RMP, Vieira LGE (2012) Improving low-temperature tolerance in sugarcane by expressing the *ipt* gene under a cold inducible promoter. Biol Plant. https://doi.org/10.1007/s10535-012-0018-1

Beznec A, Faccio P, Miralles DJ, Abeledo LG, Oneto CD, de Garibotto MB, Gonzalez G, Moreyra F, Elizondo M, Ruíz M et al (2021) Stress-induced expression of *IPT* gene in transgenic wheat reduces grain yield penalty under drought. J Genetic Eng Biotechnol 19:67

Biggs MS, Handa AK (1988) Biochemical basis of high-temperature inhibition of ethylene biosynthesis in ripening tomato fruits. Physiol Plant 72:572–578

Bortolin GS, Teixeira SB, de Mesquita PR, Ávila GE, Carlos FS, da Silva Pedroso CE, Deuner S (2020) Seed priming with salicylic acid minimizes oxidative effects of aluminum on *Trifolium* seedlings. J Soil Sci Plant Nutr 20:2502–2511

Calleja-Cabrera J, Boter M, Oñate-Sánchez L, Pernas M (2020) Root growth adaptation to climate change in crops. Front Plant Sci. https://doi.org/10.3389/fpls.2020.00544

Cao WH, Liu J, He XJ, Mu RL, Zhou HL, Chen SY, Zhang JS (2007) Modulation of ethylene responses affects plant salt-stress responses. Plant Physiol 143:707–719

Cao X, Zhu C, Zhong C, Zhang J, Wu L, Jin Q, Ma Q (2019) Nitric oxide synthase-mediated early nitric oxide burst alleviates water stress-induced oxidative damage in ammonium-supplied rice roots. BMC Plant Biol. https://doi.org/10.1186/s12870-019-1721-2

Casanova-Sáez R, Mateo-Bonmatí E, Ljung K (2021) Auxin metabolism in plants. Cold Spring Harb Perspect Biol 13:a039867

Caselles V, Casadesús A, Munné-Bosch S (2021) A dual role for abscisic acid integrating the cold stress response at the whole-plant level in *Iris pseudacorus* L. growing in a natural wetland. Front Plant Sci. https://doi.org/10.3389/fpls.2021.722525

Cebrián G, Iglesias-Moya J, García A, Martínez J, Romero J, Regalado JJ, Martínez C, Valenzuela JL, Jamilena M (2021) Involvement of ethylene receptors in the salt tolerance response of *Cucurbita pepo*. Hortic Res 8:73

Cechin I, Cardoso GS, de Fátima FT, Corniani N (2015) Nitric oxide reduces oxidative damage induced by water stress in sunflower plants. Bragantia 74:200–206

Černý M, Jedelský PL, Novák J, Schlosser A, Brzobohatý B (2014) Cytokinin modulates proteomic, transcriptomic and growth responses to temperature shocks in *Arabidopsis*. Plant Cell Environ. https://doi.org/10.1111/pce.12270

Chen Z, Zheng Z, Huang J, Lai Z, Fan B (2009) Biosynthesis of salicylic acid in plants. Plant Signal Behav 4:493–496

Chen Y, Shen H, Wang M, Li Q, He Z (2013) Salicyloyl-aspartate synthesized by the acetyl-amido synthetase GH3.5 is a potential activator of plant immunity in *Arabidopsis*. Acta Biochim Biophys Sin Shanghai 45:827–836

Chen D, Ma X, Li C, Zhang W, Xia G, Wang M (2014) A wheat aminocyclopropane-1-carboxylate oxidase gene, TaACO1, negatively regulates salinity stress in *Arabidopsis thaliana*. Plant Cell Rep 33:1815–1827

Chen T, Zhao X, Zhang C, Yang Y, Feng B, Zhang X, Fu G, Tao L (2017) Application of salicylic acid improves filling of inferior grains of Rice during late maturity under mild cold stress. Crop Sci 57:2130–2142

Chen T, Li G, Islam MR, Fu W, Feng B, Tao L, Fu G (2019) Abscisic acid synergizes with sucrose to enhance grain yield and quality of rice by improving the source-sink relationship. BMC Plant Biol. https://doi.org/10.1186/s12870-019-2126-y

Chen H, Bullock DA, Alonso JM, Stepanova AN (2021) To fight or to grow: the balancing role of ethylene in plant abiotic stress responses. Plants 33(11):33

Chen G, Zheng D, Feng N, Zhou H, Mu D, Zhao L, Shen X, Rao G, Meng F, Huang A (2022) Physiological mechanisms of ABA-induced salinity tolerance in leaves and roots of rice. Sci Rep. https://doi.org/10.1038/s41598-022-11408-0

Chen Y, Ince YÇ, Kawamura A, Favero DS, Suzuki T, Sugimoto K (2024a) ELONGATED HYPOCOTYL5-mediated light signaling promotes shoot regeneration in *Arabidopsis thaliana*. Plant Physiol 196:2549–2564

Chen Z, Zhou W, Guo X, Ling S, Li W, Wang X, Yao J (2024b) Heat stress responsive aux/IAA protein, OsIAA29 regulates grain filling through OsARF17 mediated auxin signaling pathway. Rice. https://doi.org/10.1186/s12284-024-00694-z

Cheng S, Yang Z, Wang M, Song J, Sui N, Fan H (2014) Salinity improves chilling resistance in *Suaeda salsa*. Acta Physiol Plant. https://doi.org/10.1007/s11738-014-1555-3

Cheng X, Fang T, Zhao E, Zheng B, Huang B, An Y, Zhou P (2020) Protective roles of salicylic acid in maintaining integrity and functions of photosynthetic photosystems for alfalfa (*Medicago sativa* L.) tolerance to aluminum toxicity. Plant Physiol Biochem 155:570–578

Chi C, Li X, Fang P, Xia X, Shi K, Zhou Y, Zhou J, Yu J (2020) Brassinosteroids act as a positive regulator of NBR1-dependent selective autophagy in response to chilling stress in tomato. J Exp Bot 71:1092–1106

Chi C, Xu X, Wang M, Zhang H, Fang P, Zhou J, Xia X, Shi K, Zhou Y, Yu J (2021) Strigolactones positively regulate abscisic acid-dependent heat and cold tolerance in tomato. Hortic Res. https://doi.org/10.1038/s41438-021-00668-y

Clarke SM, Cristescu SM, Miersch O, Harren FJM, Wasternack C, Mur LAJ (2009) Jasmonates act with salicylic acid to confer basal thermotolerance in *Arabidopsis thaliana*. New Phytol. https://doi.org/10.1111/j.1469-8137.2008.02735.x

Cook CE, Whichard LP, Wall ME, Egley GH, Coggon P, Luhan PA, Mcphail AT (1972) Germination stimulants. II. The structure of Strigol–a potent seed germination stimulant for Witchweed (*Striga lutea* Lour.)1,2. J Am Chem Soc. https://doi.org/10.1021/ja00772a048

Cook D, Fowler S, Fiehn O, Thomashow MF (2004) A prominent role for the CBF cold response pathway in configuring the low-temperature metabolome of *Arabidopsis*. Proc Natl Acad Sci USA. https://doi.org/10.1073/pnas.0406069101

Cooper JW, Hu Y, Beyyoudh L, Yildiz Dasgan H, Kunert K, Beveridge CA, Foyer CH (2018) Strigolactones positively regulate chilling tolerance in pea and in *Arabidopsis*. Plant Cell Environ. https://doi.org/10.1111/pce.13147

Cortleven A, Leuendorf JE, Frank M, Pezzetta D, Bolt S, Schmülling T (2019) Cytokinin action in response to abiotic and biotic stresses in plants. Plant Cell Environ 42:998–1018

Coupel-Ledru A, Tyerman SD, Masclef D, Lebon E, Christophe A, Edwards EJ, Simonneau T (2017) Abscisic acid down-regulates hydraulic conductance of grapevine leaves in Isohydric genotypes only. Plant Physiol 175:1121–1134

Cuevas JC, López-Cobollo R, Alcázar R, Zarza X, Koncz C, Altabella T, Salinas J, Tiburcio AF, Ferrando A (2009) Putrescine as a signal to modulate the indispensable ABA increase under cold stress. Plant Signal Behav 4:219–220

Dai H, Wei S, Pogrzeba M, Rusinowski S, Krzyżak J, Jia G (2020) Exogenous jasmonic acid decreased Cu accumulation by alfalfa and improved its photosynthetic pigments and antioxidant system. Ecotoxicol Environ Saf. https://doi.org/10.1016/j.ecoenv.2020.110176

Das D, Paries M, Hobecker K, Gigl M, Dawid C, Lam HM, Zhang J, Chen M, Gutjahr C (2022) PHOSPHATE STARVATION RESPONSE transcription factors enable arbuscular mycorrhiza symbiosis. Nat Commun. https://doi.org/10.1038/s41467-022-27976-8

Davies PJ (1995) The plant hormones: their nature, occurrence, and functions. In Plant hormones: physiology, biochemistry and molecular biology. Springer, Dordrecht, pp 1–12. https://doi.org/10.1007/978-94-011-0473-9_1

De Araújo ED, De Melo AS, Do Socorro Rocha M, Carneiro RF, Rocha MDM (2018) Germination and initial growth of cowpea cultivars under osmotic stress and salicylic acid. Revista Caatinga 31:80–89

De Ollas C, Hernando B, Arbona V, Gómez-Cadenas A (2013) Jasmonic acid transient accumulation is needed for abscisic acid increase in citrus roots under drought stress conditions. Physiol Plant. https://doi.org/10.1111/j.1399-3054.2012.01659.x

Dey A, Samanta MK, Gayen S, Maiti MK (2016) The sucrose non-fermenting 1-related kinase 2 gene SAPK9 improves drought tolerance and grain yield in rice by modulating cellular osmotic potential, stomatal closure and stress-responsive gene expression. BMC Plant Biol. https://doi.org/10.1186/s12870-016-0845-x

Diao QN, Cao YY, Wang H, Zhang YP, Bin SH (2022) Nitric oxide confers chilling stress tolerance by regulating carbohydrate metabolism and the antioxidant defense system in melon (*Cucumis melo* L.) seedlings. HortScience 57:1249–1256

Ding Y, Sheng J, Li S, Nie Y, Zhao J, Zhu Z, Wang Z, Tang X (2015) The role of gibberellins in the mitigation of chilling injury in cherry tomato (*Solanum lycopersicum* L.) fruit. Postharvest Biol Technol 101:88–95

Ding Y, Ma Y, Liu N, Xu J, Hu Q, Li Y, Wu Y, Xie S, Zhu L, Min L et al (2017) microRNAs involved in auxin signalling modulate male sterility under high-temperature stress in cotton (*Gossypium hirsutum*). Plant J 91:977–994

Ding F, Wang C, Xu N, Wang M, Zhang S (2021) Jasmonic acid-regulated putrescine biosynthesis attenuates cold-induced oxidative stress in tomato plants. Sci Hortic. https://doi.org/10.1016/j.scienta.2021.110373

Dobrá J, Černý M, Štorchová H, Dobrev P, Skalák J, Jedelský PL, Lukšanová H, Gaudinová A, Pešek B, Malbecka J et al (2015) The impact of heat stress targeting on the hormonal and transcriptomic response in *Arabidopsis*. Plant Sci. https://doi.org/10.1016/j.plantsci.2014.11.005

Dong H, Zhen Z, Peng J, Chang L, Gong Q, Wang NN (2011) Loss of ACS7 confers abiotic stress tolerance by modulating ABA sensitivity and accumulation in *Arabidopsis*. J Exp Bot 62:4875–4887

Du H, Wu N, Chang Y, Li X, Xiao J, Xiong L (2013) Carotenoid deficiency impairs ABA and IAA biosynthesis and differentially affects drought and cold tolerance in rice. Plant Mol Biol 83:475–488

Du H, Wu N, Cui F, You L, Li X, Xiong L (2014) A homolog of ETHYLENE OVERPRODUCER, OsETOL1, differentially modulates drought and submergence tolerance in rice. Plant J 78:834–849

Dubois V, Moritz T, García-Martínez JL (2011) Comparison of the role of gibberellins and ethylene in response to submergence of two lowland rice cultivars, Senia and Bomba. J Plant Physiol 168:233–241

Egbichi I, Keyster M, Jacobs A, Klein A, Ludidi N (2013) Modulation of antioxidant enzyme activities and metabolites ratios by nitric oxide in short-term salt stressed soybean root nodules. S Afr J Bot 88:326–333

El-Sappah AH, Rather SA, Wani SH, Elrys AS, Bilal M, Huang Q, Dar ZA, Elashtokhy MMA, Soaud N, Koul M et al (2022) Heat stress-mediated constraints in maize (*Zea mays*) production: challenges and solutions. Front Plant Sci. https://doi.org/10.3389/fpls.2022.879366

Enders TA, Strader LC (2015) Auxin activity: past, present, and future. Am J Bot 102:180–196

Erogul D, Sen F (2016) The effect of preharvest gibberellic acid applications on fruit quality of "Angelino" plums during storage. Sci Hortic 202:111–116

Esim N, Atici O, Mutlu S (2014) Effects of exogenous nitric oxide in wheat seedlings under chilling stress. Toxicol Ind Health 30:268–274

Es-sbihi FZ, Hazzoumi Z, Amrani Joutei K (2020) Effect of salicylic acid foliar application on growth, glandular hairs and essential oil yield in *Salvia officinalis* L. grown under zinc stress. Chem Biol Technol Agric 7:1–11

Fan Y, Lv Z, Li Y, Qin B, Song Q, Ma L, Wu Q, Zhang W, Ma S, Ma C et al (2022) Salicylic acid reduces wheat yield loss caused by high temperature stress by enhancing the photosynthetic performance of the flag leaves. Agronomy 12:1386

Fang P, Yan M, Chi C, Wang M, Zhou Y, Zhou J, Shi K, Xia X, Foyer CH, Yu J (2019) Brassinosteroids act as a positive regulator of photoprotection in response to chilling stress. Plant Physiol 180:2061–2076

Farber M, Attia Z, Weiss D (2016) Cytokinin activity increases stomatal density and transpiration rate in tomato. J Exp Bot 67:6351–6362

Farooq M, Basra SMA, Wahid A, Ahmad N, Saleem BA (2009) Improving the drought tolerance in rice (*Oryza sativa* L.) by exogenous application of salicylic acid. J Agron Crop Sci 195:237–246

Farooq M, Khan MA, Zhao D-D, Asif S, Kim E-G, Jang Y-H, Park J-R, Lee I-J, Kim K-M (2022) Extrinsic role of gibberellin mitigating salinity effect in different rice genotypes. Front Plant Sci 13:1041181

Fedina I, Nedeva D, Georgieva K, Velitchkova M (2009) Methyl jasmonate counteract UV-B stress in barley seedlings. J Agron Crop Sci. https://doi.org/10.1111/j.1439-037X.2008.00358.x

Feng ZT, Deng YQ, Fan H, Sun QJ, Sui N, Wang BS (2014) Effects of NaCl stress on the growth and photosynthetic characteristics of *Ulmus pumila* L. seedlings in sand culture. Photosynthetica. https://doi.org/10.1007/s11099-014-0032-y

Feng Y, Wang Y, Zhang G, Gan Z, Gao M, Lv J, Wu T, Zhang X, Xu X, Yang S et al (2021) Group-C/S1 bZIP heterodimers regulate *MdIPT5b* to negatively modulate drought tolerance in apple species. Plant J 107:399–417

Fujii H, Chinnusamy V, Rodrigues A, Rubio S, Antoni R, Park SY, Cutler SR, Sheen J, Rodriguez PL, Zhu JK (2009) In vitro reconstitution of an abscisic acid signalling pathway. Nature. https://doi.org/10.1038/nature08599

Fukao T, Bailey-Serres J (2008) Submergence tolerance conferred by Sub1A is mediated by SLR1 and SLRL1 restriction of gibberellin responses in rice. Proc Natl Acad Sci USA 105:16814–16819

Gao S, Fang J, Xu F, Wang W, Sun X, Chu J, Cai B, Feng Y, Chu C (2014) *CYTOKININ OXIDASE/DEHYDROGENASE4* integrates Cytokinin and auxin signaling to control Rice crown root formation. Plant Physiol 165:1035–1046

Garcion C, Lohmann A, Lamodière E, Catinot J, Buchala A, Doermann P, Métraux JP (2008) Characterization and biological function of the *Isochorismate Synthase2* gene of *Arabidopsis*. Plant Physiol 147:1279–1287

Gasch P, Fundinger M, Müller JT, Lee T, Bailey-Serres J, Mustropha A (2016) Redundant ERF-VII transcription factors bind to an evolutionarily conserved cis-motif to regulate hypoxia-responsive gene expression in *Arabidopsis*. Plant Cell 28:160–180

Gautam S, Singh PK (2009) Salicylic acid-induced salinity tolerance in corn grown under NaCl stress. Acta Physiol Plant 31:1185–1190

Ghorbel M, Brini F, Sharma A, Landi M (2021) Role of jasmonic acid in plants: the molecular point of view. Plant Cell Rep. https://doi.org/10.1007/s00299-021-02687-4

Gibbs DJ, Lee SC, Md Isa N, Gramuglia S, Fukao T, Bassel GW, Correia CS, Corbineau F, Theodoulou FL, Bailey-Serres J, et al (2011) Homeostatic response to hypoxia is regulated by the N-end rule pathway in plants. Nature 479:7373, 415–418

Golovatskaya IF, Kadyrbaev MK, Boyko EV, Filonova MV (2023) Salicylic acid improves cold resistance of *Solanum tuberosum* regenerants via regulation of the antioxidant system. Russ J Plant Physiol. https://doi.org/10.1134/S1021443723600319

Gomes GLB, Scortecci KC (2021) Auxin and its role in plant development: structure, signalling, regulation and response mechanisms. Plant Biol 23:894–904

Gomez-Roldan V, Fermas S, Brewer PB, Puech-Pagès V, Dun EA, Pillot JP, Letisse F, Matusova R, Danoun S, Portais JC et al (2008) Strigolactone inhibition of shoot branching. Nature. https://doi.org/10.1038/nature07271

Gonzalez-Guzman M, Rodriguez L, Lorenzo-Orts L, Pons C, Sarrion-Perdigones A, Fernandez MA, Peirats-Llobet M, Forment J, Moreno-Alvero M, Cutler SR et al (2014) Tomato PYR/PYL/RCAR abscisic acid receptors show high expression in root, differential sensitivity to the abscisic acid agonist quinabactin, and the capability to enhance plant drought resistance. J Exp Bot 65:4451–4464

Groen SC, Whiteman NK (2014) The evolution of ethylene signaling in plant chemical ecology. J Chem Ecol 40:700–716

Guo H, Ecker JR (2004) The ethylene signaling pathway: new insights. Curr Opin Plant Biol 7:40–49

Guo T, Gull S, Ali MM, Yousef AF, Ercisli S, Kalaji HM, Telesiński A, Auriga A, Wróbel J, Radwan NS et al (2022) Heat stress mitigation in tomato (*Solanum lycopersicum* L.) through foliar application of gibberellic acid. Sci Rep. https://doi.org/10.1038/S41598-022-15590-Z

Gupta S, Seth CS (2021) Salicylic acid alleviates chromium (VI) toxicity by restricting its uptake, improving photosynthesis and augmenting antioxidant defense in *Solanum lycopersicum* L. Physiol Mol Biol Plants 27:2651–2664

Haider I, Andreo-Jimenez B, Bruno M, Bimbo A, Floková K, Abuauf H, Ntui VO, Guo X, Charnikhova T, Al-Babili S et al (2018) The interaction of strigolactones with abscisic acid during the drought response in rice. J Exp Bot. https://doi.org/10.1093/jxb/ery089
Han GZ (2017) Evolution of jasmonate biosynthesis and signalling mechanisms. J Exp Bot. https://doi.org/10.1093/jxb/erw470
Hasan SA, Hayat S, Ahmad A (2011) Brassinosteroids protect photosynthetic machinery against the cadmium induced oxidative stress in two tomato cultivars. Chemosphere 84:1446–1451
Hatamzadeh A, Molaahmad Nalousi A, Ghasemnezhad M, Biglouei MH (2015) The potential of nitric oxide for reducing oxidative damage induced by drought stress in two turfgrass species, creeping bentgrass and tall fescue. Grass Forage Sci 70:538–548
Hattori Y, Nagai K, Furukawa S, Song XJ, Kawano R, Sakakibara H, Wu J, Matsumoto T, Yoshimura A, Kitano H, et al (2009) The ethylene response factors *SNORKEL1* and *SNORKEL2* allow rice to adapt to deep water. Nature 2009 460:7258 460:1026–1030
Hayat S, Hayat Q, Alyemeni MN, Wani AS, Pichtel J, Ahmad A (2012) Role of proline under changing environments. Plant Signal Behav. https://doi.org/10.4161/PSB.21949
He M, He CQ, Ding NZ (2018) Abiotic stresses: general defenses of land plants and chances for engineering multistress tolerance. Front Plant Sci. https://doi.org/10.3389/fpls.2018.01771
Hedden P, Sponsel V (2015) A century of gibberellin research. J Plant Growth Regul 34:740–760
Hedden P, Thomas SG (2012) Gibberellin biosynthesis and its regulation. Biochem J 444:11–25
Hinz M, Wilson IW, Yang J, Buerstenbinder K, Llewellyn D, Dennis ES, Sauter M, Dolferus R (2010) *Arabidopsis RAP2.2*: An ethylene response transcription factor that is important for hypoxia survival. Plant Physiol 153:757–772
Holsteens K, De Jaegere I, Wynants A, Prinsen ELJ, Van de Poel B (2022) Mild and severe salt stress responses are age-dependently regulated by abscisic acid in tomato. Front Plant Sci 13:982622
Hou ZH, Gao Y, Zheng JC, Zhao MJ, Liu Y, Cui XY, Li ZY, Wei JT, Yu TF, Zheng L et al (2024) GmBSK1-GmGSK1-GmBES1.5 regulatory module controls heat tolerance in soybean. J Adv Res. https://doi.org/10.1016/j.jare.2024.09.004
Hsu YT, Kao CH (2003) Role of abscisic acid in cadmium tolerance of rice (*Oryza sativa* L.) seedlings. Plant Cell Environ 26:867–874
Hu Y, Jiang L, Wang F, Yu D (2013) Jasmonate regulates the INDUCER OF CBF expression-C-repeat binding factor/dre binding factor1 Cascade and freezing tolerance in *Arabidopsis*. Plant Cell. https://doi.org/10.1105/tpc.113.112631
Hu Q, Zhang S, Huang B (2018) Strigolactones and interaction with auxin regulating root elongation in tall fescue under different temperature regimes. Plant Sci. https://doi.org/10.1016/j.plantsci.2018.03.008
Hu B, Deng F, Chen G, Chen X, Gao W, Long L, Xia J, Chen ZH (2020) Evolution of abscisic acid signaling for stress responses to toxic metals and metalloids. Front Plant Sci 11:547525
Huang TL, Nguyen QTT, Fu SF, Lin CY, Chen YC, Huang HJ (2012) Transcriptomic changes and signalling pathways induced by arsenic stress in rice roots. Plant Mol Biol 80:587–608
Huang J, Zhu C, Hussain S, Huang J, Liang Q, Zhu L, Cao X, Kong Y, Li Y, Wang L et al (2020a) Effects of nitric oxide on nitrogen metabolism and the salt resistance of rice (*Oryza sativa* L.) seedlings with different salt tolerances. Plant Physiol Biochem 155:374–383
Huang L, Zhang L, Zeng R, Wang X, Zhang H, Wang L, Liu S, Wang X, Chen T (2020b) Brassinosteroid priming improves peanut drought tolerance via eliminating inhibition on genes in photosynthesis and hormone signaling. Genes (Basel) 11:1–23
Huang D, Wang Q, Jing G, Ma M, Li C, Ma F (2021) Overexpression of *MdIAA24* improves apple drought resistance by positively regulating strigolactone biosynthesis and mycorrhization. Tree Physiol 41:134–146
Huang S, Zhang H, Purves RW, Bueckert R, Tar'an B, Warkentin TD (2023) Comparative analysis of heat-stress-induced abscisic acid and heat shock protein responses among pea varieties. Crop Sci 63:139–150

Husain T, Fatima A, Suhel M, Singh S, Sharma A, Prasad SM, Singh VP (2020) A brief appraisal of ethylene signaling under abiotic stress in plants. Plant Signal Behav. https://doi.org/10.1080/15592324.2020.1782051

Hwang I, Sakakibara H (2006) Cytokinin biosynthesis and perception. Physiol Plant 126:528–538

Hyoung S, Cho SH, Chung JH, So WM, Cui MH, Shin JS (2020) *Cytokinin oxidase PpCKX1* plays regulatory roles in development and enhances dehydration and salt tolerance in *Physcomitrella patens*. Plant Cell Rep 39:419–430

Iglesias-Moya J, Abreu AC, Alonso S, Torres-García MT, Martínez C, Fernández I, Jamilena M (2023) Physiological and metabolomic responses of the ethylene insensitive squash mutant *etr2b* to drought. Plant Sci 336:111853

Ignatenko A, Talanova V, Repkina N, Titov A (2019) Exogenous salicylic acid treatment induces cold tolerance in wheat through promotion of antioxidant enzyme activity and proline accumulation. Acta Physiol Plant. https://doi.org/10.1007/S11738-019-2872-3

Islam F, Yasmeen T, Arif MS, Riaz M, Shahzad SM, Imran Q, Ali I (2016) Combined ability of chromium (Cr) tolerant plant growth promoting bacteria (PGPB) and salicylic acid (SA) in attenuation of chromium stress in maize plants. Plant Physiol Biochem 108:456–467

Ito S, Nozoye T, Sasaki E, Imai M, Shiwa Y, Shibata-Hatta M, Ishige T, Fukui K, Ito K, Nakanishi H et al (2015) Strigolactone regulates anthocyanin accumulation, acid phosphatases production and plant growth under low phosphate condition in *Arabidopsis*. PLoS One. https://doi.org/10.1371/journal.pone.0119724

Jeon J, Kim J (2013) *Arabidopsis response regulator1* and *Arabidopsis histidine phosphotransfer protein2 (AHP2), AHP3*, and *AHP5* function in cold signaling. Plant Physiol. https://doi.org/10.1104/pp.112.207621

Jeon J, Cho C, Lee MR, Van Binh N, Kim J (2016) *CYTOKININ RESPONSE FACTOR2 (CRF2)* and *CRF3* regulate lateral root development in response to cold stress in *Arabidopsis*. Plant Cell. https://doi.org/10.1105/tpc.15.00909

Jespersen D, Yu J, Huang B (2015) Metabolite responses to exogenous application of nitrogen, cytokinin, and ethylene inhibitors in relation to heat-induced senescence in creeping bentgrass. PLoS One. https://doi.org/10.1371/journal.pone.0123744

Jia W, Ma M, Chen J, Wu S (2021) Plant Morphological, Physiological and Anatomical Adaption to Flooding Stress and the Underlying Molecular Mechanisms. Int J Mol Sci 22:1088

Jiang C, Gao X, Liao L, Harberd NP, Fu X (2007) Phosphate starvation root architecture and anthocyanin accumulation responses are modulated by the gibberellin-DELLA signaling pathway in *Arabidopsis*. Plant Physiol 145:1460–1470

Jiang C, Belfield EJ, Cao Y, Smith JAC, Harberd NP (2013) An *Arabidopsis* soil-salinity–tolerance mutation confers ethylene-mediated enhancement of sodium/potassium homeostasis. Plant Cell 25:3535–3552

Jianing G, Yuhong G, Yijun G, Rasheed A, Qian Z, Zhiming X, Mahmood A, Shuheng Z, Zhuo Z, Zhuo Z et al (2022) Improvement of heat stress tolerance in soybean (*Glycine max* L), by using conventional and molecular tools. Front Plant Sci. https://doi.org/10.3389/fpls.2022.993189

Jin Y, Pan W, Zheng X, Cheng X, Liu M, Ma H, Ge X (2018) OsERF101, an ERF family transcription factor, regulates drought stress response in reproductive tissues. Plant Mol Biol. https://doi.org/10.1007/s11103-018-0762-5

Jin Y, Yang P, Li J, Yang Y, Yang R, Fu H, Li J (2024) Brassinosteroids alleviate salt stress by enhancing sugar and glycine betaine in pepper (*Capsicum annuum* L.). Plants. https://doi.org/10.3390/plants13213029

Jing H, Strader LC (2019) Interplay of auxin and Cytokinin in lateral root development. Int J Mol Sci 20:486

Joo J, Lee YH, Song SI (2019) OsbZIP42 is a positive regulator of ABA signaling and confers drought tolerance to rice. Planta. https://doi.org/10.1007/s00425-019-03104-7

Joshi R, Sahoo KK, Tripathi AK, Kumar R, Gupta BK, Pareek A, Singla-Pareek SL (2018) Knockdown of an inflorescence meristem-specific cytokinin oxidase—OsCKX2 in rice reduces yield penalty under salinity stress condition. Plant Cell Environ 41:936–946

Kadioglu A, Saruhan N, Sağlam A, Terzi R, Acet T (2011) Exogenous salicylic acid alleviates effects of long term drought stress and delays leaf rolling by inducing antioxidant system. Plant Growth Regul 64:27–37

Kadlecova Z, Faltus M, Prasil I (2000) Relationship between abscisic acid content, dry weight and freezing tolerance in barley cv. Lunet. J Plant Physiol. https://doi.org/10.1016/S0176-1617(00)80050-4

Kamran M, Wang D, Alhaithloul HAS, Alghanem SM, Aftab T, Xie K, Lu Y, Shi C, Sun J, Gu W, et al (2021) Jasmonic acid-mediated enhanced regulation of oxidative, glyoxalase defense system and reduced chromium uptake contributes to alleviation of chromium (VI) toxicity in choysum (*Brassica parachinensis* L.). Ecotoxicol Environ Saf. https://doi.org/10.1016/j.ecoenv.2020.111758

Kang DJ, Seo YJ, Lee JD, Ishii R, Kim KU, Shin DH, Park SK, Jang SW, Lee IJ (2005) Jasmonic acid differentially affects growth, ion uptake and abscisic acid concentration in salt-tolerant and salt-sensitive rice cultivars. J Agron Crop Sci. https://doi.org/10.1111/j.1439-037X.2005.00153.x

Kang G, Li G, Xu W, Peng X, Han Q, Zhu Y, Guo T (2012) Proteomics reveals the effects of salicylic acid on growth and tolerance to subsequent drought stress in wheat. J Proteome Res 11:6066–6079

Karlova R, Boer D, Hayes S, Testerink C (2021) Root plasticity under abiotic stress. Plant Physiol 187:1057–1070

Kavi Kishor PB, Tiozon RN, Fernie AR, Sreenivasulu N (2022) Abscisic acid and its role in the modulation of plant growth, development, and yield stability. Trends Plant Sci 27:1283–1295

Kazan K (2015) Diverse roles of jasmonates and ethylene in abiotic stress tolerance. Trends Plant Sci 20:219–229

Kende H (1993) Ethylene biosynthesis. Annu Rev Plant Physiol Plant Mol Biol 44:283–307

Ketsa S, Chidtragool S, Klein JD, Lurie S (1999) Ethylene synthesis in mango fruit following heat treatment. Postharvest Biol Technol 15:65–72

Khalvandi M, Siosemardeh A, Roohi E, Keramati S (2021) Salicylic acid alleviated the effect of drought stress on photosynthetic characteristics and leaf protein pattern in winter wheat. Heliyon. https://doi.org/10.1016/j.heliyon.2021.e05908

Khan MIR, Iqbal N, Masood A, Per TS, Khan NA (2013) Salicylic acid alleviates adverse effects of heat stress on photosynthesis through changes in proline production and ethylene formation. Plant Signal Behav. https://doi.org/10.4161/PSB.26374

Khan MIR, Fatma M, Per TS, Anjum NA, Khan NA (2015) Salicylic acid-induced abiotic stress tolerance and underlying mechanisms in plants. Front Plant Sci. https://doi.org/10.3389/FPLS.2015.00462/FULL

Khan A, Bilal S, Khan AL, Imran M, Shahzad R, Al-Harrasi A, Al-Rawahi A, Al-Azhri M, Mohanta TK, Lee IJ (2020) Silicon and gibberellins: synergistic function in harnessing ABA signaling and heat stress tolerance in date palm (*Phoenix dactylifera* L.). Plants. https://doi.org/10.3390/PLANTS9050620

Khator K, Parihar S, Jasik J, Shekhawat GS (2024) Nitric oxide in plants: an insight on redox activity and responses toward abiotic stress signaling. Plant Signal Behav. https://doi.org/10.1080/15592324.2023.2298053

Kim YS, Choi D, Lee MM, Lee SH, Kim WT (1998) Biotic and abiotic stress-related expression of 1-Aminocyclopropane-l-carboxylate oxidase gene family in *Nicotiana glutinosa* L. Plant Cell Physiol 39:565–573

Kim H, Lee K, Hwang H, Bhatnagar N, Kim DY, Yoon IS, Byun MO, Kim ST, Jung KH, Kim BG (2014) Overexpression of PYL5 in rice enhances drought tolerance, inhibits growth, and modulates gene expression. J Exp Bot 65:453–464

Kim SH, Hussain S, Pham HTT, Kadam US, Bahk S, Ramadany Z, Lee J, Song YH, Lee KO, Hong JC et al (2024) Phosphorylation of auxin signaling repressor IAA8 by heat-responsive MPKs causes defective flower development. Plant Physiol 196:2825–2840

Ko D, Kang J, Kiba T, Park J, Kojima M, Do J, Kim KY, Kwon M, Endler A, Song W-Y et al (2014) *Arabidopsis* ABCG14 is essential for the root-to-shoot translocation of cytokinin. Proc Natl Acad Sci 111:7150–7155

Kobayashi K, Nakamura Y, Ohta H (2009) Type A and type B monogalactosyldiacylglycerol synthases are spatially and functionally separated in the plastids of higher plants. Plant Physiol Biochem. https://doi.org/10.1016/j.plaphy.2008.12.012

Koevoets IT, Venema JH, Elzenga JTM, Testerink C (2016) Roots withstanding their environment: exploiting root system architecture responses to abiotic stress to improve crop tolerance. Front Plant Sci. https://doi.org/10.3389/FPLS.2016.01335

Kondo S, Fiebig A, Okawa K, Ohara H, Kowitcharoen L, Nimitkeatkai H, Kittikorn M, Kim M (2011) Jasmonic acid, polyamine, and antioxidant levels in apple seedlings as affected by ultraviolet-C irradiation. Plant Growth Regul. https://doi.org/10.1007/s10725-010-9539-9

Kong X, Wang T, Li W, Tang W, Zhang D, Dong H (2016) Exogenous nitric oxide delays salt-induced leaf senescence in cotton (*Gossypium hirsutum* L.). Acta Physiol Plant 38:1–9

Kosová K, Prášil IT, Vítámvás P, Dobrev P, Motyka V, Floková K, Novák O, Turečková V, Rolčik J, Pešek B et al (2012) Complex phytohormone responses during the cold acclimation of two wheat cultivars differing in cold tolerance, winter Samanta and spring Sandra. J Plant Physiol 169:567–576

Kumar S, Kaur G, Nayyar H (2008) Exogenous application of abscisic acid improves cold tolerance in chickpea (*Cicer arietinum* L.). J Agron Crop Sci. https://doi.org/10.1111/j.1439-037X.2008.00335.x

Kumari R, Kapoor P, Mir BA, Singh M, Parrey ZA, Rakhra G, Parihar P, Khan MN, Rakhra G (2024) Unlocking the versatility of nitric oxide in plants and insights into its molecular interplays under biotic and abiotic stress. Nitric Oxide 150:1–17

Kuppu S, Mishra N, Hu R, Sun L, Zhu X, Shen G, Blumwald E, Payton P, Zhang H (2013) Water-deficit inducible expression of a Cytokinin biosynthetic gene *IPT* improves drought tolerance in cotton. PLoS One 8:e64190

Kuroha T, Nagai K, Gamuyao R, Wang DR, Furuta T, Nakamori M, Kitaoka T, Adachi K, Minami A, Mori Y et al (2018) Ethylene-gibberellin signaling underlies adaptation of rice to periodic flooding. Science 361:181–186

Kuromori T, Seo M, Shinozaki K (2018) ABA transport and plant water stress responses. Trends Plant Sci. https://doi.org/10.1016/j.tplants.2018.04.001

La VH, Lee BR, Zhang Q, Park SH, Islam MT, Kim TH (2019) Salicylic acid improves drought-stress tolerance by regulating the redox status and proline metabolism in *Brassica rapa*. Hortic Environ Biotechnol 60:31–40

Lamers J, Zhang Y, van Zelm E, Leong CK, Meyer AJ, de Zeeuw T, Verstappen F, Veen M, Deolu-Ajayi AO, Gommers CMM et al (2025) Abscisic acid signaling gates salt-induced responses of plant roots. Proc Natl Acad Sci USA 122:e2406373122

Larkindale J, Knight MR (2002) Protection against heat stress-induced oxidative damage in *Arabidopsis* involves calcium, abscisic acid, ethylene, and salicylic acid. Plant Physiol. https://doi.org/10.1104/pp.010320

Larkindale J, Hall JD, Knight MR, Vierling E (2005) Heat stress phenotypes of *Arabidopsis* mutants implicate multiple signaling pathways in the Acquisition of Thermotolerance. Plant Physiol 138:882–897

Lee TM (1997) Polyamine regulation of growth and chilling tolerance of rice (*Oryza sativa* L.) roots cultured in vitro. Plant Sci. https://doi.org/10.1016/S0168-9452(96)04542-6

Lefevere H, Bauters L, Gheysen G (2020) Salicylic acid biosynthesis in plants. Front Plant Sci 11:521987

Lei G, Shen M, Li ZG, Zhang B, Duan KX, Wang N, Cao YR, Zhang WK, Ma B, Ling HQ et al (2011) EIN2 regulates salt stress response and interacts with a MA3 domain-containing protein ECIP1 in *Arabidopsis*. Plant Cell Environ 34:1678–1692

Li B, Wang W (2021) Salicylic acid induces tolerance of *Vitis riparia× V. labrusca* to chilling stress by altered photosynthetic, antioxidant mechanisms and expression of cold stress responsive genes. Plant Signal Behav. https://doi.org/10.1080/15592324.2021.1973711

Li B, Zhang C, Cao B, Qin G, Wang W, Tian S (2012) Brassinolide enhances cold stress tolerance of fruit by regulating plasma membrane proteins and lipids. Amino Acids 43:2469–2480

Li S, An Y, Hailati S, Zhang J, Cao Y, Liu Y, Geng J, Hu T, Yang P (2019) Overexpression of the *Cytokinin oxidase/dehydrogenase (CKX)* from *Medicago sativa* enhanced salt stress tolerance of *Arabidopsis*. J Plant Biol 62:374–386

Li C, Li L, Reynolds MP, Wang J, Chang X, Mao X, Jing R (2021a) Recognizing the hidden half in wheat: root system attributes associated with drought tolerance. J Exp Bot 72:5117–5133

Li M, Wang C, Shi J, Zhang Y, Liu T, Qi H (2021b) Abscisic acid and putrescine synergistically regulate the cold tolerance of melon seedlings. Plant Physiol Biochem 166:1054–1064

Li Y, Zhou J, Li Z, Qiao J, Quan R, Wang J, Huang R, Qin H (2022) SALT AND ABA RESPONSE ERF1 improves seed germination and salt tolerance by repressing ABA signaling in rice. Plant Physiol 189:1110–1127

Li X, Li C, Shi L, Lv G, Li X, Liu Y, Jia X, Liu J, Chen Y, Zhu L et al (2024a) Jasmonate signaling pathway confers salt tolerance through a NUCLEAR FACTOR-Y trimeric transcription factor complex in *Arabidopsis*. Cell Rep 43:113825

Li Z, Li Z, Ji Y, Wang C, Wang S, Shi Y, Le J, Zhang M (2024b) The heat shock factor 20-HSF4-cellulose synthase A2 module regulates heat stress tolerance in maize. Plant Cell 36:2652–2667

Liao Z, Zhang Y, Yu Q, Fang W, Chen M, Li T, Liu Y, Liu Z, Chen L, Yu S et al (2023) Coordination of growth and drought responses by GA-ABA signaling in rice. New Phytol 240:1149–1161

Licausi F, Kosmacz M, Weits DA, Giuntoli B, Giorgi FM, Voesenek LACJ, Perata P, Van Dongen JT (2011) Oxygen sensing in plants is mediated by an N-end rule pathway for protein destabilization. Nature 479:7373, 419–422

Litvin AG, Van Iersel MW, Malladi A (2016) Drought stress reduces stem elongation and alters gibberellin-related gene expression during vegetative growth of tomato. J Am Soc Hortic Sci 141:591–597

Liu XX, Fu C, Yang WW, Zhang Q, Fan H, Liu J (2016) The involvement of TsFtsH8 in *Thellungiella salsuginea* tolerance to cold and high light stresses. Acta Physiol Plant. https://doi.org/10.1007/s11738-016-2080-3

Liu L, Xia W, Li H, Zeng H, Wei B, Han S, Yin C (2018) Salinity inhibits rice seed germination by reducing α-amylase activity via decreased bioactive gibberellin content. Front Plant Sci 9:333088

Liu XL, Zhang H, Jin YY, Wang MM, Yang HY, Ma HY, Jiang CJ, Liang ZW (2019) Abscisic acid primes rice seedlings for enhanced tolerance to alkaline stress by upregulating antioxidant defense and stress tolerance-related genes. Plant Soil 438:39–55

Liu Y, Zhang M, Meng Z, Wang B, Chen M (2020) Research progress on the roles of cytokinin in plant response to stress. Int J Mol Sci. https://doi.org/10.3390/ijms21186574

Liu R, Shu B, Wang Y, Feng J, Yu B, Gan Y, Liang Y, Qiu Z, Yan S, Cao B (2024) The Jasmonic acid biosynthetic genes *SmLOX4* and *SmLOX5* are involved in heat tolerance in eggplant. Plant Cell Physiol. https://doi.org/10.1093/pcp/pcae088

Lotfi R, Abbasi A, Pessarakli M, Rastogi A, Kalaji HM, Alizadeh K (2024) A comparison of Jasmonic acid and salicylic acid-induced salinity stress tolerance in safflower plants, particularly on sodium (Na) and potassium (K) nutrient contents. J Plant Nutr 47:515–528

Lou D, Wang H, Liang G, Yu D (2017) OsSAPK2 confers abscisic acid sensitivity and tolerance to drought stress in rice. Front Plant Sci. https://doi.org/10.3389/fpls.2017.00993

Lu G, Gao C, Zheng X, Han B (2009) Identification of *OsbZIP72* as a positive regulator of ABA response and drought tolerance in rice. Planta. https://doi.org/10.1007/s00425-008-0857-3

Luo MH, Yuan S, Chen YE, Liu WJ, Du JB, Lei T, Wang MB, Lin HH (2009) Effects of salicylic acid on the photosystem 2 of barley seedlings under osmotic stress. Biol Plant 53:663–669

Luo P, Wu J, Li TT, Shi P, Ma Q, Di DW (2024) An overview of the mechanisms through which plants regulate ROS homeostasis under cadmium stress. Antioxidants. https://doi.org/10.3390/antiox13101174
Lyons JM (1973) Chilling injury in plants. Annu Rev Plant Physiol 24:445–466
A. H. Mackerness S, Surplus SL, Blake P, John CF, Buchanan-Wollaston V, Jordan BR, Thomas B (1999) Ultraviolet-B-induced stress and changes in gene expression in *Arabidopsis thaliana*: role of signalling pathways controlled by jasmonic acid, ethylene and reactive oxygen species. Plant Cell Environ. https://doi.org/10.1046/j.1365-3040.1999.00499.x
Magome H, Nomura T, Hanada A, Takeda-Kamiya N, Ohnishi T, Shinma Y, Katsumata T, Kawaide H, Kamiya Y, Yamaguchi S (2013) CYP714B1 and CYP714B2 encode gibberellin 13-oxidases that reduce gibberellin activity in rice. Proc Natl Acad Sci USA 110:1947–1952
Mähönen AP, Bishopp A, Higuchi M, Nieminen KM, Kinoshita K, Törmäkangas K, Ikeda Y, Oka A, Kakimoto T, Helariutta Y (2006) Cytokinin signaling and its inhibitor AHP6 regulate cell fate during vascular development. Science 311:94–98
Mäntylä E, Lång V, Palva ET (1995) Role of abscisic acid in drought-induced freezing tolerance, cold acclimation, and accumulation of LTI78 and RAB18 proteins in *Arabidopsis thaliana*. Plant Physiol. https://doi.org/10.1104/pp.107.1.141
Mao C, He J, Liu L, Deng Q, Yao X, Liu C, Qiao Y, Li P, Ming F (2020) OsNAC2 integrates auxin and cytokinin pathways to modulate rice root development. Plant Biotechnol J 18:429–442
Marcińska I, Czyczyło-Mysza I, Skrzypek E, Grzesiak MT, Janowiak F, Filek M, Dziurka M, Dziurka K, Waligórski P, Juzoń K et al (2013) Alleviation of osmotic stress effects by exogenous application of salicylic or abscisic acid on wheat seedlings. Intl J Mol Sci 14:13171–13193
Maris D', Dempsey A, Corina Vlot A, Wildermuth MC, Klessig DF (2011) Salicylic acid biosynthesis and metabolism. Arabidopsis Book Am Soc Plant Biol 9:e0156
Maruyama K, Takeda M, Kidokoro S, Yamada K, Sakuma Y, Urano K, Fujita M, Yoshiwara K, Matsukura S, Morishita Y et al (2009) Metabolic pathways involved in cold acclimation identified by integrated analysis of metabolites and transcripts regulated by DREB1A and DREB2A. Plant Physiol. https://doi.org/10.1104/pp.109.135327
Marzec M, Daszkowska-Golec A, Collin A, Melzer M, Eggert K, Szarejko I (2020) Barley strigolactone signalling mutant hvd14.d reveals the role of strigolactones in abscisic acid-dependent response to drought. Plant Cell Environ. https://doi.org/10.1111/pce.13815
Masood A, Iqbal N, Khan NA (2012) Role of ethylene in alleviation of cadmium-induced photosynthetic capacity inhibition by Sulphur in mustard. Plant Cell Environ 35:524–533
Maurya VK, Ranjan V, Gothandam KM, Pareek S (2020) Exogenous gibberellic acid treatment extends green chili shelf life and maintain quality under modified atmosphere packaging. Sci Hortic. https://doi.org/10.1016/J.SCIENTA.2019.108934
Mazumder MK, Sharma P, Moulick D, Tata SK, Choudhury S (2022) Salicylic acid ameliorates zinc and chromium-induced stress responses in wheat seedlings: a biochemical and computational analysis. Cereal Res Commun 50:407–418
Mehdi S, Alavi N, Javad M, Khosrow A, Kalantari M, Arvin MJ, Kalantari KM (2014) Salicylic acid and nitric oxide alleviate osmotic stress in wheat (*Triticum aestivum* L.) seedlings. J Plant Interact 9:683–688
Mei W, Chen W, Wang Y, Liu Z, Dong Y, Zhang G, Deng H, Liu X, Lu X, Wang F et al (2023) Exogenous kinetin modulates ROS homeostasis to affect heat tolerance in Rice seedlings. Int J Mol Sci. https://doi.org/10.3390/ijms24076252
Metwally A, Finkemeier I, Georgi M, Dietz KJ (2003) Salicylic acid alleviates the cadmium toxicity in barley seedlings. Plant Physiol 132:272–281
Miao J, Li X, Li X, Tan W, You A, Wu S, Tao Y, Chen C, Wang J, Zhang D et al (2020) OsPP2C09, a negative regulatory factor in abscisic acid signalling, plays an essential role in balancing plant growth and drought tolerance in rice. New Phytol. https://doi.org/10.1111/nph.16670
Miki Y, Takahashi D, Kawamura Y, Uemura M (2019) Temporal proteomics of *Arabidopsis* plasma membrane during cold- and de-acclimation. J Proteome. https://doi.org/10.1016/j.jprot.2018.11.008

Min Z, Li R, Chen L, Zhang Y, Li Z, Liu M, Ju Y, Fang Y (2019) Alleviation of drought stress in grapevine by foliar-applied strigolactones. Plant Physiol Biochem. https://doi.org/10.1016/j.plaphy.2018.11.037

Min MK, Kim R, Hong WJ, Jung KH, Lee JY, Kim BG (2021) Ospp2c09 is a bifunctional regulator in both ABA-dependent and independent abiotic stress signaling pathways. Int J Mol Sci. https://doi.org/10.3390/ijms22010393

Mitchell JW, Mandava N, Worley JF, Plimmer JR, Smith M V. (1970) Brassins—a new family of plant hormones from rape pollen. Nature 225:5237, 1065–1066

Mittelheuser CJ, Van Steveninck RFM (1969) Stomatal closure and inhibition of transpiration induced by (RS)-abscisic acid. Nature 221:5177, 281–282

Mittler R, Zandalinas SI, Fichman Y, Van Breusegem F (2022) Reactive oxygen species signalling in plant stress responses. Nat Rev Mol Cell Biol. https://doi.org/10.1038/s41580-022-00499-2

Miyazono KI, Miyakawa T, Sawano Y, Kubota K, Kang HJ, Asano A, Miyauchi Y, Takahashi M, Zhi Y, Fujita Y et al (2009) Structural basis of abscisic acid signalling. Nature. https://doi.org/10.1038/nature08583

Mohan TC, Castrillo G, Navarro C, Zarco-Fernandez S, Ramireddy E, Mateo C, Zamarreño AM, Paz-Ares J, Muñoz R, Garcia-Mina JM et al (2016) Cytokinin determines thiol-mediated arsenic tolerance and accumulation in *Arabidopsis thaliana*. Plant Physiol. https://doi.org/10.1104/pp.16.00372

Mostofa MG, Fujita M (2013) Salicylic acid alleviates copper toxicity in rice (*Oryza sativa* L.) seedlings by up-regulating antioxidative and glyoxalase systems. Ecotoxicology 22:959–973

Mostofa MG, Rahman MM, Nguyen KH, Li W, Watanabe Y, Tran CD, Zhang M, Itouga M, Fujita M, Tran LSP (2021) Strigolactones regulate arsenate uptake, vacuolar-sequestration and antioxidant defense responses to resist arsenic toxicity in rice roots. J Hazard Mater. https://doi.org/10.1016/j.jhazmat.2021.125589

Mughal N, Shoaib N, Chen J, Li Y, He Y, Fu M, Li X, He Y, Guo J, Deng J et al (2024) Adaptive roles of cytokinins in enhancing plant resilience and yield against environmental stressors. Chemosphere. https://doi.org/10.1016/j.chemosphere.2024.143189

Muhammad Aslam M, Waseem M, Jakada BH, Okal EJ, Lei Z, Saqib HSA, Yuan W, Xu W, Zhang Q (2022) Mechanisms of abscisic acid-mediated drought stress responses in plants. Int J Mol Sci. https://doi.org/10.3390/IJMS23031084

Mutlu S, Karadağoğlu Ö, Atici Ö, Nalbantoğlu B (2013) Protective role of salicylic acid applied before cold stress on antioxidative system and protein patterns in barley apoplast. Biol Plant 57:507–513

Nagai K, Mori Y, Ishikawa S, Furuta T, Gamuyao R, Niimi Y, Hobo T, Fukuda M, Kojima M, Takebayashi Y et al (2020) Antagonistic regulation of the gibberellic acid response during stem growth in rice. Nature 584:109–114

Nagar P, Sharma N, Jain M, Sharma G, Prasad M, Mustafiz A (2022) OsPSKR15, a phytosulfokine receptor from rice enhances abscisic acid response and drought stress tolerance. Physiol Plant. https://doi.org/10.1111/PPL.13569

Ng LM, Melcher K, Teh BT, Xu HE (2014) Abscisic acid perception and signaling: structural mechanisms and applications. Acta Pharmacol Sin 35:567–584

Nie G, Zhou J, Jiang Y, He J, Wang Y, Liao Z, Appiah C, Li D, Feng G, Huang L et al (2022) Transcriptome characterization of candidate genes for heat tolerance in perennial ryegrass after exogenous methyl Jasmonate application. BMC Plant Biol. https://doi.org/10.1186/s12870-021-03412-9

Nishiyama R, Watanabe Y, Fujita Y, Le DT, Kojima M, Werner T, Vankova R, Yamaguchi-Shinozaki K, Shinozaki K, Kakimoto T et al (2011) Analysis of Cytokinin mutants and regulation of Cytokinin metabolic genes reveals important regulatory roles of Cytokinins in drought, salt and abscisic acid responses, and abscisic acid biosynthesis. Plant Cell 23:2169–2183

Noriega G, Caggiano E, Lecube ML, Cruz DS, Batlle A, Tomaro M, Balestrasse KB (2012) The role of salicylic acid in the prevention of oxidative stress elicited by cadmium in soybean plants. Biometals 25:1155–1165

Olszewski N, Sun T, Gubler F (2002) Gibberellin Signaling. Plant Cell 14:S61–S80

Omoarelojie LO, Kulkarni MG, Finnie JF, Pospíšil T, Strnad M, Van Staden J (2020) Synthetic strigolactone (rac-GR24) alleviates the adverse effects of heat stress on seed germination and photosystem II function in lupine seedlings. Plant Physiol Biochem. https://doi.org/10.1016/j.plaphy.2020.07.043

Pan C, Zhang H, Ma Q, Fan F, Fu R, Ahammed GJ, Yu J, Shi K (2019) Role of ethylene biosynthesis and signaling in elevated CO_2-induced heat stress response in tomato. Planta 250:563–572

Pandey P, Srivastava RK, Dubey RS (2013) Salicylic acid alleviates aluminum toxicity in rice seedlings better than magnesium and calcium by reducing aluminum uptake, suppressing oxidative damage and increasing antioxidative defense. Ecotoxicology 22:656–670

Pandey A, Khan MK, Hamurcu M, Athar T, Yerlikaya BA, Yerlikaya S, Kavas M, Rustagi A, Zargar SM, Sofi PA, et al (2023) Role of exogenous nitric oxide in protecting plants against abiotic stresses. Agronomy 13:1201

Pantoja-Benavides AD, Garces-Varon G, Restrepo-Díaz H (2022) Foliar cytokinins or brassinosteroids applications influence the rice plant acclimatization to combined heat stress. Front Plant Sci. https://doi.org/10.3389/fpls.2022.983276

Parveen A, Ahmar S, Kamran M, Malik Z, Ali A, Riaz M, Abbasi GH, Khan M, Bin SA, Rizwan M et al (2021) Abscisic acid signaling reduced transpiration flow, regulated Na+ ion homeostasis and antioxidant enzyme activities to induce salinity tolerance in wheat (*Triticum aestivum* L.) seedlings. Environ Technol Innov 24:101808

Parys E, Romanowska E, Siedlecka M, Poskuta JW (1998) The effect of lead on photosynthesis and respiration in detached leaves and in mesophyll protoplasts of Pisum sativum. Acta Physiol Plant 20:313–322

Peng Y, Yang J, Li X, Zhang Y (2021) Salicylic acid: biosynthesis and signaling. Annu Rev Plant Biol 72:761–791

Perez-Borroto LS, Guzzo MC, Posada G, Peña Malavera AN, Castagnaro AP, Gonzalez-Olmedo JL, Coll-García Y, Pardo EM (2022) A brassinosteroid functional analogue increases soybean drought resilience. Sci Rep. https://doi.org/10.1038/s41598-022-15284-6

Pizzio GA, Rodriguez L, Antoni R, Gonzalez-Guzman M, Yunta C, Merilo E, Kollist H, Albert A, Rodriguez PL (2013) The PYL4 A194T mutant uncovers a key role of PYR1-LIKE4/PROTEIN PHOSPHATASE 2CA interaction for abscisic acid signaling and plant drought resistance. Plant Physiol 163:441–455

Poór P, Nawaz K, Gupta R, Ashfaque F, Khan MIR (2021) Ethylene involvement in the regulation of heat stress tolerance in plants. Plant Cell Reports 3(41):675–698

Prerostova S, Dobrev PI, Gaudinova A, Knirsch V, Körber N, Pieruschka R, Fiorani F, Brzobohatý B, Cerný M, Spichal L et al (2018) Cytokinins: their impact on molecular and growth responses to drought stress and recovery in *Arabidopsis*. Front Plant Sci. https://doi.org/10.3389/fpls.2018.00655

Prerostova S, Dobrev PI, Kramna B, Gaudinova A, Knirsch V, Spichal L, Zatloukal M, Vankova R (2020) Heat acclimation and inhibition of Cytokinin degradation positively affect heat stress tolerance of *Arabidopsis*. Front Plant Sci. https://doi.org/10.3389/fpls.2020.00087

Prerostova S, Rezek J, Jarosova J, Lacek J, Dobrev P, Marsik P, Gaudinova A, Knirsch V, Dolezal K, Plihalova L et al (2023) Cytokinins act synergistically with heat acclimation to enhance rice thermotolerance affecting hormonal dynamics, gene expression and volatile emission. Plant Physiol Biochem. https://doi.org/10.1016/j.plaphy.2023.107683

Qin F, Shinozaki K, Yamaguchi-Shinozaki K (2011a) Achievements and challenges in understanding plant abiotic stress responses and tolerance. Plant Cell Physiol. https://doi.org/10.1093/pcp/pcr106

Qin H, Gu Q, Zhang J, Sun L, Kuppu S, Zhang Y, Burow M, Payton P, Blumwald E, Zhang H (2011b) Regulated expression of an *Isopentenyltransferase gene (IPT)* in peanut significantly improves drought tolerance and increases yield under field conditions. Plant Cell Physiol 52:1904–1914

Qiu ZB, Guo JL, Zhu AJ, Zhang L, Zhang MM (2014) Exogenous jasmonic acid can enhance tolerance of wheat seedlings to salt stress. Ecotoxicol Environ Saf. https://doi.org/10.1016/j.ecoenv.2014.03.014

Rahman MM, Mostofa MG, Keya SS, Ghosh PK, Abdelrahman M, Anik TR, Gupta A, Tran LSP (2024) Jasmonic acid priming augments antioxidant defense and photosynthesis in soybean to alleviate combined heat and drought stress effects. Plant Physiol Biochem. https://doi.org/10.1016/j.plaphy.2023.108193

Rasheed F, Mir IR, Sehar Z, Fatma M, Gautam H, Khan S, Anjum NA, Masood A, Sofo A, Khan NA (2022a) Nitric oxide and salicylic acid regulate glutathione and ethylene production to enhance heat stress acclimation in wheat involving sulfur assimilation. Plants. https://doi.org/10.3390/PLANTS11223131

Rasheed R, Ashraf MA, Ali S, Iqbal M, Zafar S (2022b) Plant metabolism adjustment in exogenously applied NO under stress. In: Singh et al. (eds) Nitric oxide in plant biology: an ancient molecule with emerging roles, pp 261–296

Raza A, Charagh S, Najafi-Kakavand S, Abbas S, Shoaib Y, Anwar S, Sharifi S, Lu G, Siddique KHM (2023) Role of phytohormones in regulating cold stress tolerance: physiological and molecular approaches for developing cold-smart crop plants. Plant Stress. https://doi.org/10.1016/j.stress.2023.100152

Reguera M, Peleg Z, Abdel-Tawab YM, Tumimbang EB, Delatorre CA, Blumwald E (2013) Stress-induced Cytokinin synthesis increases drought tolerance through the coordinated regulation of carbon and nitrogen assimilation in Rice. Plant Physiol 163:1609–1622

Ren H, Bao J, Gao Z, Sun D, Zheng S, Bai J (2023) How rice adapts to high temperatures. Front Plant Sci 14:1137923

Rezaul IM, Baohua F, Tingting C, Weimeng F, Caixia Z, Longxing T, Guanfu F (2019) Abscisic acid prevents pollen abortion under high-temperature stress by mediating sugar metabolism in rice spikelets. Physiol Plant. https://doi.org/10.1111/ppl.12759

Rivero RM, Kojima M, Gepstein A, Sakakibara H, Mittler R, Gepstein S, Blumwald E (2007) Delayed leaf senescence induces extreme drought tolerance in a flowering plant. Proc Natl Acad Sci 104:19631–19636

Robison JD, Yamasaki Y, Randall SK (2019) The ethylene signaling pathway negatively impacts CBF/DREB-regulated cold response in soybean (*Glycine max*). Front Plant Sci 10:121

Rosales MA, Maurel C, Nacry P (2019) Abscisic acid coordinates dose-dependent developmental and hydraulic responses of roots to water deficit. Plant Physiol 180:2198–2211

Ruan YL, Jin Y, Yang YJ, Li GJ, Boyer JS (2010) Sugar input, metabolism, and signaling mediated by invertase: roles in development, yield potential, and response to drought and heat. Mol Plant. https://doi.org/10.1093/mp/ssq044

Ruan J, Zhou Y, Zhou M, Yan J, Khurshid M, Weng W, Cheng J, Zhang K (2019) Jasmonic acid signaling pathway in plants. Int J Mol Sci. https://doi.org/10.3390/ijms20102479

Saeed W, Naseem S, Ali Z (2017) Strigolactones biosynthesis and their role in abiotic stress resilience in plants: a critical review. Front Plant Sci. https://doi.org/10.3389/fpls.2017.01487

Sakakibara H (2006) Cytokinins: activity, biosynthesis, and translocation. Annu Rev Plant Biol 57:431–449

Samtani H, Sharma A, Khurana JP, Khurana P (2022a) Thermosensing in plants: deciphering the mechanisms involved in heat sensing and their role in thermoresponse and thermotolerance. Environ Exp Bot. https://doi.org/10.1016/j.envexpbot.2022.105041

Samtani H, Sharma A, Khurana P (2022b) Overexpression of HVA1 enhances drought and heat stress tolerance in *Triticum aestivum* doubled haploid plants. Cells 11:912

Samtani H, Sharma A, Khurana P (2023) Ectopic overexpression of T*aHsfA5* promotes thermomorphogenesis in *Arabidopsis thaliana* and thermotolerance in *Oryza sativa*. Plant Mol Biol 112:225–243

Santosh Kumar VV, Yadav SK, Verma RK, Shrivastava S, Ghimire O, Pushkar S, Rao MV, Senthil Kumar T, Chinnusamy V (2021) The abscisic acid receptor OsPYL6 confers drought tolerance to indica rice through dehydration avoidance and tolerance mechanisms. J Exp Bot. https://doi.org/10.1093/jxb/eraa509

Sauter KJ, Davis DW, Li PH, Wallerstein IS (1990) Leaf ethylene evolution level following high-temperature stress in common bean. HortScience 25:1282–1284

Sedaghat M, Sarvestani ZT, Emam Y, Bidgoli AM, Sorooshzadeh A (2020) Foliar-applied GR24 and salicylic acid enhanced wheat drought tolerance. Russ J Plant Physiol. https://doi.org/10.1134/S1021443720040159

Sehar Z, Mir IR, Khan S, Masood A, Khan NA (2023) Nitric oxide and proline modulate redox homeostasis and photosynthetic metabolism in wheat plants under high temperature stress acclimation. Plants 12:1256

Senaratna T, Touchell D, Bunn E, Dixon K (2000) Acetyl salicylic acid (aspirin) and salicylic acid induce multiple stress tolerance in bean and tomato plants. Plant Growth Regul 30:157–161

Seo M, Koshiba T (2002) Complex regulation of ABA biosynthesis in plants. Trends Plant Sci 7:41–48

Shahzad K, Hussain S, Arfan M, Hussain S, Waraich EA, Zamir S, Saddique M, Rauf A, Kamal KY, Hano C et al (2021) Exogenously applied gibberellic acid enhances growth and salinity stress tolerance of maize through modulating the morpho-physiological, biochemical and molecular attributes. Biomolecules 11:1005

Shan C, Zhou Y, Liu M (2015) Nitric oxide participates in the regulation of the ascorbate-glutathione cycle by exogenous jasmonic acid in the leaves of wheat seedlings under drought stress. Protoplasma. https://doi.org/10.1007/s00709-015-0756-y

Sharifi P, Bidabadi SS (2020) Strigolactone could enhances gas-exchange through augmented antioxidant defense system in *Salvia nemorosa* L. plants subjected to saline conditions stress. Ind Crops Prod. https://doi.org/10.1016/j.indcrop.2020.112460

Sharma M, Gupta SK, Majumder B, Maurya VK, Deeba F, Alam A, Pandey V (2017) Salicylic acid mediated growth, physiological and proteomic responses in two wheat varieties under drought stress. J Proteome 163:28–51

Sharma L, Dalal M, Verma RK, Kumar SVV, Yadav SK, Pushkar S, Kushwaha SR, Bhowmik A, Chinnusamy V (2018) Auxin protects spikelet fertility and grain yield under drought and heat stresses in rice. Environ Exp Bot 150:9–24

Sharma A, Samtani H, Sahu K, Sharma AK, Khurana JP, Khurana P (2023) Functions of Phytochrome-interacting factors (PIFs) in the regulation of plant growth and development: a comprehensive review. Int J Biol Macromol 244:125234

Sharma A, Samtani H, Laxmi A (2024a) Molecular dialogue between light and temperature signalling in plants: from perception to thermotolerance. J Exp Bot. https://doi.org/10.1093/jxb/erae356

Sharma P, Jha AB, Dubey RS (2024b) Strigolactones: coordination with other phytohormones and enhancement of abiotic stress responses. Environ Exp Bot. https://doi.org/10.1016/j.envexpbot.2024.105782

Shemi R, Wang R, Gheith ESMS, Hussain HA, Hussain S, Irfan M, Cholidah L, Zhang K, Zhang S, Wang L (2021) Effects of salicylic acid, zinc and glycine betaine on morpho-physiological growth and yield of maize under drought stress. Sci Rep. https://doi.org/10.1038/S41598-021-82264-7

Shen J, Liu J, Yuan Y, Chen L, Ma J, Li X, Li J (2022) The mechanism of abscisic acid regulation of wild Fragaria species in response to cold stress. BMC Genomics. https://doi.org/10.1186/s12864-022-08889-8

Sheteiwy MS, Ulhassan Z, Qi W, Lu H, AbdElgawad H, Minkina T, Sushkova S, Rajput VD, El-Keblawy A, Jośko I et al (2022) Association of jasmonic acid priming with multiple defense mechanisms in wheat plants under high salt stress. Front Plant Sci. https://doi.org/10.3389/fpls.2022.886862

Shi WG, Liu W, Yu W, Zhang Y, Ding S, Li H, Mrak T, Kraigher H, Bin LZ (2019) Abscisic acid enhances lead translocation from the roots to the leaves and alleviates its toxicity in *Populus* canescens. J Hazard Mater 362:275–285

Shirani Bidabadi S, Sharifi P (2021) Strigolactone and methyl Jasmonate-induced antioxidant defense and the composition alterations of different active compounds in *Dracocephalum kotschyi* Boiss under drought stress. J Plant Growth Regul. https://doi.org/10.1007/s00344-020-10157-6

Shohat H, Cheriker H, Kilambi HV, Illouz Eliaz N, Blum S, Amsellem Z, Tarkowská D, Aharoni A, Eshed Y, Weiss D (2021) Inhibition of gibberellin accumulation by water deficiency promotes fast and long-term "drought avoidance" responses in tomato. New Phytol 232:1985–1998

Shu K, Zhou W, Chen F, Luo X, Yang W (2018) Abscisic acid and gibberellins antagonistically mediate plant development and abiotic stress responses. Front Plant Sci 9:355447

Shuai H, Meng Y, Luo X, Chen F, Zhou W, Dai Y, Qi Y, Du J, Yang F, Liu J, et al (2017) Exogenous auxin represses soybean seed germination through decreasing the gibberellin/abscisic acid (GA/ABA) ratio. Scientific Reports 7:1–11

Singh S, Parihar P, Singh R, Singh VP, Prasad SM (2016) Heavy metal tolerance in plants: role of transcriptomics, proteomics, metabolomics, and Ionomics. Front Plant Sci 6:1143

Sirhindi G, Mir MA, Sharma P, Gill SS, Kaur H, Mushtaq R (2015) Modulatory role of jasmonic acid on photosynthetic pigments, antioxidants and stress markers of *Glycine max* L. under nickel stress. Physiol Mol Biol Plants. https://doi.org/10.1007/s12298-015-0320-4

Skoog F, Armstrong DJ (1970) Cytokinins. Annu Rev Plant Physiol 21:359–384

Smoleńska-Sym G, Gawrońska H, Kacperska A (1995) Modifications of abscisic acid level in winter oilseed rape leaves during acclimation of plants to freezing temperatures. Plant Growth Regul. https://doi.org/10.1007/BF00024496

Soares C, de Sousa A, Pinto A, Azenha M, Teixeira J, Azevedo RA, Fidalgo F (2016) Effect of 24-epibrassinolide on ROS content, antioxidant system, lipid peroxidation and Ni uptake in *Solanum nigrum* L. under Ni stress. Environ Exp Bot 122:115–125

Sohag AAM, Tahjib-Ul-Arif M, Afrin S, Khan MK, Hannan MA, Skalicky M, Mortuza MG, Brestic M, Hossain MA, Murata Y (2020) Insights into nitric oxide-mediated water balance, antioxidant defence and mineral homeostasis in rice (*Oryza sativa* L.) under chilling stress. Nitric Oxide 100–101:7–16

Sorrentino G, Haworth M, Wahbi S, Mahmood T, Zuomin S, Centritto M (2016) Abscisic acid induces rapid reductions in mesophyll conductance to carbon dioxide. PLoS One 11:e0148554

Sripinyowanich S, Klomsakul P, Boonburapong B, Bangyeekhun T, Asami T, Gu H, Buaboocha T, Chadchawan S (2013) Exogenous ABA induces salt tolerance in indica rice (*Oryza sativa* L.): the role of *OsP5CS1* and *OsP5CR* gene expression during salt stress. Environ Exp Bot 86:94–105

Srivastava A, Pandey GC (2023) Role of Cytokinins and gibberellins in crops response to heat and drought stress. J Cereal Res. https://doi.org/10.25174/2582-2675/2023/136065

Srivastava S, Srivastava AK, Suprasanna P, D'Souza SF (2013) Identification and profiling of arsenic stress-induced microRNAs in *Brassica juncea*. J Exp Bot 64:303–315

Stroïnski A, Gizewska K, Zielezinska M (2013) Abscisic acid is required in transduction of cadmium signal to potato roots. Biol Plant 57:121–127

Su Y, Xia S, Wang R, Xiao L (2017) Phytohormonal quantification based on biological principles. Hormone metabolism and signaling in plants. Elsevier Inc., pp 431–470

Su Y, Huang Y, Dong X, Wang R, Tang M, Cai J, Chen J, Zhang X, Nie G (2021) Exogenous methyl Jasmonate improves heat tolerance of perennial ryegrass through alteration of osmotic adjustment, antioxidant defense, and expression of Jasmonic acid-responsive genes. Front Plant Sci. https://doi.org/10.3389/fpls.2021.664519

Sui N (2015) Photoinhibition of *Suaeda salsa* to chilling stress is related to energy dissipation and water-water cycle. Photosynthetica. https://doi.org/10.1007/s11099-015-0080-y

Sun X, Xi DH, Feng H, Du JB, Lei T, Liang HG, Lin HH (2009) The dual effects of salicylic acid on dehydrin accumulation in water-stressed barley seedlings. Russ J Plant Physiol 56:348–354

Sun J, Qi L, Li Y, Chu J, Li C (2012) Pif4-mediated activation of *YUCCA8* expression integrates temperature into the auxin pathway in regulating *Arabidopsis* hypocotyl growth. PLoS Genet. https://doi.org/10.1371/journal.pgen.1002594

Sun H, Tao J, Liu S, Huang S, Chen S, Xie X, Yoneyama K, Zhang Y, Xu G (2014a) Strigolactones are involved in phosphate- and nitrate-deficiency-induced root development and auxin transport in rice. J Exp Bot. https://doi.org/10.1093/jxb/eru029

Sun L, Zhang Q, Wu J, Zhang L, Jiao X, Zhang S, Zhang Z, Sun D, Lu T, Sun Y (2014b) Two rice authentic histidine Phosphotransfer proteins, OsAHP1 and OsAHP2, mediate Cytokinin signaling and stress responses in Rice. Plant Physiol 165:335–345

Sun C, Zhang Y, Liu L, Liu X, Li B, Jin C, Lin X (2021) Molecular functions of nitric oxide and its potential applications in horticultural crops. Hortic Res 8:71

Sun H, Rang X, Han H, Pei Z, Zhao J, Zhu Z, Li J, Zhang P, Zhao Y, Duan Y (2024) Exogenous gibberellic acid ameliorates chilling injury in peach (*Prunus persica* L.) by improving the antioxidant system. Agronomy 14:816

Surina S, Yamagami A, Miyaji T, Chagan Z, Chung K, Mitsuda N, Nishida K, Tachibana R, Zhu Z, Miyakawa T et al (2024) BIL9 promotes both plant growth via BR signaling and drought stress resistance by binding with the transcription factor HDG11. Plant Cell Physiol. https://doi.org/10.1093/pcp/pcae009

Suzuki N, Miller G, Morales J, Shulaev V, Torres MA, Mittler R (2011) Respiratory burst oxidases: the engines of ROS signaling. Curr Opin Plant Biol. https://doi.org/10.1016/j.pbi.2011.07.014

Suzuki N, Koussevitzky S, Mittler R, Miller G (2012) ROS and redox signalling in the response of plants to abiotic stress. Plant Cell Environ 35:259–270

Swamy PM, Smith BN (1999) Role of abscisic acid in plant stress tolerance. Curr Sci 76:1220–1227

Tai Z, Yin X, Fang Z, Shi G, Lou L, Cai Q (2017) Exogenous GR24 alleviates cadmium toxicity by reducing cadmium uptake in switchgrass (*Panicum virgatum*) seedlings. Int J Environ Res Public Health. https://doi.org/10.3390/ijerph14080852

Takahashi D, Gorka M, Erban A, Graf A, Kopka J, Zuther E, Hincha DK (2019) Both cold and sub-zero acclimation induce cell wall modification and changes in the extracellular proteome in *Arabidopsis thaliana*. Sci Rep. https://doi.org/10.1038/s41598-019-38688-3

Talukdar D (2013) Arsenic-induced oxidative stress in the common bean legume, *Phaseolus vulgaris* L. seedlings and its amelioration by exogenous nitric oxide. Physiol Mol Biol Plants 19:69–79

Tan M, Cheng D, Yang Y, Zhang G, Qin M, Chen J, Chen Y, Jiang M (2017) Co-expression network analysis of the transcriptomes of rice roots exposed to various cadmium stresses reveals universal cadmium-responsive genes. BMC Plant Biol 17:1–18

Taşgin E, Atici Ö, Nalbantoğlu B, Popova LP (2006) Effects of salicylic acid and cold treatments on protein levels and on the activities of antioxidant enzymes in the apoplast of winter wheat leaves. Phytochemistry 67:710–715

Tashjian AH, Gordon Sato JH, Tashjian AH, Yasumura Y, Levine L, Sato G, Parker ML (1966) Germination of Witchweed (*Striga lutea* Lour.): isolation and properties of a potent stimulant. Science (1979) 154:1189–1190

Tessi TM, Maurino VG, Shahriari M, Meissner E, Novak O, Pasternak T, Schumacher BS, Ditengou F, Li Z, Duerr J et al (2023) AZG1 is a cytokinin transporter that interacts with auxin transporter PIN1 and regulates the root stress response. New Phytol 238:1924–1941

Thomashow MF (1999) Plant cold acclimation: freezing tolerance genes and regulatory mechanisms. Annu Rev Plant Biol. https://doi.org/10.1146/annurev.arplant.50.1.571

Tilman D, Balzer C, Hill J, Befort BL (2011) Global food demand and the sustainable intensification of agriculture. Proc Natl Acad Sci USA. https://doi.org/10.1073/pnas.1116437108

Torun H (2019) Time-course analysis of salicylic acid effects on ROS regulation and antioxidant defense in roots of hulled and hulless barley under combined stress of drought, heat and salinity. Physiol Plant 165:169–182

Tsuchiya Y, Vidaurre D, Toh S, Hanada A, Nambara E, Kamiya Y, Yamaguchi S, McCourt P (2010) A small-molecule screen identifies new functions for the plant hormone strigolactone. Nat Chem Biol. https://doi.org/10.1038/nchembio.435

Tuteja N (2007) Abscisic acid and abiotic stress signaling. Plant Signal Behav 2:135–138

Umehara M, Cao M, Akiyama K, Akatsu T, Seto Y, Hanada A, Li W, Takeda-Kamiya N, Morimoto Y, Yamaguchi S (2015) Structural requirements of strigolactones for shoot branching inhibition in rice and Arabidopsis. Plant Cell Physiol 56(6):1059–1072. https://doi.org/10.1093/pcp/pcv028

Umezawa T, Okamoto M, Kushiro T, Nambara E, Oono Y, Seki M, Kobayashi M, Koshiba T, Kamiya Y, Shinozaki K (2006) CYP707A3, a major ABA 8′-hydroxylase involved in dehydration and rehydration response in *Arabidopsis thaliana*. Plant J. https://doi.org/10.1111/j.1365-313X.2006.02683.x

Uniyal S, Bhandari M, Singh P, Singh RK, Tiwari SP (2022) Cytokinin biosynthesis in cyanobacteria: insights for crop improvement. Front Genet. https://doi.org/10.3389/fgene.2022.933226

Uno Y, Furihata T, Abe H, Yoshida R, Shinozaki K, Yamaguchi-Shinozaki K (2000) *Arabidopsis* basic leucine zipper transcription factors involved in an abscisic acid-dependent signal transduction pathway under drought and high-salinity conditions. Proc Natl Acad Sci USA. https://doi.org/10.1073/pnas.190309197

Usman B, Nawaz G, Zhao N, Liao S, Liu Y, Li R (2020) Precise editing of the ospyl9 gene by rna-guided cas9 nuclease confers enhanced drought tolerance and grain yield in rice (*Oryza sativa* l.) by regulating circadian rhythm and abiotic stress responsive proteins. Int J Mol Sci. https://doi.org/10.3390/ijms21217854

Valluru R, Davies WJ, Reynolds MP, Dodd IC (2016) Foliar abscisic acid-to-ethylene accumulation and response regulate shoot growth sensitivity to mild drought in wheat. Front Plant Sci. https://doi.org/10.3389/FPLS.2016.00461

Veerasamy M, He Y, Huang B (2007) Leaf senescence and protein metabolism in creeping bentgrass exposed to heat stress and treated with cytokinins. J Am Soc Hortic Sci. https://doi.org/10.21273/jashs.132.4.467

Venzhik Y, Talanova V, Titov A (2016) The effect of abscisic acid on cold tolerance and chloroplasts ultrastructure in wheat under optimal and cold stress conditions. Acta Physiol Plant. https://doi.org/10.1007/s11738-016-2082-1

Verma RK, Santosh Kumar VV, Yadav SK, Pushkar S, Rao MV, Chinnusamy V (2019) Overexpression of ABA receptor *PYL10* gene confers drought and cold tolerance to Indica Rice. Front Plant Sci. https://doi.org/10.3389/FPLS.2019.01488

von Schwartzenberg K, Núñez MF, Blaschke H, Dobrev PI, Novák O, Motyka V, Strnad M (2007) Cytokinins in the bryophyte *Physcomitrella patens*: analyses of activity, distribution, and Cytokinin oxidase/dehydrogenase overexpression reveal the role of extracellular Cytokinins. Plant Physiol 145:786–800

Waadt R, Seller CA, Hsu PK, Takahashi Y, Munemasa S, Schroeder JI (2022) Plant hormone regulation of abiotic stress responses. Nat Rev Mol Cell Biol 23:680–694

Walia H, Wilson C, Condamine P, Liu X, Ismail AM, Close TJ (2007) Large-scale expression profiling and physiological characterization of jasmonic acid-mediated adaptation of barley to salinity stress. Plant Cell Environ. https://doi.org/10.1111/j.1365-3040.2006.01628.x

Wang YH, Irving HR (2011) Developing a model of plant hormone interactions. Plant Signal Behav 6:494–500

Wang Y, Liu C, Li K, Sun F, Hu H, Li X, Zhao Y, Han C, Zhang W, Duan Y et al (2007) *Arabidopsis* EIN2 modulates stress response through abscisic acid response pathway. Plant Mol Biol 64:633–644

Wang F, Guo Z, Li H, Wang M, Onac E, Zhou J, Xia X, Shi K, Yu J, Zhou Y (2016) Phytochrome A and B function antagonistically to regulate cold tolerance via abscisic acid-dependent jasmonate signaling. Plant Physiol 170(1):459–471. https://doi.org/10.1104/pp.15.01171

Wang J, Zhang Q, Cui F, Hou L, Zhao S, Xia H, Qiu J, Li T, Zhang Y, Wang X et al (2017a) Genome-wide analysis of gene expression provides new insights into cold responses in *Thellungiella salsuginea*. Front Plant Sci. https://doi.org/10.3389/fpls.2017.00713

Wang X, Zhuang L, Shi Y, Huang B (2017b) Up-regulation of HSFA2c and HSPs by ABA contributing to improved heat tolerance in tall fescue and *Arabidopsis*. Int J Mol Sci. https://doi.org/10.3390/ijms18091981

Wang WC, Lin TC, Kieber J, Tsai YC (2019) Response regulators 9 and 10 negatively regulate salinity tolerance in Rice. Plant Cell Physiol 60:2549–2563
Wang W, Wang X, Zhang J, Huang M, Cai J, Zhou Q, Dai T, Jiang D (2020) Salicylic acid and cold priming induce late-spring freezing tolerance by maintaining cellular redox homeostasis and protecting photosynthetic apparatus in wheat. Plant Growth Regul 90:109–121
Wang F, Niu H, Xin D, Long Y, Wang G, Liu Z, Li G, Zhang F, Qi M, Ye Y et al (2021a) OsIAA18, an Aux/IAA transcription factor gene, is involved in salt and drought tolerance in rice. Front Plant Sci. https://doi.org/10.3389/fpls.2021.738660
Wang X, Li Q, Xie J, Huang M, Cai J, Zhou Q, Dai T, Jiang D (2021b) Abscisic acid and jasmonic acid are involved in drought priming-induced tolerance to drought in wheat. Crop J. https://doi.org/10.1016/j.cj.2020.06.002
Wang X, Liu WC, Zeng XW, Yan S, Qiu YM, Wang JB, Huang X, Yuan HM (2021c) HbSnRK2.6 functions in ABA-regulated cold stress response by promoting HbiCE2 transcriptional activity in *Hevea brasiliensis*. Int J Mol Sci. https://doi.org/10.3390/ijms222312707
Wang Y, Du F, Wang J, Li Y, Zhang Y, Zhao X, Zheng T, Li Z, Xu J, Wang W et al (2021d) Molecular dissection of the gene *osga2ox8* conferring osmotic stress tolerance in rice. Int J Mol Sci 22:9107
Wang Y, Jiang H, Mao Z, Liu W, Jiang S, Xu H, Su M, Zhang J, Wang N, Zhang Z et al (2021e) Ethylene increases the cold tolerance of apple via the *MdERF1B-MdCIbHLH1* regulatory module. Plant J 106:379–393
Wang HL, Yang Q, Tan S, Wang T, Zhang Y, Yang Y, Yin W, Xia X, Guo H, Li Z (2022a) Regulation of cytokinin biosynthesis using PtRD26pro-IPT module improves drought tolerance through PtARR10-PtYUC4/5-mediated reactive oxygen species removal in Populus. J Integr Plant Biol 64:771–786
Wang W, Wang X, Lv Z, Khanzada A, Huang M, Cai J, Zhou Q, Huo Z, Jiang D (2022b) Effects of cold and salicylic acid priming on free proline and sucrose accumulation in winter wheat under freezing stress. J Plant Growth Regul 41:2171–2184
Wang X, Wu Z, Zhou Q, Wang X, Song S, Dong S (2022c) Physiological response of soybean plants to water deficit. Front Plant Sci. https://doi.org/10.3389/fpls.2021.809692
Wang X, Li Z, Shi Y, Liu Z, Zhang X, Gong Z, Yang S (2023a) Strigolactones promote plant freezing tolerance by releasing the WRKY41 -mediated inhibition of CBF / DREB1 expression. EMBO J. https://doi.org/10.15252/embj.2022112999
Wang X, Miao J, Kang W, Shi S (2023b) Exogenous application of salicylic acid improves freezing stress tolerance in alfalfa. Front Plant Sci. https://doi.org/10.3389/FPLS.2023.1091077/FULL
Wang X, Qi X, Zhuang Z, Bian J, Li J, Chen J, Li Z, Peng Y (2024) Interactions between Brassinosteroids and Strigolactones in alleviating salt stress in maize. Int J Mol Sci. https://doi.org/10.3390/ijms251910505
Wani SH, Kumar Sah S (2014) Biotechnology and abiotic stress tolerance in rice. Rice Research: Open Access. https://doi.org/10.4172/jrr.1000e105
Wassie M, Zhang W, Zhang Q, Ji K, Cao L, Chen L (2020) Exogenous salicylic acid ameliorates heat stress-induced damages and improves growth and photosynthetic efficiency in alfalfa (*Medicago sativa* L.). Ecotoxicol Environ Saf. https://doi.org/10.1016/J.ECOENV.2020.110206
Wasternack C, Hause B (2019) The missing link in jasmonic acid biosynthesis. Nat Plants. https://doi.org/10.1038/s41477-019-0492-y
Wasternack C, Song S (2017) Jasmonates: biosynthesis, metabolism, and signaling by proteins activating and repressing transcription. J Exp Bot. https://doi.org/10.1093/jxb/erw443
Wei L, Zhang J, Wang C, Liao W (2020) Recent progress in the knowledge on the alleviating effect of nitric oxide on heavy metal stress in plants. Plant Physiol Biochem 147:161–171
Wei L, Zhang J, Wei S, Hu D, Liu Y, Feng L, Li C, Qi N, Wang C, Liao W (2022) Nitric oxide enhanced salt stress tolerance in tomato seedlings, involving Phytohormone equilibrium and photosynthesis. Int J Mol Sci 23:4539

Wild M, Davière JM, Regnault T, Sakvarelidze-Achard L, Carrera E, Lopez Diaz I, Cayrel A, Dubeaux G, Vert G, Achard P (2016) Tissue-specific regulation of gibberellin signaling fine-tunes *Arabidopsis* Iron-deficiency responses. Dev Cell 37:190–200
Wilkinson S, Clephan AL, Davies WJ (2001) Rapid low temperature-induced stomatal closure occurs in cold-tolerant *Commelina communis* leaves but not in cold-sensitive tobacco leaves, via a mechanism that involves apoplastic calcium but not abscisic acid. Plant Physiol 126:1566–1578
Woeste KE, Ye C, Kieber JJ (1999) Two *Arabidopsis* mutants that overproduce ethylene are affected in the posttranscriptional regulation of 1-Aminocyclopropane-1-carboxylic acid synthase. Plant Physiol 119:521–530
Wu Y-S, Yang C-Y (2019) Ethylene-mediated signaling confers thermotolerance and regulates transcript levels of heat shock factors in rice seedlings under heat stress. Bot Stud 60:23
Wu SH, Wang C, Chen J, Lin BL (1994) Isolation of a cDNA encoding a 70 kDa heat-shock cognate protein expressed in vegetative tissues of *Arabidopsis thaliana*. Plant Mol Biol 25:577–583
Wu XX, Zhu XH, Chen JL, Yang SJ, Ding HD, Zha DS (2013) Nitric oxide alleviates adverse salt-induced effects by improving the photosynthetic performance and increasing the anti-oxidant capacity of eggplant (*Solanum melongena* L.). J Hortic Sci Biotechnol 88:352–360
Wu C, Cui K, Wang W, Li Q, Fahad S, Hu Q, Huang J, Nie L, Mohapatra PK, Peng S (2017) Heat-induced cytokinin transportation and degradation are associated with reduced panicle cytokinin expression and fewer spikelets per panicle in rice. Front Plant Sci. https://doi.org/10.3389/fpls.2017.00371
Wu Y, Liu J, Zhao L, Wu H, Zhu Y, Ahmad I, Zhou G (2024) Abiotic stress responses in crop plants: a multi-scale approach. J Integr Agric. https://doi.org/10.1016/j.jia.2024.09.003
Wybouw B, De Rybel B (2019) Cytokinin—a developing story. Trends Plant Sci 24:177–185
Xiao Y, Liu D, Zhang G, Gao S, Liu L, Xu F, Che R, Wang Y, Tong H, Chu C (2019) Big Grain3, encoding a purine permease, regulates grain size via modulating cytokinin transport in rice. J Integr Plant Biol 61:581–597
Xing J, Xu Y, Tian J, Gianfagna T, Huang B (2009) Suppression of shade- or heat-induced leaf senescence in creeping bentgrass through transformation with the *IPT* gene for cytokinin synthesis. J Am Soc Hortic Sci. https://doi.org/10.21273/jashs.134.6.602
Xu K, Xu X, Fukao T, Canlas P, Maghirang-Rodriguez R, Heuer S, Ismail AM, Bailey-Serres J, Ronald PC, Mackill DJ (2006) Sub1A is an ethylene-response-factor-like gene that confers submergence tolerance to rice. Nature 2006 442:7103 442: 705–708
Xu ZS, Xia LQ, Chen M, Cheng XG, Zhang RY, Li LC, Zhao YX, Lu Y, Ni ZY, Liu L et al (2007) Isolation and molecular characterization of the *Triticum aestivum* L. ethylene-responsive factor 1 (TaERF1) that increases multiple stress tolerance. Plant Mol Biol 65:719–732
Yaaran A, Negin B, Moshelion M (2019) Role of guard-cell ABA in determining steady-state stomatal aperture and prompt vapor-pressure-deficit response. Plant Sci. https://doi.org/10.1016/j.plantsci.2018.12.027
Yan H, Filardo F, Hu X, Zhao X, Fu DH (2016) Cadmium stress alters the redox reaction and hormone balance in oilseed rape (*Brassica napus* L.) leaves. Environ Sci Pollut Res Int 23:3758–3769
Yan Z, Wang J, Wang F, Xie C, Lv B, Yu Z, Dai S, Liu X, Xia G, Tian H et al (2021) MPK3/6-induced degradation of ARR1/10/12 promotes salt tolerance in *Arabidopsis*. EMBO Rep. https://doi.org/10.15252/embr.202152457
Yang JC, Li M, Xie XZ, Han GL, Sui N, Wang BS (2013) Deficiency of phytochrome B alleviates chilling-induced photoinhibition in rice. Am J Bot. https://doi.org/10.3732/ajb.1200574
Yang R, Yang T, Zhang H, Qi Y, Xing Y, Zhang N, Li R, Weeda S, Ren S, Ouyang B et al (2014) Hormone profiling and transcription analysis reveal a major role of ABA in tomato salt tolerance. Plant Physiol Biochem 77:23–34
Yao C, Finlayson SA (2015) Abscisic acid is a general negative regulator of *Arabidopsis* axillary bud growth. Plant Physiol 169:611–626

Ye H, Li L, Yin Y (2011) Recent advances in the regulation of Brassinosteroid signaling and biosynthesis pathways. J Integr Plant Biol 53:455–468

Yin W, Xiao Y, Niu M, Meng W, Li L, Zhang X, Liu D, Zhang G, Qian Y, Sun Z et al (2020) ARGONAUTE2 enhances grain length and salt tolerance by activating *BIG GRAIN3* to modulate Cytokinin distribution in Rice. Plant Cell 32:2292–2306

Yoo SD, Cho YH, Tena G, Xiong Y, Sheen J (2008) Dual control of nuclear EIN3 by bifurcate MAPK cascades in C2H4 signalling. Nature 2008 451:7180 451: 789–795

Yu LJ, Luo YF, Liao B, Xie LJ, Chen L, Xiao S, Li JT, Hu SN, Shu WS (2012) Comparative transcriptome analysis of transporters, phytohormone and lipid metabolism pathways in response to arsenic stress in rice (*Oryza sativa*). New Phytol 195:97–112

Yu J, Cang J, Lu Q, Fan B, Xu Q, Li W, Wang X (2020) ABA enhanced cold tolerance of wheat 'dn1' via increasing ROS scavenging system. Plant Signal Behav. https://doi.org/10.1080/15592324.2020.1780403

Zeng H, Tang Q, Hua X (2010) *Arabidopsis* Brassinosteroid mutants *det2-1* and *bin2-1* display altered salt tolerance. J Plant Growth Regul 29:44–52

Zhang Y, Li X (2019) Salicylic acid: biosynthesis, perception, and contributions to plant immunity. Curr Opin Plant Biol 50:29–36

Zhang L, Li Z, Quan R, Li G, Wang R, Huang R (2011) An AP2 domain-containing gene, *ESE1*, targeted by the ethylene signaling component *EIN3* is important for the salt response in *Arabidopsis*. Plant Physiol 157:854–865

Zhang Q, Li J, Zhang W, Yan S, Wang R, Zhao J, Li Y, Qi Z, Sun Z, Zhu Z (2012) The putative auxin efflux carrier OsPIN3t is involved in the drought stress response and drought tolerance. Plant J 72:805–816

Zhang H, Zhang Q, Zhai H, Li Y, Wang X, Liu Q, He S (2017) Transcript profile analysis reveals important roles of jasmonic acid signalling pathway in the response of sweet potato to salt stress. Sci Rep. https://doi.org/10.1038/srep40819

Zhang H, Duan W, Xie B, Wang B, Hou F, Li A, Dong S, Qin Z, Wang Q, Zhang L (2020) Root yield, antioxidant capacities, and hormone contents in different drought-tolerant sweet potato cultivars treated with ABA under early drought stress. Acta Physiol Plant. https://doi.org/10.1007/S11738-020-03116-X

Zhang Q, Li D, Wang Q, Song X, Wang Y, Yang X, Qin D, Xie T, Yang D (2021) Exogenous salicylic acid improves chilling tolerance in maize seedlings by improving plant growth and physiological characteristics. Agronomy. https://doi.org/10.3390/AGRONOMY11071341

Zhang X, Zhang L, Ma C, Su M, Wang J, Zheng S, Zhang T (2022) Exogenous strigolactones alleviate the photosynthetic inhibition and oxidative damage of cucumber seedlings under salt stress. Sci Hortic. https://doi.org/10.1016/j.scienta.2022.110962

Zhao D, Shen L, Fan B, Yu M, Zheng Y, Lv S, Sheng J (2009a) Ethylene and cold participate in the regulation of *LeCBF1* gene expression in postharvest tomato fruits. FEBS Lett 583:3329–3334

Zhao Y, Hu Y, Dai M, Huang L, Zhou DX (2009b) The *WUSCHEL-related Homeobox* gene *WOX11* is required to activate shoot-borne crown root development in Rice. Plant Cell 21:736–748

Zhao M, Liu W, Xia X, Wang T, Zhang W-H (2014) Cold acclimation-induced freezing tolerance of Medicago truncatula seedlings is negatively regulated by ethylene. Physiol Plant 152:115–129

Zhao Y, Cheng S, Song Y, Huang Y, Zhou S, Liu X, Zhou DX (2015) The interaction between Rice ERF3 and WOX11 promotes crown root development by regulating gene expression involved in Cytokinin signaling. Plant Cell 27:2469–2483

Zhao S, Ma Q, Xu X, Li G, Hao L (2016a) Tomato Jasmonic acid-deficient mutant *spr2* seedling response to cadmium stress. J Plant Growth Regul. https://doi.org/10.1007/s00344-015-9563-0

Zhao Y, Chan Z, Gao J, Xing L, Cao M, Yu C, Hu Y, You J, Shi H, Zhu Y et al (2016b) ABA receptor PYL9 promotes drought resistance and leaf senescence. Proc Natl Acad Sci USA. https://doi.org/10.1073/pnas.1522840113

Zhao J, Yu N, Ju M, Fan B, Zhang Y, Zhu E, Zhang M, Zhang K (2019) ABC transporter *OsABCG18* controls the shootward transport of cytokinins and grain yield in rice. J Exp Bot 70:6277–6291

Zhao S, Zhang Q, Liu M, Zhou H, Ma C, Wang P (2021) Regulation of plant responses to salt stress. Int J Mol Sci. https://doi.org/10.3390/IJMS22094609

Zhou Y-L, Huo S-F, Wang L-T, Meng J-F, Zhang Z-W, Xi Z-M (2018) Exogenous 24-Epibrassinolide alleviates oxidative damage from copper stress in grape (*Vitis vinifera* L.) cuttings. Plant Physiol Biochem 130:555–565

Zhou Y, Zhai H, He S, Zhu H, Gao S, Xing S, Wei Z, Zhao N, Liu Q (2020) The Sweetpotato *BTB-TAZ* protein gene, *IbBT4*, enhances drought tolerance in transgenic *Arabidopsis*. Front Plant Sci. https://doi.org/10.3389/fpls.2020.00877

Zhu B, Choi DW, Fenton R, Close TJ (2000) Expression of the barley dehydrin multigene family and the development of freezing tolerance. Mol Gen Genet. https://doi.org/10.1007/s004380000299

Zhu T, Deng X, Zhou X, Zhu L, Zou L, Li P, Zhang D, Lin H (2016) Ethylene and hydrogen peroxide are involved in brassinosteroid-induced salt tolerance in tomato. Sci Rep. https://doi.org/10.1038/srep35392

Žižková E, Dobrev PI, Muhovski Y, Hošek P, Hoyerová K, Haisel D, Procházková D, Lutts S, Motyka V, Hichri I (2015) Tomato (*Solanum lycopersicum* L.) *SlIPT3* and *SlIPT4* isopentenyltransferases mediate salt stress response in tomato. BMC Plant Biol 15:1–20

Zwack PJ, Rashotte AM (2015) Interactions between cytokinin signalling and abiotic stress responses. J Exp Bot 66:4863–4871

Chapter 10
Nature's Own Engineers: Role of Rhizosphere Microbiome to Confer Resistance to Host Plants Against Drought and Salinity

Debjyoti Bandhu Banerjee, Proharsha Dey, and Anirban Bhar

Abstract The two main abiotic factors that continuously reduce agricultural productivity are salinity and drought. The predicament is further worsened by a rapid population explosion and constantly shifting climate. A number of strategies have been considered to reduce abiotic stress and boost output. Because chemical fertilizers can increase productivity and profit, they have become overly popular; nonetheless, the extended use of these fertilizers damages the ecosystem of the soil. It is necessary to use sustainable farming methods in order to boost crop yields, reduce stress, and protect the environment. The rhizosphere is the biologically most diverse part of the ecosystem of plants that includes a wide variety of beneficial as well as pathogenic microbial communities. Both plant growth-promoting endophytic bacteria (PGPEB) and plant growth-promoting rhizobacteria (PGPR) can be used as bioinoculants and have largely been found to be the most promising candidates for regulating salinity and drought. Several processes, including phytohormone modulation, antioxidant enzyme production, and upregulation of stress-responsive genes, may contribute to the beneficial effects of PGPR-host interaction. In order to counteract various abiotic challenges, such as salinity and drought, rhizosphere engineering has become increasingly popular as an emerging modern technique. The purpose of this chapter is to examine our present knowledge of the cooperative mechanism that underlies PGPR's resistance to salinity and drought stress as well as the potential future applications of rhizosphere engineering for sustainable agriculture. In order to support further research in these areas, the chapter also highlights a few pertinent future research questions.

Keywords Abiotic stress · Drought stress · Microbiome · Rhizosphere · Rhizosphere engineering · Salinity stress

D. B. Banerjee · P. Dey · A. Bhar (✉)
Molecular Plant & Microbiology Laboratory (MPML), Department of Botany (Post Graduate), Ramakrishna Mission Vivekananda Centenary College, Kolkata, India
e-mail: anirbanbhar@rkmvccrahara.ac.in

A. Singh et al. (eds.), *Sustainable Crop Production*, Sustainability Sciences in Asia and Africa, https://doi.org/10.1007/978-981-95-7844-3_10

Editor's Notes

Roots are responsible for providing anchorage, and efficient uptake of nutrient and water; they are also the first organs to encounter soil salinity and drought. Being anchored in soil, roots are also sites of symbiotic and antagonistic interactions with the soil microflora, the microbiome. The chapter by Banerjee et al. explores the roles played by the constituents of rhizobiome, namely, plant growth-promoting rhizobacteria (PGPR) and plant growth-promoting endophytic bacteria (PGPEB), in ameliorating drought and salinity stresses. The PGPRs and PGPEBs not only reduce the demand of chemical fertilizers by facilitating availability of soil nutrition but also trigger stress responses in plants by inducing phytohormone signaling, emission of volatiles, and plant stress responsive genes. Rhizosphere engineering aims to modify the soil microflora to promote plant growth and ameliorate stress. The authors put forward a strong case to investigate microflora, both endophytic and soil-borne, develop microbial consortia, and establish efficient delivery mechanism of nano-biofertilizers to reap the complete benefit of sustainable agriculture.

Actionable Points

1. Use biofertilizers containing PGPR and PGPEB to improve productivity and stress tolerance.
2. Improve soil organic matter by applying natural soil amendments.
3. Enhance soil sustenance by encouraging crop rotation and intercropping with natural microbiome modulating crops.
4. Practice monitoring and regulating soil microbial health.
5. Adopt minimal tillage to maintain soil natural microflora and balanced root-microbe interactions.

10.1 Introduction

Plants being sessile in nature are facing constant environmental challenges with dynamic magnitudes. They are also exposed to a plethora of biotic and abiotic stress factors (Zhu 2016). These stressors alter the plant responses, such as retarded growth rates, crop yield, gene expression, and cellular metabolism (Gull et al. 2019). The continuous climate change exaggerates these abiotic stress situations and imposes damage to the crop plants. Additionally, population outbursts and xenobiotic interventions deteriorate the above abiotic stress conditions. According to the reports of the United Nations, the world population could achieve 9.7 billion by 2050 (https://www.un.org). Hence, the increasing global population, rapid climate change, and associated abiotic stress deserve more attention to tackle the situation in the future (Nowicka et al. 2018). Among many other abiotic stresses, drought and salinity are directly connected to climate change and affect global agriculture viciously. The climate change-induced water stress has challenged several drought--sensitive crop yields (Nasir and Toth 2022). On the other hand, it has been found that climate change and human interventions readily increase soil salinity jeopardizing the agroecosystem dramatically (Ondrasek et al. 2022). The adverse effect of drought negatively imparts stem proliferation, water and nutrient uptake, leaf water potential, and flowering, which ultimately kills the plant and reduce yield. Then again, salt stress affects seed germination, seedling establishment, and ion toxicity (Khan et al. 2021b). On the perception of different abiotic stress, plants also activate inherent defense signaling cascade, e.g., stimulation of stress hormones, e.g., abscisic acid (ABA), ethylene (ET), and jasmonic acid (JA), accumulation of reactive oxygen species (ROS), and modulation of several stress responsive genes to impart tolerance, etc. (Rejeb et al. 2014). Several transcription factors (*MYB*, *WRKY*, *bZIP*, *bHLH*, *AP2/ ERF*, *NAC*, etc.) act as the master regulator of different abiotic stress responses (Qian et al. 2021; Wei et al. 2021). Besides, different epigenetic modification such as DNA methylation and histone modification also modulate stress-responsive genes in plants (Boyko and Kovalchuk 2008).

Scientists are constantly trying to ameliorate the abiotic stress factors to improve crop yield. Genomic modifications are one of the profusely utilized techniques to develop abiotic stress-tolerant plants harboring different candidate genes. On the contrary, chemical and synthetic fertilizers are also profoundly utilized to improve crop yield during the past couple of decades. It has been found that phosphorus amelioration in the soil can improve calcium to phosphorus ratio and positively impact soil health along with crop rotation (Polovyy et al. 2023). On the contrary, the exponential use of chemical fertilizers to promote crop yield and food security has some adverse effects on the ecosystem also, e.g., water contamination, soil deterioration, greenhouse gas emission, etc. (Wang et al. 2018). Chemical fertilizers also have detrimental effects on soil microbial community. Instead, the combination of chemical fertilizers with organic manures is sometimes found to restore soil fertility. Chemical fertilizers, Nitrogen, Phosphorus, Potassium (NPK) when utilized along with livestock manure and wheat straw, nitrogen cycle genes (*nifH*, archaeal and bacterial *amoA*, *nirS*, *nirK*, and *nosZ*) were more strongly induced in agricultural fields (Sun et al. 2015). On the other hand, it was reported that the use of pig manure-based organic-inorganic compound fertilizer promotes soil nutrient availability, microbial biomass, and crop yield in rice-wheat cropping systems (Zhao et al. 2016). Under this scenario, it is much needed to come up with sustainable agricultural developmental strategies that along with the stress tolerance and increase in crop yield also overcome the negative effects of chemical fertilizer.

The rhizosphere is the active ecological niche of diverse microbial communities, influenced by plant root exudates (Dubey and Sharma 2021; Vyas et al. 2022). Rhizobacteria are associated with plant roots and consist of both beneficial and pathogenic bacteria which establishes a dynamic collaborative interface (Khan et al. 2021b). The rhizosphere microbial communities have a tremendous effect on plant growth, development, as well as stress response (Das et al. 2023). Plant growth-promoting rhizobacteria (PGPR) are linked with plant roots and promote plant growth and boost immunity. It also stimulates nutrient uptake from soil, thus lowering the demand for endogenous fertilizer (Yang et al. 2009). PGPR also promotes antagonistic activity to plant pathogens as well as increases nutrient availability (Pérez-Montaño et al. 2014). The rhizosphere microbial community is also capable of reducing abiotic stress by varied mechanisms of action (Caddell et al. 2019). Firmicutes, Bacteroidetes, and Actinobacteria are the major representative phyla among PGPR, and *Acetobacter*, *Arthrobacter*, *Azotobacter*, *Bacillus*, *Exiguobacterium*, *Herbaspirillum*, *Pseudomonas*, etc. are the major PGPR genera (Kour et al. 2019). Furthermore, a diversified microbial community is also found inside the plant tissue known as plant endophytes. Along with the rhizosphere microbial community, these endophytes also have a beneficial role in the regulation of plant growth and development and maintenance of the physiology of plants such as nitrogen fixation, phytohormone synthesis, and protection from various stresses. There is also a mutual correlation between the endophytic microbe and rhizospheric microbes (Chaudhary et al. 2022). Hence, the greater productivity and stress tolerance could be achieved by modulation of the rhizosphere microbial community of different crop plants. The increase in fitness of the plants under abiotic stress by employing diverse rhizosphere microbiome is an emergent field of agribiotechnology. Before achieving successful rhizosphere engineering, complete rhizosphere microbial map of different crop plants should be deciphered under varied environmental conditions. In this view, the present chapter discusses the role and mechanism of action of different rhizosphere microorganisms in combating drought and salinity stress. The comprehensive belowground interaction was elaborated, and the emerging idea of rhizosphere engineering has also been highlighted. The chapter has also emphasized some relevant future research directions to comply with abiotic stress-tolerant sustainable agriculture.

10.2 Rhizosphere Engineering (RE): An Emerging Approach to Manage Plant Abiotic Stress Response (The Challenges over Possibilities)

10.2.1 Rhizosphere

The rhizosphere is a complex and unified network of plant roots and a plethora of microbial communities such as bacteria, fungi, archaea, etc. It can be divided into three zones: (i) inner most, endo-rhizosphere constituting root cortex and endodermis, (ii) middle zone, rhizoplane, and (iii) outermost, ecto-rhizosphere. Amino acids, carbohydrates, organic acids, and secondary metabolites secreted by plants known as rhizodeposits control the growth of microorganisms surrounding the roots (Hakim et al. 2021). The rhizodeposits are principally influenced by the plant root secreted exudates with rich in carbon and nitrogenous compounds (Tian et al. 2020). The plants invest a large amount of their accumulated carbon through photosynthesis in the maintenance of their rhizosphere exudates. Along with the beneficial exudates attracting beneficial microbes and plant-derived siderophores to improve mineral acquisition, rhizodeposits also include allelochemicals responsible to counteract unwanted microbes (Patra et al. 2021). Most of these natural rhizosphere interactions are beneficial for the plants to withstand different adverse environmental conditions (Ahkami et al. 2017) and promote plant growth and development (Dries et al. 2021). The recruitment of the functional microbiota in the root surroundings is predominantly dependent on the types of plants they inhabit. The inter--organismal signaling between plant root exudates and the microbiome shapes the ecological balance in the rhizosphere microenvironment (Venturi and Keel 2016). This communication is also influenced by the neighboring plant communities (Lyu and Smith 2022). The root exudates change steadily with the developmental stages of the plants and that play a crucial role in a paradigm shift of the rhizosphere microbiome community over time (Park et al. 2023). Modification and/or introduction of new PGPR population in the rhizosphere can recover soil health and promote growth as they increase nutrient uptake, suppress pathogens, and improve fitness (Kumar and Dubey 2020). Inoculation of artificially cultured natural PGPR in the targeted agricultural fields is the most common practice, e.g., *Rhizobium*, *Azorhizobium*, *Allorhizobium*, and *Ochrobactrum* are predominantly used as inocula for legumes (Quiza et al. 2015). Free-living PGPRs like *Micrococcus*, *Azotobacter*, *Arthrobacter*, *Bacillus*, and even some actinomycetes show beneficial effects and can be used on some specific plants (Adesemoye and Egamberdieva 2013; Atsede and Fassil 2018). Moreover, in spite of using a single bacterial strain, a combination of different strains can provide better and enduring effects even in a changing environmental condition. It has been reported that coculture containing a biocontrol agent and a growth promoter strain provides excellent results in RE (Singh et al. 2024a; Checcucci et al. 2022). Nonetheless, employing natural PGPR as an inoculum may have certain drawbacks. Most natural PGPRs are confined to specific plant genera; hence, applying them on plants from different genera may produce unsatisfactory results. Newly inoculated bacteria may encounter competition with an existing dominating microbial population, putting their survival at risk (Haskett et al. 2021). At the time of working with two bacterial strains in coculture, there are still a number of ecological factors that need to be carefully taken into account. The two strains of a simple coculture can interact in six different ways: (i) commensalism; (ii) amensalism; (iii) competition; (iv) predation; (v) collaboration; and (vi) no interaction. Taking into account every possible permutation and combination, the number of possible microbial interactions increases exponentially with the increase of number of strains in coculture mix (Großkopf and Soyer 2014). Hence, it demands urgent solving and disentangling of these disseminated signals to utilize data for holistic agricultural applications. Plant biotechnology is booming in this post-omics era. Bioinformatics and gene editing technology are at the forefront of development. It is now possible to achieve desired phenotypes by changing the genotype, and this can be used to combat a wide range of biotic and abiotic stress tolerance. This state-of-the-art technology can also be applied to modify the chemistry of the soil or add amendments to it. As discussed earlier, plants release rhizodeposition like flavonoids, iso-flavonoids, amino acids, mucilage, organic acids, polysaccharides, and proteins that help to shape the rhizosphere (O'Connell et al. 1996). Additionally, researchers have identified several genes that produce particular rhizosphere exudates, and gene editing technology allows for the modification or manipulation of rhizodeposition (Ryan et al. 2009). For a better and more effective result, a cocktail of exudates can be displayed rather than changing or promoting a single distinct exudate (Herz et al. 2018). Conversely, it has been discovered that induced genes may not be able to express correctly or efficiently in some situations due to posttranscriptional modification or metabolic feedback, or they may only express a very small quantity of phenotype that is insufficient to have an impact on the rhizosphere (Ryan et al. 2009). However, once being released, induced exudate may even quickly deteriorate. Additionally, edited genes can affect other vital physiological processes unrelated to RE because of the pleiotropic effect (Hakim et al. 2021).

10.2.2 Rhizosphere Engineering

The term "rhizosphere engineering" although was first suggested long back by O'Connell (O'Connell et al. 1996); the effective and amenable application of the same in agricultural sustainability is still prevaricating. This dynamic practice with very limited work has massive potential that can lead modern agriculture to a sustainable system in changing climatic conditions and stress emergences. RE can be defined as the targeted modifications of microbes, soil, and plants and, additionally, management to alter rhizospheric process for a particular reason to foster plant growth and development, carbon sequestration, etc. The components of rhizosphere, i.e., soil, microbes, and plants, can be manipulated by chemical, physical, biological, genomic, and agronomic approaches (Dessaux et al. 2016; Wang and Kuzyakov 2024). There are three key mechanisms that have been employed to engineer the rhizosphere: (a) amendments of soil, (b) microbe-mediated alterations by utilizing upstream or downstream techniques to modify microbe genetically, and (c) modifications mediated by plants involving transgenics, genetic engineering, and gene editing of plants (Solanki et al. 2024). Soil amendment is considered as the simplest method for rhizospheric engineering. A wide variety of soil amendments are used to boost plant production. Compost addition and utilization of biochar are the two most dominant methods; besides microalgae, sewage sludge, fly ash, zeolites, and animal wastes have long been used as general soil amendments for RE (Sharma et al. 2021; Jakkula and Wani 2018). Pyrolysis of organic biomass yields biochar, a carbon-rich compound that has drawn a significant interest in mitigating climate change, environmental management, and sustainable agriculture. It promotes soil's microbial activity, nutrient retention, and water holding capacity (Waheed et al. 2025). Indigenous microbial strain can be manipulated through genetic engineering approaches, for example, CRISPR/Cas9, RNAi interference, transfer of beneficial bacteria from healthy to target site, and through inoculation of SynComs (Synthetic Microbial Communities) (Joshi et al. 2025). An efficient PGPR strain *Pseudomonas* sp. UW4 can mitigate biotic as well as abiotic stress was genetically engineered to produce more IAA. In order to transform that strain second copy of the target, IAA biosynthesis gene, i.e., *ami*, *phe*, *nit*, *nthA*, and *nthB*, was introduced. Interestingly, it was reported that transformed *Pseudomonas* sp. UW4 strain can produce 2–3 times higher IAA (Kaur et al. 2021). Similarly, a genetically modified *Rhizobium etli*, with overexpressed trehalose-6-phosphate synthase gene, induces nitrogenase activity and nodule formation in *Phaseolus vulgaris* upon inoculation (Solanki et al. 2024). Microbiome-mediated smart agriculture system (MiMSAS) is a cutting-edge method for modifying rhizosphere microbiome to combat different stressors. Development of synthetic microbial community or SynCom, constructed through bottom-up approach, also plays a significant role in mitigating different stress. The microbe-microbe interaction and host-microbe interaction are mediated by a small number of core microbiota. Therefore, it is important to identify keystone taxa at the time of constructing SynCom. A top-down approach to building effective SynComs has been made possible by the rapid developments in culture-independent analysis, gene editing technologies, next-generation sequencing, genomics, and bioinformatics tools (Bano et al. 2021). The plant microbiome can be engineered by two methods: reinoculation of exogenously enriched native beneficial microorganisms and direct inoculation of exogenous beneficial microorganisms. Several studies demonstrated that inoculation of consortia of PGPR showed positive effects in combating biotic and abiotic stresses. PGPR acts as a biocontrol agent by synthesizing ACC deaminase (1-aminocyclopropane-1- carboxylate deaminase), lytic enzymes, antibiotics, and siderophores and through induction of induced systemic resistance (ISR) (Hakim et al. 2021; Lau et al. 2022). Hence, modulation of the rhizosphere community and soil amendment has immense possibilities in RE with several missing questions to be addressed in the near future.

10.3 An Overview of Plants' Principal Responses to Drought and Salinity

Plants continuously encounter different abiotic stresses, e.g., extreme temperature, chilling, drought, salinity, etc. that hinder plant growth and development and curtail crop productivity (Zhang et al. 2022). Climate change and xenobiotic alterations due to global population outbursts further deteriorate the abiotic stress scenarios of plants. The lowering of soil water profile and increase in the salinity level is one of the major concerns across the globe (Grillakis 2019; Corwin 2021). The mathematical model-based analysis of temporal soil water regimes demonstrated that different vegetation patterns are affected differently in terms of climate change-associated global drought (Farkas et al. 2014). Agricultural land is affected most brutally due to these abrupt but continuous abiotic fraternities. The global loss of crops, drought, and associated salinity stress have been considered to be the "costliest abiotic stress" globally (Grillakis 2019; Dos Santos et al. 2022). Low relative water content, leaf osmotic potential, water potential alterations, and nutritional imbalance are some major physiological changes under drought conditions. In addition, retarded root, shoot lengths, accumulation of glycine betaine, and proline are some crucial morphological and biochemical symptoms, respectively (Yadav et al. 2020). Drought stress also negatively impacts vegetative growth, early seedling growth, and root-shoot dry weight in several crop plants, such as pea (*Pisum sativum* L.), alfalfa (*Medicago sativa* L.), and rice (*Oryza sativa* L.) (Fahad et al. 2017). Additionally, drought and salinity stress are also known to damage cellular macromolecules, e.g., DNA damage, degradation of essential protein, lipid membrane, etc. due to loss of cellular redox homeostasis (Bano et al. 2022). Water stress also leads to the modulation of hormonal cross talk and activation of intracellular channel proteins. Hyperosmotic signal in response to drought and salinity leads to the accumulation of many stress-responsive phytohormones. ABA was found to be the most prominent contributor not only by classical stomatal closure but also controls stress-induced gene expression and osmolyte synthesis, e.g., prolines, trehalose, polyhydric sugar alcohols (mannitol, sorbitol, pinitol, etc.), regulation of reactive oxygen species (ROS), etc. (Rajasheker et al. 2019). Recently, the interesting role of jasmonic acid (JA) in modulating abiotic stress tolerance particularly drought and salinity emerged with obvious interaction with ABA pathway (Riemann et al. 2015). ABA and associated hormonal cross talk are known to induce stress-responsive pathways by activating *ABA-RESPONSIVE ELEMENT BINDING PROTEINS* (*AREBs*)/ *ABRE-BINDING FACTORS* (*ABF*s). *SNF1-related protein kinases 2* (*SnRK2s*) phosphorylates the transcription factors for their activation. In the higher plants, ABA-unresponsive subclass I SnRK2s were found to activate drought resistance in ABA-independent manner (Soma et al. 2021). Similarly, other transcription factors, e.g., *NAC*, *dehydration-responsive element-binding* (*DREB*), *WRKY*, etc., are also found to be induced during drought and salinity stress. The *NAC* (*NAM* (*No Apical Meristem*), *ATAF1* (*Arabidopsis thaliana Activation Factor 1*), and *CUC2* (*Cup-shaped Cotyledon*)) is the largest group of transcription factors, and comparative analysis in pearl millet (*Pennisetum glaucum*) revealed significant induction of *NAC* TFs in response to drought and salinity stress (Dudhate et al. 2021). Similarly, *VvNAC17* was found to enhance sensitivity to ABA-mediated tolerance against drought, salinity, and freezing stress combinations (Ju et al. 2020). *CaNAC46*, from *Capsicum annuum*, was reported to reduce drought and salinity when expressed in *Arabidopsis thaliana* (Qian et al. 2021; Ma et al. 2021). Similar responses of *NAC* TFs were well documented in a wide range of plants during abiotic stress encounters (Hoang et al. 2020; Hou et al. 2020; Han et al. 2020; Li et al. 2021). *DREB* is interestingly known to interact with *DRE/CRT cis*-element during the regulation of drought response in an ABA-independent pathway (Singh and Chandra 2021). Recently, a novel inducible promoter of *DREB1G* from *Phoenix dactylifera* to the *AtRD29* gene showed enhanced tolerance to drought and salinity (Kodackattumannil et al. 2023). This could be utilized for the strategic expression of different abiotic stress-responsive genes to develop multiple abiotic stress-tolerant plants. Along with *DREB*, recently many other TFs (*AP2/ERF*, *bHLH*, *bZIP*, *GATA*, *HD-Zip*, *Homeo-box*, *MADS-box*, *MYB*, *Tri-helix*, *WHIRLY*, *WOX*, *YABBY*, *zinc finger*, etc.) are also known to modulate drought and salinity stress (Hussain et al. 2021). The rare involvement of *Dof* Tfs in drought and salinity response has been documented recently in soybean. The *GmDof41* was found to be directly interacting with promoter of *GmDREB2A* and regulates their expression (Wei et al. 2023).

Recently, it has been observed that the rhizosphere microbiome plays a pivotal role in ameliorating abiotic stress, e.g., drought and associated salinity. Hence, the application of beneficial microorganisms in plant abiotic stress amelioration comes out as the most promising approach in the future for sustainable agriculture. These beneficial microbes can alter the stress condition by different mechanisms such as upregulation of the stress-responsive and antioxidant gene, production of phytohormones, solubilization of phosphate, etc. Drought stress can significantly alter bacterial assemblages and microbial communities in rhizosphere (Wu et al. 2024). A significant increase in Actinobacteria was observed with a dramatic decrease in Firmicutes during drought stress in rice. Additionally, ABA was also released prominently in the root exudates as compared to JA and L-cystine concentrations (Li et al. 2023a). These findings laid an interesting idea of ABA-dependent alteration of root microbiome under drought stress. Similarly, *Enterobacter* sp. was found to interact with 1-aminocyclopropane-1-carboxylic acid (ACC) deaminase-producing rhizosphere bacterial population to induce salinity stress in *Cajanus cajan* (Anand et al. 2023a). Furthermore, plant growth-promoting endophytic bacteria (PGPEB) also help plants to withstand stress conditions by enhancing systemic resistance and plant immunity to abiotic stresses (Lata et al. 2018). It has been found that desert--adapted fungal endophytes *Periconia* sp. and *Neocamarosporium* sp. were capable of inducing drought and salinity resistance in crops by induction of antioxidative defense and proline synthesis (Hosseyni Moghaddam et al. 2021). The involvement of ROS signaling in endophyte-mediated abiotic stress tolerance is also evident in other studies (Verma et al. 2021). More intriguing research in elucidating the mechanism of cross talk between plants, rhizosphere microbes, and soil metabolites is required for further insight into this field (Fig. 10.1).

10.4 Drought Stress Tolerance by Rhizosphere Microbial Modulation: The Mechanistic Relationship

10.4.1 The Mechanism of Drought Stress Amelioration by Plant Growth-Promoting Rhizobacteria (PGPR)

Drought is one of the most significant abiotic stressors that impede plant growth and development. Plant breeding, selection, and genetic engineering techniques to increase drought tolerance are time-consuming and expensive. Plants employ a wide variety of adaptive strategies, i.e., biochemical, physiological, molecular, metabolic, and cellular mechanisms to combat with the negative consequence of drought stress (Chieb and Gachomo 2023). Interestingly, plant growth-promoting rhizobacteria (PGPR) can act as a promising agent to alleviate the detrimental effects of drought stress tolerance. The mechanisms through which PGPR mitigates drought

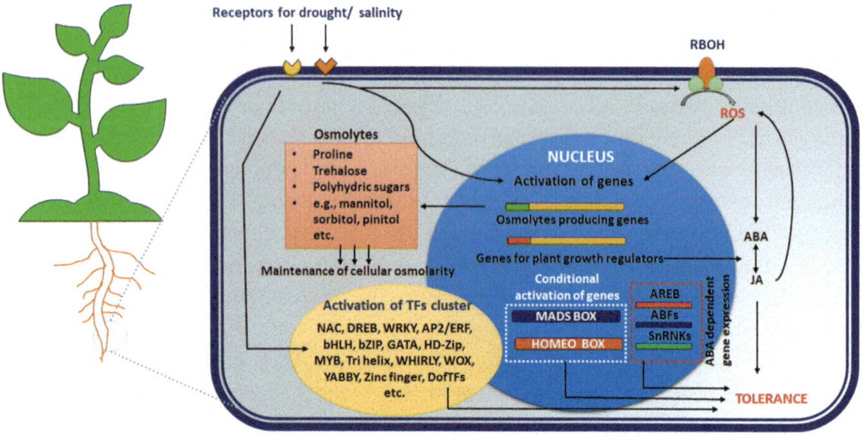

Fig. 10.1 The schematic representation of principal molecular responses of plants associated with drought and salinity stress. *NAC* (NAM, ATAF1/2, and CUC2) domain transcription factors, *DREB* dehydration-responsive element binding, *AP2/ ERF* APETALA2/Ethylene Responsive Factor, *bHLH* basic helix-loop-helix, *bZIP* basic leucine zipper, *HD Zip* homeodomain-leucine zipper, *MYB* myeloblastosis viral oncogene homolog, *WHIRLY* whirly (WHY), *WOX* WUSCHEL-related homeobox, *Dof* DNA binding with one finger, *MADS box* MINICHROMOSOME MAINTENANCE 1, *AGAMOUS* DEFICIENS and SERUM RESPONSE FACTOR, *AREB* ABA-responsive element--binding protein, *ABF* abscisic acid-responsive transcription factor, *SnRK* SNF1-related protein kinase, *ABA* abscisic acid, *JA* jasmonic acid

stress are known as rhizobacterial-induced drought endurance and resilience (RIDER) (Kaushal and Wani 2016). PGPRs can mitigate drought stress by the following methods: (i) synthesis of phytohormones like indole-3 acetic acid (IAA), gibberellic acid, cytokinin, and abscisic acid (ABA), (ii) the activity of ACC deaminase to reduce ethylene content in roots, (iii) bacterial compounds that induce systemic tolerance, and (iv) bacterial exopolysaccharides (Vurukonda et al. 2016). Moreover, it can also alleviate the detrimental effect of drought stress by modifying root morphology, altering stress-responsive genes, siderophore, osmolytes, and volatile organic compound (VOC) production (Ahmad et al. 2022). A detailed understanding of the diverse and beneficial rhizospheric microbial communities, as well as the mechanisms through which PGPR mediates drought tolerance, is essential for advancing toward a sustainable future in food security.

10.4.1.1 Modulation of Antioxidative System and Hormonal Cross Talk

Bioinoculants, including *Klebsiella* sp. IG3 and *Enterobacter ludwigii* IG 10, were introduced into a wheat plant in an experiment, and the results showed that the plant was significantly more resistant to drought stress. This was evidenced by a significant decrease in malondialdehyde content and H_2O_2 when compared to the uninoculated stressed plants (Gontia-Mishra et al. 2016). Studies have revealed that *Bacillus* sp. is the most potent drought ameliorating rhizobacteria by promoting growth and crop yield. In maize, five strains of *Bacillus* sp., isolated from a dry Caribbean region with phytates hydrolyzing, phosphate solubilizing, ACC deaminase, and indole-like compound synthesizing activity, impart drought tolerance (Moreno-Galván et al. 2020). In a separate study, *Bacillus megaterium* (MU2) and *Bacillus licheniformis* (MU8) were found to alleviate drought stress in wheat plants. These bacteria were reported to produce ACC deaminase and IAA along with antagonistic activities against plant pathogens. Besides, some novel peptides were also reported from drought-induced *Bacillus megaterium* (MU2) (Rashid et al. 2022). The application of *Bacillus subtilis* Rhizo SF 48 was reported to enhance drought resistance in tomato plants. This bacterial strain produces ACC deaminase and significantly increases proline, superoxide dismutase (SOD), and ascorbate peroxidase (APX) activity, thus reducing membrane damage in drought-affected tomato plants (Gowtham et al. 2020). A similar type of effect on the antioxidative system was observed in potatoes by the application of *Bacillus subtilis* HAS31 as PGPR (Batool et al. 2020). In *Mentha piperita*, *Pseudomonas fluorescens* WCS417 r along with *Bacillus amyloliquefaciens* GB03 reduces drought stress by increasing SOD and peroxidase (PX) along with phenolics content (Chiappero et al. 2019). Besides redox alterations, ACC deaminase activity is another potent mechanism of microbe-mediated drought stress amelioration. Enhancement of ethylene production during drought stress negatively affects the plants. ACC deaminase, a pyridoxal phosphate-dependent enzyme, found in several bacteria and fungi that disintegrate ACC the precursor of ethylene to ammonia and α-ketobutyrate, thus helps the plants to withstand drought (Singh et al. 2015). The PGPR containing ACC deaminase-producing *Pseudomonas aeruginosa* (DtM10), *Enterobacter cloacae* (DtM16), *Achromobacter xylosoxidans* (DtM29), and *Leclercia adecarboxylata* (DtM34) inoculated in drought induced maize plants remarkably modulates root-shoot length, photosynthetic pigment formation, and nutrient concentration in maize seedlings (Danish et al. 2020). The ACC deaminase-producing PGPRs containing genera, *Pseudomonas*, *Enterobacter*, and *Achromobacter*, were reported to enhance growth promotion under drought stress in grapevine (*Vitis vinifera* L.) (Duan et al. 2021). On the other hand, eight ACC deaminase-producing bacteria belonging to three genera (*Pseudomonas*, *Bacillus*, and *Serratia*) were significantly reducing drought tolerance in jujube (*Ziziphus jujuba*) by inducing antioxidative defense, hormonal induction, and growth and physiological changes (Zhang et al. 2020).

10.4.1.2 Modulation by Releasing Volatile Organic Compounds (VOCs)

Volatile organic compounds (VOCs) released from plant leaves and other parts are known for their ability to intercommunication between plants. The application of VOCs in restricting herbivores is common, but recently, their application in the improvement of different stress also attracted more studies on it (Brilli et al. 2019). VOCs that are secreted by PGPR are characterized by lipophilic, low molecular weight compounds with high vapor pressure and low boiling point (Mhlongo et al.

2022). VOCs of rhizospheric bacteria possess a crucial role in plant-bacteria interaction. It plays a significant role in plant growth promotion and development (Laller et al. 2023). Acetoin, 2,3-butanediol, and pentyl furan are some reported VOCs that promote plant growth. *Bacillus amyloliquefaciens IN937a*, *Bacillus subtilis* GB 03, and *Enterobacter cloacae JM22* are reported to produce plant growth-promoting VOCs like acetoin, 2,3-butanediol, and pentyl furan. Despite the fact that PGPR VOCs devoid of any plant growth hormones and siderophores, it mediates control of plant endogenous auxin homeostasis and root iron uptake. Moreover, it also plays a significant role in triggering induced systemic resistance (ISR) in response to several biotic and abiotic stresses (Liu and Zhang 2015; Ahmad et al. 2022; Chieb and Gachomo 2023). It was evident from transcriptional and histochemical analysis that *Bacillus subtilis* GB 03 regulates auxin homeostasis in *Arabidopsis*. Interestingly, gene expression for auxin biosynthesis upregulated in the aerial regions of the GB03 exposed plants (Zhang et al. 2007).

An interesting investigation demonstrated that priming of wheat with rhizospheric bacteria from harsh environmental conditions improved drought tolerance evidenced by reduced VOC and increased photosynthetic efficiency (Timmusk et al. 2014). These VOCs can be secreted both by plants and microbes. The plant-released VOCs can be categorized into ethylene, methanol, terpenes, methyl jasmonate, methyl salicylate, etc. (Chieb and Gachomo 2023). The PGPRs are also capable of producing some highly diffusible VOCs (alcohol, alkenes, ketones, etc.) required for their own sustenance and communication with plants (Peñuelas et al. 2014). During the rhizosphere interaction, these volatile compounds are released as indicators of stress level. The VOCs released by the *Pseudomonas pseudoalcaligenes* were observed to reduce drought stress in maize by reducing lipid peroxidation and oxidative damage. The GC-MS analysis revealed the presence of dimethyl disulfide, 2,3-butanediol, and 2-pentylfuran within VOCs, to be the reason behind the effect (Yasmin et al. 2021). In a separate study with secreted VOCs, metabolic profiling of four PGPR strains *Lysinibacillus sphaericus* (T19), *Pseudomonas koreensis* (N19), *Pseudomonas fluorescens* (N04), and *Paenibacillus alvei* (T22) is done through solid-phase micro-extraction gas chromatography-mass spectrometry (SPME-GC-MS). It revealed 121 identified metabolites; majority have the capacity to influence host metabolism includes alkanes, alkenes, terpenoids, pyrazines, furans, sulfides, ketones, aldehydes, and derivatives of salicylic acids (Mhlongo et al. 2022). VOCs 2R, 3R-butanediol is synthesized during colonization of *Pseudomonas chlororaphis* 06 in plant roots that helps to prevent water loss by closing stomata; furthermore, it also increases free salicylic acid production under drought stress in *Arabidopsis thaliana* (Ali and Khan 2021). Recent studies with plant growth-promoting drought-tolerant bacteria *Bacillus altitudinis* FD48 VOC stimulate induced systemic tolerance in rice plants under drought stress. Surprisingly, among 40 bioactive compounds identified through GC-MS-ATD analysis, dimethyl disulfide,1-hexanol, 2,3-butanediol, benzene, butanoic acid, pentadecane, acetic acid, etc. plant growth-promoting compounds are more apparent (Narayanasamy and Uthandi 2024). The coordination of VOCs and antioxidative systems was also observed in rice (Rafedzi et al. 2023). Recently, the screening of VOCs from *Paenibacillus peoriae* GXUN15128 and their effect on growth promotion activity of *Arabidopsis thaliana* revealed many novel compounds (Wang et al. 2023). Hence, VOCs are an important regulatory mechanism for controlling drought by altering PGPRs in the rhizosphere region (Tanveer et al. 2023) (Fig. 10.2).

10.4.2 The Mechanism of Drought Stress Tolerance by Plant Growth-Promoting Endophytic Bacteria (PGPEB)

Plant endophytic bacteria are beneficial groups of microbes residing within the plant tissue and secrete various phytochemicals to aid tolerance in adverse environmental conditions. The entry and transmission of these endophytes within the plant

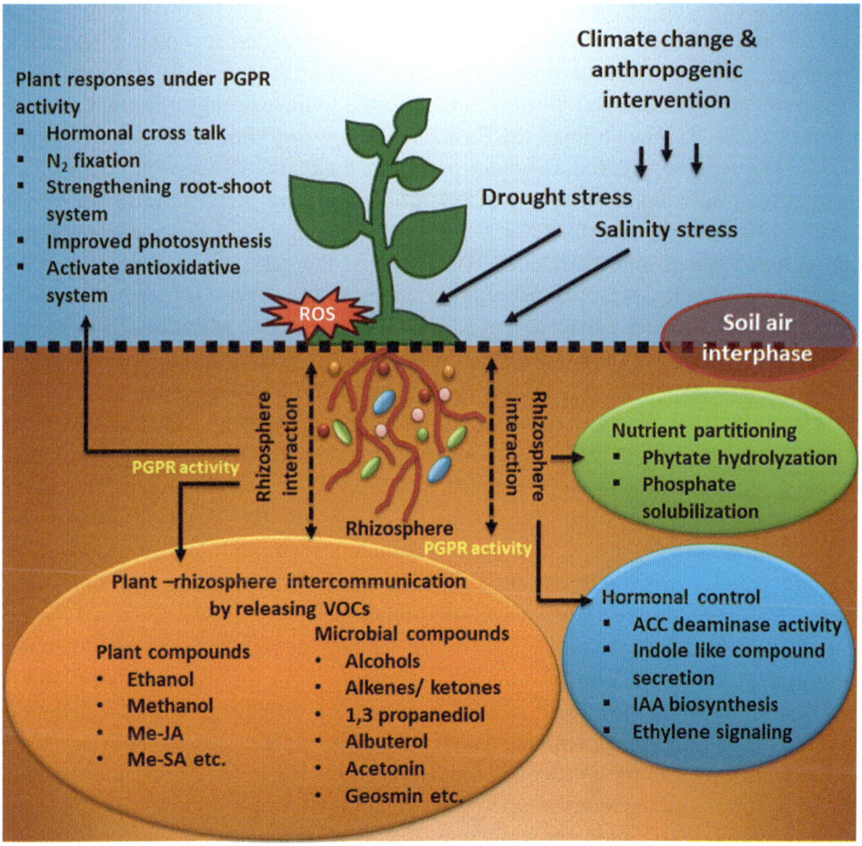

Fig. 10.2 The diagrammatic representation of plant root and rhizosphere microbial communities ameliorating drought and salinity stress. The drought- and salt stress-induced conditions can be restored by active PGPRs in the plant rhizosphere. The rhizospheric interaction includes nutrition partitioning, hormonal control, and plant-rhizosphere intercommunication by releasing volatile organic compounds (VOCs). Rhizosphere microbes were also capable of reducing oxidative stress by activating the antioxidative enzyme system, root-shoot strength, and nitrogen fixation

body follow several consecutive complex steps which starts with attachment and secretion of certain metabolites (Pathak et al. 2022). These endophytes consist of several plant growth-promoting traits like nitrogen metabolism and production of siderophore, different phytohormones, nitrogen fixation, mineral uptake phosphate solubilization, etc. (Ullah et al. 2019). Endophytic diazotrophic bacteria like *Gluconacetobacter diazotrophicus* Pal5 were reported to ameliorate drought stress tolerance in rice and sugarcane plants under low nitrogen environments. Drought and low nitrogen are the two most limiting factors for maize plants worldwide. Inoculation of *Gluconacetobacter diazotrophicus* Pal5 showed significant root and shoot growth by induction of auxin production and atmospheric N_2 fixation under moderate drought stress and low nitrogen condition. As the auxin is responsible for cell elongation and lateral root formation, induction of endogenous auxin concentration through the diazotroph promotes drought tolerance in maize plants under moderate conditions (Tufail et al. 2021). Recently, 26 endophytic bacteria and fungi were isolated from the medicinal plant *Endostemon obtusifolius* showing flavonoid, ammonia, and siderophore production capabilities. Surprisingly, *Paenibacillus polymyxa* and *Fusarium oxysporum* were reported to have drought-tolerant activity (Ogbe et al. 2023). The strain of the *Fusarium oxysporum* should be studied further, as this fungus is considered as potent pathogen for plants (Bhar et al. 2018). The drought-tolerant endophytic fungus, *Penicillium citrinum*, was reported from wheat capable to solubilize phosphate and zinc and produce extracellular enzymes and siderophores (Kaur and Saxena 2023). Recently, the genome sequence of endophyte, *Sphingomonas paucimobilis* ZJSH1, reported from roots of *Dendrobium officinale* exhibited growth enhancement and drought tolerance capabilities by inducing phytohormones and antioxidative defense system (Li et al. 2023b). The drought--tolerant endophytes (*Alcaligenes faecalis* subsp. *phenolicus*, *Pseudomonas moorei*, *Bacillus siamensis*, *Bacillus cereus*, and *Achromobacter deleyi* were isolated from grain *Amaranths* which showed increased nitrogen fixing, phosphate solubilizing, and ACC deaminase activity (Bhagyashree et al. 2022). The endophytic inoculum combination *Rossellomorea aquimaris* strain 3.13 and *Bacillus velezensis* strain 5.18 with the coinoculum were proved to induce growth and inulin production in *Helianthus tuberosus* L. (Jerusalem artichoke) under drought stress (Boonmahome et al. 2023).The drought-induced seed germination retardation was restored by the action of seed endophyte isolates in maize by downregulating oxidative damage (Siddique et al. 2022). The meta-analysis encompassing the data involving endophytic communities in growth improvement under drought stress revealed that endophytic bacteria have a higher propensity to modulate stress conditions than endophytic fungal communities (Tufail et al. 2022). More research employing a high-throughput metagenomics approach is necessary to put the missing piece in place to complete the interaction puzzle.

10.4.3 Rhizosphere Engineering for Drought Stress Reversal: PGPR and Beyond

Among the abiotic stresses, drought is one of the biggest hindrances that limit crop production worldwide as discussed earlier. A statistical analysis showed that 41.11% of total yield loss is caused by drought (Yashavanthakumar et al. 2021). It is estimated that the average world temperature will increase 1.8 to 4° C by 2100 as a result of global warming (Ozturk et al. 2021). As every crop is susceptible to drought stress, yield loss is expected to be exorbitantly high in the near future. The PGPR can play a critical role in this regard, and by manipulating the rhizosphere community, drought tolerance can be imposed.

The PGPR controls several adaptation mechanisms that help plants grow under water-deficient conditions. The PGPRs are principally soil microflora which acts as "microbial seed bank" in controlling plant stress response in a particular species--specific pattern (Xun et al. 2021). The species-specific improvements of biochemical and morphological parameters were observed in maize when treated with *Pseudomonas fluorescens*, *P. putida*, *Azospirillum lipoferum*, and *Bacillus subtilis* as PGPRs (Singh et al. 2024b). The most studied effect includes the production of indole acetic acid (IAA), ABA, and 1-aminocyclopropane-1-carboxylate (ACC) deaminase (Vurukonda et al. 2016), which have been already discussed in detail earlier. Upregulation of IAA results in an increase of adventitious root, lateral root, cell division, and shoot length, thus increasing water uptake and biomass (Khan et al. 2019). Another study on maize proved that *Azospirillum* can induce drought tolerance by producing IAA (García et al. 2017). An experiment on Mint (*Mentha pulegium* L.) demonstrated drought tolerance via the production of ABA, after inoculation with a coculture of *Azospirillum brasilense* and *Azotobacter chroococum* (Asghari et al. 2020). The DKC3014 is a drought-sensitive variety of *Zea mays* and can grow under drought stress when inoculated with a coculture of two PGPR strains *Cupriavidus necator* 1C2 and *Pseudomonas fluorescens* S3X. These PGPRs promote the production of siderophore, IAA, and ACC deaminase (Pereira et al. 2020).

10.4.3.1 Drought Tolerance Through Carbon Dot Mechanism

An interesting approach to modify the rhizosphere community to impart drought tolerance was reported in maize, where foliar sprayed carbon dots induce root exudates like pyruvic acid, succinic acid, and betaine that ultimately increase the abundance of *Pseudomonas, Nitrospira, Sphingomonas*, and *Conocybe* ranging from 122.6% to 344.4%. These altered microbial community increased nutrient availability, photosynthetic rate by elevating electron transport rate, and fresh weight of root and shoot significantly under drought stress (Yang et al. 2022). It also improves nitrogen-fixing ability and water uptake efficiency in soybean under drought stress (Wang et al. 2022). Similarly, another study demonstrated that carbon dots improved the photosynthetic ability of soybean plants under drought stress by modulating the redox imbalances and nitrogen metabolism. The expression of *GmNRT*, *GmAMT*, and *GmAQP* genes was instigated along with rhizosphere recruitment of Actinobacteria, Ascomycota, Acidobacteria, Glomeromycota, etc. (Ji et al. 2023). Carbon dots are nanomaterials with zero dimensions, consisting of amorphous carbon, hydrogen, and oxygen characterized by unique luminescence, high thermal stability and water solubility, low toxicity, and chemically inert. These relatively new techniques have several fascinating applications like biosensing, catalysis, energy conversion, etc. (Kurian and Paul 2021). The major role of carbon dots relies on their unique scavenging properties of ROS generated through diverse abiotic stresses. The structure of the carbon dots allows them to donate proton for active scavenging capabilities (Lu et al. 2019). Besides, carbon dots are also able to change the rhizosphere microbiome communities. The foliar doping of nitrogen in carbon dots (N-CDs) readily increases the organic exudations in roots of maize and increases the population of beneficial microorganisms, e.g., *Azotobacter*, *Bacillus*, *Lysobacter*, *Mucilaginibacter*, Sphingomonas, *Rhizobium*, *Pseudomonas*, *Mucilaginibacter*, *Flavisolibacter*, etc. weeks after N-CD applications (Yue et al. 2023). Therefore, the introduction of carbon dots to combat drought stress by altering the rhizospheric microbial community now seems to be one of the most promising approaches in the modern era.

10.4.3.2 Melatonin-Mediated Drought Tolerance

Additionally, application of melatonin (N-acetyl-5-methoxytryptamine) can also alter the rhizospheric microbiome that subsequently mitigates drought stress in barley root, by inducing the metabolism of nonstructural carbohydrates (Ye et al. 2022). Melatonin acts as a key signaling molecule in plant abiotic stress that serves as a scavenger for reactive oxygen species (ROS), induces an antioxidant defense system, and alters stress-responsive genes. It is mainly synthesized in the mitochondria and chloroplast in plants. It is also reported that melatonin mitigates low-temperature stress in barley, by altering the rhizospheric microbial community (Jiang et al. 2022) although the detailed mechanism of this rhizospheric microbe modulation through melatonin is yet to be deciphered. A study revealed *Streptomyces* modulates microbial community by its metabolites under severe drought stress in sugarcane (Xing et al. 2023). Poly γ-glutamic acid is a water-soluble, nontoxic, and biodegradable polymer that enhances the rhizospheric microbial community by uplifting the abundance of PGPR such as *Actinobacteria*, *Firmicutes*, Alphaproteobacteria, and Deltaproteobacteria in maize on exposure to drought, thus showing increased resistance to drought (Ma et al. 2022). Introducing *Rhodobacter sphaeroides* KE149, an IAA-producing rhizobacteria, alleviates drought tolerance in adzuki bean plants by regulating different stress-responsive phytohormone and mineral concentrations (Kang et al. 2020). Some examples of the use of PGPR bacteria in drought stress ameliorations have been summarized in Table 10.1.

Table 10.1 The list of examples of drought stress tolerance by PGPR

Plant growth-promoting rhizobacteria (PGPR)	Strain	Isolated from	Shows tolerance in	Mechanisms of action	References
Bacillus thuringiensis	AZP2	*Pinus ponderosa* roots grown on gneiss rock at Mt. Lemmon	*Triticum aestivum*	Increasing the activity of monodehydroascorbate reductase (MDHAR), superoxide dismutase (SOD) glutathione reductase (GR), and catalase (CAT) Lowering the secretion of benzaldehyde, β-pinene	Timmusk et al. (2014)
Pseudomonas jessenii *Pseudomonas synxantha* *Arthrobacter nitroguajacolicus*	R62 YB3 and YB5	–	Drought-sensitive variety of *Oryza sativa*, i.e., IR64	By upregulating APX, SOD, POD antioxidant defense system Lowering the level of H_2O_2 and MDA	Yogendra et al. (2015)
Pseudomonas putida *Bacillus amyloliquefaciens*	NBRIRA NBRISN13	–	*Cicer arietinum* L	A consortium of these two PGPR significantly improves drought stress tolerance by mineral solubilization, siderophore activity, and ACC deaminase activity Modulation of microbial diversity and soil enzymes. β-glucosidase acid/alkaline phosphatase, protease, urease	Kumar et al. (2016)
Bacillus subtilis	LDR2	–	*Triticum aestivum*	Increasing the endogenous phytohormone auxin level, whereas decreasing ABA and ACC Interacting with the CTR1 regulatory component and DREB2 transcription factor	Barnawal et al. (2017)

Table 10.1 (continued)

Plant growth-promoting rhizobacteria (PGPR)	Strain	Isolated from	Shows tolerance in	Mechanisms of action	References
Bacillus thuringiensis strain	*IAM 12077*	Rhizosphere soil mixture of different autochthonous shrub species	*Thymus vulgaris*, *Santolina chamaecyparissus* *Lavandula dentata*	Modulating nutrient acquisition, without altering rhizospheric microbial community Increasing shoot potassium(K+), calcium (ca++), and magnesium (Mg++) content	Armada et al. (2018)
Arthrobacter sp. *Pseudomonas* sp. *Microbacterium* sp. *Bacillus* sp.	–	*Saccharum officinarum*	*Saccharum officinarum*	*Pseudomonas* and *Arthrobacter* showed nitrogen-fixing ability Enhancing IAA production	Pereira et al. (2019)
Sphingomonas sp.	*Cra20*	Root surface of *Leontopodium leontopodioides* Tian Shan Mountains, China,	*Arabidopsis thaliana*	Stimulating the formation of lateral roots and root hairs Modulating native rhizospheric bacterial community	Luo et al. (2019)
Bacillus subtilis, *Bacillus thuringiensis*, *Bacillus megaterium*	–	Rhizosphere of chickpea	*Cicer arietinum L.* Punjab Noor-2009 (drought sensitive) and 93,127 (drought tolerant)	Increasing chlorophyll, sugar content Rising L-asparagine riboflavin, aspartate, glycerol, 3-hydroxy-3-methyglutarate nicotinamide in leaves	Khan et al. (2019)
Pseudomonas fluorescens *Bacillus amyloliquefaciens*	WCS417 r GB03	–	*Mentha piperita*	Significant increase in total phenolic content Reduction of proline content in inoculated plants Decrease in membrane lipid peroxidation ACC deaminase activity in GB03 strain	Chiappero et al. (2019)
Pseudomonas fluorescens *Pseudomonas putida*	–	–	*Zea mays*	Downregulating the cold-dehydrin 410 gene and WRKY 18,12-oxo--phytodienoatereductase 7 (OPR7) along with the lipoxygenase gene	Ahmad et al. (2019)

Pseudomonas lini *Serratia plymuthica*	–	*Ziziphus jujuba*	–	ACC deaminase activity to reduce the ethylene production Enhancing the antioxidant enzyme activities and decreasing the level of ABA and malondialdehyde (MDA)	Zhang et al. (2020)
Pseudomonas aeruginosa, Enterobacter cloacae, Achromobacter xylosoxidans, and Leclercia adecarboxylata	–	*Rhizosphere of Zea mays*	*Zea mays*	ACC deaminase activity that cut down ethylene production Improving NPK uptake	Danish et al. (2020)
Bacillus subtilis	HAS31	–	*Solanum tuberosum* L. Santae [drought tolerant] and PRI-red [drought sensitive]	Upregulating activity of CAT, SOD, and peroxidase (POD) Increasing production of total soluble sugar, lowering superoxide(O_2^-), H_2O_2, and MDA production	Batool et al. (2020)
Bacillus sp. Pseudomonas sp.	*WM13–24* *M30–35*	*Haloxylon ammodendron [xerophyte]*	*Lolium perenne* L.	Modulating the activities of antioxidant enzymes catalase (CAT), peroxidase (POD) Decreasing the phytohormone ABA content in leaves	He et al. (2021)
Arthrobacter endophyticus, Zobellella denitrificans *Staphylococcus sciuri*	–	–	*Pistacia vera* L.	Increasing the production of auxin and ACC deaminase Production of EPS and siderophore, moreover phosphate and zinc-solubilizing activities	Khalilpour et al. (2021)
Bacillus subtilis Pseudomonas azotoformans	FAB1 FAP3	Rhizosphere and rhizoplane of *Triticum aestivum*	*Triticum aestivum*	Improving physiological traits such as stomatal conductance, transpiration rate, photosynthetic rate, intercellular CO_2 concentration, and intrinsic water use efficiency Lowering the level of stress marker for example CAT, SOD, and MDA	Ansari et al. (2023)

(continued)

Table 10.1 (continued)

Plant growth-promoting rhizobacteria (PGPR)	Strain	Isolated from	Shows tolerance in	Mechanisms of action	References
Serratia odorifera	–	–	*Hordeum vulgare* L.	Increasing Chl-A and Chl- B content, POD, CAT, and SOD activity Increasing NPK uptake by barley plant	Gul et al. (2023)
Bacillus pseudomycoides	ARN7	–	*Zea mays*	Promoting chlorophyll, total soluble sugar, relative water content [RWC], and phenolic content Upregulation of antioxidant enzyme activity	Anbuganesan et al. (2024)
Bacillus subtilis *Curtobacterium citreum*	SRJ4 MJ1	Rhizospheric soil sample of *Zea mays*	*Zea mays*	SRJ 4 promotes antioxidant activity, RWC, and drought-responsive gene expression DHN1 Downregulation of WRKY18 gene expression	Yasmeen et al. (2024)

10.4.4 Successful Commercialization of Endophytes: The Lab--to-Land Transition

Endophytes are the inherent microbial populations residing within the plant tissues and principally comprising bacteria, actinomycetes, and fungi. Due to long-term coevolutionary assemblage with plants, endophytes help plants in many aspects, e.g., nutrient acquisition, solubilization, growth, development, as well as stress tolerance (Anand et al. 2023b). Hence, studying plant endophytes gives clues to their ecological relevance with different developmental phases of plants, as well as roles and types of endophytes in accordance with their strategies of coexistence. The functional classification of plant endophytes includes facultative, obligate, opportunistic, and passenger endophytes (Hardoim et al. 2008). The major growth-promoting and stress-tolerant endophytes belong to the obligate category, and passenger endophytes may also sometimes possess unique stress-tolerance capabilities. For instance, the genome sequence of passenger endophyte, *Enterobacter cloacae* S23, showed plant growth-promoting and stress tolerance activity in ground nut nodules (Ramakrishnan et al. 2023). Similarly, the ground nut nodule also possesses rhizobacteria (*Rhizobium phaseoli* S18 and *Rhizobium pusense* S6R2) as passenger endophytes to promote salt stress tolerance (Oviya et al. 2023). They are reported to produce plant hormones and nutrient chelators for their activities. The endophytes are also reported to produce diverse secondary molecules and alkaloids in many cases. The microbial production of azadirachtin was successfully employed against feeding locusts (*Schistocerca gregaria*) (Butterworth and Morgan 1968). The excellent insecticidal effectiveness of these bacterial antibiotics leads to many endophytes as commercial insecticides, e.g., TreeAzin, AzaMax, BioNEEM, AzaGuard, and AzaSol (Anand et al. 2023). Although the beneficial roles of microorganisms in agricultural practices have long been documented, improper scientific intervention restricts their successful implications. The specificity, lack of consistency in function, and low demand in the market grossly lead the start-ups or commercial production of endophyte-dependent agricultural products to fail. Recently, an evidence-based consistent pipeline of successful endophyte agricultural formulation was documented in the case of barley. The article beautifully describes stepwise procedures to establish consistent microbial inoculation technology for agriculture (Murphy et al. 2018). The rhizosphere endophytes are known to modulate soil health and interact with the exophytes through quorum sensing (Chaudhary et al. 2022). The rhizosphere is the primary colonization site of endophytes. The major bacteria found in the rhizosphere include Proteobacteria, Acidobacteria, Gemmatimonadetes, Actinobacteria, Chloroflexi, and Bacteroidetes. On the other hand, the fungal population encompasses Ascomycota, Mortierellomycota, Basidiomycota, Chytridiomycota, and Olpidiomycota (Shi et al. 2021). The fungal endophytes, *Aspergillus* sp., *Cladosporium* sp., *Fusarium* sp., *Penicillium* sp., *Verticillium* sp., and *Ascomycete* sp., are very commonly used in agricultural practices to induce growth and amelioration of stress (Sahoo et al. 2017). Many fungal genera, e.g., *Fusarium* sp., *Trichoderma* sp., *Penicillium* sp., *Aspergillus* sp.,

Alternaria sp., etc., were reported to combat drought stress successfully. On the other hand, salinity stress can also be successfully reduced by the induction of redox alterations by fungal endophytes (Verma et al. 2022). Along with the improvement of soil health and direct interaction, endophytes may also be capable of modulating abiotic stress tolerance by secreting secondary metabolites and other chemical compounds (Ogbe et al. 2020; Singh et al. 2022). Hence, the isolation and application of these potent secondary metabolites directly to the field condition may also be used as a good alternative to live bacterial application. The active metabolites released from the organisms are called "biopreparations," which have been popularly used in organic farming to improve crop yield. To date, many species of *Bacillus* sp., *Thiobacillus* sp., *Trichoderma* sp., *Pseudomonas* sp., *Azospirillum* sp., *Rhizobium* sp., and *Gluconacetobacter* sp. etc. were successfully used as commercial "biopreparations" (Kuźniar et al. 2019). Additionally, deciphering the mode of action of stress amelioration by the endophytic bacterial community may allow biotechnological application of the same by altering or manipulating the genetic regulations (Lata et al. 2018). Many biotechnological companies have recently emerged that utilized this microbial information judicially and transformed the "lab-to-land" transition of these tiny wonders into agricultural modulators. Some of these companies with their principal products are tabulated in Table 10.2.

10.4.5 Successful Commercialization of PGPR

PGPR may act as a source of metabolites and enzymes that also help in nutrient mobilization, detoxification of heavy metals, and stress mitigation that subsequently promote plant growth and development. Commercialization of PGPR-based bioformulation can serve as a potential alternative to hazardous synthetic chemicals (Nagrale et al. 2023). PGPR's effectiveness depends extensively on marketing and commercialization. Market analysis, proper management, and surveying are prerequisites for synthesizing appropriate microbial inoculation. Information about particular crops and climatic conditions are crucial for selecting an ideal microbial isolate (Mushtaq 2020). PGPR-based inoculant development includes isolation of bacteria from plant tissues, screening under in vitro conditions, assessing strain combination, considering management practices, optimization of the product, determining ecotoxicological effect, formulation of the product, and registration and approval by a regulatory body (Backer et al. 2018). There are legal and technical issues in the way to commercialization of plant growth-promoting microorganisms that lead to low registration especially in countries like India. Developing robust production technologies, efficient delivery methods, and streamlining bioagent registration could improve disease management, increase crop yields, and promote sustainable agricultural practices globally (Hubballi et al. 2023). *Bacillus velezensis* GB03, a gram-positive PGPR, has been reported to boost plant growth and immunity in different crop plants, e.g., wheat, cucumber, cotton, barley, soybean, and pepper. In 1998, US Environmental Protection Agency validating its safety approved

Table 10.2 List of the biotechnological companies using microorganisms as promising products for agricultural improvements and sustainability

Sl. no.	Company name	Country of origin	Principal products	Url
1.	Novozymes/ Novonesis	Denmark	Agricultural bio-solutions, microbial enzymes, biocleaning, aquaculture, bioenergy, microbial bioagriculture, microbial food technology, etc.	https://www.novozymes.com/en/solutions/bioag
2.	Vegalab	Switzerland	Microbial insecticides, fungicides, nematicides, biocontrol agents, living bacterial biocontrol agents, mycorrhizal biocontrol agents, biofertilizers, plant amendments, cellular boost for drought stress amendments, etc.	https://www.vegalab.com/
3.	UPL Limited	India	Seed treatment, microbial bio-inoculants, bio-stimulant, biofertilizers, plant growth--promoting rhizobacteria, adjuvants, post-harvest managements, etc.	https://www.upl-ltd.com/
3.	Chr. Hansen	Denmark	Microbial inoculants, biopesticides, nematicides, soil microbial inoculants, bioformulations etc.	https://www.chr-hansen.com/en
4.	Kiwa Bio-Tech	China	Biological organics, microbial fertilizers, micro-inoculants, bacterial soil modulators, microbe--based aqua-soluble fertilizers, etc.	https://kiwabiotech.com/
5.	Agri Life	India	Microbe-based soil health and stress managements, bio-composting, biostimulants, biofertilizers, microbial biopesticides, botanicals, microbial nanotechnology-based stress amelioration products, etc.	https://www.agrilife.in/
6.	Seipasa	Spain	Biopesticides, biostimulants, and biofertilizers are the major products	https://www.seipasa.com/en/
7.	Mapleton Agri Biotech Pty Ltd	Australia	Nitrogen-fixing bacterial formulation (TwinN), mixture of microbial fertilizers, and plant--based amino acids (PastureN), nutrient acquisition, and solubilization using *Bacillus* and vesicular arbuscular mycorrhizae (VAM) (CataPult SuperFine), etc.	https://www.mabiotec.com/
8.	Rizobacter	Argentina	Bioinoculants, e.g., Signum, Rizoliq, Ribol, Rilotus, etc., biostimulation agents, adjuvants, biofertilizers, etc.	https://www.rizobacter.com.ar/en

(continued)

Table 10.2 (continued)

Sl. no.	Company name	Country of origin	Principal products	Url
9.	Lallemand Inc.	USA	Bioingredients, biofuels, micro--inoculants, nutraceuticals, etc.	https://www.lallemand.com/en/
10.	Kan Biosys	India	Nutrient, plant disease and stress managements, soil amendments, etc.	https://www.kanbiosys.com/
11.	Symborg	Spain	Agricultural products, biofertilizers, biostimulants, micro-inoculants, etc.	https://symborg.com/en/
12.	Valagro/ Syngenta biologicals	Italy	Microorganism-based biostimulants, seed modulators, biofertilizers, nutrients specially, water-soluble nutrients, biocontrol agents, etc.	https://www.syngentabiologacals.com/usa/en-us/
13.	Biomax Naturals	India	Liquid biofertilizers, biostimulants, etc.	http://www.biomaxnaturals.com/
14.	LKB Bio Fertilizer	Malayasia	Products related to improve soil fertility, disease resistance, abiotic stress managements, root developments, nutrient acquisition, enhanced growth, quality and yield, etc.	https://lkb-group.com/
15.	T. Stanes & Company Limited	India	Micronutrients, organic fertilizers, botanical insecticides, biofertilizers, seed amendments, stress managements, etc.	https://tstanes.com/
16.	Agrinos	Canada/ USA	Biofertilizers, crop yield management products, soil managements, micro-inoculants, etc.	http://agrinos.com/

the GB03 strain for commercial usage under the brand name Kodiak® (Jang et al. 2023). Another fungicidal and bactericidal product named Amylo-X®· was synthesized from *Bacillus amyloliquefaciens* subsp. *plantarum* D747 authorized in Italy from 2012. Amylo-X can be regarded as a useful crop protection tool because of its wide range of activity, excellent toxicological and ecotoxicological profile, and unique method of action (Ladurner et al. 2012).

10.4.6 The Bottlenecks for Endophyte Isolation and Discovery

With immense possibilities, the discovery and subsequent utilization of endophytes in agriculture are challenging. The first major challenge is to isolate the true endophyte community as there is every possibility of contamination with the surface microbiome. Secondly, the isolated microbes sometimes are difficult to culture and sometimes viable but not culturable (Ayilara et al. 2023). The root tip culture method was utilized to identify different groups of endophytic microbiomes of bananas. It was found that *Enterobacter* sp., *Klebsiella* sp., *Ochrobactrum* sp., *Pantoe* sp., *Staphylococcus* sp., and *Bacillus* sp. were the most easily culturable forms, whereas *Brevundimonas* sp., *Methylobacterium* sp., Alcaligenes, *Ralstonia* sp., *Pseudomonas* sp., *Corynebacterium* sp., *Microbacterium* sp., *Staphylococcus* sp., and *Oceanobacillus* sp. were extensively non-culturable. On the other hand, nonfilamentous actinobacteria were cultivable after prolonged culturing (Thomas et al. 2008). In the culture-dependent method, the undesirable fraction was proliferating with abundance in competition with the desirable non-culturable fraction in the plate due to their slow growth, which is the major challenge in the prolonged culture of the cultivable bacteria (Afzal et al. 2019). During the identification of the endophytic bacteria by partial 16 s rDNA sequencing method, contamination from mitochondrial and plastid genome is the major constraint in the culture-independent method (Garcias-Bonet et al. 2012). To overcome the situation, an enrichment approach was taken to isolate soybean endophytes by fractionation and homogenization of the entire stem during the isolation process and was found to be very effective (Ikeda et al. 2009). Among the culture-independent techniques, stable isotope probing (SIP) and nanoscale secondary ion mass spectrometry techniques (NanoSIMS) were popularly utilized coupled with Raman spectroscopy (Ayilara et al. 2023). Due to the recent advancement of the omics technique, the metagenomic approach is currently popularized over all other non-culture-dependent pathways to identify endophytic holobionts of the plants (Kumar and Dubey 2020). High-throughput next-generation sequencing (NGS) and short-gun sequencing coupled with advanced bioinformatics have now been adapted in metagenomics to discover an entire microbial community with secreted metabolites (Saini et al. 2024). The endophytic communities were deciphered through metagenomic NGS (mNGS) in *Acacia mearnsii*, which demonstrated that Proteobacteria and *Bradyrhizobium* play crucial role in plant development and stress tolerance (As Shiddiq et al. 2023). Metagenomics studies are now also utilized for unveiling intra-microbial and plant microbe interaction dynamics (Singh et al. 2024). Recently, comparative shoot-root endophyte analyses by both culture-dependent classical method and culture-independent metagenomic approach highlighted significant differences in microbial flora in shoot and root tissues of two varieties of *Ocimum sanctum* (Gupta et al. 2024). Interestingly, the metagenomic analysis of soybean endophytes revealed that the leaf-residing potential bacteria were "signature microbes" for health and diseases of the plants. These bacteria can be utilized as potential biomarkers against diverse stress response in agricultural applications (Chouhan et al. 2023).

10.5 Salt Stress and Microbial Modulation: The Functional Mechanism

10.5.1 The Mechanism of Salt Stress Tolerance by Altered PGPR Circuitry

Halotolerant plant growth-promoting rhizobacteria (HT-PGPR) are the dominant microorganisms for mitigating salt stress through their secondary metabolite production. Under stress conditions, it produces certain lipophilic VOC, like 1, 3-propanediol, albuterol, dimethyl disulfide, acetoin, and geosmin that have a pivotal role in plant growth. HT-PGPR imparts salinity tolerance by several mechanisms such as upregulation of high-efficiency K^+ transporter, antioxidant enzymes, phosphate solubilization, and IAA production (Kumar et al. 2023). A similar study reported that bacteria isolated from extreme environmental conditions help in stress tolerance in crop plants, e.g., *Pseudomonas fluorescens* isolated from date palm imparts salt tolerance in maize by promoting root growth (Zerrouk et al. 2016). Root exudate from rhizospheric bacteria plays a significant role in promoting stress tolerance by modulating plant-water relationship, ion transport, and nutrient uptake (Ilangumaran and Smith 2017). So, it may be concluded that PGPR promotes drought and salinity stress tolerance through a merely similar mechanism. However, the detailed mechanisms are yet to be understood. Again, about 243 bacterial strains have been isolated from the microbiota of coastal halophytes *Limonium sinense*, out of which 15 strain of *Glutamicibacter* show PGP traits. The inoculated plant shows better stress response than the non-inoculated control plants under salt stress (Qin et al. 2018). Similarly, around 126 bacteria were isolated from the rhizosphere of *Artemisia princeps*, *Chenopodium ficifolium*, *Oenothera biennis*, and *Echinochloa crus-galli*, among which seven showed a significant plant growth-promoting trait. *Arthrobacter woluwensis* AK1 was demonstrated as the most potent candidate and helps to cope with the salinity stress in soybean plants by inducing the production of auxin and ABA, simultaneously upregulating the stress-responsive gene *GmLAX3* and *GmST1* (Khan et al. 2019a). Another group of halotolerant bacteria ALT 29 and ALT 43 imparts improved salinity stress tolerance in soybean plants by increasing IAA, chlorophyll, and salicylic acid (SA) content along with the upregulation of *GmFLD19* and *GmNARK* (Khan et al. 2021a). The strains of *Bacillus amyloliquefaciens* isolated from *Sasamorpha borealis* showed PGPR activity and a better adaptation to salt stress in *Capsicum annuum L.* (Kazerooni et al. 2021). More research in this field is required to understand the mechanism of action properly (Fig. 10.2).

10.5.2 Rhizosphere Engineering: The Emerging Tool for Combating Salinity Stress

Soil salinization is a worldwide concern that impacts 1100Mha soil, i.e., 7% of earth's land area. It is a complex phenomenon that is caused by natural geochemical processes and secondary anthropogenic activities. A substantial percentage of cultivated land has been negatively impacted by primary salinization events such as rise in the sea level, intrusion of saltwater into freshwater aquifers, etc. (Balasubramaniam et al. 2023). One of the most significant limiting factors of crop growth and productivity is salt stress. Three major impacts of salinity stress on plants are as follows: it lowers water potential, disrupts ion homeostasis, and produces toxicity. Salt stress possesses detrimental effects on crucial processes such as photosynthesis, germination, water retention capacity, growth and productivity (Parihar et al. 2015). Plants' response to salinity stress is quite complex and multidimensional. There are three reported salinity tolerance mechanisms in plants, i.e., toxic ion exclusion, compartmentalization of toxic ion, and shoot ion-independent tolerance (Negrão et al. 2017).

Salinity stress and shortage of Fe^{3+} are two significant limiting factors that plants have faced in arid soil (Sultana et al. 2021). Siderophores produced by PGPR have high Fe^{3+} affinity. It can work as a chelating agent for inorganic Fe^{3+} thus increasing Fe^{3+} uptake by the plant. Several strains of PGPR like *Bacillus* sp. and *Pseudomonas* sp. can produce siderophores and enhance tolerance in host plants (Aznar and Dellagi 2015). Nitrogen-fixing bacteria like *Azotobacter*, *Rhizobium*, and *Azospirillum* play a critical role by fixing the atmospheric N_2, especially during salinity stress. PO_4^{3-} and K^+ both are reported to be solubilized by halotolerant bacteria by acidification, ion exchange, and chelating mechanism (Etesami and Beattie 2018; Saghafi et al. 2019). A recent study proved that it is possible to increase salinity tolerance in *Zea mays* L. by inoculating the rhizosphere with *Kocuria rhizophila* Y1 strain (Li et al. 2020). The strain is responsible for the upregulation of IAA and downregulation of ABA under high salt stress. ABA is responsible for stomatal closing thus limiting water loss. Another study showed that the use of *Rhizobium leguminosarum* RhOF125, *Paenibacillus mucilaginosus* BLA7, and *Ensifer meliloti* RhOL1 strains can promote salinity tolerance in faba bean (*Vicia faba*) (Benidire et al. 2020). A detailed investigation showed an increase in *Sphingobacterium*, *Chryseobacterium*, *Stenotrophomonas*, *Agrobacterium*, and *Sinorhizobium* species and a decrease in Enterobacteriaceae under salinity stress conditions. Among 89 rhizospheric and endospheric bacteria isolated from two halophytic species, namely, *Suaeda mollis* and *Salsola tetrandra*, three strains of *Bacillus atrophaeus* thrive to reduce the detrimental effects of salinity stress during the seed germination stage in *Arabidopsis thaliana* (Kerbab et al. 2021). *Enterobacter ludwigii* B30 collected from *Paspalum vaginatum* imparts salinity tolerance in *Cynodon dactylon* by rising chlorophyll, carotenoids, SOD, and catalase concentration (Wei et al. 2022). Application of selenium (Se) along with salt-tolerant bacteria enhances plant rhizosphere microbial community, which imparts salt tolerance in soybean plants (Wang et al. 2023). Silicon (Si) and selenium (Se) both have immense potential in promoting stress tolerance in plants, integrated effects of Si^{+4} and Se^{-2}, reducing Na^{+} through root meristematic zone, and also modulating the expression of salt stress-responsive gene *ZmNHX* and *ZmSOS2*, in maize (Xu et al. 2021). Some examples of the use of PGPR bacteria in salt stress ameliorations have been summarized in Table 10.3.

10.6 The Encapsulation Technique for Better Performance of Engineered PGPR

The bioactivity of the PGPR decreases over time due to some environmental factor; encapsulation of PGPR promotes their efficiency for a longer time. In vitro culture of PGPR is majorly done while investigating their efficiency and effect, but their application in open agricultural fields and getting desirable results has a big difference. Bioactivity degradation and competition with previously dwelling PGPR are the main cause behind this (Vejan et al. 2019). Thus, the application and optimal outcome of PGPR in natural conditions need a specialized delivery method that will provide nutrition and protection and increase cell productivity of efficiency (Schoebitz et al. 2013)). In encapsulation technique, the nutritious core contains all the bacteria while the core is encapsulated by a protective outer shell or capsule. Scientists have developed several methods for encapsulation. Spray drying, extrusion, emulsification, and freeze-drying are some well-used examples of physical encapsulation techniques, whereas chemical methods contain molecular inoculation, coacervation, and co-crystallization (Herrmann and Lesueur 2013; Saberi Riseh et al. 2021). Also, different polymers are used for encapsulation depending upon the PGPR strain, host plant, and ecological factors to optimize the efficiency. Synthetic polymers like polyethylene glycol (PEG), poly(lactic-co-glycolic acid) PLGA, polyvinyl alcohol, poly(hydroxyethylmethacrylate) (PHEMA), and hydroxyethyl methacrylate-methyl methacrylate (HEMA-MMA) are used. Biopolymers could be a better replacement because of their biodegradable and eco-friendly nature. The most used one is alginate; other than this, gelatine, xanthan, whey protein, gellan gum, chitosan, and starch are also reported to be used (Verma et al. 2020). Synthetic polymers can be tailored accordingly and can be industrially produced in large amounts (Piskin 1995). Spray drying is a well-used method that includes spraying a homogenized mixture (slurry) of bacterial culture and carrier mixture into a hot chamber along with warm gas (Watanabe et al. 2002). This process is used in both agriculture and nonagricultural systems. However, exposure to high temperature (more than 140° C) can cause damage to the carried bacteria, thus resulting in less efficiency and cell production (Schoebitz et al. 2013). Another highly used method is the extrusion technique or ionic gelling. Food-graded biopolymers are used to form hydrogel beads that encapsulate the bacteria culture. This is the easiest and most effective technique, where the microbial culture is mixed with biopolymer solution (mostly sodium alginate) and is dropped into $CaCl_2$

Table 10.3 The list of examples of salinity stress tolerance by PGPR

Rhizobacteria	Strain	Isolated from	Shows tolerance in	Mechanisms of action	References
Microbacterium oleivorans, *Brevibacterium iodinum*, *Rhizobium massiliae*	KNUC7074 KNUC7183 KNUC7586	Rhizosphere of *Capsicum annum L.* plant		Increasing concentration of several antioxidant enzymes such as guaiacol peroxidase, ascorbate peroxidase, and catalase Accumulation of higher amount of proline and total soluble sugar	Hahm et al. (2017)
Klebsiella sp.	SBP-8		*Triticum aestivum*	Interacting with the antioxidative enzyme synthesis such as SOD, POX, and CAT Lowering concentration of malondialdehyde (MDA) Increase of osmolyte contents	Singh and Jha (2017)
Bacillus cereus, *Serratia marcescens*, *Pseudomonas aeruginosa*			*Triticum aestivum*	Increasing IAA, HCN synthesis along with nitrogen fixation and phosphate solubilization Upregulation of antioxidant activities	Desoky et al. (2020)
Kocuria rhizophila		Rhizosphere of *Zea mays*	*Zea mays*	By upregulating antioxidant gene *ZmGR1* and *ZmAPX1* and salt tolerance responsive *geneZmNHX1*, *ZmNHX2*, *ZmNHX3*, *ZmWRKY58*, and *ZmDREB2A* Modulation of phytohormone IAA and ABA	Li et al. (2020)
Pseudomonas fluorescens, *Bacillus pumilus* *Exiguobacterium aurantiacum*			*Triticum aestivum*	Altering the concentration of osmolytes, proline, and soluble protein at different magnitude by each of the rhizobacteria	Nawaz et al. (2020)

(continued)

Table 10.3 (continued)

Rhizobacteria	Strain	Isolated from	Shows tolerance in	Mechanisms of action	Referenc
Pseudomonas argentinensis *P. Azotoformans*,			*Brassica juncea*	Synthesis of auxin, ACC deaminase, and aminolevulinic acid	(Phour a Sindhu 2020)
Bacillus sp. *Azospirillum brasilense* *Azospirillum lipoferum* *Pseudomonas stutzeri*	*KF719179* *KJ194586* *KJ434039* *KJ685889*	Rhizospheric soil of salt-tolerant plants in Pakistan	*Wheat Triticum aestivum*	Elevating the production of chlorophyll-a, b, soluble sugar, proline, antioxidant enzymes CAT, SOD, and POX	Ilyas et a (2020)
Acinetobacter bereziniae, *Enterobacter ludwigii*, *Alcaligenes faecalis*	IG 2 IG 10 IG 27		*Pisum sativum*	Altering biochemical parameters like chlorophyll, protein, soluble sugar and expression of antioxidant enzymes Decreasing the level of electrolyte leakage	Sapre et (2022)
Bacillus [Halophilic]		Extreme environment of Qinghai--Tibetan plateau zone, China	*Triticum aestivum*	Modulation of SODA in ROS detoxification, thus reducing the detrimental effects of oxidative stress Regulation of lipid peroxidation and ABA biosynthesis	Ayaz et a (2022)
Pantoea agglomerans	FAP10,		*Triticum aestivum* var. 343	Modulating physiological traits such as stomatal conductance, transpiration rate, photosynthetic rate, intercellular CO_2 concentration, and intrinsic water use efficiency	Ansari et al. (2024)
Pseudomonas putida	R12		*Brassica juncea*	Lowering uptake of Na^+, increasing uptake of K^+ Increasing total soluble sugar, total protein, and relative water content and antioxidant production ACC deaminase activity	Alhoqail (2024)

solution to form the hydrogel beads. Then, the beads are collected, rinsed, and dried (Kosaraju 2005; Yeung et al. 2016). Interfacial polymerization is a chemical technique where microencapsulation of PGPR is done via emulsification. Two chemicals capable of emulsification are used, the PGPR is dissolved in the discontinuous inorganic solvent (mainly water), whereas an organic solvent (mainly oil) works as a continuous phase or vice versa (Vejan et al. 2019). This process involves the use of high pH and toxic chemicals like sebacoyl chloride but also provides a high active loading capacity, of up to 90% (Yeo et al. 2001).

After production, the viability of the encapsulated bacteria should be determined for sustainable use, as several biotic and abiotic factors influence the viability of PGPR (Krasaekoopt et al. 2004; Capela et al. 2006). The viable bacteria recovery level is represented by encapsulation yield (EY), which can be calculated via an equation to determine the viability:

$$\text{Percentage of } EY\left(EY\%\right) = \left(N_{\text{bead}} / N_{\text{suspension}}\right) \times 100$$

N_{bead} represents the total viable count recovered from beads and $N_{\text{suspension}}$ is the viable count escaped into suspension. It has been reported that the ET ranged between the lowest 31% and a higher 65% range (Amine et al. 2014; Vejan et al. 2019).

10.7 PGPR Delivery System and "Plant Probiotics": Where Do we Stand?

The growth-promoting and stress-ameliorating activities of PGPRs are well documented, and after the advent of high-throughput sequencing technology and metagenomics, the rhizo-microbial map of different crop plants has continuously been unveiled. Among the next major constraints in successful applicability of the rhizosphere engineering is convenient delivery system of the engineered rhizobacteria in the agricultural fields. The major factors that limit the successful delivery of the PGPRs are reduced diffusion, low permeability and target delivery, etc. Nanotechnology based nano-emulsions are proved to be very efficient delivery system for the PGPRs in the agricultural fields as they are low-cost technology and suitable for large-scale applications (Ravichandran et al. 2022). The utilization of nano-emulsions in the agricultural delivery system has further analyzed, and implications have been shouted for their controlled delivery mechanism (Tyagi et al. 2023). Nano-formulations using PGPRs are utilized for the crop production under abiotic stress tolerance (Shanmugam 2022). The nanoparticles coupled with the biofertilizers are termed as nano-biofertilizers (NBFs), which utilizes the metallic or organic, polysaccharide, chitosan-mediated delivery system used for the "green synthesis" of the plant growth-promoting molecules (PGPMs) from PGPRs. These are successfully utilized as the alternative of the chemical fertilizers as "smart agriculture" practice (Garg et al. 2023). The probiotics are the utilization of the living microorganisms in the health benefits of the hosts (either plants or animals). The probiotic application of the microorganisms in the agricultural sustenance has wide applicability (Checcucci et al. 2022). The synergistic application of nano-based probiotics in the agriculture has proven to be effective against drought and salinity tolerance in plants (Upadhayay et al. 2023). The application of PGPRs along with nanoparticles (silver, gold, zinc oxide, silica, titanium, etc.) is recently employed as potential delivery materials (Nayana et al. 2020). Titanium dioxide nanoparticles are known for the salinity stress tolerance during foliar spray in broad bean plants (Abdel Latef et al. 2018). Similarly, silicon dioxide nanoparticles have remarkable effect on drought stress tolerance in seedlings of hawthorns (Ashkavand et al. 2018). The combination of these nanoparticles with the salt- and drought-tolerant PGPRs could be able to restore the stress situation dramatically. Although recently many studies have initiated to combine nanotechnology with agri-biotechnology (Vasant et al. 2023), more research is necessary in the field of nanomaterial-based delivery system of PGPRs and "plant probiotic concept" for their judicial application in the agricultural sustainability.

10.8 Conclusion and Future Directions

Drought and salinity stress causes dramatic crop loss globally along with other abiotic stress conditions. Rapid climate change and urbanization further deteriorate the situation. Plants are constantly interacting with their surrounding environments with varied dimensions and magnitudes. Rhizosphere is the most interesting, yet complex active ecological niche that directly establishes communications with environments. The rhizosphere harbors an assemblage of diverse microbial communities. These microbiomes are involved in both positive and negative interactive networks with themselves and plants. This "plant-microbiome-environment (soil)" tripartite interaction normalizes the growth, development, as well as fitness of plants against different stresses. PGPRs are the most studied rhizosphere bacterial communities that exhibited growth promotion activities in plants. Despite the developmental effects, these microbial communities are well documented as abiotic stress modulators. Hence, modulation of the rhizosphere microbiome under stressful conditions may restore the normal growth of the plants. Rhizosphere engineering is the most emerging technique to counteract abiotic stress in plants. Although several works regarding the isolation and activity of PGPRs in plant developments have been carried out, detailed analyses of rhizosphere microbial communities under changing environmental conditions are still elusive. The detailed microbial map of different plants, especially, crop plants under different environmental conditions, is the prerequisite for their successful applications. Although this field has wide possibilities, it still warrants some future questions to be answered. The microbial map of different plants, especially crop plants under different environmental conditions, needs to be deciphered first. Once the microbial map has been deciphered, this could be analyzed and utilized for the crop improvement under varied environmental stress. The rhizosphere contains both culturable and non-culturable forms of the microbes. How the non-culturable microbial forms be utilized in crop improvements is a real challenge. The advent of metagenomic sequencing technology is capable to reveal holistic microbial compositions of the crop plants. Many of these microbes are difficult to culture. If the beneficial microbes fall into the non-culturable fraction, it will be difficult to utilize the same for the crop improvements. Hence, a judicial approach is necessary to utilize the benefits from this non-culturable microbes, e.g., metabolite sequestration. In most of the cases, PGPRs are plant specific. More research is needed to overcome this species specificity for their wide applicability. The involvement of systems biology tools to evaluate the efficacy of tailored microbial consortia may provide useful information for their future application. The advent of genome editing technology now permits to specifically silent or enhance the effect of a particular gene to enhance the effectivity of the beneficial microbes. The molecular mechanism of the "plant-rhizosphere-soil" tripartite interactome has to be deciphered to break the species barrier of the beneficial microbes to enhance their application plasticity. The adaptability and efficacy of engineered microbiomes in different crops under changing environmental conditions should be deciphered. The scalability and environmental risk assessment should also be kept in mind. Although the rhizosphere engineering is not supposed to hamper the ecology of the soil, there is a chance of disruption of native microbial flora and horizontal gene transfer due to excessive and prolonged use. Gene flow and the effect on nontarget species have also been kept in mind. Hence, a strong ethical consideration and legislation are mandatory before extensive commercial application. Along with the discovery of potential stress-tolerant PGPRs, more research should be focused on the successful delivery system of the PGPRs in the agricultural fields. Bioformulations should be improved to design "plant probiotics" for the agricultural applications. Finally, the economic viability of the technology should properly be investigated for effective future application.

Acknowledgments This research did not receive any specific grant from funding agencies in the public, commercial, or not-for-profit sectors. The authors are sincerely thankful to the respected editors of the book for the constructive suggestions and invitations.

Declaration of Competing Interest The authors declare that they have no known competing financial interests or personal relationships that could have appeared to influence the work reported in this paper.

Statements and Declarations

Funding No funding has received to perform this work.

Availability of Data and Material Not applicable.

Code Availability Not applicable.

References

Abdel Latef AAH, Srivastava AK, El-sadek MSA et al (2018) Titanium dioxide nanoparticles improve growth and enhance tolerance of broad bean plants under saline soil conditions. Land Degrad Dev 29:1065–1073. https://doi.org/10.1002/ldr.2780

Adesemoye AO, Egamberdieva D (2013) Beneficial effects of plant growth-promoting *Rhizobacteria* on improved crop production: prospects for developing economies. In: Maheshwari DK, Saraf M, Aeron A (eds) Bacteria in agrobiology: crop productivity. Springer Berlin Heidelberg, Berlin, Heidelberg, pp 45–63

Afzal I, Shinwari ZK, Sikandar S, Shahzad S (2019) Plant beneficial endophytic bacteria: mechanisms, diversity, host range and genetic determinants. Microbiol Res 221:36–49. https://doi.org/10.1016/j.micres.2019.02.001

Ahkami AH, Allen White R, Handakumbura PP, Jansson C (2017) Rhizosphere engineering: enhancing sustainable plant ecosystem productivity. Rhizosphere 3:233–243. https://doi.org/10.1016/j.rhisph.2017.04.012

Ahmad I, Zaib S, Alves PCMS, Luthe DS, Bano A, Shakeel SN (2019) Molecular and physiological analysis of drought stress responses in Zea mays treated with plant growth promoting rhizobacteria. Biol Plant 63536–63547. https://doi.org/10.32615/bp.2019.092

Ahmad HM, Fiaz S, Hafeez S et al (2022) Plant growth-promoting Rhizobacteria eliminate the effect of drought stress in plants: a review. Front Plant Sci 13:875774. https://doi.org/10.3389/fpls.2022.875774

Alhoqail WA (2024) ACC-Deaminase producing *Pseudomonas putida* RT12 inoculation: A promising strategy for improving *Brassica juncea* tolerance to salinity stress. Not Bot Horti Agrobot Cluj-Na 52(1):13550. https://doi.org/10.15835/nbha52113550

Ali S, Khan N (2021) Delineation of mechanistic approaches employed by plant growth promoting microorganisms for improving drought stress tolerance in plants. Microbiol Res 249:126771. https://doi.org/10.1016/j.micres.2021.126771

Amine KM, Champagne CP, Salmieri S et al (2014) Effect of palmitoylated alginate microencapsulation on viability of *Bifidobacterium longum* during freeze-drying. LWT - Food Sci Technol 56:111–117. https://doi.org/10.1016/j.lwt.2013.11.003

Anand G, Shrivas VL, Dubey S et al (2023a) Stress-buster Enterobacter sp. alleviates salinity stress in *Cajanus cajan* together with impacting its rhizospheric microbiome. S Afr J Bot 156:202–212. https://doi.org/10.1016/j.sajb.2023.02.040

Anand U, Pal T, Yadav N et al (2023b) Current scenario and future prospects of Endophytic microbes: promising candidates for abiotic and biotic stress management for agricultural and environmental sustainability. Microb Ecol 86:1455–1486. https://doi.org/10.1007/s00248-023-02190-1

Ansari FA, Ahmad I, Pichtel J (2023) Synergistic effects of biofilm-producing PGPR strains on wheat plant colonization growth and soil resilience under drought stress. Saudi J Biol Sci 30(6):103664. https://doi.org/10.1016/j.sjbs.2023.103664

Ansari FA, Ahmad I, Pichtel J, Husain FM (2024) *Pantoea agglomerans* FAP10: A novel biofilm-producing PGPR strain improves wheat growth and soil resilience under salinity stress. Environ Exp Bot 222:105759. https://doi.org/10.1016/j.envexpbot.2024.105759

Anbuganesan V, Vishnupradeep R, Bruno LB, Sharmila K, Freitas H, Rajkumar M (2024) Combined application of biochar and plant growth-promoting rhizobacteria improves heavy metal and drought stress tolerance in *Zea mays*. Plants (Basel). 13(8):1143

Armada E, Leite MFA, Medina A, Azcón R, Kuramae EE (2018) Native bacteria promote plant growth under drought stress condition without impacting the rhizomicrobiome. FEMS Microbiol Ecol 94(7). https://doi.org/10.1093/femsec/fiy092

As Shiddiq M, Pramudia D, Nefilinda N et al (2023) Metagenomic profile of Endophytic bacteria in *Acacia mearnsii*. IOP Conf Ser Earth Environ Sci 1182:012030. https://doi.org/10.1088/1755-1315/1182/1/012030

Asghari B, Khademian R, Sedaghati B (2020) Plant growth promoting rhizobacteria (PGPR) confer drought resistance and stimulate biosynthesis of secondary metabolites in pennyroyal (*Mentha pulegium* L.) under water shortage condition. Sci Hortic 263:109132. https://doi.org/10.1016/j.scienta.2019.109132

Ashkavand P, Tabari M, Zarafshar M et al (2018) Effect of SiO2 nanoparticles on drought resistance in hawthorn seedlings. https://doi.org/10.48538/FRP-2015-0034

Atsede M, Fassil A (2018) Isolation and screening of antibiotic producing actinomycetes from rhizosphere and agricultural soils. Afr J Biotechnol 17:700–715. https://doi.org/10.5897/AJB2017.16080

Ayaz M, Ali Q, Jiang Q, Wang R, Wang Z, Mu G, Khan SA, Khan AR, Manghwar H, Wu H, Gao X, Gu Q (2022) Salt tolerant bacillus strains improve plant growth traits and regulation of phytohormones in wheat under salinity stress. Plants 11(20):2769. https://doi.org/10.3390/plants11202769

Ayilara MS, Adeleke BS, Babalola OO (2023) Bioprospecting and challenges of plant microbiome research for sustainable agriculture, a review on soybean Endophytic bacteria. Microb Ecol 85:1113–1135. https://doi.org/10.1007/s00248-022-02136-z

Aznar A, Dellagi A (2015) New insights into the role of siderophores as triggers of plant immunity: what can we learn from animals? J Exp Bot 66:3001–3010. https://doi.org/10.1093/jxb/erv155

Backer R, Rokem JS, Ilangumaran G et al (2018) Plant growth-promoting Rhizobacteria: context, mechanisms of action, and roadmap to commercialization of biostimulants for sustainable agriculture. Front Plant Sci 9:1473. https://doi.org/10.3389/fpls.2018.01473

Balasubramaniam T, Shen G, Esmaeili N, Zhang H (2023) Plants' response mechanisms to salinity stress. Plants 12:2253. https://doi.org/10.3390/plants12122253

Bano S, Wu X, Zhang X (2021) Towards sustainable agriculture: rhizosphere microbiome engineering. Appl Microbiol Biotechnol 105:7141–7160. https://doi.org/10.1007/s00253-021-11555-w

Bano A, Gupta A, Rai S et al (2022) Mechanistic role of reactive oxygen species and its regulation *via* the antioxidant system under environmental stress. In: Hasanuzzaman M, Nahar K (eds) Physiology. IntechOpen

Barnawal D, Bharti N, Pandey SS, Pandey A, Chanotiya CS, Kalra A (2017) Plant growth-promoting rhizobacteria enhance wheat salt and drought stress tolerance by altering endogenous phytohormone levels and *TaCTR1 / TaDREB2* expression. Physiol Plant 161(4):502–514. https://doi.org/10.1111/ppl12614

Batool T, Ali S, Seleiman MF et al (2020) Plant growth promoting rhizobacteria alleviates drought stress in potato in response to suppressive oxidative stress and antioxidant enzymes activities. Sci Rep 10:16975. https://doi.org/10.1038/s41598-020-73489-z

Benidire L, El Khalloufi F, Oufdou K et al (2020) Phytobeneficial bacteria improve saline stress tolerance in *Vicia faba* and modulate microbial interaction network. Sci Total Environ 729:139020. https://doi.org/10.1016/j.scitotenv.2020.139020

Bhagyashree KB, Shivaprakash MK, Reddy MR (2022) Isolation and identification of bacterial endophytes from grain Amaranth (*Amaranthus caudatus*) for plant growth promotion. Indian J Agric Res. https://doi.org/10.18805/IJARe.A-6015

Bhar A, Chatterjee M, Gupta S, Das S (2018) Salicylic acid regulates systemic defense signaling in chickpea during *Fusarium oxysporum f. sp. ciceri* race 1 infection. Plant Mol Biol Report 36:162–175. https://doi.org/10.1007/s11105-018-1067-1

Boonmahome P, Namwongsa J, Vorasoot N et al (2023) Single and co-inoculum of endophytic bacteria promote growth and yield of Jerusalem artichoke through upregulation of plant genes under drought stress. PLoS One 18:e0286625. https://doi.org/10.1371/journal.pone.0286625

Boyko A, Kovalchuk I (2008) Epigenetic control of plant stress response. Environ Mol Mutagen 49:61–72. https://doi.org/10.1002/em.20347

Brilli F, Loreto F, Baccelli I (2019) Exploiting plant volatile organic compounds (VOCs) in agriculture to improve sustainable defense strategies and productivity of crops. Front Plant Sci 10:264. https://doi.org/10.3389/fpls.2019.00264

Butterworth JH, Morgan E (1968) Isolation of a substance that suppresses feeding in locusts. Chem Commun Lond:23–24

Caddell DF, Deng S, Coleman-Derr D (2019) Role of the plant root microbiome in abiotic stress tolerance. In: Verma SK, White JJF (eds) Seed endophytes. Springer International Publishing, Cham, pp 273–311

Capela P, Hay TKC, Shah NP (2006) Effect of cryoprotectants, prebiotics and microencapsulation on survival of probiotic organisms in yoghurt and freeze-dried yoghurt. Food Res Int 39:203–211. https://doi.org/10.1016/j.foodres.2005.07.007

Chaudhary P, Agri U, Chaudhary A et al (2022) Endophytes and their potential in biotic stress management and crop production. Front Microbiol 13:933017. https://doi.org/10.3389/fmicb.2022.933017

Checcucci A, Luise D, Pini F (2022) Editorial: optimizing probiotic applications in agriculture: exploring the role of growth and health promoter's microorganisms in plants and livestock animals. Front Microbiol 13:993075. https://doi.org/10.3389/fmicb.2022.993075

Chiappero J, Cappellari LDR, Sosa Alderete LG et al (2019) Plant growth promoting rhizobacteria improve the antioxidant status in *Mentha piperita* grown under drought stress leading to an enhancement of plant growth and total phenolic content. Ind Crops Prod 139:111553. https://doi.org/10.1016/j.indcrop.2019.111553

Chieb M, Gachomo EW (2023) The role of plant growth promoting rhizobacteria in plant drought stress responses. BMC Plant Biol 23:407. https://doi.org/10.1186/s12870-023-04403-8

Chouhan U, Gamad U, Choudhari JK (2023) Metagenomic analysis of soybean endosphere microbiome to reveal signatures of microbes for health and disease. J Genet Eng Biotechnol 21:84. https://doi.org/10.1186/s43141-023-00535-4

Corwin DL (2021) Climate change impacts on soil salinity in agricultural areas. Eur J Soil Sci 72:842–862. https://doi.org/10.1111/ejss.13010

Danish S, Zafar-Ul-Hye M, Hussain S et al (2020) Mitigation of drought stress in maize through inoculation with drought tolerant ACC deaminase containing PGPR under axenic conditions. Pak J Bot 52. https://doi.org/10.30848/PJB2020-1(7)

Das T, Bhattacharyya A, Bhar A (2023) Linking Phyllosphere and Rhizosphere Microbiome to the Plant–Insect Interplay: The New Dimension of Tripartite Interaction. Physiologia 3:129–144. https://doi.org/10.3390/physiologia3010011

Desoky ESM, Saad AM, El-Saadony MT, Merwad ARM, Rady MM (2020) Plant growth-promoting rhizobacteria: Potential improvement in antioxidant defense system and suppression of oxidative stress for alleviating salinity stress in *Triticum aestivum* (L.) plants. Biocatal Agric Biotechnol 30101878. https://doi.org/10.1016/j.bcab.2020.101878

Dessaux Y, Grandclément C, Faure D (2016) Engineering the rhizosphere. Trends Plant Sci 21(3):266–278. https://doi.org/10.1016/j.tplants.2016.01.002

Dos Santos TB, Ribas AF, De Souza SGH et al (2022) Physiological responses to drought, salinity, and heat stress in plants: a review. Stress (Amsterdam Neth) 2:113–135. https://doi.org/10.3390/stresses2010009

Dries L, Hendgen M, Schnell S et al (2021) Rhizosphere engineering: leading towards a sustainable viticulture? OENO One 55:353–363. https://doi.org/10.20870/oeno-one.2021.55.2.4534

Duan B, Li L, Chen G et al (2021) 1-Aminocyclopropane-1-carboxylate deaminase-producing plant growth-promoting Rhizobacteria improve drought stress tolerance in grapevine (*Vitis vinifera* L.). Front Plant Sci 12:706990. https://doi.org/10.3389/fpls.2021.706990

Dubey S, Sharma S (2021) Rhizospheric engineering by plant-mediated indirect selection of microbiome for agricultural sustainability. Crit Rev Plant Sci 40:379–397. https://doi.org/10.1080/07352689.2021.1959137

Dudhate A, Shinde H, Yu P et al (2021) Comprehensive analysis of NAC transcription factor family uncovers drought and salinity stress response in pearl millet (*Pennisetum glaucum*). BMC Genomics 22:70. https://doi.org/10.1186/s12864-021-07382-y

Etesami H, Beattie GA (2018) Mining halophytes for plant growth-promoting halotolerant bacteria to enhance the salinity tolerance of non-halophytic crops. Front Microbiol 9:148. https://doi.org/10.3389/fmicb.2018.00148

Fahad S, Bajwa AA, Nazir U et al (2017) Crop production under drought and heat stress: plant responses and management options. Front Plant Sci 8:1147. https://doi.org/10.3389/fpls.2017.01147

Farkas C, Gelybó G, Bakacsi Z et al (2014) Impact of expected climate change on soil water regime under different vegetation conditions. Biologia (Bratisl) 69:1510–1519. https://doi.org/10.2478/s11756-014-0463-8

García JE, Maroniche G, Creus C et al (2017) In vitro PGPR properties and osmotic tolerance of different Azospirillum native strains and their effects on growth of maize under drought stress. Microbiol Res 202:21–29. https://doi.org/10.1016/j.micres.2017.04.007

Garcias-Bonet N, Arrieta JM, De Santana CN et al (2012) Endophytic bacterial community of a Mediterranean marine angiosperm (*Posidonia oceanica*). Front Microbiol 3. https://doi.org/10.3389/fmicb.2012.00342

Garg D, Sridhar K, Stephen Inbaraj B et al (2023) Nano-biofertilizer formulations for agriculture: a systematic review on recent advances and prospective applications. Bioengineering 10:1010. https://doi.org/10.3390/bioengineering10091010

Gontia-Mishra I, Sapre S, Sharma A, Tiwari S (2016) Amelioration of drought tolerance in wheat by the interaction of plant growth-promoting rhizobacteria. Plant Biol 18:992–1000. https://doi.org/10.1111/plb.12505

Gowtham HG, Singh SB, Murali M et al (2020) Induction of drought tolerance in tomato upon the application of ACC deaminase producing plant growth promoting rhizobacterium *Bacillus subtilis* Rhizo SF 48. Microbiol Res 234:126422. https://doi.org/10.1016/j.micres.2020.126422

Grillakis MG (2019) Increase in severe and extreme soil moisture droughts for Europe under climate change. Sci Total Environ 660:1245–1255. https://doi.org/10.1016/j.scitotenv.2019.01.001

Großkopf T, Soyer OS (2014) Synthetic microbial communities. Curr Opin Microbiol 18:72–77. https://doi.org/10.1016/j.mib.2014.02.002

Gul F, Khan IU, Rutherford S, Dai ZC, Li G, Du DL (2023) Plant growth promoting rhizobacteria and biochar production from P*arthenium hysterophorus* enhance seed germination and productivity in barley under drought stress. Front Plant Sci 14. https://doi.org/10.3389/fpls.2023.1175097

Gull A, Ahmad Lone A, Ul Islam Wani N (2019) Biotic and abiotic stresses in plants. In: Bosco De Oliveira A (ed) Abiotic and biotic stress in plants. IntechOpen

Gupta RK, Singh S, Chanotiya CS et al (2024) Diversity and functional characterization of endophytes in two contrasting cultivars of *Ocimum sanctum*: insights from culture-based and metagenomic approaches. J Plant Growth Regul. https://doi.org/10.1007/s00344-024-11373-0

Hahm MS, Son JS, Hwang YJ, Duk-Kee Kwon DK, Ghim SY (2017) Alleviation of salt stress in pepper (*Capsicum annum* L.) plants by plant growth-promoting Rhizobacteria. J Microbiol Biotechnol 27(10):1790–1797. https://doi.org/10.4014/jmb.1609.09042

Hakim S, Naqqash T, Nawaz MS et al (2021) Rhizosphere engineering with plant growth-promoting microorganisms for agriculture and ecological sustainability. Front Sustain Food Syst 5:617157. https://doi.org/10.3389/fsufs.2021.617157

Han D, Du M, Zhou Z et al (2020) Overexpression of a *Malus baccata* NAC transcription factor gene MbNAC25 increases cold and salinity tolerance in Arabidopsis. Int J Mol Sci 21:1198. https://doi.org/10.3390/ijms21041198

Hardoim PR, Van Overbeek LS, Elsas JDV (2008) Properties of bacterial endophytes and their proposed role in plant growth. Trends Microbiol 16:463–471. https://doi.org/10.1016/j.tim.2008.07.008

Haskett TL, Tkacz A, Poole PS (2021) Engineering rhizobacteria for sustainable agriculture. ISME J 15:949–964. https://doi.org/10.1038/s41396-020-00835-4

He A, Niu S, Yang D, Ren W, Zhao L, Sun Y, Meng L, Zhao Q, Paré PW, Zhang J (2021) Two PGPR strains from the rhizosphere of *Haloxylon ammodendron* promoted growth and enhanced drought tolerance of ryegrass. Plant Physiol Biochem 161:74–85. https://doi.org/10.1016/j.plaphy.2021.02.003

Herrmann L, Lesueur D (2013) Challenges of formulation and quality of biofertilizers for successful inoculation. Appl Microbiol Biotechnol 97:8859–8873. https://doi.org/10.1007/s00253-013-5228-8

Herz K, Dietz S, Gorzolka K et al (2018) Linking root exudates to functional plant traits. PLoS One 13:e0204128. https://doi.org/10.1371/journal.pone.0204128

Hoang XLT, Nguyen Y-NH, Thao NP, Tran L-SP (2020) NAC transcription factors in drought and salinity tolerance. In: Hasanuzzaman M, Tanveer M (eds) Salt and drought stress tolerance in plants. Springer International Publishing, Cham, pp 351–366

Hosseyni Moghaddam MS, Safaie N, Soltani J, Hagh-Doust N (2021) Desert-adapted fungal endophytes induce salinity and drought stress resistance in model crops. Plant Physiol Biochem 160:225–238. https://doi.org/10.1016/j.plaphy.2021.01.022

Hou D, Zhao Z, Hu Q et al (2020) PeSNAC-1 a NAC transcription factor from moso bamboo (*Phyllostachys edulis*) confers tolerance to salinity and drought stress in transgenic rice. Tree Physiol 40:1792–1806. https://doi.org/10.1093/treephys/tpaa099

Hubballi M, Rajamanickam S, Mawar R et al (2023) PGPM: fundamental, bioformulation, commercialization, and success at farmer's field. In: Mawar R, Sayyed RZ, Sharma SK, Sattiraju KS (eds) Plant growth promoting microorganisms of arid region. Springer Nature Singapore, Singapore, pp 257–301

Hussain Q, Asim M, Zhang R et al (2021) Transcription factors interact with ABA through gene expression and signaling pathways to mitigate drought and salinity stress. Biomolecules 11:1159. https://doi.org/10.3390/biom11081159

Ilyas N, Mazhar R, Yasmin H, Khan W, Iqbal S, El, Enshasy H, Dailin DJ (2020) Rhizobacteria isolated from saline soil induce systemic tolerance in wheat (*Triticum aestivum* L.) against salinity stress. Agronomy 10(7):989. https://doi.org/10.3390/agronomy10070989

Ikeda S, Kaneko T, Okubo T et al (2009) Development of a bacterial cell enrichment method and its application to the community analysis in soybean stems. Microb Ecol 58:703–714. https://doi.org/10.1007/s00248-009-9566-0

Ilangumaran G, Smith DL (2017) Plant growth promoting Rhizobacteria in amelioration of salinity stress: a systems biology perspective. Front Plant Sci 8:1768. https://doi.org/10.3389/fpls.2017.01768

Jakkula VS, Wani S (2018) Zeolites: potential soil amendments for improving nutrient and water use efficiency and agriculture productivity. Sci Rev Chem Commun 8:1–15

Jang S, Choi S-K, Zhang H et al (2023) History of a model plant growth-promoting rhizobacterium, *Bacillus velezensis* GB03: from isolation to commercialization. Front Plant Sci 14:1279896. https://doi.org/10.3389/fpls.2023.1279896

Ji Y, Yue L, Cao X et al (2023) Carbon dots promoted soybean photosynthesis and amino acid biosynthesis under drought stress: reactive oxygen species scavenging and nitrogen metabolism. Sci Total Environ 856:159125. https://doi.org/10.1016/j.scitotenv.2022.159125

Jiang M, Ye F, Liu F et al (2022) Rhizosphere melatonin application reprograms nitrogen-cycling related microorganisms to modulate low temperature response in barley. Front Plant Sci 13:998861. https://doi.org/10.3389/fpls.2022.998861

Joshi D, Kaushik A, Kumar R et al (2025) Improving plant performance through microbiome manipulation: the potential role of current bioengineering approaches. Bacteria 4:12. https://doi.org/10.3390/bacteria4010012

Ju Y, Yue X, Min Z et al (2020) VvNAC17, a novel stress-responsive grapevine (*Vitis vinifera* L.) NAC transcription factor, increases sensitivity to abscisic acid and enhances salinity, freezing, and drought tolerance in transgenic Arabidopsis. Plant Physiol Biochem 146:98–111. https://doi.org/10.1016/j.plaphy.2019.11.002

Kang S-M, Adhikari A, Lee K-E et al (2020) Inoculation with Indole-3-acetic acid-producing rhizospheric *Rhodobacter sphaeroides* KE149 augments growth of adzuki bean plants under water stress. J Microbiol Biotechnol 30:717

Kaur R, Saxena S (2023) Penicillium citrinum, a drought-tolerant Endophytic fungus isolated from wheat (*Triticum aestivum* L.) leaves with plant growth-promoting abilities. Curr Microbiol 80:184. https://doi.org/10.1007/s00284-023-03283-3

Kaur R, Kaur S, Kaur G (2021) Molecular and physiological manipulations in rhizospheric bacteria. Acta Physiol Plant 43:77. https://doi.org/10.1007/s11738-021-03251-z

Kaushal M, Wani SP (2016) Plant-growth-promoting rhizobacteria: drought stress alleviators to ameliorate crop production in drylands. Ann Microbiol 66:35–42. https://doi.org/10.1007/s13213-015-1112-3

Kazerooni EA, Maharachchikumbura SSN, Adhikari A et al (2021) Rhizospheric *Bacillus amyloliquefaciens* protects *Capsicum annuum* cv. Geumsugangsan from multiple abiotic stresses via multifarious plant growth-promoting attributes. Front Plant Sci 12:669693. https://doi.org/10.3389/fpls.2021.669693

Kerbab S, Silini A, Chenari Bouket A et al (2021) Mitigation of NaCl stress in wheat by rhizosphere engineering using salt habitat adapted PGPR halotolerant bacteria. Appl Sci 11:1034. https://doi.org/10.3390/app11031034

Khalilpour M, Mozafari V, Abbaszadeh-Dahaji P (2021) Tolerance to salinity and drought stresses in pistachio (*Pistacia vera* L.) seedlings inoculated with indigenous stress-tolerant PGPR isolates. Scientia Horticulturae 289:110440. https://doi.org/10.1016/j.scienta.2021.110440

Khan MA, Ullah I, Waqas M et al (2019a) Halo-tolerant rhizospheric Arthrobacter woluwensis AK1 mitigates salt stress and induces physio-hormonal changes and expression of *GmST1* and *GmLAX3* in soybean. Symbiosis 77:9–21. https://doi.org/10.1007/s13199-018-0562-3

Khan N, Bano A, Rahman MA et al (2019b) Comparative physiological and metabolic analysis reveals a complex mechanism involved in drought tolerance in chickpea (*Cicer arietinum* L.) induced by PGPR and PGRs. Sci Rep 9(1):2097. https://doi.org/10.1038/s41598-019-38702-8

Khan MA, Sahile AA, Jan R et al (2021a) Halotolerant bacteria mitigate the effects of salinity stress on soybean growth by regulating secondary metabolites and molecular responses. BMC Plant Biol 21:176. https://doi.org/10.1186/s12870-021-02937-3

Khan N, Ali S, Shahid MA et al (2021b) Insights into the interactions among roots, rhizosphere, and Rhizobacteria for improving plant growth and tolerance to abiotic stresses: a review. Cells 10:1551. https://doi.org/10.3390/cells10061551

Kodackattumannil P, Whitley K, Sasi S et al (2023) Novel inducible promoter *DREB1G* cloned from date palm exhibits high fold expression over *AtRD29* to drought and salinity stress. Plant Cell, Tissue and Organ Culture (PCTOC) 154:367–380. https://doi.org/10.1007/s11240-023-02460-3

Kosaraju SL (2005) Colon targeted delivery systems: review of polysaccharides for encapsulation and delivery. Crit Rev Food Sci Nutr 45:251–258. https://doi.org/10.1080/10408690490478091

Kour D, Rana KL, Yadav N et al (2019) Rhizospheric microbiomes: biodiversity, mechanisms of plant growth promotion, and biotechnological applications for sustainable agriculture. Plant Growth Promot Rhizobacteria Agric Sustain Theory Pract:19–65

Krasaekoopt W, Bhandari B, Deeth H (2004) The influence of coating materials on some properties of alginate beads and survivability of microencapsulated probiotic bacteria. Int Dairy J 14:737–743. https://doi.org/10.1016/j.idairyj.2004.01.004

Kumar et al (2016) Synergistic effect of *Pseudomonas putida* and *Bacillus amyloliquefaciens* ameliorates drought stress in chickpea (*Cicer arietinum* L.). Plant Signal Behav. https://doi.org/10.1080/15592324.2015.1071004

Kumar A, Dubey A (2020) Rhizosphere microbiome: engineering bacterial competitiveness for enhancing crop production. J Adv Res 24:337–352. https://doi.org/10.1016/j.jare.2020.04.014

Kumar V, Raghuvanshi N, Pandey AK et al (2023) Role of halotolerant plant growth-promoting Rhizobacteria in mitigating salinity stress: recent advances and possibilities. Agriculture (Basel) 13:168. https://doi.org/10.3390/agriculture13010168

Kurian M, Paul A (2021) Recent trends in the use of green sources for carbon dot synthesis–a short review. Carbon Trends 3:100032. https://doi.org/10.1016/j.cartre.2021.100032

Kuźniar, Włodarczyk, Wolińska (2019) Agricultural and other biotechnological applications resulting from trophic plant-endophyte interactions. Agronomy 9:779. https://doi.org/10.3390/agronomy9120779

Ladurner E, Benuzzi M, Fiorentini F, Lucchi A (2012) Efficacy of Amylo-X®, a new fungicide/bactericide based on *Bacillus amyloliquefaciens* strain D747 for the control of microbial plant pathogens

Laller R, Khosla PK, Negi N et al (2023) Bacterial volatiles as PGPRs: inducing plant defense mechanisms during stress periods. South Afr J Bot 159:131–139. https://doi.org/10.1016/j.sajb.2023.05.041

Lata R, Chowdhury S, Gond SK, White JF (2018) Induction of abiotic stress tolerance in plants by endophytic microbes. Lett Appl Microbiol 66:268–276. https://doi.org/10.1111/lam.12855

Lau S-E, Teo WFA, Teoh EY, Tan BC (2022) Microbiome engineering and plant biostimulants for sustainable crop improvement and mitigation of biotic and abiotic stresses. Discov Food 2:9. https://doi.org/10.1007/s44187-022-00009-5

Li X, Sun P, Zhang Y et al (2020) A novel PGPR strain *Kocuria rhizophila* Y1 enhances salt stress tolerance in maize by regulating phytohormone levels, nutrient acquisition, redox potential, ion homeostasis, photosynthetic capacity and stress-responsive genes expression. Environ Exp Bot 174:104023. https://doi.org/10.1016/j.envexpbot.2020.104023

Li W, Zeng Y, Yin F et al (2021) Genome-wide identification and comprehensive analysis of the NAC transcription factor family in sunflower during salt and drought stress. Sci Rep 11:19865. https://doi.org/10.1038/s41598-021-98107-4

Li G, Wang K, Qin Q et al (2023a) Integrated microbiome and Metabolomic analysis reveal responses of rhizosphere bacterial communities and root exudate composition to drought and genotype in Rice (*Oryza sativa* L.). Rice 16:19. https://doi.org/10.1186/s12284-023-00636-1

Li J, Wu H, Pu Q et al (2023b) Complete genome of *Sphingomonas paucimobilis* ZJSH1, an endophytic bacterium from *Dendrobium officinale* with stress resistance and growth promotion potential. Arch Microbiol 205:132. https://doi.org/10.1007/s00203-023-03459-2

Liu X-M, Zhang H (2015) The effects of bacterial volatile emissions on plant abiotic stress tolerance. Front Plant Sci 6. https://doi.org/10.3389/fpls.2015.00774

Lu F, Yang S, Song Y et al (2019) Hydroxyl functionalized carbon dots with strong radical scavenging ability promote cell proliferation. Mater Res Express 6:065030. https://doi.org/10.1088/2053-1591/ab0c55

Luo Y, Wang F, Huang Y, Zhou M, Gao J, Yan T, Sheng H, An L (2019) *Sphingomonas* sp. Cra20 increases plant growth rate and alters rhizosphere microbial community structure of *Arabidopsis thaliana* under drought stress. Front Microbiol. https://doi.org/1010.3389/fmicb.2019.01221

Lyu D, Smith DL (2022) The root signals in rhizospheric inter-organismal communications. Front Plant Sci 13:1064058. https://doi.org/10.3389/fpls.2022.1064058

Ma J, Wang L, Dai J et al (2021) The NAC-type transcription factor CaNAC46 regulates the salt and drought tolerance of transgenic *Arabidopsis thaliana*. BMC Plant Biol 21:11. https://doi.org/10.1186/s12870-020-02764-y

Ma H, Li P, Liu X et al (2022) Poly-γ-glutamic acid enhanced the drought resistance of maize by improving photosynthesis and affecting the rhizosphere microbial community. BMC Plant Biol 22:11. https://doi.org/10.1186/s12870-021-03392-w

Mhlongo MI, Piater LA, Dubery IA (2022) Profiling of volatile organic compounds from four plant growth-promoting Rhizobacteria by SPME–GC–MS: a metabolomics study. Meta 12:763. https://doi.org/10.3390/metabo12080763

Moreno-Galván A, Romero-Perdomo FA, Estrada-Bonilla G et al (2020) Dry-Caribbean bacillus spp. strains ameliorate drought stress in maize by a strain-specific antioxidant response modulation. Microorganisms 8:823. https://doi.org/10.3390/microorganisms8060823

Murphy B, Doohan F, Hodkinson T (2018) From concept to commerce: developing a successful fungal endophyte inoculant for agricultural crops. J Fungi 4:24. https://doi.org/10.3390/jof4010024

Mushtaq Z (2020) PGPR: present role, mechanism of action and future prospects along bottlenecks in commercialization. EQA Int J environ Qual:9–15. https://doi.org/10.6092/ISSN.2281-4485/11103

Nagrale DT, Chaurasia A, Kumar S et al (2023) PGPR: the treasure of multifarious beneficial microorganisms for nutrient mobilization, pest biocontrol and plant growth promotion in field crops. World J Microbiol Biotechnol 39:100. https://doi.org/10.1007/s11274-023-03536-0

Narayanasamy S, Uthandi S (2024) Plant growth promoting signatory volatiles emitted by a drought-tolerant bacterium *Bacillus altitudinis* FD48 and its role in moisture stress alleviation in rice (*Oryza sativa* L.). Plant Stress 14:100621. https://doi.org/10.1016/j.stress.2024.100621

Nasir MW, Toth Z (2022) Effect of drought stress on potato production: a review. Agronomy 12:635. https://doi.org/10.3390/agronomy12030635

Nawaz A, Shahbaz M, Asadullah, Imran A, Marghoob MU, Imtiaz M, Mubeen F (2020) Potential of salt tolerant PGPR in growth and yield augmentation of wheat (*Triticum aestivum* L.) under saline conditions. Front Microbiol. https://doi.org/1110.3389/fmicb.2020.02019

Nayana AR, Joseph BJ, Jose A, Radhakrishnan EK (2020) Nanotechnological advances with PGPR applications. In: Hayat S, Pichtel J, Faizan M, Fariduddin Q (eds) Sustainable agriculture reviews 41. Springer, Cham, pp. 163–180

Negrão S, Schmöckel SM, Tester M (2017) Evaluating physiological responses of plants to salinity stress. Ann Bot 119:1–11. https://doi.org/10.1093/aob/mcw191

Nowicka B, Ciura J, Szymańska R, Kruk J (2018) Improving photosynthesis, plant productivity and abiotic stress tolerance—current trends and future perspectives. J Plant Physiol 231:415–433. https://doi.org/10.1016/j.jplph.2018.10.022

O'Connell KP, Goodman RM, Handelsman J (1996) Engineering the rhizosphere: expressing a bias. Trends Biotechnol 14:83–88. https://doi.org/10.1016/0167-7799(96)80928-0

Ogbe AA, Finnie JF, Van Staden J (2020) The role of endophytes in secondary metabolites accumulation in medicinal plants under abiotic stress. South Afr J Bot 134:126–134. https://doi.org/10.1016/j.sajb.2020.06.023

Ogbe AA, Gupta S, Stirk WA et al (2023) Growth-promoting characteristics of fungal and bacterial endophytes isolated from a drought-tolerant mint species *Endostemon obtusifolius* (E. Mey ex Benth) N E Br. Plants 12:638. https://doi.org/10.3390/plants12030638

Ondrasek G, Rathod S, Manohara KK et al (2022) Salt stress in plants and mitigation approaches. Plants 11:717. https://doi.org/10.3390/plants11060717

Oviya G, Rangasamy A, Ariyan M et al (2023) Halotolerant nodule Rhizobial and passenger endophytes alleviates salinity stress in groundnut (*Arachis hypogaea* L.). J Plant Growth Regul 42:6620–6635. https://doi.org/10.1007/s00344-023-10919-y

Ozturk M, Turkyilmaz Unal B, García-Caparrós P et al (2021) Osmoregulation and its actions during the drought stress in plants. Physiol Plant 172:1321–1335. https://doi.org/10.1111/ppl.13297

Parihar P, Singh S, Singh R et al (2015) Effect of salinity stress on plants and its tolerance strategies: a review. Environ Sci Pollut Res 22:4056–4075. https://doi.org/10.1007/s11356-014-3739-1

Park I, Seo Y-S, Mannaa M (2023) Recruitment of the rhizo-microbiome army: assembly determinants and engineering of the rhizosphere microbiome as a key to unlocking plant potential. Front Microbiol 14:1163832. https://doi.org/10.3389/fmicb.2023.1163832

Pathak P, Rai VK, Can H et al (2022) Plant-endophyte interaction during biotic stress management. Plants 11:2203. https://doi.org/10.3390/plants11172203

Patra A, Sharma VK, Chattopadhyay A et al (2021) Rhizodeposition: an unseen teaser of nature and its prospects in nutrients dynamics. In: Rakshit A, Singh SK, Abhilash PC, Biswas A (eds) Soil science: fundamentals to recent advances. Springer Singapore, Singapore, pp 85–126

Peñuelas J, Asensio D, Tholl D et al (2014) Biogenic volatile emissions from the soil. Plant Cell Environ 37:1866–1891. https://doi.org/10.1111/pce.12340

Pereira LB, Andrade GS, Meneghin SP, Vicentini R, Ottoboni LMM (2019) Prospecting plant growth-promoting bacteria isolated from the rhizosphere of sugarcane under drought stress. Curr Microbiol 76(11):1345–1354. https://doi.org/10.1007/s00284-019-01749-x

Pereira SIA, Abreu D, Moreira H et al (2020) Plant growth-promoting rhizobacteria (PGPR) improve the growth and nutrient use efficiency in maize (*Zea mays* L.) under water deficit conditions. Heliyon 6:e05106. https://doi.org/10.1016/j.heliyon.2020.e05106

Pérez-Montaño F, Alías-Villegas C, Bellogín RA et al (2014) Plant growth promotion in cereal and leguminous agricultural important plants: from microorganism capacities to crop production. Microbiol Res 169:325–336. https://doi.org/10.1016/j.micres.2013.09.011

Phour M, Sindhu SS (2020) Amelioration of salinity stress and growth stimulation of mustard (*Brassica juncea* L.) by salt-tolerant Pseudomonas species. Appl Soil Ecol 149:103518. https://doi.org/10.1016/j.apsoil.2020.103518

Piskin E (1995) Biodegradable polymers as biomaterials. J Biomater Sci Polym Ed 6:775–795. https://doi.org/10.1163/156856295X00175

Polovyy V, Yashchenko L, Lopushniak V, Kolesnyk T (2023) The influence of chemical amelioration on the structure of the phosphate Fund of Retisol of Western Polissia in Ukraine. Ecol Eng Environ Technol 24:100–110. https://doi.org/10.12912/27197050/163669

Qian Y, Zhang T, Yu Y et al (2021) Regulatory mechanisms of bHLH transcription factors in plant adaptive responses to various abiotic stresses. Front Plant Sci 12:677611. https://doi.org/10.3389/fpls.2021.677611

Qin S, Feng W-W, Zhang Y-J et al (2018) Diversity of bacterial microbiota of coastal halophyte Limonium sinense and amelioration of salinity stress damage by symbiotic plant growth-promoting Actinobacterium *Glutamicibacter halophytocola* KLBMP 5180. Appl Environ Microbiol 84:e01533–18. https://doi.org/10.1128/AEM.01533-18

Quiza L, St-Arnaud M, Yergeau E (2015) Harnessing phytomicrobiome signaling for rhizosphere microbiome engineering. Front Plant Sci 6. https://doi.org/10.3389/fpls.2015.00507

Rafedzi EAK, Bharudin I, Kamaruddin S et al (2023) Volatile organic compounds (VOCs) emission and antioxidant activities of Rice treated with a consortium of plant growth-promoting Rhizobacteria under drought conditions. Univer J Agric Res 11:916–928. https://doi.org/10.13189/ujar.2023.110516

Rajasheker G, Jawahar G, Jalaja N et al (2019) Role and regulation of Osmolytes and ABA interaction in salt and drought stress tolerance. In: Plant signaling molecules. Elsevier, pp 417–436

Ramakrishnan P, Ariyan M, Rangasamy A et al (2023) Draft genome sequence of *Enterobacter cloacae* S23 a plant Growth promoting Passenger Endophytic bacterium isolated from Groundnut Nodule possesses stress tolerance traits. Curr Genomics 24:36–47. https://doi.org/10.2174/1389202924666230403123208

Rashid U, Yasmin H, Hassan MN et al (2022) Drought-tolerant *Bacillus megaterium* isolated from semi-arid conditions induces systemic tolerance of wheat under drought conditions. Plant Cell Rep 41:549–569. https://doi.org/10.1007/s00299-020-02640-x

Ravichandran M, Samiappan SC, Rangaraj S et al (2022) Nanoemulsion formulations with plant growth promoting rhizobacteria (PGPR) for sustainable agriculture. In: Bio-based nanoemulsions for agri-food applications. Elsevier, pp 207–223

Rejeb I, Pastor V, Mauch-Mani B (2014) Plant responses to simultaneous biotic and abiotic stress: molecular mechanisms. Plants 3:458–475. https://doi.org/10.3390/plants3040458

Riemann M, Dhakarey R, Hazman M et al (2015) Exploring Jasmonates in the hormonal network of drought and salinity responses. Front Plant Sci 6. https://doi.org/10.3389/fpls.2015.01077

Ryan PR, Dessaux Y, Thomashow LS, Weller DM (2009) Rhizosphere engineering and management for sustainable agriculture. Plant Soil 321:363–383. https://doi.org/10.1007/s11104-009-0001-6

Saberi Riseh R, Ebrahimi-Zarandi M, Tamanadar E et al (2021) Salinity stress: toward sustainable plant strategies and using plant growth-promoting Rhizobacteria encapsulation for reducing it. Sustainability 13:12758. https://doi.org/10.3390/su132212758

Saghafi D, Delangiz N, Lajayer BA, Ghorbanpour M (2019) An overview on improvement of crop productivity in saline soils by halotolerant and halophilic PGPRs. 3 Biotech 9:261. https://doi.org/10.1007/s13205-019-1799-0

Sahoo S, Sarangi S, Kerry RG (2017) Bioprospecting of endophytes for agricultural and environmental sustainability. In: Patra JK, Vishnuprasad CN, Das G (eds) Microbial biotechnology. Springer Singapore, Singapore, pp 429–458

Saini P, Verma A, Tiwari H et al (2024) Omics-based approaches in studying fungal endophytes and their associated secondary metabolites. In: Singh BP, Abdel-Azeem AM, Gautam V et al (eds) Endophytic fungi. Springer, Cham, pp 209–227

Sapre S, Gontia-Mishra I, Tiwari S (2022) Plant growth-promoting rhizobacteria ameliorates salinity stress in pea (*Pisum sativum*). J Plant Growth Regul 41(2):647–656. https://doi.org/10.1007/s00344-021-10329-y

Schoebitz M, López MD, Roldán A (2013) Bioencapsulation of microbial inoculants for better soil–plant fertilization. A review. Agron Sustain Dev 33:751–765. https://doi.org/10.1007/s13593-013-0142-0

Shanmugam H (2022) An insight on developing nanoformulations suitable for delivering plant beneficial microorganisms to crops under abiotic stresses. In: Mitigation of plant abiotic stress by microorganisms. Elsevier, pp 273–297

Sharma P, Sharma MMM, Malik A et al (2021) Rhizosphere, rhizosphere biology, and Rhizospheric engineering. In: Mohamed HI, El-Beltagi HE-DS, Abd-Elsalam KA (eds) Plant growth-promoting microbes for sustainable biotic and abiotic stress management. Springer, Cham, pp 577–624

Shi Y, Yang H, Chu M et al (2021) Differentiation and variability in the rhizosphere and Endosphere microbiomes of healthy and diseased cotton (*Gossypium* sp.). Front Microbiol 12:765269. https://doi.org/10.3389/fmicb.2021.765269

Siddique S, Naveed M, Yaseen M, Shahbaz M (2022) Exploring potential of seed Endophytic bacteria for enhancing drought stress resilience in maize (*Zea mays* L.). Sustainability 14:673. https://doi.org/10.3390/su14020673

Singh RP, Jha PN (2017) Analysis of fatty acid composition of PGPR *Klebsiella* sp. SBP-8 and its role in ameliorating salt stress in wheat. Symbiosis 73(3):213–222. https://doi.org/10.1007/s13199-017-0477-4

Singh K, Chandra A (2021) DREBs-potential transcription factors involve in combating abiotic stress tolerance in plants. Biologia (Bratisl) 76:3043–3055. https://doi.org/10.1007/s11756-021-00840-8

Singh RP, Shelke GM, Kumar A, Jha PN (2015) Biochemistry and genetics of ACC deaminase: a weapon to "stress ethylene" produced in plants. Front Microbiol 6. https://doi.org/10.3389/fmicb.2015.00937

Singh S, Singh S, Singh A, Kalra A (2022) Endophytic microbes mitigate biotic-abiotic stresses and modulate secondary metabolite pathway in plants. In: Microbial syntrophy-mediated eco-enterprising. Elsevier, pp 87–124

Singh A, Neha, Yadav S et al (2024a) Metagenomics: a approach for understanding microbe-microbe and plant-microbiome-interactions. In: Microbial Technology for Agro-Ecosystems. Elsevier, pp 309–332

Singh S, Roychowdhury R, Mukherjee A, et al (2024b) Impact of species-specific plant growth-promoting Rhizobacteria (PGPR) on maize (*Zea mays*) phenotypic and biochemical diversity. https://doi.org/10.1101/2024.04.28.591576

Solanki MK, Joshi NC, Singh PK et al (2024) From concept to reality: transforming agriculture through innovative rhizosphere engineering for plant health and productivity. Microbiol Res 279:127553. https://doi.org/10.1016/j.micres.2023.127553

Soma F, Takahashi F, Yamaguchi-Shinozaki K, Shinozaki K (2021) Cellular phosphorylation signaling and gene expression in drought stress responses: ABA-dependent and ABA-independent regulatory systems. Plants 10:756. https://doi.org/10.3390/plants10040756

Sultana S, Alam S, Karim MM (2021) Screening of siderophore-producing salt-tolerant rhizobacteria suitable for supporting plant growth in saline soils with iron limitation. J Agric Food Res 4:100150. https://doi.org/10.1016/j.jafr.2021.100150

Sun R, Guo X, Wang D, Chu H (2015) Effects of long-term application of chemical and organic fertilizers on the abundance of microbial communities involved in the nitrogen cycle. Appl Soil Ecol 95:171–178. https://doi.org/10.1016/j.apsoil.2015.06.010

Tanveer S, Ilyas N, Akhtar N et al (2023) Induction of regulatory mechanisms by plant growth promoting rhizobacteria in crops facing drought stress. Crop Pasture Sci 74:856–870. https://doi.org/10.1071/CP22263

Thomas P, Swarna GK, Roy PK, Patil P (2008) Identification of culturable and originally non-culturable endophytic bacteria isolated from shoot tip cultures of banana cv. Grand Naine. Plant Cell Tissue Organ Cult 93:55–63. https://doi.org/10.1007/s11240-008-9341-9

Tian T, Reverdy A, She Q et al (2020) The role of rhizodeposits in shaping rhizomicrobiome. Environ Microbiol Rep 12:160–172. https://doi.org/10.1111/1758-2229.12816

Timmusk S, Abd El-Daim IA, Copolovici L et al (2014) Drought-tolerance of wheat improved by rhizosphere bacteria from harsh environments: enhanced biomass production and reduced emissions of stress volatiles. PLoS One 9:e96086. https://doi.org/10.1371/journal.pone.0096086

Tufail MA, Touceda-González M, Pertot I, Ehlers R-U (2021) *Gluconacetobacter diazotrophicus Pal5* enhances plant robustness status under the combination of moderate drought and low nitrogen stress in *Zea mays* L. Microorganisms 9:870. https://doi.org/10.3390/microorganisms9040870

Tufail MA, Ayyub M, Irfan M et al (2022) Endophytic bacteria perform better than endophytic fungi in improving plant growth under drought stress: a meta-comparison spanning 12 years (2010–2021). Physiol Plant 174:e13806. https://doi.org/10.1111/ppl.13806

Tyagi S, Rajput P, Pramanik A et al (2023) Plant growth-promoting Rhizobacteria Nanoemulsion and their applications. In: Husen A, Bachheti RK, Bachheti A (eds) Current trends in green Nano-emulsions. Springer Nature Singapore, Singapore, pp 123–140

Ullah A, Nisar M, Ali H et al (2019) Drought tolerance improvement in plants: an endophytic bacterial approach. Appl Microbiol Biotechnol 103:7385–7397. https://doi.org/10.1007/s00253-019-10045-4

Upadhayay VK, Chitara MK, Mishra D et al (2023) Synergistic impact of nanomaterials and plant probiotics in agriculture: a tale of two-way strategy for long-term sustainability. Front Microbiol 14:1133968. https://doi.org/10.3389/fmicb.2023.1133968

Vasant G, Bhatt S, Raghav R, Joshi P (2023) Revitalization of PGPR through integrating nanotechnology for sustainable development in agriculture. In: Plant-microbe interaction – recent advances in molecular and biochemical approaches. Elsevier, pp 227–248

Vejan P, Khadiran T, Abdullah R et al (2019) Encapsulation of plant growth promoting Rhizobacteria—prospects and potential in agricultural sector: a review. J Plant Nutr 42:2600–2623. https://doi.org/10.1080/01904167.2019.1659330

Venturi V, Keel C (2016) Signaling in the rhizosphere. Trends Plant Sci 21:187–198. https://doi.org/10.1016/j.tplants.2016.01.005

Verma A, Thakur S, Mamba G et al (2020) Graphite modified sodium alginate hydrogel composite for efficient removal of malachite green dye. Int J Biol Macromol 148:1130–1139. https://doi.org/10.1016/j.ijbiomac.2020.01.142

Verma H, Kumar D, Kumar V et al (2021) The potential application of endophytes in Management of Stress from drought and salinity in crop plants. Microorganisms 9:1729. https://doi.org/10.3390/microorganisms9081729

Verma A, Shameem N, Jatav HS et al (2022) Fungal endophytes to combat biotic and abiotic stresses for climate-smart and sustainable agriculture. Front Plant Sci 13:953836. https://doi.org/10.3389/fpls.2022.953836

Vurukonda SSKP, Vardharajula S, Shrivastava M, SkZ A (2016) Enhancement of drought stress tolerance in crops by plant growth promoting rhizobacteria. Microbiol Res 184:13–24. https://doi.org/10.1016/j.micres.2015.12.003

Vyas P, Rana AK, Kaur M (2022) Rhizospheric bacteria as soil health engineer promoting plant growth. In: Rhizosphere Engineering. Elsevier, pp 45–64

Waheed A, Xu H, Qiao X et al (2025) Biochar in sustainable agriculture and climate mitigation: mechanisms, challenges, and applications in the circular bioeconomy. Biomass Bioenergy 193:107531. https://doi.org/10.1016/j.biombioe.2024.107531

Wang C, Kuzyakov Y (2024) Rhizosphere engineering for soil carbon sequestration. Trends Plant Sci 29:447–468. https://doi.org/10.1016/j.tplants.2023.09.015

Wang Y, Zhu Y, Zhang S, Wang Y (2018) What could promote farmers to replace chemical fertilizers with organic fertilizers? J Clean Prod 199:882–890. https://doi.org/10.1016/j.jclepro.2018.07.222

Wang C, Ji Y, Cao X et al (2022) Carbon dots improve nitrogen bioavailability to promote the growth and nutritional quality of soybeans under drought stress. ACS Nano 16:12415–12424. https://doi.org/10.1021/acsnano.2c03591

Wang K, Lin Z, Dou J et al (2023) Identification and surveys of promoting plant growth VOCs from biocontrol bacteria *Paenibacillus peoriae* GXUN15128. Microbiol Spectr 11:e04346–22. https://doi.org/10.1128/spectrum.04346-22

Watanabe Y, Fang X, Minemoto Y et al (2002) Suppressive effect of saturated acyl L -ascorbate on the oxidation of linoleic acid encapsulated with Maltodextrin or gum Arabic by spray-drying. J Agric Food Chem 50:3984–3987. https://doi.org/10.1021/jf011656u

Wei Z, Ye J, Zhou Z et al (2021) Isolation and characterization of PoWRKY, an abiotic stress-related WRKY transcription factor from *Polygonatum odoratum*. Physiol Mol Biol Plants 27:1–9

Wei H, He W, Li Z et al (2022) Salt-tolerant endophytic bacterium *Enterobacter ludwigii* B30 enhance bermuda-grass growth under salt stress by modulating plant physiology and changing rhizosphere and root bacterial community. Front Plant Sci 13:959427. https://doi.org/10.3389/fpls.2022.959427

Wei J-T, Zhao S-P, Zhang H-Y et al (2023) GmDof41 regulated by the DREB1-type protein improves drought and salt tolerance by regulating the DREB2-type protein in soybean. Int J Biol Macromol 230:123255. https://doi.org/10.1016/j.ijbiomac.2023.123255

Wu C, Zhang X, Liu Y et al (2024) Drought stress increases the complexity of the bacterial network in the rhizosphere and endosphere of rice (*Oryza sativa* L.). Agronomy 14:1662. https://doi.org/10.3390/agronomy14081662

Xing Y, Dao J, Chen M et al (2023) Multi-omics reveals the sugarcane rhizosphere soil metabolism-microbiota interactions affected by drought stress. Appl Soil Ecol 190:104994. https://doi.org/10.1016/j.apsoil.2023.104994

Xu S, Zhao N, Qin D et al (2021) The synergistic effects of silicon and selenium on enhancing salt tolerance of maize plants. Environ Exp Bot 187:104482. https://doi.org/10.1016/j.envexpbot.2021.104482

Xun W, Shao J, Shen Q, Zhang R (2021) Rhizosphere microbiome: functional compensatory assembly for plant fitness. Comput Struct Biotechnol J 19:5487–5493. https://doi.org/10.1016/j.csbj.2021.09.035

Yadav S, Modi P, Dave A, et al (2020) Effect of abiotic stress on crops. Sustain Crop Prod:3

Yang J, Kloepper JW, Ryu C-M (2009) Rhizosphere bacteria help plants tolerate abiotic stress. Trends Plant Sci 14:1–4. https://doi.org/10.1016/j.tplants.2008.10.004

Yang H, Wang C, Chen F et al (2022) Foliar carbon dot amendment modulates carbohydrate metabolism, rhizospheric properties and drought tolerance in maize seedling. Sci Total Environ 809:151105. https://doi.org/10.1016/j.scitotenv.2021.151105

Yashavanthakumar KJ, Baviskar VS, Navathe S et al (2021) Impact of heat and drought stress on phenological development and yield in bread wheat. Plant Physiol Rep 26:357–367. https://doi.org/10.1007/s40502-021-00586-0

Yasmeen T, Arif MS, Tariq M, Akhtar S, Syrish A, Haidar W, Rizwan M, Hussain MI, Ahmad A, Ali S (2024) Biofilm producing plant growth promoting bacteria in combination with glycine betaine uplift drought stress tolerance of maize plant. Front Plant Sci 15. https://doi.org/10.3389/fpls.2024.1327552

Yasmin H, Rashid U, Hassan MN et al (2021) Volatile organic compounds produced by *Pseudomonas pseudoalcaligenes* alleviated drought stress by modulating defense system in maize (*Zea mays* L.). Physiol Plant 172:896–911. https://doi.org/10.1111/ppl.13304

Ye F, Jiang M, Zhang P et al (2022) Exogenous melatonin reprograms the rhizosphere microbial community to modulate the responses of barley to drought stress. Int J Mol Sci 23:9665. https://doi.org/10.3390/ijms23179665

Yeo Y, Baek N, Park K (2001) Microencapsulation methods for delivery of protein drugs. Biotechnol Bioprocess Eng 6:213–230. https://doi.org/10.1007/BF02931982

Yeung TW, Arroyo-Maya IJ, McClements DJ, Sela DA (2016) Microencapsulation of probiotics in hydrogel particles: enhancing *Lactococcus lactis subsp. cremoris* LM0230 viability using calcium alginate beads. Food Funct 7:1797–1804. https://doi.org/10.1039/C5FO00801H

Yogendra SG, Singh US, Sharma AK (2015) Bacterial mediated amelioration of drought stress in drought tolerant and susceptible cultivars of rice (*Oryza sativa* L.). Afr J Biotechnol 14(9):764–773. https://doi.org/10.5897/AJB, https://doi.org/10.5897/AJB2015.14405

Yue L, Jiao L, Tao M et al (2023) Dynamics of organic acid exudation and rhizobacteria in maize rhizosphere respond to N-CDs. Sci Total Environ 901:166500. https://doi.org/10.1016/j.scitotenv.2023.166500

Zerrouk IZ, Benchabane M, Khelifi L et al (2016) A pseudomonas strain isolated from date-palm rhizospheres improves root growth and promotes root formation in maize exposed to salt and aluminum stress. J Plant Physiol 191:111–119. https://doi.org/10.1016/j.jplph.2015.12.009

Zhang H, Kim M-S, Krishnamachari V et al (2007) Rhizobacterial volatile emissions regulate auxin homeostasis and cell expansion in *Arabidopsis*. Planta 226:839–851. https://doi.org/10.1007/s00425-007-0530-2

Zhang M, Yang L, Hao R et al (2020) Drought-tolerant plant growth-promoting rhizobacteria isolated from jujube (*Ziziphus jujuba*) and their potential to enhance drought tolerance. Plant Soil 452:423–440. https://doi.org/10.1007/s11104-020-04582-5

Zhang H, Zhu J, Gong Z, Zhu J-K (2022) Abiotic stress responses in plants. Nat Rev Genet 23:104–119. https://doi.org/10.1038/s41576-021-00413-0

Zhao J, Ni T, Li J et al (2016) Effects of organic–inorganic compound fertilizer with reduced chemical fertilizer application on crop yields, soil biological activity and bacterial community structure in a rice–wheat cropping system. Appl Soil Ecol 99:1–12. https://doi.org/10.1016/j.apsoil.2015.11.006

Zhu J-K (2016) Abiotic stress signaling and responses in plants. Cell 167:313–324. https://doi.org/10.1016/j.cell.2016.08.029

Chapter 11
Strategic Manipulation of Pathogen Recognition Receptors in Plants for Efficient Perception of Fungal Pathogens: Benefits and Risks

Sumanti Gupta, Anirban Bhar, and Debjyoti Bandhu Banerjee

Abstract The sessile nature of the plants compels them towards constant biotic and abiotic threat exposures throughout their life cycles. Increasing xenobiotic activity and rapid climate change have deteriorated the stress situation even more. Altered climatic conditions not only increase abiotic stress but also changes the behavior of existing pathogens causing emergence of new diseases. Compared to the wild plant population, crop plants are more severely impacted by these stresses, which significantly reduce potential yield. Additionally, population outburst further exacerbates the condition. In this situation, it is a major challenge for plant scientists to improve yield of crop plants to ensure food security despite all odds. Fungal invaders are the major group of pathogens responsible for substantial economic loss across the globe annually. Although many approaches have been taken to contain the spread of the diseases, e.g., the introduction of resistance genes, transgenic plants overexpressing gene of interest, crossing with the resistant varieties, targeted mutagenesis, etc., in most of these cases, resistance is either not optimum or lost with successive generations. Hence, a multifaceted approach that augments the plants' inherent immunomodulatory pathways may provide a stable resistance. Pathogen-associated molecular patterns (PAMPs) or microbe-associated molecular patterns (MAMPs) are conserved molecular structures recognized by cognate host receptors, like pattern recognition receptors (PRRs). This PAMP-PRR interaction is a prerequisite for the activation of plant immune signaling. Hence, modifying or engineering these receptors may provide an effective and robust resistance mechanism against a wide array of pathogenic groups. Although this study is in its very infancy, the present

S. Gupta
Department of Botany, Rabindra Mahavidyalaya, Hooghly, West Bengal, India

A. Bhar (✉) · D. B. Banerjee
Molecular Plant & Microbiology Laboratory (MPML), Department of Botany (Post Graduate), Ramakrishna Mission Vivekananda Centenary College, Kolkata, West Bengal, India
e-mail: anirbanbhar@rkmvccrahara.ac.in

A. Singh et al. (eds.), *Sustainable Crop Production*, Sustainability Sciences in Asia and Africa, https://doi.org/10.1007/978-981-95-7844-3_11

chapter tries to convey a comprehensive outlook for PAMP-PRR interaction, particularly for the fungal pathogens, and it also highlights their potential role in the improvement of disease resistance in plants.

Keywords Bioengineering · Danger-associated molecular pattern (DAMPs) · Fungus-host interaction · Host · Fungal pathogen · Pathogen- or microbe-associated molecular patterns (PAMPs/MAMPs) · Pattern recognition receptors (PRRs)

Editor's Notes

Fungal pathogens are a major cause of yield losses. Pathogen recognition receptors or pattern recognition receptors (PRRs) in plants are crucial for accurately recognizing pathogens and mounting a defense response. PRRs recognize pathogen-encoded pathogen-associated molecular patterns (PAMPs) and microbe-associated molecular patterns (MAMPs), and fungal infection-driven DAMPs (Danger-Associated Molecular Patterns). This precise interaction initiates pattern triggered immunity (PTI), the first line of downstream immune response, which also constitutes trans-generational memory. The chapter by Gupta et al. presents an elaborate overview of the PRR-molecular pattern interactions. A detailed account of the types of pathogen-encoded molecular patterns and host-encoded PRRs and co-receptors in a variety of crops allows readers to understand the complexity of the host-pathogen identification and evasion race. The authors propose that the specific nature of molecular interaction can form bases for bioengineering of PRRs. Although a nascent field, research scholars must devise creative strategies to manipulate the PRRs system for triggering PTI and achieving an amplified defense response.

Actionable Points

1. Fungal PAMPs and/or fungal pathogen-induced host DAMPs should be precisely identified for specific host-fungal interaction.
2. Host PRRs for a particular plant-fungal interaction model should be determined and the epitopes of PAMP-PRR interaction pinpointed for carrying out downstream bioengineering experiments.
3. The environmental parameters of disease development should be well understood to evaluate the successful bioengineered plants under greenhouse conditions prior to field releases.
4. The circadian parameters should also be properly known for raising effective engineered crops against fungal pathogens.
5. Specific host-fungal interacting secretome of either host or fungal origin should also be identified to depict appropriate ecological conditions for promoting the sustainable growth of new engineered PRR bearing crop hosts against fungal pathogens.

11.1 Introduction

Fungi are known to have played a major role in the establishment of terrestrial habitat and evolution of land plants (Selosse et al. 2015). Various fungal species are known to establish saprophytic, symbiotic, or antagonistic relationships with host plants (Cheek et al. 2020). Evolutionary studies have highlighted the emergence of pathogenic plant-fungal interactions as an outcome of gross quantitative imbalance in resource allocation between the interacting partners (Priyashantha et al. 2023). About 8000 fungal species are known to be capable of infecting plants (Fisher et al. 2012). Among the huge number of harmful organisms that pose incessant threats to plants, fungal pathogens form a large share, causing serious loss of agricultural yield, which amounts to almost 20–40% of annual global harvest (Stukenbrock and Gurr 2023).

The capacity of discretely distinguishing self from non-self and normal-self from damaged-self is considered to be a unique feature of the plant world. Thus, plants are universally equipped with sophisticated surveillance systems that readily sense the presence of foreign molecules or damaged-self molecules within the surrounding vicinity or internally (Sanabria et al. 2010). Fungal pathogens usually invade the host plants through roots or other aerial parts (Li et al. 2022). They are classified on the basis of their mode of invasion, colonization, and establishment within the host (Doehlemann et al. 2017). Interestingly, previous studies highlighted the structural specificity of the host-fungal interface, which is possibly said to govern the outcome of the resistance/virulence of the interaction in a case-specific manner (Lacaze and Joly 2020). However, almost all fungal pathogens, irrespective of their tissue specificity, are known to deploy some common strategies in penetrating the host. The role of the secretome of both pathogen and host in controlling the fungal penetration has attracted the central focus of plant scientists. Prior to invasion, both partners are known to secrete specialized signaling metabolites, proteins, and peptides in the apoplastic region that act as an important sensor for initiating and also establishing the pathogenic interaction and/or triggering downstream defense responses. The cargos, as well as the exportation mechanism carrying the signaling molecules at the apoplastic region, are again diverse for different interacting host-fungal partners (Vincent et al. 2020).

Well-dissected plant-fungal interaction studies over the past few decades have identified diverse fungal pathogen or microbe-associated molecular patterns (PAMPs/MAMPs) or fungal infection–driven danger-associated molecular patterns (DAMPs) of host origin (Guo and Cheng 2022). These PAMPs/MAMPs/DAMPs are perceived by pattern recognition receptors (PRRs) of host origin, which are known to act as the first line of defense, termed pattern-triggered immunity (PTI). In a large number of cases, PTI is reported to limit further fungal invasion. However, a few opportunistic fungal pathogens are reported to secrete specialized molecules called "effectors" that dodge the host's surveillance to successfully invade the host, causing effector-triggered susceptibility (ETS). As a counter-defense, the host deploys resistance (R) proteins that activate effector-triggered immunity (ETI) and prevent further fungal establishment within the host (Tripathi et al. 2024). Although the plant-fungal interaction studies have made significant advances over the past few decades, the loss of economically important agricultural crops caused by fungal pathogens is still on the rise. Therefore, a better understanding of the fungal-crop interaction and possible biotechnological intervention that could restrict the entry of fungal pathogens right at the doorstep of the host plant still remain the need of the hour. The present chapter discusses the types of fungal pathogens affecting crop yield and elaborates on the details of fungal PAMPs and host PRRs and the resulting downstream defense signaling. It also highlights the different aspects of bioengineering that could help in the molecular redesign of PAMP-PRR interaction models and make a measurable decline in annual crop losses. Factors that could regulate the bioengineering process shall also be discussed in brief. Hence, taken together, the present study is expected to shed light on the prospects and challenges of manipulating the plant-fungal PTI for better agricultural outcomes for the future.

11.2 Fungal Pathogens: A Recurring Threat for Crops

Fungi are the diverse group of chlorophyll-less eukaryotic organisms that have distinct cytological and chemical features. Although eukaryotic in nature, fungi have unique life cycle patterns and cell wall chemistry. Chitin present in the cell wall of fungal mycelia serves as a principal element of pathogen/microbial associated molecular patterns (PAMPs/MAMPs), which can directly be identified by the cognate host pattern recognition receptors (PRRs). This not only instigates plant immune responses but also generates transgenerational memory (Bhar et al. 2021a). The enormous utility of fungi spanned from industrial production of various chemicals, pharmaceutical applications, decomposers to human– and plant disease–causing agents spurred scientists' interest to study "fungal biology" intriguingly (Lange 2010). The agricultural production is directly connected to many fungal pathogens. The historical occurrence of loss of agricultural crops and associated human mortality due to starvation is well documented. The potato blight in Ireland (Irish famine or Great Hunger, 1845–1849) caused by *Phytophthora infestans*, the epidemic of downy mildews of vine in 1870 caused by *Plasmopara viticola* in central Europe, and wheat stem rust epidemic at Dakotas and Minnesota islands of United States caused by *Puccinia graminis tritici* during 1935 are some of the major examples. The "Great Bengal famine" of 1943 is an example of combined effect of agricultural loss and shortage of food supply during World War II. Along with administrative failure and massive import disruption due to war situation, another major cause behind the great Bengal famine was massive curtail in production due to outbreak of brown spot disease in rice caused by *Cochliobolus miyabeanus* (*Helminthosporium oryzae*). Pathogenic fungi still cause drastic reduction in crop yield even in recent times. Fungi have the largest numbers of plant pathogens that are responsible for a variety of crop diseases primarily by killing the plant cells or evoking multiple stress responses in plants (Gupta et al. 2013, 2017; Bhar et al. 2018). The primary sources of infection are soil or seed; it may also infect healthy plants through crop debris, cross infection, or even through infected agricultural instruments. The spores of the fungal pathogens spread through wind or water splashes or by released mycotoxins (Martín et al. 2022; Fleurat-Lessard 2017). The pathogen after infection usually enters the host through natural openings, e.g., stomata, hydathodes, or cracks or wounds on plant surfaces (Bhar et al. 2021b). A diverse group of insects may also act as vectors for different fungal pathogens (Wielkopolan et al. 2021). The incidence of fungal diseases also depends on change in climatic conditions. The climate change primarily affects the change in global temperature, precipitation, humidity, and carbon footprint. These factors are capable of changing the nature of a pathogen including fungal pathogens. The impact of climate change alters the pathogen evolution, pathogenicity level, virulence, and nature of plant-fungal interaction, hence capable of giving rise to new infectious pathogens (Lahlali et al. 2024). In addition to the above risks, the host ranges of the pathogen may change, or due to the massive shifts of the climatic conditions, pathogens can spread into new areas damaging agroecosystem completely (Singh et al. 2023). The prominence of emerging new fungal pathogens in relation to climatic disruption is challenging for both crop plants and humans (Fisher et al. 2020). The *Puccinia striiformis*, heteroecious macrocyclic fungi, is responsible for stripe rust disease of wheat worldwide. This disease is old, but reemergence with the new host ranges was also documented recently. The detailed investigation on life cycle has revealed wide hosts among grass family, but their exact role in crop epidemics is still unknown (Chen 2020). Similarly, many fungal diseases of the tree species are reportedly increasing due to climatic alterations. The white pine blister rust disease caused by fungal pathogen *Cronartium ribicola* shifts its occurrence toward upper elevations due to altered weather in the southern Sierra Nevada area (Dudney et al. 2021). The combination of temperature and humidity is the most crucial factor for any infectious diseases, which is now utilized for analysis and prediction of global occurrence of infectious plant diseases including fungal origin, e.g., ProMED server (https://promedmail.org/) (Romero et al. 2022). In this view, fungal pathogens can also serve as indicators of climate change, and this is now used by the global disease databases in different predictive models to assess and analyze the impact of climate change on manifestation of different diseases (Garrett et al. 2021). The emergence of artificial intelligence (AI) and Internet of Things (IoT) has revolutionized the agricultural disease predictions through machine learning (ML) and deep understanding (Jafar et al. 2024). The integrative approach of AI, IoT, ML, and associated algorithm development can successfully be implemented for smart agricultural system, but in-depth validation of the predictive model is necessary for the robust disease forecasting. The integration of publicly available data with the predictive model is another challenge for disease forecasting (Delfani et al. 2024). Further in-depth studies are required to integrate and superposing the climate change prediction model with disease prediction model for accurate sustainable projections.

11.3 Host-Fungal Encounter: A Detailed Study

11.3.1 General Pattern Molecules of Fungi

Fungal PAMPs/MAMPS, also referred to as fungal elicitors, can be broadly classified into two groups such as saccharide type and protein type, based on their chemical nature (Guo and Cheng 2022). Saccharide elicitors consist of fungal cell wall component chitin and its deacetylated form chitosan. Host plants are known to have chitin- and chitosan-degrading enzymes, which lyse the fungal cell wall and release chitin/chitosan oligosaccharides, which subsequently activate downstream immune responses in the host plant (Pusztahelyi 2018). Apart from chitins and chitosans, two types of glucans (alpha and beta), made up of glucose molecules linked with glycosidic bonds, also act as potent fungal elicitors and trigger plant immune responses (Rebaque et al. 2021). Protein elicitors also form a diverse group. The cell wall–degrading enzymes (CWDEs) belonging to the polysaccharide lyase (PL), glycoside hydrolase (GH), and carbohydrate esterase (CE) families, which are known for degrading host cell walls, are readily recognized as PAMPs by the host surveillance system (Li et al. 2020). The necrosis- and ethylene-inducing peptide 1 (Nep1)-like proteins (NLPs), containing the conserved peptide nlp20, are a major class of pathogenic fungal elicitors. Heat-stable, glycine-rich, and cysteine-free hairpin protein and glycoprotein elicitors of *Magnaporthe oryzae* and *Alternaria tenuissima* are other types of protein elicitors (Kulye et al. 2012; Nie et al. 2022). Glycosylphosphatidyl inositol (GPI)-anchored protein that are products of post-translational modifications also serve as fungal PAMPs and stimulate host PTI, which in turn is reported to restrict other pathogenic fungal infections (Song et al. 2021). Additionally, some secreted protein of unknown function also acts as fungal PAMPs (Xu et al. 2022) (Fig. 11.1 and Table 11.1).

11.3.2 The Host Surveillance and Fungal Perception System: PRRs

Although the host immune response comprises of both PTI and ETI, the present section elaborates only the host PRRs that are reported to participate in PAMP/MAMP recognition and activation of PTI. The cognate host PRRs of PAMPs/MAMPs are primarily cell membrane bound receptors such as receptor-like protein kinases (RLKs) and receptor-like proteins (RLPs). RLKs are made up of a transverse helix, an extracellular domain that contains leucine-rich regions and a cytoplasmic kinase domain. RLPs also bear similar structure except for the replacement of cytosolic kinase domain with a cytosolic tail (Lee et al. 2021). However, in many cases, the involvement of co-receptors in the recognition of PAMAP/MAMP-PRR has been established (Ma et al. 2016). *BRI1-associated receptor kinase 1* (*BAK1*) is one such receptor that is known to recognize fungal ethylene-induced xylanase

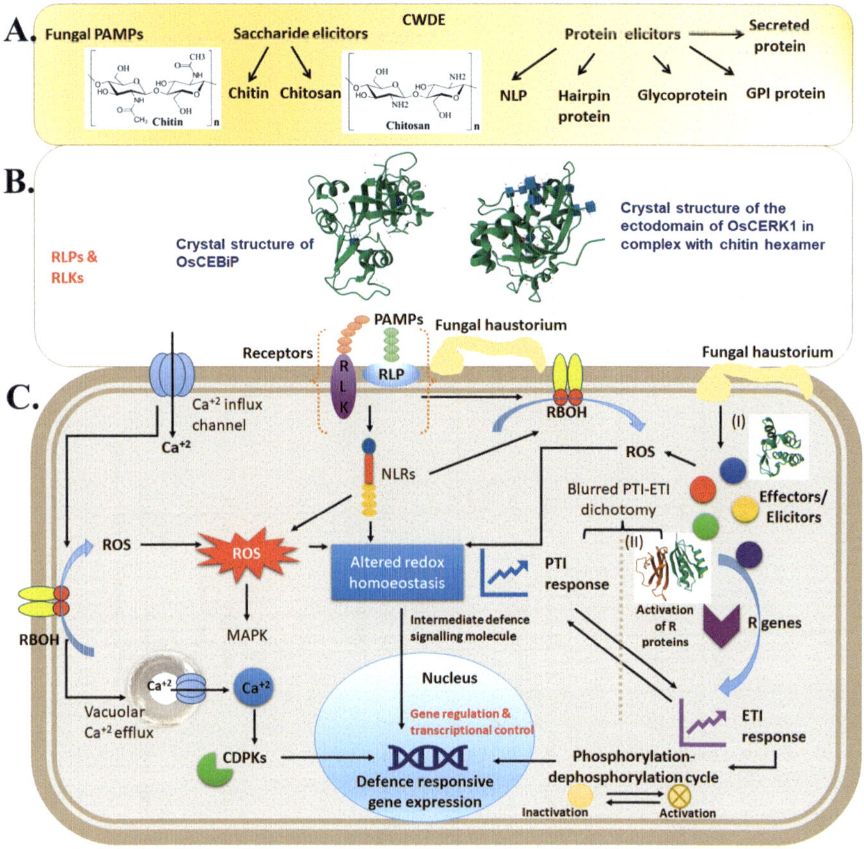

Fig. 11.1 The schematic diagram shows (**a**) principal types of fungal PAMPs, e.g., chitin and chitosan, the activity of cell wall-degrading enzymes (CWDEs) to generate DAMPs, and different types of fungal elicitors; (**b**) the RLPs and RLKs specific for fungal pathogens, e.g., the crystal structure of rice (*Oryza sativa* chitin receptors) OsCEBiP (PDB id: 5JCD) and OSCERK1 (PDB id: 7VS7) as representatives; (**c**) the detailed resistance pathway operated during plant pathogen interaction with pattern-triggered immunity (PTI) and effector-triggered immunity (ETI) during fungal ingression. (i) Representative crystal structure of fungal elicitors (PDB id: 1BEG), (ii) complex of rice blast (*Magnaporthe oryzae*) effector protein AVR-Pia with the HMA domain of Pikp-1 from rice (*Oryza sativa*) (PDB id: 6Q76) as representative structure

(EIX) as a co-receptor of *LeEIX1* (*Lycopersicon esculentum* ethylene-induced xylanase1) (Bar et al. 2010). Similarly, *OsCERK1* (an *Oryza sativa* lysin motif LysM-containing RLK) is also a known co-receptor of *OsCEBiP*, which recognizes fungal chitin (Shimizu et al. 2010). However, in *Arabidopsis thaliana*, the *CERK1* was found to be a co-receptor of *LYK5* (lysine motif receptor kinase 5), instead of *OsCEBiP* homologue *LYM2*, during chitin perception, suggesting that the PAMP-PRR interaction and downstream PTI grossly differ in case of plant-fungal interaction in a case-specific manner (Erwig et al. 2017) (Fig. 11.1 and Table 11.1).

Table 11.1 Types of fungal PAMPs/MAMPs and their corresponding PRRs in different plants

Serial no.:	Type of fungal pathogen (biotroph/ hemibiotroph/ necrotroph)	Pathogen name	PAMP/ PAMP-triggered DAMP	Nature of PAMP/ PAMP-assisted DAMP	Plant host	Invading tissue	Ligand PRR	Nature of PRR	References
1.	Hemibiotroph	*Phytophthora infestans*	Elicitin INF1	Proteinaceous elicitor	*Solanum microdontum*	Leaf	SmLRR-RLP ELICITIN RESPONSE ELR (SmELR)	Receptor-like protein (RLP)	Du et al. (2015)
2.	Necrotroph	*Botrytis cinerea*	Nep1-like proteins	Proteinaceous elicitor	*Arabidopsis thaliana*	Leaf	Leucine-rich repeat receptor protein (LRR-RP) RLP23	Receptor-like protein (RLP)	Albert et al. (2015)
3.	Biotroph	*Phytophthora sojae*	XEG1	Glycoside hydrolase	*Nicotiana benthamiana*	Leaf	LRR receptor-like protein response to XEG1 (RXEG1)	Receptor-like protein (RLP)	Wang et al. (2018)
4.	Hemibiotroph	*Pseudomonas syringae pv. tomato*	Flagellin	Bacterial flagellin	*Solanum lycopersicum*	Leaf	Flagellin	Receptor-like protein (RLP)	Zipfel (2014)
5.	Necrotroph	*Pythium periplocum*	Elicitin PpEli2	Proteinaceous elicitor	*Nicotiana benthamiana*	Root ecosystem	Receptor-like protein (REli)	Receptor-like protein (RLP)	Yang et al. (2023)

6.	Necrotroph	*Valsa mali*	VmE02	–	*Nicotiana benthamiana*	Vascular tissue	Receptor-like protein RE02	Receptor-like protein (RLP)	Nie et al. (2021)
7.	Biotroph	*Magnaporthe oryzae*	MoCel12A and MoCel12B	Endoglucanases	*Oryza sativa*	Leaf	OsCERK1		Yang et al. (2021)
8.	Hemibiotroph	*Fusarium* spp.	–	–	*Arabidopsis thaliana*	–	MDIS1-interacting receptor-like kinase 2 (MIK2)	Leucine-rich repeat receptor-like kinase	Coleman et al. (2021)
9.	Hemibiotroph	*Verticillium dahliae*	Ve1	–	*Solanum lycopersicum*	Vascular tissue	eLRR-RLP	Extracellular leucine-rich repeats	Song et al. (2020)
10.	Necrotroph	*Fusarium solani*	–	–	*Piper nigrum L*	Root tissue	Cf-9	–	Lau et al. (2020)
11.	Biotroph	*Blumeria graminis*	–	Chitin	*Haynaldia villosa*	Leaf	*CERK1-V*	LysM receptor kinase	Fan et al. (2022)
12.	Hemibiotrophic	*Phytophthora sojae*	XEG1	Glycoside hydrolase	*Glycine max; Nicotiana benthamiana*	Roots and leaves	RXEG1	Protein	Ma et al. (2015) and Lu and Tsuda (2021)

(continued)

Table 11.1 (continued)

Serial no.:	Type of fungal pathogen (biotroph/hemibiotroph/necrotroph)	Pathogen name	PAMP/PAMP-triggered DAMP	Nature of PAMP/PAMP-assisted DAMP	Plant host	Invading tissue	Ligand PRR	Nature of PRR	References
13.	Biotroph at early phase necrotroph in late phase	*Fusarium graminearum*	Chitin	Polysaccharide	*Arabidopsis* sp.	Natural openings of plants (e.g., stomata) and wounds	APEX; PIP1; RLK7	Protein	Manes et al. (2021)
14.	Necrotroph	General fungal elicitor, e.g., *Botrytis cinerea*	Xyn11/eix	Xylanase activity	*Prunus persica* and *Prunus avium*		PaEIX2; PpEIX2–2	Protein	Alvarez et al. (2024)
15.	Biotroph	*Puccinia striiformis* f. sp. *tritici* (*Pst*)	Chitin	Homopolymer of N-acetyl-d-glucosamine (GlcNAc) linked by β-(1,4) bond	*Triticum aestivum* L.	Leaf tissue	LysM-RLK, LysM-RLP, CERK1	Protein	Su et al. (2021) and Gong et al. (2020)
16.	Biotroph	*Blumeria graminis*	β-Glucans, chitin	Non-starch homopolysaccharide	*Triticum aestivum* L.		Receptor-like kinases (*TaRLK1*, *TaRLK2*), malectin-like/LRR-RLK (*RLK-V*), lectin type RLK (*LecRK-V*)	Protein	Mapuranga et al. (2022)

17.	Necrotroph	*Botrytis cinerea*	Chitin, β-glucans, β-1,2-glucan trisaccharide (B2G3)	Non-starch homopolysaccharide	*Arabidopsis thaliana*	Usually inflorescence, fruits and also stems, buds and tubers	Glucan-binding protein (GBP), chitin elicitor receptor kinase 1 (*At*CERK1)	Protein	Van Boerdonk et al. (2024) and Fuertes-Rabanal et al. (2024)
18.	Necrotroph	*Sclerotinia sclerotiorum* (Lib.)	Chitin, β-glucan as PAMPs, cutin monomer as DAMPS	Homopolysaccharides and biopolyester composed of hydroxyl and epoxy fatty acids	Wide host ranges, e.g., *Oryza sativa, Triticum aestivum, Zea mays, Hordeum vulgare, Avena sativa,* etc.	Primarily root and root-shoot junction	Lipid transfer proteins (LTPs) for cutin monomer	Protein	Hossain et al. (2023) and Kim et al. (2008)

11.3.3 *The Downstream PTI Signal Transduction Following* PAMP-PRR *Interaction*

Invasion and establishment of pathogenic fungi within their host generally follow these sequential steps: (1) adhesion to the host surface, (2) formation of infection structures, (3) penetration of the host cell wall, (4) colonization at the site of penetration, and finally (5) subsequent expansion into the host. Additionally, some pathogenic fungi secrete toxic metabolites that facilitate the infection process from initial adhesion to stable establishment within the host (Li et al. 2022). Like any plant-pathogen interaction, infection caused by pathogenic fungi is also known to induce some universal defense responses within the host plants. These responses include accumulation of reactive oxygen species (ROS) at the site of infection, often culminating in an oxidative burst and restriction of the pathogen at the site of penetration. Other responses include accumulation of phytoalexins, production of pathogenesis-related proteins (PR proteins), and activation of systemic acquired resistance (SAR) (Jones and Dangl 2006).

Following the binding of fungal PAMP/MAMP with its cognate host PRR, the resultant PTI is known to activate the mitogen-activated protein kinase (MAPK) pathway. Initially, the MAPKs induce the expression of host defense–related genes. For example, MEKK1-MKK4/MKK5-MPK3/MPK6 cascade pathway activates WRKY transcription factors during the fungal attack in *Arabidopsis* (Asai et al. 2002). Besides, MAPK also regulates the production of antimicrobial compounds, mediates cell wall strengthening, and increases callose accumulation at the site of infection (Shiu and Bleecker 2001). MAPKs are also known to modulate abscisic acid (ABA)-mediated stomatal opening and closure in *Pisum* guard cells during fungal infection (Su et al. 2017). In *Arabidopsis thaliana*, MAPKs are reported to control the synthesis of ethylene against *Botrytis cinerea* infection (Schweighofer et al. 2007). Additionally, calcium-dependent protein kinases (CDPKs) that are the largest family of calcium sensitive protein kinases containing a calcium ion–binding EF hand (predominant helix-loop-helix) motif are also important players of PTI regulation by mediating ROS burst and arresting fungal invasion in *Arabidopsis thaliana* (Dubiella et al. 2013) (Fig. 11.1).

11.4 Preventing the Invader at the Doorstep; Bioengineering PRRs for Enhanced Resistance

Successful restriction of pathogenic entry at the doorstep appears to be a hallmark achievement in agriculture that is yet to be achieved. Since the initiation of every plant-pathogen encounter is mediated by PAMP-PRR recognition, manipulating the interacting epitopes of the PAMP-PRR interaction complex by means of bioengineering tools and techniques resulting in de-recognition or uncoupling of the PAMP-PRR complex is believed to be highly needed for crop protection against harmful fungi. Reports suggest that such manipulation approaches have been undertaken for effector-R gene interaction models, where nucleotide-binding leucine-rich regions (NLRs) of R genes are altered to change ETI (Marchal et al. 2022). However, studies of similar manipulations of cell surface receptors that can modify PTI and prevent pathogen entry are still in their infancy. The tools and techniques of changing NLR receptors can be tested for their applicability in surface PRRs as well. Targeted manipulation of PTI may include the following approaches: (1) manipulating specificity by target site swapping with intra- or inter-species PRRs; (2) expanding specificity using targeted or random mutagenesis within PRRs; (3) modifying protease cleavage sites, referred to as decoy engineering, to extend specificity within intra- or inter-species PRRs; and (4) changing the binding sites of PRRs by ID exchange with effector targets, ID switching by nanoparticle introgression, and/or targeted mutagenesis at the binding sites of PRRs (Boutrot and Zipfel 2017; Dodds et al. 2024). Although the abovementioned approaches hold good promise for altering PTI and restraining fungal invaders at the early stages of infection, the experimental validation of any of the above approaches is still pending.

11.5 Factors Affecting PAMP-PRR Relation

Considering the fact that plant-fungal interaction is initiated by the recognition of fungal PAMPs by host PRRs, all other factors (abiotic and biotic) that contribute to the successful recognition process of PAMP-PRR are also considered to be equally important to trigger downstream defense responses (Priyashantha et al. 2023). Abiotic factors such as temperature undulations are known to alter both physiological and metabolic performances of the host and also alter the behavioral patterns of attacking fungi (Romero et al. 2022). On the other hand, humidity, known to act as an associated cofactor of temperature, directly influences host and pathogen physiology and modifies disease dynamics at spatial and temporal scales (Lahlali et al. 2024). Photoperiodism and light intensities have been reported to have a profound role in modulating innate immune responses of the host by regulating phytohormonal homeostasis and also contributing to SAR (Roden and Ingle 2009). Availability of adequate water is indispensable for any life form. In the case of plant-fungal interaction both below and above ground, water accessibility is known to act as an important driver of host-fungal interaction. Apart from having a direct impact on the physiology and metabolism of the host, water is primarily responsible for shaping microbial structure and community, which not only influences the fungal virulence but also controls the functionality of the surrounding microbiome (Aung et al. 2018). Researchers suggest that the carbon dioxide concentration also has a complex effect on the outcome of plant-fungal disease dynamics, which shows differential effects comprising positive, negative, and neutral effects on innate immunity of the host (Smith and Luna 2023). Apart from abiotic factors, the role of microbiota being directly involved with the plant-fungal arm race is a fact that has received scientific endorsement in recent times. The host-fungal interacting partners make constant attempts to simultaneously communicate and adjust their microbiome to function according to their individualistic advantage and impose a control on the outcome of the interaction (Mesny et al. 2024). Thus, manipulating the PAMP-PRR interaction to produce altered PTI responses under dynamic climatic conditions appears to be quite challenging and demands extensive experimentation based on realistic model predictions of natural environments.

11.6 Future Prospects and Challenges

The above sections summarize the fact that fungal-induced annual crop losses still remain a matter of deep concern for plant scientists worldwide. Thus, undertaking appropriate preventive measures to curtail the losses is of utmost importance. The new-age technological advancements can be used in designing apposite models to manipulate PAMP-PRR interaction and deliver altered PTI responses and successful restriction of fungal pathogens at the host doorstep. But, on the contrary, the feasibility of such depicted models in natural environments puts forth the following questions:

1. Will the manipulated PAMP-PRR interaction and resultant altered PTI be able to restrict fungal pathogens under altered climatic conditions that can change the entire disease triangle?
2. The role of the microbiome in having profound control over the disease triangle is well established. How will the bioengineered PTI responses explain the altered relation with their surrounding microenvironment under unpredictable natural changes?
3. Since manipulating the PRRs is believed to have a thorough effect on the host metabolism, how will the effect be passed on to its following generations in the form of transgenerational memory?

Addressing all the above questions is therefore necessary to carry out the study of manipulating host PRRs to prevent sustainable crop losses caused by fungal pathogens.

Acknowledgments This research did not receive any specific grant from funding agencies in the public, commercial, or not-for-profit sectors. The authors are sincerely thankful to the respected editors of the book for the constructive suggestions and invitations.

Declaration of Competing Interest The authors declare that they have no known competing financial interests or personal relationships that could have appeared to influence the work reported in this chapter.

Statements and Declarations

Funding No funding has been received to perform this work.

Declaration of Competing Interest Not applicable.

Availability of Data and Material Not applicable.

Code Availability Not applicable.

References

Albert I, Böhm H, Albert M, Feiler CE, Imkampe J, Wallmeroth N, Brancato C, Raaymakers TM, Oome S, Zhang H, Krol E, Grefen C, Gust AA, Chai J, Hedrich R, Van Den Ackerveken G, Nürnberger T (2015) An RLP23–SOBIR1–BAK1 complex mediates NLP-triggered immunity. Nat Plants 1(10):15140. https://doi.org/10.1038/nplants.2015.140

Alvarez A, Aceituno-Valenzuela U, Leibman-Markus M, Muñoz D, Rubilar C, Figueroa F, Pinto M, Latorre M, Stange C, Avni A, Bar M, Pizarro L (2024) Discovery of functional PRRs for the fungal elicitor Xyn11/eix in Prunus fruit trees. Plant Stress 14:100567. https://doi.org/10.1016/j.stress.2024.100567

Asai T, Tena G, Plotnikova J, Willmann MR, Chiu W-L, Gomez-Gomez L, Boller T, Ausubel FM, Sheen J (2002) MAP kinase signalling cascade in *Arabidopsis* innate immunity. Nature 415(6875):977–983. https://doi.org/10.1038/415977a

Aung K, Jiang Y, He SY (2018) The role of water in plant–microbe interactions. Plant J 93(4):771–780. https://doi.org/10.1111/tpj.13795

Bar M, Sharfman M, Ron M, Avni A (2010) BAK1 is required for the attenuation of ethylene-inducing xylanase (Eix)-induced defense responses by the decoy receptor LeEix1: BAK1 in Eix signaling. Plant J 63(5):791–800. https://doi.org/10.1111/j.1365-313X.2010.04282.x

Bhar A, Chatterjee M, Gupta S, Das S (2018) Salicylic acid regulates systemic defense signaling in chickpea during *Fusarium oxysporum* f. sp. ciceri race 1 infection. Plant Mol Biol Report 36(2):162–175. https://doi.org/10.1007/s11105-018-1067-1

Bhar A, Chakraborty A, Roy A (2021a) Plant responses to biotic stress: old memories matter. Plants 11(1):84. https://doi.org/10.3390/plants11010084

Bhar A, Jain A, Das S (2021b) Soil pathogen, *Fusarium oxysporum* induced wilt disease in chickpea: a review on its dynamicity and possible control strategies. Proc Indian Natl Sci Acad 87(2):260–274. https://doi.org/10.1007/s43538-021-00030-9

Boutrot F, Zipfel C (2017) Function, discovery, and exploitation of plant pattern recognition receptors for broad-spectrum disease resistance. Annu Rev Phytopathol 55(1):257–286. https://doi.org/10.1146/annurev-phyto-080614-120106

Cheek M, Nic Lughadha E, Kirk P, Lindon H, Carretero J, Looney B, Douglas B, Haelewaters D, Gaya E, Llewellyn T, Ainsworth AM, Gafforov Y, Hyde K, Crous P, Hughes M, Walker BE, Campostrini Forzza R, Wong KM, Niskanen T (2020) New scientific discoveries: plants and fungi. Plants People Planet 2(5):371–388. https://doi.org/10.1002/ppp3.10148

Chen X (2020) Pathogens which threaten food security: *Puccinia striiformis*, the wheat stripe rust pathogen. Food Sec 12(2):239–251. https://doi.org/10.1007/s12571-020-01016-z

Coleman AD, Maroschek J, Raasch L, Takken FLW, Ranf S, Hückelhoven R (2021) The *Arabidopsis* leucine-rich repeat receptor-like kinase MIK2 is a crucial component of early immune responses to a fungal-derived elicitor. New Phytol 229(6):3453–3466. https://doi.org/10.1111/nph.17122

Delfani P, Thuraga V, Banerjee B, Chawade A (2024) Integrative approaches in modern agriculture: IoT, ML and AI for disease forecasting amidst climate change. Precis Agric 25(5):2589–2613. https://doi.org/10.1007/s11119-024-10164-7

Dodds PN, Chen J, Outram MA (2024) Pathogen perception and signaling in plant immunity. Plant Cell 36(5):1465–1481. https://doi.org/10.1093/plcell/koae020

Doehlemann G, Ökmen B, Zhu W, Sharon A (2017) Plant pathogenic fungi. Microbiol Spectr 5(1):5.1.14. https://doi.org/10.1128/microbiolspec.FUNK-0023-2016

Du J, Verzaux E, Chaparro-Garcia A, Bijsterbosch G, Keizer LCP, Zhou J, Liebrand TWH, Xie C, Govers F, Robatzek S, Van Der Vossen EAG, Jacobsen E, Visser RGF, Kamoun S, Vleeshouwers VGAA (2015) Elicitin recognition confers enhanced resistance to *Phytophthora infestans* in potato. Nat Plants 1(4):15034. https://doi.org/10.1038/nplants.2015.34

Dubiella U, Seybold H, Durian G, Komander E, Lassig R, Witte C-P, Schulze WX, Romeis T (2013) Calcium-dependent protein kinase/NADPH oxidase activation circuit is required for rapid defense signal propagation. Proc Natl Acad Sci USA 110(21):8744–8749. https://doi.org/10.1073/pnas.1221294110

Dudney J, Willing CE, Das AJ, Latimer AM, Nesmith JCB, Battles JJ (2021) Nonlinear shifts in infectious rust disease due to climate change. Nat Commun 12(1):5102. https://doi.org/10.1038/s41467-021-25182-6

Erwig J, Ghareeb H, Kopischke M, Hacke R, Matei A, Petutschnig E, Lipka V (2017) Chitin-induced and *CHITIN ELICITOR RECEPTOR KINASE1 (CERK1)* phosphorylation-dependent endocytosis of *Arabidopsis thaliana* LYSIN MOTIF-CONTAINING RECEPTOR-LIKE KINASE5 (LYK5). New Phytol 215(1):382–396. https://doi.org/10.1111/nph.14592

Fan A, Wei L, Zhang X, Liu J, Sun L, Xiao J, Wang Y, Wang H, Hua J, Singh RP, Wang Z, Wang X (2022) Heterologous expression of the *Haynaldia villosa* pattern-recognition receptor CERK1-V in wheat increases resistance to three fungal diseases. Crop J 10(6):1733–1745. https://doi.org/10.1016/j.cj.2022.02.005

Fisher MC, Henk Daniel A, Briggs CJ, Brownstein JS, Madoff LC, McCraw SL, Gurr SJ (2012) Emerging fungal threats to animal, plant and ecosystem health. Nature 484(7393):186–194. https://doi.org/10.1038/nature10947

Fisher MC, Gurr SJ, Cuomo CA, Blehert DS, Jin H, Stukenbrock EH, Stajich JE, Kahmann R, Boone C, Denning DW, Gow NAR, Klein BS, Kronstad JW, Sheppard DC, Taylor JW, Wright GD, Heitman J, Casadevall A, Cowen LE (2020) Threats posed by the fungal kingdom to humans, wildlife, and agriculture. MBio 11(3):e00449-20. https://doi.org/10.1128/mBio.00449-20

Fleurat-Lessard F (2017) Integrated management of the risks of stored grain spoilage by seedborne fungi and contamination by storage mould mycotoxins. J Stored Prod Res 71:22–40. https://doi.org/10.1016/j.jspr.2016.10.002

Fuertes-Rabanal M, Largo-Gosens A, Fischer A, Munzert KS, Carrasco-López C, Sánchez-Vallet A, Engelsdorf T, Mélida H (2024) Linear β-1,2-glucans trigger immune hallmarks and enhance disease resistance in plants. J Exp Bot 75(22):7337–7350. https://doi.org/10.1093/jxb/erae368

Garrett KA, Nita M, De Wolf ED, Esker PD, Gomez-Montano L, Sparks AH (2021) Plant pathogens as indicators of climate change. In: Climate change. Elsevier, pp 499–513

Gong B-Q, Wang F-Z, Li J-F (2020) Hide-and-seek: chitin-triggered plant immunity and fungal counterstrategies. Trends Plant Sci 25(8):805–816. https://doi.org/10.1016/j.tplants.2020.03.006

Guo J, Cheng Y (2022) Advances in fungal elicitor-triggered plant immunity. IJMS 23(19):12003. https://doi.org/10.3390/ijms231912003

Gupta S, Bhar A, Chatterjee M, Das S (2013) *Fusarium oxysporum* f. sp. ciceri race 1 induced redox state alterations are coupled to downstream defense signaling in root tissues of chickpea (*Cicer arietinum* L.). PLoS One 8(9):e73163. https://doi.org/10.1371/journal.pone.0073163

Gupta S, Bhar A, Chatterjee M, Ghosh A, Das S (2017) Transcriptomic dissection reveals wide spread differential expression in chickpea during early time points of *Fusarium oxysporum* f. sp. ciceri Race 1 attack. PLoS One 12(5):e0178164. https://doi.org/10.1371/journal.pone.0178164

Hossain MDM, Sultana F, Li W, Tran L-SP, Mostofa MG (2023) *Sclerotinia sclerotiorum* (Lib.) de Bary: insights into the pathogenomic features of a global pathogen. Cells 12(7):1063. https://doi.org/10.3390/cells12071063

Jafar A, Bibi N, Naqvi RA, Sadeghi-Niaraki A, Jeong D (2024) Revolutionizing agriculture with artificial intelligence: plant disease detection methods, applications, and their limitations. Front Plant Sci 15:1356260. https://doi.org/10.3389/fpls.2024.1356260

Jones JDG, Dangl JL (2006) The plant immune system. Nature 444(7117):323–329. https://doi.org/10.1038/nature05286

Kim TH, Park JH, Kim MC, Cho SH (2008) Cutin monomer induces expression of the rice *OsLTP5* lipid transfer protein gene. J Plant Physiol 165(3):345–349. https://doi.org/10.1016/j.jplph.2007.06.004

Kulye M, Liu H, Zhang Y, Zeng H, Yang X, Qiu D (2012) Hrip1, a novel protein elicitor from necrotrophic fungus, *Alternaria tenuissima*, elicits cell death, expression of defence-related genes and systemic acquired resistance in tobacco. Plant Cell Environ 35(12):2104–2120. https://doi.org/10.1111/j.1365-3040.2012.02539.x

Lacaze A, Joly DL (2020) Structural specificity in plant–filamentous pathogen interactions. Mol Plant Pathol 21(11):1513–1525. https://doi.org/10.1111/mpp.12983

Lahlali R, Taoussi M, Laasli S-E, Gachara G, Ezzouggari R, Belabess Z, Aberkani K, Assouguem A, Meddich A, El Jarroudi M, Barka EA (2024) Effects of climate change on plant pathogens and host-pathogen interactions. Crop Environ 3(3):159–170. https://doi.org/10.1016/j.crope.2024.05.003

Lange L (2010) The importance of fungi for a more sustainable future on our planet. Fungal Biol Rev 24(3–4):90–92. https://doi.org/10.1016/j.fbr.2010.12.002

Lau ET, Khew CY, Hwang SS (2020) Transcriptomic analysis of pepper plants provides insights into host responses to *Fusarium solani* infestation. J Biotechnol 314–315:53–62. https://doi.org/10.1016/j.jbiotec.2020.03.014

Lee D-H, Lee H-S, Belkhadir Y (2021) Coding of plant immune signals by surface receptors. Curr Opin Plant Biol 62:102044. https://doi.org/10.1016/j.pbi.2021.102044

Li Y, Han Y, Qu M, Chen J, Chen X, Geng X, Wang Z, Chen S (2020) Apoplastic cell death-inducing proteins of filamentous plant pathogens: roles in plant-pathogen interactions. Front Genet 11:661. https://doi.org/10.3389/fgene.2020.00661

Li L, Zhu X-M, Zhang Y-R, Cai Y-Y, Wang J-Y, Liu M-Y, Wang J-Y, Bao J-D, Lin F-C (2022) Research on the molecular interaction mechanism between plants and pathogenic fungi. IJMS 23(9):4658. https://doi.org/10.3390/ijms23094658

Lu Y, Tsuda K (2021) Intimate association of PRR- and NLR-mediated signaling in plant immunity. MPMI 34(1):3–14. https://doi.org/10.1094/MPMI-08-20-0239-IA

Ma Z, Song T, Zhu L, Ye W, Wang Y, Shao Y, Dong S, Zhang Z, Dou D, Zheng X, Tyler BM, Wang Y (2015) A *Phytophthora sojae* glycoside hydrolase 12 protein is a major virulence factor during soybean infection and is recognized as a PAMP. Plant Cell 27(7):2057–2072. https://doi.org/10.1105/tpc.15.00390

Ma X, Xu G, He P, Shan L (2016) SERKing coreceptors for receptors. Trends Plant Sci 21(12):1017–1033. https://doi.org/10.1016/j.tplants.2016.08.014

Manes N, Brauer EK, Hepworth S, Subramaniam R (2021) MAMP and DAMP signaling contributes resistance to *Fusarium graminearum* in Arabidopsis. J Exp Bot 72(18):6628–6639. https://doi.org/10.1093/jxb/erab285

Mapuranga J, Chang J, Yang W (2022) Combating powdery mildew: advances in molecular interactions between *Blumeria graminis* f. sp. tritici and wheat. Front Plant Sci 13:1102908. https://doi.org/10.3389/fpls.2022.1102908

Marchal C, Pai H, Kamoun S, Kourelis J (2022) Emerging principles in the design of bioengineered made-to-order plant immune receptors. Curr Opin Plant Biol 70:102311. https://doi.org/10.1016/j.pbi.2022.102311

Martín I, Gálvez L, Guasch L, Palmero D (2022) Fungal pathogens and seed storage in the dry state. Plants 11(22):3167. https://doi.org/10.3390/plants11223167

Mesny F, Bauer M, Zhu J, Thomma BPHJ (2024) Meddling with the microbiota: fungal tricks to infect plant hosts. Curr Opin Plant Biol 82:102622. https://doi.org/10.1016/j.pbi.2024.102622

Nie J, Zhou W, Liu J, Tan N, Zhou J, Huang L (2021) A receptor-like protein from *Nicotiana benthamiana* mediates VmE02 PAMP-triggered immunity. New Phytol 229(4):2260–2272. https://doi.org/10.1111/nph.16995

Nie Y, Li G, Li J, Zhou X, Zhang Y, Shi Q, Zhou X, Li H, Chen X-L, Li Y (2022) A novel elicitor MoVcpo is necessary for the virulence of *Magnaporthe oryzae* and triggers rice defense responses. Front Plant Sci 13:1018616. https://doi.org/10.3389/fpls.2022.1018616

Priyashantha A, Dai D-Q, Bhat D, Stephenson S, Promputtha I, Kaushik P, Tibpromma S, Karunarathna S (2023) Plant–fungi interactions: where it goes? Biology 12(6):809. https://doi.org/10.3390/biology12060809

Pusztahelyi T (2018) Chitin and chitin-related compounds in plant–fungal interactions. Mycology 9(3):189–201. https://doi.org/10.1080/21501203.2018.1473299

Rebaque D, Del Hierro I, López G, Bacete L, Vilaplana F, Dallabernardina P, Pfrengle F, Jordá L, Sánchez-Vallet A, Pérez R, Brunner F, Molina A, Mélida H (2021) Cell wall-derived mixed-linked β-1,3/1,4-glucans trigger immune responses and disease resistance in plants. Plant J 106(3):601–615. https://doi.org/10.1111/tpj.15185

Roden LC, Ingle RA (2009) Lights, rhythms, infection: the role of light and the circadian clock in determining the outcome of plant–pathogen interactions. Plant Cell 21(9):2546–2552. https://doi.org/10.1105/tpc.109.069922

Romero F, Cazzato S, Walder F, Vogelgsang S, Bender SF, Van Der Heijden MGA (2022) Humidity and high temperature are important for predicting fungal disease outbreaks worldwide. New Phytol 234(5):1553–1556. https://doi.org/10.1111/nph.17340

Sanabria NM, Huang J-C, Dubery IA (2010) Self/non-self perception in plants in innate immunity and defense. Self/Nonself 1(1):40–54. https://doi.org/10.4161/self.1.1.10442

Schweighofer A, Kazanaviciute V, Scheikl E, Teige M, Doczi R, Hirt H, Schwanninger M, Kant M, Schuurink R, Mauch F, Buchala A, Cardinale F, Meskiene I (2007) The PP2C-type phosphatase AP2C1, which negatively regulates MPK4 and MPK6, modulates innate immunity, jasmonic acid, and ethylene levels in *Arabidopsis*. Plant Cell 19(7):2213–2224. https://doi.org/10.1105/tpc.106.049585

Selosse M, Strullu-Derrien C, Martin FM, Kamoun S, Kenrick P (2015) Plants, fungi and oomycetes: a 400-million year affair that shapes the biosphere. New Phytol 206(2):501–506. https://doi.org/10.1111/nph.13371

Shimizu T, Nakano T, Takamizawa D, Desaki Y, Ishii-Minami N, Nishizawa Y, Minami E, Okada K, Yamane H, Kaku H, Shibuya N (2010) Two LysM receptor molecules, CEBiP and OsCERK1, cooperatively regulate chitin elicitor signaling in rice: LysM receptors for rice chitin signaling. Plant J 64(2):204–214. https://doi.org/10.1111/j.1365-313X.2010.04324.x

Shiu S-H, Bleecker AB (2001) Plant receptor-like kinase gene family: diversity, function, and signaling. Sci STKE 2001(113):re22. https://doi.org/10.1126/stke.2001.113.re22

Singh BK, Delgado-Baquerizo M, Egidi E, Guirado E, Leach JE, Liu H, Trivedi P (2023) Climate change impacts on plant pathogens, food security and paths forward. Nat Rev Microbiol 21(10):640–656. https://doi.org/10.1038/s41579-023-00900-7

Smith F, Luna E (2023) Elevated atmospheric carbon dioxide and plant immunity to fungal pathogens: do the risks outweigh the benefits? Biochem J 480(22):1791–1804. https://doi.org/10.1042/BCJ20230152

Song R, Li J, Xie C, Jian W, Yang X (2020) An overview of the molecular genetics of plant resistance to the verticillium wilt pathogen *Verticillium dahliae*. IJMS 21(3):1120. https://doi.org/10.3390/ijms21031120

Song T, Zhang Y, Zhang Q, Zhang X, Shen D, Yu J, Yu M, Pan X, Cao H, Yong M, Qi Z, Du Y, Zhang R, Yin X, Qiao J, Liu Y, Liu W, Sun W, Zhang Z, Wang Y, Dou D, Ma Z, Liu Y (2021) The N-terminus of an *Ustilaginoidea virens* Ser-Thr-rich glycosylphosphatidylinositol-anchored protein elicits plant immunity as a MAMP. Nat Commun 12(1):2451. https://doi.org/10.1038/s41467-021-22660-9

Stukenbrock E, Gurr S (2023) Address the growing urgency of fungal disease in crops. Nature 617(7959):31–34. https://doi.org/10.1038/d41586-023-01465-4

Su J, Zhang M, Zhang L, Sun T, Liu Y, Lukowitz W, Xu J, Zhang S (2017) Regulation of stomatal immunity by interdependent functions of a pathogen-responsive MPK3/MPK6 cascade and abscisic acid. Plant Cell 29(3):526–542. https://doi.org/10.1105/tpc.16.00577

Su Y, Chen Y, Chen J, Zhang Z, Guo J, Cai Y, Zhu C, Li Z, Zhang H (2021) Effectors of *Puccinia striiformis* f. sp. tritici suppressing the pathogenic-associated molecular pattern-triggered immune response were screened by transient expression of wheat protoplasts. IJMS 22(9):4985. https://doi.org/10.3390/ijms22094985

Tripathi A, Pandey VK, Jha AK, Srivastava S, Jakhar S, Vijay SG, Rustagi S, Malik S, Choudhary P (2024) Intricacies of plants' innate immune responses and their dynamic relationship with fungi: a review. Microbiol Res 285:127758. https://doi.org/10.1016/j.micres.2024.127758

Van Boerdonk S, Saake P, Wanke A, Neumann U, Zuccaro A (2024) β-Glucan-binding proteins are key modulators of immunity and symbiosis in mutualistic plant–microbe interactions. Curr Opin Plant Biol 81:102610. https://doi.org/10.1016/j.pbi.2024.102610

Vincent D, Rafiqi M, Job D (2020) The multiple facets of plant–fungal interactions revealed through plant and fungal secretomics. Front Plant Sci 10:1626. https://doi.org/10.3389/fpls.2019.01626

Wang Y, Xu Y, Sun Y, Wang H, Qi J, Wan B, Ye W, Lin Y, Shao Y, Dong S, Tyler BM, Wang Y (2018) Leucine-rich repeat receptor-like gene screen reveals that *Nicotiana* RXEG1 regulates glycoside hydrolase 12 MAMP detection. Nat Commun 9(1):594. https://doi.org/10.1038/s41467-018-03010-8

Wielkopolan B, Jakubowska M, Obrępalska-Stęplowska A (2021) Beetles as plant pathogen vectors. Front Plant Sci 12:748093. https://doi.org/10.3389/fpls.2021.748093

Xu Q, Hu S, Jin M, Xu Y, Jiang Q, Ma J, Zhang Y, Qi P, Chen G, Jiang Y, Zheng Y, Wei Y (2022) The N-terminus of a *Fusarium graminearum* -secreted protein enhances broad-spectrum disease resistance in plants. Mol Plant Pathol 23(12):1751–1764. https://doi.org/10.1111/mpp.13262

Yang C, Liu R, Pang J, Ren B, Zhou H, Wang G, Wang E, Liu J (2021) Poaceae-specific cell wall-derived oligosaccharides activate plant immunity via OsCERK1 during *Magnaporthe oryzae* infection in rice. Nat Commun 12(1):2178. https://doi.org/10.1038/s41467-021-22456-x

Yang K, Wang Y, Li J, Du Y, Zhai Y, Liang D, Shen D, Ji R, Ren X, Peng H, Jing M, Dou D (2023) The *Pythium periplocum* elicitin PpEli2 confers broad-spectrum disease resistance by triggering a novel receptor-dependent immune pathway in plants. Hortic Res 10(2):uhac255. https://doi.org/10.1093/hr/uhac255

Zipfel C (2014) Plant pattern-recognition receptors. Trends Immunol 35(7):345–351. https://doi.org/10.1016/j.it.2014.05.004

Chapter 12
Plant Immunity Inducers: Discovery and Delivery for Crop Protection

Manjima Mohanan, Febina Fernandez, and Anish Kundu

Abstract The global food security in the twenty-first century is concerning due to the uncertain climate change, pest outbreaks, declining soil fertility, and rising population. Since healthy and productive plants are indispensable to meet the global food demands, these factors negatively influence crop productivity. Though plants have their own immunity, they cannot thrive in certain conditions. To mitigate this, it is necessary to develop innovative and sophisticated strategies to enhance defense responses in plants. "Plant immunity induction" which is also called "plant priming" is one of the well-known strategies of crop protection that is environment-friendly as well as sustainable. Priming brings the plants into an "alarm state" by exposing them into minor stressor that can be natural or synthetic compounds, which help them to react more efficiently and rapidly in future, mirroring the induced innate immunity in the mammalian cell through vaccination. Various complex, interconnected biochemical and genetic processes are involved in creating stress memories in plants. Multiple studies showed that primed plants exhibit improved growth, reduced yield losses, and lower dependence on chemical pesticides and fertilizers. By incorporating plant priming into agricultural practices, it is possible to boost crop productivity while promoting environmental sustainability. This chapter sheds light on the concept, types, and molecular mechanisms of plant priming and its practical applications in agriculture. In addition to this, we addressed the limitations and challenges in optimizing priming responses in plants under diverse environmental conditions. Combining plant's innate immune capabilities with this innovative approach is undoubtedly helpful in ensuring resilient agricultural systems that can overcome current and future challenges.

Authors Manjima Mohanan and Febina Fernandez have equally contributed to this chapter.

M. Mohanan · F. Fernandez · A. Kundu (✉)
Plant Biotechnology and Disease Biology Division, BRIC-Rajiv Gandhi Centre for Biotechnology (BRIC-RGCB), Thiruvananthapuram, Kerala, India
e-mail: anishkundu@rgcb.res.in

A. Singh et al. (eds.), *Sustainable Crop Production*, Sustainability Sciences in Asia and Africa, https://doi.org/10.1007/978-981-95-7844-3_12

Keywords Plant immunity · Defense priming · Stress memory · Crop protection · Effectors · Plant-induced immunity

Editor's Notes

Mohanan et al. introduce the readers to the exciting field of plant defense priming leading to the creation of stress-memory. This interesting idea, similar to vaccination-based immunity in animals, can become a key strategy for imparting crop protection in future. Defense priming can be achieved by way of chemical agents, pathogenic and beneficial microbes, natural and synthetic elicitors such as phytohormones, and derivatives thereof. Multiple molecular regulators mediate defense priming including MAPK-WRKY signal cascades, PAMPs, HAMPs, and PTMs of histone proteins. Epigenetics associated chromatin remodeling and histone modification derived defense priming mechanisms are identified as key players for trans-generational stress-memory. A novel approach for priming defense is based on RNAi and use of specific dsRNA targeting pathogen or host genes. The authors explain the advantages of specificity, environmental safety, and rapid adaptability of RNAi-based approaches and encourage the scholars to identify the determinants of specificity and effectiveness. Particularly, the authors draw the attention of policy-makers and scientists to a range of experimental and regulatory hurdles that must be addressed. Prospects of defense priming, as a means to achieve sustainable agricultural practices, must be explored by future scholars.

Actionable Points

1. Ensuring the use of effective biological (e.g., AMF, PGPR), chemical (e.g., BABA, SA), and natural compounds tailored to specific crops and stress conditions.
2. Using different distribution methods by seed priming, foliar spray, or root dipping, optimized with precise dosages and timing to achieve reliable field results.
3. Tracking the molecular effects of priming as hormone profiling and gene expression during implementation of functional field experiments.
4. Integrating priming into more general crop management techniques such as organic farming and integrated pest management.
5. Formulating policies and their commercialization to increase their uptake.

12.1 Introduction

The current state of global food security and the equilibrium between the world's expanding population's demand for food and agricultural production worldwide, along with discrepancies between supply and demand at the local, national, and regional levels, are concerning. Moreover, changes in global climate along with elevating temperature, pollution, droughts, changing pattern of pest, and pathogen invasion are becoming a threat to the agriculture. In order to meet the increasing food demands, plant protection, in general, and crop protection against plant diseases, in particular, are necessary. Pathogens, animals, and weeds collectively result in direct yield losses between 20% and 40% of global agricultural productivity losses (Kubiak et al. 2022). Every year, a significant amount of overall crops deal with damage in the field due to diseases, pests, and other abiotic stresses. Several stresses, both abiotic and biotic, that crop plants encounter in the environment limit the output of crops overall, leading to significant losses (Majrashi 2022).

Throughout evolution, plant genomes have adapted to a number of biotic stresses, involving a wide range of phytopathogens, as well as abiotic stresses like variation in light, heat, drought, frost, and strong winds. Genomic stress adaptation has several advantages, especially if the stress is frequent (Horne et al. 2014). However, many stressful events happen sporadically and temporarily. Genetic and phenotypic plasticity are two ways plants might adapt to the unpredictability and inconsistencies of stress occurrence (Hilker and Schmülling 2019).

Scientists have developed a number of strategies in recent decades to adapt crops to the changing climate and protect the crops from various biotic and abiotic stresses, which include generating transgenic crops, resistant crop varieties, and chemical fertilizers. These approaches help protect crops and reduce farmers' financial losses (Oberč and Arroyo Schnell 2020). Crop production mostly relies on the use of chemicals to combat infections when crops lack genetic resilience on their own. Despite their efficiency, chemical-based plant defense has adverse environmental effects, puts at risk the environment as a whole, and is not sustainable. However, modern synthetic chemicals are not always cost-effective and only accessible through sophisticated agricultural production systems (Urech 2000). Furthermore, similar to antibiotics, it can be difficult to discover new chemicals to control plant disease, and widespread usage of existing agents may lead to the selection of pesticide-tolerant pathobiota. Recent advancement in plant immunity research in the creation of plant's memory against stresses has found that epigenetic modifications and modified genetic rearrangements create short-term or long-term memory in the plant cells, which enable the plant to tolerate stresses in the current as well as subsequent generations, and this phenomenon is called "plant immunity induction" or "priming of plants" (Thakur et al. 2020; Wang et al. 2020; Fan et al. 2018; Liu et al. 2022). In this chapter, we have discussed the overall aspects of plant priming including its molecular mechanisms, priming agents, and applications for resilience in agriculture.

12.2 Plant Priming

Priming is an immunity-inducing method that has a vital role in enhancing plants' defense system. Priming can be inherited in the progenies of the plants. Plant priming heritability defines the degree to which a plant's capacity to "memorize" a stress and react more successfully in the future can be transferred to its progeny through genetic or epigenetic modifications. It clarifies the extent to which this priming ability is hereditary rather than environmental. The generation of priming-related memory in plant cells in response to environmental stresses can be categorized into two types, i.e., mitotic stress memory and meiotic stress memory. Mitotic stress memories is responsible for somatic memory which generally resides for short term and is initiated by an intense stress; however, it can be reactivated within a limited period (i.e., a few hours or few days) (Liu et al. 2021). On the contrary, meiotic stress memory not only sustains for long term but also directly transmits from the parents to the first-generation progeny and thus is referred to as transgenerational memory (Lämke and Bäurle 2017; Oberkofler et al. 2021; Molinier et al. 2006). These memories are generally created during interactions with the environments. To deal with environmental stress, plants adopted an extensive array of resistance enhancement approaches. Their survival is aided via constitutive physical and chemical defense systems such as cell walls, waxy cuticles, lignin depositions, and the production of secondary metabolites (Agrios 2005). Plants can also protect themselves more specifically against pathogens by using extracellular or intracellular proteins known as pattern recognition receptors (PRRs) and resistance proteins. These receptors can identify and perceive specific pathogen-derived molecules, which include effectors, damage-associated molecular patterns (DAMPs), and pathogen-associated molecular patterns (PAMPs). These molecules result in PAMP-triggered immunity (PTI) and effector-triggered immunity (ETI) (Mauch-Mani et al. 2017). To counteract attacks, plants rely on inducible defense responses, and pathogens attempt to inhibit these defense processes. PTI and ETI typically have distinct trends and intensities and are triggered by different activation pathways. According to recent research, PTI and ETI complement one another in the integrated plant immunity, and they converge into same downstream signaling pathways (Ngou et al. 2021).

These abiotic and biotic stresses can have divergent degrees of effect, either by themselves or in combination with other stresses, which significantly challenge the plant system. As a result, plant must allocate the energy for development and reproduction to defense and stress signaling systems, which drastically decrease the plant's fitness. Even though little stress might trigger a plant to activate defense mechanisms, it is not a big burden for a healthy plant; therefore, the fitness loss might not be noticeable. Furthermore, in the long term, it might be beneficial. Mechanisms such as priming, hormesis, or adaptive responses function in a way that is somewhat analogous to vaccination in humans and animals (Tiwari and Singh 2021).

"Priming" is a crop protection strategy that enhances plant's immune responses by leveraging its natural defense mechanisms. This approach improves the plant's ability to react to potential threats based on previous encounters (Pastor et al. 2013). This is a novel method for promoting sustainable crop growth. Similar to human vaccination, where mild doses of inactivated pathogens are used to build immunity, plants can be immunized by exposing them to low levels of stress (either biotic or abiotic) as an initial trigger. This process helps plants to recognize and activate their defense systems. Since plants lack specialized immune cells like animals, they possess an effective defense system, and, although they do not have memory cells, they can retain a memory of previous stress exposure and become more sensitive to similar future threats due to past experiences. This crop improvement strategy is sustainable, profitable, sophisticated, and does not cause any damage to the surrounding environment. Priming is essentially bringing a plant into a "conscious" or "alarm" mode by intentionally exposing it to a minor biotic or abiotic stress known as the initial stimulus. The plant reacts more rapidly and effectively, thus exhibiting enhanced defense, when it is subsequently subjected to pathogen attack, insects, or any other environmental factor known as the secondary stimulus. Being in a primed condition enables the plant to respond more effectively to threats (Tiwari and Singh 2021).

In addition to chemicals and environmental stress signals, pathogens and beneficial microorganisms can also operate as cautionary signals that initiate priming (Mauch-Mani et al. 2017; Singh et al. 2017). Primary stimuli, such as the signals commenced by the microbes, alert the plant to danger and prime it by activating the early defense responses. Natural elicitors like plant hormones and their derivatives, such as salicylic acid (SA) and jasmonic acid (JA), methyl jasmonate (MeJA) as well as other signaling elicitors like pipecolic acid, benzothiadiazol (BTH), and β-amino butyric acid (BABA), are examples of priming agents (Oostendorp et al. 2001). Abiotic stimuli encompass stressors like heat, cold, salinity, ultraviolet light, and ozone.

12.3 Molecular Mechanism of Priming

Based on existing knowledge, plants can enhance their defense against biotic and abiotic stresses through plant priming. Elucidating the molecular mechanisms involved in the plant priming can lead to advancing sustainable agriculture practices. Multiple complex and multifaceted biochemical and genetic processes take part in plant priming; however, plant defense priming follows certain key molecular mechanisms: dormant signaling protein accumulation, modifications of histones, molecular patterns perception, primary metabolism alterations, and systemic signaling. Plant's effective and swift responses to biotic and abiotic stresses are obtained through the collective actions of these molecular mechanisms.

Mitogen-activated protein kinases (MPKs) are dormant proteins that are integral to cellular signal amplifications. These proteins remain in an inactive state and become activated upon exposure to stress signals, resulting in a quick and robust defense response. When a primed plant encounters with a stressor, these pre-accumulated dormant proteins become active, enhancing the plant's defense response more rapidly than the non-primed plant. Also, these MPKs can be activated by other MPK kinases (MPKKs) through phosphorylation, and further these activated MPKs activate transcription co-activators and various transcription factors, such as WRKY22 and WRKY29, which are crucial for defense-related gene expression (Fig. 12.1). This cascade amplifies the signals generated during the exposure to stressors. This gene expression modulation is required to mount an effective defense response against infections and stressors (Conrath 2011).

MAPKs are vital in integrating diverse signaling pathways, enabling plants to respond effectively to various stimuli. They help in coordinating the reactions to stresses such as pathogen attacks, environmental fluctuations, and physical damage, ensuring a unified and robust defense response.

In the case of pathogen infection, plants recognize pathogen-associated molecular patterns (PAMPs) and pathogen effectors, while herbivore infestation is perceived by the plant through herbivore-associated molecular patterns (HAMPs). A priming response will be triggered during this recognition, enhancing the plant's potential to activate defense mechanisms upon subsequent encounters with these

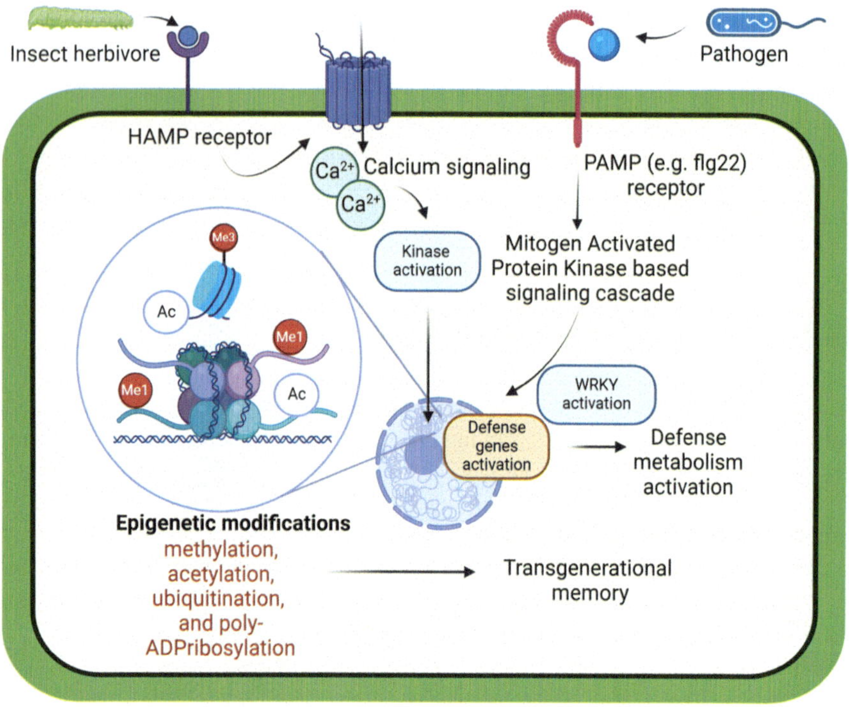

Fig. 12.1 Schematic representations of significant molecular mechanisms of priming. External stimulus or immunity boosters such as elicitors (e.g., PAMPs, HAMPs) or effector molecules can activate plant's immunity by commencing defense-associated signaling. HAMPs from herbivores is perceived by specific receptors through calcium ion signaling and various kinase activations. These molecules initiate activation of transcription factors of defense genes which further induces defense metabolism. PAMPs and effector molecules are perceived by receptors and activate MAP kinase-mediated signaling cascade including the activation of WRKY proteins. Moreover, these inducers result in epigenetic modifications that alter the accessibility of DNA for transcription. These alterations can result in an increased activation of defense genes with subsequent stress exposure. Epigenetic modifications are major factors for transgenerational priming. (Figure is generated through Biorender (www.biorender.com))

stimuli (Fig. 12.2). For example, the bacterial PAMP flagellin or its conserved N-terminal 22-amino-acid epitope flg22 is recognized by the PRR FLS2 in *Arabidopsis thaliana,* triggering an MPK cascade (MEKK1, MEK4/5, and MPK3/6) in leaf protoplasts. This cascade amplifies the flagellin signal, activating defense genes like WRKY22 and WRKY29, which are crucial for plant immunity (Colaianni et al. 2021). Therefore, mitogen-activated protein kinases are crucial for plant defense, regulating and amplifying stress signals to ensure effective immunity. Their role highlights their potential as a candidate to be targeted for priming for improving crop resilience in agriculture.

The priming process may also include epigenetic alterations, such as histone modifications (i.e., methylation, acetylation, ubiquitination, and poly-ADP ribosylation), which affect the accessibility of DNA for transcription. These alterations can result in an increased activation of defense genes with subsequent stress exposure (Fig. 12.1). Studies reveal that treating barley plants with immunity activators like trigonelline, *Bacillus subtilis* (B50) culture filtrate, or isonicotinic acid methyl ester

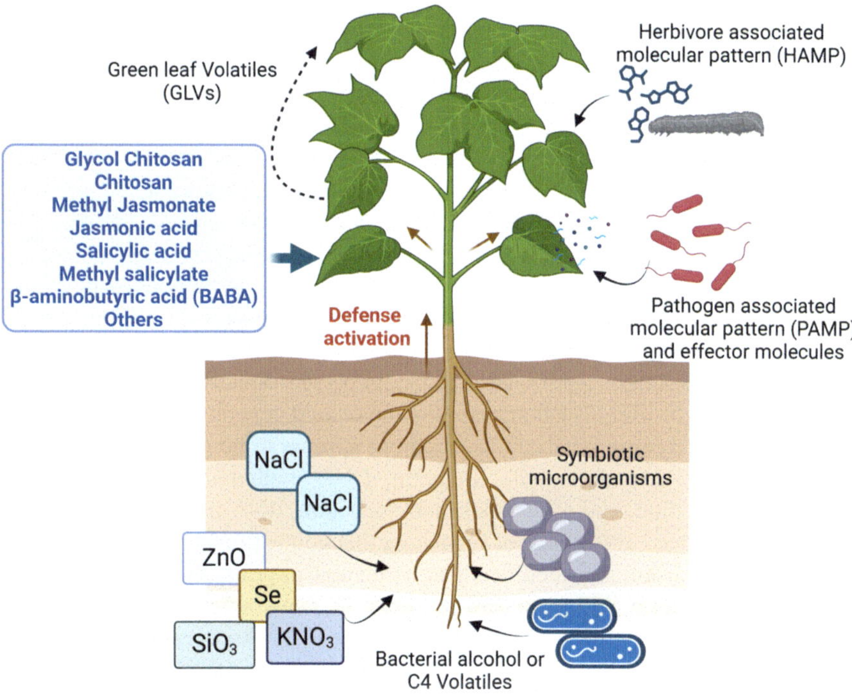

Fig. 12.2 Schematic representations of major known biotic and abiotic "priming" agents that induce immunity in plants. Biotic inducers are living organisms (such as insects and pathogens) and the chemical substances they produce are included in biotic inducers (e.g., PAMP, HAMP, phytohormones and their derivatives, chitosan and its derivative, GLVs, bacterial alcohols, C4 volatiles, mutualistic microbes). Abiotic inducers are generally synthesized molecules (e.g., BABA, NaCl, zinc oxide, selenium (Se), KNO_3, SiO_2) (Figure is generated through Biorender (www.biorender.com))

leads to slight but notable reductions in DNA cytosine methylation, linked to powdery mildew resistance. These reports highlight a connection between chromatin modifications and plant immunity. Additionally, the histone variant H2A.Z has been shown to play a crucial role in regulating SA-responsive genes in *A. thaliana* (Zhang et al. 2022; Berr et al. 2012; Conrath 2011; Yang et al. 2015). On the other hand, it was hypothesized that modified chromatin can prime the defense genes for faster and robust activation (Widmann et al. 1999).

Novel discoveries and approaches for enhancing resilience of crops have recently been made possible by developments in our knowledge of the genetic and epigenetic mechanisms of plant stress memory (Liu et al. 2021; Racette et al. 2020). Plants undertake genome-wide reorganization and fundamentally guided epigenetic changes following a previous stress event and thus provide a stress memory (Liu et al. 2022). Stress priming can improve responsiveness and promote the expression of desired characteristics in crops by exposing themselves to eliciting events early in their development or in parent generations. Genetic assimilation may take place when epigenetic variation is preserved over generations, and this will probably allow for stable meiotic stress memory phenotypes in the course of time (Miryeganeh and Saze 2020).

As mentioned earlier, stress memory has been primarily classified in two types: meiotic stress memory, also known as transgenerational memory, which affect future generations; and mitotic stress memory also referred as somatic memory, which alters present generation (Avramova 2015; Iwasaki and Paszkowski 2014). According to Bäurle, multiple generation stress memory is known as transgenerational memory (Fig. 12.1), whereas single-generation stress memory is known as intergenerational memory (Lämke and Bäurle 2017). Although somatic memory is typically temporary and is triggered by sudden stress, it can be revived in a brief period of time, generally hours to days. Nonetheless, somatic memory can also be persisted for an extended period of time across several phases of development of current life cycle. Intergenerational memory is passed directly to first-generation progeny, but maybe diminished or erased in the second generation.

As evidenced by prior study, stress memory appears to be primarily epigenetic and involves the remodeling of chromatin, DNA methylation, localization of nucleosomes, modification of histones, and noncoding RNA-mediated processing. Through the retained abundance of regulators that control functional gene expressions via post-translational modifications, structural imprints on chromatin, transcription factor accumulation, signaling metabolites and stress-response proteins are either promoted or supressed. This in turn influences subsequent gene transcription events after the initial stress onset (Avramova 2015).

An emerging strategy to enhance plant resistance involves the exogenous application of double-stranded RNA (dsRNA) molecules targeting essential pathogen genes or modulating plant immune responses. This approach leverages RNA interference (RNAi) mechanisms to suppress pathogen virulence genes or to prime the plant immune system, providing a non-transgenic method of disease control. Studies have demonstrated that foliar application or soil drenching with pathogen-specific dsRNA can lead to the uptake of these molecules by plant tissues, ultimately resulting in the silencing of targeted pathogen genes upon infection (Das and Sherif 2020; Koseoglou et al. 2022). This method offers several advantages, including specificity, environmental safety, and the ability to rapidly adapt dsRNA sequences to emerging pathogen strains.

Furthermore, dsRNA molecules can be designed to target plant susceptibility (S) genes, downregulating their expression to reduce disease susceptibility. For instance, targeting susceptibility genes involved in pathogen entry or defense suppression could impede pathogen establishment or virulence. The effectiveness of exogenous dsRNA delivery depends on factors such as stability of the molecules, mode of application, and plant species, but recent advancements in nanoparticle carriers and formulation techniques have improved delivery efficiency (Koseoglou et al. 2022).

12.4 Plant Priming Agents

Unlike animals, plants have no prominent immune system to defend themselves from their enemies. Instead, they rely on several signal transduction pathways (Jones and Dangl 2006). The plants can attain adaptive immunity through priming. Plant priming is an immunity boosting that enhances the plants' defensive capacity against subsequent stressors after pre-conditioning them. It resembles the plant's "memory," so it can improve the defense in the future based on the memory of previous exposure. The substances and treatments that boost plant immunity are known as plant priming agents (Balmer et al. 2015; Tripathi et al. 2024; Conrath et al. 2006). Numerous studies about priming agents were conducted to assess their effectiveness in addressing various stresses (Lephatsi et al. 2022; Cecchini et al. 2015; Tripathi et al. 2024; Pare et al. 2005).

12.4.1 Biotic and Abiotic Priming Agents

Plant priming agents were categorized into biotic and abiotic inducers. Living organisms themselves and the chemical substances they produce are included in biotic inducers. Arthropods, pathogens, and beneficial microbes are included in biotic inducers. Plant priming researchers have demonstrated that *Bacillus* and *Trichoderma* species can induce plant priming. The plant volatiles emissions and the jasmonic acid-related defenses are triggered by some herbivore-associated molecular patterns (HAMPs), like volicitin and inceptin. These are the oral secretions of caterpillars, which also act as biotic priming agents. Pathogen-associated molecular patterns (PAMPs) such as flg22 (a peptide from bacterial flagellin) and chitin (a component of fungal cell walls) can also act as the biotic priming agents (Pieterse et al. 2014).

Arbuscular mycorrhizal fungi (AMF) play a pivotal role in the nutrient acquisition and stress tolerance; in addition, they can aid in the systemic priming of plant defenses, which enables a quicker and more powerful reaction to pathogens or pest attack. A plant growth-promoting root-specific endophytic fungi, *Piriformospora indica*, can induce chlorophyll and proline content in rice, which helps tolerate salt stress (Jogawat et al. 2013). By intensifying the production of defense-associated secondary metabolites, some chemical substances can resist various stresses to plants that are specifically called elicitors. Elicitors that are effective against drought and biotrophic pathogens primarily activate the SA pathway but also inadvertently increased systemic vulnerability to chewing insects (Fig. 12.2) (Venegas-Molina et al. 2020).

By using certain abiotic factors such as chemical treatments and environmental factors, we can prime the plants. These include synthetic elicitors such as beta-glucosidase (Mattiacci et al. 1995), β-amino butyric acid (BABA) (Jakab et al. 2001), benzothiadiazol (BTH), probenazol, and pipecolic acid (Návarová et al. 2012; Tripathi et al. 2024) and natural elicitors such as vitamins, plant hormones, and hormonal derivatives (jasmonic acid, methyl jasmonic acid, salicylic acid, methyl salicylate, etc.). Glycol chitosan, a derivative of chitosan, has also efficient priming activity in *Piper nigrum* that protects it from *Phytophthora capsici* (Indu et al. 2022).

12.4.2 Insect-Derived Priming Agents

Insect herbivores produce various molecules called herbivore-associated molecular pattern (HAMP), which can act as a stimulus of defense in plants and has the potential to prime plant's immunity against second attack of herbivory. Volicitin (N-(17-hydroxylinolenoyl)-l-glutamine), a classical example of HAMP, is a fatty acid-amino acid conjugate discovered from the oral secretion (OS) beet armyworm caterpillars and is a crucial component in plant recognition of damage from insect herbivory (Alborn et al. 1997). It is a robust insect-derived elicitor that can induce volatile organic compounds (VOCs) in plants (Fig. 12.2). As a result, it enhances the indirect defense mechanism in plants by captivating natural enemies of plant infesting insect herbivores (Alborn et al. 1997; Paré et al. 1998). Volicitin's chemical structure directly impacts how well it elicits reactions in plants. It has been demonstrated that synthetic analogues of volicitin differ in their capacity to produce VOCs, suggesting that volicitin's precise structure is essential to its biological function (Truitt et al. 2004). In addition to its action as an elicitor, volicitin also helps prime plant defenses, making it possible to react to herbivore attacks more quickly and efficiently. When the plant is attacked again, this priming effect may cause defensive mechanisms to activate more rapidly and powerfully. Volicitin has the ability to alter the transcriptional rearrangements in plants, which induces defense-related gene expressions in plants (Truitt et al. 2004). Volicitin, as a biotic elicitor, activates plant chemical defense upon recognition by specific plant receptors and triggers the signaling pathways involving the plant defense-related hormones, also promotes the VOC production. This diverse strategy enables plants to improve their overall resistance and react to herbivore challenges in an efficient manner. *Pieris brassicae* is an herbivore in cabbage plants, which secretes elicitor β-glucosidase that can induce the defense response of cabbage plants to herbivore injury; through that, they can incite the emission of volatiles that are used by parasitoids of the herbivore to locate their victims (Mattiacci et al. 1995). Other examples of HAMPs are *N*-linolenoyl-L-glutamic acid from *Manduca sexta* OS fed on tobacco, other FAC derivatives having an alteration in fatty acid chain or amino acid moieties (Mithöfer and Boland 2008; Lait et al. 2003), caeliferins from *Schistocerca americana* OS fed on maize (Alborn et al. 2007), inceptins from *Spodoptera frugiperda* OS fed on cow pea/maize (Schelmz et al. 2006), and bruchins from oviposition fluid of *Bruchus pisorum* (Doss et al. 2000).

12.4.3 Green Leaf Volatiles

Green leaf volatiles (GLVs) are generated from the C_{18} oxylipin pathway and from certain bacteria. These elicitors can induce cellular defense responses in plants. (Z)-3-hexenal, (Z)-3-hexenol, and (Z)-3-hexenyl acetate are some known examples of green leaf volatiles that activate defense-related genes, while (E)-2-hexenal mainly triggers genes linked to abiotic stress responses, like heat and oxidative stress. GLVs act as airborne signals that quickly trigger defense-related genes and may serve as self-produced DAMPs in injured leaves (Fig. 12.2) (Yamauchi et al. 2018).

12.4.4 Bacterial Alcohol

Specific plant growth-promoting rhizobacteria (PGPR) produce specific volatile compounds known as the bacterial alcohols. These compounds can induce primed state in plants, promoting their resistance to pathogens. C4 bacterial volatiles enhance defense signaling components, priming the plant for more robust pathogen responses.

12.4.5 Immune-Specific Priming Agents

Immune-specific priming agents include ETI-eliciting effectors that can sensitize plants to resist pathogens. It has been reported that the ETI landscape of tomato is more limited than in the model plant *A. thaliana*. *A. thaliana* shows better ETI responses against pathogens. For example, among 21 type III effectors belonging to the Hop and Avr families that are reported to elicit ETI in *A. thaliana*, 17 have also been shown to trigger the ETI in tomato (Lonjon et al. 2024). However, some effectors are exclusively found to be active only in tomatoes. Tomato plants can resist *Pseudomonas syringae* infection through these ETI eliciting effectors; this happens when a non-virulent strain, either before or concurrently with a virulent strain, delivers the effector (Lonjon et al. 2024). The ETI responses generated by these effectors can provide substantial disease resistance when utilized as priming agents, indicating that these responses can be adequately retained and activated when needed.

12.4.6 Mutualistic Microbes as Priming Agents

Pseudomonas, *Bacillus*, *Trichoderma*, and mycorrhiza species are beneficial root symbionts; they can modify the plant immune system for enhanced defense without directly activating expensive defenses (Fig. 12.2) (Pieterse et al. 2014). *Piriformospora indica* can activate induced resistance in plants. This *P. indica*-involved resistance relies on signaling of JA and the cytosolic activity of the non-expressor of pathogenesis-related genes 1 (*NPR1*). It is a master regulator protein acting as SA majorly taking part in SA signaling.

12.4.7 Synthetic Priming Agent

Various synthetic abiotic molecules can be used to prime plants by enhancing defense. For example, β-aminobutyric acid (BABA) has a vital role in plant sensitization. In *A. thaliana*, aspartyl tRNA synthetase (IBI1) is responsible for BABA perception. Multiple immune responses in *A. thaliana* are activated by this perception, including callose-associated cell wall defenses that are under control by abscisic acid (ABA). The IBI1-VOZ transcription factor module directs pathogen-triggered ABA signaling toward strengthening cell wall defenses while also down-regulating genes involved in responses to abiotic stress (Schwarzenbacher et al. 2020; Van der Ent et al. 2009).

Priming agents are a revolutionary way to increasing plant resilience, providing a sustainable and environmentally friendly alternative to traditional stress management strategies, and these agents also bridge the gap between basic research and agricultural uses by activating the plant's intrinsic defense mechanisms. Emerging technologies, such as nanotechnology and synthetic biology, have significant potential for developing next-generation priming agents by increasing stability and delivery potential. As science and technology improve, priming agents are poised to become a key component of future agricultural methods, creating a resilient and sustainable food system (Table 12.1).

Table 12.1 Major known priming agents and their sources

Type of priming agent	Priming stimuli	Sources as well as effective against	References
Biotic agents	Volicitin	Beet armyworm caterpillars	Alborn et al. (1997) and Truitt et al. (2004)
Biotic agents	Green leaf volatiles	Plants	Yamauchi et al. (2018)
	Bacterial alcohols	*Pseudomonas koreensis* (N19), *Ps. fluorescens* (N04), *Lysinibacillus sphaericus* (T19), and *Paenibacillus alvei* (T22)	Mhlongo et al. (2022)
Biotic agents	ETI-eliciting effectors	*Pseudomonas*	Lonjon et al. (2024)
Biotic agents	*Piriformospora indica*, *Pseudomonas*, *Bacillus*, *Trichoderma*, mycorrhiza	Mutualistic microorganisms	Jogawat et al. (2013), Molitor and Kogel (2009), and Tsai et al. (2020)
Biotic agents	Jasmonic acid	Plants	Tsai et al. (2020)
Biotic agents	Salicylic acid	Plants	Tsai et al. (2020)
Biotic agents	*β*-Glucosidase	*Pieris brassicae*	Mattiacci et al. (1995)
Abiotic agents	*β*-Aminobutyric acid	Synthesized	Schwarzenbacher et al. (2020), Van der Ent et al., (2009)
Abiotic agents	Probenazol	Synthesized	Návarová et al. (2012)
Abiotic agents	Benzothiadiazol	Synthesized	Návarová et al. (2012)
Abiotic agents	Zinc oxide (ZnO) and selenium (Se), KNO_3, SiO_2, NaCl	Synthesized and purified	Hussain et al. (2020), Ahmad et al. (2024), Ali et al. (2021) and García-Locascio et al. (2024)

12.5 Methods for Discovering and Confirming Novel Priming Agents

The identification of novel elicitors for plant priming involves utilizing advanced methods that enable researchers to effectively identify and evaluate potential candidates. A harmonious application of these technologies will ensure the exploitation of new inducers for more detailed and time-saving exploration into innovative elicitors that will help evolve more potent plant priming strategies in the future. The cutting-edge techniques used to accelerate the discovery of novel elicitors that can enhance plant defense responses are discussed below.

12.5.1 High-Throughput Screening (HTS)

HTS involves rapidly testing large numbers of compounds or potential inducers for their ability to trigger plant defense responses (Schmitz et al. 2021). It enables the identification of effective priming agents in a cost- and time-efficient manner. For breeders, it offers powerful tools to accelerate trait selection and improve crop development efficiency. For farmers, it leads to innovative, sustainable solutions that boost productivity and reduce input costs. HTS can be applied to test libraries of small molecules, natural products, or synthetic chemicals, and by using automated systems, researchers can evaluate thousands of compounds on plant cells or whole plants to observe their effects on defense-related gene expression, pathogen resistance, or stress tolerance. This method allows for the discovery of novel inducers by systematically testing a wide variety of substances under controlled conditions. Some examples of HTS are as follows:

(a) Natural product screening: Earlier, the elicitors had been a source of natural products originating from botanical sources, microorganisms, and fungi. Through screening of large natural product libraries, researchers discovered certain compounds that can trigger plant defense mechanisms. HTS methods can be applied to screen natural product libraries, followed by targeted isolation and characterization of active compounds. This approach is particularly useful in discovering novel elicitors that may have evolved to interact with plant defense systems.
(b) Synthetic chemical library screening: Screening large chemical libraries enables the discovery of synthetic or semisynthetic molecules that can induce plant defense responses. This approach focuses on testing a wide range of chemicals, from small organic compounds to synthetic peptides, to identify potential priming agents. Along with HTS, QSAR modeling and molecular docking could serve as tools to predict new plant defense inducers by designing them in silico and optimizing their activity beforehand.
(c) Bioprospecting: This approach involves exploring diverse ecosystems (e.g., soil, marine, or forest environments) to discover new biological sources of inducers. It can also involve testing combinations of inducers to identify synergistic effects that enhance priming. Synergistic screening may involve pairing elicitors from different natural sources (e.g., plant extracts combined with microbial metabolites) to maximize their effectiveness in priming plant defenses (Sekurova et al. 2019).

12.5.2 "Omics" Technologies

(a) Genomics: The sequencing and analysis of plant genomes provide insights into the genes and genetic pathways involved in plant defense mechanisms. Identifying genes related to immunity can help pinpoint potential targets for priming.

(b) Transcriptomics: This technique involves measuring gene expression levels under various conditions (e.g., exposure to potential inducers). By comparing gene expression profiles before and after treatment with different compounds, researchers can identify defense-related target genes for priming.
(c) Proteomics: Proteomic approaches help analyze changes in the protein levels and posttranslational modifications of the targets in response to priming agents. Understanding how proteins involved in plant defense are regulated can guide the identification of effective inducers.
(d) Metabolomics: Global metabolomics approach is extremely useful to find out changes in plant metabolites (e.g., secondary metabolites, hormones) in response to exposure to potential inducers. This also helps identify the plant's new bioactive compounds that may trigger priming responses.
(e) Phenomics: Phenomics can be a useful tool when integrates with HTS. By using imaging technologies, phenomics measures the phenotypic changes in plants (e.g., growth, disease resistance, stress tolerance) after treatment with different immunity inducers. It provides real-time, large-scale data on plant responses to various compounds (Balmer et al. 2015; Tugizimana et al. 2018).

12.5.3 CRISPR/Cas9 and Gene Editing

CRISPR/Cas9 technology can be used to modify plant genomes and identify genes involved in defense signaling pathways. By knocking out specific genes or introducing new ones, researchers can better understand the genetic basis of priming responses and identify key pathways that can be targeted for priming induction. Through gene editing, plants can be engineered to become more responsive to priming agents or more resistant to stress, helping identify the most effective inducers and mechanisms (Dey 2021).

12.5.4 Artificial Intelligence (AI) and Machine Learning (ML)

AI and ML can be used to analyze large datasets generated from omics technologies, identification of gene classes, protein classes, and compound chemical classes. Majorly, this can be used for the prediction of the functional group responsible for priming activity in the molecule. These tools help predict the specific structural pattern of priming molecules, which may help the scientist to synthesize novel priming agents by chemical modification. AI and ML can also predict which compounds or genetic targets might be most effective for inducing plant defense responses (Reel et al. 2021).

12.6 Applications of Priming

The most effective strategy for safeguarding plants and improving agricultural output in order to achieve sustainable global food security is defense priming. Priming is an important approach to deal with solving the problems of food scarcity worldwide. This method has the potential to protect the crop mostly without causing a yield penalty, which can improve agriculture. Priming's transgenerational potential can boost farming's profitability, rendering it economically viable as well as sustainable. The physiological mechanisms of priming-induced stress tolerance have been uncovered in several researches, which could serve as a conceptual basis for enhancing crop protection against various stresses. It is still uncertain how priming causes stress memory at the molecular level. As mentioned above, studies at the chromatin level will be especially critical in unraveling the mechanisms driving stress memory. Gaining more knowledge on the mechanisms underlying priming will promote the discovery of novel and efficient priming agents and their targets, which can replicate the benefits of priming in increasing crop production. Additionally, since priming application may vary depending on the species, it is important to optimize them for use in crop production in the future.

Research studies have demonstrated that various distinct priming agents have been identified and that they can be applied to mitigate different kinds of stresses as well as plant development and growth promotion. "Seed priming" is an important strategy in agriculture to improve germination, growth, and stress tolerance in crops. In this method, seeds are pretreated with specific priming agents to boost biochemical changes that help the seeds in water uptake, cell division, and ATP production. It also helps seedlings to manage oxidative and other stresses (Devika et al. 2021). A study has shown the role of nonthermal plasma (NTP) in priming (Pastor et al. 2013; Whitehead 2016, which is a partially ionized gas comprising neutral atoms or molecules, charged particles, and reactive species such as reactive oxygen species (ROS) and reactive nitrogen species (RNS). NTP functions as a moderate stress that can cause plants to respond adaptively. By responding in such a manner, plants become more resilient to upcoming environmental stresses like salinity, drought, or disease infestations, which lead to the overexpression of stress-related genes (Misra et al. 2019). They also increase the production of 2016 ROS-enhanced antioxidant activity and improve nutrient uptake (Lu et al. 2016). It has been demonstrated that pretreating seedlings with NTP is an appropriate priming agent for mitigating the toxicity of zinc oxide (ZnO) and selenium (Se) in *Melissa officinalis* (Babajani et al. 2019). Polyethylene glycol (PEG)-mediated reduction of drought stress was observed when tomato NTP priming was applied, indicating the importance of H_2O_2 and NO signaling in plant defense (Adhikari et al. 2020).

Herbivore-induced plant volatile (HIPV)-mediated systemic signaling has been demonstrated in several plant species including sagebrush, lima beans, poplar, and blueberry (Karban et al. 2006; Frost et al. 2007; Heil and Bueno 2007). In these cases, HIPVs released from an attacked part of the plant primed the healthy parts of the same plant to respond more strongly, and it is released faster than other inducible volatiles (Frost et al. 2007). Methylated forms of plant hormones, including methyl jasmonic acid and methyl salicylic acid, have been identified as HIPV signals (Heil and Bueno 2007; Frost et al. 2008).

A study on rice found that KNO_3, SiO_2, and SA priming greatly increased seedling growth under drought stress (Ali et al. 2021). The seed's emergence, growth, and vigor were enhanced by priming them with these compounds, which triggered pre-germination metabolic processes. These processes included increased water imbibition, cell division and elongation, repair of damaged nucleic acids, and activation of reserve mobilizing enzymes such as dehydrogenase, α-amylase, β-amylase, and acid phosphatase. Under terminal dryness, rice priming with selenium markedly enhanced shoot and root length as well as seedling dry biomass, whereas wheat priming with $CaCl_2$ and hydropriming boosted plant height, leaf area, panicle length, and grain yield (Khaliq et al. 2015).

Plant defense responses against a variety of harmful organisms can be triggered by chitosan or glycol-chitosan. A study revealed the connection between the pathogen-induced basal defense pathways functioning in a resistant plant host and the orchestration pattern of defense pathways in chitosan-primed ginger. The findings indicate the potential of using chitosan for the sustainable management of ginger soft rot and show that it can prepare standing ginger plants against the disease (Varghese and Thomas 2023). In addition, glycol chitosan treatment on black pepper exhibited resistance against *Phytopthora* (Indu et al. 2022).

Selenium nanoparticles serve as useful nano-priming agents that can improve tomato growth and germination quality. Tomato seedlings that were nano-primed grew the best in terms of total chlorophyll content and antioxidant capacity in aerial tissues and also showed significant increase in the total length of the plant (García-Locascio et al. 2024). ZnO nanoparticles, both foliar and priming, may enhance spinach growth performance in a study involving salinity stress. By increasing the amount of chlorophyll, gas exchange parameters, and antioxidant enzyme activity in saline soil, the combination of ZnO nanoparticles and biochar greatly enhanced growth performance (Ahmad et al. 2024). Hydropriming outperformed NaCl priming in a study on Canola (*Brassica napus L.*), in terms of enhancing germination, growth, and yield characteristics (BiBi et al. 2024).

At low concentrations, NaCl can function as a priming agent, successfully breaking dormancy and encouraging water uptake by seeds, which also helps activate the enzymatic mechanisms involved in germination. Plants frequently boost chlorophyll production as a compensatory mechanism to maximize photosynthetic efficiency under mild stress circumstances, such as those brought on by low concentrations of NaCl. Increased chlorophyll concentration makes it easier for plants to absorb and use light, which in turn encourages carbon assimilation and biomass buildup, which eventually results in a higher plant seed yield (Hussain et al. 2020). In terms of chlorophyll concentration, hydropriming encourages seeds to be as hydrated as possible, which activates metabolic pathways that are involved in the manufacture of chlorophyll (Farooq et al. 2010). This increases light absorption and photosynthetic efficiency, which eventually raises chlorophyll levels and boosts plant vigor.

In greenhouses and fields, bacterial and viral diseases have been controlled with a variety of biological and chemical agents. Treatment with beneficial microbes decreases plant symptoms and disease severity following pathogen infection, according to extensive investigations using beneficial bacteria and fungi associated with roots and leaves (Pieterse et al. 2014). It was previously reported that foliar applications of *Pseudozyma aphidis* produced systemic resistance against the gray mold disease, *Botrytis cinerea* on *A. thaliana*, the powdery mildew fungus *Podosphaera xanthii* on cucumber, and the bacterial canker pathogen *Clavibacter michiganensis* on tomato (Barda et al. 2015; Gafni et al. 2015). In a novel study, significant protection against *X. axonopodis* and, surprisingly, against cucumber mosaic virus, pepper mottle virus, pepper mild mottle virus, and broad bean wilt virus was provided by newly isolated leaf-colonizing yeasts from the leaves of field-grown pepper plants (Lee et al. 2017).

12.7 Limitations and Challenges in Future Perspective

Despite the potential benefits of plant priming, several challenges and limitations need to be addressed for its widespread application in the future:

(a) Consistency and Effectiveness: The efficacy of priming can differ between various plant species, environmental states, and stress conditions. Hence, ensuring fruitful results in diverse environment is challenging. Factors such as age of the plant, genotype, and developmental stage influence their response to priming, making it more laborious to optimize and standardize protocols.
(b) Understanding Mechanisms: The exact molecular and physiological mechanisms involved in priming are still not clear. It is essential to delve more on this area to unravel how various elicitors activate plant defense responses.
(c) Environmental Variability: Abiotic factors, particularly temperature, soil conditions, and humidity influence the effectiveness of priming response. Uncertainty in the weather and climatic changes may make the priming treatment and standardization more complex.
(d) Toxicity and Plant Health: Overpriming or inappropriate application of priming agents may have a harmful impact on plant health, development, and productivity. This leads to the accumulation of toxicity, stunted growth, or unexpected interactions with other chemicals and agricultural practices. Utmost care should be taken to ensure safe and optimal dosage and to prevent crop harming.
(e) Scalability and Cost: The recent advanced delivery systems of priming agents such as nano-priming or encapsulation are more expensive, and their implementation is more complex when it comes to commercial agriculture. Cost-effectiveness and scalability are significant challenges for farmers with small-scale setups.
(f) Regulatory and Public Perception Issues: Discovery and implementation of novel priming methods involving synthetic chemicals face regulatory obstacles

or public skepticism. Thorough risk assessment, adherence to local, national, and international regulations, good laboratory and manufacturing practices, and transparent usage instructions can address these regulatory challenges, and compliance with these standards ensures the safety and efficacy for crop protection. Obtaining trust of consumers will be crucial for the successful implementation of priming strategies.

(g) Integration with Other Crop Management Practices: Priming methods have to be integrated with other well-known agricultural practices like the management of pests and soil health to improving its effectiveness. It is essential to find complementary farming methods that synergistically work with priming strategy.

(h) Long-Term Sustainability: The careful evaluation of long-term impact of priming on soil health, plant growth, balanced crop productivity, and ecosystem wellness is crucial as overpriming on repeated manner could lead to unexpected ecological consequences.

In the United States, the Environmental Protection Agency (EPA) approved *Bacillus thuringiensis* (Bt)-based biopesticides following thorough toxicological, environmental, and efficacy evaluations. Similarly, in India, the Central Insecticides Board and Registration Committee (CIBRC) approved Bt products after data submissions on shelf life, bioefficacy, and nontarget effects. Also, salicylic acid-based biostimulants have been evaluated by the European Food Safety Authority (EFSA) and European Union (EU) to ensure they meet safety, efficacy, and labeling standards.

Aligning with safety standards can be made more feasible by leveraging publicly available resources such as the FAO's Guidelines for the Registration of Biopesticides, which outline minimum data requirements and streamlined approval processes. Collaboration with local research institutions and agricultural extension services can also provide technical support in fulfilling regulatory requirements. Furthermore, tools such as simplified data collection templates, regional pilot studies, and cooperative regulatory frameworks can lower barriers to compliance. These additions not only clarify the regulatory pathway but also make it more accessible and actionable for diverse stakeholders in resource-limited settings.

12.8 Conclusion

Despite significant advancements in the use of defensive priming for crop protection over the last decade, some elusive areas in priming remain before we completely comprehend the plant immune memory and how it might be used in sustainable agriculture. In this regard, the establishment and maintenance processes of plant defense priming and the sustainable molecular approaches for priming in crop plants are not fully understood. Additionally, several priming mechanisms including seed priming have been extensively established for protecting crops from harmful microorganisms, but their efficacy against herbivores has received less attention. With the advancement of priming methods and the growing understanding of plant immune memory, taking advantage of priming would open up new possibilities for improved crop protection in future agriculture. As discussed above, various challenges are still remaining to properly exploit the benefit of priming strategies. Addressing these challenges will require continued elaborative and in-depth research, innovation, and collaboration between scientists, farmers, and policymakers to optimize plant priming for sustainable agriculture in the future.

Acknowledgments We acknowledge the Department of Biotechnology (DBT, India) and Rajiv Gandhi Centre for Biotechnology, Thiruvananthapuram, India, for supporting this work with intramural funding (RGCB/2023/00661). MM duly acknowledges CSIR, India, and FF duly acknowledge UGC, India, for research fellowships.

Competing Interests Authors have no relevant financial or nonfinancial competing interest to disclose.

References

Adhikari B, Adhikari M, Ghimire B, Adhikari BC, Park G, Choi EH (2020) Cold plasma seed priming modulates growth, redox homeostasis and stress response by inducing reactive species in tomato (*Solanum lycopersicum*). Free Radic Biol Med 156:57–69

Agrios GN (2005) How plants defend themselves against pathogens. In: Plant pathology, pp 207–248. https://doi.org/10.1016/B978-0-08-047378-9.50012-9

Ahmad S, Khan Sehrish A, Hussain A, Zhang L, Owdah Alomrani S, Ahmad A, Sarker PK (2024) Salt stress amelioration and nutrient strengthening in spinach (*Spinacia oleracea* L.) via biochar amendment and zinc fortification: seed priming versus foliar application. Sci Rep 14(1):15062

Alborn HT, Turlings TCJ, Jones T, Stenhagen G, Loughrin JH, Tumlinson JH (1997) An elicitor of plant volatiles from beet armyworm oral secretion. Science 276(5314):945–949

Alborn HT, Hansen TV, Jones TH, Bennett DC, Tumlinson JH, Schmelz EA, Teal PE (2007) Disulfooxy fatty acids from the American bird grasshopper *Schistocerca americana*, elicitors of plant volatiles. Proceedings of the National Academy of Sciences, 104(32), 12976–12981

Ali LG, Nulit R, Ibrahim MH, Yien CYS (2021) Efficacy of KNO_3, SiO_2 and SA priming for improving emergence, seedling growth and antioxidant enzymes of rice (*Oryza sativa*), under drought. Sci Rep 11(1):3864

Avramova Z (2015) Transcriptional 'memory' of a stress: transient chromatin and memory (epigenetic) marks at stress-response genes. Plant J 83(1):149–159

Babajani A, Iranbakhsh A, Oraghi Ardebili Z, Eslami B (2019) Seed priming with non-thermal plasma modified plant reactions to selenium or zinc oxide nanoparticles: cold plasma as a novel emerging tool for plant science. Plasma Chem Plasma Process 39:21–34

Balmer A, Pastor V, Gamir J, Flors V, Mauch-Mani B (2015) The 'prime-ome': towards a holistic approach to priming. Trends Plant Sci 20(7):443–452

Barda O, Shalev O, Alster S, Buxdorf K, Gafni A, Levy M (2015) *Pseudozyma aphidis* induces salicylic-acid-independent resistance to *Clavibacter michiganensis* in tomato plants. Plant Dis 99(5):621–626

Berr A, Ménard R, Heitz T, Shen WH (2012) Chromatin modification and remodelling: a regulatory landscape for the control of *Arabidopsis* defense responses upon pathogen attack. Cell Microbiol 14(6):829–839

BiBi R, Elahi NN, Danish S, Alahmadi TA, Ansari MJ (2024) Enhancing germination and growth of canola (*Brassica napus L.*) through hydropriming and NaCl priming. Sci Rep 14(1):14026
Cecchini NM, Steffes K, Schläppi MR, Gifford AN, Greenberg JT (2015) *Arabidopsis* AZI1 family proteins mediate signal mobilization for systemic defense priming. Nat Commun 6(1):7658
Colaianni NR, Parys K, Lee HS, Conway JM, Kim NH, Edelbacher N et al (2021) A complex immune response to flagellin epitope variation in commensal communities. Cell Host Microbe 29(4):635–649
Conrath U (2011) Molecular aspects of defense priming. Trends Plant Sci 16(10):524–531
Conrath U, Beckers GJ, Flors V, García-Agustín P, Jakab G, Mauch F, Newman MA, Pieterse CM, Poinssot B, Pozo MJ, Pugin A (2006) Priming: getting ready for battle. Mol Plant-Microbe Interact 19(10):1062–1071
Das PR, Sherif SM (2020) Application of exogenous dsRNAs-induced RNAi in agriculture: challenges and triumphs. Front Plant Sci 11:946
Devika OS, Singh S, Sarkar D, Barnwal P, Suman J, Rakshit A (2021) Seed priming: a potential supplement in integrated resource management under fragile intensive ecosystems. Front Sustain Food Syst 5:654001
Dey A (2021) CRISPR/Cas genome editing to optimize pharmacologically active plant natural products. Pharmacol Res 164:105359
Doss RP, Oliver JE, Proebsting WM, Potter SW, Kuy S, Clement SL, Williamson RT, Carney JR DeVilbiss ED (2000) Bruchins: insect-derived plant regulators that stimulate neoplasm formation. Proceedings of the National Academy of Sciences, 97(11), pp.6218–6223
Fan Y, Ma C, Huang Z, Abid M, Jiang S, Dai, T, Zhang, W, Ma, S, Jiang, D and Han X, (2018) Heat priming during early reproductive stages enhances thermo-tolerance to post-anthesis heat stress via improving photosynthesis and plant productivity in winter wheat (*Triticum aestivum* L.). Frontiers in plant science, 9, p.805
Farooq M, Wahid A, Ahmad N, Asad SA (2010) Comparative efficacy of surface drying and redrying seed priming in rice: changes in emergence, seedling growth and associated metabolic events. Paddy Water Environ 8:15–22
Frost CJ, Mescher MC, Carlson JE, De Moraes CM (2008) Why do distance limitations exist on plant-plant signaling via airborne volatiles? Plant Signal Behav 3(7):466–468
Gafni A, Calderon CE, Harris R, Buxdorf K, Dafa-Berger A, Zeilinger-Reichert E, Levy M (2015) Biological control of the cucurbit powdery mildew pathogen *Podosphaera xanthii* by means of the epiphytic fungus *Pseudozyma aphidis* and parasitism as a mode of action. Front Plant Sci 6:132
García-Locascio E, Valenzuela EI, Cervantes-Avilés P (2024) Impact of seed priming with Selenium nanoparticles on germination and seedlings growth of tomato. Sci Rep 14(1):6726
Heil M, Silva Bueno JC (2007) Within-plant signaling by volatiles leads to induction and priming of an indirect plant defense in nature. Proc Natl Acad Sci 104(13):5467–5472
Hilker M, Schmülling T (2019) Stress priming, memory, and signalling in plants. Plant Cell Environ 42(3):753–761
Horne SD, Chowdhury SK, Heng HH (2014) Stress, genomic adaptation, and the evolutionary trade-off. Front Genet 5:92
Hussain T, Koyro HW, Zhang W, Liu X, Gul B, Liu X (2020) Low salinity improves photosynthetic performance in *Panicum antidotale* under drought stress. Front Plant Sci 11:481
Indu M, Meera B, Sivakumar KC, Mahadevan C, Shafi KM, Nagarathnam B, Sowdhamini R, Sakuntala M (2022) 'Priming' protects *Piper nigrum* L. from *Phytophthora capsici* through reinforcement of phenylpropanoid pathway and possible enhancement of Piperine biosynthesis. Front Plant Sci 13:1072394
Iwasaki M, Paszkowski J (2014) Epigenetic memory in plants. EMBO J 33(18):1987–1998
Jakab G, Cottier V, Toquin V, Rigoli G, Zimmerli L, Métraux JP, Mauch-Mani B (2001) β-Aminobutyric acid-induced resistance in plants. Eur J Plant Pathol 107:29–37
Jogawat A, Saha S, Bakshi M, Dayaman V, Kumar M, Dua M et al (2013) *Piriformospora indica* rescues growth diminution of rice seedlings during high salt stress. Plant Signal Behav 8(10):e26891

Jones JDG, Dangl J (2006) The plant immune system. Nature 444:323–329
Karban R, Shiojiri K, Huntzinger M, McCall AC (2006) Damage-induced resistance in sagebrush: volatiles are key to intra-and interplant communication. Ecology 87(4):922–930
Khaliq A, Aslam F, Matloob A, Hussain S, Geng M, Wahid A, ur Rehman H (2015) Seed priming with selenium: consequences for emergence, seedling growth, and biochemical attributes of rice. Biol Trace Elem Res 166:236–244
Koseoglou E, van der Wolf JM, Visser RG, Bai Y (2022) Susceptibility reversed: modified plant susceptibility genes for resistance to bacteria. Trends Plant Sci 27(1):69–79
Kubiak A, Wolna-Maruwka A, Niewiadomska A, Pilarska AA (2022) The problem of weed infestation of agricultural plantations vs. the assumptions of the European biodiversity strategy. Agronomy 12(8):1808
Lait CG, Alborn HT, Teal PE, Tumlinson JH III (2003) Rapid biosynthesis of N-linolenoyl-L-glutamine, an elicitor of plant volatiles, by membrane-associated enzyme (s) in *Manduca sexta*. Proc Natl Acad Sci 100(12):7027–7032
Lämke J, Bäurle I (2017) Epigenetic and chromatin-based mechanisms in environmental stress adaptation and stress memory in plants. Genome Biol 18:1–11
Lee G, Lee SH, Kim KM, Ryu CM (2017) Foliar application of the leaf-colonizing yeast *Pseudozyma churashimaensis* elicits systemic defense of pepper against bacterial and viral pathogens. Sci Rep 7(1):39432
Lephatsi M, Nephali L, Meyer V, Piater LA, Buthelezi N, Dubery IA, Opperman H, Brand M, Huyser J, Tugizimana F (2022) Molecular mechanisms associated with microbial biostimulant-mediated growth enhancement, priming and drought stress tolerance in maize plants. Sci Rep 12(1):10450
Liu H, Able AJ, Able JA (2021) Small RNAs and their targets are associated with the transgenerational effects of water-deficit stress in durum wheat. Sci Rep 11(1):3613
Liu H, Able AJ, Able JA (2022) Priming crops for the future: rewiring stress memory. Trends Plant Sci 27(7):699–716
Lonjon F, Lai Y, Askari N, Aiyar N, Bundalovic-Torma C, Laflamme B, Wang PW, Desveaux D, Guttman DS (2024) The effector-triggered immunity landscape of tomato against *Pseudomonas syringae*. Nat Commun 15(1):5102
Lu X, Naidis GV, Laroussi M, Reuter S, Graves DB, Ostrikov K (2016) Reactive species in non-equilibrium atmospheric-pressure plasmas: generation, transport, and biological effects. Phys Rep 630:1–84
Majrashi AA (2022) Preliminary assessment of weed population in vegetable and fruit farms of Taif, Saudi Arabia. Braz J Biol 82:e255816
Mattiacci L, Dicke M, Posthumus MA (1995) Beta-glucosidase: an elicitor of herbivore-induced plant odor that attracts host-searching parasitic wasps. Proc Natl Acad Sci 92(6):2036–2040
Mauch-Mani B, Baccelli I, Luna E, Flors V (2017) Defense priming: an adaptive part of induced resistance. Annu Rev Plant Biol 68:485–512
Mhlongo MI, Piater LA, Dubery IA (2022) Profiling of volatile organic compounds from four plant growthpromoting rhizobacteria by SPME–GC–MS: A metabolomics study. Metabolites, 12(8), 763
Miryeganeh M, Saze H (2020) Epigenetic inheritance and plant evolution. Popul Ecol 62(1):17–27
Misra NN, Yepez X, Xu L, Keener K (2019) In-package cold plasma technologies. J Food Eng 244:21–31
Mithofer A, Boland W (2008) Recognition of herbivory-associated molecular patterns. Plant Physiology, 146(3), 825–831
Molinier J, Ries G, Zipfel C, Hohn B (2006) Transgeneration memory of stress in plants. Nature 442(7106):1046–1049
Molitor A, Kogel KH (2009) Induced resistance triggered by *Piriformospora indica*. Plant Signal Behav 4(3):215–216
Návarová H, Bernsdorff F, Döring AC, Zeier J (2012) Pipecolic acid, an endogenous mediator of defense amplification and priming, is a critical regulator of inducible plant immunity. Plant Cell 24(12):5123–5141

Ngou BPM, Ahn HK, Ding P, Jones JD (2021) Mutual potentiation of plant immunity by cell-surface and intracellular receptors. Nature 592(7852):110–115
Oberč BP, Arroyo Schnell A (2020) Approaches to sustainable agriculture. Exploring the pathways towards the future of farming. IUCN, Brussels, p 486
Oberkofler V, Pratx L, Bäurle I (2021) Epigenetic regulation of abiotic stress memory: maintaining the good things while they last. Current opinion in plant biology, 61, 102007
Oostendorp M, Kunz W, Dietrich B, Staub T (2001) Induced disease resistance in plants by chemicals. Eur J Plant Pathol 107(1):19–28
Paré PW, Alborn HT, Tumlinson JH (1998) Concerted biosynthesis of an insect elicitor of plant volatiles. Proc Natl Acad Sci 95(23):13971–13975
Pare PW, Farag MA, Krishnamachari V, Zhang H, Ryu CM, Kloepper JW (2005) Elicitors and priming agents initiate plant defense responses. Photosynth Res 85:149–159
Pastor V, Luna E, Mauch-Mani B, Ton J, Flors V (2013) Primed plants do not forget. Environ Exp Bot 94:46–56
Pieterse CM, Zamioudis C, Berendsen RL, Weller DM, Van Wees SC, Bakker PA (2014) Induced systemic resistance by beneficial microbes. Annu Rev Phytopathol 52(1):347–375
Racette K, Zurweller B, Tillman B, Rowland D (2020) Transgenerational stress memory of water deficit in peanut production. Field Crop Res 248:107712
Reel PS, Reel S, Pearson E, Trucco E, Jefferson E (2021) Using machine learning approaches for multi-omics data analysis: a review. Biotechnol Adv 49:107739
Schmelz EA, Carroll M J, LeClere S, Phipps SM, Meredith J, Chourey PS, Alborn HT Teal PE (2006) Fragments of ATP synthase mediate plant perception of insect attack. Proceedings of the National Academy of Sciences, 103(23), pp.8894–8899
Schwarzenbacher RE, Wardell G, Stassen J, Guest E, Zhang P, Luna E, Ton J (2020) The IBI1 receptor of β-aminobutyric acid interacts with VOZ transcription factors to regulate abscisic acid signaling and callose-associated defense. Mol Plant 13(10):1455–1469
Sekurova ON, Schneider O, Zotchev SB (2019) Novel bioactive natural products from bacteria via bioprospecting, genome mining and metabolic engineering. Microb Biotechnol 12(5):828–844
Singh P, Dave A, Vaistij FE, Worrall D, Holroyd GH, Wells JG, Roberts MR (2017) Jasmonic acid-dependent regulation of seed dormancy following maternal herbivory in *Arabidopsis*. New Phytol 214(4):1702–1711
Thakur A, Sharma KD, Siddique KH, Nayyar H (2020) Cold priming the chickpea seeds imparts reproductive cold tolerance by reprogramming the turnover of carbohydrates, osmo-protectants and redox components in leaves. Scientia Horticulturae, 261, 108929
Tiwari M, Singh P (2021) Plant defense priming: a new tool for sustainable global food security. In: Agricultural innovations and sustainability. Agrobios Research, pp 133–153
Tripathi DK, Bhat JA, Antoniou C, Kandhol N, Singh VP, Fernie AR, Fotopoulos V (2024) Redox regulation by priming agents toward a sustainable agriculture. Plant Cell Physiol 65(7):1087–1102
Truitt CL, Wei HX, Pareé PW (2004) A plasma membrane protein from *Zea mays* binds with the herbivore elicitor volicitin. Plant Cell 16(2):523–532
Tsai HJ, Shao KH, Chan MT, Cheng CP, Yeh KW, Oelmüller R, Wang SJ (2020) *Piriformospora indica* symbiosis improves water stress tolerance of rice through regulating stomata behavior and ROS scavenging systems. Plant Signal Behav 15(2):1722447
Tugizimana F, Mhlongo MI, Piater LA, Dubery IA (2018) Metabolomics in plant priming research: the way forward. Int J Mol Sci 19(6):1759
Urech P (2000) Sustainable agriculture and chemical control: opponents or components of the same strategy? Crop Prot 19(8–10):831–836
Van der Ent S, Van Hulten M, Pozo MJ, Czechowski T, Udvardi MK, Pieterse CM, Ton J (2009) Priming of plant innate immunity by rhizobacteria and β-aminobutyric acid: differences and similarities in regulation. New Phytol 183(2):419–431
Varghese L, Thomas G (2023) Chitosan triggers tolerance to *Pythium myriotylum* infection in ginger (*Zingiber officinale*) by modulating multiple defense signaling pathways. Physiol Mol Plant Pathol 125:101983

Venegas-Molina J, Proietti S, Pollier J, Orozco-Freire W, Ramirez-Villacis D, Leon-Reyes A (2020) Induced tolerance to abiotic and biotic stresses of broccoli and *Arabidopsis* after treatment with elicitor molecules. Sci Rep 10(1):10319

Wang X, Zhang J, Song J, Huang M, Cai J, Zhou Q, Dai T, Jiang D (2020). Abscisic acid and hydrogen peroxide are involved in drought priming-induced drought tolerance in wheat (*Triticum aestivum* L.). Plant Biology, 22(6), pp.1113–1122.

Widmann C, Gibson S, Jarpe MB and Johnson GL (1999) Mitogen-activated protein kinase, conservation of a three-kinase module from yeast to human. Physiol. Rev. 79, 143–180. https://doi.org/10.1152/physrev.1999.79.1.143

Whitehead JC (2016) The chemistry of cold plasma. In: Cold plasma in food and agriculture. Academic, pp 53–81

Yamauchi Y, Matsuda A, Matsuura N, Mizutani M, Sugimoto Y (2018) Transcriptome analysis of *Arabidopsis thaliana* treated with green leaf volatiles: possible role of green leaf volatiles as self-made damage-associated molecular patterns. J Pestic Sci 43(3):207–213. Applications of plant priming

Yang L, Li B, Zheng XY, Li J, Yang M, Dong X, He G, An C, Deng XW (2015) Salicylic acid biosynthesis is enhanced and contributes to increased biotrophic pathogen resistance in *Arabidopsis* hybrids. Nat Commun 6(1):7309

Zhang Y, Tang M, Huang M, Xie J, Cheng J, Fu Y, Jiang D, Yu X, Li B (2022) Dynamic enhancer transcription associates with reprogramming of immune genes during pattern triggered immunity in *Arabidopsis*. BMC Biol 20(1):165

Chapter 13
Engineering Artificial miRNAs to Prime Defense Responses Against Plant Pests and Pathogens

Tilahun Rabuma and Neeti Sanan-Mishra

Abstract MicroRNAs (miRNAs) are endogenously produced, small ribonucleotides that switch off gene expression to regulate the typical development and response to stress in plants. Artificial miRNAs (amiRNAs) are synthetic molecules bioengineered in the laboratory for "turning off" gene expression. These can be used for functional genomics to understand the biological role of specific genes, as well as for improving specific plant traits like enhancing resistance to pathogens and pests. This chapter starts with a basic understanding of the plant miRNA biogenesis pathway in light of recent developments and discusses the role of miRNAs in protecting plants from pathogens. It then provides a comprehensive update on the engineering and processing of amiRNAs and their potential application in protecting crops from the attack of pests and pathogens. The challenges in incorporating amiRNAs as endogenous cellular components for employing the miRNA biogenesis pathways to reprogram synthetic circuits are also discussed.

Keywords miRNA · Design strategy · Tools · Applications · Crop improvement

T. Rabuma
Department of Biotechnology, College of Natural and Computational Science, Wolkite University, Wolkite, Ethiopia

Plant RNAi Biology Group, International Centre for Genetic Engineering and Biotechnology, New Delhi, India

N. Sanan-Mishra (✉)
Plant RNAi Biology Group, International Centre for Genetic Engineering and Biotechnology, New Delhi, India
e-mail: neeti@icgeb.res.in

A. Singh et al. (eds.), *Sustainable Crop Production*, Sustainability Sciences in Asia and Africa, https://doi.org/10.1007/978-981-95-7844-3_13

Abbreviations and Acronyms

AGO1	Argonaute 1
AO	L-Ascorbate Oxidase
CBP80/20	Cap-Binding Protein 80/20
CCSD	Copper Chaperone or Superoxide Dismutase
CGMV	Cucumber Green Mosaic Virus
CHR2	Chromatin Remodeling Factor 2
CLF	Curly Leaf
CMV	Cucumber Mosaic Virus
CPL1	C-Terminal Domain Phosphatase-Like 1
CSD	Copper-Zinc Superoxide dismutase
D-body	Dicing Body
DCL1	Dicer-Like 1
DDL	Dawdle
DRB1	Double-Stranded RNA-Binding Protein 1
ELP2/5	Elongator Protein 2/5
FISH	Fluorescent In Situ Hybridization
GCN5	General Control Non-repressed Protein 5
HIGS	Host-Induced Gene Silencing
HSP90	Heat Shock Protein 90
HST	Hasty
HR	Hypersensitivity Response
HYL1	Hyponastic Leaves 1
ILP1	Polyploidy1-1d
MED	Mediator
NOT 2a/2b	Negative on TATA-Less
NTR1	NTC-Related Protein 1
NUP1	Nuclear Pore Complex 1
PKL	PICKLE
PP4	Protein Phosphatase 4
Pri-miRNA	Primary microRNA
Pre-miRNA	Precursor miRNA
PSTVd	Potato Spindle Tuber Viroid
RACK1	Receptor of Activated C Kinase 1
RBSDV	Rice Black-Streaked Dwarf Virus
RISC	RNA-Induced Silencing Complex
RNA POL II	RNA Polymerase II
ROS	Reactive Oxygen Species
RSV	Rice Stripe Virus
RTBV	Rice Tungro Bacilliform Virus
SAR	Systemic Acquired Resistance
SE	Serrate
SEAP1	Serrate-Associated Protein 1
SMA1	SMALL1
SMEK1	Suppressor of MEK 1

SMV	Soybean Mosaic Virus
SOD	Superoxide Dismutase
STA1	Stabilized 1
STV1	Short Valve 1
SWI/SNF	Switch/Sucrose Non-fermentable
SWN	Swinger
TCP	Teosinte Branched 1/Cycloidea/PCF
TGH	Tough
THP1	Tho2/Hpr2 Phenotype 1
ToLCNDV	Tomato Leaf Curl New Delhi Virus
ToLCV	Tomato Leaf Curl Virus
TREX-2	Transcription-Coupled Export 2
TSWV	Tomato Spotted Wilt Virus
TuMV	Turnip Mosaic Virus
TYLCLV	Tomato Yellow Leaf Curl Virus
TYMV	Turnip Yellow Mosaic Virus
VAT	Virus Aphid-Mediated Transmission
WDV	Wheat Dwarf Virus

Editor's Notes

Artificial miRNA-based gene silencing is one of the most powerful technologies applied extensively in functional genomics. As is the case with any technology, knowledge about the operational principles are key to optimal amiRNA design and deployment. The authors of this chapter, Rabuma and Sanan-Mishra, provide a comprehensive overview of the field of miRNAs and amiRNAs and their potential in conferring defense response. The chapter explains miRNA biogenesis, structural and processing features, and precise target-pairing properties. Knowledge rooted in the principles of miRNA biology has led to identification of parameters that have been implemented in popular computational tools for user convenience. The authors argue that computational genomics approaches are key to successful design and deployment of amiRNA against specific targets. Stakeholders have an option to choose a range of gene silencing tools including genome editing and amiRNA technology to develop stress-resilient crops.

Actionable Points

1. The amiRNAs can be designed with nearly perfect sequence complementarity and specificity with their target to achieve high efficiency in gene silencing. The efficiency of amiRNA in inducing gene silencing is due to minimizing the off-target effects and potential for missing variants or mutants.
2. Transgenic lines or CRISPR-based miRNA editing can be used to alter levels of key defense-associated miRNAs in crops. The strategy can include overexpressing beneficial miRNAs that enhance immunity or silencing negative regulators that increase susceptibility. Host-Induced Gene Silencing (HIGS) has shown success in controlling infections caused by RSV, TuMV, ToLCNDV, and other viruses.

3. Breeders can select or design crop varieties where these miRNA-target modules are fine-tuned for optimized resistance gene expression without compromising plant fitness, e.g., miR482 that regulates Resistance (R) genes and miR528 that regulates defense-related transcription factor, *OsSPL9* to enhance durability of resistance against viruses and fungi.

13.1 Introduction

The noncoding RNA (ncRNA) comprises a large family, which plays a vital role in negatively regulating gene expression at the posttranscriptional and transcriptional levels (Li and Peng 2021). This phenomenon is collectively known as RNA interference or gene silencing (Raja and Tennyson 2023). The ncRNAs are categorized according to their length as small (20–24 nucleotides; nt) and long (≥200 nt) (Axtell 2013; Imaduwage and Hewadikaram 2024). They serve as key regulators of the genetic networks that control plant growth and defense response.

Rapid advancement in RNAi research has unlocked novel opportunities for employing this technology for precisely improving pest/pathogen resistance in plants, as well as for a wide range of agricultural and basic biology innovations. Small ncRNAs can be broadly divided into two groups, namely, small interfering RNA (siRNA) and microRNA (miRNA). Among these, the hairpin precursor-based miRNAs play an important role in a variety of biological and metabolic processes in plants to control nearly all facets of growth and adaptation to the environment (Jeena et al. 2022). They offer tremendous potential for crop improvement by enabling accurate regulation of specific gene expression, thereby improving resistance to various environmental stresses.

The miRNAs represent a broad class of small ncRNA (Lau et al. 2001; Lee and Ambros 2001) that can downregulate expression of genes to trigger posttranscriptional gene silencing (PTGS) or transcriptional gene silencing (TGS). During PTGS, they exert their effect by pairing with complementary sequences in the mRNA coding region, causing either transcript cleavage or suppression of translation (Zhang et al. 2022). If the target gene promotes undesirable traits, then the approach of miRNA or artificial miRNA (amiRNA) expression is employed to silence the pathway. In contrast, for genes regulating beneficial or desirable traits, the approach involves generating miRNA cleavage-resistant mRNA or utilizing artificial target mimics to evade the regulation by miRNA.

Despite the theory and application-based support for applying miRNA-mediated silencing in plants, the biotechnological scope of amiRNA and target mimics for improving crop plants have not yet been fully explored. Hence, this chapter highlights the current progress in understanding and applying amiRNA technology, emphasizing its potential in enhancing plant defense to pests and pathogens. It also details the principles of target identification and discusses several parameters critical for amiRNA selection and design.

13.2 Biogenesis Pathway of microRNA

The amiRNA technology harnesses the natural biogenesis pathways of miRNA in plants to generate artificial or engineered, functional miRNAs (Tiwari et al. 2014; Bhagwat et al. 2016; Yasir et al. 2022). Designing and implementing effective amiRNA-based gene-silencing techniques in agricultural crops thus requires a thorough understanding of plant miRNA biogenesis and function.

The miRNAs are biologically processed from *pri-miRNA* transcripts that are typically synthesized from defined genomic loci by RNA POL II (Sanchita et al. 2018). In plants, these transcripts are processed in the D-body in the nucleus, which ensures efficient and accurate transcription (Dong et al. 2008; Liu et al. 2012; Zhu et al. 2013). The *pri-miRNAs* have a cap at their 5′-end and a poly-A tail at their 3′-end. Their transcription initiation and co-transcriptional processing require POL II (Hajheidari et al. 2012). POL II recruitment to the *MIR* gene promoters is facilitated by MED20A, MED17, and MED18 (Kim and Chen 2011) and supported by other proteins like THP1 and Yeast SAC3 Homolog A (core subunits of the TREX-2 complex) and NOT 2a/2b, ELP2/5, SMA1, and STV1 to initiate the transcription of *MIR* genes (Wang et al. 2013; Fang et al. 2015; Li et al. 2018; Zhang et al. 2020). *MIR* gene transcription is also regulated by factors controlling histone modification and chromatin remodeling, like GCN5, SWN, CLF, PKL, CHR2, and the ATPase subunit of the SWI/SNF chromatin-remodeling complex (Kim et al. 2009; Wang et al. 2017, 2019).

Spliceosome-associated proteins, such as ILP1, NTR1, and CWC15, also bind to the promoters of *MIR* and increase the occupancy of POL II, thereby favorably regulating *pri-miRNA* transcription. Several other splicing-related factors like CBP80/20 (Kim et al. 2008; Raczynska et al. 2009), STA1 (Chaabane et al. 2012), SMA122 (Li et al. 2018), PP4 regulatory subunit 3 (Wang et al. 2019), THP1 in the TREX-2 complex (Zhang et al. 2020), SEAP1 (Li et al. 2021), WD40 repeat protein RBV21 (Liang et al. 2022), AAR2 (Fan et al. 2022), and histone deacetylase 9 (Park et al. 2023) also regulate the process. HST, an ortholog of Exportin 5, also plays a role in promoting *MIR* gene transcription (Cambiagno et al. 2021).

The *pri-miRNA* undergoes stepwise enzymatic processing into hairpin-loop pre-miRNA and functional miRNA duplexes within the D-body to ensure stability and efficient processing (Stroynowska et al. 2014). DCL1 within the microprocessor complex catalyzes the dual steps of processing of *pri-miRNA* to generate the pre-miRNA and the further processing of the stem region of the pre-miRNA to produce a mature miRNA duplex consisting of guide and passenger (miRNA*) strands (Tiwari et al. 2014). Another important player in this process is a phosphoprotein, HYL1, also known as DRB1, which is dephosphorylated by CPL1 and PP4/SMEK1 (Koiwa et al. 2002; Yang et al. 2014). Dephosphorylated HYL1 binds to the folded region of *pri-miRNAs* through the dsRBD and serves as a molecular anchor to guide accurate strand selection and processing by DCL1 (Manavella et al. 2012; Su et al. 2017). Localization of HYL1 to D-bodies is mediated by the MOS4-associated complex (Li et al. 2020).

Additional components of the microprocessor complex regulating *pri-miRNA* processing include SE, a zinc finger protein; CWC15 (Zhou et al. 2024); RACK1, a partner protein of the SE (Speth et al. 2013); TGH, a G-patch domain-containing protein (Ren et al. 2012); and PRL1 and DDL. PRL1 is a conserved WD40 protein, and DDL is a forkhead domain-containing protein; both stabilize the *pri-miRNA* to enhance DCL1 activity (Yu et al. 2008; Zhang et al. 2014).

After processing, the mature miRNA duplexes are methylated by enzyme HEN1 and are loaded onto AGO1. The miRNA-AGO complex is exported to the cytoplasm via the TREX-2 complex. At the nuclear envelope, THP1 interacts with NUP1 to promote export of miRNA to the cytoplasm (Bologna et al. 2018; Zhang et al. 2020). Although HST was previously thought to mediate nuclear-cytoplasmic transport of plant miRNAs (Cambiagno et al. 2021), recent studies suggest that AGO1 shuttling is required for mature miRNA export (Bologna et al. 2018), while HST facilitates the transcription of *MIR* genes (Cambiagno et al. 2021). In addition, HSP90 was proposed to associate with AGO1 to promote miRNA loading and removal of the passenger strand (Iki et al. 2010; Bologna et al. 2018). In the cytoplasm, the AGO1-RISC recognizes the complementary target mRNA, resulting in its cleavage or translational inhibition.

13.3 The Role of miRNA in Plant Protection Against Pathogens

Pathogens, including bacteria, viruses, and fungi, can cause disease in plants by disrupting the functions of host cells and suppressing immune defenses. Plants have a variety of sophisticated defense mechanisms, like structural and chemical defenses, HR and SAR (Wang et al. 2019). Another line of resistance is through miRNA-driven silencing of host and pathogen mRNA (Zhang et al. 2011a; Yang et al. 2021; Rabuma et al. 2022).

miRNAs regulate gene function in diverse organisms, including bacteria, fungi, viruses, insects, and nematodes. This broad regulatory capacity underlines their potential to contribute to plant protection against various pathogens. During host-pathogen interactions, specific miRNAs play a vital role in modulating the disease response in plants. In various plants, multiple miRNAs have been identified that either positively or negatively regulate resistance to fungal, bacterial, and viral pathogens by regulating the expression of resistance (R) genes. In *Arabidopsis*, miR163 expression was upregulated after infection with *Pseudomonas syringae pv. tomato* DC3000. It targets methyltransferase genes like *PXMT1* and *FAMT* to increase the plant's sensitivity to the bacterial pathogen (Chow and Ng 2017).

Several studies involving overexpression, knockout, or knockdown of miRNAs in a variety of plant species have confirmed their involvement in regulating pathogen infections (Basso et al. 2019). Overexpression of miR1918 in tomato plants increased the susceptibility to late blight disease, indicating its role as a negative regulator in the defense against *Phytophthora infestans* (Luan et al. 2016).

miR319a was shown to accumulate during *Sclerotinia sclerotiorum* infection in *Brassica rapa* plants and increase sensitivity to the pathogen. In plant leaves, miR319a controls cell division arrest by negatively regulating the expression of the *TCP* gene family. It also acts to repress the *BraTCP4-1* gene, which encodes a protein that regulates plant immunity by changing the expression levels of *WRKY* genes (Dong et al. 2021).

The perception of bacterial flagellin peptide, FLG22, leads to a notable increase in the levels of miR393, which negatively regulates the expression of transcripts encoding the F-box auxin receptors, *TIR1, AFB2,* and *AFB3*. Consequently, repression of the auxin signaling pathways significantly enhances pattern-triggered immunity (PTI), thereby bolstering resistance to *Pseudomonas syringae* pv. tomato DC3000 in *Arabidopsis* plants (Navarro et al. 2006). Likewise, overexpression of miR393 suppresses auxin signaling, which increases bacterial resistance. This finding implies that miRNAs play a crucial role in the plant-induced immune response by modulating the phytohormones (Navarro et al. 2006; Peláez and Sanchez 2013). The miR393*, the complementary strand of miR393, has also been shown to play an important role in antibacterial immunity by inhibiting the expression of *MEMB12 (SNARE),* which encodes a protein involved in membrane fusion, thereby facilitating the exocytosis of pathogenesis-related protein, PR1 (Zhang et al. 2011b).

miR160a has been demonstrated to positively regulate the callose deposition process induced by pathogen-associated molecular patterns (PAMPs) during defense against *Magnaporthe oryzae* in rice (Li et al. 2014). In contrast, two other miRNAs, namely, miR398b and miR773, have been found to exert a negative regulatory effect on PAMP-induced callose deposition and influence the plant's overall resistance to bacterial pathogens (Li et al. 2010). miR398b mediates an integrative regulatory module by targeting genes that encode members of the *SOD* family, including *CCSD* and *CSD* (Li et al. 2010, 2019). These interactions suggest that miR398b may have a pivotal function in modulating the expression of these genes, which could ultimately influence the plant immune response to bacterial invasions. Tomato miR482, which regulates specific NBS-LRR genes, was downregulated in the defense reaction against bacteria and viruses (Shivaprasad et al. 2012). These studies reveal a complex interplay between miRNA regulation and plant innate immunity mechanisms, emphasizing the importance of understanding the role of miRNA in plant defense strategies and their responses to pathogenic challenges.

miRNAs also provide an extra layer of inherent defense against viral infection by influencing plant gene expression (Kumar et al. 2022). This process involves a complex network of signaling pathways and defense responses, which are regulated by miRNAs to enhance the plant's resilience against invading viruses. The viruses can modulate the plant miRNAs in favor of their infection. For instance, miR168, which targets *AGO1*, a core component of the RISC, is upregulated by several viruses to dampen the RNA silencing defense mechanism (Liu et al. 2017). Similarly, RTBV upregulates miR6024 to bring down the levels of *NBS-LRR resistance* genes for compromising the plant's defense.

The miRNA-mediated regulation of auxin significantly contributes to the VAT resistance against *Aphis gossypii* in melon crops (Yang et al. 2021). The aphid-infestation-mediated induction of miR393 represses *auxin receptor* genes in VAT^{+}-resistant plants (Sattar et al. 2015). Likewise, it was demonstrated that TuMV infection triggered miR1885-dependent silencing of the photosynthesis-related gene *BraCP24*, enhancing floral transition. In contrast, TuMV-induced BraTNLI expression overcomes the miR1885-mediated turnover of the *R* gene in *Brassica* (Cui et al. 2019).

It has been demonstrated that Osa-miR528 has a role in defense against RSV (Kumar et al. 2022). In cleavage-defective AGO18 mutants, it is sequestered away from AGO1 upon RSV infection, preventing the formation of a functional RISC. These events result in an increased expression of the downstream *AO* gene and trigger a defense response against RSV by building up ROS (Wu et al. 2017). Furthermore, it was demonstrated that the OsSPL9 transcription factor modulates the expression of the Osa-miR528-AO module by adhering to the promoter of Osa-miR528, leading to downregulation of Osa-miR528, which results in increased resistance against RSV through *AO* expression (Yao et al. 2019).

Thus, regulation of gene expression by miRNAs facilitates the plant response to biotic stress (Yang et al. 2021). These findings underscore the importance of understanding the molecular mechanisms by which miRNAs operate, as they may serve as valuable tools in developing innovative approaches to enhance bacterial and viral pathogen resistance in crops, ultimately leading to better agricultural sustainability and food security.

13.4 Artificial microRNA Design and Synthesis

The native miRNA duplex in the stem region of hair-pin looped pre-miRNA can be replaced with artificially designed miRNA-miRNA* sequences using oligonucleotide-directed mutagenesis (Schwab et al. 2006; Niu et al. 2006; Bhagwat et al. 2016; Redhu et al. 2023). The pre-miRNAs containing the amiRNA duplex when delivered inside the plants cells are recognized as endogenous precursors and processed to produce the functional amiRNA duplex, which assembles with RISC to cleave their target gene in the same way as the natural miRNA (Redhu et al. 2023). The mature amiRNAs are highly specific and have several potential applications in providing resistance to abiotic stressors and protection from virus and other pathogens (Carbonell et al. 2019; Jiang et al. 2023).

13.4.1 Features for miRNA Selection and Design

Designing amiRNAs involves identification of a sequence with nearly perfect sequence complementarity within the target gene to achieve precise and efficient gene silencing (Bhagwat et al. 2016). The effectiveness of amiRNA in inducing gene silencing also depends on the choice of precursor backbone, identification of the specific target sequence to be silenced, strategies to minimize off-target effects, and approaches for not missing variants or mutants (Yusuf et al. 2021). Depending on the intended application, amiRNAs can be designed to recognize and silence multiple genes simultaneously by targeting conserved regions within gene families, or they can be designed to recognize and silence only a single target (Zhang et al. 2018). Therefore, careful consideration of the parameters for amiRNA design is important, and these aspects are elaborated below.

13.4.1.1 Target Site Identification

The effectiveness of gene silencing using the amiRNA approach is determined by identification of the correct target gene sequence to be knocked down. This process of identifying the target gene integrates computational analysis and experimental validation with biological knowledge to confirm the specificity and effectiveness of the designed amiRNA. Bioinformatics tools and databases play a crucial role in identifying and analyzing the ORFs containing the target sequence of the designed amiRNA. The specific target gene sequences can be obtained from publicly accessible plant genome repositories such as EMBL, PGDB, GenBank, and miRBase (Van and Do 2017). A thorough understanding of the biological role of the gene is crucial for selecting the most relevant target genes when tailoring amiRNA for a specific functional or agricultural outcome.

13.4.1.2 Selection Parameters for Candidate amiRNA

Several parameters are considered to design and select the best candidate amiRNA using diverse bioinformatics tools. The basic parameters include several key features like amiRNA length and binding affinity between amiRNA and its target sequence.

The ideal guide amiRNA sequence should be 21 nucleotides long, contain a uracil (U) residue at the first position from the 5′ end, and preferably have an adenine (A) residue at the tenth position. The guide strand should be designed to have lower thermodynamic stability at its 5′ end. This asymmetry in thermodynamic stability favors the selective loading of the guide strand into RISC, enhancing gene silencing efficiency (Khvorova et al. 2003; Schwarz et al. 2003; Schwab et al. 2006; Ossowski et al. 2008).

Only a few mismatched nucleotides are acceptable between the amiRNA and the target sequences, mostly at the 3′ ends of the amiRNA. An amiRNA, which has a 5′ mismatch with amiRNA* but has perfect complementarity to the target gene sequence, serves as a very effective gene silencing agent (Park et al. 2009). The binding strength between miRNA and its target is also determined by free energy calculations (Gu et al. 2021). Stronger binding is indicated by lower free energy, suggesting that the transcript is more likely to be the actual target of the amiRNA (Yue et al. 2009).

13.4.1.3 Bioinformatics Tools Used for Designing amiRNA

Several Web-based bioinformatics tools have been developed to design potential amiRNA with high specificity and perfect or near-perfect complementarity with target genes or gene groups across diverse monocot and eudicot species. These tools automate the prediction process by incorporating critical design parameters modeled on endogenous miRNA features. These tools enable the generation of a highly specific pool of 21-nt amiRNA sequences, with advanced options for filtering off-target sites. The tools popularly used include Web MicroRNA Designer 3 (WMD3), PsRNAtarget, MicroRNA Designer, amiRNA Designer, Plant Small RNAMaker Site (P-SAMS), and AMIRdesigner (Schwab et al. 2006; Ossowski et al. 2008; Dai and Zhao 2011; Bhagwat et al. 2016; Mickiewicz et al. 2016; Fahlgren et al. 2016; Yasir et al. 2022).

WMD3, developed by Prof. Weigel's group, was the first Web-based tool that enabled the design of amiRNA for target genes in species with comprehensive genomic or EST resources. WMD3 supports various parameters for the selection of an effective amiRNA and provides primers for PCR amplification resulting in the generation of amiRNA-amiRNA* nucleotide segments via site-directed mutagenesis within any of the three natural pre-miRNA backbones, each available through their respective vector. These three vectors are as follows: pRS300 contains *preMIR319a from Arabidopsis thaliana*, pNW55 contains *pre-MIR528 from Oryza sativa*, and pChalmiRNA2 and 3 contain *pre-MIR1157 from Chlamydomonas reinhardtii*. The amiRNA-amiRNA* could be incorporated into a natural pre-miRNA backbone by replacing the miRNA-miRNA*, using inverse PCR amplification (Dhakate et al. 2014). Due to its robustness and compatibility across plant species, WMD3 has been widely adopted (Schwab et al. 2006; Ossowski et al. 2008). For example, WMD3 was used to engineer highly specific amiRNA targeting three separate target genes (*Pds*, *Spl11* and *Eui1/CYP714D1*) in Nipponbare (japonica) and IR64 (Indica) rice varieties (Warthmann et al. 2008, 2013). Similarly, the WMD3 program was used to generate amiR-bnashp1 and amiR-bnashp2 from pRS300 in *N. benthamiana* to silence *shp1/shp2* gene (Dhakate et al. 2014). Using the WMD3 program, Zhang et al. (2018) evaluated 21 nt sequences for their suitability as amiRNAs candidates and eliminated those predicted to cause off-target effects. Building on this foundation, Mickiewicz et al. (2016) introduced amiRNA Designer, which considers the free energy decomposition of miRNA-target duplexes. This software allowed users to analyze the genes of specific interest to engineer amiRNA and filter out candidates with potential off-target effects.

This was followed by the development of the microRNA designer program that designed the potential amiRNA for integration in pre-miR168a backbone using single-step inverse PCR amplification (Bhagwat et al. 2016). Their approach involved predicting candidate amiRNAs as done in WMD3; however, their final selection was based on the free energies (ΔG) of both the 5′-ends in a potential amiRNA-amiRNA duplex. Another Web-based tool, P-SAMS amiRNA Designer, provides two design modules, amiRNA Designer and syntasiRNA Designer (Fahlgren et al. 2016), which predict the amiRNA as well as the oligonucleotides required for cloning into BsaI/ccdB vectors containing At-pre-miR390a or Os-pre-miR390 (Carbonell 2019). Recently, the AMIRdesigner tool was developed to design and process oligos for the development of amiRNAs using pre-miR171 and pre-miR164 backbones (Yasir et al. 2022). Most importantly, the amiRNA designed by the AMIRdesigner tool could be assembled into miRNA clusters capable of targeting multiple sites in a transcript.

13.4.1.4 Improving amiRNA Efficacy

Despite the advancements in bioinformatics tools, the current computational programs for target prediction do not discriminate between efficient and inefficient amiRNA based on seed sequence, suggesting that strong binding between amiRNA and target mRNA is essential for target recognition and repression (Arroyo et al. 2014). Therefore, the bioinformatics prediction alone is not sufficient for the selection process, and experimental validations are needed to evaluate the response of the amiRNA and confirm that the predicted transcripts are targeted.

The programs designed to predict amiRNA sequences produce an extensive list of many potential amiRNA candidate sequences for the target gene required to be silenced. The effectiveness of these amiRNAs in vivo remains uncertain, so the engineered amiRNA plasmid constructs need to be screened individually in planta using techniques like quantitative droplet digital PCR to assess the target mRNA levels for selecting the effective amiRNA (Cabrera et al. 2024).

The inconsistent efficacy of amiRNAs has been a significant barrier to their widespread adoption in plant systems (Li et al. 2014). As a result, enhancing the effectiveness of amiRNA-based gene silencing has become a key research focus. Arroyo et al. (2014) demonstrated that amiRNAs with matches in the canonical seed sequence are more effective in simultaneous repression of multiple target genes when compared with those with mismatches or bulges. Other factors, such as accessibility of target (Duan et al. 2008; Jiang et al. 2011) and intracellular amiRNA accumulation (Lafforgue et al. 2011), have been identified as critical determinants of silencing efficiency in plant cells.

To address these issues and improve silencing outcomes, researchers have explored strategies like expressing multiple amiRNAs from a single miRNA backbone to increase the gene silencing efficiency compared to a single amiRNA with only one hit in the transcript region. Altering the configuration and sequence of amiRNAs has also been shown to enhance their silencing potential (Redhu et al. 2023). However, these approaches failed to quantify the efficiency of amiRNA when the targeted transcript was subject to translation inhibition rather than transcript cleavage. Therefore, to address this issue, Li et al. (2014) developed an epitope-tagged protein-based amiRNA (ETPamir) screening system, enabling direct evaluation of the silencing efficiency of amiRNA. This approach allowed for the accurate selection of highly effective amiRNA from a large pool of computationally predicted amiRNA sequences to achieve complete gene silencing in plants.

Additionally, it was noted that the pre-miRNA backbone used to design and integrate the amiRNA may have an impact on how effectively it silences target genes. Bally et al. (2020), for instance, examined the effects of five distinct pre-miRNA hairpin backbones and found that larvae fed on transgenic *N. benthamiana* expressing anti-ACE2amiR from precursor backbones At-miR159, Tc-ba, Dm-ba, Dv-miR279, and Dv-miR1 resulted in 30%, 30%, 50%, and 60% mortality rate, respectively.

Experimental studies with plant viruses revealed that mutations within viral transcripts can significantly compromise the effectiveness of amiRNA-mediated silencing. To overcome this challenge, Lafforgue et al. (2013) tested two strategies aimed at reducing the likelihood of escape mutants and improving the efficacy of amiRNA. They were able to improve the resistance against TuMV in *A. thaliana* by designing a pair of amiRNAs targeting distinct and highly conserved regions in the TuMV genome. These amiRNAs were designed to be complementary to highly conserved RNA motifs in the viral genome. Teotia et al. (2023) also showed that dual (two-hit) amiRNAs were more effective than single (one-hit) amiRNAs in silencing *GFP* and endogenous *PDS* reporter genes in *A. thaliana*. Extending this approach, Khalid et al. (2023) engineered seven clusters of amiRNAs (pAMIN14 construct), each targeting distinct viral genes to effectively combat TYLCLV in transgenic tomato lines.

13.4.2 Expressing amiRNA in Plant Cell

The amiRNA-expressing constructs can be introduced into plant cells via *Agrobacterium*-mediated transformation or gene-gun based direct delivery methods. Warthmann et al. (2013) provided comprehensive protocols for using amiRNA in targeted gene silencing, covering steps from cloning amiRNA precursors to *Agrobacterium*-mediated transformation and subsequent plant regeneration. This process involves cloning the desired amiRNA sequence into a suitable plant expression vector followed by its introduction into the plant genome. Commonly used plasmids like pCAMBIA, which express the amiRNA containing gene cassette under a constitutive promoter and terminator, have been utilized to express the amiRNA in plants (Li et al. 2014). Within the plant cell, the endogenous miRNA processing machinery recognizes the amiRNA precursor and facilitates the generation of the functional amiRNA/amiRNA* duplex. For example, amiRCP-1 and amiRCP-2 were engineered in the pre-miR319a of *Arabidopsis* and subsequently introduced into grapevine somatic embryos following *A. tumefaciens* co-cultivation. Successful expression of amiRNA has been achieved in several plant species. Moreover, several studies have demonstrated the promise of amiRNA in enhancing plant traits and boosting productivity (Table 13.1). Nonetheless, challenges remain in ensuring efficient delivery, stable integration, and controlled expression of amiRNA in plant cells. In addition, the potential risk of off-target effects of amiRNA warrants careful evaluation and mitigation to prevent unintended consequences.

Table 13.1 Examples of amiRNAs employed for enhancing plant tolerance to different abiotic and biotic stresses

S. no.	Backbone	Designed amiRNA	Target gene(s)	Traits improved	Crop plant	Reference
A. Abiotic tolerance: drought and salt tolerance						
1	Ath-pre-miR319a	amiR-ProDH1	*StProDH1* and *StProDH2* genes	Transgenic plants have high relative water, chlorophyll and proline contents under drought stress.	*Solanum tuberosum* L.	Li et al. (2020)
2		amiR-CBP80	*CBP80* gene	Changes in morphology and physiology resulting in drought tolerance	*Solanum tuberosum* L.	Pieczynski et al. (2013)
3	Osa-pre-miR528	amiR-*DST*	*Drought and salt tolerance* (*DST*) gene	*DST* downregulated rice plants (T_0) showed Vigorous growth and longer panicles Higher stress tolerance and better yields	*Oryza sativa*	Faisal et al. (2017)
B. Biotic tolerance: virus infection						
4	Ath-pre-miR319a	amiR-P1	*P1* gene	Induce resistance to soybean mosaic virus (SMV)	*Nicotiana benthamiana*	Latif et al. (2024)
5		AC1-amiR1 AC1-amiR2	*AC1* gene (ATP/GTP-binding domain) of ToLCNDV	Transgenic lines exhibited reduced disease symptoms and a high degree of resistance to tomato leaf curl New Delhi virus (ToLCNDV)	*Solanum lycopersicum* (Pusa Ruby)	Sharma and Prasad (2020)

(continued)

Table 13.1 (continued)

S. no.	Backbone	Designed amiRNA	Target gene(s)	Traits improved	Crop plant	Reference
6		Nine amiRs	Genes for inclusion protein (CI), nuclear inclusion protein (NIa, NIb), CP of PVY	amiRs that targeted the *Nib* and *CP* displayed higher silencing efficiency than those that targeted *CI* and *Nia* of *Potato virus Y* (PVY^N)	*Nicotiana benthamiana* and *N. tabacum* (NC89)	Song et al. (2014)
7		amiRCP-1, amiRCP-2	*CP* of (GFLV)	Reduction in the *grapevine fan leaf virus* (GFLV) tagged with GUS when co-expressed with amiRNA	*Vitis vinifera* (Chardonnay)	Jelly et al. (2012)
8	Ath-pre-miR 159	amiR-159P25, amiR-159HCPro amiR-159Brp1	*P25*, *HC-Pro* and *Brp1* of PVX, PVY and PSTVd	Transgenic potato lines exhibited resistance to coinfection with *potato virus X* (PVX), *potato virus Y* (PVY), *potato spindle tuber viroid* (PSTVd).	*Solanum tuberosum* (Youjin)	Jiang et al. (2023)
9	Ath-pre-miR159a	amiR-2a/b, amiR-3′UTR	*2a* and *2b* genes and the 3′UTR of CMV	Transgenic tomato plants expressing amiRNA demonstrated effective resistance to *cucumber mosaic virus* (CMV) infection alone and on co-infection with TMV and TYLCV	*Solanum lycopersicum* L. (Huafan A101)	Zhang et al. (2011a)

(continued)

Table 13.1 (continued)

S. no.	Backbone	Designed amiRNA	Target gene(s)	Traits improved	Crop plant	Reference
10	Ath-pre-miR159b	amiR derived from virulence-modulating region (VMR)	*StTCP23* gene (3′ untranslated region) and VMR of potato spindle tuber viroid	Downregulation of *StTCP23 in* transgenic potatoes disrupts the GA signaling pathways and leads to stunted plant growth and small, spindle-shaped tubers	*Solanum tuberosum* L.	Bao et al. (2019)
11	Ath-pre-miR164 Ath-pre-miR171	amiR-bC1, amiR-CP, amiR-V2, amiR-C3, amiR-C4, amiR-C1, amiR-C2	Seven virus genes encoding CP, V2, C1, C2, C3, and C4 and the β satellite gene encoding βC1	Resistance to *tomato yellow leaf curl virus* (TYLCLV) and its β satellite is conferred in tomato	*Solanum lycopersicum* (A57)	Khalid et al. (2023)
12	Ath-pre-miR390a	amiZYMV_HC-115s, amiZYMV_HC-1162s, amiZYMV_HC-182as, amiZYMV_HC-196as	Virus-encoded *CP* gene	Reduction in the titer of *Zucchini yellow mosaic virus* (ZYMV) titer. Transgenic plants showed 16–42.5% resistance to ZYMV compared to the controls	*Cucurbita pepo* L.	Berbati et al. (2023)
13		amiR-TSWV-L, amiR-TSWV-M, amiR-TSWV-S	*L, M,* or *S* segments of TSWV	Silencing of *tomato spotted wilt virus* (TSWV) RNA leads to resistance	*Solanum lycopersicum*	Carbonell et al. (2019)
14	Ath-pre-miR156, Ath-pre-miR164 Ath-pre-miR171	amiR1-CP, amiR4-MP amiR6-Rep	*CP, MP, Rep* genes of CGMMV	Suppression of *cucumber green motile mosaic virus* (CGMMV) replication and disease in virus-infected plants	*Nicotiana benthamiana*	Liang et al. (2019)

(continued)

Table 13.1 (continued)

S. no.	Backbone	Designed amiRNA	Target gene(s)	Traits improved	Crop plant	Reference
C. Biotic tolerance: fungus infection						
15	Ath-pre-miR319a	amiCmMLO17 amiCmKIC	*CmMLO17* and *CmKIC* genes	Silencing of *CmMLO17* by amiRNA reduced susceptibility of *Chrysanthemum morifolium* to *A. alternata* infection	*Chrysanthemum morifolium*	Xin et al. (2021)
16	Osa-pre-miR528	amiRNA1, amiRNA2, amiRNA3, amiRNA4	*TaAAED1* (wheat thioredoxin-like gene)	*TaAAED1* negatively regulates wheat resistance to pathogens and silencing of this gene enhances resistance. Two transgenic lines (L12 and L17) expressing amiRNA1 demonstrated great resistance to WYMV and *Puccinia striiformis f.* sp. *tritici* infection.	*Triticum aestivum*	Liu et al. (2021)
17	At-pre-miR164b	5 amiRNAs	5 different regions of the *Avr3a* gene of *Phytophthora infestans* RXLR	Resistant lines had lower pathogen load and Avr3a transcript levels than the non-transgenic control	*Solanum tuberosum* L.	Thakur et al. (2015)
D. Biotic tolerance: insect pest infestation						
1	Pre-miR159a	Plant amiRNA vector	*MpAce1* (*Myzus persicae* acetyl cholinesterase 1) gene	Significant reduction in aphid colonies fed on T1 transgenic plants of both cultivars, due to downregulation of the aphid *Ace1*	*Solanum lycopersicum* L. (Jamila and Tomaland)	Faisal et al. (2021)

(continued)

Table 13.1 (continued)

S. no.	Backbone	Designed amiRNA	Target gene(s)	Traits improved	Crop plant	Reference
2	Four insect pre-miRs and at-pre-miR159	amiR-Dm-ba, amiR-Tc-ba, amiR-Dv-miR1, amiR-Dv-miR279	*Ace2* (acetyl cholinesterase 2) gene of *Helicoverpa armigera*	*H. armigera* fed on transgenic *N. benthamiana* leaves showed increased mortality, developmental defects, and slower growth rates	*Nicotiana benthamiana*	Bally et al. (2020)
3	Ath-pre-miR164b	HaAce1-amiR1	*Ace1* gene of *H. armigera*	The transgenic tobacco-treated insect group showed 25% larval mortality and 20% adult deformities	*Nicotiana tabacum* L. (Petit Havana)	Saini et al. (2018)
4	Pre-miR let-7a (mse-let-7a) of *Manduca sexta*	amiR-*HaEcR*-5p, amiR-*HaEcR*-3p	*EcR* (Ecdysone receptor) gene of *H. armigera*	Increased larval mortality and decreased oogenesis of the insect pest. Checked by using *E. coli* HT115 strain	No transgenic crop development	Yogindran and Rajam (2016)

13.4.3 Analysis of amiRNA Gene Expression in Plants

The analysis of amiRNA expression in plants is an important step in validating their accurate processing and effective function. Analyzing the domains of amiRNA function in transgenic plants is also critical to prevent their unexpected impacts on plant growth and development (Bhagwat et al. 2016). To overcome these challenges, assessing amiRNA expression levels is an essential step to understand their functional implication.

To assess the presence and activity of selected amiRNAs, the technique involves their integration into plant expression vectors. Expression analysis can subsequently be carried out using transient expression systems or through the development of stable transgenic lines in model plants such as *Nicotiana tabacum*. Transient expression of amiRNA is commonly monitored via reporter genes like *luciferase*, *GFP*, and *GUS*. GUS activity is visually detectable following treatment of the tissue with X-gluc, which yields an insoluble blue precipitate due to β-glucuronidase activity, while GFP expression can be visualized under UV light using fluorescence microscopes (Jefferson et al. 1987; Jeoung et al. 2002).

A combined approach using both *GUS* and *GFP* reporter genes simultaneously was employed to detect the expression of amiRNA during transient transformation in strawberry (*Fragaria vesca*) fruits. In this strategy, an amiRNA construct was designed by replacing the mature miR166 sequences with amiR-GUS, amiR-GFP, amiR-NE, and amiR390 (Li et al. 2019). The production of amiR-GUS and amiR-GFP led to a reduction in *GUS* and *GFP* expression levels, while amiR390 was able to efficiently target the endogenous *TAS3* gene and increase the ta-siRNA3 biosynthesis. Molecular confirmation of amiRNA expression and activity can be achieved using several techniques that precisely detect the presence and expression levels of a particular amiRNA and/or its target transcript. Although the Northern blotting technique remains a traditional method, it requires radiolabeling and is time-consuming and labor-intensive (Dave et al. 2019). Consequently, more sensitive and high-throughput techniques including qRT-PCR, FISH, nuclease protection assays, and microarray analysis are often preferred.

Liang et al. (2019) utilized transient expression assays along with Northern blot and qRT-PCR to quantify the amiRNA designed to target the CGMV. They observed that the *amiR4-MP* gene was the most highly expressed in *N. benthamiana* infiltrated with the constructs. Western blotting confirmed the reduction in translated protein levels due to posttranscriptional gene silencing. Arroyo et al. (2014) employed a *luciferase* assay and immunoblotting to demonstrate that the designed amiRNA successfully silenced the *pyruvate carboxylase* (*PC*) and *glutaminase* (*GLS*) transcripts having a 3′ UTR fused to a *luciferase* reporter gene. The expression of amiRNAs, pAMIN14, and pAMIE14, which were designed to target seven genes of TYLCLV, was validated with transient GFP silencing assays in *N. tabacum* (Khalid et al. 2023). Latif et al. (2024) conducted a transient expression assay using an SMV clone with a *GFP* cistron to evaluate amiR-P1 efficacy. In their study, an *N. benthamiana* transgenic line expressing amiR-P1 challenged with SMV-GFP exhibited minimal accumulation of SMV-GFP, confirming functional amiRNA activity. Reporter gene-based assays have also shown that amiRNA expression efficiency is influenced by tissue type and structure, age of the transgenic plant, nature of the reporter gene, and the transformation method, as environmental factors like temperature can limit *Agrobacterium*-mediated transformation. Von Born et al. (2018) further reported that the effectiveness of amiRNA-mediated gene silencing decreases with plant age in a temperature-dependent manner. Moreover, experiments conducted on rice plants revealed that low-temperature growth, the application of surfactant and wounding, and vacuum infiltration increased the expression efficiency, while utilizing young seedlings enhanced visualization and shortened the experiment duration (Burman et al. 2020).

13.5 Improving Plant Traits with Artificial microRNA

amiRNA-based gene silencing is an effective approach for silencing the expression of specific genes to improve traits in plants (Park et al. 2009). It is a cost-efficient and effective method of improving plant health and yield. amiRNA can be utilized to silence particular susceptibility genes in plants or virulence genes of pathogens. This targeted approach not only improves plant capacity to resist disease but also lowers reliance on agrochemicals or pesticides to support environment-friendly agriculture practices.

The amiRNAs are often used to silence or regulate gene expression without changing the genetic code; their effects are typically reversible and can be controlled, allowing researchers to fine-tune gene expression. Since miRNAs do not involve direct genome alteration, the risk of off-target mutations or permanent changes to the DNA is lower compared to other genome engineering techniques. However, more investigation is needed to fully harness the capabilities of these molecules.

Numerous studies have described the use of amiRNA in enhancing plant resistance against viral pathogens, though a few studies have been performed on fungal infections, insects, and nematodes (Table 13.1). However, the potential of amiRNA in improving plant defense against bacterial infections or broadening their use to a wider range of pathogens is still limited.

13.5.1 amiRNA for Targeting Viral Pathogens

Several investigations have demonstrated that amiRNA can be utilized to silence the expression of plant genes required for viral replication and pathogenicity. The approach involves integrating pre-amiRNA constructs into plant genomes so that the cellular miRNA biogenesis machinery can be utilized to express the specific amiRNA in the transgenic plants. The amiRNA-RISC ultimately silences essential virus genes and prevents their replication in plant cells. This strategy is categorized as HIGS and has shown promising results for combating viral infections in plants.

The amiRNAs have been used to silence genes of RSV and RBSDV in *O. sativa* (Sun et al. 2016; Zhou et al. 2022), ToLCV in tomato (Sharma and Prasad 2020), PSTVd (Carbonell and Daròs 2017) and TSWV in *N. benthamiana* (Carbonell et al. 2019), TuMV (Lafforgue et al. 2013) and CMV in *A. thaliana* (Duan et al. 2008), *P69* gene of TYMV and *HC-Pro* gene of TuMV (Niu et al. 2006) in *A. thaliana*, and *WDV* gene in barley (Kis et al. 2016). The amiRNA targeting the ToLCNDV *AC1* gene, which is a pathogenicity determinant and plays a key role in replication of the virus, resulted in imparting virus resistance and minimizing loss in crop yield (Sharma and Prasad 2020).

13.5.2 amiRNA for Targeting Insect Pests

The amiRNAs have been extensively used to silence insect genes using HIGS-based approaches (Table 13.1). amiRNAs against *Helicoverpa armigera* have been expressed in transgenic *N. tabacum* L. cv. Petit Havana (Saini et al. 2018) and *N. benthamiana* (Bally et al. 2020), while amiRNAs against *Myzus persicae* have been expressed in *S. lycopersicum* L. (Faisal et al. 2021). The bacterially expressed amiRNAs have been used to silence essential genes in *H. armigera* (Yogindran and Rajam 2016). A key limitation in applying amiRNAs directly against insects through HIGS lies in the structural differences between pre-miRNA backbones of insects and plants. These differences can greatly reduce the efficiency of amiRNA processing and, in some cases, lead to an almost complete failure in processing the designed amiRNA (Mann et al. 2023).

13.5.3 amiRNA for Targeting Fungal Pathogens

A few reports are available on HIGS approaches utilizing amiRNAs to combat other pathogens like fungi (Mann et al. 2023). Therefore, the use of amiRNA approaches for silencing critical genes of a fungal pathogen needs to be investigated thoroughly (Thakur et al. 2015; Mann et al. 2023) (Table 13.1). amiRNA have been expressed in *Solanum tuberosum* L. to target *RXLR* in *Phytophthora infestans* (Thakur et al. 2015), in *Triticum aestivum* to target specific genes in *Puccinia striiformis f.* sp. *tritici* (Liu et al. 2021), and in *Chrysanthemum morifolium* to target *Alternaria alternata* infection (Xin et al. 2021).

13.6 Summary and Perspectives

The amiRNA harnesses the cellular miRNA pathways to selectively silence genes, offering a powerful strategy to enhance plant tolerance to a broad spectrum of biotic and abiotic stress conditions. Recent developments in understanding miRNA biogenesis and target specificity have contributed to the design of more precise and effective amiRNAs, which have broadened their application in crop improvement.

The careful selection of suitable target genes, precise design parameters, and efficient delivery techniques are some of the crucial elements that affect how effectively amiRNA induces gene silencing. Therefore, engineering of amiRNA requires a systematic approach that combines computational predictions, experimental validation, and biological knowledge. Designing effective amiRNA to enhance disease resistance in crops marks a major breakthrough in RNA silencing technology with immense potential to tackle challenges associated with food security. In the case of rapidly evolving pathogens like viruses, utilizing single or multiple amiRNA to target conserved genomic regions of viruses can significantly improve silencing efficiency and durability of resistance.

Delivery of amiRNA through non-transgenic approaches is emerging as a viable alternative to control pathogens or manage insect pests. Promising approaches under investigation include foliar application of stabilized small RNAs in model or crop plants (Joga et al. 2016; McLoughlin et al. 2018), use of nano-carriers like chitosan-, silicon- and carbon-based nanoparticles, and clay nanosheets along with cross-linkers like tripolyphosphate, dextran sulfate, and poly-D-glutamic acid (Cunningham et al. 2018). These strategies offer environmentally friendly and regulatory-compliant avenues for pest and pathogen control.

Moreover, developments in CRISPR-Cas-mediated plant genome editing have provided an alternative means to knockout or knockin *MIR* genes. However, editing noncoding miRNA loci is more challenging than targeting protein-encoding genes due to the presence of homologous, orthologous, or paralogous *MIR* gene families and highly folded secondary structures of pre-miRNA regions (Barrangou et al. 2015). Additional constraints such as the intricacy of regulatory networks, criteria for gRNA design, overlapping transcripts, and the genomic loci of *MIR* genes may limit the editing process (Barrangou et al. 2015; Basak and Nithin 2015). Despite these hurdles, the accuracy and plasticity of amiRNA technology provide a promising adjunct strategy for functional genomics and crop engineering.

The integration of amiRNA into crop improvement pipelines holds great promise for developing stress-resilient and high-yield cultivars. While many studies have demonstrated their potential, widespread agricultural application of amiRNA for crop improvement remains underexplored. amiRNA-mediated gene silencing has also shown tremendous capacity in nutritional improvement of crops (Liu and Chen 2010), with potential applications in increasing the uptake and accumulation of essential mineral and vitamin content to address malnutrition, especially in resource-poor regions. However, the application of amiRNAs for modifying traits like flavor, fragrance, and color of horticultural crops has received limited attention so far (Kumar et al. 2024).

The amiRNA has demonstrated considerable potential in enhancing crop resilience to various abiotic stresses (Table 13.1), including drought, salinity, cold, osmotic pressure, and heat (Cedillo-Jiménez et al. 2016; Zhou et al. 2017; as reviewed by Patil et al. 2021). For example, the *DST* gene negatively regulates stomatal closure in rice leaves, resulting in increased loss of water through transpiration. Silencing of the *DST* gene through amiRNA not only improved drought and salt stress tolerance but also enhanced yield (Faisal et al. 2017). Additional studies showed that loss of function of the *DST* gene enhanced stomatal closure and reduced stomatal density, resulting in improved drought and salt tolerance in rice (Huang et al. 2009). Mutation in the rice *HST1* gene, which encodes the DST transcription factor, resulted in higher heat tolerance, indicating the role of *DST* in negatively regulating heat tolerance (Ding et al. 2023). In potatoes, amiRNA-mediated silencing of *StProDH1* improved drought tolerance, with transgenic lines showing healthier phenotypes under drought stress (Li et al. 2020). The transgenic potato plants accumulated more proline and sugars during rehydration than the non-transgenic plants. Proline is known to function as an antioxidant and regulates cellular osmotic adjustment, indicating that it also scavenges ROS and quenches singlet oxygen (Matysik et al. 2002; Hayat et al. 2012).

Further studies are required to fully exploit the potential of amiRNA technology. Continued research and biotechnological advancements in optimizing the construct design, delivery methods, and functional characterization across a wider range of crop species will enable their application for a variety of biotic or abiotic stress conditions. Future advancements in biotechnology, coupled with traditional breeding and genome editing approaches, will help usher in amiRNAs as next-generation tools for targeted and durable gene silencing.

References

Arroyo JD, Gallichotte EN, Tewari M (2014) Systematic design and functional analysis of artificial microRNAs. Nucleic Acids Res 42(9):6064–6077

Axtell MJ (2013) Classification and comparison of small RNAs from plants. Annu Rev Plant Biol 64(1):137–159

Bally J, Fishilevich E, Doran RL, Lee K, De Campos SB, German MA, Narva KE, Waterhouse PM (2020) Plin-amiR, a pre-microRNA-based technology for controlling herbivorous insect pests. Plant Biotechnol J 18(9):1925–1932

Bao S, Owens RA, Sun Q, Song H, Liu Y, Eamens AL, Feng H, Tian H, Wang M, Zhang R (2019) Silencing of transcription factor encoding gene *StTCP23* by small RNAs derived from the virulence modulating region of potato spindle tuber viroid is associated with symptom development in potato. PLoS Pathog 15(12):e1008110

Barrangou R, Birmingham A, Wiemann S, Beijersbergen RL, Hornung V, Smith A (2015) Advances in CRISPR-Cas9 genome engineering: lessons learned from RNA interference. Nucleic Acids Res 43:3407–3419

Basak J, Nithin C (2015) Targeting non-coding RNAs in plants with the CRISPR-Cas technology is a challenge yet worth accepting. Front Plant Sci 6:1–8

Basso MF, Ferreira PCG, Kobayashi AK, Harmon FG, Nepomuceno AL, Molinari HBC, Grossi-de-Sa MF (2019) MicroRNAs and new biotechnological tools for its modulation and improving stress tolerance in plants. Plant Biotechnol J 17(8):1482–1500

Berbati M, Kaldis A, Voloudakis A (2023) Efficient artificial microRNA-mediated resistance against zucchini yellow mosaic virus in zucchini via agroinfiltration. J Virol Methods 321:114805

Bhagwat B, Chi M, Han D, Tang H, Tang G, Xiang Y (2016) Design, construction and validation of artificial microRNA vectors using *Agrobacterium*-mediated transient expression system. In: Fett-Neto A (ed) Biotechnology of plant secondary metabolism, Methods in molecular biology, vol 1405. Humana Press, New York

Bologna NG, Iselin R, Abriata LA, Sarazin A, Pumplin N, Jay F, Grentzinger T, Dal Peraro M, Voinnet O (2018) Nucleo-cytosolic shuttling of ARGONAUTE1 prompts a revised model of the plant microRNA pathway. Mol Cell 69(4):709–719.e5

Burman N, Chandran D, Khurana JP (2020) A rapid and highly efficient method for transient gene expression in rice plants. Front Plant Sci 11:584011

Cabrera GT, Meijboom KE, Abdallah A et al (2024) Artificial microRNA suppresses *C9ORF72* variants and decreases toxic dipeptide repeat proteins *in vivo*. Gene Ther 31:105–118

Cambiagno DA, Giudicatti AJ, Arce AL, Gagliardi D, Li L, Yuan W, Lundberg DS, Weigel D, Manavella PA (2021) HASTY modulates miRNA biogenesis by linking pri-miRNA transcription and processing. Mol Plant 14(3):426–439

Carbonell A (2019) Design and high-throughput generation of artificial small RNA constructs for plants. Methods Mol Biol 1932:247–260

Carbonell A, Darròs JA (2017) Artificial microRNAs and synthetic trans-acting small interfering RNAs interfere with viroid infection. Mol Plant Pathol 18(5):746–753

Carbonell A, López Del Rincón C, Daròs J (2019) Fast-forward identification of highly effective artificial small RNAs against different tomato spotted wilt virus isolates. Mol Plant-Microbe Interact 32(2):142–156

Cedillo-Jiménez CA, Hernández-Salazar M, Escobar-Feregrino T, Caballero-Pérez J, Arteaga-Vázquez M, Cruz-Ramírez A, Torres-Pacheco I, Guevara-González R, Cruz-Hernández A (2016) MicroRNAs sequencing for understanding the genetic regulation of plant genomes. In: Plant genomics. InTech eBooks

Chaabane SB, Liu R, Chinnusamy V, Kwon Y, Park JH, Kim SY, Zhu JK, Yang SW, Lee BH (2012) STA1, an *Arabidopsis* pre-mRNA processing factor 6 homolog, is a new player involved in miRNA biogenesis. Nucleic Acids Res 41(3):1984–1997

Chow H, Ng DK (2017) Regulation of miR163 and its targets in defense against *Pseudomonas syringae* in *Arabidopsis thaliana*. Sci Rep 7:46433

Cui C, Wang J, Zhao J, Fang Y, He X, Guo H, Duan C (2019) A *Brassica* miRNA regulates plant growth and immunity through distinct modes of action. Mol Plant 13(2):231–245

Cunningham FJ, Goh NS, Demirer GS, Matos JL, Landry MP (2018) Nanoparticle-mediated delivery towards advancing plant genetic engineering. Trends Biotechnol 36:882–897

Dai X, Zhao PX (2011) psRNATarget: a plant small RNA target analysis server. Nucleic Acids Res 39(Web Server issue):155–159

Dave VP, Ngo TA, Pernestig A, Tilevik D, Kant K, Nguyen T, Wolff A, Bang DD (2019) MicroRNA amplification and detection technologies: opportunities and challenges for point of care diagnostics. Lab Investig 99(4):452–469

Dhakate P, Shivaraj SM, Singh A (2014) Design of artificial miRNA for redundant silencing of *Brassica SHP1* and *SHP2*:transient assay-based validation of transcript cleavage from polyploid Brassicas. Acta Physiol Plant 36:2125–2135

Ding Y, Zhou M, Wang K, Qu A, Hu S, Jiang Q, Yi K, Wang F, Cai C, Zhu C, Chen Z (2023) Rice DST transcription factor negatively regulates heat tolerance through ROS-mediated stomatal movement and heat-responsive gene expression. Front Plant Sci 14:1068296

Dong Z, Han M, Fedoroff N (2008) The RNA-binding proteins HYL1 and SE promote accurate in vitro processing of pri-miRNA by DCL1. Proc Natl Acad Sci USA 105(29):9970–9975

Dong W, Ren W, Wang X, Mao Y, He Y (2021) MicroRNA319a regulates plant resistance to *Sclerotinia* stem rot. J Exp Bot 72(10):3540–3553

Duan CG, Wang CH, Fang RX, Guo HS (2008) Artificial MicroRNAs highly accessible to targets confer efficient virus resistance in plants. J Virol 82(22):11084–11095

Fahlgren N, Hill ST, Carrington JC, Carbonell A (2016) P-SAMS: a web site for plant artificial microRNA and synthetic trans-acting small interfering RNA design. Bioinformatics 32(1):157–158

Faisal AM, Biswas S, Zerin T, Rahman T, Seraj ZI (2017) Downregulation of the DST transcription factor using artificial microRNA to increase yield, salt and drought tolerance in rice. Am J Plant Sci 08(09):2219–2237

Faisal M, Abdel-Salam EM, Alatar AA (2021) Artificial microRNA-based RNA interference and specific gene silencing for developing insect resistance in *Solanum lycopersicum*. Agronomy 11(1):136

Fan L, Gao B, Xu Y, Flynn N, Le B, You C, Li S, Achkar N, Manavella PA, Yang Z, Chen X (2022) *Arabidopsis* AAR2, a conserved splicing factor in eukaryotes, acts in microRNA biogenesis. Proc Natl Acad Sci USA 119(41):e2208415119

Fang X, Cui Y, Li Y, Qi Y (2015) Transcription and processing of primary microRNAs are coupled by Elongator complex in *Arabidopsis*. Nat Plants 1:15075

Gu T, Xie M, Barbazuk WB, Lee J (2021) Biological features between miRNAs and their targets are unveiled from deep learning models. Sci Rep 11(1):23825

Hajheidari M, Farrona S, Huettel B, Koncz Z, Koncz C (2012) CDKF1 and CDKD protein kinases regulate phosphorylation of serine residues in the C-terminal domain of *Arabidopsis* RNA polymerase II. Plant Cell 24(4):1626–1642

Hayat S, Hayat Q, Alyemeni MN, Wani AS, Pichtel J, Ahmad A (2012) Role of proline under changing environments. Plant Signal Behav 7(11):1456–1466

Huang X, Chao D, Gao J, Zhu M, Shi M, Lin H (2009) A previously unknown zinc finger protein, DST, regulates drought and salt tolerance in rice via stomatal aperture control. Genes Dev 23(15):1805–1817

Iki T, Yoshikawa M, Nishikiori M, Jaudal MC, Matsumoto-Yokoyama E, Mitsuhara I, Meshi T, Ishikawa M (2010) In vitro assembly of plant RNA-Induced silencing complexes facilitated by molecular chaperone HSP90. Mol Cell 39(2):282–291

Imaduwage I, Hewadikaram M (2024) Predicted roles of long non-coding RNAs in abiotic stress tolerance responses of plants. Mol Hortic 4(1):20

Jeena GS, Singh N, Shikha, Shukla RK (2022) An insight into microRNA biogenesis and its regulatory role in plant secondary metabolism. Plant Cell Rep 41(8):1651–1671

Jefferson RA, Kavanagh TA, Bevan MW (1987) GUS fusions: beta-glucuronidase as a sensitive and versatile gene fusion marker in higher plants. EMBO J 6(13):3901–3907

Jelly NS, Schellenbaum P, Walter B, Maillot P (2012) Transient expression of artificial microRNAs targeting Grapevine fanleaf virus and evidence for RNA silencing in grapevine somatic embryos. Transgenic Res 21(6):1319–1327

Jeoung JM, Krishnaveni S, Muthukrishnan S, Trick HN, Liang GH (2002) Optimization of sorghum transformation parameters using genes for green fluorescent protein and beta-glucuronidase as visual markers. Hereditas 137(1):20–28

Jiang F, Song Y, Han Q, Zhu C, Wen F (2011) The choice of target site is crucial in artificial miRNA-mediated virus resistance in transgenic *Nicotiana tabacum*. Physiol Mol Plant Pathol 76(1):2–8

Jiang L, Mu R, Wang Z, Liu S, Lu D (2023) Silencing P25, HC-Pro and BRP1 of potato virus (Viroid) using artificial microRNA confers resistance to PVX, PVY and PSTVD in transgenic potato. Potato Res 66(1):231–244

Joga MR, Zotti MJ, Smagghe G, Christiaens O (2016) RNAi efficiency, systemic properties and novel delivery methods for pest insect control: what we know so far. Front Physiol 7:553

Khalid A, Zhang X, Ji H, Yasir M, Farooq T, Dai X, Li F (2023) Large artificial microRNA cluster genes confer effective resistance against multiple tomato yellow leaf curl viruses in transgenic tomato. Plants 12(11):2179

Khvorova A, Reynolds A, Jayasena SD (2003) Functional siRNAs and miRNAs exhibit strand bias. Cell 115(2):209–216

Kim YJ, Chen X (2011) The plant mediator and its role in noncoding RNA production. Front Biol 6(2):125–132

Kim S, Yang JY, Xu J, Jang IC, Prigge MJ, Chua NH (2008) Two cap-binding proteins CBP20 and CBP80 are involved in processing primary MicroRNAs. Plant Cell Physiol 49(11):1634–1644

Kim VN, Han J, Siomi MC (2009) Biogenesis of small RNAs in animals. Nat Rev Mol Cell Biol 10(2):126–139

Kis A, Tholt G, Ivanics M, Várallyay É, Jenes B, Havelda Z (2016) Polycistronic artificial miRNA-mediated resistance to Wheat dwarf virus in barley is highly efficient at low temperature. Mol Plant Pathol 17(3):427–437

Koiwa H, Barb AW, Xiong L, Li F, McCully MG, Lee B, Sokolchik I, Zhu J, Gong Z, Reddy M, Sharkhuu A, Manabe Y, Yokoi S, Zhu J, Bressan RA, Hasegawa PM (2002) C-terminal domain phosphatase-like family members (AtCPLs) differentially regulate *Arabidopsis thaliana* abiotic stress signaling, growth and development. Proc Natl Acad Sci USA 99(16):10893–10898

Kumar K, Mandal SN, Neelam K, De Los Reyes BG (2022) MicroRNA-mediated host defense mechanisms against pathogens and herbivores in rice: balancing gains from genetic resistance with trade-offs to productivity potential. BMC Plant Biol 22(1):351

Kumar M, Panwar V, Chaudhary V, Kumar R (2024) Artificial miRNAs: a potential tool for genetic improvement of horticultural crops. Sci Hortic 331:113160

Lafforgue G, Martínez F, Sardanyés J, De La Iglesia F, Niu QW, Lin SS, Elena SF (2011) Tempo and mode of plant RNA virus escape from RNA interference-mediated resistance. J Virol 85(19):9686–9695

Lafforgue G, Martínez F, Niu Q, Chua N, Daròs J, Elena SF (2013) Improving the effectiveness of artificial microRNA (amiR)-mediated resistance against turnip mosaic virus by combining two amiRs or by targeting highly conserved viral genomic regions. J Virol 87(14):8254–8256

Latif MF, Tan J, Zhang W, Yang W, Zhuang T, Lu W, Qiu Y, Du X, Zhuang X, Zhou T, Kundu JK, Yin J, Xu K (2024) Transgenic expression of artificial microRNA targeting soybean mosaic virus P1 gene confers virus resistance in plant. Transgenic Res 33(3):149–157

Lau NC, Lim LP, Weinstein EG, Bartel DP (2001) An abundant class of tiny RNAs with probable regulatory roles in *Caenorhabditis elegans*. Science 294(5543):858–862

Lee RC, Ambros V (2001) An extensive class of small RNAs in *Caenorhabditis elegans*. Science 94(5543):862–864

Li X, Peng J (2021) The epitranscriptome of small non-coding RNAs. Non-coding RNA Res 6(4):167–173

Li Y, Zhang Q, Zhang J, Wu L, Qi Y, Zhou JM (2010) Identification of microRNAs involved in pathogen-associated molecular pattern-triggered plant innate immunity. Plant Physiol 152(4):2222–2231

Li J, Zhang D, Sheen J (2014) Epitope-tagged protein-based artificial miRNA screens for optimized gene silencing in plants. Nat Protoc 9(4):939–949

Li S, Xu R, Li A, Liu K, Gu L, Li M, Zhang H, Zhang Y, Zhuang S, Wang Q, Gao G, Li N, Zhang C, Li Y, Yu B (2018) SMA1, a homolog of the splicing factor Prp28, has a multifaceted role in miRNA biogenesis in Arabidopsis. Nucleic Acids Res 46(17):9148–9159

Li H, Dong X, Mao W, Guan Y, Zhang Z (2019) An effective artificial microRNA vector based on Fv-miR166 precursor from strawberry. Sci Hortic 256:108643

Li S, Zhang N, Zhu X, Ma R, Yang J, Tang X, Si H (2020) Enhanced drought tolerance with artificial microRNA-mediated StProDH1 gene silencing in potato. Crop Sci 60(3):1462–1471

Li M, Yu H, Liu K, Yang W, Zhou B, Gan L, Li S, Zhang C, Yu B (2021) Serrate-associated Protein 1, a splicing-related protein, promotes miRNA biogenesis in Arabidopsis. New Phytol 232(5):1959–1973

Liang C, Hao J, Li J, Baker B, Luo L (2019) Artificial microRNA-mediated resistance to cucumber green mottle mosaic virus in *Nicotiana benthamiana*. Planta 250(5):1591–1601

Liang C, Cai Q, Wang F, Li S, You C, Xu C, Gao L, Cao D, Lan T, Zhang B, Mo B, Chen X (2022) *Arabidopsis* RBV is a conserved WD40 repeat protein that promotes microRNA biogenesis and ARGONAUTE1 loading. Nat Commun 13(1):1217

Liu Q, Chen Y (2010) A new mechanism in plant engineering: the potential roles of microRNAs in molecular breeding for crop improvement. Biotechnol Adv 28(3):301–307

Liu C, Axtell MJ, Fedoroff NV (2012) The helicase and RNase IIIa domains of *Arabidopsis* Dicer-Like1 modulate catalytic parameters during microRNA biogenesis. Plant Physiol 159(2):748–758

Liu SR, Zhou JJ, Hu CG, Wei CL, Zhang JZ (2017) MicroRNA-mediated gene silencing in plant defense and viral counter-defense. Front Microbiol 8:1801

Liu P, Zhang X, Zhang F, Xu M, Ye Z, Wang K, Liu S, Han X, Cheng Y, Zhong K, Zhang T, Li L, Ma Y, Chen M, Chen J, Yang J (2021) A virus-derived siRNA activates plant immunity by interfering with ROS scavenging. Mol Plant 14(7):1088–1103

Luan Y, Cui J, Wang W, Meng J (2016) MiR1918 enhances tomato sensitivity to *Phytophthora infestans* infection. Sci Rep 6:35858

Manavella PA, Hagmann J, Ott F, Laubinger S, Franz M, Macek B, Weigel D (2012) Fast-forward genetics identifies plant CPL phosphatases as regulators of MIRNA processing factor HYL1. Cell 151(4):859–870

Mann CWG, Sawyer A, Gardiner DM, Mitter N, Carroll BJ, Eamens AL (2023) RNA-based control of fungal pathogens in plants. Int J Mol Sci 24(15):12391

Matysik J, Alia BB, Mohanty P (2002) Molecular mechanisms of quenching of reactive oxygen species by proline under stress in plants. Curr Sci 82(5):525–532

McLoughlin AG, Wytinck N, Walker PL, Girard IJ, Rashid KY, De Kievit T, Fernando WGD, Whyard S, Belmonte MF (2018) Identification and application of exogenous dsRNA confers plant protection against *Sclerotinia sclerotiorum* and *Botrytis cinerea*. Sci Rep 8(1):7320

Mickiewicz A, Rybarczyk A, Sarzynska J, Figlerowicz M, Blazewicz J (2016) AmiRNA designer—new method of artificial miRNA design. Acta Biochim Pol 3(1):71–77

Navarro L, Dunoyer P, Jay F, Arnold B, Dharmasiri N, Estelle M, Voinnet O, Jones JDG (2006) A plant miRNA contributes to antibacterial resistance by repressing auxin signaling. Science 312(5772):436–443

Niu QW, Lin SS, Reyes JL et al (2006) Expression of artificial microRNAs in transgenic *Arabidopsis thaliana* confers virus resistance. Nat Biotechnol 24(11):1420–1428

Ossowski S, Schwab R, Weigel D (2008) Gene silencing in plants using artificial microRNAs and other small RNAs. Plant J 53(4):674–690

Park W, Zhai J, Lee J (2009) Highly efficient gene silencing using perfect complementary artificial miRNA targeting *AP1* or heteromeric artificial miRNA targeting *AP1* and *CAL* genes. Plant Cell Rep 28(3):469–480

Park J, Giudicatti AJ, Bader ZE, Han MK, Møller C, Arce AL, Xu Z, Yang SW, Manavella PA, Yun D (2023) The high expression of osmotically responsive gene 15–histone deacetylase 9 complex associates with HYPONASTIC LEAVES 1 to modulate microRNA expression in response to abscisic acid signaling. Plant Cell 35(8):2910–2928

Patil S, Joshi S, Jamla M, Zhou X, Taherzadeh MJ, Suprasanna P, Kumar V (2021) MicroRNA-mediated bioengineering for crop climate-resilience. Bioengineered 12(2):10430–10456

Peláez P, Sanchez F (2013) Small RNAs in plant defense responses during viral and bacterial interactions: similarities and differences. Front Plant Sci 4:343

Pieczynski M, Marczewski W, Hennig J, Dolata J, Bielewicz D, Piontek P, Wyrzykowska A, Krusiewicz D, Strzelczyk-Zyta D, Konopka-Postupolska D, Krzeslowska M, Jarmolowski A, Szweykowska-Kulinska Z (2013) Down-regulation of *CBP80* gene expression as a strategy to engineer a drought-tolerant potato. Plant Biotechnol J 11(4):459–469

Rabuma T, Gupta OP, Chhokar V (2022) Recent advances and potential applications of cross-kingdom movement of miRNAs in modulating plant disease response. RNA Biol 19(1):519–532

Raczynska KD, Simpson CG, Ciesiolka A, Szewc L, Lewandowska D, McNicol J, Szweykowska-Kulinska Z, Brown JWS, Jarmolowski A (2009) Involvement of the nuclear cap-binding protein complex in alternative splicing in *Arabidopsis thaliana*. Nucleic Acids Res 38(1):265–278

Raja I, Tennyson J (2023) Small RNA—regulator of biotic stress and pathogenesis in food crops. In: Plant small RNA in food crops. Elsevier eBooks, pp 233–269

Redhu SN, Yashveer S, Taunk J, Banga P, Singh V, Tokas J, Grewal S, Arora I (2023) Plant microRNAs: unexplored biogenesis, prediction tools and applications. Gene Rep 32:101799

Ren G, Xie M, Dou Y, Zhang S, Zhang C, Yu B (2012) Regulation of miRNA abundance by RNA binding protein TOUGH in *Arabidopsis*. Proc Natl Acad Sci USA 109(31):12817–12821

Saini RP, Raman V, Dhandapani G, Malhotra EV, Sreevathsa R, Kumar PA, Sharma TR, Pattanayak D (2018) Silencing of *HaAce1* gene by host-delivered artificial microRNA disrupts growth and development of *Helicoverpa armigera*. PLoS One 13(3):e0194150

Sanchita TR, Asif MH, Trivedi PK (2018) Dietary plant miRNAs as an augmented therapy: cross-kingdom gene regulation. RNA Biol 15(12):1433–1439

Sattar S, Addo-Quaye C, Thompson GA (2015) miRNA-mediated auxin signalling repression during Vat-mediated aphid resistance in *Cucumis melo*. Plant Cell Environ 39(6):1216–1227

Schwab R, Ossowski S, Riester M, Warthmann N, Weigel D (2006) Highly specific gene silencing by artificial microRNAs in *Arabidopsis*. Plant Cell 18(5):1121–1133

Schwarz DS, Hutvágner G, Du T, Xu Z, Aronin N, Zamore PD (2003) Asymmetry in the assembly of the RNAi enzyme complex. Cell 115(2):199–208

Sharma N, Prasad M (2020) Silencing AC1 of Tomato leaf curl virus using artificial microRNA confers resistance to leaf curl disease in transgenic tomato. Plant Cell Rep 39(11):1565–1579

Shivaprasad PV, Chen H, Patel K, Bond DM, Santos BA, Baulcombe DC (2012) A MicroRNA superfamily regulates nucleotide binding site–Leucine-rich repeats and other mRNAs. Plant Cell 24(3):859–874

Song Y, Han Q, Jiang F, Sun R, Fan Z, Zhu C, Wen F (2014) Effects of the sequence characteristics of miRNAs on multi-viral resistance mediated by single amiRNAs in transgenic tobacco. Plant Physiol Biochem 77:90–98

Speth C, Willing E, Rausch S, Schneeberger K, Laubinger S (2013) RACK1 scaffold proteins influence miRNA abundance in *Arabidopsis*. Plant J 76(3):433–445

Stroynowska-Czerwinska A, Fiszer A, Krzyzosiak WJ (2014) The panorama of miRNA-mediated mechanisms in mammalian cells. Cell Mol Life Sci 71:2253–2270

Su C, Li Z, Cheng J, Li L, Zhong S, Liu L, Zheng Y, Zheng B (2017) The protein phosphatase 4 and SMEK1 complex dephosphorylates HYL1 to promote mIRNA biogenesis by antagonizing the MAPK cascade in *Arabidopsis*. Dev Cell 41(5):527–539.e5

Sun L, Lin C, Du J, Song Y, Jiang M, Liu H, Zhou S, Wen F, Zhu C (2016) Dimeric artificial microRNAs mediate high resistance to RSV and RBSDV in transgenic rice plants. Plant Cell Tissue Org Cult 126:127–139

Teotia S, Wang X, Zhou N, Wang M, Liu H, Qin J, Han D, Li C, Li CE, Pan S, Tang H, Kang W, Zhang Z, Tang X, Peng T, Tang G (2023) A high-efficiency gene silencing in plants using two-hit asymmetrical artificial MicroRNAs. Plant Biotechnol J 21(9):1799–1811

Thakur A, Sanju S, Siddappa S, Srivastava N, Shukla PK, Pattanayak D, Sharma S, Singh B (2015) Artificial microRNA mediated gene silencing of *Phytophthora infestans* single effector Avr3a gene imparts moderate type of late blight resistance in potato. Plant Pathol J 14(1):1–12

Tiwari M, Sharma D, Trivedi PK (2014) Artificial microRNA mediated gene silencing in plants: progress and perspectives. Plant Mol Biol 86:1–18

Van Vu T, Do VN (2017) Customization of artificial MicroRNA design. In: Rani S (ed) MicroRNA profiling, Methods in molecular biology, vol 1509. Humana Press, New York

Von Born P, Bernardo-Faura M, Rubio-Somoza I (2018) An artificial miRNA system reveals that relative contribution of translational inhibition to miRNA-mediated regulation depends on environmental and developmental factors in *Arabidopsis thaliana*. PLoS One 13(2):e0192984

Wang L, Song X, Gu L, Li X, Cao S, Chu C, Cui X, Chen X, Cao X (2013) NOT2 proteins promote polymerase II-dependent transcription and interact with multiple MicroRNA biogenesis factors in *Arabidopsis*. Plant Cell 25(2):715–727

Wang X, Lee RS, Alver BH, Haswell JR, Wang S, Mieczkowski J, Drier Y, Gillespie SM, Archer TC, Wu JN, Tzvetkov EP, Troisi EC, Pomeroy SL, Biegel JA, Tolstorukov MY, Bernstein BE, Park PJ, Roberts CW (2017) SMARCB1-mediated SWI/SNF complex function is essential for enhancer regulation. Nat Genet 49(2):289–295

Wang J, Chen S, Jiang N, Li N, Wang X, Li Z, Li X, Liu H, Li L, Yang Y, Ni T, Yu C, Ma J, Zheng B, Ren G (2019) Spliceosome disassembly factors ILP1 and NTR1 promote miRNA biogenesis in *Arabidopsis thaliana*. Nucleic Acids Res 47(15):7886–7900

Warthmann N, Chen H, Ossowski S, Weigel D, Hervé P (2008) Highly specific gene silencing by artificial miRNAs in rice. PLoS One 3(3):e1829

Warthmann N, Ossowski S, Schwab R, Weigel D (2013) Artificial MicroRNAs for specific gene silencing in rice. In: Yang Y (ed) Rice protocols, Methods in molecular biology, vol 956. Humana Press, Totowa

Wu J, Yang R, Yang Z, Yao S, Zhao S, Wang Y, Li P, Song X, Jin L, Zhou T, Lan Y, Xie L, Zhou X, Chu C, Qi Y, Cao X, Li Y (2017) ROS accumulation and antiviral defence control by microRNA528 in rice. Nat Plants 3(1):16203

Xin J, Liu Y, Li H, Chen S, Jiang J, Song A, Fang W, Chen F (2021) CmMLO17 and its partner CmKIC potentially support *Alternaria alternata* growth in *Chrysanthemum morifolium*. Hortic Res 8(1):101

Yang X, Ren W, Zhao Q, Zhang P, Wu F, He Y (2014) Homodimerization of HYL1 ensures the correct selection of cleavage sites in primary miRNA. Nucleic Acids Res 42(19):12224–12236

Yang X, Zhang L, Yang Y, Schmid M, Wang Y (2021) miRNA mediated regulation and interaction between plants and pathogens. Int J Mol Sci 22(6):291

Yao S, Yang Z, Yang R, Huang Y, Guo G, Kong X, Lan Y, Zhou T, Wang H, Wang W, Cao X, Wu J, Li Y (2019) Transcriptional regulation of *MIR528* by *OsSPL9* orchestrates antiviral response in rice. Mol Plant 12(8):1114–1122

Yasir M, Motawaa M, Wang Q, Zhang X, Khalid A, Cai X, Li F (2022) Simple webserver-facilitated method to design and synthesize artificial miRNA gene and its application in engineering viral resistance. Plants 11:2125

Yogindran S, Rajam MV (2016) Artificial miRNA-mediated silencing of ecdysone receptor (EcR) affects larval development and oogenesis in *Helicoverpa armigera*. Insect Biochem Mol Biol 77:21–30

Yu B, Bi L, Zheng B, Ji L, Chevalier D, Agarwal M, Ramachandran V, Li W, Lagrange T, Walker JC, Chen X (2008) The FHA domain proteins DAWDLE in Arabidopsis and SNIP1 in humans act in small RNA biogenesis. Proc Natl Acad Sci USA 105(29):10073–10078

Yue D, Liu H, Huang Y (2009) Survey of computational algorithms for MicroRNA target prediction. Curr Genomics 10(7):478–492

Yusuf NHM, Latip MA, Kumar VS (2021) Artificial microRNA derived from the precursors of *Ananas comosus*, *Arabidopsis thaliana* and *Oryza sativa* effectively silences endogenous genes in MD2 pineapple. Plant Gene 26:100289

Zhang X, Li H, Zhang J, Zhang C, Gong P, Ziaf K, Xiao F, Ye Z (2011a) Expression of artificial microRNAs in tomato confers efficient and stable virus resistance in a cell-autonomous manner. Transgenic Res 20(3):569–581

Zhang X, Zhao H, Gao S, Wang WC, Katiyar-Agarwal S, Huang HD, Raikhel N, Jin H (2011b) *Arabidopsis* Argonaute 2 regulates innate immunity via miRNA393(∗)-mediated silencing of a Golgi-localized SNARE gene, MEMB12. Mol Cell 42(3):356–366

Zhang S, Liu Y, Yu B (2014) PRL1, an RNA-Binding protein, positively regulates the accumulation of miRNAs and SIRNAs in *Arabidopsis*. PLoS Genet 10(12):e1004841

Zhang N, Zhang D, Chen SL, Gong BQ, Guo Y, Xu L, Zhang XN, Li JF (2018) Engineering artificial microRNAs for multiplex gene silencing and simplified transgenic screen. Plant Physiol 178(3):989–1001

Zhang B, You C, Zhang Y, Zeng L, Hu J, Zhao M, Chen X (2020) Linking key steps of microRNA biogenesis by TREX-2 and the nuclear pore complex in *Arabidopsis*. Nat Plants 6(8):957–969

Zhang F, Yang J, Zhang N, Wu J, Si H (2022) Roles of microRNAs in abiotic stress response and characteristics regulation of plant. Front Plant Sci 13:919243

Zhou J, Deng K, Cheng Y, Zhong Z, Tian L, Tang X, Tang A, Zheng X, Zhang T, Qi Y, Zhang Y (2017) CRISPR-Cas9 based genome editing reveals new insights into microRNA function and regulation in rice. Front Plant Sci 8:1598

Zhou L, Yuan Q, Ai X, Chen J, Lu Y, Yan F (2022) Transgenic rice plants expressing artificial mirna targeting the rice stripe virus *MP* gene are highly resistant to the virus. Biology (Basel) 11(2):332

Zhou B, Yu H, Xue Y, Li M, Zhang C, Yu B (2024) The spliceosome-associated protein CWC15 promotes miRNA biogenesis in *Arabidopsis*. Nat Commun 15(1):2399

Zhu H, Zhou Y, Castillo-González C, Lu A, Ge C, Zhao YT, Duan L, Li Z, Axtell MJ, Wang XJ, Zhang X (2013) Bidirectional processing of pri-miRNAs with branched terminal loops by *Arabidopsis* Dicer-like1. Nat Struct Mol Biol 20(9):1106–1115

Chapter 14
Genctic and Molecular Basis of Cytoplasmic Male Sterility and Fertility Restoration in Crop Plants

Shalini Pareek, Alok Das, and Abhishek Bohra

Abstract Crop researchers have developed a number of hybrids and varieties exhibiting superiority in desired traits (e.g., yield, quality, and resilience) via exploitation of heterosis in breeding programs. Controlled pollination has always been a Herculean challenge in the development of crop hybrids as well as field-scale production of hybrid seeds. In this context, cytoplasmic male sterility (CMS) has served as a robust genetic mechanism to control pollination in heterosis breeding and hybrid seed production in crop plants. CMS, manifested as production of nonfunctional pollen, results from compromised harmony between the organelle and nuclear genomes. Research has established mitochondrial ORF (Open Reading Frame) as causal genes for inducing CMS in several plant species. The restorer of fertility (*Rf*) genes residing in the nucleus overcomes the action of CMS-causing mitochondrial genes, thus resuming pollen fertility in the F1 hybrids. Thus, the hybrid production using CMS technology entails three components: CMA line (A) line, maintainer (B) line, and restorer (R) line. Genetic mapping and cloning of *Rf* genes in combination with evolving landscape of omics technologies in plants have provided new insights into CMS-Rf system. More recently, sequence-specific nucleases combined with mitochondrial targeted signals (mito-TALENs) have been deployed to edit mitochondrial genomes in plants, thus offering concrete evidence supporting mitochondrial role in CMS occurrence. This is notable given the fact that evidence linking mitochondrial genes with CMS was previously lacking in crop plants. Knowledge of the genetic and molecular basis of CMS and fertility restoration is imperative to understand the nucleo-cytoplasmic crosstalk and for breeding modern crop hybrids with better yield, quality and resilience.

Keywords Cytoplasm · Fertility restoration · Heterosis · Male sterility · Mitochondria · Pollen

---

S. Pareek · A. Das · A. Bohra (✉)
ICAR-Indian Institute of Pulses Research (IIPR), Kanpur, India

A. Singh et al. (eds.), *Sustainable Crop Production*, Sustainability Sciences in Asia and Africa, https://doi.org/10.1007/978-981-95-7844-3_14

Editor's Notes
Male sterility forms the bedrock of heterosis breeding in crop plants. Among the multitude of mechanisms, cytoplasmic male sterility based on nucleo-cytoplasmic incompatibility is most used. Pareek et al. in this chapter provide a comprehensive overview of complex genetics and the phenotypic manifestations of diverse male sterile and fertility restorer systems across plants. Understanding the molecular genetics of cytoplasmic male sterility and fertility restorer system is essential for large-scale commercial applications for heterosis breeding. Even though the genetics of heterosis has been explained based on dominance and overdominance hypothesis, molecular and mechanistic details supporting these hypotheses are incomplete meriting careful dissection. Researchers and scholars will benefit from the enriching discussions on numerous CMS and restorer systems. The authors encourage the researchers to generate deeper functional understanding of these systems by integrating modern technologies such as genome editing.

Actionable Points
1. Cytoplasmic male sterility (CMS) has facilitated hybrid breeding and seed production in a number of crops.
2. The action of mitochondrial genes leading to the formation of nonfunctional pollen is countered by nuclear genes or Rf genes.
3. CMS-Rf in plants presents a unique system to deepen our understanding of cytoplasmic-nuclear crosstalk.
4. Molecular marker technology has contributed to hasten the progress of hybrid breeding in crop plants.
5. Exciting opportunities have emerged for functionally validating CMS-associated genes through gene editing.

14.1 Introduction

Failure to produce functional anthers, pollen, or male gametes in flowers that are typically hermaphroditic is termed as male sterility (MS). It is often seen as an incidental developmental anomaly. However, in nature, a variety of mechanisms give rise to MS in plants (Bohra et al. 2021a, b, c). Many hermaphroditic plants have evolved strategies to promote cross-pollination, such as dichogamy, chasmogamy, heterostyly, self-incompatibility (SI), and unisex sterility. Unisex sterility can be controlled by genetic factors (nuclear, cytoplasmic, or both) or environmental influences.

Researchers have provided detailed classifications and molecular insights into naturally occurring MS mechanisms in plants (Budar and Pelletier 2001; Kaul 1988). From a technical perspective, various methods can induce MS in plants. In maize, for instance, mechanical removal of male flower organs, or de-tasseling, is a straightforward method (Smith et al. 2004). Alternatively, chemical hybridizing agents (CHAs) can be applied to inhibit pollen development (Tu and Banga 1998) MS, first noted by German botanist Joseph Gottlieb Kölreuter in 1763, has been reported in over 610 plant species (Mayr 1986; Kaul 1988). It includes cytoplasmic male sterility (CMS), caused by mitochondrial genes and linked nuclear genes (Mishra and Bohra 2018), and genic male sterility (GMS), which is caused by nuclear genes alone (Vedel et al. 1994). CMS technology has been widely used for the development of superior hybrids in a number of crops (Bohra et al. 2016, 2024).

MS mutants can disrupt the development of anther tissues, either at the sporophytic or at the gametophytic stage. Sporophytic mutants primarily affect the tapetum and meiocytes, leading to pollen abortion or sterility, while gametophytic mutants mainly impair microspore or pollen grain development (Guo and Liu 2012). Mitochondrial mutations that cause CMS, or from nuclear-encoded genes or transgenes that lead to pollen ablation, are known as nuclear-encoded male sterility (NMS) (Chase et al. 2010). The Texas (T) cytoplasm (cms-7) in maize, first described in the Golden June line (Rogers and Edwardson 1952), carries the CMS trait. This discovery was significant for geneticists and breeders, as the absence of functional pollen eliminated the need for the costly detasseling process in hybrid corn production. The failure of anther exertion and pollen abortion has been reported in CMS-T, while female fertility remains unaffected, allowing seed production if viable pollen is provided. The maize CMS-T gained attention during the 1969–1970 due to Southern corn leaf blight epidemic (Williams and Levings 1992). Prior to the epidemic, CMS-T had become the primary method of pollen control in US hybrid corn, accounting for 85% of hybrid corn acreage by 1970 (Wych 1988). Subsequent research identified the mitochondrial gene *T-urf13* responsible for causing CMS, with evidence suggesting that this gene also increases susceptibility to the disease (Dewey et al. 1985, 1987). The CMS-T was phased out in 1971, serving as a warning about the risks of relying on a single source of CMS for large-scale seed production. At the same time, this incident underscored the importance of diversifying CMS sources within the hybrid seed industry. MS systems, owing to their ability to eliminate the need for manual emasculation, have contributed significantly to accelerating crop improvement through harnessing the heterosis or hybrid vigor (Bohra et al. 2016, 2017a; Dong et al. 2012). Besides pollination control, MS also plays a crucial role in ensuring seed purity in large-scale seed production programs, which is challenging to achieve through conventional means such as manual emasculation (Bohra et al. 2024; Liu et al. 2013). Nuclear-encoded RF genes play a key role in restoring fertility in the presence of cytoplasmic male sterility (CMS). When a CMS line is crossed with a restorer line that carries these *Rf* genes, the *Rf* genes counterbalance the sterility effects of CMS, resulting in fertile pollen and enabling seed production. This restoration of fertility is crucial for hybrid breeding programs, as it facilitates the exploitation of hybrid vigor (heterosis), which can boost yield and enhance crop quality (Melonek et al. 2021). The objective of this chapter is to provide a detailed insight into the genetic and molecular mechanism behind cytoplasmic male sterility and fertility restoration in crop plants.

14.2 Types of Male Sterility

The application of GMS (originates from mutations in the nuclear genome that might be sensitive to environmental conditions) and CMS (occurs due to the interaction of mitochondrial and nuclear genes) in hybrid seed production involves two main systems: the two line and three-line systems, respectively.

1. *Two-line system:* In this system, an environment-sensitive genic male sterility (EGMS) mutant (MT) line is used (Liu et al. 2023). EGMS line fails to produce functional pollen under restrictive conditions (e.g., long-day for photoperiod-sensitive GMS, short photoperiod length for reverse PGMS, humidity sensitive HGMS, or high-temperature for temperature-sensitive TGMS). The same line behaves as male fertile, allowing self-pollination under permissive conditions (short-day for PGMS, long-day for reverse PGMS or low temperature for TGMS). The EGMS line is cross-pollinated with pollen from a wild-type (WT) line for commercial production of hybrids. Unlike CMS, most GMS mutants are unsuitable for hybrid seed production due to the difficulty in maintaining pollen sterility as a substantial proportion (50%) of the population needs to be discarded, a procedure which remains quite demanding in terms of labor and time and causes a significant increase in the cost of seed production (Bohra et al. 2024). However, the discovery of EGMS mutants has enabled the use of certain GMS traits in hybrid crop breeding (Virmani and Ilyas-Ahmed 2001). EGMS lines exhibit pollen fertility changes in response to environmental factors like day length and temperature. The first photoperiod-sensitive GMS (PGMS) mutant, Nongken 58S (NK58S), was identified in 1973 in japonica rice (*Oryza sativa* ssp. *japonica*). NK58S is male sterile under long days condition but produces functional pollen under short days condition (Shi 1985). The temperature-sensitive GMS (TGMS) mutant *Annong S-1* was discovered in indica rice (*O. sativa* ssp. *indica*) in 1988. It is male sterile at high and male fertile at low temperature (Deng et al. 1999). HGMS is reported in rice and *Arabidopsis* (Aarts et al. 1995; Xue et al. 2018). Unlike CMS, the fertility transition in PGMS and TGMS systems enables hybrid seed production with the two lines only, thus bypassing the cumbersome steps of identification and application of a third line, i.e., R-line. Also, the MS trait of the lines can be maintained simply by growing them under permissive conditions (short-day or low temperature) without the need for crossing with the B line (Chen and Liu 2014). Spontaneous *thermo-sensitive female sterility 1* (*tfs1*) point mutation in ARGONAUTE7 (AGO7) is responsible for complete sterility and partial fertility under regular/high temperature and low temperature, respectively. AGO7 and miR390 form RNA-induced silencing complex to activate biogenesis of siRNA from *TRANS-ACTING3* (*TAS3*). These siRNAs repress auxin response factor genes. The mutant TFS1 protein loses the potential to load the miR390/miR390* duplex and eject miR390* during RISC formation (Li et al. 2022). The mechanism underlying EGMS involves multiple regulatory components, including receptor-like kinases (RLKs); UbL40 (ubiquitin-60S ribosomal protein L40) and

noncoding RNAs in TGMS; transcription factors such as CSA and CSA2 in PGMS, and MYB33 and MYB65 in TGMS; plant homeodomain (PHD) finger proteins in both P/TGMS; pollen coat–associated factors in HGMS; and the regulatory roles of phytohormones, particularly auxin and jasmonic acid, in EGMS (Liu et al. 2023).

2. *Three-line system (ABR system):* This requires three lines (Chen and Liu 2014):
   (a) CMS line (A-line): Contains sterile cytoplasm (S) and a recessive, nonfunctional nuclear restorer gene (*rf*)
   (b) Maintainer line (B-line): Has normal cytoplasm (N) and the same nuclear genome as the CMS line
   (c) Restorer line (R-line): Has either normal (N) or sterile (S) cytoplasm and a functional nuclear dominant restorer gene (*Rf*)

The CMS line is maintained by crossing with the cognate maintainer or B-line (Bohra et al. 2017). The genetically pure seeds of B- and R-lines are obtained through self-pollination and can be maintained through selfing and sib-mating. A cross between A- and R-lines leads to the production of hybrid seeds. The mitochondrial genome controlled, trait typically follows a maternal (non-Mendelian) inheritance in most angiosperm species (Hagemann 2004). Various CMS systems have been identified in various crops, including maize (S-, C-, or T-systems; Laughnan and Gabay-Laughnan 1983), *Brassica napus* (Ogura System; Ogura 1968), rye (Pampa [P]-cytoplasm; Geiger and Schnell 1970), rice (Wang et al. 2005), wheat (G-type, Sv-type, msH1-system; Wilson and Ross 1962; Mukai and Tsunewaki 1979; Martín et al. 2008), pigeonpea (A2 and A4 cytoplasm; Bohra et al. 2016, 2022), and chilli (S-type cytoplasm, Shifriss and Frankel 1971). The pollen sterility in these MS systems can be reversed by the action of fertility restorer (*Rf*) genes specific to each CMS system (Kempe and Gils 2011). For instance, in *Oryza rufipogon*, FA-CMS, OsRF19 protein cleaves the sterility causing *FA182* transcript (Jiang et al. 2022) while, in WA-CMS, PPR coded by *Rf20* contend with *WA352* for binding to COX11, which results in ROS scavenging to restore fertility (Song et al. 2024). In maize, restoration complex formed by restorer gene *ZmRF5* with *RS31A* and *MORF8* produces enough ATP6C for normal assembly of complex V, leading to fertility restoration (Lin et al. 2024). In T-CMS of wheat, Melonek et al. (2021) reported that *Rf1* and *Rf3* cleave *orf279* transcript. In oilseed rape, Ogura CMS is restored by the Rfo gene, which encodes the PPR-B protein that inhibits translation of the mitochondrial orf138 (Wang et al. 2021b). In tobacco, *ORFs orf82, orf103, orf115a, orf91, orf115b,* and *orf100* are reported to be molecular markers for the detection of sua-CMS (Zheng et al. 2018). Allotopic expression of *atp1* gene was found responsible for CMS and abortive seed trait in tobacco (Dewey et al. 2023). Similarly, *ms-1, ms-2, ms-3, ms-4, ms-5* are the *CMS* genes identified in muskmelon (McCreight and Elmstrom 1984). For hybrid breeding, these restoration systems can act as the diploid sporophyte or the haploid gametophyte (Hanson and Conde 1985; Schnable and Wise 1998). For example, MS of maize T-CMS is countered by dominant alleles of the two loci *rf1* and *rf2* in a sporophytic mechanism (Laughnan and Gabay-Laughnan 1983). A gametophytic restoration pattern was discovered in hybrid rice plant involving two nonallelic nuclear restorer genes, *Rf5* and *Rf6* (Huang et al. 2012). The F1 hybrid plants containing either *Rf5* or *Rf6* displayed 50% normal pollen grains, while those with both *Rf5* and *Rf6* displayed 75% normal pollens. Both stable CMS mutants and restoration system in each crop species pave the way for efficient hybrid seed production provided that the CMS mutation is not associated with yield penalties or other undesirable traits (Bohra et al. 2024). However, CMS systems are often sensitive to environmental factors like temperature and photoperiod (Kaul 1988), and instability or to disease and insects/pests susceptibility has also been observed in crops such as *B. napus* (Havey 2004), maize (Laughnan and Gabay-Laughnan 1983). In pearl millet, the positive association of sterile cytoplasm with ergot susceptibility was indicated by the results demonstrating that hybrids made from A X R were significantly more susceptible than their corresponding B X R hybrids (Thakur et al. 1989). Similarly, in sorghum, Dhillon et al. (2005) explored the relative susceptibility of different male sterile cytoplasm to shoot fly *Atherigona soccata* and found that oviposition and deadheart formation was significantly lower in the maintainer lines as compared to their respective sterile lines. Table 14.1 shows major sources of male sterility and fertility restoration used for hybrid development in crop plants.

14.3 Diversity of CMS Phenotypes in Plants and Underlying Mechanisms

Defects inherited through the maternal parent in the development of male flower organs inducing CMS provide evidence in support of the mitochondrial role in regulating flower formation (Bohra et al. 2016). MADS-box transcription factors play crucial roles in flower development (Masiero et al. 2011). Concerning the diversity of the MS phenotypes, carpelloid CMS and petaloid CMS refer to the conditions where the stamens convert into carpels and petals, respectively, without affecting the normal function of female organs.

(a) *Sterile pollen/anther deformities*

Anterograde regulation in plants involves mitochondria-targeted products coded by nuclear mitochondrial-sorting gene (MSG), including RF proteins, tissue-specific regulatory factors (TSRFs), and subunits of the mitochondrial electron transfer chain (mtETC) complexes. TSRFs may control the accumulation of CMS proteins in male flower organs (anthers) through translational or posttranslational mechanisms, ensuring male specificity. These CMS proteins can interact with mtETC subunits, influencing their function, redox status, or ATP production, which in turn generates retrograde signals such as reactive oxygen species (ROS) and cytochrome c release, interfering with the regular process of programmed cell death (PCD) in the tapetum or microspores (Chen and Liu 2014). Research on CMS systems in various crops has identified four key mechanisms through which pollen sterility is manifested in plants: the cyto-

Table 14.1 Major sources of male sterility and fertility restoration used for hybrid development in crop plants

Crop	Name of hybrid	A line	B line	R line	References
Rice	Rajlaxmi	CRMS32A	–	IR42266-29-4R	Rao et al. (2013)
	Pusa RH10	–	Pusa6B	PRR78	Basavaraj et al. (2010)
	DRRH 2	IR68897A	IR68897B	DR 714	Kumari et al. (2023)
	PRH 10	PUSA6A	PUSA6B	PRR 78	Kumari et al. (2023)
	DRRH 3	APMS6A	APMS6B	RPHR1005	Kumari et al. (2023)
	CR Dhan 703	CRMS 32A	–	CRL123R	Verma et al. (2022)
	CR Dhan 704	PMS 17A	–	CR546	Verma et al. (2024)
	DRRH 4	APMS-6A	–	AR 9–18	Senguttuvel et al. (2023)
	KRH-2	IR-58025A	IR-58025B	MR-3R	Rajanna et al. (2011)
Sorghum	Hybrid HJH 1513	ICSA465	–	HC308	Kumari et al. (2023)
Pearl millet	NBH 4903 (MH 2035)	NB 105A	–	NB 98R	Satyavati et al. (2018)
	Mahabeej-1005 (MH 1852)	MBJ-2A	–	MBJ-1R	Satyavati et al. (2018)
	PBH 306 (MH 1962)	PB 501A	–	PB 89R	Satyavati et al. (2018)
	Bio 448 (MH 1671)	13A	–	R 210	Satyavati et al. (2018)
	Bio 70 (MH 1632)	11A	–	R 207	Satyavati et al. (2018)
	NBH 5767 (MH 1785)	NB 345A	–	NB 97R	Satyavati et al. (2018)
	NBH 5061 (MH 1812)	NB 66A	–	NB 98R	Satyavati et al. (2018)
	KBH 108 (MH 1737)	KMPS-74A	–	KPR-9288R	Satyavati et al. (2018)
Sunflower	KBSH-88	CMS-1103A	–	GKVK-763	https://icar-iior.org.in/node
	Tilhan Tech-SUNH-1	MJA 5	–	MJR 1	https://icar-iior.org.in/sites/default/files/i
Pigeonpea	GTH-1	GT 288A	–	GTR-11	Saxena and Nadarajan (2010)
	IPH 15–03	PA 163A	–	AK 250189R	Bohra et al. (2017b)
	IPH 09–5	PA 163A	–	AK 261322R	Bohra et al. (2015)

toxicity model, energy deficiency model, aberrant programmed cell death (PCD) model, and retrograde regulation model. These models are outlined below:

Cytotoxicity

In this model, CMS 10–35 kDa transmembrane proteins disrupt mitochondrial function in the anther cells, leading to male sterility. The first identified CMS protein, URF13 in maize CMS-T, is toxic to both *Escherichia coli* and eukaryotic cells (Dewey et al. 1987; Korth et al. 1991; Korth and Levings 1993; Levings 1993). In rice with Boro II cytoplasm, an abnormal mitochondrial *orf79* is cotranscribed with a duplicated *atp6* gene and encodes a cytotoxic peptide responsible for CMS. Additionally, two PPR proteins encoded by the *Rf-1* locus in the Boro II system inhibit the production of this cytotoxic peptide through distinct mechanisms (Wang et al. 2006). Similarly, Jing et al. (2012) identified a new *CMS* gene, *orf288*, which is co-transcribed with atp6 and is responsible for the hau CMS in *Brassica juncea*. They found that *orf288* exhibits cytotoxicity to *E. coli* and induces male sterility in *Arabidopsis thaliana* when expressed from the nuclear genome, both with and without mitochondrial targeting presequences.

Deficiency in ATP supply

Interference in the normal functioning of mitochondria prevents cells from meeting the high energy demands during male reproductive development. Chen and Liu (2014) described that mitochondria, being crucial for ATP production via the mtETC, are particularly important in anther cells, which require more energy than other tissues. To meet these demands, cells may increase mitochondrial number or activity. Lee and Warmke (1979) noted rapid mitochondrial division during maize CMS-T anther development, suggesting the *CMS* gene disrupts mitochondrial function, limiting the supply of ATP for male development. The sequences and structures of several CMS proteins suggest a molecular basis for the energy deficiency model. Many CMS proteins, such as URF13 (maize CMS-T), ORF138 (*Brassica* CMS-Ogu), ORF79 (rice CMS-BT), ORFH79 (rice CMS-HL), and preSatp6 (sugar beet CMS-Owen), are mitochondrial transmembrane proteins (Dewey et al. 1987; Uyttewaal et al. 2008; Kazama and Toriyama 2003; Yamamoto et al. 2005). These proteins may integrate into the inner mitochondrial membrane, disrupting the proton gradient and impairing ATP synthesis (Rhoads et al. 1995).

Impaired ROS scavenging and tapetal PCD

In plants, during PCD, cytochrome c release from mitochondria and ROS overproduction take place (Yao et al. 2002). According to Luo et al. (2013), coordinated interactions between the sporophytic (anther wall) and gametophytic (microspore) cells lead to the PCD of the tapetum, the innermost layer of the anther wall. This process is required for male gametophyte development. Premature or delayed PCD results in male sterility (Kawanabe et al. 2006). CMS-PET1 cytoplasm in sunflower initiates early tapetal PCD, resulting in cytochrome c release from mitochondria. Similarly, ORF522 affects the activities of F1Fo-ATPase and cytochrome *c* oxidase (Balk and Leaver 2001; Sabar et al. 2003). In rice CMS-WA lines, the CMS protein WA352 accumulates in tapetal cells, interacts with OsCOX11, and disrupts peroxide metabolism, leading to early cytochrome c release and ROS production (Luo et al. 2013).

Retrograde regulation

In rice, the *Rf17* gene in the cytoplasm of CMS-CW encodes a mitochondrial acyl-carrier protein synthase like-domain protein called *RETROGRADE-REGULATED MALE STERILITY* (RMS). RNAi-mediated gene silencing of RMS was found to restore fertility in the CMS plant, while its overexpression in the fertility restorer line induced pollen abortion (Fujii and Toriyama 2009). Similarly, Chinese-wild (CW)-type CMS, CWA, in rice mitochondria upregulated the expression of RMS (Suketomo et al. 2020). The elevated expression of RMS results in the suppression of pollen germination, ultimately causing gametophytic sterility (Fujii and Toriyama 2009).

(b) *Carpelloid*

The carpelloid CMS phenotypes closely resemble those of homeotic mutants. The expression of two *MADS-box* genes responsible for regulating whorls 2 and 3 in carrot flowers is suppressed in carpelloid CMS lines. This suggests that retrograde signaling from mitochondria plays a key role in regulating the expression of these nuclear *MADS-box* genes, ultimately driving the organ conversion seen in carpelloid CMS (Linke et al. 2003). *LFY, UFO,* and *ASK1 in B. juncea and AP3-like, PI-like, and AG-like genes* in *B. napus* were the genes responsible for carpelloidy (Yang et al. 2008; Teixeira et al. 2005). Similar genes in carrot are *GLOBOSA* and *DEFICIENS* (Linke et al. 2003). In a related context, protoplast fusion has been explored as a method for generating novel CMS systems in *B. napus* (inap CMS). This was achieved by creating somatic hybrids with *Isatis indigotica* (Chinese woad) through recurrent backcrossing. In this case, CMS resulted from the conversion of tetradynamous stamens into carpelloid structures, which led to the development of female reproductive organs and the absence of pollen, causing male sterility. Notably, following pollination with *B. napus*, *I. indigotica* retained its normal morphology (Kang et al. 2017).

(c) *Petaloid*

The development of petaloid male-sterile plants from carrot cell cultures was reported by Wright et al. (1996). Petaloid CMS carrots show an overexpression of the mitochondrial *atp9* genes, which is linked to distinct characteristics in the organization and expression of these gene sequences (Szklarczyk et al. 2014). In a recent work by Singh and Karmakar (2021), male sterility was successfully introgressed into two carrot varieties, 'Kashi Arun' and 'Kashi Krishna', as well as 11 advanced tropical carrot lines through backcrossing, utilizing naturally occurring petaloid CMS plants from open-pollinated populations.

MADS-box genes are key transcription factors that regulate flowering and flower development. Flowering is controlled by a complex network of signals and genes (Wils and Kaufmann 2017). The ABCDE model, based on studies of *A. thaliana* and *Antirrhinum majus* (Coen and Meyerowitz 1991), outlines how specific gene classes govern each floral whorl. According to this model, the development of sepals, petals, stamens, carpels, and ovules is controlled by A + E genes, A + B + E genes, B + C + E genes, C + E genes, and C + D+ E genes, respectively, in which class A genes and class C genes antagonize each other (Zhang et al. 2023a). Except for AP2, all genes in this model belong to the MADS-box gene family (Chandler 2018; Kieffer and Davies 2001; Lu et al. 2019). Class A gene is *APETALA1* (*AP1*), class B includes *APETALA3* (*AP3*) and *PISTILLATA* (*PI*), class C is *AGAMOUS* (*AG*), class D genes contain *STK*, *SHATTERPROOF1* (*SHP1*), and *SHATTERPROOF2* (*SHP2*), and lastly class E genes include *SEPALLATA1* (*SEP1*), *SEP2*, *SEP3*, and *SEP4* (Zhang et al. 2023a). Qiao et al. (2024) discovered a *Lagerstroemia indica* mutant, 'Xiang Yun', with a longer flowering period and inactive pollen. Transcriptome analysis revealed significantly reduced expression of *MADS* genes involved in flower development compared to the wild type 'Hong Ye'. The sterility of 'Xiang Yun' was characterized by inactive pollen, anther dehiscence disorder, and an abnormal tapetum layer, which resulted in embryo sac abortion. Similarly, onion CMS lines exhibited abnormal floral organ development. Lou et al. (2023) used RACE to clone, the full-length cDNAs of five *MADS-box* genes—*AcFUL, AcDEF, AcPI, AcAG*, and *AcSEP3*—representing A, B, C, and E classes. Phylogenetic analysis of these sequences revealed that CMS in onions was linked to the downregulation of *AcDEF, AcPI*, and *AcAG*, upregulation of *AcSEP3*, and no change in *AcFUL* expression. Overexpression of MADS-box genes *TaAG-A* and *TaAG-B* causes deformities in wheat, while silencing these genes in a fertile line via virus-induced gene silencing (VIGS) results in plants with striped leaves, underdeveloped spikes, and reduced seed set. These findings suggest that *TaAG-A* and *TaAG-B* may contribute to male sterility in the wheat CMS line (Yang et al. 2017).

14.4 Genetic Basis of Male Sterility and Fertility Restoration

14.4.1 Probe Hybridization

To improve pepper breeding efficiency, 176 key pepper parent materials were re-sequenced, and the PepperSNP50K liquid-phase breeding chip, containing 51,172 markers, was developed. The integration of *CaRf* functional characterization with the PepperSNP50K chip enabled the creation of a high-quality red pepper hybrid (Tang et al. 2024). A point mutation in the *OsTMS19* gene was identified as the cause of the TGMS phenotype, confirmed by in situ hybridization of *OsTMS19* transcripts in wild-type rice anthers at various development stages (Zhou et al. 2024). *ADT1* specific probe was used to determine mutation in the *abnormal degraded tapetum 1* (*ADT1*) gene caused excessive ROS accumulation, leading to premature tapetal PCD and pollen abortion. ADT1 encodes a protein crucial for regulating ROS homeostasis and the proper timing of tapetum PCD during anther development (Liu et al. 2024). Cho et al. (2006) identified cytoplasmic male sterility

(CMS) factors in onion plants using PCR-RFLP, a method that distinguishes between male-fertile (N) and male-sterile (S) cytoplasm. The PCR-RFLP marker was based on an amplicon from the *psbA* gene located in the chloroplast genome. By digesting the amplified products from onion varieties with different cytoplasm using the restriction enzyme *MspI*, researchers found that plants with N-cytoplasm contained a functional *MspI* recognition site (CCGG), which was cleaved by the enzyme. In contrast, S-cytoplasm plants exhibited a mutation at this site (CTGG), preventing *MspI* from recognizing and cutting the amplicon. This difference in restriction enzyme digestion patterns enabled the identification of the cytoplasm type in onion plants. Using RFLP-PCR, Yasumoto et al. (2008) determined the frequency of ORF687 in wild radish with an *Rf* gene. Their findings indicated that 14.5% of 207 studied plants showed an identical RFLP pattern to the known *Rf* type of *orf687* mainly collected from regions of Southern Japan. Similarly, PCR-RFLP analysis using the *SspI* enzyme was conducted to determine the presence of the ORF687 (*Rfo*) sequence and the *rforfo* genotype in radish plant DNA. Results showed that the sterile plant had the *rforfo* genotype and lacked the *Rft* allele. Among 32 F_2 progeny, segregation was observed with 18 plants having the *Rforfo* genotype and 14 plants with the *rforfo* genotype (Yamagishi et al. 2024). The PCR-RFLP analysis of the *rbc*L-ORF106 region of chloroplast DNA (using *RsaI*) and the V7 region of mitochondrial DNA (using *Scr*FI) in the BC_5 lines revealed that all progenies displayed restriction patterns identical to those of the cytoplasm donor, *Solanum aethiopicum* Gilo Group. These findings provide strong evidence for the maternal inheritance of both cpDNA and mtDNA in the backcross progenies, confirming the successful cytoplasmic substitution during the development of the male-sterile eggplant line (Khan et al. 2023).

14.4.2 Gene Editing for CMS and Fertility Restoration

Recent research has demonstrated the immense utility of gene editing (GEd) for improving reproductive traits in crops, contributing significantly toward enhancing agricultural yields and safeguarding food security targets (Kumar et al. 2023). GEd can produce male-sterile/restorer lines, a crucial step in hybrid seed production (Table 14.2). For example, TALENs have been used in rapeseed to elucidate their roles, with successful fertility restoration via homologous recombination (Kazama et al. 2019). Similarly, CRISPR/Cas9-based gene editing of *GhCYP703A2* in cotton revealed its role in fertility, leading to the development of male-sterile lines (Ma et al. 2022). MitoCRISPR/Cas9 targeting the *mtatp9* gene in tobacco also demonstrated the link between mitochondrial genes and CMS (Chang et al. 2023).

Kumar and Purty (2023) reported the successful fertility restoration in male-sterile *barnase* lines using *barstar* restorer lines generated through two completely different tapetum-specific promoters for hybrid rice seed production. The major genes *TGW6* and *Wx*, controlling 1000 grain weight and amylose content, were edited in a rice B-line using CRISPR/Cas9 technology with 90% mutation

Table 14.2 List of genes for the gene editing studies used for CMS/restoration

Crop	Name of *CMS/Rf* gene edited	Gene editing technique	Mutation frequency (%)	References
Cotton	*GhCYP703A2*	CRISPR/Cas9	–	Ma et al. (2022)
Tobacco	*mtatp9*	MitoCRISPR/Cas9	3.38 (in petaloid stamen type), 8.02 (abortive stamen type)	Chang et al. (2023)
Rice	*TGW6 and Wx*	CRISPR/Cas9	90%	Han et al. (2018)
Indica rice	*TMS5*	CRISPR/Cas9	–	Barman et al. (2019)
Chinese rice	*RMS*	CRISPR/Cas9	28.50 (G6 site), 11.11(G7 site)	Suketomo et al. (2020)
Wheat	*Ms45*	CRISPR/Cas9	16%	Singh et al. (2018)
Maize	*Rf4*	CRISPR/ Cas9 + HDR editing	42%	Jaqueth et al. (2020)
Maize	*ZmRf5*	CRISPR/Cas9	–	Lin et al. (2024)
Maize	*CENH3*	CMS cyto-swapping	2.3–5.6%	Bortiri et al. (2024)
Soybean	*GmAMS1*	CRISPR/Cas9	25%	Chen et al. (2021)
Brassica napus	*BnDAD1*	CRISPR/Cas9	–	Tu et al. (2024)
Brassica oleracea	*CENH3*	CRISPR/Cas9	14.29 (for sgRNA1), 52.38 (for sgRNA2)	Han et al. (2024)

frequency. This approach facilitated the development of high-yielding, glutinous CMS rice lines (Han et al. 2018). Similarly, the YK17S, created using CRISPR/Cas9 gene editing technology to modify the *TMS5* gene, has been identified as a novel TGMS line with enhanced yield potential. This line shows great promise for widespread application in two-line hybrid breeding of indica rice (Barman et al. 2019). The role of the *Ms45* gene in wheat was investigated by inducing mutations in the A, B, and D homeologs using CRISPR-Cas9. Mutations were recovered in all three homeologs, including a plant with mutations in all three from two different genotypes. Genetic analysis revealed that all three *Ms45* homeologs contribute to male fertility, and triple homozygous mutants are necessary to achieve male sterility by aborting pollen development (Singh et al. 2018). In maize, positional cloning identified Rf_4 as a bHLH transcription factor with a nuclear localization signal. Unexpectedly, Rf_4 also identified as the variant of male sterility gene *Ms23*, crucial in early tapetal development, with CMS-C fertility restoration likely a downstream event in this pathway. CRISPR/Cas9 coupled with homology-dependent repair (HDR) editing generated isogenic variants with an F → Y substitution, revealing that RF4-F187 is linked to fertility restoration, while RF4-Y187 is associated with non-restoration. This substitution, located in the four-helix bundle core, stabilizes dimer conformation (Jaqueth et al. 2020). Stable male-sterile lines were generated for the first time by targeted editing of AMS (ABORTED MICROSPORES gene, a bHLH transcription factor that affects tapetal development) homologs in soybean using CRISPR/Cas9 technology. Targeted editing of *GmAMS1* resulted in a male-sterile phenotype (Chen et al. 2021). Fertility was restored in genome-edited CMS plants with reduced *RETROGRADE-REGULATED MALE STERILITY* (RMS) expression and unchanged PPR2 expression when mutations were introduced in the RMS promoter regions, either inside or outside the PPR2 CDS. No fertility restoration occurred when the mutation was within the RMS CDS. These findings showed that PPR2 is not involved in fertility restoration, which is instead driven by reduced RMS expression, offering a new fertility restorer line of Chinese wild rice for agricultural use (Suketomo et al. 2020). In a study by Tu et al. (2024), a novel approach utilizing CRISPR/Cas9 technology was introduced to disrupt *defective in anther dehiscence1* (BnDAD1) in *B. napus* to generate male-sterile lines. *BnDAD1*, a crucial gene in the jasmonic acid biosynthesis pathway, is predominantly expressed in the stamens of rapeseed flower buds. The targeted disruption of this gene resulted in reduced levels of α-linolenic acid and jasmonate in the double mutants, leading to impaired anther dehiscence and incomplete pollen maturation. By crossing these male-sterile double mutants with male-fertile lines, researchers successfully demonstrated a two-line hybrid system, facilitating the production of F_1 seeds. Lin et al. (2024) identified the nuclear restorer gene *ZmRf5*, which encodes a P-type pentatricopeptide repeat (PPR) protein, regulating the maize CMS-C trait. Overexpression of *ZmRf5* restored fertility in CMS-C plants, while CRISPR/Cas9-mediated editing of *ZmRf5* caused pollen abortion in the restorer line. Recent advances include the creation of CRISPR/Cas9 multiplex vectors targeting multiple traits in rice (He et al. 2024), the development of *Brassica oleracea* CMS lines via CENH3 mutagenesis (Han et al. 2024), and a CMS cyto-swapping method in maize (Bortiri et al. 2024). These breakthroughs highlight CRISPR/Cas9's growing role in creating hybrid vigor and male sterility across diverse crops, offering an efficient alternative to traditional breeding methods.

14.4.3 Fertility Restoration

14.4.3.1 Genomic Level

The fertility restoration is often linked to mitochondrial DNA rearrangements, sub-stoichiometric shifts, or loss of specific mtDNA regions, such as the pvs sequence in *Phaseolus vulgaris* (Chase and Ortega 1992). These shifts can be triggered by nuclear genes like *Fr*, marking the first known example of an *Rf* gene restoring fertility at the genomic level through substoichiometric adjustment (Abad et al. 1995). In maize, the location of *ZmRf5* was initially identified through bulked sergeant analysis (BSA) and QTL mapping in a small population. Linkage markers, *umc1736* and *bnlg1520*, were developed during primary mapping, while fine-mapping employed markers from the B73-ref and Mo17 genome sequences (Lin et al. 2024). Similarly, using QTL-seq, Zheng et al. (2020) identified a novel QTL, *qRf8-1*, associated with fertility restoration in maize CMS-C. Through BSA-seq analysis of mixed DNA pools from sterile and fertile F_2 buds, a 6.25 Mb candidate interval was identified. In maize, Rf_{12}, a strong fertility restorer for C-type CMS, was mapped to chromosome 2, and the candidate gene, *Zm00001d007531*, was identified as encoding a p-type PPR protein (Zhang et al. 2023b). In *B. napus* L., the restorer gene *Rfw* was found to suppresses the expression of the male sterility gene via transcriptional, translational, or other mechanisms, thereby ensuring the normal functioning of mitochondrial electron transport chain (*mtETC*) genes (Yang et al. 2021). Similarly, in *B. juncea*, the fertility-restoring gene *Rfo* was localized to the distal region of chromosome A09 within a large C genome translocation (Gudi et al. 2020). In hybrid rice, Song et al. (2024) cloned Rf_{20}, a restorer of fertility identified through a GWAS study of core parental lines. Structural variations in Rf_{20} were linked to the restoration of fertility in wild-abortive CMS types. In *Capsicum annuum*, genome re-sequencing and recombination analysis revealed *CA00g82510* as a novel candidate gene for the fertility restoration locus *CaRf032* (Zhang et al. 2020). Feng et al. (2020) developed the InDel-1892 marker in cotton, enabling simultaneous identification of Rf_2 and Rf_1, which are associated with fertility restoration in CMS-D8 and CMS-D2, respectively. Genome-wide association study (GWAS) of CMS PET1 of sunflower identified a 7.72 Mb region on chromosome 13 containing 21 candidate genes for Rf_1, including 20 pentatricopeptide repeat (*PPR*) genes and one probable aldehyde dehydrogenase gene (Goryunov et al. 2019). Using map and sequence-based chromosome walking, Ma et al. (2021) identified that Rf_5 is linked to R_{11} (a rust resistance gene) in sunflower. Restorer genes countering the respective CMS types in different crops are shown in Table 14.3.

14.4.3.2 Transcriptional Level/Post-transcriptional Level

CMS-associated transcripts usually undergo posttranscriptional modifications like cleavage, polyadenylation, capping, and splicing. Zhang et al. (2021) used high-throughput sequencing and degradome analyses to identify microRNAs (miRNAs) that control gene expression at transcriptional level and their targets in a soybean CMS line (JLCMS9A) and its maintainer line (JLCMS9B). Of the 280 miRNAs, 30, 23, and 21 belonged to the miR166, miR156, and miR171 families, respectively. Overexpression of soybean *miR2119b* in *Arabidopsis* was found to control male fertility via HT-ROS signaling pathway under high-temperature stress (Ding et al. 2021). Rf_1 and Rf_3 were shown to bind the mitochondrial *orf279* transcript inducing cleavage and preventing expression of the CMS in wheat (Melonek et al. 2021). An RNA-seq mapping strategy, validated using chromosome engineering and physical mapping, identified a putative fertility restoration region on 6HchS in *Hordeum chilense* with six candidate genes, including the barley restorer gene *Rfm1* (Rodríguez-Suárez et al. 2022). In maize, ZmRF5 was identified as being targeted

Table 14.3 Restorer genes countering the respective CMS types in different crops

Crop	CMS type	Restorer loci	Protein encoded	Restoration mode	References
Rice	WA	*Rf3 (Chr. 1), Rf4 (Chr. 10)*	Unknown	Sporophytic	Lin and Yuan (1980) and Zhao et al. (2023)
	HL	*Rf5 (Chr. 10), Rf6 (Chr. 8)*	PPR	Gametophytic	Rao (1988)
	BT	*RF1A Rf1B (Chr. 10)*	PPR	Gametophytic	Shinjyo (1969)
	LD	*Rf2 (Chr. 2)*	Glycine-rich repeats	Gametophytic	Watanabe (1971)
	CW	*Rf 17 (Chr. 4)*	Acyl carrier protein synthase	Gametophytic	Katsuo and Mizushima (1958)
	RT-102	*Rf 102*	Unknown	–	Okazaki et al. (2013)
	RT-98	*Rf 98 (Chr. 10)*	Unknown	–	Igarashi et al. (2013)
Maize	C (Charrua)	*Rf4 (Chr. 8S)*	Unknown	Sporophytic	Beckett (1971)
	S (USDA)	*Rf3 (Chr. 2)*	Unknown	Gametophytic	Xiao et al. (2020)
	T(Texas)	*Rf1(Chr. 3), Rf2 (Chr. 9)*	Aldehyde dehyrdogenase	Sporophytic	Rogers and Edwardson (1952)
B. napus	Ogu	*Rfo (Chr. C03)*	Unknown	Sporophytic	Ogura (1968)
	Pol	*Rfp (Chr. C09)*	Unknown	Sporophytic	Wang et al. (1995)
	Napus	*Rfn (Chr. C09)*	Unknown	Sporophytic	Li et al. (1998)
B. juncea	Hau	–	–	Sporophytic	Heng et al. (2018)
	CMS-orf220	–	–	–	Yang et al. (2010)
Sorghum	A3	*Rf3*	Unknown	Gametophytic	Schertz (1993) and Tang et al. (1999)
	A1	*Rf1(Chr. SBI-08), Rf2 (Chr. SBI-02), Rf5 (Chr. SBI-05), Rf6 ((Chr. SBI-04)*	PPR	Gametophytic	Stephens and Holland (1954) and Radchenko et al. (2024)
Sunflower	PET1	*Rf1 (Chr. 13)*	–	Gametophytic	Leclercq (1969)
Pigeonpea	A4	*CcLG08*	–	–	Sinha et al. (2015)

(continued)

Table 14.3 (continued)

Crop	CMS type	Restorer loci	Protein encoded	Restoration mode	References
Tomato	MSA1				Shikanai et al. (1996)
	O				Kuwabara et al. (2021)
	P				Kuwabara et al. (2021)
Radish	Kosena	*Rfk1 (=Rfo or PPR-B) (Chr. 9)*	PPR		Iwabuchi et al. (1999) and Koizuka et al. (2003)
	DCGMS	*Rfd*			Lee et al. (2008)
	Ogura	*Rfo (=Rfk1 or PPR-B) (Chr. A09)*	PPR		Desloire et al. (2003) and Brown et al. (2003)
Pearl millet	A1				Burton (1958)
	A2 and A3 cytoplasm				Rai et al. (2001) and Rao et al. (1997)
	A4 CMS				Hanna (1989)
	A5 CMS				Rai (1995)
Wheat	T-cytoplasm	*Rf1 (Chr. 1A) and Rf3 (Chr. 1B)*	PPR		Wilson and Ross (1962), Melonek et al. (2021) and Brownfield (2021)
	K-type	*RF9 (Chr. 6AS)*			Liu et al. (2011) and Wu et al. (2023)
	YA-type	*RF9 (Chr. 6AS)*			Liu et al. (2006)
Secale cereale	Pampa	*QRfp-4R (Chr. 4R)*	PPR		Niedziela et al. (2021)
Petunia	PCF	*Rf-PPR592*	PCF		Bentolila et al. (2002)
Sugar beet	O-type	*Rf1*	PPR		Arakawa et al. (2020)

(continued)

Table 14.3 (continued)

Crop	CMS type	Restorer loci	Protein encoded	Restoration mode	References
Soybean	ZD	*Rf (Chr. 16)*			Zhao et al. (1998), Wang et al. (2010) and Dong et al. (2012)
	N21566				Zhao and Gai (2006)
	RN	*Rf1, Rf3*			Wang et al. (2010), Dong et al. (2012), Guo et al. (2022) and Sun et al. (2022)
	N8855	*Rf (Chr. 16)*	GmPPR576		Ding et al. (2002) and Wang et al. (2021)
	N23661				Nie et al. (2017)
	SXCMS5A			Gametophytic	Bai et al. (2022)
	NJCMS			Gametophytic	He et al. (2021) and Nie et al. (2017)
	XX				Zhao et al. (1998) and Sun et al. (2003)
Common bean	CMS-Sprite	Fr and Fr2			Mackenzie et al. (1988)

to the mitochondria, where it interacts with the splicing factor RS31A through MORF8 to form a cleavage/restoration complex. This complex cleaves CMS-associated atp6c transcripts, shifting the primary cleavage site from the 480th nucleotide to the 344th nucleotide. This adjustment facilitates rapid transcript degradation while preserving atp6c RNA for efficient protein translation. As a result, sufficient ATP6C is produced to support the assembly of complex V, ultimately restoring male fertility (Lin et al. 2024).

14.4.3.3 Posttranslational Level

In CMS systems, fertility restoration may be regulated by translational or posttranslational mechanisms. Moreover, in sugar beet, posttranslational interaction between preSATP6 and bvORF20 was found to change the higher order structure of preSATP6, which, in turn, may lead to fertility restoration (Kitazaki et al. 2015). In *Brassica, Rf* genes have been shown to restore male fertility by suppressing atp6/

orf288 through a posttranslational mechanism (Heng et al. 2018). Likewise, Wang et al. (2021a) demonstrated that fertility restoration in the Ogura system from radish operates through a specific blockade of ribosome progression along the *orf138* CMS transcript. RFO showed a translation inhibition mechanism in which RFO binding within the CDS of *orf138* to inhibit its elongation in *B. oleracea* (Xing et al. 2024). In maize, a PPR protein RfZm5 facilitates the translation of *atp6c* to overcome the shortage of ATP6C (Lin et al. 2024). Restorer gene, *Rfo*, encodes a P-type PPR protein, ORF687, which suppresses the translational of ORF138 in radish (Yamagishi et al. 2021).

14.5 Heterosis or Hybrid Breeding

Two key models are commonly discussed in the context of heterosis. The first is the "dominance" model, which suggests that recessive alleles at different loci are complemented in the hybrid. The second is the "overdominance" model, which proposes that interactions between different alleles in the hybrid contribute to an increase in vigor (Birchler et al. 2010).

- *Dominance hypothesis*

In this model, one parent carries gene copies absent in the other parent, resulting in the hybrid possessing more genes than either of the parents (Fu and Dooner 2002). According to the dominance hypothesis, heterosis arises from masking of harmful or inferior recessive alleles contributed by one parent by the beneficial dominant alleles from the other parent, leading to heterozygous genotypes at different loci (Davenport 1908; Bruce 1910; Jones 1917).

- *Overdominance hypothesis*

The overdominance hypothesis explains heterosis as the enhanced fitness of heterozygous genotypes compared to homozygous genotypes at a single locus (East 1908; Shull 1908). A variation of this model, known as "pseudo-overdominance," suggests that complementation occurs between different recessive alleles located in close linkage on opposite homologs, creating the appearance of overdominance. Li et al. (2008) found that in two recombinant inbred (RI) populations of rice hybrids, heterosis resulted from a complex interplay of partial-to-complete dominance, overdominance, and epistasis.

14.6 Hybrid Seed Production

The natural cytoplasmic male sterility (CMS) systems developed or identified in various crop species play a critical role in producing hybrid seeds, particularly for hybrid vigor (Bohra et al. 2025). Over time, advances in biotechnology have led to the creation of genetically engineered CMS systems, further enhancing hybrid seed production strategies.

A well-studied example is the CMS system associated with *Texas* cytoplasm in maize, first identified in the 1940s and widely adopted in hybrid maize production in the United States during the 1960s. However, this system became problematic due to its association with Southern corn blight, caused by the phytopathogen *Bipolaris maydis* (Tatum 1971; Ullstrup 1972). The outbreak of this disease in the early 1970s caused significant maize yield losses in the United States, leading to the discontinuation of the CMS-T (Texas) system for hybrid maize production. Maize was among the first crops in which cytoplasmic male sterility (CMS) systems were utilized for hybrid seed production.

In radish (*Raphanus sativus*), the Ogura CMS system, first discovered in a Japanese radish cultivar in 1968, is one of the most stable and widely used CMS systems for hybrid production. Additionally, CMS systems have been developed in *B. juncea* through somatic hybridization with *Moricandia arvensis* and subsequent backcrossing (Kirti et al. 1992; Prakash et al. 1998). In rice, several CMS systems are employed, including CMS-BT (Boro-Taichung), Honglian (HL), Lead rice (LD), Chinese wild rice (CW), Wild Abortive (WA), and Fujian Abortive (FA). Among these, CMS-WA has been particularly important in hybrid rice breeding (Kim and Zhang 2018; Toriyama 2021). CRISPR/Cas9 has been used in rice to induce alternative splicing of the *OsMADS1* gene, which significantly enhances hybrid vigor. This modification has been shown to increase grain yield by approximately 34.1–47.5% compared to the corresponding inbred rice lines (Wang et al. 2024).

The Owen-CMS system has been predominantly used in hybrid beet seed production. The Rf_1 gene locus, associated with fertility restoration in Owen-CMS, has been shown to contain a cluster of genes (*bvORF18–bvORF21*), with bvORF20 being particularly effective in restoring partial fertility in transgenic sugar beets (Matsuhira et al. 2012). In maize, the CMS-C system is commonly used for hybrid seed production, with the major restorer gene, Rf_4, playing a key role in fertility restoration. Studies on maize nested association mapping (NAM) lines have shown that Rf_4 requires the presence of PPR153 for successful restoration of CMS-C across diverse maize germplasm. This finding provides valuable insights into identifying inbreds suitable for CMS-C-based hybrid production (Jaqueth et al. 2024).

Moreover, recent advancements in maize breeding include using CENH3 and MATRILINEAL (*MTL*) gene, also known as ZmPLA1 as haploid inducers to produce haploids with male sterility (Kelliher et al. 2017; Bortiri et al. 2024). These haploids inherit the cytoplasm from the female parent and are marked by a dominant anthocyanin color marker, making identification easier. This CMS system, combined with a cyto-swapping method, offers a faster and more cost-effective alternative to conventional trait introgression methods for hybrid seed production (Bortiri et al. 2024).

14.7 Conclusion

The latest research on CMS and fertility restoration in plants has provided novel insights into the communication between the nuclear and mitochondrial genomes. New breeding targets have been uncovered by comparative analysis of CMS-associated genes/orfs following sequencing of whole mitochondria in combination with genetic mapping and cloning of *Rf* genes in plants. Gene editing involving mitoTALENs has demonstrated to be a robust tool for validation of CMS-associated genes in plants. A deeper understanding of CMS-Rf systems will drive innovations in heterosis breeding and seed production in crop plants.

References

Aarts MG, Keijzer CJ, Stiekema WJ, Pereira A (1995) Molecular characterization of the *CER1* gene of *Arabidopsis* involved in epicuticular wax biosynthesis and pollen fertility. Plant Cell 7(12):2115–2127

Abad AR, Mehrtens BJ, Mackenzie SA (1995) Specific expression in reproductive tissues and fate of a mitochondrial sterility-associated protein in cytoplasmic male-sterile bean. Plant Cell 7(3):271–285

Arakawa T, Matsunaga M, Matsui K, Itoh K, Kuroda Y, Matsuhira H, Kitazaki K, Kubo T (2020) The molecular basis for allelic differences suggests restorer-of-fertility 1 is a complex locus in sugar beet (*Beta vulgaris* L.). BMC Plant Biol 20(1):503

Bai Z, Ding X, Zhang R, Yang Y, Wei B, Yang S, Gai J (2022) Transcriptome analysis reveals the genes related to pollen abortion in a cytoplasmic male-sterile soybean (*Glycine max* (L.) Merr.). Int J Mol Sci 23(20):12227

Balk J, Leaver CJ (2001) The PET1-CMS mitochondrial mutation in sunflower is associated with premature programmed cell death and cytochrome c release. Plant Cell 13(8):1803–1818

Barman HN, Sheng Z, Fiaz S, Zhong M, Wu Y, Cai Y, Wang W, Jiao G, Tang S, Wei X, Hu P (2019) Generation of a new thermo-sensitive genic male sterile rice line by targeted mutagenesis of *TMS5* gene through CRISPR/Cas9 system. BMC Plant Biol 19(1):109

Beckett JB (1971) Classification of male-sterile cytoplasms in maize (*Zea mays* L.). Crop Sci 11:724–727

Bentolila S, Alfonso AA, Hanson MR (2002) A pentatricopeptide repeat-containing gene restores fertility to cytoplasmic male-sterile plants. Proc Natl Acad Sci USA 99(16):10887–10892

Birchler JA, Yao H, Chudalayandi S, Vaiman D, Veitia RA (2010) Heterosis. Plant Cell 22(7):2105–2112

Bohra A, Singh IP, Yadav AK, Pathak A, Soren KR, Chaturvedi SK, Nadarajan N (2015) Utility of informative SSR markers in the molecular characterization of cytoplasmic genetic male sterility-based hybrid and its parents in pigeonpea. Natl Acad Sci Lett 38:13–19

Bohra A, Jha UC, Adhimoolam P, Bisht D, Singh NP (2016) Cytoplasmic male sterility (CMS) in hybrid breeding in field crops. Plant Cell Rep 35(5):967–993

Bohra A, Jha A, Singh IP, Pandey G, Pareek S, Basu PS, Chaturvedi SK, Singh NP (2017a) Novel CMS lines in pigeonpea [*Cajanus cajan* (L.) Millspaugh] derived from cytoplasmic substitutions, their effective restoration and deployment in hybrid breeding. Crop J 5:89–94

Bohra A, Jha R, Pandey G, Patil PG, Saxena RK, Singh IP, Singh D, Mishra RK, Mishra A, Singh F, Varshney RK, Singh NP (2017b) New hypervariable SSR markers for diversity analysis, hybrid purity testing and trait mapping in pigeonpea [*Cajanus cajan* (L.) Millspaugh]. Front Plant Sci 8:377

Bohra A, Gandham P, Rathore A, Thakur V, Saxena RK, Satheesh Naik SJ, Varshney RK, Singh NP (2021a) Identification of microRNAs and their gene targets in cytoplasmic male sterile and fertile maintainer lines of pigeonpea. Planta 253:59

Bohra A, Rathore A, Gandham P, Saxena RK, Satheesh Naik SJ, Dutta D, Singh IP, Singh F, Rathore M, Varshney RK, Singh NP (2021b) Genome-wide comparative transcriptome analysis of the A4-CMS line ICPA 2043 and its maintainer ICPB 2043 during the floral bud development of pigeonpea. Funct Integr Genomics 21:251–263

Bohra A, Prasad G, Rathore A, Saxena RK, Satheesh Naik SJ, Pareek S, Jha R, Pazhamala L, Datta D, Pandey G, Tiwari A, Maurya AK, Soren KR, Akram M, Varshney RK, Singh NP (2021c) Global gene expression analysis of pigeonpea with male sterility conditioned by A2 cytoplasm. The Plant Genome 14:e20132

Bohra A, Satheesh Naik SJ, Tiwari A, Maurya AK, Tyagi S, Yadav V (2022) Discovery and application of male sterility systems in pigeonpea. In: Bohra A, Parihar AK, Satheesh Naik SJ, Chandra A (eds) Plant male sterility systems for accelerating crop improvement. Springer, Singapore, pp 149–166

Bohra A, Lamichaney A, Parihar AK, Satheesh Naik SJ, Karmakar P (2024) Hybrid seed production: application, challenges and opportunities. In: Lamichaney A, Parihar AK, Bohra A, Karmakar P, Satheesh Naik SJ (eds) Hybrid seed production for boosting crop yields: applications, challenges and opportunities. Springer Nature Singapore

Bohra A, Tiwari A, Pareek S, Joshi R, Satheesh Naik SJ, Kumari K, Verma RL, Parihar AK, Patil PG, Dixit GP (2025) Past and future of cytoplasmic male sterility and heterosis breeding in crop plants. Plant Cell Rep 22:33

Bortiri E, Selby R, Egger R, Tolhurst L, Dong S, Beam K, Meier K, Fabish J, Delaney D, Dunn M, McNamara D, Setliff K, Castro Miranda Lunny R, Gergen S, Dawe RK, Kelliher T (2024) Cyto-swapping in maize by haploid induction with a *cenh3* mutant. Nat Plants 10(4):567–571

Brown GG, Formanová N, Jin H, Wargachuk R, Dendy C, Patil P, Laforest M, Zhang J, Cheung WY, Landry BS (2003) The radish *Rfo* restorer gene of Ogura cytoplasmic male sterility encodes a protein with multiple pentatricopeptide repeats. Plant J 35(2):262–272

Brownfield L (2021) Plant breeding: revealing the secrets of cytoplasmic male sterility in wheat. Curr Biol 31:R724–R726

Bruce AB (1910) The Mendelian theory of heredity and the augmentation of vigor. Science 32(827):627–628

Budar F, Pelletier G (2001) Male sterility in plants: occurrence, determinism, significance and use. C R Acad Sci III 324(6):543–550

Burton GW (1958) Cytoplasmic male sterility in pearl millet [*Pennisetum glaucum* (L.) R. Br]. Agron J 50:230–231

Chandler JW (2018) Class VIIIb APETALA2 Ethylene response factors in plant development. Trends Plant Sci 23(2):151–162

Chang Y, Liu B, Jiang Y, Cao D, Liu Y, Li Y (2023) Induce male sterility by CRISPR/Cas9-mediated mitochondrial genome editing in tobacco. Funct Integr Genomics 23(3):205

Chase CD, Ortega VM (1992) Organization of ATPA coding and 3′ flanking sequences associated with cytoplasmic male sterility in *Phaseolus vulgaris* L. Curr Genet 22(2):147–153

Chase CD, Ribarits A, Heberle-Bors E (2010) Male sterility. In: Pua E, Davey M (eds) Plant developmental biology—biotechnological perspectives. Springer, Berlin/Heidelberg, pp 437–457

Chen L, Liu YG (2014) Male sterility and fertility restoration in crops. Annu Rev Plant Biol 65:579–606

Chen X, Yang S, Zhang Y, Zhu X, Yang X, Zhang C, Li H, Feng X (2021) Generation of male-sterile soybean lines with the CRISPR/Cas9 system. Crop J 9(6):1270–1277

Cho KS, Yang TJ, Hong SY, Kwon YS, Woo JG, Park HG (2006) Determination of cytoplasmic male sterile factors in onion plants (*Allium cepa* L.) using PCR-RFLP and SNP markers. Mol Cell 21(3):411–417

Coen ES, Meyerowitz EM (1991) The war of the whorls: genetic interactions controlling flower development. Nature 353(6339):31–37

Davenport CB (1908) Degeneration, albinism and inbreeding. Science 28(718):454–455
Deng HF, Shu FB, Yuan DY (1999) An overview of research and utilization of Annong S-1. Hybrid Rice 14:1–3
Desloire S, Gherbi H, Laloui W, Marhadour S, Clouet V, Cattolico L, Falentin C, Giancola S, Renard M, Budar F, Small I, Caboche M, Delourme R, Bendahmane A (2003) Identification of the fertility restoration locus, Rfo, in radish, as a member of the pentatricopeptide-repeat protein family. EMBO Rep 4(6):588–594
Dewey RE, Levings CS, Timothy DH (1985) Nucleotide sequence of ATPase subunit 6 gene of maize mitochondria. Plant Physiol 79(3):914–919
Dewey RE, Timothy DH, Levings CS (1987) A mitochondrial protein associated with cytoplasmic male sterility in the T cytoplasm of maize. Proc Natl Acad Sci USA 84(15):5374–5378
Dewey RE, Selote D, Griffin HC, Dickey AN, Jantz D, Smith JJ, Matthiadis A, Strable J, Kestell C, Smith WA (2023) Cytoplasmic male sterility and abortive seed traits generated through mitochondrial genome editing coupled with allotopic expression of atp1 in tobacco. Front Plant Sci 14:1253640
Dhillon MK, Sharma HC, Reddy BVS, Singh R, Naresh JS, Kai Z (2005) Relative susceptibility of different male-sterile cytoplasms in sorghum to shoot fly, *Atherigona soccata*. Euphytica 144(3):275–283
Ding D, Gai J, Cui Z, Qiu J (2002) Development of a cytoplasmic-nuclear male-sterile line of soybean. Euphytica 124:85–91
Ding X, Chen L, Guo J, Gai J, Yang S (2021) A small RNA of miR2119b from soybean CMS line acts as a negative regulator of male fertility in transgenic *Arabidopsis*. Plant Physiol Biochem 167:210–221
Dong DK, Li Z, Yuan FJ, Zhu SL, Chen P, Yu W, Yang QH, Fu XJ, Yu XM, Li BQ, Zhu DH (2012) Inheritance and fine mapping of a restorer-of-fertility (*Rf*) gene for the cytoplasmic male sterility in soybean. Plant Sci 188-189:36–40
East EM (1908) Inbreeding in corn. In: Reports of the Connecticut Agricultural Experiment Station for Years 1907–1908, pp 419–428
Feng J, Zhu H, Zhang M, Zhang X, Guo L, Qi T, Tang H, Wang H, Qiao X, Zhang B, Shahzad K, Xing C, Wu J (2020) Development and utilization of an InDel marker linked to the fertility restorer genes of CMS-D8 and CMS-D2 in cotton. Mol Biol Rep 47(2):1275–1282
Fu H, Dooner HK (2002) Intraspecific violation of genetic colinearity and its implications in maize. Proc Natl Acad Sci USA 99(14):9573–9578
Fujii S, Toriyama K (2009) Suppressed expression of retrograde-regulated male sterility restores pollen fertility in cytoplasmic male sterile rice plants. Proc Natl Acad Sci USA 106(23):9513–9518
Geiger HH, Schnell FW (1970) Cytoplasmic male sterility in rye (*Secale cereale* L.). Crop Sci 10(5):590–593
Goryunov DV, Anisimova IN, Gavrilova VA, Chernova AI, Sotnikova EA, Martynova EU, Boldyrev SV, Ayupova AF, Gubaev RF, Mazin PV et al (2019) Association mapping of fertility restorer gene for CMS PET1 in sunflower. Agronomy 9(2):49
Gudi S, Atri C, Goyal A, Kaur N, Akhtar J, Mittal M, Kaur K, Kaur G, Banga SS (2020) Physical mapping of introgressed chromosome fragment carrying the fertility restoring (*Rfo*) gene for Ogura CMS in *Brassica juncea* L. Czern & Coss. Theor Appl Genet 133(10):2949–2959
Guo JX, Liu YG (2012) Molecular control of male reproductive development and pollen fertility in rice. J Integr Plant Biol 54(12):967–978
Guo FL, Lin CJ, Wang PN, Yang XL, Wu Z, Peng B, Zhao LM, Zhang CB (2022) Fine mapping of a restorer-of-fertility gene *GmRf1* for the cytoplasmic male sterility in soybean. J Plant Genet Resour 23:518–526
Hagemann R (2004) The sexual inheritance of plant organelles. In: Daniell H, Chase C (eds) Molecular biology and biotechnology of plant organelles. Springer, Dordrecht, pp 93–113
Han Y, Luo D, Usman B, Nawaz G, Zhao N, Liu F, Li R (2018) Development of high yielding glutinous cytoplasmic male sterile rice (*Oryza sativa* L.) lines through CRISPR/Cas9 based mutagenesis of Wx and TGW6 and proteomic analysis of anther. Agronomy 8(12):290

Han F, Zhang X, Liu Y, Liu Y, Zhao H, Li Z (2024) One-step creation of CMS lines using a *BoCENH3*-based haploid induction system in *Brassica* crop. Nat Plants 10(4):581–586

Hanna WW (1989) Characteristics and stability of a new cytoplasmic-nuclear male-sterile source in pearl millet. Crop Breed Genet Cytol 29:1457–1459

Hanson MR, Conde MF (1985) Functioning and variation of cytoplasmic genomes: lessons from cytoplasmic-nuclear interactions affecting male fertility in plants. Int Rev Cytol 94:213–267

Havey MJ (2004) The use of cytoplasmic male sterility for hybrid seed production. Springer, Dordrecht, pp 623–634

He T, Ding X, Zhang H, Li Y, Chen L, Wang T, Yang L, Nie Z, Song Q, Gai J, Yang S (2021) Comparative analysis of mitochondrial genomes of soybean cytoplasmic male-sterile lines and their maintainer lines. Funct Integr Genomics 21:43–57

He D, Zhou R, Huang C, Li Y, Peng Z, Li D, Duan W, Huang N, Cao L, Cheng S, Zhan X, Sun L, Wang S (2024) Improvement of flowering stage in Japonica rice variety jiahe212 by using CRISPR/Cas9 system. Plants (Basel) 13(15):2166

Heng S, Gao J, Wei C, Chen F, Li X, Wen J, Yi B, Ma C, Tu J, Fu T, Shen J (2018) Transcript levels of *orf288* are associated with the *hau* cytoplasmic male sterility system and altered nuclear gene expression in *Brassica juncea*. J Exp Bot 69(3):455–466

Huang W, Hu J, Yu C, Huang Q, Wan L, Wang L, Qin X, Ji Y, Zhu R, Li S, Zhu Y (2012) Two non-allelic nuclear genes restore fertility in a gametophytic pattern and enhance abiotic stress tolerance in the hybrid rice plant. Theor Appl Genet 124(5):799–807

Igarashi K, Kazama T, Motomura K, Toriyama K (2013) Whole genomic sequencing of RT98 mitochondria derived from *Oryza rufipogon* and northern blot analysis to uncover a cytoplasmic male sterility-associated gene. Plant Cell Physiol 54:237–243

Iwabuchi M, Koizuka N, Fujimoto H, Sakai T, Imamura J (1999) Identification and expression of the Kosena radish (*Raphanus sativus* cv. Kosena) homologue of the Ogura radish CMS-associated gene, orf138. Plant Mol Biol 39:183–188

Jaqueth JS, Hou Z, Zheng P, Ren R, Nagel BA, Cutter G, Niu X, Vollbrecht E, Greene TW, Kumpatla SP (2020) Fertility restoration of maize CMS-C altered by a single amino acid substitution within the Rf4 bHLH transcription factor. Plant J 101(1):101–111

Jaqueth JS, Li B, Vollbrecht E (2024) Pentatricopeptide repeat 153 (PPR153) restores maize C-type cytoplasmic male sterility in conjunction with RF4. PLoS One 19(7):e0303436

Jiang H, Lu Q, Qiu S, Yu H, Wang Z, Yu Z, Lu Y, Wang L, Xia F, Wu Y, Li F, Zhang Q, Liu G, Song D, Ma C, Ding Q, Zhang X, Zhang L, Zhang X, Li X, Zhang J, Xiao J, Li X, Wang N, Ouyang Y, Zhou F, Zhang Q (2022) Fujian cytoplasmic male sterility and the fertility restorer gene *OsRf19* provide a promising breeding system for hybrid rice. Proc Natl Acad Sci USA 119(34):e2208759119

Jing B, Heng S, Tong D, Wan Z, Fu T, Tu J, Ma C, Yi B, Wen J, Shen J (2012) A male sterility-associated cytotoxic protein ORF288 in *Brassica juncea* causes aborted pollen development. J Exp Bot 63:1285–1295

Jones DF (1917) Dominance of linked factors as a means of accounting for heterosis. Genetics 2(5):466–479

Kang L, Li P, Wang A, Ge X, Li Z (2017) A novel cytoplasmic male sterility in *Brassica napus* (inap CMS) with carpelloid stamens via protoplast fusion with Chinese woad. Front Plant Sci 8:529

Katsuo K, Mizushima U (1958) Studies on the cytoplasmic difference among rice varieties, *Oryza sativa* L. 1. On the fertility of hybrids obtained reciprocally between cultivated and wild varieties. Jpn J Breed 8:1–5

Kaul MLH (1988) Male sterility in higher plants. Springer, Berlin, pp 15–96

Kawanabe T, Ariizumi T, Kawai-Yamada M, Uchimiya H, Toriyama K (2006) Abolition of the tapetum suicide program ruins microsporogenesis. Plant Cell Physiol 47(6):784–787

Kazama T, Toriyama K (2003) A pentatricopeptide repeat-containing gene that promotes the processing of aberrant atp6 RNA of cytoplasmic male-sterile rice. FEBS Lett 544(1-3):99–102

Kazama T, Okuno M, Watari Y, Yanase S, Koizuka C, Tsuruta Y, Sugaya H, Toyoda A, Itoh T, Tsutsumi N, Toriyama K, Koizuka N, Arimura SI (2019) Curing cytoplasmic male sterility via TALEN-mediated mitochondrial genome editing. Nat Plants 5(7):722–730

Kelliher T, Starr D, Richbourg L, Chintamanani S, Delzer B, Nuccio ML, Green J, Chen Z, McCuiston J, Wang W, Liebler T, Bullock P, Martin B (2017) MATRILINEAL, a sperm-specific phospholipase, triggers maize haploid induction. Nature 542(7639):105–109

Kempe K, Gils M (2011) Pollination control technologies for hybrid breeding. Mol Breed 27(4):417–437

Khan MMR, Arita T, Iwayoshi M, Ogura-tsujita Y, Isshiki S (2023) Development of a male-sterile line of eggplant utilizing the cytoplasm of *Solanum aethiopicum* Gilo. Group. Environ Control Biol 61(4):63–67

Kieffer M, Davies B (2001) Developmental programmes in floral organ formation. Semin Cell Dev Biol 12(5):373–380

Kim YJ, Zhang D (2018) Molecular control of male fertility for crop hybrid breeding. Trends Plant Sci 23(1):53–65

Kirti PB, Narasimhulu SB, Prakash S, Chopra VL (1992) Somatic hybridization between *Brassica juncea* and *Moricandia arvensis* by protoplast fusion. Plant Cell Rep 11(5-6):318–321

Kitazaki K, Arakawa T, Matsunaga M, Yui-Kurino R, Matsuhira H, Mikami T, Kubo T (2015) Post-translational mechanisms are associated with fertility restoration of cytoplasmic male sterility in sugar beet (*Beta vulgaris*). Plant J 83(2):290–299

Koizuka N, Imai R, Fujimoto H, Hayakawa T, Kimura Y, Kohno-Murase J, Sakai T, Kawasaki S, Imamura J (2003) Genetic characterization of a pentatricopeptide repeat protein gene, *orf687*, that restores fertility in the cytoplasmic male-sterile Kosena radish. Plant J 34(4):407–415

Korth KL, Levings CS 3rd (1993) Baculovirus expression of the maize mitochondrial protein URF13 confers insecticidal activity in cell cultures and larvae. Proc Natl Acad Sci USA 90(8):3388–3392

Korth KL, Kaspi CI, Siedow JN, Levings CS 3rd (1991) URF13, a maize mitochondrial pore-forming protein, is oligomeric and has a mixed orientation in *Escherichia coli* plasma membranes. Proc Natl Acad Sci USA 88(23):10865–10869

Kumar P, Purty RS (2023) Successful fertility restoration in male-sterile Barnase line by optimal expression of Barstar gene for hybrid-rice seed production. J Crop Improv 38(1):9–24

Kumar M, Prusty MR, Pandey MK, Singh PK, Bohra A, Guo B, Varshney RK (2023) Application of CRISPR/Cas9-mediated gene editing for abiotic stress management in crop plants. Front Plant Sci 14:1157678

Kumari S, Chakrabarty SK, Paul D, Singh Y, Bhowmick PK, Prasad ASH (2023) Variability and heterosis of seed vigour traits in hybrid rice (*Oryza sativa* L.). Indian J Genet Plant Breed 83(02):168–175

Kuwabara K, Harada I, Matsuzawa Y, Ariizumi T (2021) Organelle genome assembly uncovers the dynamic genome reorganization and cytoplasmic male sterility associated genes in tomato. Hortic Res 8:1–11

Laughnan JR, Gabay-Laughnan S (1983) Cytoplasmic male sterility in maize. Annu Rev Genet 17:27–48

Leclercq P (1969) Unesterilite male chez le tournesol. Ann Amel Plantes 19:99–106

Lee SLJ, Warmke HE (1979) Organelle size and number in fertile and T-cytoplasmic male-sterile corn. Am J Bot 66(2):141–148

Lee YP, Park S, Lim C, Kim H, Lim H, Ahn Y, Sung SK, Yoon MK, Kim S (2008) Discovery of a novel cytoplasmic male-sterility and its restorer lines in radish (*Raphanus sativus* L.). Theor Appl Genet 117(6):905–913

Levings CS 3rd (1993) Thoughts on cytoplasmic male sterility in cms-T maize. Plant Cell 5(10):1285–1290

Li XQ, Jean M, Landry BS, Brown GG (1998) Restorer genes for different forms of *Brassica* cytoplasmic male sterility map to a single nuclear locus that modifies transcripts of several mitochondrial genes. Proc Natl Acad Sci USA 95(17):10032–10037

Li L, Lu K, Chen Z, Mu T, Hu Z, Li X (2008) Dominance, overdominance and epistasis condition the heterosis in two heterotic rice hybrids. Genetics 180(3):1725–1742

Li H, You C, Yoshikawa M, Yang X, Gu H, Li C, Cui J, Chen X, Ye N, Zhang J, Wang G (2022) A spontaneous thermo-sensitive female sterility mutation in rice enables fully mechanized hybrid breeding. Cell Res 32(10):931–945

Lin SC, Yuan LP (1980) Hybrid rice breeding in China. In: IRRI (ed) Innovative approaches to rice breeding. IRRI, Manila, pp 35–51

Lin Y, Yang H, Liu H, Lu X, Cao H, Li B, Chang Y, Guo Z, Ding D, Hu Y, Xue Y, Liu Z, Tang J (2024) A P-type pentatricopeptide repeat protein ZmRF5 promotes 5′ region partial cleavages of atp6c transcripts to restore the fertility of CMS-C maize by recruiting a splicing factor. Plant Biotechnol J 22(5):1269–1281

Linke B, Nothnagel T, Borner T (2003) Flower development in carrot CMS plants: mitochondria affect the expression of MADS box genes homologous to *GLOBOSA* and *DEFICIENS*. Plant J 34(1):27–37

Liu CG, Hou N, Liu LK, Liu JC, Kang XS, Zhang AM (2006) A YA-type cytoplasmic male sterile source in common wheat. Plant Breed 125:437–440

Liu H, Cui P, Zhan K, Lin Q, Zhuo G, Guo X, Ding F, Yang W, Liu D, Hu S, Yu J, Zhang A (2011) Comparative analysis of mitochondrial genomes between a wheat K-type cytoplasmic male sterility (CMS) line and its maintainer line. BMC Genomics 12:163

Liu C, Ma N, Wang PY, Fu N, Shen HL (2013) Transcriptome sequencing and *de novo* analysis of a cytoplasmic male sterile line and its near-isogenic restorer line in chili pepper (*Capsicum annuum* L.). PLoS One 8(6):e65209

Liu D, Shi J, Liang W, Zhang D (2023) Molecular mechanisms underlying plant environment-sensitive genic male sterility and fertility restoration. Seed Biol 2(1):13

Liu J, Ye Q, Jiang W, Liu S, Wu Z, Hu X, Wang X, Zhang Z, Guo D, Chen X, He H, Hu L (2024) *Abnormal Degraded Tapetum 1* (*ADT1*) is required for tapetal cell death and pollen development in rice. Theor Appl Genet 137(7):170

Lou H, Huang Y, Zhu Z, Xu Q (2023) Cloning and expression analysis of onion (*Allium cepa* L.) MADS-Box genes and regulation mechanism of cytoplasmic male sterility. Biochem Genet 61(5):2116–2134

Lu H, Liu Z, Lan S (2019) Genome sequencing reveals the role of MADS-box gene families in the floral morphology evolution of orchids. Hortic Plant J 5(6):247–254

Luo D, Xu H, Liu Z, Guo J, Li H, Chen L, Fang C, Zhang Q, Bai M, Yao N, Wu H, Wu H, Ji C, Zheng H, Chen Y, Ye S, Li X, Zhao X, Li R, Liu YG (2013) A detrimental mitochondrial-nuclear interaction causes cytoplasmic male sterility in rice. Nat Genet 45(5):573–577

Ma H, Wu Y, Lv R, Chi H, Zhao Y, Li Y, Liu H, Ma Y, Zhu L, Guo X, Kong J, Wu J, Xing C, Zhang X, Min L (2022) Cytochrome P450 mono-oxygenase CYP703A2 plays a central role in sporopollenin formation and ms5ms6 fertility in cotton. J Integr Plant Biol 64(10):2009–2025

Mackenzie SA, Pring DR, Bassett MJ, Chase CD (1988) Mitochondrial DNA rearrangement associated with fertility restoration and cytoplasmic reversion to fertility in cytoplasmic male sterile *Phaseolus vulgaris* L. Proc Natl Acad Sci USA 85(8):2714–2717

Martín AC, Atienza SG, Ramírez MC, Barro F, Martín A (2008) Male fertility restoration of wheat in *Hordeum chilense* cytoplasm is associated with 6HchS chromosome addition. Aust J Agric Res 59(3):206–213

Masiero S, Colombo L, Grini PE, Schnittger A, Kater MM (2011) The emerging importance of type I MADS box transcription factors for plant reproduction. Plant Cell 23(3):865–872

Matsuhira H, Kagami H, Kurata M, Kitazaki K, Matsunaga M, Hamaguchi Y, Hagihara E, Ueda M, Harada M, Muramatsu A, Yui-Kurino R, Taguchi K, Tamagake H, Mikami T, Kubo T (2012) Unusual and typical features of a novel restorer-of-fertility gene of sugar beet (*Beta vulgaris* L.). Genetics 192(4):1347–1358

Mayr E (1986) Joseph Gottlieb Kölreuter's contributions to biology. *Osiris* 2:135–176

McCreight JD, Elmstrom GM (1984) A third male sterile gene in muskmelon. Hortic Sci 19:268–270

Melonek J, Duarte J, Martin J, Beuf L, Murigneux A, Varenne P, Comadran J, Specel S, Levadoux S, Bernath-Levin K, Torney F, Pichon JP, Perez P, Small I (2021) The genetic basis of cytoplasmic male sterility and fertility restoration in wheat. Nat Commun 12(1):1036

Mishra A, Bohra A (2018) Non-coding RNAs and plant male sterility: current knowledge and future prospects. Plant Cell Rep 37(2):177–191

Mukai Y, Tsunewaki K (1979) Basic studies on hybrid wheat breeding: VIII. A new male sterility-fertility restoration system in common wheat utilizing the cytoplasms of *Aegilops kotschyi* and *Ae. variabilis*. Theor Appl Genet 54(4):153–160

Nie Z, Zhao T, Yang S, Gai J (2017) Development of a cytoplasmic male-sterile line NJCMS 4A for hybrid soybean production. Plant Breed 136:516–525

Niedziela A, Brukwiński W, Bednarek PT (2021) Genetic mapping of pollen fertility restoration QTLs in rye (*Secale cereale* L.) with CMS Pampa. J Appl Genet 62:185–198

Ogura H (1968) Studies on the new male-sterility in Japanese radish with special reference to the utilization of this sterility towards the practical raising of hybrid seeds. Mem Fac AgricKagoshima Univ 6:36–78

Okazaki M, Kazama T, Murata H, Motomura K, Toriyama K (2013) Whole mitochondrial genome sequencing and transcriptional analysis to uncover an RT102-type cytoplasmic male sterility-associated candidate gene derived from *Oryza rufipogon*. Plant Cell Physiol 54:1560–1568

Prakash S, Kirti P, Bhat S, Gaikwad K, Kumar VD, Chopra V (1998) A *Moricandia arvensis*–based cytoplasmic male sterility and fertility restoration system in *Brassica juncea*. Theor Appl Genet 97:488–492

Qiao Z, Deng F, Zeng H, Li X, Lu L, Lei Y, Li L, Chen Y, Chen J (2024) MADS-Box family genes in *Lagerstroemia indica* and their involvement in flower development. Plants (Basel) 13(5):709

Radchenko EE, Anisimova IN, Ryazanova MK, Kibkalo IA, Alpatieva NV (2024) Newly developed restorer lines of sorghum [*Sorghum bicolor* (L.) Moench] resistant to greenbug. Plants 13:425

Rai KN (1995) A new cytoplasmic-nuclear male sterility system in pearl millet. Plant Breed 114:445–447

Rai K, Kumar AK, Andrews D, Rao AS (2001) Commercial viability of alternative cytoplasmic-nuclear male-sterility systems in pearl millet. Euphytica 121:107–114

Rajanna CM, Narayanswamy R, Choudhury MM, Silva JT, Prasad R (2011) Characterization of rice (*Oryza sativa* L.) hybrids and their parental lines based on the seed, seedling, and plant morphological traits. Seed Sci Biotechnol 5(1):36–41

Rao YS (1988) Cytohistology of cytoplasmic male sterile lines in hybrid rice. In: Smith WH, Bostian LR, Cervantes EP (eds) Hybrid rice. IRRI, Manila, pp 115–128

Rao AS, Mengesha MH, Reddy KN, Reddy CR (1997) Identification of new male sterile lines in pearl millet germplasm. Ind J Plant Genet Resour 10(1):17–25

Rao RN, Ratho SN, Pand D, Rao KS, Sarkunan, Behera KS, Patnaik SSC, Mohanty SK, Das S, Swain P, Das KM (2013) Production technology for hybrid rice-Rajalaxmi, NRRI Research Bulletin No. 59. ICAR-National Rice Research Institute, Cuttack, pp 1–2

Rhoads DM, Levings CS 3rd, Siedow JN (1995) URF13, a ligand-gated, pore-forming receptor for T-toxin in the inner membrane of cms-T mitochondria. J Bioenerg Biomembr 27(4):437–445

Rodriguez-Suarez C, Requena-Ramirez MD, Hornero-Mendez D, Atienza SG (2022) The breeder's tool-box for enhancing the content of esterified carotenoids in wheat: from extraction and profiling of carotenoids to marker-assisted selection of candidate genes. Methods Enzymol 671:99–125

Rogers JS, Edwardson JR (1952) The utilization of cytoplasmic male-sterile inbreds in the production of corn hybrids. Agron J 44(1):8–13

Sabar M, Gagliardi D, Balk J, Leaver CJ (2003) ORFB is a subunit of F1F(O)-ATP synthase: insight into the basis of cytoplasmic male sterility in sunflower. EMBO Rep 4(4):381–386

Satyavathi CT, Khandelwal V, Rajpurohit BS, Supriya A, Beniwal BR, Kamlesh K, Sushila B, Shripal S, Mahesh CK, Yadav SL (2018) Pearl Millet-hybrid and varieties. ICAR-All India Coordinated Research Project on Pearl Millet, Mandore, pp 1–142

Saxena KB, Nadarajan N (2010) Prospects of pigeonpea hybrids in Indian agriculture. Electron J Plant Breed 1(4):1107–1117

Schertz KF (1993) Male-sterility in sorghum: its characteristics and importance. In: Use of molecular markers in sorghum and pearl millet breeding for developing countries: proceedings of an ODA plant science research programme conference on genetic improvement. Plant Sci 9:568–570

Schnable PS, Wise RP (1998) The molecular basis of cytoplasmic male sterility. Trends Plant Sci 3:175–180

Senguttuvel P, Hari Prasad AS, Sundaram RM, Revathi P, Kemparaju KB, Sruthi K, Subba Rao LV et al (2023) DRRH-4 (IET 27937)—world's first public bred aerobic rice hybrid. J Rice Res 16(1):105–106

Shi MS (1985) The discovery and study of the photosensitive recessive male-sterile rice (*Oryza sativa* L. ssp. *japonica*). Sci Agric Sin 2:44–48

Shifriss C, Frankel R (1971) New sources of cytoplasmic male sterility in cultivated peppers. J Hered 62:254–256

Shikanai T, Nakata S, Harada K, Watanabe K (1996) Analysis of the heterogeneous transcripts of the highly edited orf206 in tomato mitochondria. Plant Cell Physiol 37(5):692–696

Shinjyo C (1969) Cytoplasmic-genetic male sterility in cultivated rice, *Oryza sativa* L. Jpn J Genet 44:149–156

Shull GH (1908) The composition of a field of maize. J Hered 4(1):296–301

Singh BK, Karmakar P (2021) Introgression of cytoplasmic male sterility (CMS) in tropical carrots (*Daucus carota* subsp. *sativus* Schubl. & Martens). Veg Sci 48(2):203–208

Singh M, Kumar M, Albertsen MC, Young JK, Cigan AM (2018) Concurrent modifications in the three homeologs of *Ms45* gene with CRISPR-Cas9 lead to rapid generation of male sterile bread wheat (*Triticum aestivum* L.). Plant Mol Biol 97(4-5):371–383

Sinha P, Saxena KB, Saxena RK, Singh VK, Suryanarayana V, Sameer Kumar CV, Katta MAVS, Khan AW, Varshney RK (2015) Association of nad7a gene with cytoplasmic male sterility in pigeonpea. Plant Genome 8(2):eplantgenome2014.11.0084

Smith CW, Betran J, Runge ECA (2004) Corn: origin, history, technology, and production. Wiley, Hoboken

Song S, Li Y, Qiu M, Xu N, Li B, Zhang L, Li L, Chen W, Li J, Wang T, Qiu Y, Gong M, Yu D, Dong H, Xia S, Pan Y, Yuan D, Li L (2024) Structural variations of a new fertility restorer gene, *Rf20*, underlie the restoration of wild abortive-type cytoplasmic male sterility in rice. Mol Plant 17(8):1272–1288

Stephens JC, Holland RF (1954) Cytoplasmic male-sterility for hybrid sorghum seed production. Agron J 46:20–23

Suketomo C, Kazama T, Toriyama K (2020) Fertility restoration of Chinese wild rice-type cytoplasmic male sterility by CRISPR/Cas9-mediated genome editing of nuclear-encoded *RETROGRADE-REGULATED MALE STERILITY*. Plant Biotechnol (Tokyo) 37(3):285–292

Sun H, Zhao L, Wang S, Wang Y, Li J (2003) Progress on the utilization of heterosis in soybean. Chin J Oil Crop Sci 25:92–96

Sun Y, Zhang Y, Jia S, Lin C, Zhang J, Yan H, Peng B, Zhao L, Zhang W, Zhang C (2022) Identification of a candidate restorer-of-fertility gene *Rf3* encoding a pentatricopeptide repeat protein for the cytoplasmic male sterility in soybean. Int J Mol Sci 23:5388

Szklarczyk M, Szymanski M, Wojcik-Jagla M, Simon PW, Weihe A, Borner T (2014) Mitochondrial atp9 genes from petaloid male-sterile and male-fertile carrots differ in their status of heteroplasmy, recombination involvement, post-transcriptional processing as well as accumulation of RNA and protein product. Theor Appl Genet 127(8):1689–1701

Tang HV, Chen W, Pring DR (1999) Mitochondrial orf107 transcription, editing, and nucleolytic cleavage conferred by the gene *Rf3* are expressed in sorghum pollen. Sex Plant Reprod 12(1):53–59

Tang B, Yang H, Yin Q, Miao W, Lei Y, Cui Q, Cheng J, Zhang X, Chen Y, Du J, Xie L, Tang S, Wang M, Li J, Cao M, Chen L, Xie F, Li X, Zhu F, Wang Z, Xiong C, Dai X, Zou X, Liu F (2024) Fertility restorer gene *CaRf* and PepperSNP50K provide a promising breeding system for hybrid pepper. Hortic Res 11(10):223

Tatum LA (1971) The southern corn leaf blight epidemic. Science 171(3976):1113–1116

Teixeira RT, Farbos I, Glimelius K (2005) Expression levels of meristem identity and homeotic genes are modified by nuclear-mitochondrial interactions in alloplasmic male-sterile lines of *Brassica napus*. Plant J 42(5):731–742

Thakur RP (1989) Ergot susceptibility in relation to cytoplasmic male sterility in pearl millet. Plant Dis 73(8):676–678

Toriyama K (2021) Molecular basis of cytoplasmic male sterility and fertility restoration in rice. Plant Biotechnol (Tokyo) 38(3):285–295

Tu ZP, Banga SK (1998) Chemical hybridizing agents. Narosa Publishing House, New Delhi

Tu M, Wang R, Guo W, Xu S, Zhu Y, Dong J, Yao X, Jiang L (2024) A CRISPR/Cas9-induced male-sterile line facilitating easy hybrid production in polyploid rapeseed (*Brassica napus*). Hortic Res 11(7):uhae139

Ullstrup A (1972) The impacts of the southern corn leaf blight epidemics of 1970–1971. Annu Rev Phytopathol 10:37–50

Uyttewaal M, Arnal N, Quadrado M, Martin-Canadell A, Vrielynck N, Hiard S, Gherbi H, Bendahmane A, Budar F, Mireau H (2008) Characterization of *Raphanus sativus* pentatricopeptide repeat proteins encoded by the fertility restorer locus for Ogura cytoplasmic male sterility. Plant Cell 20(12):3331–3345

Vedel F, Pla M, Vitart V, Gutierres S, Chetrit P, De Paepe R (1994) Molecular basis of nuclear and cytoplasmic male sterility in higher plants. Plant Physiol Biochem 32:601–608

Verma R, Jena D, Rout D, Hembram B, Katara JL, Sarkar S, Samantaray S, Patra SD, Mohapatra SD, Mukherjee AK, Nayak AK, Singh ON (2022) Production technology for hybrid rice-CR Dhan-703. NERRI Technology Bulletin-199, pp 1–4

Verma RL, Sah RP, Meher J, Marndi BC, Bose LK, Chattopadhyay K, Dash SK, Singh P, Mohanty MR, Jena D, Rout D, Kuanr L, Kar MK, Samantaray S, Nayak AK (2024) NRRI Research Bulletin No. 55. ICAR-National Rice Research Institute, Cuttack, p 60

Virmani SS, Ilyas-Ahmed M (2001) Environment-sensitive genic male sterility (EGMS) in crops. Adv Agron 72:139–195

Wang HM, Ketela T, Keller WA, Gleddie SC, Brown GG (1995) Genetic correlation of the *orf224/atp6* gene region with Polima CMS in Brassica somatic hybrids. Plant Mol Biol 27(4):801–807

Wang Y, Xue Y, Li J (2005) Towards molecular breeding and improvement of rice in China. Trends Plant Sci 10(12):610–614

Wang Z, Zou Y, Li X, Zhang Q, Chen L, Wu H, Su D, Chen Y, Guo J, Luo D, Long Y, Zhong Y, Liu YG (2006) Cytoplasmic male sterility of rice with boro II cytoplasm is caused by a cytotoxic peptide and is restored by two related PPR motif genes via distinct modes of mRNA silencing. Plant Cell 18(3):676–687

Wang Y, Zhao L, Wang X, Sun H (2010) Molecular mapping of a fertility restorer gene for cytoplasmic male sterility in soybean. Plant Breed 129:9–12

Wang T, He T, Ding X, Zhang Q, Yang L, Nie Z, Zhao T, Gai J, Yang S (2021) Confirmation of *GmPPR576* as a fertility restorer gene of cytoplasmic male sterility in soybean. J Exp Bot 72:7729–7742

Wang C, Lezhneva L, Arnal N, Quadrado M, Mireau H (2021a) The radish Ogura fertility restorer impedes translation elongation along its cognate CMS-causing mRNA. Proc Natl Acad Sci USA 118(35):e2105274118

Wang H, Xiao Q, Wei C, Chen H, Chen X, Dai C, Wen J, Ma C, Tu J, Fu T, Shen J, Yi B (2021b) A mitochondria-localized pentatricopeptide repeat protein is required to restore *hau* cytoplasmic male sterility in *Brassica napus*. Theor Appl Genet 134(5):1377–1386

Wang C, Wang Z, Cai Y, Zhu Z, Yu D, Hong L, Wang Y, Lv W, Zhao Q, Si L, Liu K, Han B (2024) A higher-yield hybrid rice is achieved by assimilating a dominant heterotic gene in inbred parental lines. Plant Biotechnol J 22(6):1669–1680

Watanabe Y (1971) Establishment of cytoplasmic and genetic male- sterile lines by means of Indica-Japonica cross. Oryza Cuttack 8(Suppl 2):9–16

Williams ME, Levings CS (1992) Molecular biology of cytoplasmic male sterility. In: Plant breeding reviews. Wiley, New York, pp 23–51

Wils CR, Kaufmann K (2017) Gene-regulatory networks controlling inflorescence and flower development in *Arabidopsis thaliana*. Biochim Biophys Acta Gene Regul Mech 1860(1):95–105

Wilson J, Ross W (1962) Male sterility interaction of *Triticum aestivum* nucleous and *Triticum timopheevii* cytoplasm. Wheat Inform Service 14:29–30

Wright J, Reilley A, Labriola J, Kut S, Orton T (1996) Petaloid male-sterile plants from carrot cell cultures. Hortic Sci 31(3):421–425

Wu B, Xia Y, Zhang G, Wang Y, Wang J, Ma S, Song Y, Yang Z, Ma L, Niu N (2023) Transcriptomics reveals a core transcriptional network of K-type cytoplasmic male sterility microspore abortion in wheat (*Triticum aestivum* L.). BMC Plant Biol 23:618

Wych RD (1988) Production of hybrid seed corn. In: Sprague GF, Dudley JW (eds) Corn and corn improvement, 3rd edn. American Society of Agronomy, Madison, pp 565–607

Xiao S, Zang J, Pei Y, Liu J, Liu J, Song W, Shi Z, Su A, Zhao J, Chen H (2020) Activation of mitochondrial orf355 gene expression by a nuclear-encoded DREB transcription factor causes cytoplasmic male sterility in maize. Mol Plant 13(9):1270–1283

Xing M, Xu Y, Lu Y, Yan J, Zeng A (2024) Analysis of Rfo-mediated network in regulating fertility restoration in *Brassica oleracea*. Int J Mol Sci 25(22):12026

Xue Z, Xu X, Zhou Y, Wang X, Zhang Y, Liu D, Zhao B, Duan L, Qi X (2018) Deficiency of a triterpene pathway results in humidity-sensitive genic male sterility in rice. Nat Commun 9(1):604

Yamagishi H, Jikuya M, Okushiro K, Hashimoto A, Fukunaga A, Takenaka M, Terachi T (2021) A single nucleotide substitution in the coding region of Ogura male sterile gene, *orf138*, determines effectiveness of a fertility restorer gene, *Rfo*, in radish. Mol Gen Genomics 296(3):705–717

Yamagishi H, Hashimoto A, Fukunaga A, Takenaka M, Terachi T (2024) Identification and variation of a new restorer of fertility gene that induces cleavage in *orf138* mRNA of Ogura male sterility in radish. Theor Appl Genet 137(10):231

Yamamoto MP, Kubo T, Mikami T (2005) The 5′-leader sequence of sugar beet mitochondrial atp6 encodes a novel polypeptide that is characteristic of Owen cytoplasmic male sterility. Mol Gen Genomics 273(4):342–349

Yang JH, Qi XH, Zhang MF, Yu JQ (2008) MADS-box genes are associated with cytoplasmic homeosis in cytoplasmic male-sterile stem mustard as partially mimicked by specifically inhibiting mtETC. Plant Growth Regul 56:191–201

Yang J, Liu X, Yang X, Zhang M (2010) Mitochondrially-targeted expression of a cytoplasmic male sterility-associated *orf220* gene causes male sterility in *Brassica juncea*. BMC Plant Biol 10:231

Yang W, Lou X, Li J, Pu M, Mirbahar AA, Liu D, Sun J, Zhan K, He L, Zhang A (2017) Cloning and functional analysis of MADS-box Genes, TaAG-A and TaAG-B, from a wheat K-type cytoplasmic male sterile line. Front Plant Sci 8:1081

Yang Q, Nong X, Xu J, Huang F, Wang F, Wu J, Zhang C, Liu C (2021) Unraveling the genetic basis of fertility restoration for cytoplasmic male sterile line WNJ01A originated from *Brassica juncea* in *Brassica napus*. Front Plant Sci 12:721980

Yao N, Tada Y, Sakamoto M, Nakayashiki H, Park P, Tosa Y, Mayama S (2002) Mitochondrial oxidative burst involved in apoptotic response in oats. Plant J 30(5):567–579

Yasumoto K, Matsumoto Y, Terachi T, Yamagishi H (2008) Restricted distribution of *orf687* as the pollen fertility restorer gene for Ogura male sterility in Japanese wild radish. Breed Sci 58(2):177–182

Zhang Z, Zhu Y, Cao Y, Yu H, Bai R, Zhao H, Zhang B, Wang L (2020) Fine mapping of the male fertility restoration gene *CaRf032* in *Capsicum annuum* L. Theor Appl Genet 133(4):1177–1187

Zhang C, Fu F, Lin C, Ding X, Zhang J, Yan H, Wang P, Zhang W, Peng B, Zhao L (2021) MicroRNAs involved in regulatory cytoplasmic male sterility by analysis of RNA-seq and small RNA-seq in Soybean. Front Genet 12:654146

Zhang Y, Zhao T, Wang Y, Yang R, Li W, Liu K, Sun N, Hussian I, Ma X, Yu H, Zhao K, Chen J, Yu X (2023a) Expression characterization of ABCDE Class MADS-Box genes in *Brassica rapa* with different pistil types. Plants (Basel) 12(11):2218

Zhang P, Zhao Z, Zheng M, Liu Y, Niu Q, Liu X, Shi Z, Yi H, Yu T, Rong T, Cao M (2023b) Fine mapping and candidate gene analysis of a novel fertility restorer gene for C-type cytoplasmic male sterility in maize (*Zea mays* L.). Theor Appl Genet 136(11):234

Zhao T, Gai J (2006) Discovery of new male-sterile cytoplasm sources and development of a new cytoplasmic-nuclear male-sterile line NJCMS3A in soybean. Euphytica 152:387–396

Zhao L, Sun H, Huang M (1998) Breeding of ZA type cytoplasmic male sterile line and its preliminary study. Soybean Sci 17:268–270

Zhao Z, Ding Z, Huang J, Meng H, Zhang Z, Gou X, Tang H, Xie X, Ping J, Xiao F, Liu YG, Xie Y, Chen L (2023) Copy number variation of the restorer *Rf4* underlies human selection of three-line hybrid rice breeding. Nat Commun 14(1):7333

Zheng Y, Liu Z, Sun Y, Liu G, Yang A, Li F (2018) Characterization of genes specific to sua-CMS in *Nicotiana tabacum*. Plant Cell Rep 37(9):1245–1255

Zheng M, Yang T, Liu X, Lu G, Zhang P, Jiang B, Zhou S, Lu Y, Lan H, Zhang S, Li C, Rong T, Cao M (2020) qRf8-1, a Novel QTL for the fertility restoration of maize CMS-C identified by QTL-seq. G3 (Bethesda) 10(7):2457–2464

Zhou L, Mao YC, Yang YM, Wang JJ, Zhong X, Han Y, Zhang YF, Shi QS, Huang XH, Meyers BC, Zhu J, Yang ZN (2024) Temperature and light reverse the fertility of rice P/TGMS line ostms19 via reactive oxygen species homeostasis. Plant Biotechnol J 22(7):2020–2032

Chapter 15
Space Plant Biology: Growing Crops Beyond Planet Earth

Mohana Sai Akula and Penna Suprasanna

Abstract The study of plant biology in environments beyond Earth's gravity emerges as a crucial scientific pursuit as humankind prepares for establishing a permanent presence on the Moon and Mars. Migrating from short durations in space to creating permanent habitats to facilitate life around celestial bodies elicits strong research focus on space food security. Space agriculture has extended beyond survival to include mental health, cultural identity, and the shaping of habitation in the future. Plants not only are vital for sustenance but also play a key role in oxygen production, water recycling, and psychological well-being, which are essential for prolonged human space missions. Transitioning from the gravitational conditions of Earth (1 g) to the altered gravity environments of space travel poses fundamental challenges for plant growth, morphology, cellular orientation, and biochemical processes. Space plant biology, thus, is not only about growing crops on planets but also about nurturing life systems while navigating the cosmos. This chapter explores a spectrum of phenomena, from the influence of gravity in evolutionary processes on Earth to plant adaptations under conditions of microgravity and hypergravity. We assemble information from space experiments conducted on the International Space Station (ISS), BIOS-3, Veggie, and China's Tiangong and augment it with ground-based studies using clinostats, RPMs, and centrifuges. This chapter also summarizes the evolution of human food systems in space, from simple, packaged foods to complex, self-sustaining, culturally integrating bioregenerative systems. Combined with knowledge gained from experiments on board the ISS and Tiangong, and in the EDEN International Space Station (EDEN-ISS) – and Mars Desert Research Station (MDRS) – simulated environments, crops grown in environmentally challenging habitats are more desirable for space food security. In addition, this chapter addresses the broader relevance of agricultural technology for space applications.

M. S. Akula
Amity Institute of Biotechnology, Amity University Maharashtra, Mumbai, India

P. Suprasanna (✉)
Amity Center for Nuclear Biotechnology, Amity Institute of Biotechnology, Amity University Maharashtra, Mumbai, India
e-mail: psuprasanna@mum.amity.edu

A. Singh et al. (eds.), *Sustainable Crop Production*, Sustainability Sciences in Asia and Africa, https://doi.org/10.1007/978-981-95-7844-3_15

Keywords Space plant biology · Microgravity · Crops · Plant growth · Plant adaptations

Editor's Notes

As the planet reaches its tipping point, meeting the demands of an ever-exploding population, and, growing food in space is capturing the imagination of scientists. Inter-galactic space missions and establishing human settlements beyond planet Earth seem to be gaining form inspired by science fiction as space technologies progress. The ability to grow crops and idea of *"space food security"* is becoming a reality in what may be defined as "Astro-Anthropocene" era.

The chapter on Astrobiology by Akula and Suprasanna is an extremely valuable contribution intended to inspire students, scholars, researchers, policymakers, and general public. Beyond testing fundamental tenets of agriculture, growing crops in space will require assessment of basic biological principles such as growth and development in microgravity, establishment of polarity, adaptive features, to list a few. Progress in astrobiology is also intrinsically linked to sustainable developmental goals (SDGs) and food sovereignty. Successful astrobiological experiments of growing crops in space and beyond planet earth is also a testament to technological marvels and AI-powered crop production. The authors hope that the chapter stimulates the minds of all stakeholders.

Actionable Points

1. Testing and optimization of modular greenhouses, automated hydroponic systems, and artificial intelligence (AI)-power crop growth monitoring
2. Establishing facility for growing rapid cycling crop varieties, with high nutritional value, in compact spaces with greater resource use efficiency
3. Testing Closed-Loop Systems for recycling oxygen, CO_2, and water and waste materials in Bioregenerative Life Support System (BLSS)
4. Optimizing soil-less space farming utilizing hydroponics, aeroponics, aquaponics
5. Undertaking experiments in microgravity and variable gravity for investigating developmental and adaptive processes such as stress responses, establishment of polarity, hormone gradients, gravitropism, plant architecture
6. Incorporating concepts of living architecture such as greenhouses with photosynthetically active walls composed of mosses or algae for air purification; glass houses that permit natural light, and crop beds that act as humidity regulators

15.1 Introduction

Space exploration has been focusing on propulsion systems, life support, and scientific instrumentation; however, now that human missions are evolving from short stays to permanent residences on the Moon, Mars, and beyond, food, which was previously just a logistical requirement, is now being considered something that is essential for living, autonomy, and mental health in off-world living. In this new era of interplanetary exploration, ensuring food security in space goes beyond a mere technical requirement; it poses a sociocultural and ecological challenge intricately related to terrestrial food systems (Meinen et al. 2018; Watkins et al. 2022). Since the first days of human space travel, astronauts have relied on vacuum-packed, shelf-stable foods engineered for nutrition, energy, and efficiency. Although well practiced in early space missions, this method is becoming progressively inadequate for future missions. Extended human missions to Mars and the emerging possibility of permanent lunar settlements highlight the limitations of the stored foods, which include deterioration of nutrients, the absence of freshness, and mental fatigue (Zeidler et al. 2021; Shaw and Soma 2022). Also, a human community that cannot produce its own food defies the basic idea of space self-sufficiency. Thus, a new idea arises, centering on food production rather than food transportation, which gradually leads to food sovereignty in space and on planetary surfaces. Astronauts cannot just eat, but they will have to grow life, play a role in waste management, purify water, and eat the food they grow. These systems must be robust, compact, automated, and ecologically harmonious. Deep space food security is closely connected to planetary sustainability on Earth. Innovations pioneered for growing food in outer space, including hydroponics, aeroponics, energy-efficient lighting, and closed-loop nutrient systems, are rapidly transforming food production methods everywhere from cities to deserts to regions with difficult climates (Agritecture 2023). Agriculture beyond this world is no longer just for astronaut nutrition, but for environmentally sound food production back on Earth encompassing research into future food security in space through the lenses of plant biological innovations, environmental engineering, psychological fortitude, and global sustainability.

15.2 Rethinking Food Security Beyond Earth

As humans continue to explore beyond our planet, through long-term missions on the Moon, Mars, and deep space, the ability to sustainably have food resources is a growing priority in mission design. The magnitude of the term "food" in space is changing from a supplied consumable to a living, self-sustaining system ensuring nutritional value, emotional well-being, cultural familiarity, and ecological balance (Douglas et al. 2021). Pre-cooked foods that have been a mainstay of astronaut diets are found to be lacking for missions to come (Tang et al. 2021). Freeze-dried and thermostabilized foods offer a bare minimum of caloric intake but decline in nutritional value with storage time, especially in essential vitamins C and B1 (Zeidler et al. 2021). In addition, long-term reliance on repetitive, unappetizing meals leads to appetite fatigue, reduced food intake, and possible emotional detachment from regular meal times. For missions to Mars of 3 years and beyond, relying solely on Earth rations is a logistical problem and a nutritional hazard. Findings of studies conducted on the International Space Station (ISS) and the European Space Agency

(ESA)'s Concordia station indicate that astronauts lose decreasing interest in food as sense qualities degrade, which then influences their mood and overall performance (Poulet et al. 2021). In addition, systematic review of spaceflight nutrition established that preloaded food fails to meet recommended daily allowance for such micronutrients as vitamin D, iron, and folate, resulting in long-term health problems such as bone demineralization, anemia, and immune suppression (Ross et al. 2023). In situ fresh crop production, therefore, is a prime requirement for deep space travel. Regenerative food systems must be brought into future habitat planning in order to meet space travelers' evolving needs—nutritionally, emotionally, and operationally.

In isolated environments such as space, the cultivation, care, and harvest of plants serve as therapeutic activities, often described as "green time", that help reduce stress and enhance resilience (Kordyum and Hasenstein 2021). Crews have been shown to have higher morale with the ability to grow herbs or leafy greens, with findings from analog missions like HI-SEAS and CHAPEA, suggesting the impact of plant growing on crew morale (Ferl and Paul 2016; Wheeler 2017). Beyond cultural identity, onboard food cultivation encourages functional independence. Oxygen is produced, carbon dioxide is absorbed, and water is regenerated by transpiration of crops grown in closed-loop systems. The role of plant modules as integrated components of more complex life support systems, such as bioregenerative systems similar to those currently developed by NASA (Veggie), the European Space Agency (ESA—MELiSSA), and Russia (BIOS-3), exemplifies the success of such designs, which increase not only survival rates but also the agency and involvement of the crew (Wheeler 2010; Carillo et al. 2020; De Micco et al. 2026).

15.3 Space Farming, Global Food Security, and the SDGs

Agriculture in space might instantly reply to challenges beyond the home planet, but it also offers flexible solutions to major Earthly problems. Innovations initially developed for outer space agriculture such as hydroponics, vertical farming, closed-loop irrigation, and efficient LED systems have since been applied to food-insecure and climate-vulnerable regions on Earth (Agritecture 2023). This integration brings together the focus on our outer space surroundings with the United Nations Sustainable Development Goals (SDGs). Space spin-offs in agriculture promote the sustainable, decentralized, and resilient food production systems necessary for achieving SDG 2 (Zero Hunger), SDG 6 (Clean Water and Sanitation), and SDG 13 (Climate Action). One of the most notable activities includes the creation of a greenhouse by EDEN-ISS in Antarctica, which not only acts as an Earth-based analog to extraplanetary agriculture but also creates a framework for terrestrial food production in extreme or isolated areas (Schubert et al. 2021). Such advances underscore that food security on Earth and in space are not disparate pathways but inextricable destinies. Valuable knowledge from space exploration will guide how we tackle problems such as drought, desertification, and malnutrition in the coming times.

15.4 The Shift from "Food Transport" to "Food Sovereignty" in Orbit and in Outer Space

Food sovereignty is about the rights and the ability to produce and manage food systems according to the values, needs, and self-determination of the people involved. This principle acquires an especially crucial character within the framework of space exploration. As we aim to establish permanent outposts on other celestial bodies, like the Moon or Mars, dependence on Earth-based supply chains is both inefficient and vulnerable to disruption. Hence, space missions are starting to implement in situ crop production, water recycling systems, and composting technologies that use microbial activity to recover key nutrients (Meinen et al. 2018; Watkins et al. 2022). Current innovations being tested for space applicability include modular greenhouses, automated hydroponic systems, and artificial intelligence (AI)-powered crop monitoring (Romano and Aronne 2021). The future demands that crews of astronauts will not just eat food, but they will also grow plant ecosystems for sustenance. Additionally, this cultural shift enables dietary practices to be tailored to the sociocultural, physiological, and psychological aspects of crew members. Space agriculture thus becomes not just a logistical solution but a means of preserving human dignity, diversity, and adaptability, even in the extraterrestrial environments.

15.5 The Role of Plants in Sustained Space Missions

Plants offer one potential source of fresh, nutrient-rich food, when traditional food supply chains collapse. Growing crops in space not only is for immediate nutritional requirement but also is an integral part of the well-being of astronauts. Green leafy vegetables, legumes, and fruit-bearing crops supply vitamins, fibers, and phytochemicals that are essential to support the health of astronauts, their immune system, and for metabolic balance (Wheeler 2010, 2017). Fresh produce availability attenuates the dietary decay observed with prolonged reliance on pre-packaged space food, leading to improved health and life during extended space missions (Monje et al. 2003; Meinen et al. 2018). There have been vegetables, food crops, leafy plants which have been grown on ISS (Johnson et al. 2021), which highlight the features and potential benefits that can be accrued in space (Fig. 15.1). The prime focus is on relative ease of growing plants in space (compact spaces), securing high source of nutrients, and harvesting in less time periods with greater efficiency.

Zabel et al. (2021) note that the plant systems must not only be physically incorporated into space environments but also be genetically acclimated to their limits. Plant functional traits such as low canopy height, high harvest index, increased efficiency in nutrient acquisition, and consistent photoperiodic response (Fig. 15.2) will have to be selectively bred to render plants more compatible with bounded and

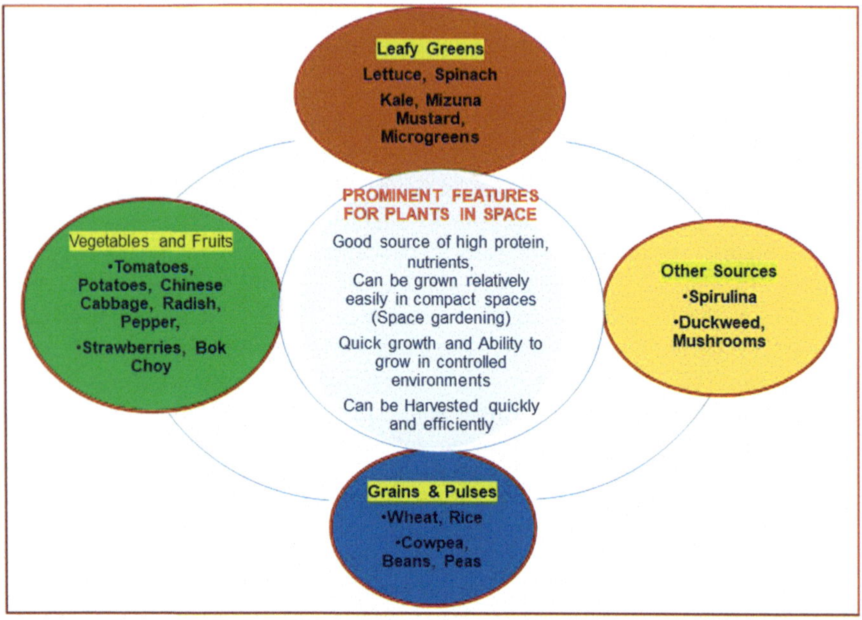

Fig. 15.1 Plant and beneficial features for space farming

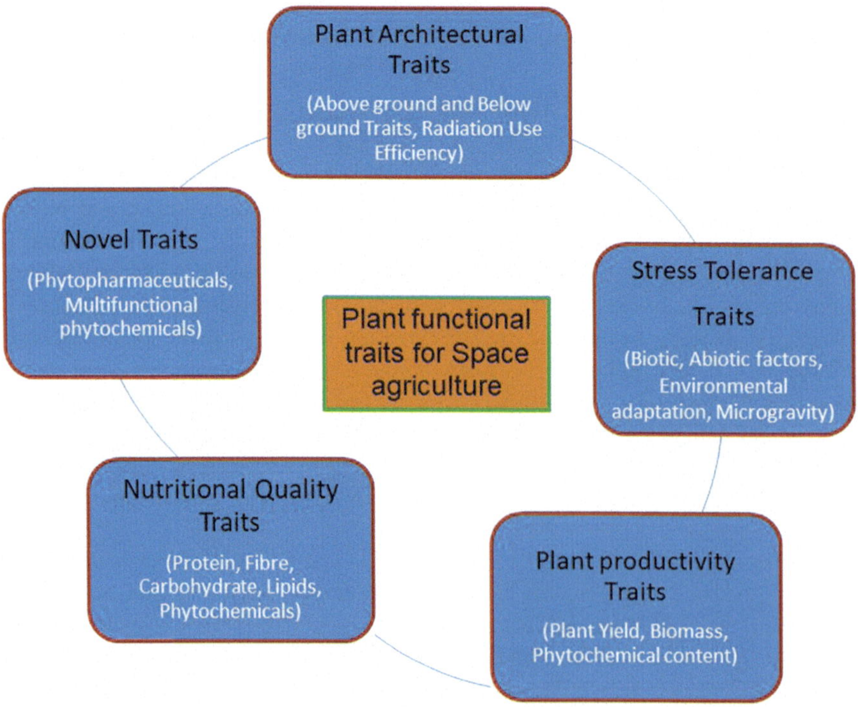

Fig. 15.2 Plant functional traits for space agriculture

energy-constrained environments. Seeds of several seed-propagated crops (*Arabidopsis*, pea, wheat, brassica) and propagules from vegetatively propagated plants (potato and sweet potato) have been tested for growth and development in microgravity under diverse experimental conditions, suggesting that plants can be grown productively in space (Mohanta et al. 2021; Kordyum and Hasenstein 2021). Further, it was observed that *Arabidopsis*, pea, wheat, and brassica were grown in orbit, including two generations and four generations in the case of pea. It was also interesting to see from the growth data that there was no significant difference in growth, development, and propagation, but yield and seed development were affected under microgravity conditions (Kordyum and Hasenstein 2021).

In the Indian context, the addition of agriculture to the successes of space exploration highlights a strategic perspective. This is a unique convergence between India's agricultural profundity and its visionary space program, where ancient farming wisdom meets cutting-edge technology. With over half the population involved in farming and increasing attention being paid to food security, transposing terrestrial agricultural techniques to non-terrestrial environments seems both plausible and progressive (ISRO 2025). The Indian Space Research Organisation (ISRO) has been embarking upon low-cost missions like Chandrayaan and Mangalyaan. In the preparations for independent space stations and crewed missions to other planets, the role of Bharatiya Krishi in enhancing the national space ambitions can hardly be exaggerated (Pandith et al. 2023). The agricultural resourcefulness of India will certainly shape the human space missions, be it through hydroponics, aeroponics, or bioregenerative systems.

As we embark on longer beyond-Earth expeditions to the Moon or onto Mars, the essentiality of long-lived life support systems that are both compact and self-sustaining, and that are intimately linked to biological processes, becomes significant. In this scenario, plants fulfill a key, yet irreplaceable, function. A biotechnology-based approach to overcome these problems is using a Bioregenerative Life Support System (BLSS) (Fig. 15.3), which utilizes biological components, mainly plants and microorganisms, to regenerate air, water, and waste materials (Ferl et al. 2002; Wheeler 2017; De Micco et al. 2023). However, there are several physical and energy constraints that need to be considered for BLSS (Carillo et al. 2020). Unlike traditional physico-chemical systems that use mechanical filter, BLSS leverages natural process found on Earth like photosynthesis to actively recycle the carbon dioxide output by astronauts using as input for such process, producing oxygen. Additionally, the water released through the transpiration of plants can be harvested and reused to improve the reclamation process in closed habitats (Zabel et al. 2016).

15.6 Gravity as a Determinant in Terrestrial Plant Development

Gravity is a key and universal environmental signal that drives the overall plant's response and directional growth of terrestrial plants. It provides an essential directional cue controlling tissue polarity, dictating nutrient and hormone spread, and aligning

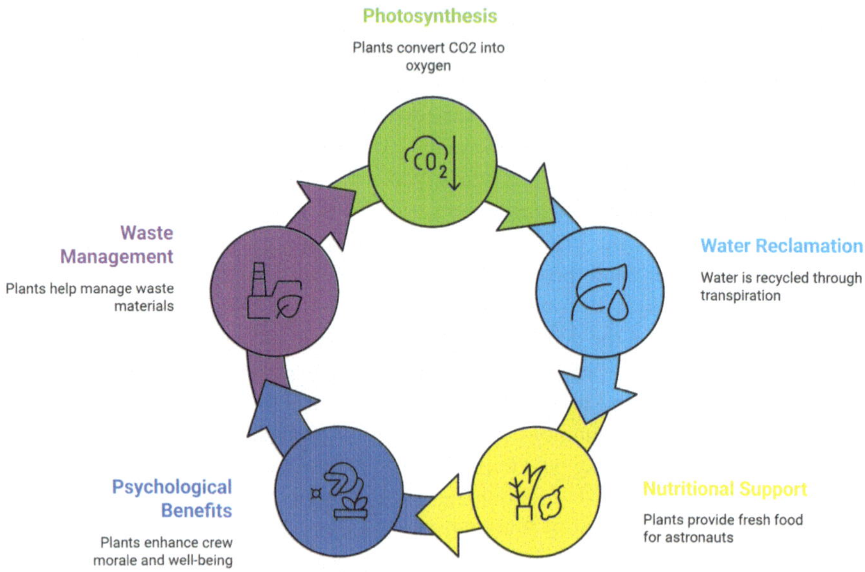

Fig. 15.3 Features of bioregenerative life support systems

major developmental events from seed germination through maturity (Ferl and Paul 2010; Farooq et al. 2024). Studying how plants perceive and respond to gravity is essential to understand the biological dysregulations that occur in microgravity and other altered-gravity-induced abnormalities observed during space exploration missions. One example of the biological influence of gravity is seed germination, during which seeds activate a cascade of metabolic and physiological processes leading to radicle emergence and seedling establishment. Gravity causes the radicle to grow positively gravitropically (downward) anchoring the plant in the soil, while the plumule or shoot grows negatively gravitropically (upward) to access light for photosynthesis (Paul et al. 2013; Hoson 2014). This root-shoot polarity is essential for a properly oriented seedling and is important for spatial orientation and resource acquisition at the full-grown plant level.

Statocytes are specialized cells in the root cap and shoot endodermis that mediate the perception of gravity. In these cells, amyloplasts, dense organelles rich in starch, settle down in accordance with the pull of gravity. This mechanical sedimentation activates mechanosensitive ion channels that elicit intracellular signaling cascades of calcium influx, ROS, and changes in auxin distribution. By regulating differential cell elongation, these gradients of auxin lead to the bending of roots and shoots depending on the specific tropic response exhibited (Kiss et al. 2019). Apart from giving directionality, gravity is essential to vascular transport systems (Farooq et al. 2024). It affects xylem tension and phloem loading, allowing the one-directional transport of water and minerals from the roots to the shoots as well as the downward flow of photosynthates. Disruption of these transport systems in microgravity or clinorotation settings has resulted in abnormal nutrient transport, diminishment of growth rates, and morphological abnormalities (Zabel et al. 2016; Zheng et al. 2015). Similarly, gravity controls the spatial distribution of plant hormones (gibberellins, cytokinins, ethylene, abscisic acid (ABA)) and their sensitivity. In normally grown plants, these hormones together regulate plant structure, flowering, aging, and responses to environmental stresses. In space, however, altered auxin transport and altered hormone signaling lead to abnormal growth patterns such as skewed shoot orientation, looping roots, and delayed vascular tissue formation (Monje et al. 2003).

Another prime consequence of changed gravity forces appears in cell morphology. Microtubules and actin filaments form the cytoskeleton, which dynamically adapts to gravitational forces. Elevated levels of reactive oxygen species, inducible nitric oxide synthase, and mislocalized markers of specific cellular compartments are all consequences of microgravity-induced disturbance of cytoskeletal organization, irregular deposition of the cell wall, and atypical orientation of cell division. This has notable effects such as changes in cell shape, tissue thickness, and even organ formation, which are often indicative of the alteration of internal positional cues in the tissues (Kirk et al. 2022). In conclusion, gravity is not just a passive background signal, but rather a key regulatory signal for plant development (Muthert et al. 2020). Its influence ranges from molecular signaling to gross whole-organism morphology. In microgravity or weightless conditions, such as those experienced during spaceflight, plants are forced to make complex physiological and developmental adaptations (Maffei et al. 2024), which we will learn more about in the following sections of this chapter.

15.7 Altered Gravity Conditions Encountered in Spaceflight

The constant 1 *g* force of gravity on Earth is an important directional cue that plays a critical role in the development of plants. The adaptive ability of plants is influenced by the various gravitational conditions, including microgravity, hypergravity, and partial gravity, that they are subjected to during an ensuing spaceflight (Muthert et al. 2020).

15.7.1 Microgravity (0–10^{-3} g): Explorations in the ISS and Deep Space

Microgravity (μ*g*) refers to a condition of weak gravitational force, usually less than 10^{-3} *g* of the normal gravitational force on Earth (Zheng et al. 2015). This is a regular experience aboard ships such as the International Space Station (ISS) and even on deep space missions. The microgravity environment lacks a dominant gravitational vector, which disrupts normal processes critical for plants, including gravitropism, fluid transport, and cellular organization (Vorselen et al. 2014; Ferl and Paul 2016). Sedimentation of amyloplasts in statocytes, which are essential for gravity perception, is disrupted in microgravity, resulting in modified redistribution of auxin. This change leads to aberrant root orientation, curling shoots, and low polarity (Kiss et al. 2019). Additionally, water transport changes from gravity driven to capillary dependent, which can complicate systems such as hydroponics and root zone wetting (Monje et al. 2003). Studies of plant growth in the ISS, particularly within the framework of the NASA Veggie project and the ESA Plant Habitat experiments, have revealed various phenomena, including diminished root anchorage, delayed flowering, and altered gene expression after prolonged exposure to microgravity (Paul et al. 2013; Zabel et al. 2016).

15.7.2 Hypergravity (>1 g*): Learnings from Centrifuge-Based Platforms*

Hypergravity environments describe situations in which gravitational acceleration is greater than Earth 1 *g*, usually accomplished via centrifuges or suborbital launch trajectories. The same hypergravity conditions serve as useful models to simulate stress conditions of biological systems and also provide information on the sensitivity of plants to changes in gravity levels (Zheng et al. 2015; Farooq et al. 2024). Plants have been found to exhibit shorter hypocotyl lengths (i.e., pushing stems) and increased lignification in increased gravity experiments conducted on centrifuge platforms (Wolverton and Kiss 2009), while improved root growth orientation also increases (Swamy et al. 2021), which could indicate higher gravity strengthens tropic responses. Hypergravity affects cytoskeletal organization at the molecular level; it promotes the accumulation of reactive oxygen species (ROS) and changes hormonal levels (auxin, gibberellin, etc.). Simulated hypergravity platforms are of great use not only in basic research but also for preparing biological specimens to better endure deformation during the stresses of a space launch and to understand how plants adapt to changing gravity conditions and stresses, situations that might be encountered in a reentry vehicle or with planetary centrifuge systems.

15.7.3 Partial Gravity: Moon (0.16 g*) and Mars (0.38* g*)*

Partial gravity, which means a gravity force greater than microgravity but lower than the Earth's gravity level, is closely related to future agriculture in the extraterrestrial environment (Farooq et al. 2024). The Moon (0.16 *g*) and Mars (0.38 *g*) are environments that provide gravity and can be reduced. These environments serve as permanent habitats where plants must grow, thrive, and recycle resources efficiently for the success of the mission (Maffei et al. 2024). Partial gravity may coordinate statocyte settling, auxin transport, and directional growth, but not as efficiently as on Earth. Partial gravity is expected to have intermediate effects on biological responses relative to microgravity and full Earth gravity and may yield unique expression profiles, growth patterns, and hormonal signaling pathways (Zabel et al. 2016).

Due to the prolonged timescales of surface missions, it is vital to explore partial gravity farming systems that address a variety of design components, including root stability, nutrient segregation, and structural support (Kordyum and Hasenstein 2021). These systems will also need to manage issues related to lower transpiration rates and uneven nutrient distribution that may limit the growth of the plants. Gravity, fundamentally, not only is a gradation of environments but also presents different challenges to plant growth in space. An in-depth understanding of plant responses across this gravity continuum is essential for the development of resilient crops and growth techniques applicable to future off-planet farming.

15.8 Tropic Responses and Spaceflight-Induced Modifications

On planet Earth, plant growth is impacted by external variables like gravity, light, water, and touch. Examples of important growth responses are gravitropism and phototropism. For example, gravitropism leads plants to grow their roots downward and absorb nutrients from the soil, and phototropism helps a plant grow its shoots toward the light for photosynthesis (Muthert et al. 2020; Luigi et al. 2022). These responses, though, are profoundly altered in space conditions, especially microgravity. A comprehensive knowledge of the mechanisms that underlie how these tropisms are modified in the absence or perturbation of gravity is essential for engineering successful agricultural systems for long-duration space exploration missions (Reed and Vanden Bosch 2023).

15.8.1 Gravitropism: Orientation of Root and Shoot in Microgravity

Generally, roots show positive gravitropism, whereas shoots show negative gravitropism. Consequently, roots grow toward important soil resources and shoots toward light, both of which are critical for plant survival and growth. The microgravity conditions on board the International Space Station (ISS), where gravitational orientation is largely a limiting factor, lead to a lack of sedimentation of amyloplasts in statocytes and interference with gravity reception (Ferl and Paul 2016). This leads to changes in auxin gradients, irregular curvature of organs, and loss of directional growth, which ultimately results in roots growing in random directions and shoots failing to orient themselves consistently. In fact, both through ISS experiments in microgravity conditions and via ground-based clinostat studies, impaired gravitropic responses were observed in a number of species, including *Arabidopsis* and wheat, leading to reduced total root elongation (Kiss et al. 2019). The described changes also make it difficult to anchor plants and efficiently uptake nutrients, which could present major challenges to the agriculture systems of interplanetary missions.

15.8.2 Phototropism on Altered Gravity Environment

Under microgravity, phototropism commonly supersedes the gravitotropic stimuli when there is no or minimal influence from the primary stimulus of gravity (Luigi et al. 2022). Plants grown in space environments have adapted to rely more on light-related cues for their growth-regulating needs. A relationship between light dynamics and plant growth regulators that is taken for granted here on Earth is particularly influential in the controlled environment agriculture systems that are presently developed for the International Space Station (ISS) under LED components to control how plant structures are shaped. In space, where there is no gravity, receptors of blue light or phototropins are required to assist in determining shoot direction (Zabel et al. 2016). But this adaptation has limits, as, for instance, in root systems, which typically respond weakly to light signals. It is an interesting phenomenon, as studies have suggested that phototropic responses can be enhanced when plants are exposed to microgravity (Paul et al. 2013), due to the plant's need to establish spatial coordinates. This finding opens up a new direction to ultimately design light-driven directional systems to sculpt plant architecture in a space habitat.

15.9 Effect on Auxin Transportation, Signaling, and Oriented Growth

Plants show growth responses to environmental stimuli, including gravity and light, which are controlled by auxin movement and signaling pathways (Casal and Estevez 2021). Auxin, one of the key plant hormones, establishes growth polarity by differential gravity- or light-induced accumulation across a tissue, leading to differential cell elongation and curvature (Mroue et al. 2018). Molecular and spatial reorganization of auxin transport under microgravity conditions leads to misregulated signals and incorrect growth trajectories. For example, when exposed to new gravitropic signals, targeting of auxin transport mechanisms fails, leading to erroneous growth responses (Monje et al. 2003). Lack of amyloplast sedimentation impairs the proper localization of PIN (PIN-FORMED) proteins (responsible for auxin efflux) and decreases the plant's potential to coordinate its structure. Additionally, microsatellite gravity has been shown to influence different aspects of cellular signaling, including cytoskeletal integrity, calcium flux, and environmental sensing-related gene expression (Khodadad et al. 2011). Such perturbations can delay developmental timelines, reduce biomass production, and compromise plant health. Incorporating phototropism as a possible second type of directional growth is promising but not sufficient. If we are to develop truly resilient space agriculture, engineers must consider not only light orientation and auxin manipulation but also substrate design and feedback-controlled growth environments in compensation for the lack of gravitation input.

15.10 Physiological and Biochemical Adaptations in Microgravity

Even under microgravity and without extreme mechanical stress, plants are under considerable stress (Baba et al. 2022). The absence of consistent gravitational orientation provokes a range of physiological and biochemical responses that influence plant function at organismal, tissue, and molecular levels. These modifications include changes in stress signaling pathways and structural changes in important organelles, shedding light on how plants can adapt to stressful environments (Zheng et al. 2015). In this section, we outline the various physiological changes that plants undergo as a result of microgravity, based on the results obtained from crops grown in space and ground-based studies.

15.11 Stress-Induced Biochemical Shifts (e.g., ROS, Antioxidants, and Proline Accumulation)

Plants are subject to oxidative stress in low-gravity environments. This is due to alterations in cell communication and reduced metabolic capacity of organelles to produce large amounts of ROS species like superoxide anions, hydrogen peroxide, and hydroxyl free radicals (Khodadad et al., 2011; Mantri et al. 2012; Suprasanna et al. 2021). The excess ROS can damage proteins, lipids, and nucleic acids and disrupt normal physiology. Plants respond to this by upregulating their antioxidant defense systems, including enzymes (e.g., superoxide dismutase (SOD), catalase, and peroxidases) and nonenzymatic antioxidants (e.g., flavonoids and carotenoids). Moreover, as an osmoprotectant and stress marker, proline is accumulated in a comparatively higher concentration under low-gravitational environment (Zabel et al. 2016), thus helping in osmoprotection and free radical scavenging. These biochemical changes reflect the efforts in restoring stability to the plant and serve as effective indicators for evaluating the adaptability and survivability of crops in space environments. Arena et al. (2014) compared plant and mammalian cellular defense responses to space radiation and concluded that plants invoke sophisticated defense responses like DNA repair mechanisms (Geraskin 2024), osmoprotectant accumulation, and controlled programmed cell death. Interestingly, plant cells were more tolerant to radiation damage, a characteristic that can be utilized in space crop selection and genetic engineering protocols.

15.12 Cellular Reorganization, Gene Expression Changes, and Altered Metabolic Pathways

Microgravity conditions lead to massive reorganization of internal structures at the cellular level with reorganization of cytoskeletal elements (microtubules, actin filaments) on the most significant scale (Lopez Garzon et al. 2025). Such disruption disturbs the spatial organization of organelles like the nucleus, mitochondria, and Golgi bodies, affecting cellular polarity and intraorganellar transport (Kirk et al. 2022). Experiments investigating gene expression from space-grown *Arabidopsis*, wheat, and lettuce have revealed extensive transcriptional reprogramming. Genes involved in cell wall remodeling, photosynthesis, auxin signaling, and stress signaling pathways are commonly up- or downregulated, indicating that the plant adapts its metabolism (Paul et al. 2013). Significantly, the modified levels of gravity modulate auxin transport, which then impacts growth and morphogenesis. In addition, adjustments in both primary and secondary metabolism are observed, reflected as reprogrammed sugar biosynthesis, amino acid pathways, and secondary metabolite production (Tuominen et al. 2009; Hou et al. 2025). These changes could affect not only plant development but also nutritional content and flavor profiles of edible crops grown in space.

15.13 Structural Changes in Chloroplasts, Cell Walls, and Root Hairs

Under microgravity conditions, organelles like chloroplasts undergo structural changes (Takemura et al. 2017). These changes include disorganization of grana stacking, increased chlorophyll content, and uneven distribution of chlorophyll within mesophyll cells. Such adaptations can negatively impact the physiology of photosynthesis and light energy capture in some of the ancient branching groups of plant species that are sensitive to fluctuations in light intensity (Ferl et al. 2002). In a similar way, changes in the expression of cellulose synthase complexes and wall-associated kinases result in rearrangements in the composition of cell wall architecture when plants are exposed to an altered gravitational environment. These result in thinner and less strongly desired walls, making the plant more weakened structurally and more sensitive to pressure of the environment. In many plants grown under microgravity, root hairs are observed to be fewer in number and shorter in length, which lead to insufficient nutrient absorption and inadequate anchorage in substrate systems (Kiss et al. 2019). These structural inexactitudes pose significant barriers to closed-loop sustainable agricultural production systems, which require complete manifestation of all growth parameters in order to obtain the best yield.

15.14 Notable Space-Based Cultivation Experiments: A Global Perspective

Over the past 60 years, a range of space agencies across the globe have executed cutting-edge plant-growing experiments in a microgravity environment to assess the potential of crops for long-duration human spaceflight (Kordyum and Hasenstein 2021). Each of these experimental systems offers valuable insights, right from the early life support systems developed in Russia to the modern era with the rotary orbital greenhouses on both the ISS and the Tiangong station. This contributes to understanding how plants adapt, and whether biomass can be produced, and if closed-loop bioregenerative systems are feasible. This section focuses on key programs, those that have made notable contributions to space agriculture.

15.15 BIOS-3 (Russia): Early Bioregenerative Systems

The BIOS-3 project, an early example and a truly ambitious attempt to build a closed-loop ecological life support system, was developed in the former Soviet Union during the 1960s and 1970s (Salisbury et al. 1991, 1997; Polyakov et al. 2009; Liu et al. 2021a, b; De Micco et al. 2023). Located in Siberia, BIOS-3 consisted of enclosed growth chambers in which humans lived alongside crops. Plants were integral within the chambers because they would regenerate oxygen, absorb carbon dioxide, and deliver sustenance (Massa et al. 2016). Wheat, beet, carrots, and onions were grown hydroponically with artificial lighting. However, even though BIOS-3 was floor based, it closely replicated the obstacles to human habitation in space and proved that human life could be sustained solely through bioregenerative systems for long periods, up to 180 days. This trailblazing mission established the groundwork for plant-human symbiosis in space, providing valuable data on recycling resources, maintaining oxygen levels, and revealing the details of space farm logistics.

15.16 Veggie (NASA): Leafy Crops on the ISS

The Veggie system is one of several projects for growing plants in space and is currently the largest one on the International Space Station (ISS). Launched in 2014, Veggie utilizes simple and flexible hydroponics technology on pillows and red, blue, and green light-emitting diodes to promote photosynthesis (Zabel et al. 2016). Veggie has successfully grown a number of leafy greens, including red romaine lettuce, mizuna mustard, and zinnia flowers (Massa et al. 2017). Not only did these crops serve as a supplement to astronauts' diets, but they also provided psychological comfort on long spaceflights (Monje et al. 2003). The Veggie experiments showed that microgravity does affect the orientation of plant roots and distribution of water, but careful control of environmental conditions and the manner of delivering light can produce edibles of high quality and nutritional value. Analysis of microbes revealed that properly nurtured plants were safe to eat in space, so astronauts dined on their first official space "salad" in 2015.

15.17 Lada Greenhouse (ISS) and Chinese Tiangong Station: Plant Growth Modules

Lada greenhouse: This Russian-developed facility has been established as an advanced plant cultivation system akin to the ISS. It has been used in the cultivation of pea, wheat, and barley while collecting data on root morphology, water use efficiency, and photosynthetic responses in a microgravity environment (Ferl and Paul 2016). The Tiangong Station of China has taken its first steps toward space growth experiments with rice and *Arabidopsis* cultivated on board Tiangong-2. The stations seek to better understand plants as they grow and flower and produce seeds for future deep-space journeys and lunar outposts. Information from both platforms is vital to plant growth in space, nutrient recycling, and practical issues in keeping crops in the pressurized environment of orbital modules.

15.18 Analog and Ground-Based Simulation Models for Gravitational Research

The study of plant responses under altered gravity is intrinsically limited by the logistical, scientific, and temporal complexities common to all space missions (De Pascale et al. 2021). As a result, ground-based analogs and simulation models have become critical tools in space biology studies. These models allow researchers to mimic aspects of microgravity, hypergravity and space-related conditions in controlled Earth environments, allowing for long-term, replicable, and ethically sound experimentation. This section explores the environments and tools that provide insight into plant biology in systems that mimic some aspects of space.

15.19 Use of Clinostats, RPMs, and Centrifuges

Clinostats, one of the oldest and most widely used methods to mimic microgravity, are devices that create a lack of static gravitational force (Borstand Van Loon 2009) by rotating the platform to change the sample orientation continuously. Unlike a 2D clinostat, which has a rotation around just one axis, more sophisticated 3D random positioning machines (RPMs) rotate on two independent axes to average gravitational vectors to an approximate near-zero (Kiss et al. 2019). These controls are especially useful to study early plant growth, for example, seed germination, root bending, and hormone redistribution. Though not a theoretical mirror of microgravity (i.e., lacking fluid dynamics and rendering no GCR, which means inferring interactions from ground), they offer insights into gravity perception and signaling pathways, for example, involving auxin gradients and statocyte behavior. In contrast, centrifuges utilize high rotary speeds to create hypergravity, subjecting the samples to amplified gravitational force. These platforms allow scientists to study how plants respond to increased gravitational forces, showcasing opposite views to what was presented in microgravity research. It is noteworthy that hypergravity has been shown to promote cell wall thickening, lignification, and root architecture adaptations that are essential for agricultural systems on planetary surfaces with higher gravity compared to Earth or while in launch and reentry environments in space travel (Wolverton and Kiss 2009).

15.20 Ground-Based Analog Stations

Analog stations, also working beyond hardware devices, simulate the environmental, operational, and psychological aspects of space missions. These have evolved from anthropocentric facilities of recent times to bioregenerative testbeds in the case of space agriculture. Lunares, located in Poland, is a simulated lunar and Martian habitat in which systems for plant growth are tested in limited light, limited volume, and isolation of crew. The combined capabilities allow the integration of crew with controlled environment agriculture systems, monitoring, and resource restrictions, thus yielding a comprehensive simulation of planetary mission-like environments. The ASTRA-LM (Analog Station for Terrestrial Research in Astronautics—Lunar & Martian) at Ladakh, India, aims to be the highest-altitude analog research station in the world (ITSD 2024). The Lunar Soil Simulant (LSS) will house modules for space farming, EVA training, and human factors and provide a ground for the evaluation of plant-soil systems (LSS + cocopeat, for example) subjected to extreme temperature variations, low-pressure atmospheres, and exposure to high-intensity UV radiation that closely mimic the conditions on the surfaces of the Moon and Mars. The facility will also include centrifugal gravity chambers that will allow simulating partial *g* forces. These analogue habitats support long-term investigations surrounding plant viability in closed-loop life support systems, seed-to-seed life cycles, and nutritional sustainability in environments designed to replicate conditions at planetary bases.

15.21 Comparative Value of Analog vs Orbital Experimentation

True microgravity and real-time spaceflight conditions provided by orbital experiments (e.g., aboard the ISS and the Tiangong) helped unlock our understanding of astrobiology, but challenges in accessibility, sample return, equipment, and costs are encountered. On the other hand, analog and simulation models offer the advantages of always-on accessibility, real-time observability, and continuous flexibility that are critical for early-stage prototyping and validation. Testing time-space on an analog mission can also help test integrated systems, including plant-microbe interactions, crew participation in the cultivation of crops, and use of the resource utilization (Poulet et al. 2021). However, these missions cannot simulate all of the conditions in an orbital environment, such as cosmic radiation, microgravity for long periods of time, and the thermal properties of space bourn platforms. Hence, a complementary approach is essential. Ground models will be of great help to cross-verify the experimental results from space. Such a synergistic approach can ensure that space agriculture research progresses systematically and scientifically toward the establishment of viable space-based food production technologies.

15.22 Challenges in Studying Microgravity-Mediated Plant Responses

Detailed understanding of the plants in microgravity demands a large number of experimental studies to be done. Due to its complex interactions with different environmental signals and physiological mechanisms, gravity represents a challenging element to examine in isolation. The limited availability of long-term simulation platforms, the perpetual presence of additional stress factors present in spaceflight, and the lack of standardization in experimental methods between missions all hinder our ability to derive universals about plant responses to this dynamic in the environment.

The limited spaceflight studies conducted on the entire gravity-plant-coupled system emphasize the need for long-time-persistent partial gravity platforms as one of the significant difficulties in the present gravitational plant biology. Microgravity experiments can be performed aboard the International Space Station (ISS), and centrifuges can generate hypergravity; however, no facility is currently available that can simulate long-duration stable lunar (0.16 g) or Martian (0.38 g) gravity. Existing methods to mimic partial gravity through centrifugal means or machine parabolic flights are generally limited to such short timescales that vibration, Coriolis effects, and exposure variability render them limited; most of what we know about how plants behave in partial gravity environments comes from short-duration tests and mathematical models. This knowledge gap very much complicates the development of future long-duration space missions, where the full life cycles of plants are expected to operate under such gravitational conditions. It is challenging to predict how tropic responses, nutrient absorption, reproductive success, and biomass accumulations will be affected in habitats analogous to those available on the Moon or Mars without a reliable tool that assists such a prediction.

15.23 Difficulty in Isolating Gravity Effects from Other Spaceflight Stressors

One of the main issues for space biology is the complicated and multifaceted space environment. In addition to the consequences of changed gravity, space-grown plants here are faced with cosmic radiation, variable temperature, atmospheric pressure variations, low humidity, and gas exchange limitations. These are various stressors that not only interact with each other but also frequently mask the individual contributions of gravity. As an example, radiation exposure mediates oxidative damage, affects cell division, and alters gene expression—factors also reported with microgravity (Khodadad et al. 2011). Under the lack of highly controlled conditions, it is therefore difficult to causally link observed phenotypic or biochemical changes only to changes in gravity. This problem is especially pronounced during orbital experiments when there are differences in factors like shielding, ventilation, and lighting among modules and missions. Additionally, microgravity causes transport of water and nutrients to violate gravitational nature on Earth, forcing movement to rely on capillary forces. Over time, this dependency can cause fluid stratification or anoxic conditions in the rhizosphere. These alterations would further confound the separation of effects attributed to mechanobiology from responses related to changed hydration dynamics or hypoxia (Monje et al. 2003).

15.24 Standardization Issues in Biological Data Across Missions

The largest obstacle to gravity-based plant research is a lack of standardization between the experiments in space. Missions differ in crop types, length, light settings, growth substrates, and data presentation. These discrepancies pose challenges for comparative studies, hinder meta-analytical approaches, and slow the establishment of general conclusions (Paul et al. 2013). Reproducibility is also challenged by biological diversity, which includes differences in seed batches, climates of growth chambers, and sampling methods. This has led to divergence in crew interactions, harvest schedules, and environmental management, despite efforts from organizations such as NASA and ESA to standardize equipment (e.g., Veggie and Advanced Plant Habitat). Mitigation involves using model organisms such as *Arabidopsis*, automated sensing technologies, and AI-supported phenotyping platforms. Plants must be treated as an integral component of the gravity biology of all organisms, and the field of plant gravity biology will remain disjointed unless globally harmonized experimental protocols and data repositories are adopted.

15.25 Farming Without Soil: Experimental Systems for Space Farming

Gravity-based farming will not yield results in microgravity, partial gravity, or enclosed habitats in the long term when human colonies are established in space, making it unsuitable for use in life support systems in outer space. Given the stringent resource limitations associated with the extraterrestrial environment, the sustainability of food systems likely hinges on soilless cultivation technologies that are optimized for extremely limited volume and resource availability (De Pascale et al. 2021). For example, NASA has already successfully demonstrated these technologies aboard the International Space Station (ISS) and similar environments like the EDEN-ISS greenhouse in Antarctica. More importantly, they are the foundation on which future agriculture could be built on Mars and the Moon. There have been several plant systems (tuber crops, cereals, fruit, and leafy vegetables) and approaches for use in space farming. Because of its unique features, such as higher harvest index, tuber yield, and carbohydrate richness coupled with simple food processing needs, potato has become a good candidate for space farming. Liu et al. (2021a, b) adopted potato as a candidate for the Whole-Body Edible and Elite Plant (WBEEP) strategy (Fig. 15.4), which could pave the way for future space agriculture, highlighting that crops with high yield and more edible tissue, enriched with essential nutrients, could be suitable for space farming.

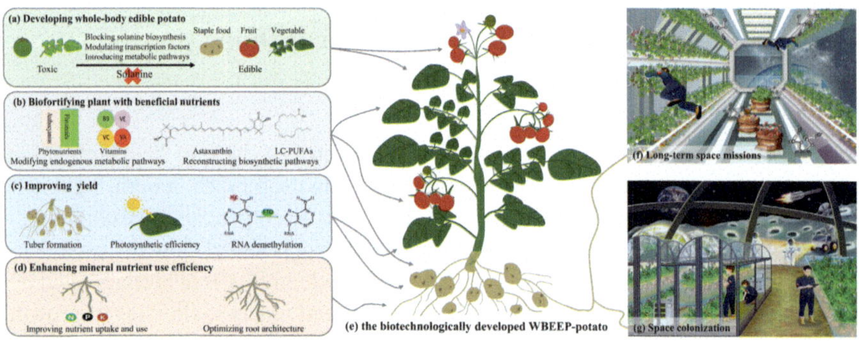

Fig. 15.4 Potato as an example of whole-body edible and elite plant strategy (Liu et al. 2021a, b)

15.26 Overview of Soilless Systems: Hydroponics, Aeroponics, and Aquaponics

In the past few decades, different soilless systems have become available for use in agriculture, and these have now been adopted in space agriculture (Fig. 15.5). Hydroponics has been developed mainly as a soilless method used in space travel and defines the development of plants in a water and nutrient fluid solution. This provides high-density plant development with controlled delivery of nutrients and significantly reduced water utilization and is suitable for use in a space environment. The ISS Veggie system has been able to use hydroponics to cultivate lettuce, radish, mustard, and zinnia, having sufficient control over nutrient and pH levels to develop healthy biomass in space (Zabel et al. 2016). Aquaponics could provide not only vegetable food for plants but also protein supplies for space crews and maximize recycling efficiency (Ferl and Paul 2016).

Aeroponics goes a step further, suspending roots of plants in the air and misting them with a highly calculated nutrient solution. Not only does this conserve a great deal of water (potentially up to 98% compared to soil vs soil-based cultivation), but it also provides enhanced oxygenation of the root zone, a significant requirement in closed systems with inadequate gas exchange (Wheeler 2010). NASA's Advanced Plant Habitat (APH) started with aeroponics experiments to grow such crops as dwarf wheat and tomatoes in microgravity simulations. One of the recent initiatives of the International Atomic Energy Agency has confirmed that seeds that are irradiated in space are more resistant to stress and have possibly better germination characteristics. ISS seeds were irradiated by cosmic rays and subsequently cultivated to assess mutagenic characteristics that support adapting crops to space agriculture environments (FAO 2023).

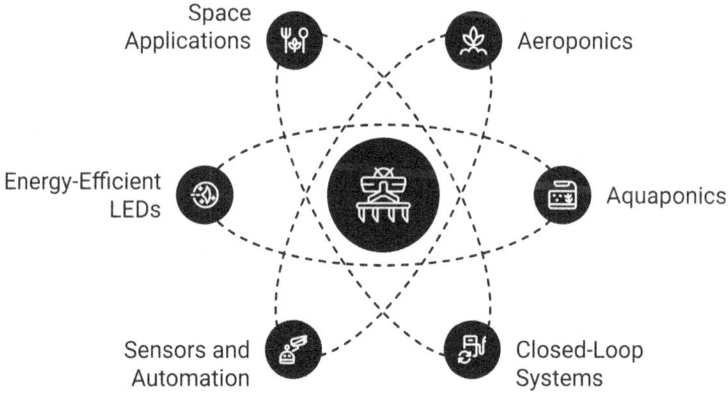

Fig. 15.5 Multifarious dimensions of space agriculture technology

15.27 Closed-Loop Systems (Oxygen, CO_2, and Water Recycling)

The implementation of soilless agriculture as part of closed biological systems represents a central component of sustainable habitable spaces in the cosmos. Hydroponically or aeroponically grown plants constitute food sources as well as play a key role in Bioregenerative Life Support Systems (BLSS). Through photosynthesis, these plants volatilize back the carbon dioxide exhaled by astronauts and emit oxygen, and their transpiration helps with water recycling and humidity control (Zeidler et al. 2021). For instance, in the MELiSSA (Micro-Ecological Life Support System Alternative) framework developed by ESA, vegetation is part of a five-module cycle involving microbial waste processing, oxygen extraction, and water recovery. These designs have demonstrated the potential for self-sustaining plant growth chambers that maintain some atmosphere independence. The water used in hydroponic or aeroponic (growing plants in nutrient-rich water) systems is continuously recycled through UV sterilization, mechanical filtration, and the capture of evapotranspiration. A close oversight essentially ensures maximum functionality with the least possible investment of resources—an even more important factor during long missions to the Moon or Mars or deep space.

15.28 Role of Sensors, Automation, and Energy-Efficient LEDs

Space agriculture requires the use of live observation, automation, and adaptive feedback loops. Modern space-based plant-growing facilities are outfitted with sensors for humidity, CO_2 levels, temperature, and pH, all linked to microcontrollers or AI-driven management systems (Moffatt et al. 2019). For example, the EDEN-ISS project employs a comprehensive suite of environmental sensors and remote control systems that allow operators to adjust light, nutrient content, and humidity from the Earth or on terminals available to onboard crew (Schubert et al. 2021). This ensures continuous plant growth, despite the delays in communications or the system malfunctions. Lighting proves another influential element. Under conditions where sunlight is sparse, or under glass on the surface of Mars with spectra filtered by its atmospheric gases, plants require very carefully optimized spectra of light. Using energy-efficient LED arrays optimally blends red, blue, and far-red wavelengths—designed for driving photosynthesis and saving energy (Poulet et al. 2021). Due to their modularity and low heat emission, crucial for temperature regulation, LEDs have perfect characteristics for both orbital and planetary agriculture.

15.29 Application in Space Stations and Future Mars Habitats

Hydroponic systems have already demonstrated their capability and safety in space with technologies like Veggie and the Advanced Plant Habitat (on board the International Space Station [ISS]) (Martimer and Gilliham 2022). These facilities delivered the first batch of freshly grown and eaten produce in orbit, with over 15 different plant varieties cultivated to date. This has benefited the crew nutritionally and psychologically (Zeidler et al. 2021). For future Mars missions, several space agencies (NASA, ESA, CNSA) are now studying inflatable or rigid greenhouses. A few models even have regolith layers covering greenhouses for radiation protection. The greenhouses would likely be using hydroponic or aeroponic delivery systems in pressurized modules. The crops, which include potatoes, wheat, soybeans, and kale, were selected based on their caloric density and the speed at which they grow. Further to this, other research projects are underway in developing seed tapes that may not require conventional substrates, completely autonomous harvesting equipment, and machine learning models for projecting crop failures, therefore minimizing human effort. These advances are not just theoretical. Experimental systems such as the Lunar Greenhouse at the University of Arizona and the GreenHab module built at MDRS by the Mars Society are already testing new formats that will be central to upcoming worlds.

15.30 Space Farming Infrastructure: From Orbit to Outpost

To establish semipermanent or long-term bases on the Moon and Mars, space farming infrastructure must evolve from small orbital growth units to flexible and integrated farming habitats (Nguyen et al. 2023). This evolution is essential to supporting the demands of plants, but also meeting ecological, architectural, and psychological needs in a closed loop. The design must balance technological efficiency with biological flexibility and crew involvement so that space farming is genuinely contributing to life support off-Earth in a beneficial and sustainable manner.

15.31 Evolution from ISS Veggie to Planned Lunar Greenhouses

NASA's Veggie, or Vegetable Production System, was installed on the International Space Station for the first time in 2014, making the operational debut of controlled plant growing using microgravity. Its revolutionary, compact, modular, pillow-based design featured adjustable LED lighting, hydroponic nutrient delivery, and a soft, resizable enclosure for optimal conditions to grow crops, including lettuce, zinnias, radishes, and mustard greens. Specifically, Veggie supremely demonstrated that plants can germinate, grow, and be consumed safely in microgravity and thus represented the first example of fresh food in the human spaceflight diet (Khodadad et al. 2020; Zeidler et al. 2021). Following the success of Veggie, NASA developed an Advanced Plant Habitat (APH) and Plant Habitat-02 that enabled greater control over variables such as CO_2 levels, temperature, humidity, light cycles, and root zone moisture. They included automated sensor arrays, allowing terrestrial teams to remotely monitor crop health and adjust environmental conditions in real time.

As we look to the future of lunar exploration, different agencies are exploring concepts for pressurized lunar greenhouses with radiation shielding, thermal management systems, and robotic capabilities. Greenhouses of the lunar variety, such as the projects developed at the University of Arizona and embedded within ESA's MELiSSA framework, are being designed to facilitate the seamless integration into modular surface habitats, allowing for long-term sustainability of food, oxygen, and stable humidity levels in the face of the omnipresent harsh conditions outside their idyllic walls (Wheeler 2010).

15.32 Modular Growth Chambers, Robotic Maintenance, and Bioreactors

A key feature of agricultural infrastructure in space is its focus on modularity. Cultivation spaces need to scale and adapt, providing both food sustainability and environmental controls. Many candidate exploration architectures treat farms as interchangeable modules that would be added like "plug-in" systems to crew habitations, power plants, and waste management. Also, we cannot underestimate the power of automation. Deep space or reduced gravity environments offer little opportunity for human involvement. As a result, robotic controls are being developed to achieve planting, pruning, harvesting, and even pest surveillance. As an example, the EDEN-ISS project uses remotely controlled growth chambers tested in Antarctic isolation to replicate environmental conditions forecasted in Martian missions (Schubert et al. 2021). Another groundbreaking domain is bioreactors. These arrangements enable the growth of algae, fungi, or microbial biomass in small quarters, making it possible to recycle nutrients efficiently and produce proteins, vitamins, and bioactive agents (Ferl and Paul 2016). Through the integration of such systems into plant-based dwellings, dependence on planet Earth for diversity of diet and micronutrient supplements by crew members could be significantly reduced.

15.33 Examples from Gateway, Tiangong, and Analog Greenhouses

NASA's planned Lunar Gateway, a collection of modules designed to orbit the Moon, contains both logistics modules and payload bays which could be used for growing plants. Although primarily intended for docking and scientific endeavors, the Gateway is also set to play a key role in research related to plant resilience, radiation tolerance, and partial gravity effects in the deep space environment. China's Tiangong Space Station has also begun adding plant modules to its longer missions. Experiments aboard Tiangong-2 found that rice and *Arabidopsis* could thrive, with future iterations expected to include integrated agricultural laboratories to provide sustenance and life support. These international efforts contribute to a better understanding of the potential of space agriculture and support future advancements in extraterrestrial crop cultivation (Romano and Aronne 2021). Analog habitats such as the Mars Desert Research Station (MDRS) in Utah or the Lunares in Poland host operational greenhouses (e.g., the "GreenHab") where astronaut trainees perform research on crop cultivation under challenging, resource-constrained conditions. These platforms do not just mimic growth conditions, but they also consider the relationship between crops and the crew's dietary needs, which is essential to trying out architectural concepts in space farming.

15.34 The Concept of "Living Architecture"

Space agriculture is not just growing crops within confined spaces, but it combines life support systems and building elements into living structures (Living Architecture). It is all about incorporating walls composed of mosses or algae to purify the air, glass houses that also act as natural light, and crop beds that act as humidity regulators. Reprogramming of plant physiological status can be orchestrated genetically or agronomically to allow for the eventual prospect of building "engineered photosynthetic units" as architectural components. Future Lunar or Martian residences could include regolith-insulating greenhouses, photosynthetically active walls (Gunawardena and Steemers 2019), and modulating floor systems that respond to the demands of an active crew or crop.

Within the clinical and high-stakes environment of a space mission, food is not only a biological necessity, but it also becomes a psychological anchor, a symbol of familiarity, a well of comfort, and a projection of an individual's personality. As space agriculture progresses due to innovations in nutrient efficacy, closed-loop mechanisms, and automation, it is also vital to consider the human experience of food, with its textures, aromas, customs, and sentimental meaning. A plant ecosystem that may sustain physical life and also support emotional well-being may enhance the overall morale, group cohesion, and ultimately mission success. Specifically, Zabel et al. (2021) propose that future plant systems are genetically modified not only for growth rate or yield but also for biochemical traits for mood stabilization. Tryptophan-rich foods, serotonin precursors, or calming phytochemicals could be taken simultaneously as food and medicine to foster astronaut neuropsychological well-being in long-duration isolation.

The mere presence of living flora has also been shown to improve emotional health in tightly packed, extreme conditions. Studies performed aboard the International Space Station (ISS), Concordia Station, and analog settings such as HI-SEAS and Lunares repeatedly indicate that adding green elements mitigates stress, increases sleep length, and maintains cognitive clarity (Ferl and Paul 2016). Astronauts often say that tending to plants is "therapeutic" and that it gives them a real connection to Earth, to life, and to the passage of time. Crew members involved in the Veggie project on the ISS reported not just eating the produce but also feeling a better mood and lower anxiety after being engaged with plant care. The simple progression of watching a living organism grow, the germination of a seed transforming into rich green life, inspires feelings of empowerment, forward motion, and nurturing vital elements for isolated environments.

15.35 Spin-Offs from Space AgTech

Though there appears to be a chasm separating the problems of nourishing astronauts while in deep space and those faced by soil-based agriculture, space agriculture has actually acted as a catalyst for change on the ground. The principles of efficiency, minimization of waste, and focus on environmental management in space farming technologies are now being found transferable to urban food systems, climate-resilient agriculture, and sustainable infrastructure (Fussy and Papenbrock 2022). These unintended side benefits underscore how innovations such as vertical farming, sustainable nutrient cycling, and low-water tech created for space missions can help tackle some of the biggest challenges facing our planet when it comes to food security and climate change.

One major outcome of space agricultural technology has been the proliferation of vertical farms, organized indoor farms that use hydroponics, aeroponics, or aquaponics. Originally designed for microgravity (for which conventional soil culture is not applicable and the volume for growing crops is limited), these systems maximize yield per square meter and use 90–95% less water than conventional farming methods (Zabel et al. 2016). In countries dealing with water scarcity or soil degradation, such as those in the Middle East and certain parts of sub-Saharan Africa, vertical farming has become especially important. The recycling of organic waste into valuable compost or biofertilizers has been accelerated due to the implementation of closed-loop nutrient cycling, one of the fundamental principles of Bioregenerative Life Support Systems (BLSS) in space. Not only does this reduce dependence on chemical inputs, but it also improves the health of soil in the long run. Greenhouses not only have been equipped with advanced soil-free cultivation technologies, including hydroponics, aquaponics, and real-time nutrient monitoring, but also affirm the terrestrial application of these systems, as demonstrated by experimental life support platforms like EDEN-ISS (Schubert et al. 2021) on Earth.

15.36 Conclusions and Prospects

The future of human space exploration is evolving ambitiously, with visions of living and working off-planet on orbital outposts, lunar bases, and even Martian colonies. Our phasing is shifted from contingency preparations to tackling fundamental mission planning. The evolution of space food systems will create resilient, independent, and biointegrated networks that will be critical, not only for emergency supplies and efficient delivery of payloads but for the successful completion of long-term exploration missions. These are systems which are expected to not just feed people but to sustain ecosystems, renew air and water, stabilize psychological health, and maintain cultural practices.

As we aim to leave the safe confines of Earth orbit to establish lunar outposts and Martian colonies, the question of how crops respond to gravity, or the absence of it, has never been more pressing. This chapter has traversed the gravitational topography of plant science, from the steady state of gravity on Earth to the lack of gravitational force when in orbit, and the differential responses of plants as a function of changed gravitational fields.

Gravity from the Earth acts as a fundamental environmental cue that integrates a variety of biochemical processes, such as guiding seed placement, mediating root-soil contact, facilitating nutrient movement, and forming plant components. Plants respond to microgravity or partial gravity by undergoing comprehensive physiological and biochemical changes, which include changes of auxin redistribution and cell wall structures, stress markers, and alterations of metabolic gene expression, an indication of changes occurring at the level of whole plant physiology. In response, plants use a range of adaptation processes, including phototropic response, architectonic reorientation, and increased biochemical stress tolerance via antioxidant defenses. Innovative studies involving space missions, clinostats, centrifuges, and simulated conditions continually uncover the resilience in plants to respond and redefine homeostasis in outer space. The obtained results belong to the pillars of science and the foundations of bioregenerative life support systems (BLSS).

Cultivation of gravity-tolerant crops is crucial for long-duration missions and distant colonies. They need to grow well in off-world growing environments, work under constraints of more limited resources, respond to changed fluid dynamics, and be suited for new mass through gravity manipulation. Such crops ideally ought to be biomass-rich, nutrient-efficient high, fast-maturing, and possessed of root-shoot system flexibility and hydraulic potential; the ability to stabilize their metabolism under fluctuating forces would also be advantageous. Space agriculture in the years to come will thus rely on the ability of crops to complete their lifespan within a controlled, artificial biosphere, needing the least amount of soil inputs and possessing the highest regenerative performance. Such cultivars for growing off-Earth will be necessary to further advancements in crop genetic manipulation, omics-informed selection, and closed-loop agronomic systems. India, a country with an expansive yet innovative agricultural footprint, remains uniquely positioned to emerge as a world leader in the space agriculture domain. This is also contributing to a broadened national space enterprise that includes the fields of space bioscience that have a history of being in the domain of the aerospace industry, which are seeing rapid shifts as time moves when organizations like ISRO are making headway to undertake manned missions and with more private sector and academic collaborations. India is well-placed to become a global leader in bridging natural and astronautic resources, particularly through innovations such as upcycling exo-grain grown using ISRU (In-Situ Resource Utilization) farming strategies: space analogue space habitat to ISRU on Moon/Mars up to "bioregenerative systems." The journey from soil to satellite is not a metaphorical one—it is an actual mission in development.

References

Agritecture (2023) Space farming and sustainable development: what do they have in common? Retrieved from https://www.agritecture.com/blog/space-farming

Arena C, De Micco V, Macaeva E, Quintens R (2014) Space radiation effects on plant and mammalian cells. Acta Astron 104:419–431. https://doi.org/10.1016/j.actaastro.2014.05.005

Baba AI, Mir MY, Riyazuddin R, Cséplő Á, Rigó G, Fehér A (2022) Plants in microgravity: molecular and technological perspectives. Int J Mol Sci 23(18):10548. https://doi.org/10.3390/ijms231810548

Borst AG, Van Loon JJWA (2009) Technology and developments for the random positioning machine, RPM. Microgravity Sci Technol 21:287–292

Carillo P, Morrone B, Fusco GM, De Pascale S, Rouphael Y (2020) Challenges for a sustainable food production system on board of the International Space Station: a technical review. Agronomy 10(5):687. https://doi.org/10.3390/agronomy10050687

Casal JJ, Estevez JM (2021) Auxin-environment integration in growth responses to forage for resources. Cold Spring Harb Perspect Biol 13(4):a040030. https://doi.org/10.1101/cshperspect.a040030

De Micco V, Amitrano C, Mastroleo F, Aronne G, Battistelli A, Carnero-Diaz E, De Pascale S, Detrell G, Dussap CG, Ganigué R, Jakobsen ØM, Poulet L, Van Houdt R, Verseux C, Vlaeminck SE, Willaert R, Leys N (2023) Plant and microbial science and technology as cornerstones to bioregenerative life support systems in space. NPJ Microgravit 9(1):69. https://doi.org/10.1038/s41526-023-00317-9

De Micco V, Amitrano C, Pannico A, De Pascale S (2026) Integrating plant biology and crop science for sustainable farming in space exploration and earth applications. In: Penna S, Varshney RK (eds) Improving crops for future sustainability and climate resilience. Sustainability Sciences in Asia and Africa (SAFS). Springer, Singapore, pp 309–330. https://doi.org/10.1007/978-981-95-2754-0_9

De Pascale S, Arena C, Aronne G, De Micco V, Pannico A, Paradiso R, Rouphael Y (2021) Biology and crop production in space environments: challenges and opportunities. Life Sci Space Res 29:30–37. https://doi.org/10.1016/j.lssr.2021.02.005

Douglas GL, Wheeler RM, Fritsche RF (2021) Sustaining astronauts: resource limitations, technology needs, and parallels between spaceflight food systems and those on Earth. Sustainability 13(16):9424. https://doi.org/10.3390/su13169424

FAO (2023) Crop seeds return from space in IAEA/FAO project to help feed a warming world https://www.fao.org/newsroom/detail/FAO-IAEA-crop-seeds-return-space-project-to-help-feed-a-warmingworld/en
Farooq M, Ali S, Khan M, Jang Y-H, Kim E-G, Zhao D-D, Kim K-M (2024) Investigating plant responses to microgravity and adaptations in gravisensitive environments. Environ Sci Eur 36:28. https://doi.org/10.1186/s12302-024-00851-7
Ferl RJ, Paul AL (2010) Lunar plant biology–a review of the Apollo era. Astrobiology 10:261–274
Ferl RJ, Paul AL (2016) The biology of spaceflight: what it takes to grow plants in space. Gravit Space Res 4(2):47–61
Ferl RJ, Zupanska AK, Spinale C, Paul AL (2002) Plants in space: the final frontier. Gravit Space Biol Bull 15(2):7–18
Fussy A, Papenbrock J (2022) An overview of soil and soilless cultivation techniques-chances, challenges and the neglected question of sustainability. Plants (Basel, Switzerland) 11(9):1153. https://doi.org/10.3390/plants11091153
Geras'kin S (2024) Plant adaptation to ionizing radiation: mechanisms and patterns. Sci Total Environ 916:170201. https://doi.org/10.1016/j.scitotenv.2024.170201
Gunawardena K-C, Steemers K (2019) Living walls in indoor environments. Build Environ 148:478–487
Hoson TJL (2014) Plant growth and morphogenesis under different gravity conditions: relevance to plant life in space. Life 4:205–216
Hou T, Zheng B, Peng F, Jiang Z, Zhang W, Wang Y (2025) Under simulated microgravity and gravity, anthocyanin is regulated by DcaWRKY2 in *Dendrobium catenatum* leaves. Front Plant Sci 15:1505199. https://doi.org/10.3389/fpls.2024.1505199
ITSD-India Today Science Desk (2024) ISRO launches analogue mission in Ladakh to simulate life outside Earth. India Today. Retrieved from https://www.indiatoday.in/science/story/isro-launches-analogue-mission-in-ladakh-to-simulate-life-outside-earth-2626272-2024-11-01
ISRO (2025). https://www.isro.gov.in/CROPS.html
Johnson CM, Boles HO, Spencer LE, Poulet L, Romeyn M, Bunchek JM, Fritsche R, Massa GD, O'Rourke A, Wheeler RM (2021) Supplemental food production with plants: a review of NASA research. Front Astron Space Sci 8:734343. https://doi.org/10.3389/fspas.2021.734343
Khodadad CLM, Hummerick ME, Spencer LE, Dixit AR, Richards JT, Romeyn MW, Massa GD (2011) Microgravity effects on seed germination and plant growth. Astrobiology 11(3):311–323
Khodadad CLM, Hummerick ME, Spencer LE, Dixit AR, Richards JT, Romeyn MW, Smith TM, Wheeler RM, Massa GD (2020) Microbiological and nutritional analysis of lettuce crops grown on the International Space Station. Front Plant Sci 11:199. https://doi.org/10.3389/fpls.2020.00199
Kirk JT, Johnson CM, Smith BE (2022) Cytoskeletal dynamics and cell orientation in microgravity. Space Life Sci Rep 5(1):45–56
Kiss JZ, Wolverton C, Wyatt SE (2019) Plant tropisms in spaceflight environments. Adv Space Res 64(6):1231–1240
Kordyum E, Hasenstein KH (2021) Plant biology for space exploration—building on the past, preparing for the future. Life Sci Space Res 29(2021):1–7
Liu H, Yao Z, Fu Y, Feng J (2021a) Review of research into bioregenerative life support system(s) which can support humans living in space. Life Sci Space Res 31(2021):113–120. https://doi.org/10.1016/j.lssr.2021.09.003
Liu Y, Xie G, Yang Q (2021b) Biotechnological development of plants for space agriculture. Nat Commun 12:5998. (2021). https://doi.org/10.1038/s41467-021-26238-3
López Garzón NA, Pinzón-Fernández MV, Saavedra TJS, Nati-Castillo HA, Arias-Intriago M, Salazar-Santoliva C, Izquierdo-Condoy JS (2025) Microgravity and cellular biology: insights into cellular responses and implications for human health. Int J Mol Sci 26(7):3058. https://doi.org/10.3390/ijms26073058

Luigi GI, Leone ER, Lucius W, Muthert F, Maurizio I, Capozzi F, Manzano A, Ciska M, Herranz R, Medina FJ, Kiss JZ, van Loon JJWA, Aronne G (2022) Interaction of gravitropism and phototropism in roots of *Brassica oleracea*. Environ Exp Bot 193:104700. https://doi.org/10.1016/j.envexpbot.2021.104700

Maffei ME, Balestrini R, Costantino P, Lanfranco L, Morgante M, Battistelli A, Del Bianco M (2024) The physiology of plants in the context of space exploration. Commun Biol 7:1311. https://doi.org/10.1038/s42003-024-06989-7

Mantri N, Patade V, Penna S, Ford R, Pang E (2012) Abiotic stress responses in plants: present and future. In: Ahmad P, Prasad M (eds) Abiotic stress responses in plants. Springer, New York, pp 1–19. https://doi.org/10.1007/978-1-4614-0634-1_1

Massa GD, Wheeler RM, Morrow RC, Levine HG (2016) Growth chambers on the international space station for large plants. Acta Hortic 1134:215–222. https://doi.org/10.17660/actahortic.2016.1134.29

Massa GD, Newsham G, Hummerick ME, Morrow RC, Wheeler RM (2017) Preparation for the veggie plant growth system on the international space station. Gravit Space Res 5:24–34

Meinen E, Dueck T, Kempkes F, Stanghellini C (2018) Growing fresh food on future space missions: environmental conditions and crop management. Sci Hortic 235:270–278. https://doi.org/10.1016/j.scienta.2018.03.002

Moffatt SA, Morrow RC, Wetzel JP (2019) Astro GardenTM aeroponic plant growth system design evolution. In: 49th international conference on environmental systems 7–11 July, 2019, Boston, Massachusetts, vol 195, pp 1–13

Mohanta TK, Mishra AK, Mohanta YK, Al-Harrasi A (2021) Space breeding: the next-generation crops. Front Plant Sci 12:771985. https://doi.org/10.3389/fpls.2021.771985

Monje O, Stutte GW, Goins GD, Porterfield DM, Bingham GE (2003) Lettuce biomass production in a hybrid hydroponic and porous tube nutrient delivery system. Life Support Biosph Sci 9(3):211–221

Mortimer JC, Gilliham M (2022) SpaceHort: redesigning plants to support space exploration and on-earth sustainability. Curr Opin Biotechnol 73:246–252. https://doi.org/10.1016/j.copbio.2021.08.018

Mroue S, Simeunovic A, Robert HS (2018) Auxin production as an integrator of environmental cues for developmental growth regulation. J Exp Bot 69(2):201–212. https://doi.org/10.1093/jxb/erx259

Muthert LWF, Izzo LG, van Zanten M, Aronne G (2020) Root tropisms: investigations on earth and in space to unravel plant growth direction. Front Plant Sci 10:1807. https://doi.org/10.3389/fpls.2019.01807

Nguyen MTP, Knowling M, Tran NN, Burgess A, Fisk I, Watt M, Escribà-Gelonch M, This H, Culton J, Hessel V (2023) Space farming: horticulture systems on spacecraft and outlook to planetary space exploration. Plant Physiol Biochem 194:708–721. https://doi.org/10.1016/j.plaphy.2022.12.017

Pandith JA, Neekhra S, Ahmad S, Sheikh RA (2023) Recent developments in space food for exploration missions: a review. Life Sci Space Res 36:123–134. https://doi.org/10.1016/j.lssr.2022.09.007

Paul A-L, Amalfitano CE, Ferl RJ (2013) Plant development under spaceflight conditions: effects of gravitational changes on the physiology and gene expression of plants. Gravit Space Biol 27(1):2–12

Polyakov Y, Musaev I, Polyakov S (2009) Application of space technology to the design of sustainable settlements in hot deserts: bioregenerative life support systems. Nat Prec. https://doi.org/10.1038/npre.2009.3926.1

Poulet L, Schubert D, Zabel P (2021) Crew time in a space greenhouse using data from analog missions and veggie. Life Sci Space Res (Amst) 31:101–112. https://doi.org/10.1016/j.lssr.2021.08.002

Reed DW, Vanden Bosch CA (2023) Engineering perspectives of growing plants in space. J Indian Inst Sci 103:797–805. https://doi.org/10.1007/s41745-023-00369-6

Romano LE, Aronne G (2021) The world's smallest plants (*Wolffia* sp.) as potential species for bioregenerative life support systems in space. Front Plant Sci 12:647. https://doi.org/10.3389/fpls.2021.718160

Ross C, Sablani S, Tang J (2023) Preserving ready-to-eat meals using microwave technologies for future space programs. Foods (Basel, Switzerland) 12(6):1322. https://doi.org/10.3390/foods12061322

Salisbury FB (1991) Lunar farming: achieving maximum yield for space colonies. Bioscience 41(9):626–634

Salisbury FB, Gitelson JI, Lisovsky GM (1997) Bios-3: Siberian experiments in bioregenerative life support: attempts to purify air and grow food for space exploration in a sealed environment began in 1972. Bioscience 47(9):575–585. https://doi.org/10.2307/1313164

Schubert D, Zabel P, Bamsey M, Zeidler C (2021) Lessons learned from EDEN ISS: plant health monitoring and data analysis of a greenhouse in extreme environment. Acta Hortic 1306:83–90. https://doi.org/10.17660/ActaHortic.2021.1306.11

Shaw R, Soma T (2022) To the farm, Mars, and beyond: technologies for growing food in space, the future of long-duration space missions, and earth implications in English news media coverage. Front Commun 7:1007567. https://doi.org/10.3389/fcomm.2022.1007567

Suprasanna P, Negi P, Vinay K (2021) Genomics and epigenetics of plant abiotic stress. In: Fink G (ed) Handbook of stress: stress: genetics, epigenetics, and genomics. Elsevier Science, San Diego, pp 49–56

Swamy BK, Hosamani R, Sathasivam M, Chandrashekhar SS, Reddy UG, Moger N (2021) Novel hypergravity treatment enhances root phenotype and positively influences physiobiochemical parameters in bread wheat (*Triticum aestivum* L.). Sci Rep 11(1):15303. https://doi.org/10.1038/s41598-021-94771-8

Takemura K, Kamachi H, Kume A, Fujita T, Karahara I, Hanba YT (2017) A hypergravity environment increases chloroplast size, photosynthesis, and plant growth in the moss *Physcomitrella patens*. J Plant Res 130:181–192. https://doi.org/10.1007/s10265-016-0879-z

Tang H, Rising HH, Majji M, Brown RD (2021) Long-term space nutrition: a scoping review. Nutrients 14(1):194. https://doi.org/10.3390/nu14010194

Tuominen LK, Levine LH, Musgrave ME (2009) Plant secondary metabolism in altered gravity. In: Jain SM, Saxena PK (eds) Protocols for in vitro cultures and secondary metabolite analysis of aromatic and medicinal plants, Methods in molecular biology, vol 547. Humana Press, Totowa. https://doi.org/10.1007/978-1-60327-287-2_30

Vorselen D, Roos WH, MacKintosh FC, Wuite GJL, van Loon JJWA (2014) The role of the cytoskeleton in sensing changes in gravity by nonspecialized cells. FASEB J 28:536–547. https://doi.org/10.1096/fj.13-236356

Watkins P, Hughe JS, Gamage TV, Knoerzer K,. Ferlazzo ML, Banati RB (2022) Long term food stability for extended space missions: a review. Life Sci Space Res 32:79–95. https://doi.org/10.1016/j.lssr.2021.12.003

Wheeler RM (2010) Plants for human life support in space: from Myers to Mars. Gravit Space Res 23(2):25–35

Wheeler RM (2017) Agriculture for space: people and places paving the way. Open Agric 2(1):14–32. https://doi.org/10.1515/opag-2017-0002

Wolverton C, Kiss JZ (2009) Gravitropic and phototropic signaling interactions in plants under variable g conditions. Planta 230(1):105–113

Zabel P, Zeidler C, Schubert D, Bamsey M (2016) Review on plant cultivation technologies and their influence on space crop production. Acta Astronaut 126:324–334. https://doi.org/10.1016/j.actaastro.2016.05.022

Zabel F, Müller C, Elliott J, Minoli S, Jägermeyr J, Schneider JM, Franke JA, Moyer E, Dury M, Francois, L., Folberth, C., Liu, W., Pugh, T.A.M., Olin, S., Rabin, S.S., Mauser, W, Hank T, Ruane AC, Asseng S (2021) Large potential for crop production adaptation depends on available future varieties. Glob Change Biol 27:3870–3882. https://doi.org/10.1111/gcb.15649

Zeidler C, Zabel P, Schubert D (2021) Bioregenerative life support systems for future space missions and their application to Earth-based systems. Acta Astronaut 180:139–150. https://doi.org/10.1016/j.actaastro.2020.12.008

Zheng HQ, Han F, Le J (2015) Higher plants in space: microgravity perception, response, and adaptation. Microgravity Sci Technol 27:377–386. https://doi.org/10.1007/s12217-015-9428-y

The manufacturer's authorised representative in the EU is Springer Nature Customer Service Centre GmbH, Europaplatz 3, 69115 Heidelberg, Germany. If you have any concerns regarding our products, please contact ProductSafety@springernature.com

Printed and bound by CPI Group (UK) Ltd, Croydon, CR0 4YY
21/09/2026
02220013-0013